PROCEEDINGS OF THE THIRD INTERNATIONAL SYMPOSIUM
ON ARTIFICIAL RECHARGE OF GROUNDWATER – TISAR 98
AMSTERDAM / NETHERLANDS / 21-25 SEPTEMBER 1998

Artificial Recharge of Groundwater

Edited by
JOS H. PETERS ET AL.
Kiwa Research and Consultancy, Nieuwegein, Netherlands

A.A. BALKEMA / ROTTERDAM / BROOKFIELD / 1998

The texts of the various papers in this volume were set individually by typists under the supervision of each of the authors concerned.

Published by
A.A. Balkema, P.O. Box 1675, 3000 BR Rotterdam, Netherlands
Fax: +31.10.413.5947; E-mail: balkema@balkema.nl; Internet site: www.balkema.nl
A.A. Balkema Publishers, Old Post Road, Brookfield, VT 05036-9704, USA
Fax: 802.276.3837; E-mail: info@ashgate.com

ISBN 90 5809 017 5

Printed in the Netherlands

Artificial Recharge of Groundwater, Peters et al. (eds) © 1998 Balkema, Rotterdam, ISBN 90 5809 017 5

Table of contents

1.7 *Hydrological aspects*

1.8 *Well recharge case studies*

1.9 *Pollutants and basin recharge*

1.10 *Other types of recharge*

Artificial Recharge of Groundwater, Peters et al. (eds) © 1998 Balkema, Rotterdam, ISBN 90 5809 017 5

Preface

Artificial recharge has often been a theme at international workshops, seminars, congresses and symposia. A few examples are the gatherings in Haifa (Israel) in 1967, Dortmund (Germany) in 1979, Akersloot (Netherlands) in 1993, Helsinki (Finland) in 1996 and Schwerte (Germany) in 1997. With the goal of offering a truly global platform to experts in this field from all over the world, the American Society of Civil Engineers (ASCE) organized a symposium from 23 to 27 August 1988, in Anaheim, California. The proceedings of this symposium were edited by A. Ivan Johnson and Donald J. Finlayson, and published by the ASCE (ISBN 0-87262-713-6). The second symposium was held in Orlando, Florida, from 17 to 22 July 1994. A. Ivan Johnson and R. David Pyne edited the proceedings, and ASCE was once again the publisher (ISBN 0-7844-0091-1).

In Orlando, July 1994, the ASCE appealed to other organizations to take the initiative for the third symposium in the series. Theo Olsthoorn, Ed Roosma and myself took up the challenge. The ASCE agreed that Kiwa Research and Consultancy, and the Dutch water supply companies that are part of the Kiwa Artificial Recharge Working Group should be the guest organizations for the third symposium in the series: TISAR98, the Third International Symposium on Artificial Recharge of groundwater. It was Ed's idea to hold TISAR at the RAI International Congress and Exhibition Centre, Amsterdam, at the same time as the biennial Aquatech international exhibition on water technology. The idea has been further developed by an Organizing Committee, and the symposium will take place from 21 to 25 September 1998.

The document now in your hands contains the planned proceedings for TISAR 98. I greatly appreciate it that authors and reviewers have put in so much effort, and have collaborated with me in publishing these proceedings in advance. Our goal throughout has been to make the planned proceedings available to all participants before the start of the symposium. I'm proud to say – as things stand at the moment – that we're going to succeed.

The selection of papers was made by a Scientific Committee, on the basis of abstracts that authors submitted in response to a call for papers in 1997. This publication contains those papers that will be presented orally and in poster-form at the symposium. In August 1998, just before the symposium takes place, a final programme will be made. For all kinds of reasons, the actual proceedings often differ from the planned ones. Despite the care that has gone into the preparations, TISAR 98 will probably not be an exception to the rule.

Over two hundred participants are expected. They will have the opportunity of examining the roughly 80 contributions, of which 50 will be presented orally and 30 in the form of posters. The papers originate from 30 countries or USA-states. Prior to the symposium, on Sunday, 20 September, a short course on the basics of artificial recharge of groundwater will be given by notable and experienced professionals in the field. On Wednesday, 23, and Friday, 25 September, as part of the technical excursions, there will be visits to recharge facilities along the North Sea coast.

The Organizing Committee consists of Jef P.P.M. Ernes, A. Ivan Johnson (USA), Hein G.de Jonge, Hans G. Nowee and Jos H. Peters; and is chaired by the Symposium President, Theo Olsthoorn. On the Scientific Committee are Kees van den Akker, Herman Bouwer (USA), Jesús Carrera (Spain), Peter Dillon (Australia), Karsten H. Jensen (Denmark), Jos H. Peters, Uwe Schöttler (Germany) and Pieter Stuyfzand. The members of the Scientific Committee, together with Hein de Jonge, Ed Roosma, Theo Olsthoorn and Ivan Johnson, did an excellent job of reviewing the manuscripts. All these co-editors will also act as moderators or chairpersons for sessions on the topics below:

- Operation of basin recharge,
- Water management in arid regions,
- Clogging,
- Recharge of treated waste water,
- Pollutants and bank filtration,
- Microbiological aspects,
- Hydrological aspects,
- Well recharge case studies,
- Pollutants and basin recharge,
- Other types of recharge,
- Storage and recovery efficiency, and
- Chemical reactions and well recharge.

You will find the poster presentations in this publication, after the papers to be presented orally. They are somewhat shorter and are divided into blocks on the following topics:

- Water management in arid regions,
- Field and desk studies, plans and cases,
- Artificial groundwater recharge in Stockholm (Sweden),
- Behaviour of contaminants,
- Predicting and modelling groundwater quality,
- Clogging of recharge systems, and
- Biological treatment and removal of organic matter.

The Symposium President, members of the Organizing and Scientific committees, lecturers for the short course, co-editors, all co-operating organizations, and last but not least, Mariske Kruize-Ooms of the Symposium Secretariat, all receive my heartfelt thanks. Finally, I want to acknowledge the generosity of the many financial sponsors of TISAR 98, without whose contributions neither the symposium nor this publication would be possible. The main sponsors are listed on a separate page.

I look forward to meeting the authors of the papers and posters, the members of both committees, and all other TISAR 98 participants, in Amsterdam in September 1998. I also expect the initiative to be taken there for the fourth symposium in the series; a series that started in 1988 in Anaheim. Who will accept the challenge to organize the next gathering on this fascinating and important subject, somewhere on our planet at the beginning of the next millennium?

Editor
Jos H. Peters, Kiwa Research and Consultancy
Nieuwegein, the Netherlands
5 July 1998

Artificial Recharge of Groundwater, Peters et al. (eds) © 1998 Balkema, Rotterdam, ISBN 90 5809 017 5

Water in Europe

Th.G. Martijn
Director International Relations VEWIN, and Member Board of Directors EUREAU

1. Water does not respect our national boundaries in Europe. Consequently it is not surprising that, since a number of years, the European Commission is quite busy with European water problems. The first water directive, the one concerning the quality required of surface water intended for the production of drinking water, appeared in 1975.

2. The basis of the EU Water Policy can be found in the EU Environmental Policy. The latter has two objectives:
   1. It has to contribute to the progressive integration of the EU Member States, and
   2. It has to try to achieve the objectives fixed in Article 130 R of the European Union Treaty, namely:
   - To protect and improve the quality of the environment,
   - To contribute towards protecting human health, and
   - To assure a prudent and rational utilisation of natural resources.

3. An important Directive for water suppliers is Directive 80/778 of 15 July, 1980, relating to the quality of water intended for human consumption. This so-called Drinking Water Directive of 1980 was based on scientific knowledge of the late 1970's. It was too inflexible and did not take into account the wide variety in water conditions that exist in different parts of Europe. Proposals for a revised Drinking Water Directive were adopted by the European Commission in January 1995 and submitted to the European Parliament. It is foreseen that the revision of the Drinking Water Directive will be completed this year.

4. Water is also a pre-condition for a sustained economic development. Industrial, commercial and agricultural sectors cannot flourish without the availability of this vital source. Appreciating the vital role of water for the well-being and the economic development of a society, we observe that water is at the same time subject to conflicting interests. Many sectors put their demands on scarce available water resources. Effective water management, embodied in legislation, policies and planning, becomes especially crucial in environments that are characterised by high population densities and economic development.

5. Recently, the European Commission produced a proposal for a Framework Directive for Community Action in the field of Water Policy. The proposed new Directive provides a major step forward and comprises a number of key features that are warmly welcomed. These key features, which will determine to a large extent the future of our water industry, include:
   - Integrated water management planning on a river basin basis,

- The holistic management of water resources, both surface and groundwater in quality and quantity,
- The reference to both the intended use of water resources and the intrinsic protection of aquatic ecosystems,
- Preventive action, which means damage to be rectified at source,
- Reference to the polluter-pays-principle, full cost recovery and the principle of subsidiarity,
- Looking at all sources of pollution: Point, diffuse and accidental.

6. Artificial Recharge of groundwater systems plays and will play an important role in a majority of EU-countries to reach different objectives of the proposed Framework Directive. AR-schemes contribute to effective quantitative water resource management, safety and continuity in water supply. In addition, generally, it improves water quality and has environmental benefits, if applied in a sustainable way. This symposium will show again the benefits of AR and hopefully influences in a positive way European thinking of AR in the scope of the Framework Directive.

Artificial Recharge of Groundwater, Peters et al. (eds) © 1998 Balkema, Rotterdam, ISBN 90 5809 017 5

The role of artificial recharge of groundwater for the water supply of Amsterdam

G.ter Horst
Alderman of Public Utilities of Amsterdam

Clearly, the Netherlands are a wet country with numerous rivers, canals and lakes. Although, therefore, there never was a shortage of water, there has long been a shortage of good quality water. The groundwater in the low-lying western part of the country is essentially brackish, unsuitable as a source for drinking water.

For centuries, citizens of Amsterdam had to resort to water collected from the roofs and to water shipped in daily from a remote less polluted river, called Vecht. This water was sold by the bucket, which only the well-to-do could afford.

This situation changed in 1853, when Amsterdam became the first city in the Netherlands to receive piped drinking water. This water was taken from the coastal dunes south of Zandvoort, which proved to be filled with fresh water.

The raw groundwater was collected by canals dug through the dune area. It was then transported over about 30 km by pipeline to the city, and at first, sold by the bucket for the price of one cent. Nowadays, the water price is about double, a little over 2 cents per bucket. But for this money, the drinking water undergoes a sophisticated treatment and is delivered directly into each and every house.

The system immediately proved the importance of safe drinking water. Water-borne epidemics like cholera and typhoid virtually disappeared from the city after this water became available.

Like everywhere else, the water demand of Amsterdam grew exponentially. The canal system in the dunes was thus continuously extended until, by the end of the last century, the natural yield of the dunes was reached.

Then, from 1903 onwards, wells, screened beneath the extended clay layer, were drilled. By tapping and mining a deeper aquifer, these wells permitted to increase the extraction to far beyond the yearly natural replenishment. Severe intrusion of brackish and salt water resulted. In the fifties, the fresh-salt water interface has risen an average of 60 m and so severely endangered the continuation of the drinking water supply.

To sustain drinking water production, the dunes would have to be artificially replenished with water brought in from elsewhere. The only viable resource turned out to be surface water from a branch of the river Rhine at 55 km from the dunes.

A separate company was founded in 1950 with the objective to extract this river Rhine water, purify it and pump it to the dunes of North-Holland as well as a number of big industries. By 1956 the first river water arrived at the dunes and was artificially recharged into the new ponds. In the same period other waterworks also realised artificial recharge schemes.

The artificial recharge system has since been the backbone of the drinking water production for Amsterdam, and has proven reliable. To date about 75% of the drinking water for Amsterdam stems from the dune area. Only 15% of it is natural groundwater, 85% being artificially recharged, prepurified river water. In total about two thirds of the drinking water of the capital stems from Rhine water.

The AR system has withstood many problems, which are intrinsic to the use of water from an international river used by millions of people and thousands of industries.

The system provides a protected stock of raw water, which allows to shut off river water intake after accidental pollutions.

The AR-system, in which the river water resides for about three months, smoothes quality variations and temperature of the river water. It thus allows for a stable drinking water quality.

The underground offers an ecological and natural way to remove viruses, bacteria and protozoa and thus provides hygienic safety.

These are all sustainable facts. The potentially less sustainable aspects, like pollution of the underground by toxic chemicals, has long been attacked by extension of the purification of the river. Since 1974 it consists of coagulation and rapid sand filtration, and so removes not only sediment but also especially phosphate and heavy metals before recharging the dunes.

On the other hand, based on intensive river-basin wide quality measurements, for many decades now, continuous international efforts are taken to have governments, authorities, cities and industries reduce the pollution of the river. This has led to large improvements compared with the situation in the late sixties and seventies. Especially heavy metals went down by a factor of 10 or more. These international efforts also led to stabilising of the salt concentration in the river. Pesticides and some other toxic organic compounds remain points of concern, however.

The quality of the water to be recharged is also continuously analysed. Moreover, it is subjected to severe national and provincial standards, a number of which being even more severe than those of drinking water. Point of departure of the current regulations is, that no pollution is allowed in the recharge water that might endanger the quality of the underground and the dune ecology.

Ever since the middle of the last century the water companies have protected the coastal dunes from all kinds of urban and economic developments. It is this protection and nothing else which has preserved the dunes for the current and future generations as one of the most important nature reserves of the country. Much of the eco-hydrological research done today aims at sustaining drinking water production while maximising the ecological values.

During its 145 years of existence, the Water Supply of Amsterdam has overcome many challenges on the basis of sound investigations, many of which were carried out in-house. This allowed us to continue to produce and distribute a good tasting, good quality drinking water to the capital and surrounding communities. This water is softened and produced and distributed safely without chlorine and is highly appreciated by the population.

New technologies will be developed aiming at safe, extensible and ecological sustainable drinking water production. Among these are artificial recharge of groundwater by means of wells and direct purification of surface water by means of membranes. Even so, in the foreseeable future, the current artificial recharge scheme in the dunes continues and will continue to play an essential role for a reliable drinking water production for our citizens.

Artificial Recharge of Groundwater, Peters et al. (eds) © 1998 Balkema, Rotterdam, ISBN 90 5809 017 5

Organization

ORGANIZER
Kiwa Research and Consultancy, Netherlands

CO-OPERATING ORGANIZATIONS
American Society of Civil Engineers, ASCE, USA
Dune Water Company of South Holland, Netherlands
Amsterdam Water Supply, Netherlands
Limburg Water Supply, Netherlands
International Association of Hydrogeologists (IAH)
    Artificial Recharge Working Group
International Association of Hydrological Sciences (IAHS)
    Groundwater Committee
American Society for Testing and Materials (ASTM)

MAIN SPONSORS
Kiwa Research and Consultancy
Dune Water Company of South Holland
Limburg Water Supply
Amsterdam Water Supply
PWN North-Holland Water Supply Company
Ballast Nedam-Heerema
AQUATECH International Exhibition on Water Technology
European Commission, Directorate-General XII, Science, Research and Development

kiwa

wml

Ballast Nedam

HEEREMA

AQUA
TECH
98

EUROPEAN COMMISSION
DIRECTORATE-GENERAL XII
SCIENCE, RESEARCH AND DEVELOPMENT

1 Oral presentations

1.1 Operation of basin recharge

Artificial Recharge of Groundwater, Peters et al. (eds) © 1998 Balkema, Rotterdam, ISBN 90 5809 017 5

Seawater intrusion reversed through artificial recharge beneath the Oxnard Plain, California

S.G. Berger & F.J. Gientke
United Water Conservation District, Santa Paula, Calif., USA

ABSTRACT: The Oxnard Plain is a 142.5 square km area located along the Pacific Coast in Central California. Since the late 1800's the area has proven to be an abundant agricultural area with deep tertiary deposits of alluvium which make the area rich in groundwater resources. Early wells drilled through the overlying clay cap resulted in artesian pressures sufficient to raise water to the second floor of residences as well as support agriculture. By the 1940's the groundwater levels has progressively lowered, fluctuating in response to tidal changes. By early 1950's seawater moved into near-coastal wells and for the next 35 years continued to move further inland making hundreds of wells useless. Accordingly, United Water initiated a 3 Phase Program to restore the groundwater basins. As a result, overdraft appears to have been eliminated in the upper aquifer and seawater intrusion reversed.

1 PHASE I

In Response to seawater intrusion, United Water Conservation District (UWCD) constructed a 6096 cm high dam to create a 107.3 million m^3 reservoir. In addition, 2 spreading grounds were constructed in the forebay, adjacent the Oxnard Plain (Figure 1).
Several other features of the 1955 project were 1) construct an 12.9 km long pipeline to deliver river water 9.9 million m^3 to a large agricultural area and thus reduce the amount of plumage from this region, and 2) construct a 16.2 m^3 potable water line to several municipal areas from a recharge /pump back facility at El Rio. However, the water district lacked a permanent diversion structure on the lower Santa Clara River and was unable to divert water during high runoff periods. As a result, large volumes of divertable runoff flowed to the ocean rather than being beneficially used for arti-ficial recharge. Unfortunately, these projects were insufficient to overcome the overdraft and seawater intrusion worsened.

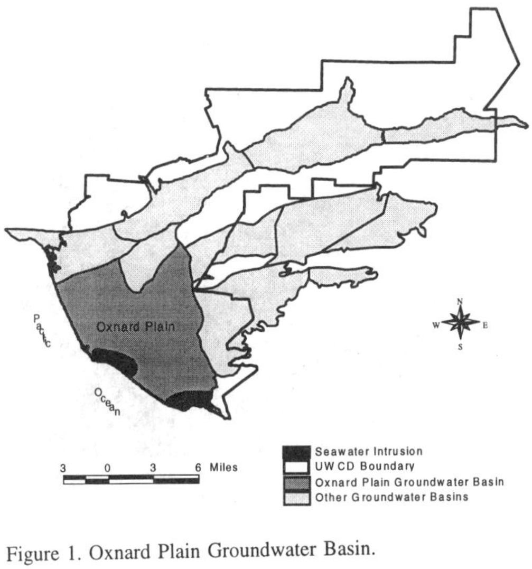

Figure 1. Oxnard Plain Groundwater Basin.

1.1 Santa Felicia Dam

In 1955, United Water completed the construction of Santa Felicia Dam about 9.7 km north of the Santa Clara River on Piru Creek. Within the 107.3 million m^3 reservoir, high winter runoff is stored and later released down to the Santa Clara River where the water is diverted into the Saticoy and El Rio Spreading Grounds. As the water flows down the 40.2 km to the diversion point, significant quantities sink through the alluvial channel and replenish the upstream valley groundwater basins. Up to 50% or more of the flow can disappear through the river channel depending on the antecedent moisture conditions, and rate of release of water from Santa Felicia Dam. Santa Felicia Dam and two Oxnard Plain Spreading Grounds (shown in Figures 2, 3 and 4).

Figure 2.

Figure 3.

Figure 4.

1.2 Pleasant Valley Pipeline

Part of the water diverted near Saticoy is sent via pipeline to the Pleasant Valley agricultural area in the southwest part of the district. Delivered water is used to offset pumping from the aquifers and thus, is considered in-lieu recharge of the groundwater basins. During the last 43 years of deliveries, about 308.3 million m^3 have been delivered to the Pleasant Valley County Water District.

1.3 Oxnard-Hueneme Pipeline

As a result of highly mineralized groundwater in the vicinity of Oxnard and Pt. Hueneme, United Water constructed this minicipal, potable water line which also serves to reduce pumping from near-coastal wells. In the last 43 years, about 498.5 million m^3 have been delivered.

1.4 Earthen diversion dike

Although United Water has invested considerable money to construct the above mentioned facilities, the District lacked a permanent, year-round diversion structure at Saticoy. Because of that, large flows invariably washed out the earthen embankment which could not be restored until flows in the river subsided and equipment could move into the river to re-establish the dike. During the interval from breach of the dike to its re-establishment, significant quantities of water continued to flow to the ocean and not to the spreading grounds.

Studies in the early 1980's showed that a permanent structure could increase the diversion by over 25% and accordingly, the District began to develop plans for a structure which would then supplement its capability to eliminate the overdraft.

2 PHASE II

To eliminate the seawater intrusion, United Water, with the local county government, constructed several more projects in the 1980's to provide additional water supplies to the overdrafted area. These projects shifted pumping from the Upper Aquifer System to the Lower, and increased diversions by 25% from the Santa Clara River to the two nearby artificial recharge basins. Also, a groundwater management agency was formed and placed restrictions on pumpage and the construction of new wells. A second surface water delivery system was constructed to bring additional river water to a second agricultural area, and growers also installed millions of dollars worth of high efficiency irrigation systems which caused pumpage to drop by about 40% to 123.4 million m^3 annually. As a result of these combined efforts, groundwater levels rose, recharge volumes increased markedly, and in 1996, widespread artesian conditions occurred throughout the Oxnard Plain. For the first time in decades, ground-water levels along the coast rose rapidly while associated chloride levels dropped dramatically (Fig. 5). Through 1996, approximately 3.207 billion m^3 of water has been artificially spread by United Water. Despite the immediate benefits to groundwater replenishment from the construction of Santa Felicia Dam and the Lower River Project, the overdraft was not eliminated, and seawater intrusion continued to move further inland. Hydrology studies estimated that the overdraft was about 30.8 million m^3 per year and that additional water supply projects were needed. The State of California also became concerned over the continued overdraft and encouraged local water officials to solve the problem. Accordingly, plans were developed by the County of Ventura and United Water to construct a permanent diversion (called the Freeman Diversion) across the river near the entrance to the Oxnard Plain; establish a groundwater management agency, now known as the Fox Canyon Groundwater Management Agency, (GMA); and construct a surface water delivery pipeline to a 2428 ha agricultural area which was experiencing heavy pumping and created a localized pumping trough.

2.1 Freeman Diversion

Literally, within a few days after completion of the Freeman Diversion in February, 1991, a prolonged rainstorm with significant runoff occurred and marked the beginning of an above-normal rainfall/runoff period which has continued through the Spring of 1998. As a result, artificial replenishment was increased significantly, and groundwater levels throughout the district rose.

2.2 Groundwater Management Agency

At the same time, the efforts of the Fox Canyon GMA to reduce pumpage and increase an awareness of water conservation resulted in a dramatic drop in agricultural water consumption from about 197.3 to 123.3 million m^3 per year.

2.3 Agricultural surface water supply

Also, the newly constructed pipeline, known as the Pumping Trough Pipeline or PTP, began supplying

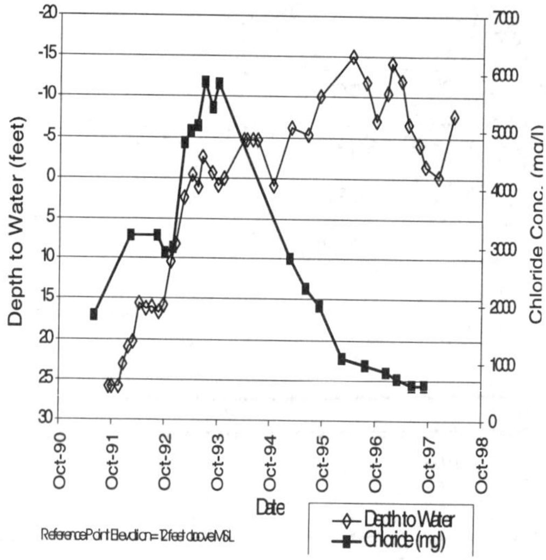

Figure 5. Monitoring Well A1-195 at Port Hueneme.

an average of about 10.5 million m^3 annually to the agricultural area. With the implementation of the three projects, chloride levels began to drop dramatically within near-coastal wells as shown in Figure 5.

Subsequent mathematical modeling of the groundwater under the Oxnard Plain confirmed that the overdraft in the Upper Oxnard Aquifer was probably eliminated and that in fact, a surplus groundwater supply was in existence. One of the final elements to be constructed in Phase II was the acquisition of a 57.5 ha shallow gravel pit adjacent to the Saticoy Spreading Grounds. This feature was incorporated into the operation with the purpose of storing large quantities of water when the two nearby spreading grounds became saturated and replenishment rates dropped.

Nonetheless, water well soundings and the math-ematical model confirmed that a residual overdraft with encroaching seawater was occurring in the vicinity of Pt. Mugu, in the southwest corner of the District, and demonstrated the need for additional supply projects discussed in Phase III below.

3 PHASE III

Finally, to overcome residual seawater intrusion and lower aquifer overdraft, United Water has identified several new projects to bring river water to agricultural areas near the coast which still rely on groundwater. This surface delivery would allow the aquifer systems to repressurize and push out the remaining seawater. However, lateral migration of intruded seawater, parallel to the coastline, may preclude the rejection of all the seawater. Also as part of this, neighboring agencies are constructing aquifer storage and retrieval projects; considering reuse and injection of highly treated wastewater, and construction of a diversion dam to bring river water from nearby drainage basin. The projects are depicted in Figure 6, and summarized below.

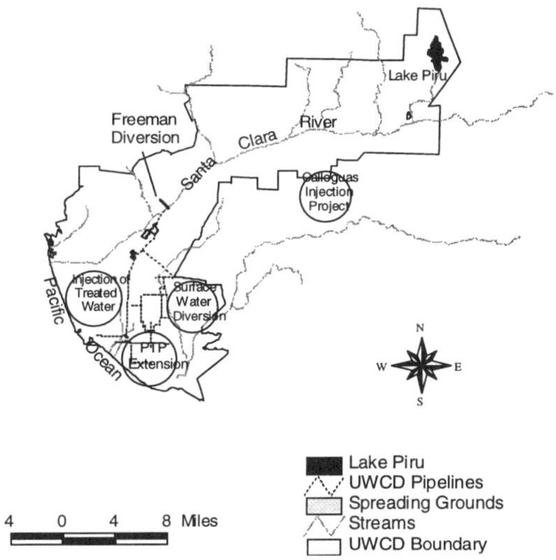

Figure 6. Seawater intrusion Reversal Project.

3.1 PTP extension

As shown in Figure 6, the remaining unresolved seawater intrusion area in the Fox Canyon Aquifer underlies the area around Pt. Mugu, California, USA in the southwest corner of the District. Although this area is distant from United's surface water supplies, plans are underway to extend the Pumping-Trough-Pipeline (PTP) southward to provide surface water from the Santa Clara River to the individuals and companies that pump from the lower aquifer. Simply stated, if about 6.2 million m^3 can be provided, then the overdraft of several thousand acre-feet not only will be overcome, but additional supplies will enable the basin to more quickly repressurize, recover, and perhaps push out the sea-water.

3.2 Major injection project

The aquifer storage and recovery project is being undertaken by United's neighboring water district, Calleguas Municipal Water District, (CMWD). Engineers from Calleguas have begun construction of a project which will store up to 370 million m^3 of water in the Las Posas Basin which among other benefits will reduce pumping near the coast during drought periods, and thus further reduce the long-term stress on the Fox Canyon and other basins.

3.3 Injection of treated wastewater

The City of Oxnard is contemplating a project to inject the Oxnard Plain, up to 24.7 million m^3 of highly treated waste water. Plans are being developed to inject water into the Fox Canyon Basin as part of a pilot project.

3.4 Surface water diversion

Lastly, Pleasant Valley County Water District is working with CMWD to divert several thousand acre-feet of tertiary treated waste water from the neighboring City of Thousand Oaks, California, USA via Calleguas Creek for use in their agricultural area. As with the above project, this effort will also reduce pumping in the Fox Canyon Aquifer.

4 CONCLUSIONS

The successes in overcoming overdraft and seawater intrusion has been a 50 year struggle from identification of seawater intrusion in the 1940's, gathering public support and financing, and implementing a comprehensive set of projects. These projects involved the introduction of intense water conservation methods; use of treated waste water; aquifer injection-recovery projects; and construction of sur-face water diversions and delivery systems. Clearly this water conservation success can be attributed to multi-year patience, strong local support, pro-active management and adequate water resources to harvest.

Artificial Recharge of Groundwater, Peters et al. (eds) © 1998 Balkema, Rotterdam, ISBN 90 5809 017 5

Sustainable groundwater management using artificial recharge in the Paris region

H. Haeffner
Suez-Lyonnaise des Eaux, CIRSEE, Le Pecq, France

M. Detay
Lyonnaise des Eaux Philippines, Ortigas Center, Metro Manila, Philippines

J.L. Bersillon
Suez-Lyonnaise des Eaux, CIRSEE, LEM ENSG, Vandoeuvre, France

ABSTRACT : Through two field examples, the authors present several results on artificial recharge basins, based on state of the art informations describing site procedures, experiments carried out and numerical approaches. A very precise monitoring of infiltration capacity has been set up to optimize recharge cycles. With other analysis, it allowed to assess the role of algae in clogging processes and removal of nutrients. Seasonal variations of clogging and of quality changes in an artificial recharge system are discussed.

1 INTRODUCTION

Lyonnaise des Eaux has used artificial recharge (AR) techniques for more than 35 years in the Paris area to achieve sustainability in overdrawn alluvial aquifers. Artificial recharge applies to the two largest groundwater treatment plants supplying West of Paris area : Croissy-sur-Seine (300,000 m^3/day) and Flins-Aubergenville (150,000 m^3/day). The recharge process has been in operation in Croissy-sur-Seine since 1959, allowing a large number of quality, quantity and ecological surveys to be carried out.

Surface water is drawn from the Seine river and undergoes pretreatment including clarification, settling and filtration. It is then sent to 15 hectares (37 acres) of infiltration basins where it undergoes cascade aeration before gradually reaching the aquifer. Infiltration rates range from 0,1 to 3 m/day. Altogether 54 million m^3 reach the aquifer every year.

2 OPERATION OF AR BASINS, MONITORING OF INFILTRATION CAPACITY

2.1 *Site procedures and monitoring methods*

Maintenance of infiltration basins is necessary to avoid clogging. Flooding and drying periods are used with cleaning of basins surface during dry periods. Cleaning consists in drying the basin by closing water inflow and waiting approximatively 1 week, removing the organic layer of macrophytes and algae when necessary and harrowing the top 30 cm of the soil to break the clogged structure. Flooding periods range from 5 weeks to 1 year and have been optimized to increase the total volume infiltrated by each basin (Detay 1994).

Recent works using continuous logging of inflow rates and water level in basins allowed precise real-time measurement of "infiltration capacity" for each basin. Infiltration rate (in m/day) is easily calculated by estimating surface and volume of the basin according to eq (1) and (2) and calculating over a short period of time [t1,t2] the infiltrated volume $Vi_{[t1,t2]}$ and obtaining infiltration rate $IR_{[t1,t2]}$ according eq (3) and (4)

$$S = S_0 + psH \tag{1}$$

$$V = S_0 H + \frac{psH^2}{2} \tag{2}$$

Where H = measured head in the basin (average of head during [t1,t2] period), p = perimeter of the basin, S_0 = Surface of the basin when H=0, s = slope coefficient of basin border

$$Vi_{[t1,t2]} = V(t1) - V(t2) + Q_{[t1,t2]} \tag{3}$$

$$IR_{[t1,t2]} = \frac{Vi_{[t1,t2]}}{S(t2-t1)} \tag{4}$$

Where $Q_{[t1,t2]}$= injected volume of water during period of time [t1,t2].

IR depends basically of applied head H in the basin and of soil permeability which is decreasing with time due to clogging. Theoretical calculation of infiltration rate requires resolution of Richards equation as flow is unsaturated between the basin and the aquifer. A simple approach consists in using Darcy's law to calculate infiltration rate, considering an equivalent system (Fig. 1) that is controlled by a clogged layer at a depth H_0 under the basin according to equation (5)

$$IR = IC \times \left(H + H_0\right) \tag{5}$$

IC is the increase in infiltration rate per meter of head in the basin. In the paper IC is referred as "infiltration capacity". It is independent from head variations in the basin and is representative of clogging in the basin. H_0 can be determined graphically by plotting IR and H with different values of H on a short period of time. H_0 is in general between 0,2 and 1 m under the bottom of the basin. Then "infiltration capacity" IC is calculated with eq (5).

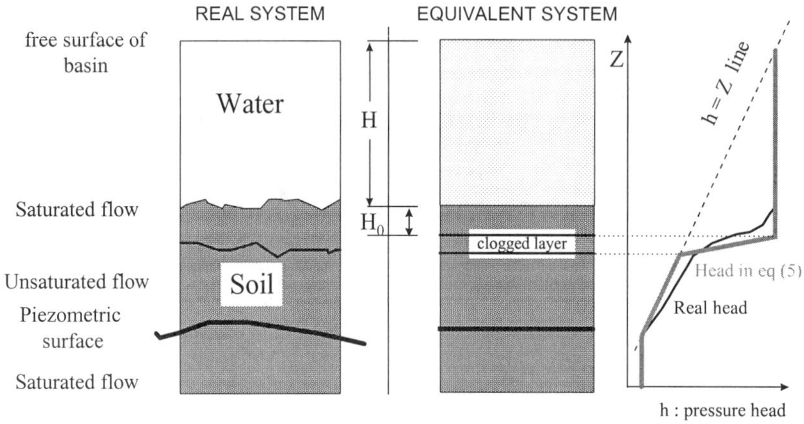

Figure 1. Considering an equivalent system to calculate infiltration rate according to equation (5).

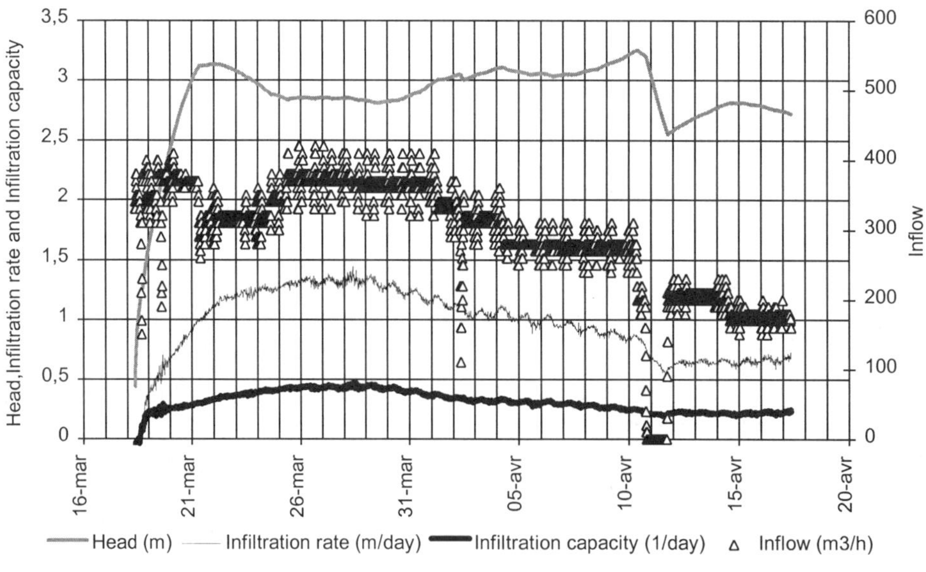

Figure 2. Example of continuous logging.

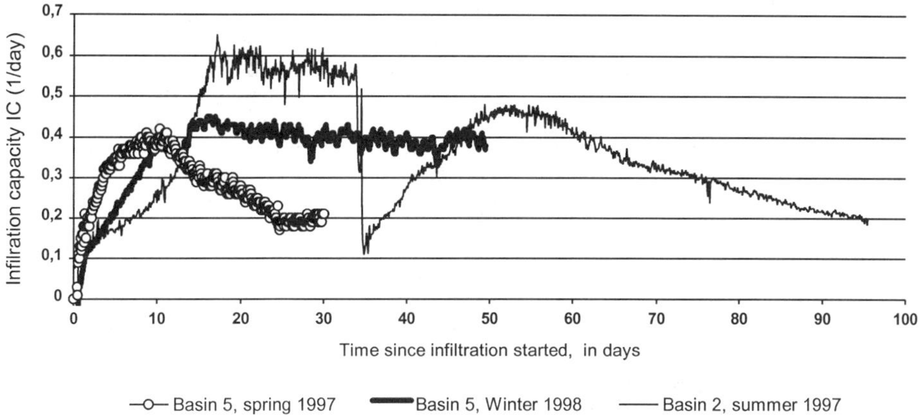

Figure 3. Three selected examples of calculations of infiltration capacity.

2.2 *Results and discussion*

Figure 2 shows a continuous logging of H and inflow rate in a basin and corresponding calculations of IR and IC. Infiltration capacity is independent from head but not infiltration rate. Infiltration capacity has the dimension of the inverse of a duration (day^{-1}). If Ho is neglected, 1/IC is equal to residence time of the free water in the basin.

Figure 3 shows several examples of calculations of infiltration capacity. For all basins, degasing of unsaturated soil is responsible for a first phase of low infiltration rate at the beginning of recharge lasting between 1 to 4 weeks. For basin 2, a short drying of the basin around day 34 resulted in a new degasing phase lasting about 18 days. As no treatment of the basin was performed, clogging continued to occur and initial infiltration capacity was not recovered after the second degasing. For basin 5, the infiltration capacity immediately after degasing is around 0,42 day^{-1} for both cycles. In the cycle of spring 1997, important clogging occurs reducing IC of more than 50 % in 25 days whereas in the cycle of winter 1998, almost no clogging happens (reduction of IC is less than 15 % in 50 days).

2.3 *Cycle time optimization*

The recharge basins are used with a variable height comprised between 0 and 2-4 m of water. Contrarily to shallow or constant level basins, the infiltration flow in each basin is not only controlled by clogging of the basin but also by the head of water applied.

Under the assumption that basin clogging is depending of time and not of infiltrated volume (which is generally valid for biological and chemical clogging and not for physical clogging by suspended matter), the maximum average infiltrated volume is obtained by operating a basin at its maximum head (H_m) and optimizing cycle time as described below in a method already described as AARC method (Dickenson 1994) :

Assuming a drying time d and flooding cycle time CT (in days), average infiltrated water per surface unit and per day $\overline{IR}$ over a whole cycle is calculated according to eq (6)

$$\overline{IR} = \overline{IC} \times (H_m + H_0)$$

$$\overline{IR} = \left(\frac{\sum_{t=0}^{t=CT} IC(t) \times \Delta t}{CT + d} \right) (H_m + H_0) \qquad (6)$$

An optimal cycle time CT is calculated with a spreadsheet from IC data and presented graphically. An example of optimal CT is given in figure 4 for basin 5 in spring 1997. Optimal cycle time in figure 4 is 22 days of infiltration if 7 days of drying time are necessary.

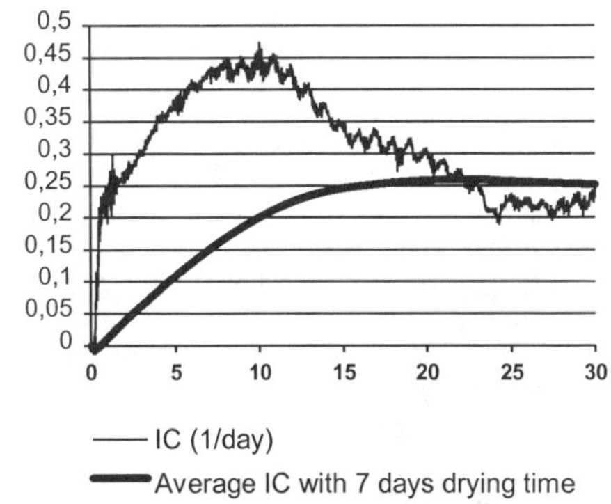

Figure 4. Optimal cycle time calculation.

Heavy clogging and short drying time (7 days) result in short cycle time and high maintenance costs. In general, longer cycle times are used (5 weeks to 1 year) as drying time is longer than 1 week and clogging is not as severe as in Figure 4. As noted by Dickenson, cycle time can be increased by a few percent past optimal time without losing much efficiency of the infiltration basin.

3 CLOGGING PHENOMENA, QUALITY CHANGES OF WATER

The artificial recharge water is containing in average around 1 mg/l of total suspended solids (see Table 1). This ensures little mechanical clogging compared to other litterature data.

Table 1. Average quality data for AR Water on some relevant parameters at the two recharge sites.

AR water concentrations	le Pecq-Croissy	Flins-Aubergenville
Total susp. solids mg/l	1,1	0,55
Turbidity NTU	0,41	0,24
Calcium mg/l Ca	93	98
Nitrate mg/l NO_3	24	29
Ammonium mg/l NH_4	0,08	0,67
Orthophosphate µg/l P	32,5	151
Conductivity µS	496	559

Sediments of AR basins have been collected in Le Pecq-Croissy using a permeable filter placed on the bottom of the basin during a recharge cycle. This allowed to separate settling particles during recharge from existing sand in the basin. Chemical analysis of this sediment showed that it is mainly constituted of calcite and also of silica.

Table 2. Composition of AR sediments in Le Pecq Croissy.

	Sample 1	Sample 2
Calcite ($CaCO_3$)	56 %	63 %
Silica (SiO_2)	23 %	22 %
Organic matter (in CH_2O)	19 %	10 %
Other	4 %	4 %
Total	102 %	99 %

Electron microscopy of basin sediments showed that calcite crystals are forming a very fine grained cement around larger particles (shells of small mussels, sand grains and diatoms). As algae growth is responsible for change in carbonate equilibrium and precipitation of calcite, the deposited calcite is believed to be precipitated in the water column when photosynthesis occurs.

Considering that organic matter, calcite and part of the silica (diatoms) is derived from biological activity in the basin, mechanical clogging is believed to be negligible compared to biological clogging (algae and bacteria) and chemical clogging (precipitation of calcite and diatoms). Additional evidence of the leading role of algae in clogging is the difference in infiltration capacity between winter and spring or summer conditions without significant change in suspendend matter in the recharge water. For recharge basin S5 in Flins-Aubergenvile, infiltration capacity is almost constant in winter whereas in spring or summer a decrease of more than 50 % can occur in several weeks (Fig. 3). Other authors (Schuh 1990) had already observed changes in clogging phenomena between seasons but without being able to separate mechanical clogging and chemical clogging due to algae.

For rapid infiltrating basins (1-3 m/day), quality changes for dissolved species are small between pretreated water and aquifer water (only nitrification, aerobic respiration of biodegradable carbon and oxygen depletion are significantly observed). Acidification of pretreated water to prevent calcite precipitation is considered in ongoing studies as a new technique to reduce clogging associated with algae growth and lenghten flooding cycles.

For slow infiltrating basins (0,05 - 0,5 m/day), uptake by algae during spring and summer is responsible for an important decrease in concentrations of silica, orthophosphate, total inorganic nitrogen and indirectly of calcium (Fig. 5, 6). In spring and summer, orthophosphate has been identified as the limiting factor for algae growth as orthophosphate is below detection limit. (Fig. 6d)

In water well P1, which receives almost only water from S8 basin, nitrogen removal is in average 1,9 mg/l N (Fig. 6a). In summer, this removal is less than nitrogen uptake by algae in the basin. Aerobic respiration and release of nitrogen from dead algae is a possible explanation. In winter, uptake by algae is very small as little orthophosphate is available in recharge water and temperature is low. The removal

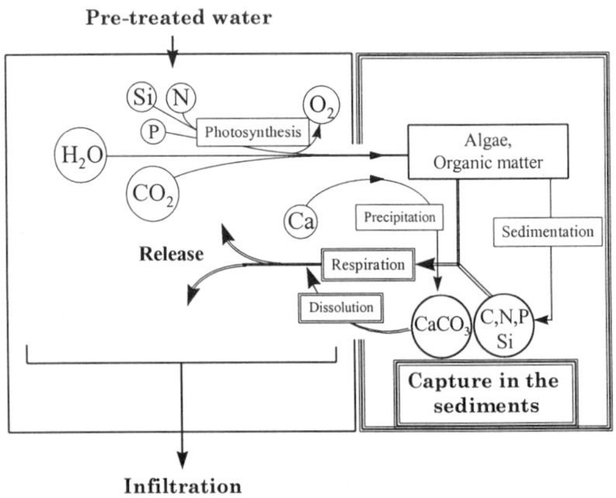

Figure 5. Element cycling in AR basins.

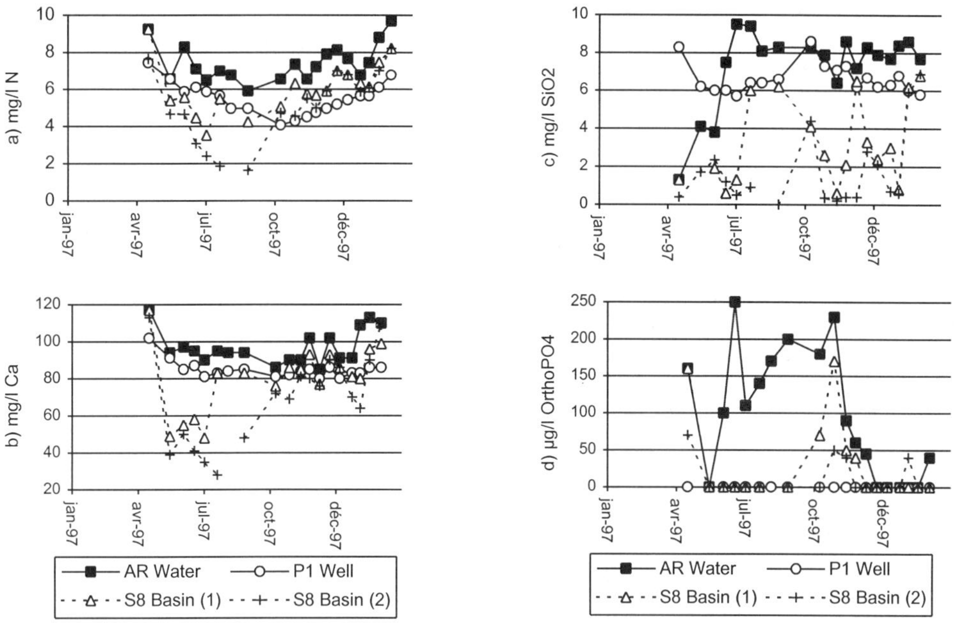

Figure 6 a, b, c, d. Evolution of water quality due to activity of algae in a slow infiltrating basin. (Basin S8, IR = 0,1m/day). P1 Well is strongly influenced by S8 infiltration basin. S8 (1) and (2) are sampling points in the basin.

of nitrogen in the basin that is observed can be the result of denitrification in the basin. Organic matter for this denitrification is available due to settling of dead algae and macrophytes in the basin. Figure 5 shows element cycling in AR Basins.

Activity of algae has a significant impact on water quantity and quality in AR basins. A good level of understanding of biological phenomena is important to operate infiltration basins properly, to take advantage of natural removal of elements and to maintain high infiltration capacity.

Other results dealing with trace metals, pathogens (virus, bacteria and parasites) and pesticides are available on the two recharge sites. A good removal of metals and pathogens has been observed because of elimination of suspended matter in the pretreatment and also of biological elimination in the infiltration to the aquifer. Sand of recharge basins has no significant enrichment in trace metals due to recharge water. Pesticides are not eliminated unless powder activated carbon is added to the pre-treatment.

4 ACTIVE GROUNDWATER MANAGEMENT

AR is a major tool for active groundwater management. Based on hydrogeologic modelling and linear programming, an optimization tool has been developed to determine monthly volumes to be infiltrated in each basin as well as cost effective pumping strategy for each well in order to face peak demand throughout the year. An example of optimized use of AR is given in Figure 7.

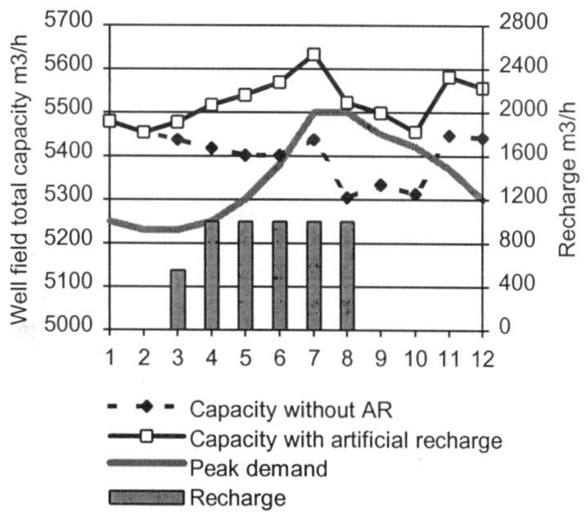

Figure 7. Example of output from the aquifer management tool developed to optimize the use of artificial recharge in Flins-Aubergenville. Optimal recharge is calculated to face peak demand.

5 CONCLUSION

Artificial Recharge is an indispensable technique for water resources management in the Paris region. It ensures both sustainable water quantity and quality in overdrawn alluvial aquifers where bank filtered water is limited in quantity and quality. In the aquifers of Flins Aubergenville and Croissy sur Seine, high ammonia concentrations from bank filtration water have been prevented in the aquifer by using artificial recharge. Accurate monitoring of infiltration capacity allow a perspective for long term studies on recharge capacity (loss or gain of infiltration capacity, assessment of maintenance techniques for the basins). Nutrient removal in artificial recharge is also a key issue in aquifers with high inorganic nitrogen content as it is a cost effective mean of reducing pollution.

ACKNOWLEDGEMENTS

Agence de l'Eau Seine-Normandie has brought financial support for part of this work.

REFERENCES

Detay, M. 1994, Rational groundwater reservoir management : the role of artificial recharge. *Artificial recharge of Groundwater II - Proc. of 2nd inter-national symposium on artificial recharge of groundwater. 17-22 july 1994* 231-240. American society of Civil engineeers, New York, New York 10017-2398

Dickenson, J.M. & Bachman S.B. 1994, The optimization of spreading ground operations. *Artificial recharge of Groundwater II - Proc. of 2nd international sym-posium on artificial recharge of groundwater. 17-22 july 1994* 630-639. American society of Civil engeneeers, New York, New York 10017-2398

Schuh, W.M. 1990. Seasonal variation of clogging of an artificial recharge basin in a northern climate. *Journal of Hydrology* 121:193-215

Artificial Recharge of Groundwater, Peters et al. (eds) © 1998 Balkema, Rotterdam, ISBN 90 5809 017 5

Dynamic simulation model for water management of a large-scale artificial recharge system

M.J.M.Mosch
Amsterdam Water Supply, Netherlands

ABSTRACT: The artificial recharge and recovery system in the Amsterdam Water Supply dune area is complex due to the large number of recharge ponds (40) and recovery canals (about 20). Next to this there is a large drainage system (9 km) and about 15 pumping stations. The management of this system is subject to many ecological, hydrological and water management boundary conditions, which all have to be satisfied simultaneously. Generally, it is difficult to predict the behaviour of this system as an entity, as well as the effects of changes of the system. To improve the management of this system, a dynamic continuous simulation model was made using the MATLAB/SIMULINK©® software. The most important components of the system are included as linked modules. This paper describes the model and its use to increase the insight into the behaviour of the artificial recharge and recovery system. It will be shown how the model is continuously updated with actual data and how it is applied in a weekly schedule.

1 INTRODUCTION

The Amsterdam Water Supply (AWS) has two production plants, in which about a 100 million m^3 of drinking water is produced each year. The plant located in Vogelenzang (Production West) produces about 70 million m^3/year. The purification to drinking water from river water can be divided into three logical steps. At first, prepurification after intake of the water from the Lek canal (rapid sand filtration and coagulation) and transport to the recharge area. Secondly, artificial recharge in the dune area (the AWD) regaining this water after two or three months and transport to the purification plant. Thirdly, final purification: aeration, rapid sand filtration, ozonisation, softening, activated carbon filtration, slow sand filtration. Finally the water is transported to the customers.

The provincial authorities required an extended study in order to optimise the competitive objectives of nature conservation and drinking water production. The AWS also had to give a description of the water management written down into a plan, which includes the actions and measurements to be taken under a variety of possible circumstances. With these tasks in mind (1/ to optimise the complex system, 2/ writing a working plan), it was the choice of the AWS to develop a simulation model of the entire artificial recharge and recovery system.

This article is about the computer model which has been developed in order to manage the artificial recharge and recovery system. The model was developed by Amsterdam Water Supply and written in the software MATLAB/SIMULINK (The Mathworks 1996).

2 THE RECHARGE SYSTEM

The recharge system of AWS is a very old one. The first recovery canal was dug in 1853. At the beginning of the 20th century deep wells were made to satisfy the increasing demand for drinking water. In 1957 the system was enlarged with an artificial recharge system. The prepurified river water is transported to the AWD, where it flows by open canals to the recharge areas. In the recharge area the water infiltrates into the sandy soil. The recharge system, of about 975 hectares, comprises 40 artificial recharge ponds and a drainage system with a length of 9 km. The discharge from this recharge area happens with the mentioned drainage system and with the open air recovery canals, all facilitated by gravity. There are about 20 recovery canals. About 240 deep wells are present, which can produce upto 4000 m^3/hour, half the production capacity. Natural dune water is pumped from two independent canals. In 1997 63 million m^3 water was infiltrated in the dunes to produce about 68 million m^3.

The main functions of the system are : 1. levelling quality fluctuations of the intake water by difference in detention time; 2. storage of water for periods in which the sources are out of use (maintenance or temporarily exceeding of standards. The system has a storage of at least 2 months); 3. natural purification and degradation of human viruses, bacteria and protozoa.

3 MANAGEMENT

The applied water management has qualitative and quantitative objectives. Quantity is managed by daily adjustments of weirs, pumping stations, deep wells and valves. Each canal has his own preferred water level. We have introduced a dependency to seasons for the most important preference levels in order to fulfil the ecological objectives. The yearly use of deep wells and certain recovery canals is limited in quantity by the permit.

The quality boundaries of the system output are derived from the drinking water standards. For example we guarantee that the concentrations of chloride and sodium do not exceed 150, respective 120 mg/l.

With respect to the quality boundaries of the artificial recharge water we take account of the Governmental recharge resolution. This resolution has been set up to protect the soil from pollution.

Quality management is carried out by selective intake of river water (water intake when chloride concentration is low) and the use of the different water quality sources within the system (natural dune water, deep ground water, river water).

4 THE MODEL

In order to model the system we make a distinction between the most important hydrological and functional parts of it. These different parts are: infiltration areas with drains, recovery canals and recharge ponds, precipitation areas from where natural dune water flows to the canals, pumping stations, storage canals, the collection reservoir and the precipitation-evaporation model. The infiltration areas Ia, Ib, II, IIIa, IIIb and IV (Fig. 1) can be seen as autonomous subsystems. The storage canal in infiltration area V makes this area dependent of all other systems. The water balance of the system as a whole is a sum of all autonomous system parts (Q_{asp}), minus the systems output (prescribed discharge at the collection reservoir Q_{so}) plus the change in volume of the collection reservoir (Q_{cr}) and the change in volume of the storage canals (Q_{sc}); The formula of the balance is:

$$Q_{sc} = Q_{asp} - Q_{so} + Q_{cr}$$

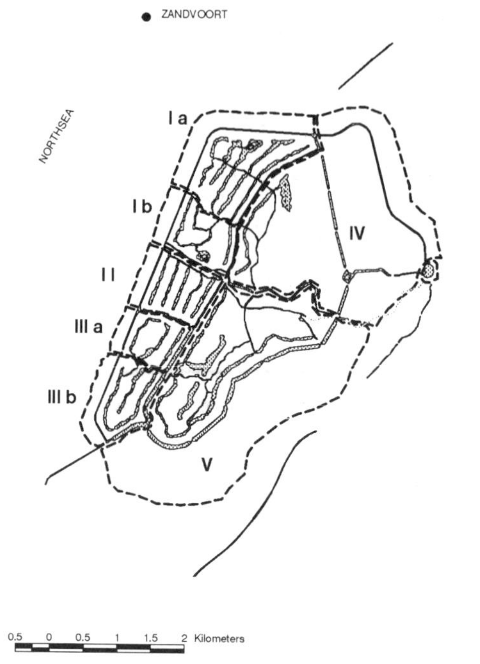

Figure 1. Overview of the Amsterdam Water Supply dune area, with its seven subsystems.

The residual part of the water balance, in the model and in reality, will be discharged from the storage canals.

The functions of the developed decision supporting water management model (BVM) are 1. Decision support in daily water management, this is comprised of calculation of the different options in water management for the next week(s) in order to optimise the management. 2. System analysis, during and after a calamity the model must give answers to management questions and allow analysis to be made. 3. Provide an up-to-date overview of quality and quantity of the AWD.

Functionally this means the model has to calculate quantity variables (discharges and water levels) and the water quality variables (chloride, hydrocarbonate) in the different parts of the system. In order to fulfil these requirements we have chosen to build a reservoir oriented model. This means the model is build with reservoirs (Fig. 2). The model consists of a large number (170) of connected reservoirs. In each of them the water and mass balance are dynamically computed. In the next subparagraphs some interesting and characteristic parts of the model will be presented.

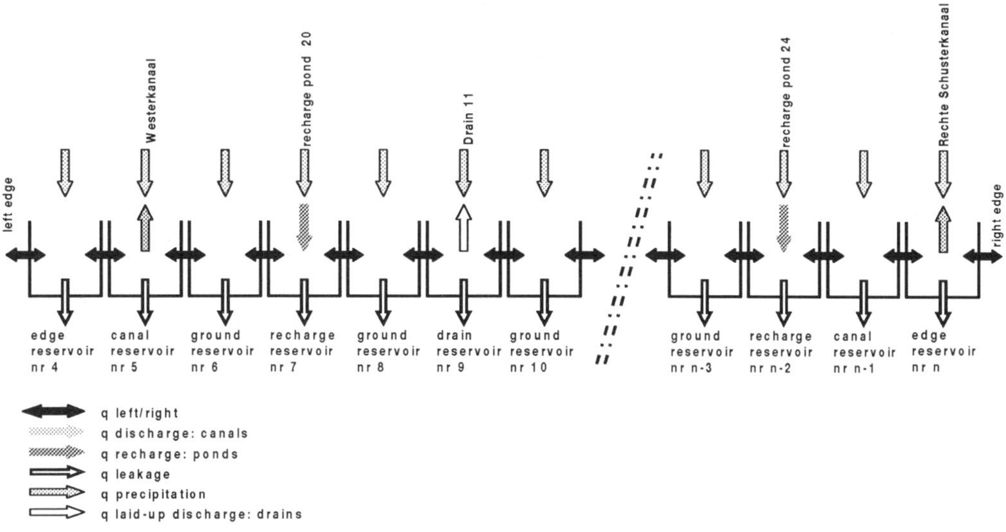

Figure 2. West- east cross-section of the recharge area II, modelled with reservoirs.

4.1 *Target function*

To optimise interaction with the real system, a target function is computed after each simulation, which contains the weighted values of the various ecological, hydrological and management objectives. The target function measures the difference between the actual situation (according to the model) and the preferred situation.

The target function is to be minimised in order to optimally fulfil the total set of objectives in this function.

The target function value (Tfv) reads

$$\overline{Tfv} = \sum \overline{Pf} \times \left| \frac{\overline{Sp} - \overline{Mr}}{\overline{Ss}} \right| \qquad (1)$$

Where Sp are the setpoints, Mr are the calculated values and Ss are the standard spreads. In the target function the normalised deviation is calculated by taking the standard spreads (Ss) into account. This happens for every element. The deviations are weighted one to each other by priority factor (Pf). The sum of these priority factors is chosen to be one.

For each element a value less than unity means that the deviation between preferred and calculated value is less than the standard. The target function is used to compare calculated scenarios with each other.

4.2 *Precipitation discharge modelling*

Precipitation, evapotranspiration, discharge and open water evaporation are important processes in the system and in its model. About 15% of the system output is natural dune water. During calibration of the model, it became clear that the precipitation-discharge relation had to be separated into two relations (Fig. 3).

The first relation is precipitation minus open water evaporation. It is used for open water parts of the model.

The second relation is more complex. For this relation a special module was developed. It holds the relation of precipitation, percolation (transport through the unsaturated zone), and discharge. It takes care of a slow changing season depending flow and is superimposed to the groundwater reservoirs of the model.

4.3 *Calibration*

The calibration was worked out by changing most relevant parameters like reservoir bottom level, permeability, porosity, hydraulic resistance in artificial recharge ponds and canals.

Calibrating the model, several system specific facts were encountered.

First, it became clear that several time series did have an inaccuracy, which made it impossible to use them. The measured drain extraction from the

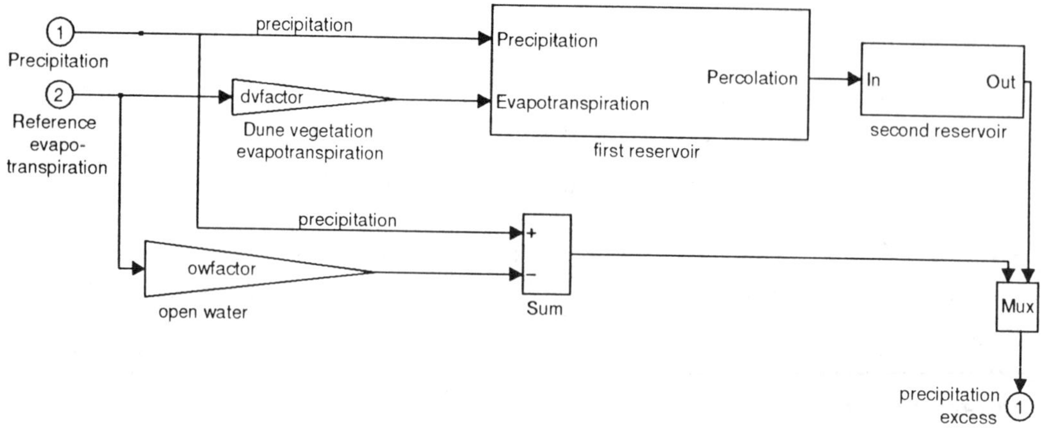

Figure 3. Simulink scheme of the used precipitation-discharge relations.

artificial recharge areas and the measured amount of recharge water lead to unacceptable water levels, which could not be reached by acceptable changes in parameters values. The main problem being that 40% of the discharge and 100% of the recharge to the hydrological subsystem was superimposed. This problem was solved in two steps. Step one: the superimposed discharge from the drains are regulated in relation with model-calculated (ground)water levels using fuzzy logic controllers (Fig. 4).

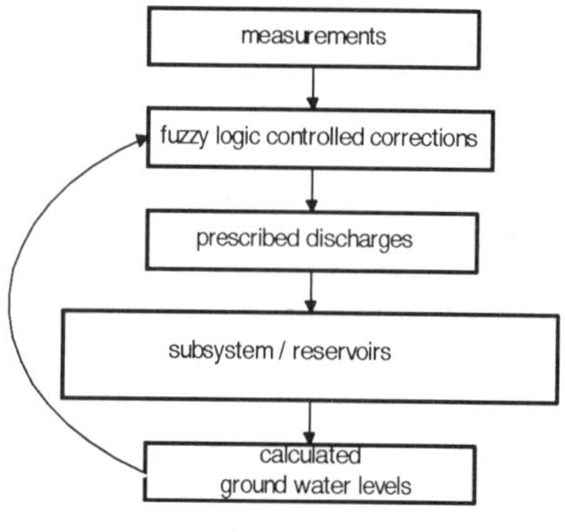

Figure 4. Measurements are used but the uncertainty in it is taken in account.

So if measurements were wrong for some periods, (ground)water levels do not exceed the maximum and minimum allowable levels. Fuzzy logic control allows a less rigid use of measured values.
Step two: during calibration the amount of recharge water was calculated using the measured water levels instead of the measured flow rates. We minimised the error between the measured and the calculated values by changing the temperature depending entry resistance of the recharge ponds.

5 PRACTICAL USE

In normal situations, the model is used weekly to support the current management (decision support). This is done by monitoring the actual situation and by predicting the consequences of operational actions for the next two or three weeks. With the model several operational values are calculated, like recharge reservoir levels, drain-discharges, target-function values, groundwater levels, concentrations of chloride, retention times and abstracted natural dune water.

In exceptional situations the model is used mainly to predict long term (one a two month) consequences at the decision points.
So when a problem occurs and the model has to be run, it has to be ready for use.

Important questions are: How can we keep, as long as possible, the most important (ground)water levels at preferred level? Must we use deep groundwater? What are the consequences for the system output? Results of the calculations are discussed with the nature department.

Predictions may start at the stopping time of the last run or earlier, i.e. at any moment for which model states were stored in previous runs. To be able to use the model quickly under exceptional situations, and in weekly management it is necessary to update the model states frequently. We have chosen to update them every week, which means we have to

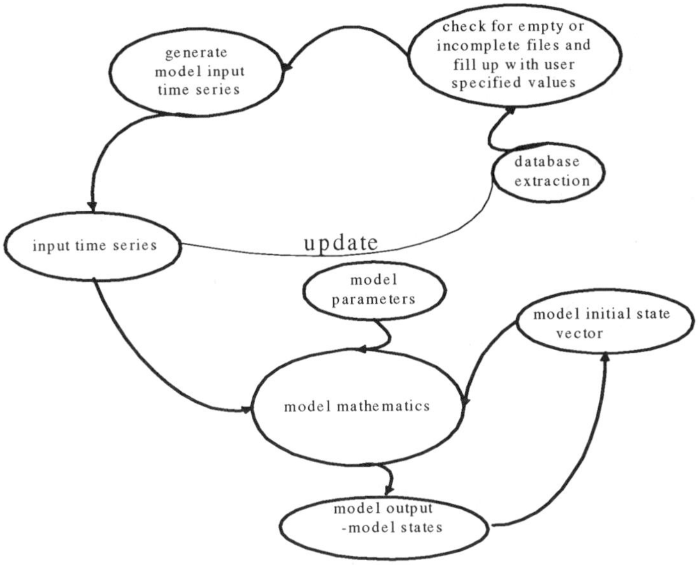

Figure 5. To run the model it needs input time series, model parameter values and model initial conditions. The input time series can be automatically updated.

update the model input with the measurements of last week.

The model input consists of time series of water flows with their concentrations of chloride. It further consists of precipitation and evaporation, water levels and water temperature. These values are measured weekly (or several times a week) and are stored in the central database of the production plant. After extraction, these raw data are processed to model input and combined with earlier time series into single time series (Fig. 5). Depending on the case, the raw data needs interpolation, gap filling, averaging with other data etc. The last step of updating is appending the database update files to the already existing input files and to run the model for at least the update period, in order to create the new model state vectors. The update of model input from the database and the calculation of the model states has been automated completely.

With the model it is possible to predict the consequences of certain management actions. In order to do this, the model must be run with user-specified input. Also certain "prediction scenarios" have to be set up. External boundary conditions like the demand of water, the amount of precipitation, and evaporation are important parts of these scenarios. The use of internal "system" boundary conditions is partly guided by user instructions, which are laid down in a working plan.

With the automated update routine it is also possible to create model input files with generated values in the future. To generate such future values the user has to choose from: 1. mean values of previous time series, 2. constant user specified values, 3. same values as last year, or 4. continuation of the last measured value. In general the user specifies for any time series with lacking data, by which method these must be replenished.

To run the model the user selects the start date from a list of state vectors stored during previous model runs and enters a stop date. He or she further chooses from several run options. One of them is to let free or to superimpose the amount of artificial recharge water to the different areas. Other calculation options are: Use setpoints strictly or adjust them depending on other system values. Calculate drain productions in relation with (ground)water levels or use superimposed time series. With respect to these time series the user must choose between the use of measurements or use of preferred discharges.

6 REMARKS

The model described in this paper is the third version. The first version had only three reservoirs, the second about twenty. With this version we think we reached the accuracy we wanted. With the total amount of 170 reservoirs and about 1000 model parameters the calibration becomes the limiting factor. Calibration is only possible when enough measurement values are available. In this case the most important measurements were water levels and chloride concentrations. We completed these time series ever since 1984.

A built graphical user interface, a GUI, makes it possible to handle the large amount of model input and output.

REFERENCES

Dijk, H.J.A.A. 1996. Infiltratiegebieden Amsterdamse Waterleidingduinen in MLAEM. (Artificial recharge in the dune area of Amsterdam Water Supply, modelled in MLAEM, in Dutch). Amsterdam Water Supply.

Geenen, A. 1994. Voortgangsnotitie Bedrijfsvoeringsmodel (Report on the progress Water Management Model, in Dutch). Amsterdam Water Supply.

Kamps, P.T.W.J. & M.J.M. Mosch 1995. Berekeningen t.b.v. het bedrijfsplan met het AMWADU model (Computations for the Water Management Model using the Groundwater Model AMWADU, in Dutch). Amsterdam Water Supply.

Mosch, M.J.M., J.H.N. Moorman & T.N. Olsthoorn 1993. Definitiestudie Bedrijfsvoeringsmodel (Definition study Water Management Model, in Dutch). Amsterdam Water Supply.

Mosch, M.J.M. 1996. Beschrijving van het BVM infiltratie en winning. Bedrijfsplan voor infiltratie en winning in de Amsterdamse waterleidingduinen. (Description of the artificial recharge and recovery Management Model in the dune area, in Dutch). Amsterdam Water Supply.

Olsthoorn, T.N. & M.J.M. Mosch 1994. Rapportage van enkele onderdelen van het Bedrijfsvoeringsmodel (Report of some characteristic parts of the Water Management Model, in Dutch). Amsterdam Water Supply.

The Mathworks, 1996. *Matlab, The language of technical computing*, version 5.

The Mathworks, 1996. *Simulink, Dynamic system simulation for Matlab*, version 2.

Artificial Recharge of Groundwater, Peters et al. (eds) © 1998 Balkema, Rotterdam, ISBN 90 5809 017 5

Maaskant Recharge Project: Monitoring and operation of the recharge system

M.W.J.Eck & S.M.L.Verheijden
N.V. Waterleidingmaatschappij Oost-Brabant (Water Supply Company of Oost-Brabant), 's-Hertogenbosch, Netherlands

A.J.Vogelaar
Kiwa Research and Consultancy, Nieuwegein, Netherlands

ABSTRACT: The Maaskant Recharge Project (PIM) will be the first large-scale recharge project outside the dune area in the Netherlands to use open recharge. The recharge and recovery will be managed by means of a specially installed monitoring system. There are both hydrological and qualitative aspects to this. The design of this monitoring system is based on an information analysis of the following management tasks: permit compliance, control and research. The measurement network consists of flowmeters, observation wells and water-level gauges. The observation wells are situated on different direction lines inside and outside the recharge area, their purpose being to collect information on hydraulic heads, groundwater leakage, travelling times, changes in water quality during aquifer passage, and clogging of pond beds and recovery wells. For the automated operation, some information will be collected continuously on line. In the future, the operation will be adjusted to meet new requests, requirements and conditions. In this way, the measurement system will become an evolving and integrated part of PIM.

1 INTRODUCTION

N.V. Waterleidingmaatschappij Oost-Brabant (WOB) is a regional water supply company in the south-east of the Netherlands, in the province of Noord Brabant. Each year, WOB produces 115 Mm^3 of drinking water from groundwater. In the future, WOB will also use surface water as raw water source for the preparation of drinking water.

The provincial authority in Noord Brabant has produced a water management policy specifying that the volume of groundwater abstracted in the east of the province must be reduced before the end of the year 2000. The aim of this is to prevent any further loss of wetlands and to reverse groundwater decline in the nature reserves. The Maaskant Recharge Project (PIM) is part of this provincial policy.

PIM consists of the following parts: intake reservoir, pretreatment, recharge area, posttreatment, and drinking water pumping station (fig. 1).

The intake reservoir holds approximately 9 Mm^3 and, to prevent stratification, will be provided with two bubble aerators. In the first phase, only water from the river Meuse can be taken in. In the next phase, a second intake will be created for water from the river Waal. Lime milk will be added to the reservoir to soften the water. Pretreatment consists of the following four treatment steps: flocculation, flotation, rapid sand filtration and activated carbon filtration. This pretreatment should rid future recharged water of all impurities. The posttreatment comprises conventional groundwater treatment for the removal of iron and manganese and active carbon filtration.

The recharge area consists of a border canal, infiltration canals, pumping wells (30 in the first phase) and nature areas (Fig. 2). The total area is approximately 180 ha, of which around 20% is open water. Recharged water remains at least 60 days in the soil for disinfection. The function of the border canal is hydrological isolation of the recharge area from the

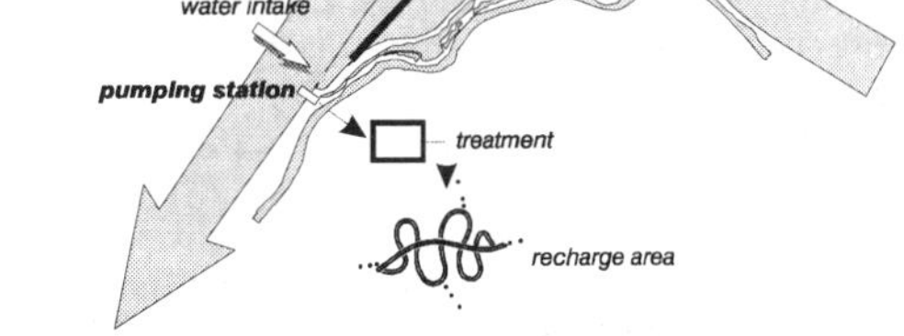

Figure 1. View of PIM showing intake from the rivers Waal and Maas (Meuse), reservoir, treatment facilities and recharge area.

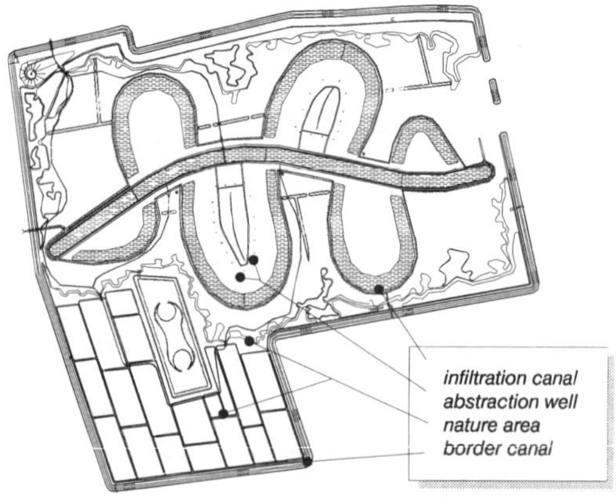

Figure 2. View of the recharge area

polder. The nature areas comprise meadows and marshes. They are being cultivated in such a way that the original vegetation, from before the time of agricultural use, can return to this river area.

There are five reasons why PIM is a unique project in the Netherlands:

1. In the final stage, PIM will be able to use two independent sources, the river Meuse and the river Waal.
2. The recharge canals are in shallow groundwater, in contrast to the open recharge systems in the dunes.
3. A clay and peat covering layer in the polder area will be confining the fully anaerobic aquifer. This contrasts with the open recharge in the (sandy) dunes, which are always partly aerobic.
4. Beginning with the design, water recovery and ecological aspects are intertwined.
5. The preconditions for drinking water production allow animals and plants to inhabit the recharge canals undisturbed.

PIM is the first large-scale recharge project in the Netherlands to use surface speading and open ponds outside the dune areas. The intended recharge area lies in a rural, agricultural polder area (approximately 2.5 m above the Amsterdam Ordnance Level (NAP)). The location is planned to be mainly in the protection area of the existing Lith groundwater pumping station. Beginning in 2000, WOB wants to produce 12.5 Mm^3 of drinking water per year in the first phase and stop the pumping of groundwater by the Lith pumping station. Eventually, the capacity of PIM will reach 50 Mm^3/a.

This paper deals with the monitoring system design for the recharge area concerning hydrological management, that is, the management of water from the pretreatment up to and including the pumping-well stage. There are quantitative and qualitative aspects to this. Although ecological goals are not discussed in this article, these preconditions arise indirectly as part of the quantitative and qualitative aspects of hydrological management.

2 MANAGEMENT AND OPERATION OF THE RECHARGE AREA

The recharge area has to be managed. Primarily, this means adherence to the terms of the permit so that there is no impact on the environment. Subject to the permit's conditions and to ecological development, the operation must run as economically as possible. This applies to management of both the groundwater and the surface water. A prerequisite for this is information. Figure 3 shows the water management information system for the project.

'External influences' in figure 3 refers to the development of WOB and relevant social changes. It is expected that these will influence the water management of PIM. Three different management tasks can be distinguished (Vogelaar et al. 1996):

- permit compliance,
- control, and
- research.

Control concerns operational management tasks, whereas research is more concerned with strategic management tasks, for example, increasing the capacity from 12.5 Mm^3/a to 25 and later to 50 Mm^3, and optimization of the process. The information needs of the three management tasks are considered in more detail below.

2.1 *Permit compliance*

In connection with hydrological isolation, information is needed on the water table in the covering clay layer and on the hydraulic head in the aquifer outside the recharge area. Water amounts must be in balance every three months, meaning the amounts of water recharged and recovered must balance. The amount and quality of surface water that flow to the surrounding polder must be monitored, and the quality of the recharge water must be safe-guarded. Finally, the spatial distribution of recharged water in the subsoil needs to be monitored.

2.2 *Control*

In general more information is needed for control than for permit compliance. The reason lies in the different time frames required for the completion of certain management tasks.

The control system for the recharge area has to satisfy a whole range of conditions. To get a good balance between the settings of the various parameters (inflow to and outflow from the recharge area), the control system receives recommendations from supporting systems. Hence, although the water amounts must be in balance every three months, an imbalance within this period is allowed. The permitted degree of imbalance is highly dependent on the time at which the imbalance occurs in the three-month period. Towards the end of the period, there is little time left for corrections, so the size of the permitted imbalance is smaller here than at the beginning.

The gathering of information on hydrological isolation must be frequent enough to prevent side-effects outside the recharge area. Hydrological isolation is necessary for more than just the groundwater permit. In the interest of operating economically, WOB does not want to see expensive pretreated recharged water leaking away.

In connection with well clogging, signals should indicate when there is a need for maintenance of the pumping wells. This has control implications for pumping-well switching. Moreover, in connection with possible clogging of infiltration canals, there must be signals that indicate this.

2.3 *Research*

PIM proceeds in phases. During the first phase, the capacity remains at 12.5 Mm^3/a, and research is conducted into the functioning of the different elements of the recharge area. Here the research questions are:
- Do the elements of the recharge area function as expected?
- Does the functionality decline?
- How do the different elements interact?
- Is fine-tuning of the different elements possible?
- What is the minimum residence time to guarantee hygienic safety, and how long is the residence time in practice?
- Does the groundwater still contain pesticides from the time of agricultural use, that is, before the setting up of the recharge area?

The change in the quality of water during the aquifer passage is a major research topic. To monitor and predict the course of this process, information is needed on the composition of the recharge water and the reactive soil elements that are present. The future hardness of the water pumped up is an important

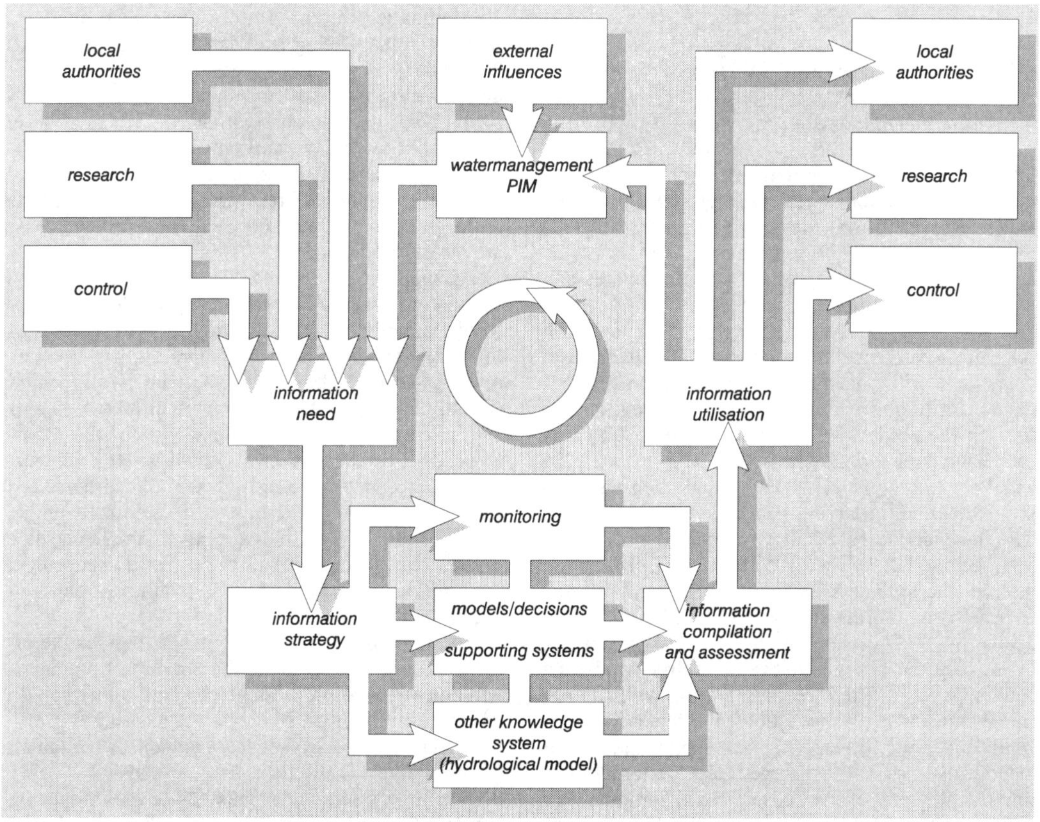

Figure 3. Water management information system for PIM (Claessen, 1997)

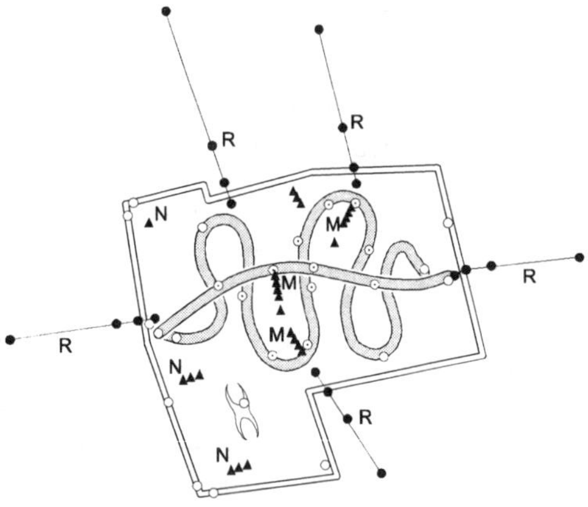

Figure 4. The measurement network in the recharge area and polder. R = polder direction line, M= meander basin direction line and N = measuring point in nature area.

parameter in this.

Research was one of the reasons for building a hydrological model, and information obtained during installation of the recharge area will be used to optimize it.

3 THE MEASUREMENT SYSTEM

3.1 *Measurement network design*

The final measurement network consists of flowmeters, observation wells (for groundwater) and water-level measuring devices (for surface water). Figure 4 shows all the measurement points of the final measurement network in the recharge area and the surrounding polder.

Of the existing observation wells, 17 have been selected as reference measurement wells, in order to show there will indeed be no impact on the environment. The 20 observation wells on the five direction lines in the polder (fig. 4, indicated by an 'R') will each have three monitoring filters at depths of 12, 25 and 40 m-below the NAP. The observation wells on these direction lines lying alongside the border canal will have additional shallow monitoring filters at 3 m below NAP. The three direction lines lying between the meanders (indicated by an 'M') will each have four monitoring filters at 12, 25, 40 and 60 m below NAP. These direction lines are meant for monitoring the changes in water quality during the aquifer passage. The three observation wells alongside the meander will have 'angled' filters lying horizontally at approximately 1 meter under the bed of the meander basin. These three filters are for monitoring the clogging of the beds of the infiltration. The observation wells in the nature areas (indicated by an 'N') will have two filters. All monitoring filters for PIM will have the same diameter, large enough for submersible pumps.

For investigating the nature and extent of well clogging, the 30 pumping wells that provide the capacity of 12.5 Mm^3/a will have at least four monitoring filters. Four pumping wells will have on-line pressure sensors in the monitoring filters.

3.2 *Monitoring strategy for water quantity*

Monitoring the water quantity comprises the following:
- collection of information on the soil composition during installation;
- on-line-measurement of water levels, hydraulic heads and flow rates;
- manual measuring of hydraulic heads;
- precipitation and evaporation.

During installation, information on soil composition has to be obtained. This means reserving time for data collection during the installation of the different wells. Thought must be given not only to drilling journals, actual borehole logging and flow measurements, but also to the cleaning of the recharge pond beds.

Automatic pressure sensors measure signals online, the signals being read and recorded hourly. All 27 surface water level measurement points have automatic readers. The filters on the R-lines at 12 m-below NAP, and the observation-well filters at 40 m below-NAP in the two north direction lines are automatically recorded. In addition, all inflows to and outflows from the different surface water parts of the recharge area and the groundwater (pumping wells) are measured on-line.

As well as these automatic measurements, the 17 existing observation wells in the polder are monitored manually every two months. The water level of the Maas greatly affects the hydraulic heads in the recharge area. This effect has been quantified (Bunnik 1997), so that future manual measurements can be corrected for the current level of the Maas. Depending on the corrected ground-water level, certain actions may have to be taken. If the corrected groundwater level falls outside its permitted range, the first action is to measure again the same day. Should this second one also be outside the range, then an investigation must be started into the cause of this.

For drawing up the water balance, besides the inflow and outflow of water, the amount of precipitation and evaporation must also be determined. To do this, precipitation gauges are installed and evaporation data obtained from a local station of the Dutch Meteorological Institute.

The three filters under the beds of the infiltration canals signal the occurrence of any clogging. Based on this information, the decision can be made whether or not to (partially) clean a bed.

The state of the pumping wells is monitored by taking annual well clogging measurements of each pumping well. The four pumping wells fitted with automatic pressure sensors will in addition be checked monthly for signs of clogging. A study has been started into the added value of this more frequent measuring of well clogging. Should the results prove positive, then future pumping wells (in phase 2) will have the same construction as the four mentioned above.

3.3 *Monitoring strategy for water quality*

Monitoring the water quality comprises the following:
- collecting soil samples for analysis during construction of the recharge system;
- analyzing the quality of the surface water before recharge and the quality of the groundwater during and after its aquifer passage.

The observation wells to be used for sampling will be made by percussion drilling, which allows highly accurate collection of soil samples for analysis. The three direction lines between the meanders (fig. 4, letter 'M') will be drilled by this method and sampled approximately every five metres. This is necessary since the subsoil is extremely heterogeneous (Meekes & de Lange 1997).

The surface water quality is analyzed at four locations in the water production process. These are pretreatment, infiltration canals, the border canal, and nature areas. The groundwater is evaluated on quality immediately after recharge. This is carried out at different depths on direction lines between the meanders, at the boundary between the recharge area and the polder, in the pumping wells, in the filters of pumping wells, and in the nature areas.

The sampling frequency varies. For parts directly related to drinking water production, sampling is every four weeks. For ecological aspects, sampling is made every three months, because water quality impacts on the ecology are slow.

For analysis concerning drinking water standards, the set for impacts is more extensive than the set used for analysis concerning ecological standards. The latter set comprises mainly nutrients, such as phosphate, sulphate, nitrate, potassium, calcium and a few other salts. In relation to drinking water productions the following are measured: dispersion of recharged drinking water in the soil, quality changes during aquifer passage, and leaching of any pesticides.

3.4 *Evaluation and reports*

The frequency of evaluation varies. Hydraulic head reports for determining impacts on the environment will be produced with the same frequency as those used for groundwater abstractions by WOB, namely, once a year. The functioning of the border canal will be given in a one-off report. The water balance must be reported once a year, but it must be in balance every quarter. Finally, there must be a yearly report on the quality of the recharged water, including the dispersion of the recharged water in the subsoil. The water board must receive a yearly report on the volume and quality of the overflow to the surrounding polder area.

No reports are necessary concerning control. However, it is important that the monitoring of the control is evaluated annually in the first few years so that adjustments can be made in time. Corrections may be required for permit or operational reasons. The need for an annual evaluation will decrease in the future.

4 CONCLUSION

Only the future can say whether the design will lead to a well-functioning monitoring system. The method has increased the insight into the different management tasks of PIM, as well as their interrelationships. The approach has produced a measurement system that can be managed and evaluated, making it an evolving part of PIM, a part that grows with PIM and is adjustable should the need arise.

REFERENCES

Bannink, A. 1992. 'Bestrijdingsmiddelen Lith: het voorkomen van bestrijdingsmiddelen in het grondwaterbeschermingsgebied Lith.' WOB, 's-Hertogenbosch 1992 (in dutch).

Bunnik, J. 1997. Waterhuishouding nulsituatie PIM. Kiwa, Nieuwegein. Reportnr. KOA 97.219a (in dutch).

Meekes, J.A.C. & F.D. de Lange 1997. Hoge Resolutie Seismiek Lith. Nederlands Instituut voor Toegepaste Geowetenschappen TNO, Delft (in dutch).

Claessen, F.A.M. 1997. *European Water Pollution Control*, (7.4). Comparing monitoring of surface and groundwater systems'.

Verheijden, S.M.L., A.J. Vogelaar & P.J. Stuyfzand 1997. Experiences with pilot-recharge in the Maaskant area: clogging, ecological engineering and changes in water quality. DVGW, Bonn. *DVGW/Kiwa-Schriftenreihe Wasser* Nr. 90.

Vogelaar, A.J., A.F.M. Meuleman & M. Zwamborn 1996. Definitief Ontwerp Infiltratiegebied PIM, Monitoringsplan. Kiwa, Nieuwegein. Rapportnummer KOA 96.120 (in dutch).

VEWIN/WOB, 1997 Brochure nr. 7 van VWPprojectgroep Stimuleringsbeleid Landbouw en Ecologie, Project infiltratie Maaskant: meanderen tussen drinkwater en natuur. VEWIN, Rijswijk (in dutch).

1.2 Water management in arid regions

Artificial Recharge of Groundwater, Peters et al. (eds) © 1998 Balkema, Rotterdam, ISBN 90 5809 017 5

Role of artificial recharge in the water resources management of Kuwait

M.N.Viswanathan & M.N.Al-Senafy
Water Resources Division, Kuwait Institute for Scientific Research, Safat, Kuwait

ABSTRACT: In arid countries like Kuwait with no natural water storages and limited groundwater resources, artificial recharge and storage of water in aquifers, can play a major role in the water resources management. Because of the absence of large storage reservoirs, the sea water desalination plants producing freshwater for urban use are forced to operate at sub-optimal conditions. Artificial recharge and storage of freshwater is required for meeting seasonal, long term and crisis conditions. Depletion of groundwater of brackish quality is another major problem facing Kuwait. This problem can be overcome by the reuse of wastewater. One of the best forms of reuse of wastewater is to recharge it using infiltration ponds. Such an arrangement improves the quality of wastewater to a level that it can be used for unrestricted landscaping and gardening purposes in urban areas.

1 INTRODUCTION

1.1 Water resources in Kuwait

Kuwait is an arid country with a total population of about 2 million people, spread over an area of about 17,780 km^2. The annual average rainfall is about 110 mm. and summer temperatures are in excess of 45^0 C. Average urban demand for fresh water in 1995 was about 880 Megalitres per day. Most of urban demand for fresh water is met by desalination of sea water. Average urban demand for brackish water in 1995 was about 200 Megalitres per day. Brackish water with total dissolved solids of about 3000 parts per million is mostly used for landscaping and gardening. Urban demand for brackish water is met by groundwater. Increasing population, higher standards of living, higher quality of water required for urban and irrigation use, and requirements for uninterrupted supply are making water resources management a daunting task.

1.2 *Problems of water resources management*

There are two major problems facing water resources management in Kuwait. One of them is lack of natural storages and the other is the declining groundwater levels in brackish water well fields.

Lack of major storages creates several problems with respect to the operation of water desalination plants and distribution system. Figure 1 shows seasonal variation in water consumption during the year 1995. The average demand for fresh water in summer increases to about 850 Megalitres per day and declines to about 600 Megalitres per day in winter months. The installed capacity of desalination plants should be in excess of 850 Megalitres per day to meet the varying fresh water demands.

Large variations in fresh water demands between summer and winter months result in the sub-optimal operation of desalination plants, adding to the cost of fresh water production. Fresh water storages of suitable size will help to even out the variations in fresh water demands and result in the optimum operation of desalination plants.

Uncertainties regarding the long term predictions for fresh water demand result in excess standby reserve capacities of desalination plants.

Figure 2 shows the annual average consumption of fresh water and installed capacities of desalination plants from 1970 to 1995. Higher installed capacities are also required because of the desalination plant unreliability. With increasing annual consumption, the amount of reserve capacity carried also increases as can be seen in figure 2, particularly between the years 1984 and 1995. Large fresh water storage capacity will reduce the need to carry excess desalination plant capacities to overcome uncertainties in demand forecasting and plant reliability.

Demand for brackish groundwater for landscaping and gardening has increased steadily since 1960. Increased consumption of brackish groundwater has resulted in the rapid decline of groundwater levels in

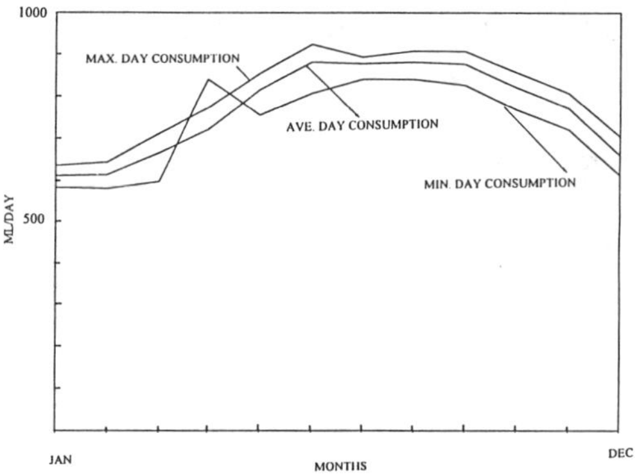

Figure 1. Seasonal variations of fresh water consumption during 1995.

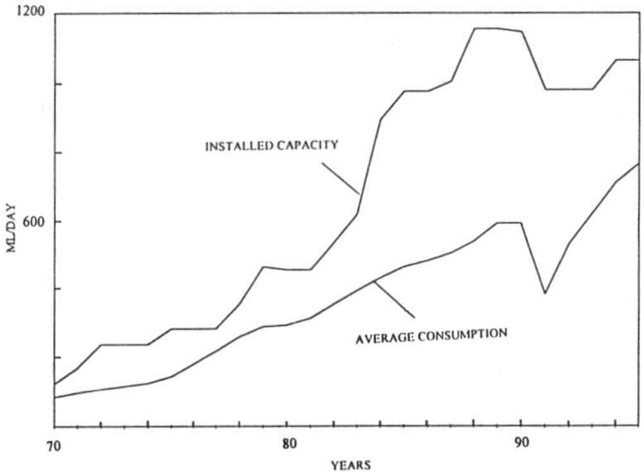

Figure 2. Desalination plant installed capacity and fresh water consumption in Kuwait between the years 1970 and 1995.

most of the brackish groundwater fields. Figure 3 shows some of the major brackish groundwater fields in Kuwait. Depleting brackish groundwater is forcing many residents to switch over to fresh water for their gardening and landscaping requirements. This in turn has placed additional burden on the desalination plants.

Artificial recharge and storage of fresh water in aquifers is likely to overcome some of the problems associated with the water resources management in Kuwait.

2 STORAGE REQUIREMENTS

2.1 Seasonal operations

The difference between summer demand and winter demand for fresh water is about 250 Megalitres per day (Fig. 1). This excess plant capacity in winter months could be used to produce additional water for artificial recharge and storage in aquifers. The excess water stored in winter months can be used in summer months, thereby reducing the plant output in summer months. By this arrangement, the desalination plants could be operated in near optimum conditions for every day of the year. An examination of figure 1 shows that a total of about 10,000 Megalitres of fresh water is available during winter months. Hence, a seasonal storage of about 10,000 Megalitres is required.

2.2 Long term storage

In 1995, an excess installed plant capacity of about 200 Megalitres per day was available. This excess capacity is required for uncertainties in long term water demands and plant unreliability. This excess amount of water could be stored in aquifers, for use during period of plant failures and demand uncertainties. In general, a period of one year storage is required for the construction of additional desalination plants and hence, minimum storage volume of water to meet one years demand will be required. Average daily water demand in Kuwait is about 750 Megalitres per day and hence a total volume of water required in storage will be about 300,000 Megalitres. At an excess installed plant capacity of about 200 Megalitres per day, it will take about 4 years to store this volume.

2.3 Water storage for crisis management

Crisis management refers to conditions where a minimum water supply is required to be maintained during periods of major emergencies. Major emergencies in arid countries and in particular Kuwait, would mean either partial or total loss of desalination plants either due to vandalism or political unrest. In such cases, the demand for drinking and cooking requirements should be met for a duration long enough to construct the damaged desalination plants. Water for bathing and toilets could come from brackish groundwater fields. Minimum per capita requirement for drinking and cooking is about 20 litres per day (Flack 1982). Assuming a total population of about 2 million people, a total of about 15,000 Megalitres of water will be required to be stored. This will meet drinking and cooking demands of 2 million people for a period of about one year. This is the minimum storage volume that should be available at all times.

3 ARTIFICIAL RECHARGE OF FRESH WATER

3.1 Storage volumes

In the previous section, three different storages of water were identified for the successful management of water resources in Kuwait. These are 1. 10,000

Megalitres for seasonal variations in water demand 2. 300,000 Megalitres for long term demand uncertainties and 3. 15,000 Megalitres for crisis conditions. The most economical form of storage of fresh water is in aquifers by artificial recharge. In addition to economics, storage of water in aquifers makes it safe from pollution, vandalism and destruction during crisis periods. One of the problems facing aquifer recharge and storage is the recovery. Unlike above ground reservoirs, where nearly all the water stored could be recovered, the recovery from aquifers is limited by the properties of the aquifer. Hence one of the primary requirements of artificial recharge is to estimate the recovery efficiency. The recovery efficiency depends on hydraulic gradients of water tables, dispersivity, aquifer transmissivity, porosity, and time of underground storage (Merritt 1985). In addition, the storage that will be used to meet seasonal variations in water demand should be in close proximity to urban areas to reduce pumping costs. Storage volumes for meeting long term uncertainties, and crisis conditions should be located at areas where the hydraulic gradients are not significant to prevent the movement and subsequent loss of stored water.

3.2 Locations for artificial recharge

Figure 3 shows three major areas of aquifers. One is the brackish water fields supplying water for the urban region. This includes aquifers at Sulaibiyah, Shigaya and Umm Gudair. These aquifers are brackish in nature The water tables in these aquifers are declining due to the excessive abstraction of groundwater for urban landscaping and gardening. Considering the close proximity of these aquifers to urban regions, these are particularly suitable for the storage of water required to meet seasonal variations of water demand.

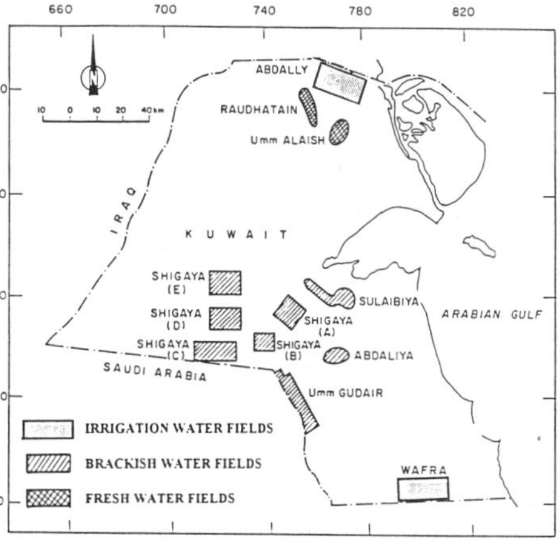

Figure 3. Locations of groundwater fields in Kuwait.

Recovery efficiency of stored water will be low, probably about 20% during the first cyclic storage and recovery (Mukhopadhyay et al. 1994). This recovery efficiency is expected to increase with subsequent cycles. Aquifers at Sulaibiyah and Shigaya (A) are particularly suited for this type of storage.

Aquifers at Wafra in the south and Abdally in the north are also brackish groundwater aquifers supplying water for irrigation. These aquifers are unsuitable for recharge and storage, because groundwater from these aquifers is constantly pumped for irrigation. In addition, the remoteness from urban regions makes them unsuitable for any form of storage.

Aquifers at Raudhatain and Umm Al-Aish in the north of Kuwait are fresh water lenses formed by the infiltration of rainwater. These lenses are floating on highly saline groundwater. In spite of low annual rainfalls, these lenses are formed due to unique lithological and geomorphological conditions. The unsaturated thickness in these aquifers vary between 20 and 40 m. If fresh water is stored in these aquifers on top of the existing lenses, the recovery efficiency will be very high because of almost negligible trans- ition zone between the stored water and native groundwater (Senay 1977). Although these aquifers are far away from the urban regions, water for long term and crisis conditions could be stored in these aquifers, because water need not be pumped in and out of these aquifers at frequent intervals.

4 ARTIFICIAL RECHARGE OF WASTEWATER

The water levels in brackish groundwater fields are declining due to the over pumping of groundwater to meet landscaping and gardening requirements. One of the options for reducing this dependence on brackish groundwater is to reuse treated wastewater. Wastewater is a resource that could be used to replace brackish groundwater for landscaping and gardening purposes. One of the major problems of reuse of wastewater is the cost of treatment to a level, that it could be used for unrestricted use without any major health hazards. The other problem is the psychological barrier that prevents people from using wastewater. Figure 4 shows the average flows of wastewater in Kuwait during the year 1995.

A total of 251 Megalitres per day of wastewater was produced during the year 1995. Of these, about 119 Megalitres per day was used for irrigation and the rest was disposed off to the sea. At present wastewater reuse is limited to areas well outside the urban regions. A reuse of this type may not overcome the problems faced in urban regions. If the generated wastewater could be used within the urban region without any restrictions on the usage, then the dependence on brackish groundwater for landscaping and gardening purposes could be completely eliminated. To achieve this objective, two conditions must be met. One is the level of treatment and the other to remove the stigma attached to the reuse of wastewater.

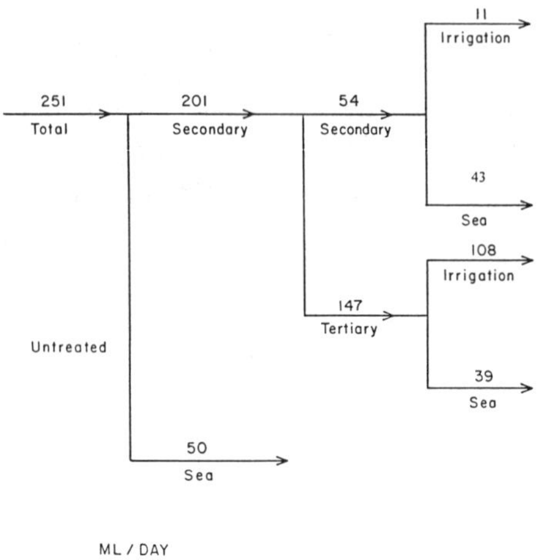

Figure 4. Wastewater flows in the year 1995.

4.1 Wastewater recharge

One of the methods to improve the wastewater quality is to recharge aquifers by infiltration ponds (Idelovitch & Michail 1984). A field study was undertaken in Sulaibiyah to investigate the improvement in waste water quality as wastewater infiltrates through the unsaturated zones of the aquifer. The investigation consisted of an experimental pond of area 1m x 1 m. Tertiary treated wastewater was allowed to infiltrate over a period of several months. The depth to water table is about 27 m. Wastewater quality as it reached the water table was studied by collecting water samples from water table by monitoring wells. The water samples were collected by bailers. Table 1 presents the wastewater quality before and after infiltration.

Wastewater quality improves as it moves down the unsaturated region. The increase in total dissolved solids is due to the leaching of salts from the unsaturated zones of the aquifer. The processes responsible for the improvement of quality are: adsorption, filtration, biological and chemical reactions. Increase of total dissolved solids is due to the leaching of salts from the unsaturated zones. The wastewater quality after infiltration is suitable for

Table 1. Wastewater quality before and after infiltration.

Parameters	Wastewater quality before infiltration (mg per litre)	Wastewater quality after infiltration (mg per litre)
Total Dissolved Solids	1400	2050
Chemical Oxygen Demand	40	10
Biochemical oxygen Demand	8	1
Ammonia as N	17	0.1
Nitrate as N	0.5	5.1
Phosphate as P	11	0.4
Total coliform per 100 ml.	4000	2

unrestricted irrigation (Idelovitch & Michael 1984). The loading rate at the experimental pond was about 1.5 m per day. At this loading rate, a total area of about 17 hectares will be required to recharge the aquifer with wastewater. The ideal site for the location of the recharge ponds will be at Sulaibiyah (Fig. 3).

5 CONCLUSIONS

1. In arid countries like Kuwait with no natural storages and limited groundwater, artificial recharge can play a major role in the management of water resources.
2. For Kuwait, three different types of storages were identified. These are seasonal, long term and crisis storages. Artificial recharge and storage in deep aquifers is ideal for seasonal storages. For long term and crisis conditions, storages as fresh water lenses will be ideal.
3. Depletion of brackish groundwater could be arrested by the reuse of treated wastewater. One of the best forms for reusing water is to recharge aquifers using infiltration ponds and storing the treated wastewater for later use.
4. Field trials undertaken in Kuwait proved that the wastewater quality could be improved by the recharge of wastewater using infiltration ponds.

REFERENCES

Flack, J.E. 1982. Urban water conservation. *American Society of Civil Engineers*: New York.

Idelovitch, E. & M. Michail. 1984. Soil aquifer treatment, a new approach to an old method wastewater reuse. *Journal WPCF* 56(8):936-943.

Merritt, M. L. 1985. Sub-surface storage of fresh water in South Florida. *US Geological Report Water Supply Paper 2261.*

Mukhopadhyay, A. F.Szekely & A. Senay 1994. Artificial groundwater recharge experiments in carbonate clastic aquifers of Kuwait. *Water Resources Bulletin.* 30:(6):1091-1107.

Senay, Y. 1977. Groundwater resources and artificial recharge in Raudhatain water field. *MEW Unpublished Report*, Kuwait: 43.

Artificial Recharge of Groundwater, Peters et al. (eds) © 1998 Balkema, Rotterdam, ISBN 90 5809 017 5

The management of surface and groundwater resources in Lebanon

S. Sadek & M. El Fadel
Department of Civil Engineering, American University of Beirut, Lebanon

ABSTRACT: Groundwater resources constitute an important component of the total water balance of Lebanon, and are especially important in the highly populated coastal zone. Increasing water demand, coupled with the near breakdown of governmental authority during the civil war, led to a lack of water management activities. This resulted in the deterioration of the groundwater quality from seawater intrusion (due to excessive pumping), and contamination with sewage and industrial wastes. Reconstruction plans currently in effect, have placed water management high on the agenda. The process involves groundwater remediation and maximization of available water resources. Artificial recharge offers a dual-purpose option for the storage of the excess surface water and improving ground water quality. This paper draws on the limited Lebanese experience in artificial recharge and on the current knowledge of the Lebanese hydrogeology, to discuss the potential of this management option.

1 INTRODUCTION

The Middle East region faces today a greater challenge than ever. The freshwater potential in nine out of 14 countries is currently less than 1,000m^3 of water per person per year, making it one of the most concentrated regions of water scarcity in the world (Postel 1993, & ESCWA 1996a). With a doubling of the population projected in the next 25 years, accompanied with the expansion of irrigated agriculture, the tapping of the groundwater sources is rapidly escalating, reaching unsustainable levels (Jaber 1995, & Jaafar 1996). This reality is in contradiction with long held views that Lebanon is water rich and that it can, and should, share its excess resources with its neighbors (US Army Corps of Engineers 1991, & Schiffer 1996).

Although Lebanon is endowed, at least in some parts, with a fair share of rainfall, the issue of groundwater quality and quantity is very acute. There has been in the past few years, widespread reports of water table decline due to over-pumping, especially in areas where irrigation relies almost totally on groundwater. In the coastal zone, the drop in the water table has caused the sea water to intrude into the aquifers, adding the problem of salinity to that of scarcity (Peaudecerf, 1971). Elsewhere in the country there were reports of contamination with toxic chemicals and with industrial and domestic wastewater.

Information about the hydrology of Lebanon is scarce; yet, it appears that with the increase in demand and the lack of appropriate legislative measures, the water budget in many regions is becoming badly unbalanced. The total disregard of basic environmental issues is adding to the pervasiveness of the problem. Deforestation and unplanned urban expansion are causing the degradation of critical watersheds. This is resulting in an increased rate of runoff and less aquifer recharge. All tends to indicate that the current water use in Lebanon is probably unsustainable. Thus, the need for management techniques and policies such as rainwater harvesting and artificial recharge of the groundwater, is today more pressing than ever.

Artificial groundwater recharge has long been acknowledged as a major sustainable management option. The Lebanese experience has been limited in this field. The process offers a dual-purpose solution for the storage of the excess surface water while remediating degraded aquifers. This paper presents a review of relevant aspects of artificial recharge in the context of its application to Lebanon.

2 LEBANON: ITS CLIMATE, MORPHOLOGY, GEOLOGY, AVAILABLE WATER RESOURCES AND PROJECTED NEEDS

2.1 *Climate*

The geographic location of Lebanon results in a temperate climate. While winters are short and relatively dry, sunny conditions dominate the rest of the year. The annual rainfall varies from 500mm/year in the Bekaa valley to 1400mm year in the mountains. The total rainfall is estimated at 9.5 billion m^3 per year and occurs mainly between the months of October and April. Climatic conditions are not uniform throughout the country as they are affected by the presence of the two major Lebanese mountain chains: the "Lebanon Mountains" and the "Anti-Lebanon Mountains", which run predominantly in a north-south direction (Fig.1). The high peaks of both mountain chains are covered with deep snow during the winter season, and con- tribute to the overall surface and groundwater flow as these caps melt.

2.2 *Morphology*

Lebanon could be divided into four main regions. They are from East to West (Fig.1):

1. A relatively flat and narrow coastal strip running from North to South.
2. The "Mount Lebanon" chain, which is parallel to the coastline reaching maximum elevations of upwards of 3000m.
3. The Bekaa Plateau, east of the "Lebanon Mountains" chain. Its elevation is about 900m above sea level.
4. The "Anti-Lebanon Mountains" chain which also runs in a predominantly North-South direction East of the Bekaa Plateau.

2.3 *Geology*

Geologic formations in Lebanon are mainly composed of fractured Karstic Limestone, with the presence of volcanic formations in the extreme north of the country. The identifiable geologic formations range in age from the Jurassic (oldest identifiable units) to the most recent Quaternary deposits.

The Jurassic contribution is a sedimentary carbonate rock, which is fissured and forms one of the two major aquifer systems. The next dominant formations belong to the Cretaceous period and can be divided into three groups The *Lower-Cretaceous* layers which are composed predominantly of clayey layers (Gres de Base) and form a relatively impervious zone separating the Jurassic and *Middle-Cretaceous* formations. The *Middle-Cretaceous* (Cenomanian and Turonian) is represented by an enormous mass of carbonate rock (yellowish-white fissured limestone) ranging in thickness from about 600 to 1000m, and which represents the second important aquifer system. *The Upper-Cretaceous* is predominantly a marly limestone, which is overlain by the most recent Neogene and Quaternary deposits of flood plains, and wind deposited dunes, etc.

In summary, the majority of the Lebanese territory is constituted by two calcareous formations of Jurassic and Cenomanian age. An important and relatively impervious zone separates them.

2.4 *Water resources*

The total flow from rainfall and snow is about 9.0 billion m^3/year. Of this total an estimated average of about 4.0 billion m^3/year remains as surface flow, and 600 million m^3/year, or 7% of the total precipitation go to aquifer recharge. The rest is lost to evapo-transpiration, and surface and subsurface flows to Syria (Orontes and El-Kabir rivers) and Northern Israel.

The above numbers are of course estimates and are based on an average year; they may vary significantly in the event of a dry period. Furthermore, the values presented do not take into account the recent and in some areas drastic changes in land use and developments that affect infiltration rates and disrupt the natural recharge of the aquifers.

2.5 *Actual and projected water needs*

Three types of water needs exist: Domestic, industrial and agricultural.

Although no clear and dependable records for domestic water use were kept during the years of the civil strife, reasonable estimates (Table 1) have been made on its magnitude (Camp Dresser & McKee 1980, & Khatib & Alami 1994), and can offer some idea as to the future needs.

Table 1. Projected domestic water need in Lebanon

Year	Consumption rate (liter/capita/day)	Total domestic consumption (million m^3/year)
1990	165	300
2000	215	550
2015	260	900

The industrial and agricultural water needs are estimated at 725 million m^3/year. They are expected to rise to about 1500 million m^3/year in 2015.

In summary, the total annual water needs are put actually (1990) at 1025 million m^3, and are projected to reach 2400 million m^3 in 2015. Although this number does not exceed the potentially available water in an average year, it is expected to exceed it in dry years. It is interesting to note here that the only major artificial water storage facility in the country is the Qaraoun Dam, with a maximum storage capacity of 220 million m^3 (Fig.1).

3 GROUNDWATER USE: ISSUES OF QUANTITY AND QUALITY

The water situation in Lebanon has two major constraining factors: the uneven spatial and temporal distributions of the water resources. The temporal constraint is important in reference to recharge issues.

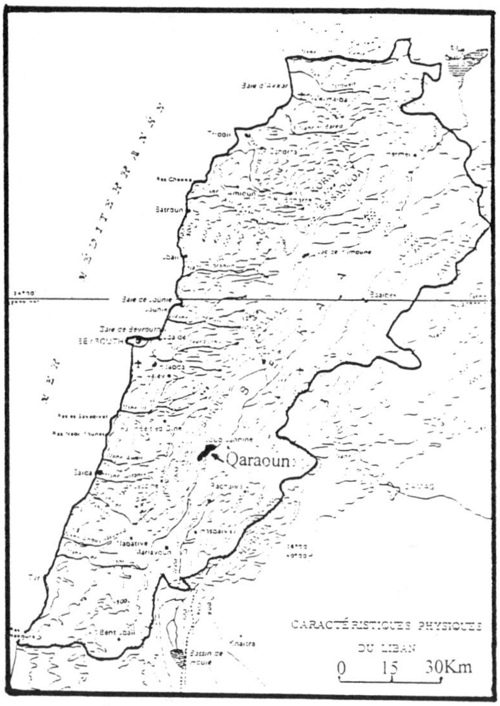

Figure 1. Map of Lebanon

The short rainy season during which most of the surface water flow occurs, implies that most of the water is lost to the sea, while the need for water during the balance of the year cannot be met. At present 80% of the government water supply is from surface sources. However this is a misleading number when one realizes that the majority of the new developments built in the last 15 years have tapped underground sources for their water, rather than rely on the unpredictable governmental supply. Very few of these new wells are regulated and assigned withdrawal rates. This situation, aggravated by nearly two decades of upheaval, has resulted in a number of problems of ground water quality deterioration. Contamination in a number of localities resulted from salt-water intrusion, direct waste water and refuse dumping into wells, and industrial leaks and surface dumping (ESCWA, 1996a).

3.1 *Salt water intrusion*

The problem of the quality of the groundwater in the coastal regions and particularly in the capital Beirut is not a new concern. Contamination with seawater and the extension of the salt/fresh water interface inland, has been in evidence since the 1960s in some areas. This problem however has reached a new scale of severity in the last two decades. The forced population movements during the war, coupled with natural growth, have strained water resources of the capital and its surroundings. New developments have all in part or in full relied on underground water supplies. This has exacerbated the intrusion problems and, if continued unabated, will contaminate even larger regions.

3.2 *Waste disposal*

The problem of ineffective septic systems particularly in mountainous areas, resulted in contamination of the groundwater. In recent years, development grew in areas where sewer systems do not-exist. Building owners have resorted to drilled wells as a source for water, but also and more alarmingly, they drilled deeper and larger "wells" for varied disposal purposes including domestic wastewater. The ensuing pollution is being detected in downstream wells resulting in great pollution concerns. These concerns are more acute in Karstic terrains given the quasi-free travel of the various pollutants through large cracks and cavities, affecting large water volumes.

3.3 *Industrial pollution*

Two separate categories of industrial pollution can be identified in Lebanon. The first is from surface and subsurface dumping of heavy metal and organic chemicals laden waters, which find their way through the highly pervious and fractured Karsts to the groundwater. The second and more widespread source of industrial pollution is from leaky underground gasoline tanks, and the uncontrolled surface dumping of petroleum by-products and residues.

4 ARTIFICIAL RECHARGE: A POSSIBLE SOLUTION?

4.1 *Advantages of artificial recharge in the Lebanon case*

It is evident from the preceding that the water situation in Lebanon faces a number of problems and challenges abound. The need for a countrywide water resources planning and management plan is essential. The Lebanese government has been in the process of elaborating a multi-phase Master Plan for water resources in the reconstruction and rehabilitation drive. The country is blessed with potentially sufficient water resources, however it is clear that a near total reliance on surface waters does not present an effective and practical solution to the water needs. The short winters and the peak river flows occurring in the months of least need, suggest that diverting some of the excess water, if not all of it, to surface or sub-surface storage ponds and aquifers would present a useful solution. Artificial recharge, whether done through injection or percolation from unlined ponds, will help ease the salt water intrusion problem by depressing the interface between salt and fresh water, and in the long term allow for a reasoned approach to ground water use.

4.2 *Practical problems and complicating factors*

The Lebanon limestone is a formation that exhibits all the typical properties of Karsts: highly jointed and fractured, with dissolution cavities and channels varying in sizes and extent. This gives the hydrologic system a degree of heterogeneity and unpredictability that could render the task of mapping it almost impossible. This of course should affect the ability of using it to manage the water resources through artificial recharge. As an example, recent subsurface investigations in the Beirut area, in preparation for the implementation of the down town reconstruction plan, have identified a number of underground cavities and conduits in what is predominantly carbonate rock.

Nonetheless, the above considerations should not prevent exploring artificial recharge as a tool for water management in Lebanon. This should be reinforced by the limited but positive experience detailed in the section that follows. Despite the common perception that water management in Karst is complicated and may present special problems (White 1988), a useful body of experience in this particular aspect of recharge in Karstic systems exists (Morozov & Lushchik 1984). Those applications suggest that artificial recharge in Karstic systems can be relatively effective if extensive information is collected about the system itself, and clean water is used in the recharge.

5 ARTIFICIAL RECHARGE: A CASE STUDY

Early artificial recharge experiments were performed (Massaad, 1969) for the purpose of exploring the possibility of storing the excess surface water in the wet winter months for use in the dry season, and the reduction of the sea water intrusion which developed from excessive exploitation of the groundwater resources. The experiments were carried out in the aquifer underlying the southern suburbs of Beirut, which has the advantage of being a closed underground reservoir (Fig.2). The aquifer was bounded on the sides by semi-permeable formations and seaward by a semi-permeable aquitard over a large depth. The aquifer is fed by natural recharge in the rainy winter season and is believed to discharge to the sea at large depths from underneath the semi-permeable formations.

Pumping test analyses yielded the following average aquifer characteristics:

- Transmissivity $T=2.4\ m^2/s$
- Storativity $S=0.03$
- Hydraulic gradient $I = 10^{-4} - 10^{-5}\ m/m$

Four artificial recharge tests were performed by well injection at a rate of less than 100 l/s. The effect of recharging was felt throughout the whole system. This showed that the aquifer system behaved as a continuous porous medium rather than a fractured Karstic medium. This was contrary to the general expectation, since artificial recharge in Karstic zones face the common problem of channeling the injected water through some underground conduits. However, the rise of the piezometric surface above a certain level could not be achieved for a long time due to the

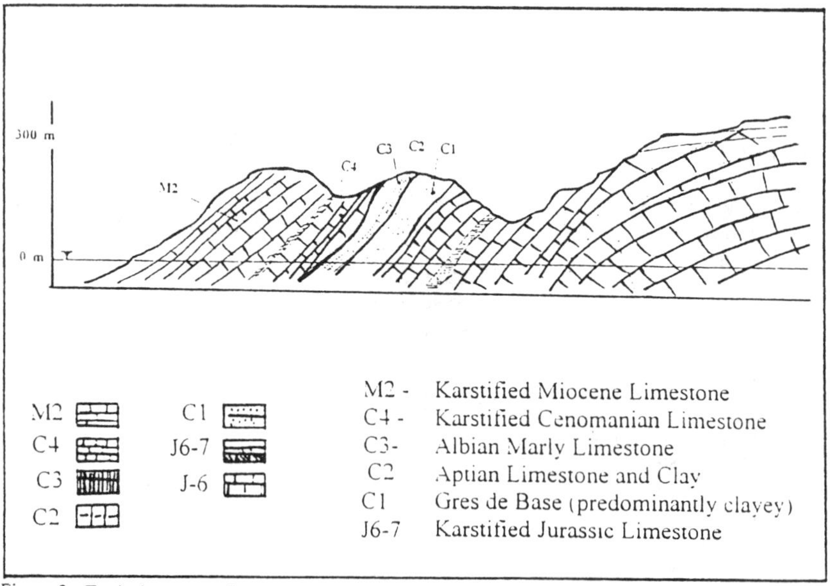

Figure 2. Typical cross-section across the test area

apparent discharge through some underground conduits or through mixing with saline water at great depths. The injected water seemed to push the saline water downward while the piezometric surface remained unaffected after a short period. Since the hydraulic gradient is small, the regional groundwater flow is weak (0.5 m/d) and unable to transport all of the injected water downstream. Mounding due to recharge should increase the lateral flow. The preliminary results showed that the injected fresh water stored in the aquifer could be recovered in the dry season without much loss to the sea. However, due to the over-exploitation of the aquifer, seawater intrusion persisted. Hence, the quality of the injected water has been significantly affected. Unless aquifer exploitation is decreased, the original objective of using artificial recharge to store fresh water cannot be met.

6 CONCLUSIONS

In this paper, an overview of the hydrogeologic characteristics and water resources of Lebanon were presented. A number of problems and challenges face the country in the future of its water sources. The artificial recharge of some of the aquifers could present an attractive solution to some of the problems.

Although early experiments in artificial recharge were initially encouraging, much remains to be done. There has been an extensive over-exploitation of the coastal aquifer in the past few years, which has increased seawater intrusion to a dramatic scale. Artificial recharge can reverse the seawater intrusion movement and safeguard the coastal irrigated land from adverse salinity effects if excessive pumping in coastal aquifers is minimized. Such a task requires large volumes of fresh water for a long time in order to clean up all the fractures and dead zones present in this Karstic zone. The proportion of the total injected water for the clean up might become prohibitively high especially since water is only available during three months.

Future work should concentrate on the study of the hydrodynamics of the groundwater flow to delineate the domain and the sources of inflows and outflows in the major aquifers. The development of a numerical model to simulate the flow for use as a management tool could be of great utility. Such a task is quite difficult to achieve due to the complexities of the common Karstic flow regime: preferential flow paths, heterogeneity and significant anisotropy. The difficulties are supplemented by the variation of salinity over large areas, which necessitates the use of a density-dependent model.

REFERENCES

Camp, Dresser & Mc Kee. 1980. Waste Water Master Plan. *Report prepared by Camp Dresser and Mc Kee for the Lebanese Council for Development and Reconstruction.*

ESCWA. 1996a. Water legislation in selected ESCWA countries, *Report E-ESCWA-ENR-1996-WG.1-WP.3*

ESCWA. 1996b. Water resources assessment in the ESCWA region using remote sensing and GIS techniques. *Report E-ESCWA-ENR-1996-7.*

Jaafar, H. 1996. Legal and administrative context for water management in Lebanon. *UN ESCWA Report E-ESCWA-ENR-1996-WG.1-CP.7, Amman Meeting.*

Jaber, B. 1995. Lebanese republic–water status paper. *UN-ESCWAE-ESCWA-ENR-1995-WG.1-CP.4 Report, Amman Meeting*

Khatib & Alami Consolidated Engineering Company. 1994. Lebanon's staged wastewater program. *Report prepared for the Lebanese Ministry for the Environment.*

Massaad, B. 1969. Les Resources en Eau Souterraines de la Region de Beyrouth. M.S. *Thesis, Lebanese University.*

Morosov, V. & A. Lushchik 1984. Artificial Recharge of Karst Water in the Flat Crimea. *Hydrogeology of Karstic Terrains V.1* (Edited by Burger, A. & Dubertret, L.), UNESCO.

Peaudecerf, P. 1971. Etude de l'invasion par les eaux marines de la nappe de Hadeth-Beyrouth, Liban. *Rapport de Mission LEB-70-01500, Organization des Nations Unies.*

Postel, S. 1993. Water Scarcity Spreading. *Vital Signs 1993* (Edited by Kane, H. & E. Ayres), Worldwatch Institute, Norton & Co., New York.

Schiffer, M. 1996. International water allocation in the Middle East. Workshop on Strategies for Intersectoral Water Management in Developing Countries, Challenges and Consequences for Agriculture. German Development Institute.

United Nations. 1970. Liban: Etudes des Eaux Souterraines. *United Nations Report DP-SF-UN-44-Lebanon, New York.*

US Army Corps of Engineers. 1991. Water in the sand: a survey of Middle East water issues. *Preliminary Study.*

White, W.B. 1988. *Geomorphology and hydrology of Karst terrains.* Oxford University Press, Oxford, UK.

Artificial Recharge of Groundwater, Peters et al. (eds) © *1998 Balkema, Rotterdam, ISBN 90 5809 017 5*

Successful operation of a large aquifer storage facility for a desert community

M.R. Lluria
Salt River Project, Phoenix, Ariz., USA

ABSTRACT: The Phoenix Metropolitian Area, the fastest growing community in the USA, is located in a semiarid region with mean annual precipitation of 180mm. Its water supply consists of limited surface water from two watersheds and groundwater. Overdraft of the regional aquifer is the result of several decades of extensive agriculture and more recently rapid urbanization. In the late 1980s Colorado River Water imported via the Central Arizona Project Aqueduct was added to the water supply. This included the development of aquifer storage facilities since new surface storage was not physically, economically nor environmentally feasible. A large underground storage facility was constructed in partnership by all the municipalities of the area. It has operated successfully since 1994, banking surplus water for future use and improving the management of peak water demands.

1 INTRODUCTION

1.1 *Background*

The Granite Reef Underground Storage Project (GRUSP) with a recharge capability of 240 million cubic meters per year commenced its operation in May of 1994 and has stored in excess of 350 million cubic meters of water up to March of 1998. This represents a large part of the imported Colorado River water surplus which if not used by the state of Arizona or stored would be taken by California or flow to the Pacific Ocean. Municipalities also use the GRUSP facility for seasonal storage and recovery. This has proven an excellent water resource management tool for the efficient handling of this precious and expensive commodity in this semiarid region of North America.

The technical performance of the total recharge system has exceeded design expectations. Recharge rates have been maintained above 1 meter per day using a full time wet basin operation with no measurable decrease. Although, algae mats have developed during the hot summer months, a biological self cleaning process spontaneously developed which has not required drying of the basins nor scarification. This has been assisted by the use of white amur fish in the delivery system to the recharge facility which has reduced algae and aquatic moss load of the inflows. The design of the surface system was based on the site conditions (Lluria & Fisk 1995). This has minimized operation and maintenance costs resulting in very low water storage unit cost. Recently, electronic devices for measuring inflows, and surface and groundwater levels have been added to the recharge operation. Theses devices are linked to a satellite based network for remote transmission via the Internet to distant terminals. This has provided the capability of real time evaluation of recharge rates and hydraulic impacts and has resulted in operational cost savings by reducing field time of monitoring crews. This remote data acquisition has also improved the accounting of aquifer storage credits mandated by regulatory agencies.

1.2 *Recharge facility setting and components*

The GRUSP is located in the eastern part of the Phoenix Metropolitian area within the Salt River Pima Maricopa Indian Reservation. The climate is semiarid with mean annual precipitation of 180 millimeters. The recharge facility is in the channel of the Salt River, the largest of the three rivers that form the Salt River Valley. The principal surface water supply of Phoenix had come from various reservoirs in these rivers. In 1988, an additional surface water source was added. Approximately 670 million cubic meters of water from the distant Colorado River are now delivered to Phoenix via the 550 kilometers long Central Arizona Project (CAP) Aqueduct. This CAP water is the principal source of aquifer storage.

The GRUSP facility stores water in an alluvial aquifer contained in a deep intermountain basin formed by tectonic process from 25 to 15 million years ago (Lluria 1987). The site was selected for its very favorable geohydrologic characteristics. The receiving aquifer has large storage capacity and the alluvial sediments that form the unsaturated zone are very permeable resulting in rapid recharge rates. The site also offered very favorable environmental conditions with absence of any surface or subsurface contamination. The surrounding area to the north, east and west is undeveloped land of the Indian reservation. To the south it is predominantly agricultural lands. To the north and south of the project site there is extensive archeological evidence of the prehistoric Hohokam civilization (300 BC to 1400 AD). This was the most concerning environmental issue during the construction phase and when locating the monitor wells.

Of important economic consequence is the existence, very near the recharge site, of an adequate water conveyance system. Water from the CAP Aqueduct and from the reservoirs of the Salt and Verde Rivers is delivered to GRUSP by an existing high capacity canal. Delivery gates with their remote control system were also in place. Water to be stored in the aquifer using the GRUSP recharge facility is delivered from the South Canal of the Salt River Project (SRP). The flow is measured by a broadcrested weir at the gates of the canal and flows along a natural drainage and a delivery channel to the four recharge basins (Lluria et al. 1993). The flow from the South Canal to all the recharge basins is by gravity. A network of monitor wells (Fig. 1)

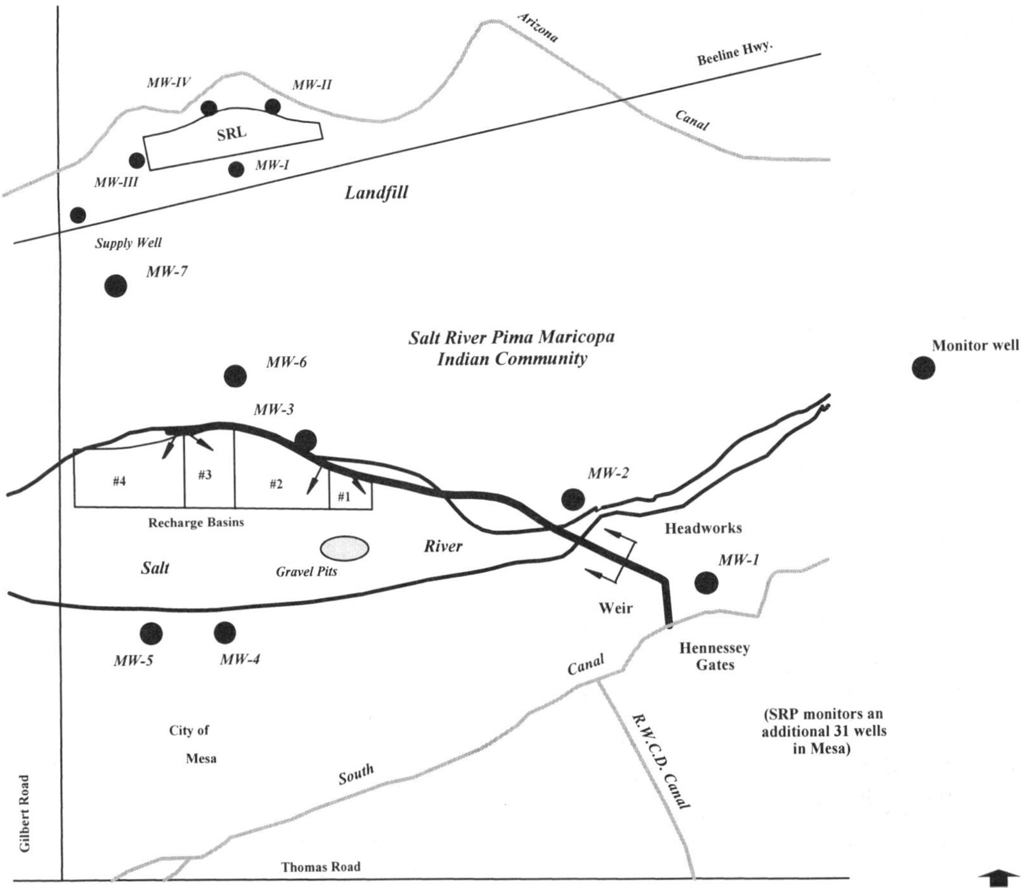

Figure 1. Components of the GRUSP and location of landfill and gravel mining

surrounding the recharge site provide the necessary groundwater hydraulic and groundwater quality information to guide the operation, manage potential impacts and fulfill the regulatory requirements. Recovery of the stored water is carried out by pumping the large capacity supply wells of the various municipalities in the Phoenix Metropolitan area.

2 TECHNICAL OPERATIONS

2.1 *Principles and mechanisms*

Two principles guide the operation of the GRUSP: technical efficiency and cost effectiveness. The end result has to be the on-time delivery and storage of water at the lowest possible unit cost. The recharge system has to be operated efficiently so that it has low operation and maintenance costs. The GRUSP recharge facility is designed to handle a wide range of inflows from 1 to 10 cubic meters per second. Proper selection of the basins, based on the inflows, is important because both their surface areas and their recharge rates are different (Table 1).

Table 1

Basin	Area (ha)	Recharge rate (m/d)
1	6.5	1.2
2	23.1	0.6
3	13.8	1.6
4	27.1	1.0

Recharge rates are also dependent on water temperature and on depth to the groundwater. Therefore, seasonal variations as well as changes due to prolonged recharge that will cause mounding have to be considered. Losses by evaporation during summer are less than 2 cm/day. The basins with the smaller surface area (Basins 1 and 2) are preferentially used, if the prevailing inflows permit their use. At the GRUSP, all basins are operated at no more than 0.6 meter of hydraulic head (depth). Use has shown that larger water depths reduce infiltration rate, possibly caused by the compression of bottom sediments (Bouwer & Rice, 1989).

A filamentous type of algae has been observed to develop on the bottom of the basins during certain times in the summer. The algal mat formed breaks up and floats to the surface. It appears to pick up finer sediments from the bottom and spread them by wave action. No measurable decrease in infiltration rates result from these periods of algal mat formation. At the GRUSP, this avoids the costly wet/dry cycling necessary at other recharge facilities, especially those that recharge treated municipal effluent.

The water conveyance system of the GRUSP allows flows to each basin from the delivery channel and also from the basins immediatly upstream by means of a gate. This adds operational flexibility and maintains the 0.6 meters depth in the basins during the large inflows. Visual observation guides the field operators to maintain the proper water level of all basins which results in maximum recharge rates.

2.2 *Maximization and control of recharge*

The GRUSP facility is a closed system. For this type, all the inflow, minus the evaporation, has to be recharged. A hydraulic balance between the surface water delivered and the water being stored in the aquifer has to be established. When the volumes delivered have wetted all of the infiltration surface while the recharge rates remain unchanged, rapid surface water buildup could occur. At GRUSP, water levels of each basin and their trends are carefully examined at least three times daily and the necessary gate adjustments carried out. When inflows are near the maximum facility capacity the inflows are gradually reduced until the hydraulic equilibrium is reestablished. At low inflows, when there are dry basins, a shifting of flow to those basins is undertaken.

Mounding of the recharged water occurs even in a highly transmissive aquifer system such as the one at the GRUSP site. This has been observed to decrease the recharge rate and may also cause hydraulic impacts to neighboring facilities. At GRUSP a depth to water of 15 meters is used as the maximum permissible groundwater level buildup under the basins. An optimization model for the operation of the GRUSP facility is under development. This model will provide guidance to maximize the total recharge while minimizing the hydraulic impacts. The abundant surface and groundwater data collected during the three years of project operation is being used for calibration.

3 MANAGEMENT OF IMPACTS

After the construction of the GRUSP two facilities were located in its vicinity. At 1.6 kilometers to the north, a sanitary landfill was constructed. In the channel of the Salt River, south of the recharge

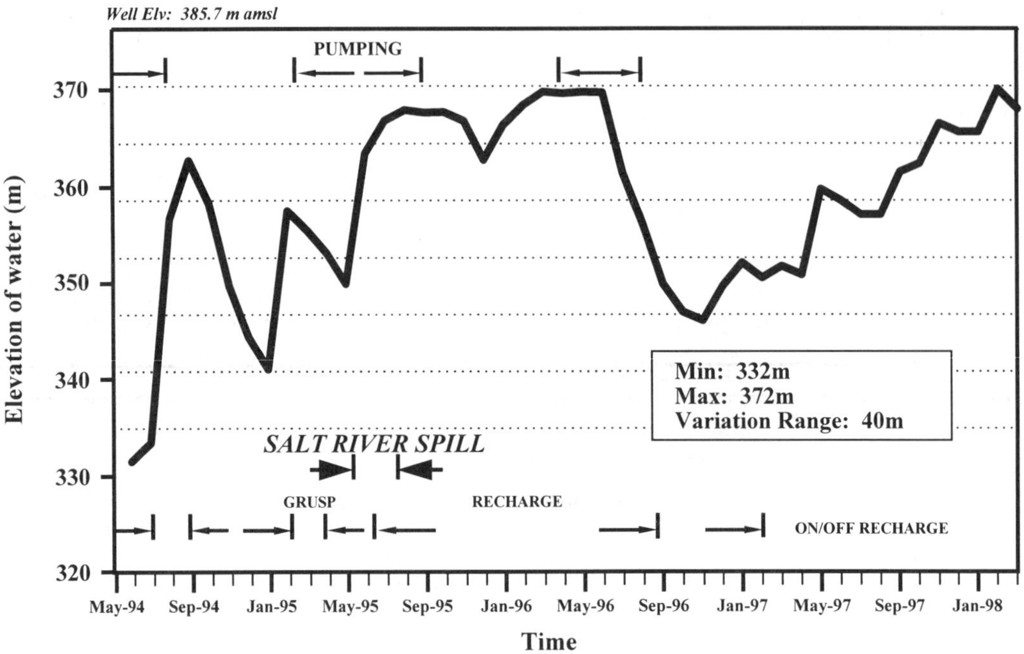

Figure 2. Hydrograph of monitor well gmw-3 at recharge facility

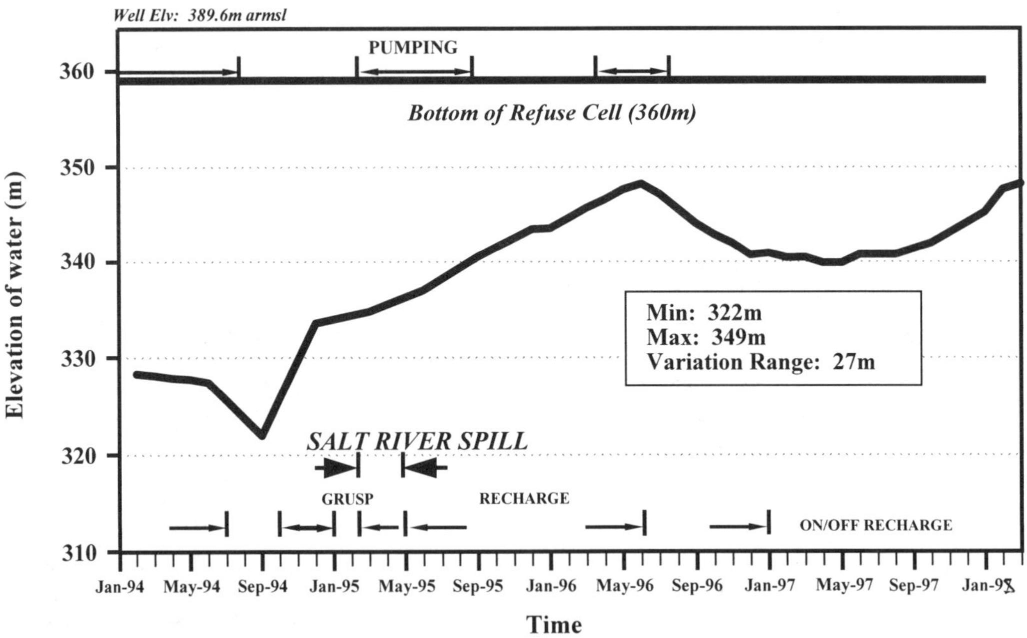

Figure 3. Hydrograph of monitor well smw-1 at landfill

basins, an extensive and deep gravel mining operation is under way.

3.1 *The Salt River Landfill*

The Salt River Landfill is a state-of-the-art facility adequately equipped with bottom liners. However, concerns that groundwater levels may reach the refuse cells and cause contamination of the aquifer has prompted additional monitoring and a plan to manage mounding caused by GRUSP recharge operations. Figure 2 is a hydrograph of a monitor well which reflects groundwater level changes below the basins since the start of the GRUSP operation. In addition to the mounding effects of GRUSP, incidental recharge of storm water in the Salt River results in considerable groundwater level rise. These effects are noticeable below the landfill (Fig. 3). A plan of operation has been developed to manage the mounding that may impact the landfill. A groundwater action level program determines when inflow reductions at GRUSP are necessary. Shifting of recharge to more distant basins is also used as a mitigation measure. Weather forecasting as well as water releases from reservoirs to the Salt River are also integrated into this plan of operation.

3.2 *The Aztec gravel mining operation*

Minor shallow water seepages from GRUSP into a gravel mining pit located 0.5 kilometers to the south occurred in 1995. These were caused by preferential flow along fractures in the alluvium. A study of the shallow river sediments in the area between the recharge and the mining operation identified an area of hydrualic connection between one of the recharge basins and the gravel pit. A small area within the recharge basin was isolated by soil berms. This measure placed an immediate stop to the seepages.

4 CONCLUSIONS

The operation of the GRUSP has succeeded in storing a considerable volume of water in an overdrafted aquifer for use during drought periods in a semiarid urban area of the USA. It has also contributed to the improvement of the management of the limited water resources of several municipalities of the Phoenix Metropolitan Area. A flexible operation of the facility during its three years of life has improved its technical efficiency and cost effectiveness. Impacts which resulted from its operation have been addressed and successfully managed and have not produced any delays or losses to the aquifer storage operation.

REFERENCES

Bouwer, H. & R.C. Rice 1989. Effects of water depth in groundwater recharge basins on infiltration rate. *J. Irrig. and Drain. Eng.* 115(4):556-568.

Lluria, M.R. 1987. Design of an artificial groundwater recharge system to an alluvial aquifer of the Basin and Range province based on geologic determinants. *Geological Society of American Annual Meeting Proceedings*: 748.

Lluria, M.R., T.L. Gorey & M. Fisk 1993. The Granite Reef Underground Storage Project: Site Selection, technical feasibility, engineering design and construction. *Proceedings of the Sixth Biennia Symposium on Artificial Recharge of Groundwater, Theniversity of Arizona.*

Lluria, M.R. & M. Fisk 1995. A large aquifer storage facility for the Phoenix area: *Proceedings of the Second Symposium on Artificial Recharge of Groundwater, American Society of Civil Engineers*: 129-138.

Artificial Recharge of Groundwater, Peters et al. (eds) © 1998 Balkema, Rotterdam, ISBN 90 5809 017 5

The role of artificial recharge in integrated water management in Egypt

Fatma Attia & Madiha Moustafa
Research Institute for Groundwater, Kanater, Egypt

Theo Olsthoorn
Amsterdam Water Supply, Netherlands

Ebel Smidt
IWACO B.V., Netherlands

ABSTRACT: Groundwater resources management in Egypt aims at optimal and sustainable use of this precious resource in the framework of overall integrated water resources management. Major concern of the management is controlled and sustainable groundwater abstraction, protection of groundwater resources from various sources of pollution and seawater intrusion, and prevention of waterlogging. Controlled artificial recharge of groundwater is an important tool for proper groundwater management within the wider objective of Egypt's Ministry of Public Works and Water Resources to manage surface and groundwater resources in an integrated manner.

Experience with uncontrolled, diffuse infiltration of irrigation excess water in the Western Delta fringes and controlled basin infiltration at the Bustan Extension Area is used to evaluate the applicability of artificial recharge in Egypt. Rainwater and treated urban sewage water are considered as alternative sources for artificial groundwater recharge. The integration of artificial recharge in Egypt's water resources management is presented in the paper.

1 WATER RESOURCES IN EGYPT

Egypt's main water resource is the Nile river. The Nile agreement with Sudan allocates 55.5 BCM/yr to Egypt, which amount is secured by the multi-year regulatory capacity provided by the Aswan High Dam. The second water source is groundwater in the Western Desert region and in Sinai in non-renewable aquifer systems. The annual extraction from the main aquifer in these areas, the Nubian Sandstone aquifer system, is about 0.5 BCM/yr. Rainfall in the coastal zones of Egypt reaches about 1.5 BCM/yr but is a rather unreliable source due to its spatial and temporal variability.

Groundwater in the Nile aquifer system and desert fringes is not a resource in itself as it is replenished from the river Nile by seepage from canals and deep percolation from irrigation application. The annual groundwater extraction in the Nile aquifer system and fringes is about 4.8 billion m^3. Groundwater extraction is expected to increase considerably in the forthcoming years, thus increasing the relative importance of groundwater in the national water resources management (Fig. 1).

Although the international community has acknowledged Egypt's efforts on birth control, the present population of 62 million is still continuing to increase. Increased population and higher living standards resulted in a situation where Egypt is to face the challenge of a water scarcity. Degradation of the land and water resources due to socio-economic exploitation adds up to the water scarcity problem. Therefore the overall objective of the national water policy for the 21st century is to utilize the available convential and non-convential water resources to meet the socio-economic and environmental needs of the country. The formulated policy focuses on three major aspects: demand management, resources development and environmental protection (Abu Zeid 1997). Although the details of the policy are still under preparation, a preliminary assessment of strategies to augment the supply of water, to increase the efficiency of the water use or to control the demand of water has been made. Some of the most promising strategies are (Attia 1997):

- Conservation projects in the Upper Nile Basin, expected to increase Egypt's share with 2 BCM/yr by 2010 (Junglei Canal, Phase I) and another 7.5 BCM/yr on the long term (Junglei Phase 2, Bahr El-Ghazal and Machar swamps projects)
- Recycling Nile aquifer water: the maximum sustainable yield is about 7.5 BCM/yr (see fig. 1)

- Recycling drainage water: actually 4 BCM/yr is recycled, while an increase of another 3 BCM/yr could be added
- Recycling wastewater after treatment: if public health aspects are properly taken into consideration the present reuse of 0.5 BCM/yr could be increased to 2.5 BCM/yr by 2010.
- Utilization of desert groundwater: an increase of 3.5 BCM/yr to a total of 4 BCM/yr is expected to be viable with a reasonable economic return.
- Utilization of rainfall: part of the total rainfall can be retained behind small, low-cost dams.
- Cost recovery as a special case of water pricing: generation of revenues for operation and maintenance of the irrigation and drainage network based on owned area or cultivated area and type of crop
- Improvement of water quality by enforcing the law concerning water quality protection
- Limitation of land reclamation
- Raising the public awareness on issues of water use and conservation.

These strategies can only be implemented when all water resources are managed in an integrated and an interactive manner. In this paper the question will be addressed which role controlled artificial recharge of groundwater can play in the implementation of the Egyptian water policy.

2 HISTORY OF ARTIFICIAL RECHARGE IN EGYPT

2.1 *Uncontrolled artificial recharge*

Large scale irrigation projects have been initiated during the sixties and seventies as a result of the construction of the High Aswan Dam, enabling the expansion of agricultural land from the central parts of the Nile Valley and Delta to its fringes. In the early stages of these developments the traditional irrigation method of flooding was practised. This method was adequate in the old land of the Nile Valley where clayey topsoil and semi-confined aquifer conditions prevail. As the topsoils in the new irrigation areas are sandy and the aquifers phreatic, considerable amounts of percolated excess irrigation caused a rapid rise of the groundwater levels, resulting locally in water logging and salinization. An example of this uncontrolled artificial recharge is the Western Delta fringes area (Fig. 2). During the first 15 years after reclamation, when adequate drainage was absent, the main rise of the groundwater table was observed (maximum 20 meter). In this period more than 150 mm/yr groundwater recharge occurred, resulting in a total estimated groundwater recharge of 6.5 BCM, of which about 4.3 BCM is stored in the groundwater mound (Attia et al., 1995). As the salt content of the irrigation water is low (normally less than 500 ppm) and high percolation rates (50-75%) are common practise, the salt content of the infiltrated water has not exceeded 1,000 ppm, which is of the same magnitude as the groundwater. Due to improved irrigation techniques and drainage systems the annual groundwater recharge in the area is reduced to about 50 mm/yr, while the groundwater outflow to the surrounding areas is about 100 MCM/yr. The effect of this built up can be observed at distances till tens of kilometres, for example in the area of controlled recharge experiments, which is approximately 30 km south-east of the northern irrigation areas and 15 km south-west of the Old Bustan Area. Groundwater monitoring during 1995 – 1998 indicated an annual groundwater rise in the area of 0.3 – 0.4 m (Fig. 3).

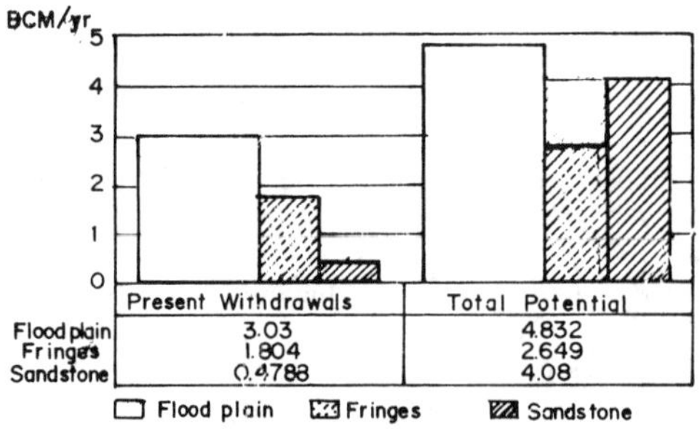

Figure 1. Groundwater use and potential in Egypt (1996)

2.2 *Controlled artificial recharge*

The application of controlled artificial recharge in Egypt has been triggered by the drought period in Upper Nile catchment areas at the end of the eighties, which renewed discussions on the need of storage mechanisms additionally to the High Aswan Dam (Attia, 1993, NEI/IWACO, 1992 and Romijn, 1993). In 1993 Egypt's Ministry of Public Works and Water Resources launched the Artificial Recharge Programme, executed by the Research Institute for Groundwater (RIGW). The programme is part of the Environmental Management of Ground Water Resources Project, executed by RIGW in the framework of cooperation between the Egyptian Ministry of Public Works and Water Resources and the Netherlands' Directorate General of International Cooperation. The project lasts from 1993-1998.

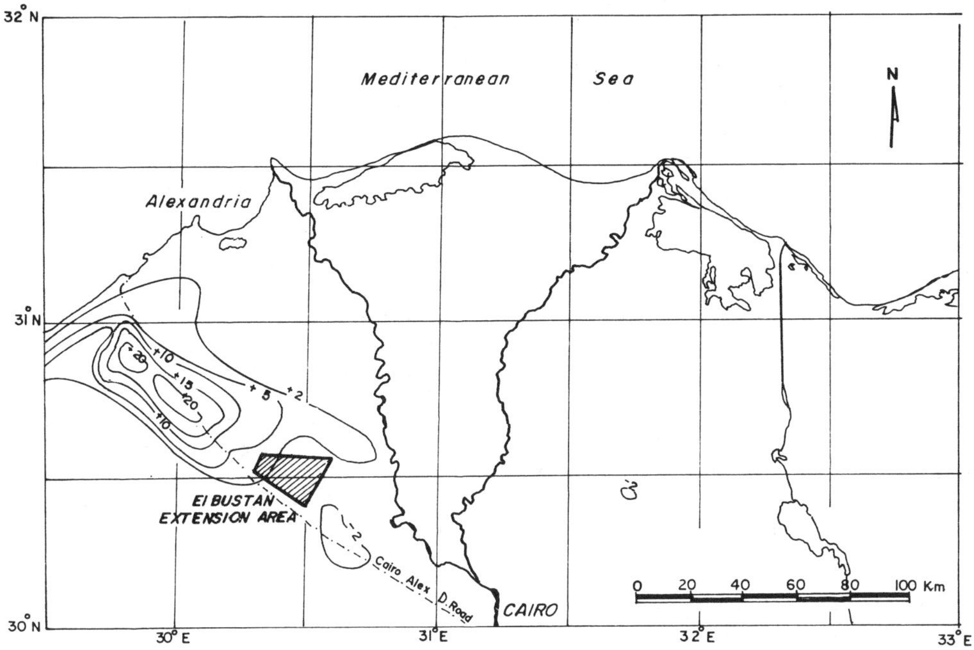

Figure 2. Groundwater level rise (in meter) between 1960 and 1990 in the Western Nile Delta.

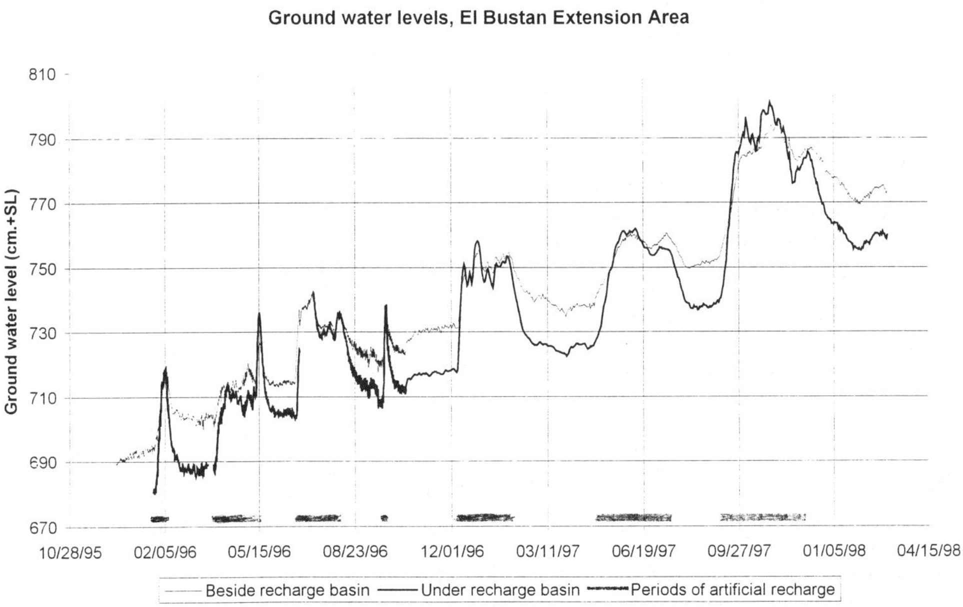

Figure 3. Groundwater levels 1995 – 1998, infiltration basin in Bustan Extension Area.

3 ARTIFICIAL RECHARGE PROGRAMME

From 1994 experimental work has been conducted in two sites: Bustan Extension, at the desert fringe about 150 km Northwest of Cairo and Bahig Hammam Area in the coastal zone about 50 km west of Alexandria. In Bustan Extension Area an infiltration basin of 135 x 155 m with an average depth of 2.5 m has been constructed. The deep excavation resulted from the need to remove the most loamy and clayey part of the low permeable top layer (about 10 m thick) covering the gravelly and sandy aquifer . In the same location also an injection well has been constructed, while in Bahig Hammam Area a scavenger well (doublet type) has been constructed which will be tested in 1998.

In Bustan Extension Area untreated Nile water is being provided by gravity through a gate at a side branch of the Bustan canal. A monitoring programme is executed for detecting the fluctuation of the groundwater by nine screens at different depths at the center of the basin, and 12 observation wells within a distance of 1 km from the basin. At larger distance other observation wells are used to monitor the natural groundwater fluctuations. An abstraction well will be constructed in 1998, enabling the recovery of infiltrated water.

Three infiltration tests have been executed in the 1996. Infiltration rates varied from 0.35 m/day to 0.08 m/day. Gradual clogging of the basin has been observed caused by the settlement of the suspended load of the untreated Nile water used for the infiltration. Settlement was eased by the low flow velocity in the basin due to the large water depth in the basin. Simulation of the recharge experiment with a linked saturated - unsaturated flow model gave good comparison between observed and calculated groundwater levels and infiltration rates. Further experiments are being executed in 1998, mainly focusing on enhancement of the infiltration rates. Details of the experiments and the numerical modelling are presented in these proceedings (Moustafa, 1998).

The conclusions from the executed experiments are:

- During controlled experimentation in 1996 0.24 MCM has been infiltrated in the artificial recharge basin in Bustan Extension Area. Scraping of the basin before new tests allowed for an average infiltration rate of 0.13 m/day.
- During three prolonged infiltration periods in 1997-1998 approximately 0.8 MCM has been infiltrated.
- Maximum groundwater level rise at the center of the basin side was 0.35 - 0.55 m, normally observed within 5 days after starting the infiltration. The temporary and locally built up groundwater mound appeared to be superposed on a regional trend of groundwater level increase, already observed before the start of the first experiment in January 1996 (see above and Figure 3).
- Cleaning of the basin, especially after dust storms, is required to prevent clogging
- The infiltration area in Bustan Extension Area has been selected as the best site after intensive hydrogeological investigations in a strip of about 25 km along a branch canal of Bustan Canal. Therefore only low to moderate infiltration rates (about 0.1 m/day) should be expected in the Western Delta fringes. As the topsoil and the aquifer in the Eastern Delta fringes is less clayey and calcareous, higher infiltration rates can be found in these areas (approximately 0.2-0.4 m/d).
- Leaching of the unsaturated zone in the beginning of the experiment resulted in temporary high TDS values of the recharged water. This process will influence the efficiency of recharge schemes, but is not considered impeding the application of infiltration basins in the area.
- The developed numerical code can be applied to test the feasibility of artificial recharge systems in different parts of Egypt, with varying hy-draulic characteristics. The maximum infil-tration rates can be forecasted.

4. FUTURE ROLE OF ARTIFICIAL RECHARGE IN EGYPT'S WATER MANAGEMENT

Most promising applications of future artificial re-charge schemes are:

Strategy of recycling Nile aquifer water

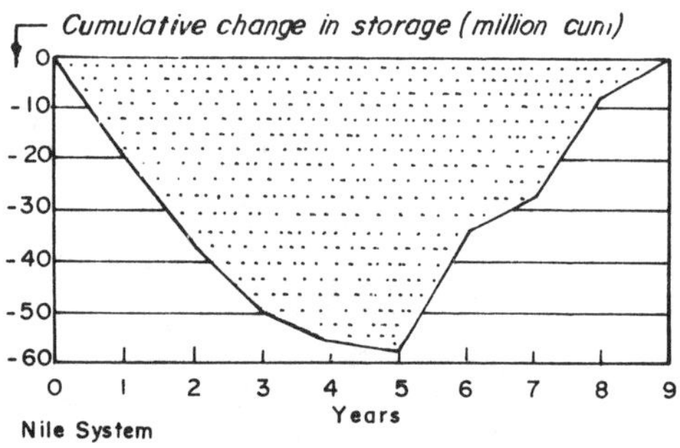

Figure 4. Groundwater management for drought mitigation - change in groundwater storage in part of the Nile Valley (RIGW, 1992).

- *Controlled use of the Nile aquifer system as a storage and treatment reservoir*. Figure 4 shows an example of multi-year groundwater management for drought mitigation. Review of the existing infiltration rates from canals, excess irrigation etc. is required to decide on the need for new infiltration schemes. In an integrated approach of the surface and groundwater systems, lining of main irrigation water carriers because of infiltration losses should be compared with the benefits of infiltration for the groundwater system.

Strategy of recycling wastewater after treatment
- *Surface infiltration of treated sewage water.* Reuse of treated sewage water is expected to increase from 0.5 BCM/yr at present to 2.5 BCM/yr by 2010. A preliminary selection of suitable sites of sewage based irrigation has been made for the Western Delta fringes (Figure 5) (Fekry and Attia, 1997). The groundwater reservoirs can be used as a storage and treatment reservoir. The order of magnitude of individual schemes will be tens to hundreds of MCM/yr.

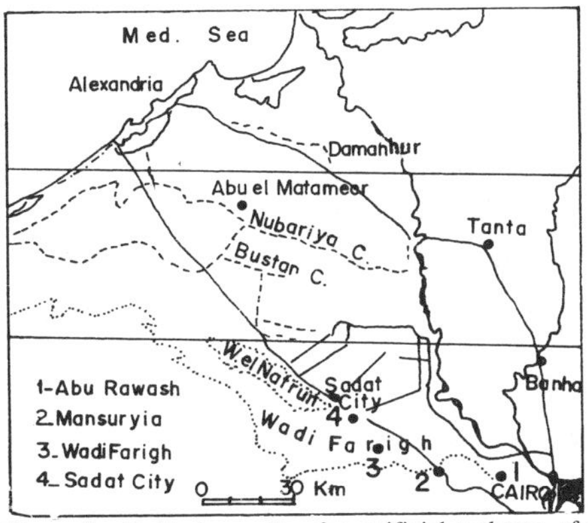

Figure 5. Preliminary sites for artificial recharge of treated sewage water in the Western Delta fringes (Fekry and Attia, 1997).

Strategy of recycling drainage water
(see strategy of improvement of water quality)

Strategy of utilization of rainwater
- *Diversion channels and small dams* aiming at decreasing direct losses to the sea or inflow to the Nile river (Northern coastal area, Golf of Suez and Aqaba, wadis discharging to the Nile valley); order of magnitude: tens of MCM/yr;
- *Overland spreading of Nile water originating from rainfall* (Nile Delta fringes); order of magnitude: tens to maximum about hundred MCM/yr

Strategy of improvement of water quality
- *Infiltration or injection of fresh water in the coastal areas* suffering from sea water intrusion (Northeast Delta); order of magnitude: tens to hundreds of MCM/yr originating from Nile flow, treated sewage water or drainage water
- *Surface infiltration, deep well injection or bank infiltration of Nile water* for public water consumption objectives. As the allowable costs for water are relatively high for public and industrial water supply, the highest economic return from artificial recharge schemes is expected from these schemes (new cities like Sadat City and 10th of Ramadan City, and old cities along the Nile or main carriers). The order of magnitudes of individual schemes is expected to be maximum some tens of MCM/yr.

5 CONCLUDING REMARKS

Large scale uncontrolled artificial recharge has occurred since the sixties in Egypt as a result of an agricultural expansion policy in combination with a water policy giving mainly attention to surface water resources. From a modern integrated water resources management point of view the groundwater mounds formed in this period can be considered strategic reserves. Appropriate management of these reserves and proper application of controlled artificial recharge in the forthcoming decades can contribute considerably to the successful implementation of the 21st century Water Policy of Egypt.

REFERENCES

Abu Zeid, M., 1997. *Egypt's water policy for the 21st century. Water Resources Outlook for the 21st century: Conflicts and Opportunities. Special Session on Water management under scarcity conditions: The Egyptian Experience. IXth World Water Congress of IWRA, Montreal, Canada, pp. 1-8.*
Attia, B.B., 1997. *Water resources policies in Egypt: Options and Evaluation. Water Resources Outlook for the 21st century: conflicts and Opportunities. Special Session on Water management under scarcity conditions: The Egyptian Experience. IXth World Water Congress of IWRA, Montreal, Canada, pp. 9-26.*
Attia, F.A.R., 1993. *Groundwater recharge in Egypt. Possibilities and implication. Regional seminar on the Potential of Artificial Recharge of Groundwater. 13-15 December 1993, University of Jordan, Amman.*
Attia, F.A.R., A. Fekry & E. Smidt, 1995. *Artificial recharge of groundwater: a tool for Egypt's future water management. Proceedings of the IWRA international conference on water*

resources management in arid countries, Muscat, 12-16 March 1995. Pp. 183-189.

Attia, F.A.R., E.H. Smidt, A. Fekry, M. Mostafa, 1995. *Artificial recharge of groundwater in Egypt: two cases of future possible application.*

Fekry, A. & F.A.R. Attia, 1997. *Preliminary selection of suitable sites for sewage based irrigation in Egypt. Proceedings of WTSA, third Gulf water conference, Muscat. Oman.*

Moustafa, M. 1998. *Execution and results of Artificial Recharge Experiments in Bustan Extension Area (Egypt). TISAR, Amsterdam.*

NEI/Iwaco, 1992. *Lake Burullus environmental and economic resources assessment study. Nile water conservation.*

RIGW, 1992. *Water security project: Groundwater Resources and Protection of Groundwater Development.*

1.3 Clogging

Artificial Recharge of Groundwater, Peters et al. (eds) © 1998 Balkema, Rotterdam, ISBN 90 5809 017 5

A conceptual and numerical model to characterize clogging

A. Pérez-Paricio & J. Carrera
Universitat Politècnica de Catalunya, Barcelona, Spain

ABSTRACT: Clogging is one of the most troublesome phenomena concerning Artificial Recharge. Different approaches exist in dealing with clogging, such as the measurement of specific parameters and determination of correlations, as well as empirical models. More sophisticated models have also been suggested in order to overcome the limitations of more simplistic methods. However, because clogging is affected by several mechanisms, a comprehensive model was not previously available.

A new generic model is presented in this paper, which includes five fundamental processes: accumulation of suspended sediments, bacterial growth, chemical reactions (precipitation/dissolution), generation of gas, and compaction. As well as the mathematical framework, a brief description on the numerical implementation is given. An application of the model to a synthetic example, where a limestone confined aquifer is recharged by more acidic water that contains suspended sediments, is presented. This example shows how two processes, both of a very different nature, can modify the aquifer's effective porosity and, hence its permeability. Other issues, such as the validity of the conceptual model, its limitations and future work, are briefly discussed.

1 INTRODUCTION

Clogging is a ubiquitous and may cause serious difficulty in Artificial Recharge systems, irrespective of specific conditions such as the water quality, aquifer properties or type of recharge system. This is why many studies have been conducted to improve our knowledge of this problem (Pérez-Paricio 1998).

But at least two main problems remain unsolved. First, the basic mechanisms are not properly described from a quantitative point of view, especially with regards the combined effects of different processes. Second, its prediction is a very complex task that extends beyond existing capabilities, as recognised by Pyne (1995). However, promising techniques are under continuous experimentation.

In any case, a new approach is needed. This paper summarises some of the basic aspects of a comprehensive clogging model which has been developed within the framework of the European Union Project entitled *Artificial Recharge of Groundwater*. The conceptual model is based on a previous thorough literature review (Pérez-Paricio 1998), which served to identify the main clogging processes and several mathematical expressions, as well as to suggest interesting alternatives.

Five elementary processes can be identified: accumulation and detachment of suspended particles, bacterial growth and die-off, chemical precipitation and dissolution, gas generation, and compaction of the clogging layer.

Very interesting models are currently available to account for individual processes. This is certainly the case for physical clogging (caused by suspended particles) or bacterial clogging (Pérez-Paricio 1998).

Among the former, Iwasaki (1937) introduced a filtration parameter to relate the spatial distribution of the concentration of particles with the concentration itself. Herzig et al. (1970) used it to describe the dynamics of deep bed filtration. More recent models can be found, but a good example is given by Adin & Rebhun (1987). An attractive analytical approach is described in Osei-Bonsu (1996).

As for bacterial clogging, recent trends have involved the development of numerous models. Three types can be distinguished: strictly macroscopic, microcolonies-based, and biofilm-type models. It is beyond the scope of this paper to describe them, but Taylor & Jaffe (1990b) and the analysis performed by Vandevivere et al. (1995) are good examples.

Although the five-process classification is very useful for qualitative purposes, in reality these are effectively inter-connected. For instance, Bouwer and Rice (1984) found that denitrification gases were responsible for clogging during basin management operations. Therefore, for many situations it is not possible to consider only the impact of independent processes, as has been done up to now.

From this discussion it is evident that much work has been done to achieve a satisfactory quantitative description of clogging, but, unfortunately, current approaches have significant limitations. Here, a new comprehensive model is presented. Whilst limitations still exist, as will be described, this model should represent a notable advance.

The structure of this paper is as follows: first, the mathematical framework is briefly described; second, some comments are made concerning its numerical implementation; third, application to a synthetic example is presented. Finally, we discuss these results and future trends.

2 MATHEMATICAL FRAMEWORK

The basic equations are introduced below, although more detailed information shall be presented in future publications. In general terms, it is important to note the main features of the model:

1. Three phases are considered: liquid, gas and solid, with a set of possible species.
2. Flow of liquid and gas phases determined by Darcy's Law.
3. Deformation of the solid matrix. Solid displacements are calculated for a given (user-selected) mechanical constitutive model.
4. Transport of suspended particles, chemical species (solutes) and heat using the advection-dispersion equation. Bacteria are not transported, but grow attached to the medium at the present stage of the model.
5. Kinetic approach for attachment/detachment of particles, for precipitation/dissolution of species (also equilibrium) and for bacterial growth attached to the matrix. Both heterogeneous (between different phases) and homogeneous reactions are dealt with.
6. Intrinsic permeability changes are a consequence of porosity variations. These, in turn, are originated by the five processes introduced earlier. The Kozeny-Carman equation was adopted to relate intrinsic permeability to porosity.

2.1 *General equations*

Prior to a description of the specific equations that control kinetics of suspended particles, solutes and bacteria, the general equations governing flow and transport are given.

Liquid and gas pressures are obtained by performing mass balances of water and air (note that balances are made for the species not for the phases) within the porous medium. For water, this can be expressed as follows:

$$\frac{\partial}{\partial t}\left[\rho_l w_l^w S_l\phi+\rho_g w_g^w S_g\phi+\rho_s w_s^w(1-\phi)\right]+\nabla\cdot\begin{bmatrix}\rho_l w_l^w \mathbf{q}_l+\rho_l w_l^w S_l\phi\mathbf{u}-\mathbf{D}_l^w\nabla w_l^w+\\ \rho_g w_g^w \mathbf{q}_g+\rho_g w_g^w S_g\phi\mathbf{u}-\mathbf{D}_g^w\nabla w_g^w+\\ \rho_s w_s^w(1-\phi)\mathbf{u}-\mathbf{D}_s^w\nabla w_s^w\end{bmatrix}=R^w \qquad (1)$$

where superscript refer to the species (w for water, in this case) and subscript to the phase (l: liquid; g: gas; s: solid). The coefficients are density (ρ :: ML^{-3}), mass of species per unit mass of phase (w :: dimensionless), degree of saturation (S :: dimensionless), Darcy velocity ($\mathbf{q}$ :: LT^{-1}), porosity (ϕ :: dimensionless), diffusion coefficient times fluid density (D :: $ML^{-1}T^{-1}$), displacement velocity of the aquifer matrix ($\mathbf{u}$ :: LT^{-1}) and sink/source term (R :: ML^3T^{-1}).

By analogy, a similar expression could be written for the air balance. As for the balance of momentum for the porous medium, this reduces to the equation of stress equilibrium if inertial terms are neglected (Olivella et al. 1994):

$$\nabla\cdot\boldsymbol{\sigma}+\mathbf{b}=\mathbf{0} \qquad (2)$$

where $\boldsymbol{\sigma}$ is the stress tensor ($ML^{-1}T^{-2}$) and $\mathbf{b}$ is the vector of body forces ($ML^{-2}T^{-2}$). By providing an adequate constitutive model, equation (2) can be transformed into an expression in terms of the solid velocities, fluid pressures and temperature. Here it is only interesting to note that such a model leads to a compatibility condition between the total strain rate and solid velocities ($\mathbf{u}$).

Heat transport is studied by means of an energy balance in the system, for the three phases. Here this is mathematically written in terms of internal energy. Provided that temperature is the same for all phases, a unique equation for total energy is required:

$$\frac{\partial}{\partial t}\left[E_l\rho_l S_l\phi+E_g\rho_g S_g\phi+E_s\rho_s(1-\phi)\right]+\nabla\cdot\begin{bmatrix}-\lambda\cdot\nabla T+\mathbf{j}'_{El}+E_l\rho_l S_l\phi\mathbf{u}+\mathbf{j}'_{Eg}+\\ E_g\rho_g S_g\phi\mathbf{u}+\mathbf{j}'_{Es}+E_s\rho_s S_s(1-\phi)\mathbf{u}\end{bmatrix}=R^E \qquad (3)$$

where E is the specific internal energy corresponding to each phase, λ is the thermal conductivity coefficient (Fourier's law), T is absolute temperature (K), and $\mathbf{j}'$ represents the advective energy fluxes.

The final type of equations is given by mass balances of the different constituents, in order to obtain the temporal and spatial evolution of their concentrations. If the flow equation is used to simplify the transport expression, a generic mass balance equation can be written:

$$\frac{\partial}{\partial t}\left[\rho_l S_l \phi c\right] = \nabla\cdot\left[\mathbf{D}_l^c \nabla c\right] - \nabla\cdot\left[\rho_l \mathbf{q}_l c\right] + R^w c_{i/o} + r(c) \tag{4}$$

where c is the concentration of a given constituent (in moles per unit mass of liquid), $c_{i/o}$ is the external (recharge) concentration, and r(c) is a sink/source term that must be explicitly given. Precisely, this term allows us to take into account different kinetic processes, such as the accumulation and detachment of suspended particles and the precipitation of chemical species. The adopted equations are now given.

2.2 *Specific kinetic terms: Suspended particles*

The sink/source term for suspended particles has received much attention during the last decades, as filtration is used widely in science and engineering. A detailed literature review suggests that a definitive expression is not currently available, but empirical approaches are usually undertaken. There are various explanations for this, but, in general, it is due to the high complexity of the problem as a result of natural heterogeneities, failure of size-averaging approaches and microscopic uncertainties. Interesting models have been developed. Some of them, for instance, have tried to obtain a correlation between microscopic magnitudes and the macroscopic observations (Yao et al. 1971, Rajagopalan & Chu 1981).

We have opted for the Adin & Rebhun (1987) equation, making important modifications in the attachment and detachment coefficients.
Their original equation is given by:

$$r(c) = K_1 q c (F - a) - K_2 a J \tag{5}$$

where K_1 is the attachment coefficient (L^{-1}), K_2 is the detachment coefficient (T^{-1}), a is the concentration of retained particles, F is the filter capacity of the medium (dimensionless) and J is the hydraulic gradient. It must be pointed out that all the concentrations in equation (5) are dimensionless.

One of the problems with this equation lies with the coefficients. In order to best represent the actual physical situation, we have selected expressions for them based on the relevant processes.

Therefore, retention of particles is assumed to occur by surface forces (van der Waals and electrostatic), Brownian diffusion, and volumetric interactions (inertia, sedimentation and interception). In accordance with previous research (Herzig et al. 1970), surface forces and Brownian diffusion are only relevant for colloids (size lower than 1 μm). On the other hand, large particles (diameter larger than 30 μm) are only affected by volumetric interactions. Intermediate particles can be subject to all of them. Table 1 summarises these aspects, whilst making a theoretical distinction between the transport and the attachment mechanisms.

Our modified equation can be written as follows:

$$r(c) = K_a c - K_d a \tag{6}$$

where K_a is the new attachment coefficient (T^{-1}) and K_d is the new detachment coefficient (T^{-1}).

Table 1. Processes affecting retention of particles.

Transport mechanisms	Relevant for:
Sedimentation	$d_p > 10$ μm
Inertia	$d_p > 10$ μm
Brownian diffusion	$d_p < 1$ μm
Attachment mechanisms	Relevant for:
Interception	$d_p > 1$ μm
Electrokinetic forces	$d_p < 10$ μm
London-van der Waals	$d_p < 10$ μm

Previous work on the filtration coefficient for each physical process (Boyd and Ghosh 1974) allows us to write the attachment coefficient as a sum of different terms. Each term represents the transport and attachment mechanisms, which depend on various parameters and on the unknowns of the system (state variables):

Attachment = Interception + Inertia + Sedimentation + Brown + van der Waals + Electrokinetic (7a)

$$K_a \propto \left(\frac{d_p}{d_g}\right)^2 + d_p^{3/2} + \frac{g\left(\rho_p - \rho_l\right) d_p^2}{\mu_l q} + \left(\frac{K_B T}{\mu_l d_p d_g q}\right)^{2/3} + d_p^2 + \frac{2.5\cdot 10^{16} Q_p \left(Q_V d_g - 6 Q_S\right)}{\mu_l q d_p} \tag{7b}$$

where the following nomenclature applies:

d_p	(L)	Diameter of particle
d_g	(L)	Diameter of grains (aquifer)
ρ_p	(ML^{-3})	Density of particle
μ_l	($ML^{-1}T^{-1}$)	Viscosity of liquid
K_B	($ML^2T^{-2}K^{-1}$)	Boltzmann's constant
Q_p	(cul)	Charge on suspended particle
Q_V	($cul\cdot L^{-3}$)	Volumetric charge density
Q_S	($cul\cdot L^{-2}$)	Surface charge density

These terms have been obtained by several authors. Boyd and Ghosh (1974) made use of previous work to express interception, sedimentation, Brownian diffusion and electrokinetic forces.

With regards to the detachment coefficient (K_d), a different approach has been adopted. Most previous works tend to neglect it or to suggest linear relationships with the attachment coefficient (K_a). However, detachment of retained particles can be very relevant, especially in order to restore the performance of a clogged system. That is why we have implemented a new expression based on our conceptual understanding of the process. Detachment is enhanced if chemical changes (reduction of ionic strength, increase of pH value) or hydrodynamic variations (higher hydraulic gradients) occur. Thus:

$$K_d = 1 - A_I \Delta I + A_{pH} \Delta pH + A_J \Delta J \quad (8)$$

where I is the ionic strength at a given time, and J represents the hydraulic gradient. Relative changes (Δ) are considered, and three constants (A coefficients) have been introduced for practical reasons.

2.3 *Specific kinetic terms: Bacteria*

The sink/source term for bacteria-catalysed precipitation of minerals is described below. Among all the mathematical possibilities, we have opted by a multiple Monod equation, as this seems to be more appropriate to interpret real data.

Let x be the mass of a species of bacteria per unit volume of aquifer. Then, the mass balance for bacteria in the medium is described by

$$\frac{\partial x}{\partial t} + \nabla \cdot [x\mathbf{u}] = r(x)$$

$$\left\{\frac{\partial x}{\partial t} + \mathbf{u}\nabla \cdot x\right\} = -[\nabla \cdot \mathbf{u}]x + r(x) \quad (9)$$

$$\frac{Dx}{Dt} = -[\nabla \cdot \mathbf{u}]x + r(x)$$

The first equation includes a term to account for deformation of the solid (second from the left). The final equation is obtained by defining the material derivative (left term of second equation).

From these equations it is necessary to derive a mathematical relationship for the sink/source term, r(x). A multiple Monod approach serves this purpose:

$$r(x) = \mu_{max} \frac{[De^-]}{[De^-] + g_{De^-}} \cdot \frac{[Ae^-(0)]}{[Ae^-(0)] + g_{Ae^-}} \cdot \left\{\prod_{j=1}^{m} \frac{g_{Ae^-}(j)}{[Ae^-(j)] + g_{Ae^-}(j)}\right\} x - \mu_{dec} x \quad (10)$$

where the following notation applies:

μ_{max}	T^{-1}	Maximum growth rate constant
[De-]	ML^{-3}	Concentration of electron donor
g_{De^-}	ML^{-3}	Saturation value for electron donor
$[Ae^-]$	ML^{-3}	Concentration of electron acceptor
g_{Ae^-}	ML^{-3}	Saturation value of electron acceptor
m	L^0	Number of limiting electron acceptors (if they exist)
μ_{dec}	T^{-1}	Decay rate constant (die-off)

3 NUMERICAL IMPLEMENTATION

A new 3-D finite elements model, named CLOG has been developed based on two existing models: CODE_BRIGHT (Olivella & García-Molina 1996) and RETRASO (Saaltink et al. 1997), which were produced for different purposes at the Universitat Politècnica de Catalunya.

The former model accounts for coupled deformation of the medium and transport of brine, gas and heat. It is a 3-D code that deals with saline environments. The latter is a 2-D code capable of treating the reactive transport of solutes, with equilibrium and kinetic precipitation of minerals. Both of them resort to Newton-Raphson iterations at each time step to come to the solution of the system.

CLOG has been generated by integrating them. This involved implementing equations (6) to (10), as well as modifying the reading and chemical initialisation modules. Further information will be presented in future publications.

4 APPLICATION: CALCITE EXAMPLE

An illustrative synthetic example of CLOG is presented here. It considers a confined aquifer where groundwater was formerly in equilibrium with the limestone matrix. Then, recharge water which is acidic is added to the system, inducing a parallel flow pattern. As a result, dissolution of calcite takes place within the zone of the aquifer closest to the recharge zone.

The simulated domain is vertical 2-D, although it is physically unidimensional, as the vertical hydraulic conductivity is 1000 times larger than the horizontal one. Recharge takes place at the left boundary (2 m^3/d per metre of aquifer thickness), while a constant head condition is imposed on the right boundary. Table 2 compiles the most interesting issues, namely the geometry, hydraulic properties and chemical speciation.

Several scenarios were considered, including recharging with a 100 mg/l concentration of suspended solids whose diameter was 10 μm. In all cases the concentrations of both initial and boundary waters remained unchanged.

These scenarios are summarised in Table 3, but we will refer only to the most relevant results here. Run 2 served to obtain the steady state Darcy velocities for the chemical calculations. Run 3 corresponded to transient flow but assuming that porosity was con-

Table 2. Characteristics of the 'calcite' example.

GEOMETRY			
Domain:	2-D, vertical profile		
Dimensions:	Length	1000 m	
	Height	2 m	
Number of:	Nodes	42	
	Elements	40	
Element size:	50 m x 2 m (equal triangles)		
Time steps:	50 (2 days)		
MODEL PARAMETERS			
Conductivity:	Horizontal = 1 m/d		
	Vertical = 1000 m/d		
Specific storage:	4.E-07		
Porosity:	0.10		
Mean grain diameter:	d_g= 0.01 m		
Diffusion coefficient:	0.		
Dispersivity:	Longitud. = 100 m		
	Vertical = 1000 m		
CHEMISTRY		initial	boundary
Primary species:	H_2O	1.0E+01	1.0E+01
	H^+	1.0E-08	3.0E-06
	HCO_3^-	1.0E-03	3.0E-05
	Ca^{2+}	1.0E-03	4.0E-05
Secondary species:	OH^-		
	CO_2		
	CO_3^{2-}		
	$CaCO_3$		
	$Ca(HCO_3)_2$		
Minerals:	$CaCO_3$		
Suspended particles:	Mean particles d_p =10 μm		

Table 3. Summary of studied scenarios.

RUN	FLOW	TSS	Δϕ (DIS)	Δϕ (TSS)
2	Steady	No	No	No
3	Transient	No	No	No
4	Transient	No	Yes	No
5	Transient	Yes	No	No
6	Transient	Yes	No	Yes
7	Transient	Yes	Yes	Yes

Δϕ (DIS) = porosity changing by calcite dissolution
Δϕ (DIS) = porosity changing by suspended solids (TSS)
Note that the chemical conditions remained always constant.

stant. Therefore, calcite dissolution was not taken into account from the hydraulic point of view. Figure 1 shows the temporal evolution of liquid pressure, which reaches the steady state before 1 day time.

Run 4 considered porosity changes due to calcite dissolution, by means of a simple linear equation. Obviously, the final head build-up was smaller than in previous simulations, because the hydraulic conductivity was updated for every time step as a function of the current porosity value. The final porosity at the recharge zone was 0.14.

Run 5 included suspended particles, but with an almost negligible attachment rate. This implies that porosity remained constant, because the suspended sediments behaved like non-reactive species.

Run 6 took into account porosity changes caused only by attachment of suspended particles, neglecting the (most important, in our example) dissolution

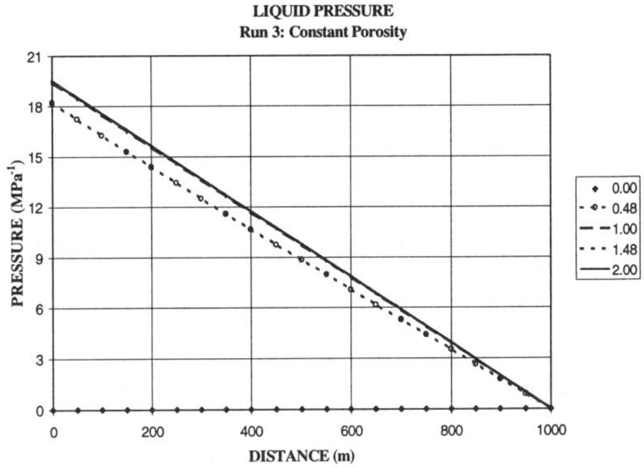

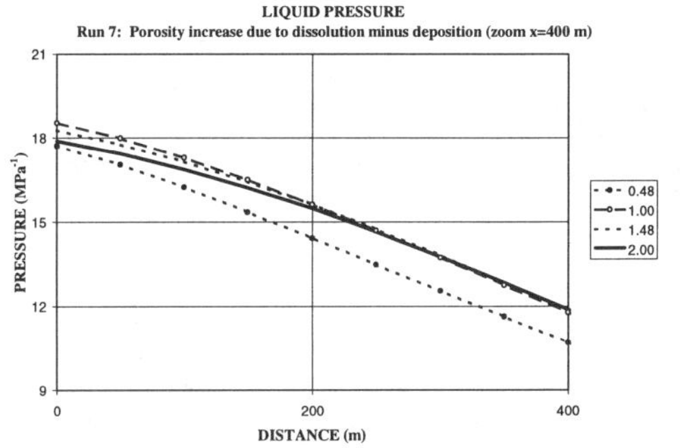

Figure 1. Pressure evolution at given times (right legend). Plot for run 7 has been zoomed, as physical changes take place only near the recharge zone. Times are expressed in days.

Table 4. Final properties at the recharge boundary.

RUN	ϕ_0	ϕ_f	K_0 (m/d)	K_f (m/d)	P_{l0} (MPa)	P_{lf} (MPa)
2	0.10	0.10	1	1	0	19.6
3	0.10	0.10	1	1	0	19.5
4	0.10	0.14	1	2.9	0	17.4
5	0.10	0.10	1	1	0	19.5
6	0.10	0.09	1	0.8	0	20.3
7	0.10	0.13	1	2.3	0	17.9

subscript 0 indicates initial value; f, the final value

term. Final porosity was 0.09, but it induced a significant increase of 0.74 MPa in liquid pressure (Table 4). Finally, run 7 merged both influences, yielding a final porosity of 0.13. Temporal evolution of liquid pressure indicates that the maximum build-up occurred at around 1 day, and then it gradually decreased due to the calcite dissolution (Figure 1).
Figure 2 depicts the temporal evolution of porosity for Run 6, where there is a reduction due to the deposition of sediments, and for Run 7, where calcite dissolution out the negative effect of particles.

5. DISCUSSION AND CONCLUSIONS

A new code (CLOG) has been developed that is capable of dealing with complex phenomena. CLOG

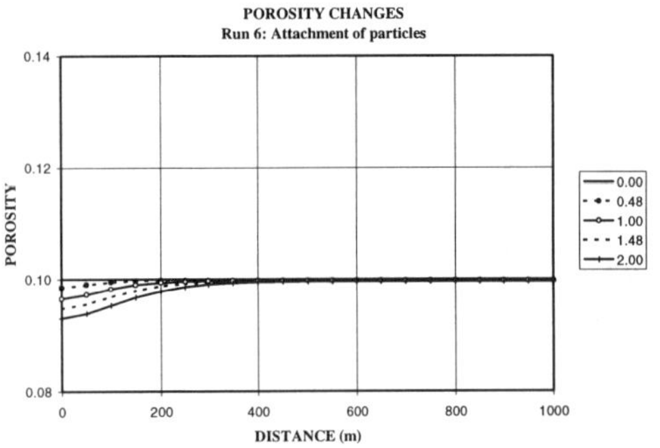

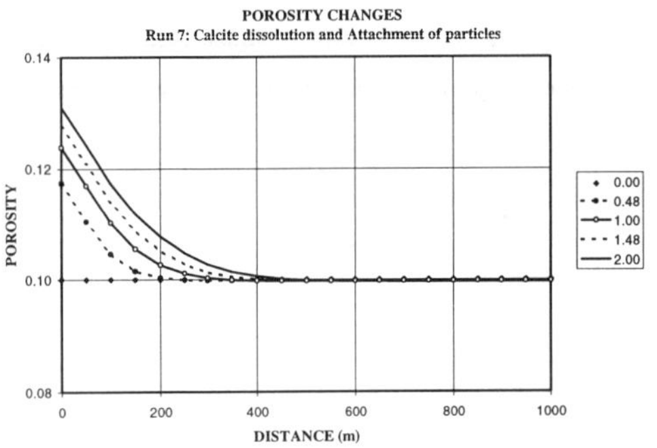

Figure 2. Effects of the different scenarios on aquifer porosity.

represents an important step in the understanding of the physical, chemical and biological processes within aquifers, especially under complicated recharge regimes (i.e. Artificial Recharge).

The synthetic example described in this paper provides some indication of the features of CLOG. Whilst the cases modelled are hypothetical, there are, in fact, interesting parallels with what has been observed in field recharge studies (Pavelic et al. 1998). This demonstrates the interest of CLOG, and, furthermore, its applicability to experimental data.

Uncertainties at field scales are still numerous, as some microbiological and geochemical parameters and reactions are poorly understood. In this respect our work is ongoing. The use of a quantitative, process-based, tool like CLOG is essential in determining the sustainability of recharge schemes.

ACKNOWLEDGEMENTS

We are indebted to the European Union for funding this work through the Project ARGW (Artificial Recharge of Groundwater).

Thanks are also due to the Generalitat de Catalunya (Catalan Government, Spain) for supporting a temporary stay of the first author at the Adelaide (South Australia) laboratory of the Centre for Groundwater Studies.

REFERENCES

Adin, A.& M. Rebhun 1987. Deep bed filtration: accumulation-detachment model parameters. *Chem. Engg. Sci.* 42(5): 1213-1219.

Bouwer, H. & R.C. Ricc 1989. Effect of water depth in groundwater recharge basins on infiltration. *J. of Irrigation and Drainage Engng.* 115(4): 556-567.

Issues in Artificial Recharge. *Proc. of the Int. Symp. on Groundwater Management*, San Antonio (TX, USA): 265-270.

Boyd, R.H. & M.M. Ghosh 1974. An investigation on the influence of some physicochemical variables on porous-media filtration. *AWWA J.* 66(2): 94-98.

Herzig, J.P., D.M. Leclerc & P. Le Goff 1970. Flow of suspensions through porous media -Application to deep filtration. *Industr. and Engng. Chemistry.* 62(5): 8-35.

Iwasaki, T. 1937. Some notes on sand filtration. *AWWA J.* 29(10): 15-91.

Olivella, S., J. Carrera, A. Gens & E.E. Alonso. 1994. Nonisothermal multiphase flow of brine and gas through saline media. *Transport in Porous Media.* 15: 271-293.

Olivella, S. & A. García-Molina 1996. CODE_BRIGHT: User's Guide. Universitat Politècnica de Catalunya.

Osei-Bonsu, K. 1996. Clogging by sediments in injected fluid flowing radially in a confined aquifer. Ph.D. Thesis, *Flinders University of South Australia*, Adelaide. 253 pp.

Pavelic, P., P.J. Dillon, K.E. Barry, A.L. Herczeg, K.J. Rattray, P. Hekmeijer & N.Z. Gerges 1998. Well clogging effects determined from mass balances and hydraulic response at a stormwater ASR site. *Proc. 3rd Int. Symp. on Artificial Recharge of Groundwater*, Amsterdam (these Proceedings).

Pérez-Paricio, A. 1998. Clogging and Artificial Recharge of Groundwater: Fundamental aspects. *M.Sc. Thesis Universitat Politècnica de Catalunya,* Barcelona. 170 pp.

Pyne, R.D.G. 1995. Groundwater recharge and wells: a guide to Aquifer Storage and Recovery. *CRC Press, Inc. (Lewis Publ.),* Boca Raton (FL, USA).

Rajagopalan, R., & R.Q. Chu 1981. Dynamics of adsorptionof colloidal particles in packed beds. *J. of Colloid and Interface Sci.* 86(2): 299-317.

Saaltink, M.W., C. Ayora & I. Benet 1997. RETRASO: User's Guide. Universitat Politècnica de Catalunya.

Taylor, S.W. & P.R. Jaffe 1990. Biofilm growth and the related changes in the physical properties of a porous medium. 1. Experimental investigation. *Water Resour. Res..* 26(9): 2153-2159.

Vandevivere, P., P. Baveye, D. Sanchez de Lozada & P. DeLeo 1995. Microbial clogging of saturated soils and aquifer materials: evaluation of mathematical models. *Water Resour. Res.* 31(9): 2173-2180.

Yao, K-M, M.T. Holsibian & C.R. O'Melia 1971. Water and wastewater filtration: concepts and applications. *Envir. Sci. Technology.* 5(11): 1105.

Artificial Recharge of Groundwater, Peters et al. (eds) © *1998 Balkema, Rotterdam, ISBN 90 5809 017 5*

Well clogging effects determined from mass balances and hydraulic response at a stormwater ASR site

P. Pavelic, P.J. Dillon, K.E. Barry & A.L. Herczeg
Centre for Groundwater Studies and CSIRO Land and Water, Glen Osmond, S.A., Australia

K.J. Rattray & P. Hekmeijer
Centre for Groundwater Studies and Flinders University of South Australia, S.A., Australia

N.Z. Gerges
Primary Industries and Resources, S.A., Australia

ABSTRACT: The nature and extent of well clogging at an experimental Aquifer Storage and Recovery (ASR) site in South Australia, where urban stormwater has been opportunistically injected into a confined carbonaceous aquifer was studied over four years. Mass balances of injected and extracted sediment and microorganisms were determined from flow rate, concentration and size distribution data, and the quantity of aquifer materials dissolved inferred from a geochemical mass balance model derived from a series of groundwater samples. Piezometric head response between the injection well and a nearby observation well allowed a comparison between years where injectant quantities and qualities differ. Our results reveal no significant clogging by either coliform, Pseudomonas or heterotrophic iron bacteria, but there is some clogging by injected sediment, especially where sediment concentrations are high. However, most of the clogging is believed to be offset by increased secondary porosity through calcite dissolution, which also raises the issue of rearrangement and mobilisation of residual aquifer sediments. The artificial recharge capacity at this site has been shown to be sustained over the four years, but issues related to the fate of injected sediments and the stability of the injection well remain unanswered.

1 INTRODUCTION

Clogging of injection wells is a complex problem. It can be physical (from accumulation of particulate matter in the injectant), chemical (from precipitations reaction in the aquifer), or biological (from bacterial slime production) (Olsthoorn 1982; Osei-Bonsu 1996). Clogging is characterised by a reduction in the pore space of the aquifer, which can reduce the aquifer's capacity to store and transmit water. In some circumstances this may limit the viability of the artificial recharge operation. Therefore, any study on artificial recharge must at least consider the potential for clogging. For instance, a recent survey of current and historic artificial recharge practices by injection (ASR) has shown that 70% of injection wells report some degree of clogging (Pavelic & Dillon 1997). This survey also showed that most studies on clogging deal with waters of high quality, typically at or near potable standards. In addition, studies tend to be of a relatively brief duration, typically as little as one injection event or season, and those that are more extended rarely take the measurements necessary to quantify the rates of clogging or the success of any remedial or preventative measures.

This study attempts to challenge this trend by considering ASR with relatively low quality stormwater, by considering hydraulic, chemical and microbiological data collected from a field site over four years. Its aims are to examine the nature and extent of clogging, and to determine the success of planned (as well as unplanned) measures to unclog the aquifer. The sustainability of ASR for this particular system will also be addressed.

2 SITE DESCRIPTION AND EXPERIMENTAL PROGRAM

This study commenced in April 1993 at the experimental research site known as Andrews Farm in South Australia. Its broad objectives are to determine the technical viability and economic sustainability of injecting urban stormwater during wet months into a brackish aquifer for the purpose of recovering this water in dry months for irrigation supplies. The site has been a focus for a research program conducted by the Centre for Groundwater Studies in Adelaide (Dillon & Pavelic 1998).

The aquifer targeted for injection is a confined

Tertiary carbonaceous sand, intersected at a depth of 105 metres. A transmissivity of 180 m^2 day^{-1} and storage coefficient is $5x10^{-4}$ have been measured through aquifer testing (Gerges et al. 1995). Three observation wells were drilled at distances of 25, 65 and 325 metres down-gradient of the injection well to examine changes in water quality and piezometric head. Ephemeral stormwater runoff from a peri-urban catchment passes through a small constructed wetland and is stored in a pond before being pumped via a screen into the injection well.

Four seasons of injection took place between August 1993 and October 1996 (Fig. 1). Intermittent injection occurred during the winter to spring period. Volumes injected were limited by the amount of stormwater available. The first season included both mainswater and stormwater, whilst the second, third and fourth were stormwater only. A net total of 248 ML of water has been added to aquifer storage with little abstraction up to March 1997, apart from a number of small redevelopment events and a refill of the wetland in 1993/94. Redevelopment occurs by airlifting whenever rates of recharge decline to levels that the site operators (Primary Industries and Resources, South Australia), consider to be inadequate or when piezometric heads approach being artesian. A major extraction phase has since commenced, but at the time of writing was incomplete and will not be discussed here. In all cases, the average rates of injection varied between 15 and 20 Ls^{-1}.

Characteristically, the stormwater is colder, has higher concentrations of dissolved oxygen, suspended solids, organic carbon, coliforms and higher pH and turbidity, but has lower concentrations of dissolved solids than the ambient groundwater which is brackish (Table 1).

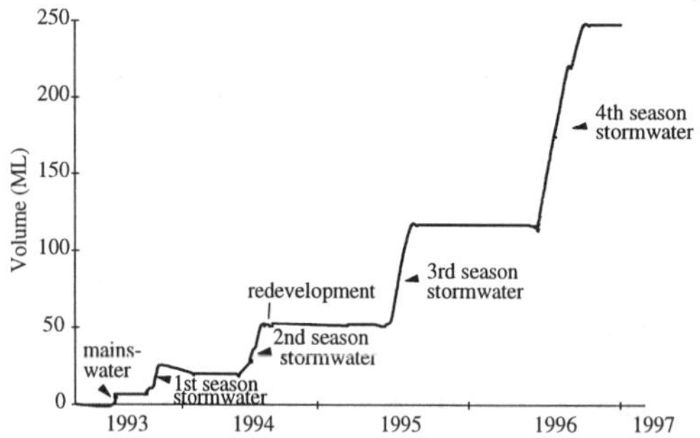

Figure 1. Cumulative net volume of injected water. *The focus for the mass balance calculations is the second season. Intervals of pumping can be inferred by a decrease in volume.*

Table 1. Composition of typical stormwater and ambient groundwater

Parameter[I]	Stormwater[II]	Groundwater[IV]
pH (-)	8.2	7.7
Temperature (°C)	10.1	24.0
Dissolved Oxygen	9.7	0.7
Turbidity (NTU)	50	13
Suspended solids	29	13
Total dissolved solids	190	2340
Bicarbonate	118	228
Calcium	23.6	237
Dissolved Organic Carbon	4.3	1.2
Total coliforms (cfu /100mL)	620[III]	0

[I] mg L^{-1} unless otherwise stated [II] mean and [III] median concentration in 1994 [IV] 65 m observation well (pre-injection)

The key processes that we will focus on here include accumulation of injected particulates, as well as the potential for microbial growth and production of their extracellular polymers (polysaccharides). Calcite dissolution may also have the potential to impact on aquifer permeability. Mass balances for each of these parameters are determined for one season, the second (1994), in which 32.7 ML was injected and 28 days later redevelopment through airlifting extracted 0.6 ML. Piezometric head information are used to compare between years.

3 METHODS

3.1 *Suspended sediment and microorganisms*

Fluxes of particulates (including both suspended sediment and microorganisms) injected and recovered from the injection well were determined. The flux is the product of concentration and flow rate, and for injection this is relatively easy to determine. The flux extracted from the aquifer during redevelopment can vary significantly with time, and requires careful integration of time-series concentrations, particularly at the beginning of any withdrawal period. In the case of suspended sediment it is necessary to distinguish between the contributions derived from the two different sources: the injectant and the aquifer. This is made feasible in this case because the two sources have very different and characteristic particle size distributions. In the case of microorganisms, partitioning is not necessary as the reference microorganism used here (coliform bacteria), was not present in the groundwater prior to injection.

3.2 *Carbonate reactions*

From the perspective of well clogging in a carbonate aquifer system, the most important chemical reaction to consider when oxygenated low salinity stormwater containing substantial amounts of organic matter is injected into an anaerobic brackish aquifer is calcite dissolution or precipitation. This is described by the reaction: $CaCO_3 + CO_2 + H_2O \leftrightarrow 2HCO_3^- + Ca^{2+}$. Of course, the direction this reaction takes can either alleviate or contribute to clogging.

The mass balance approach relied on determining the amount of calcite added to, or subtracted from, the groundwaters at the injection well and nearest observation well as a result of geochemical reactions through measurement of the bicarbonate concentration. Mixing processes (as distinct from geochemical processes), would influence the bicarbonate concentration through diffusion/ dispersion, but could be quantified and accounted for from chloride data, as this ion behaves conservatively. A more extensive description of this aspect, and the geochemical processes that operate at this site in general, can be found in Rattray et al. (1996).

3.3 *Piezometric head response*

Differences in the rate of piezometric rise between the injection well and a nearby observation well provide one means for evaluating temporal changes in clogging (Olsthoorn 1982). This, so called "water level difference" method is based on the Theis equation, and so is subject to standard assumptions concerning aquifer homogeneity and uniform injection rate. According to the Theis equation, during an injection event, the rate of piezometric head rise between these two wells should be equal if there is no clogging after some quasi steady-state is reached. Therefore, any deviations from this behaviour could be attributed to clogging. This approach provides a measure of the "clogging rate", and comparisons between seasons must be normalised to account for variations in injection rate and water temperature (which affect velocity around the injection well and viscosity respectively). Of course, this clogging rate does not provide direct information on the type of clogging, nor its precise position in the aquifer.

Water level differences between the 25 and 65 m observation wells were used. The 25 m well was chosen instead of the injection well because there was a greater consistency of data. It provides an excellent surrogate, as Gerges et al. (1995) has shown, due to the preferential hydraulic path created during the initial well development between these two wells. Paths such as these do not constitute a substantial proportion of the aquifer's volume, and so in general, the flow regime may be considered to be that of radial flow, albeit with an expanded well radius.

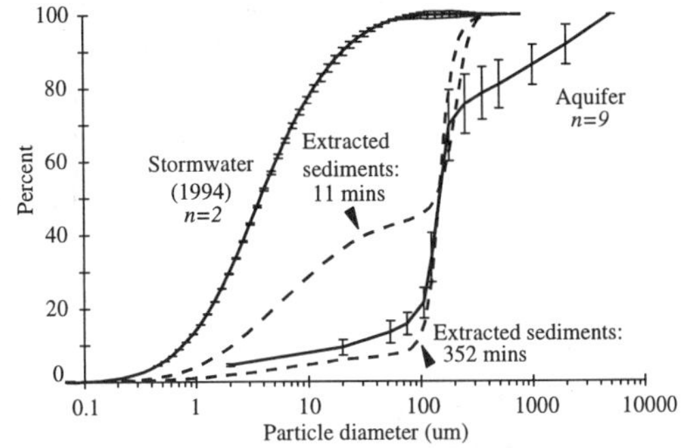

Figure 2. Cumulative particle size distributions of aquifer, stormwater and extracted sediments. *Standard deviations are shown by bars for the aquifer and stormwater.*

4 RESULTS AND DISCUSSION

4.1 *Sediment balance*

The cumulative particle size distribution curves for the aquifer and stormwater (second season) differ markedly, with little overlap between the smaller stormwater particles and the larger aquifer particles (Fig. 2). Their median particle sizes are ~4 μm and ~150 μm respectively. The uniformity of the injectant's distribution in that season is reflected by the small standard deviations that are superimposed on the plot.

During the airlifting event which followed the second season of injection, the total suspended solids concentration in the extracted water peaked at 12,500 mg L^{-1} almost immediately after the water within the well casing had been purged, and tailed-off thereafter towards its lowest value at the end of the event (Fig. 3a). This peak value was almost three orders of magnitude greater than the average concentration injected (29 mg L^{-1}). The particle size distribution at 11 minutes (Fig. 2) clearly shows a combination of both stormwater and injected sediments. Initially the extracted waters contain a higher proportion of stormwater sediments when suspended solids concentrations are high, whilst both the suspended solids concentration and stormwater contributions decline with time.

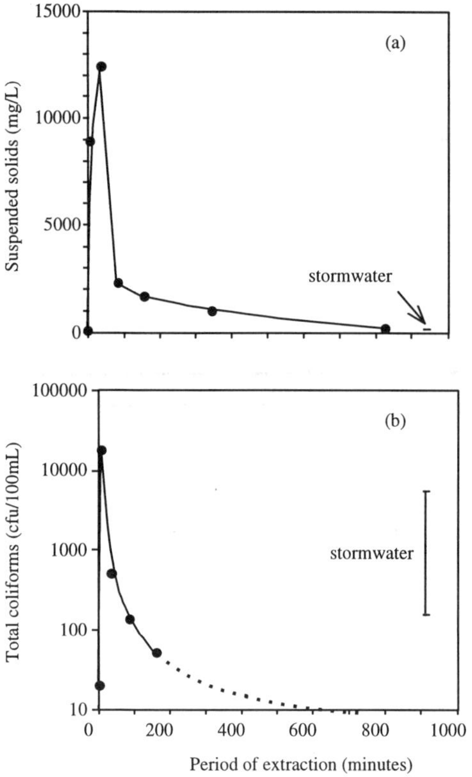

Figure 3. Changes in the concentration of (a) suspended solids and (b) coliforms in extracted waters during airlifting. *The rate of airlift pumping was constant at 13 Ls^{-1}.*

The mass balance results are summarised by Table 2. This shows that around 950 kg of sediment was injected into the aquifer, and a slightly higher quantity of sediment was extracted. However, by far the greatest proportion of sediments are derived from the aquifer (~99%), with only ~1% of sediments derived from stormwater. This figure represents the flux-weighted mean percentage, whilst over the course of the airlift percentages varied from around 4.9 to 0.5%.

As a consequence, this means that virtually all of the injected sediments are retained to eventually settle within the porespace of the aquifer. The data also suggests that these injected sediments do not accumulate around the well, but penetrate further into the aquifer to distances where airlifting (and perhaps any mechanical method), is ineffective in their removal. The mechanism leading to the mobilisation and recovery of aquifer sediments will be described later.

Table 2. Mass balance for injected and recovered waters in the second season

	Flux in or out	
SEDIMENTS	(kg)	(%)
Injected sediments	948	100.0
Extracted sediments	988	104.0
- from stormwater	7	1.2
- from aquifer	981	98.8
COLIFORMS	(cfu)	(%)
Injected coliforms	2.0×10^{11}	100.0
Extracted coliforms	3.3×10^{9}	1.6
CALCITE	(kg)	(%)
total dissolution	2300	n/a

4.2 *Microbial balance*

Total coliform bacteria were also measured during the first 163 minutes of the airlift, and Figure 3b shows that concentrations peaked at 17,000 cfu per 100 mL within a few minutes of pumping (Fig. 3b). During injection concentrations ranged from 160 to 5,800 cfu per 100 mL, with a median value of 620 cfu per 100 mL. Extrapolating the curve from 200 to 832 minutes (a reasonable approximation given the relatively low concentrations over this range) allows the total quantity recovered to be estimated at 3.3×10^{9} cfu, or ~2% of that injected.

It is interesting to note the correlation between the suspended solids and coliforms in Figure 3 (R^2 = 0.54 on a semi-log basis). Their extracted fluxes were also similar. This is not altogether surprising given that microorganisms tend to be associated with the solid phase. Hence it does not appear that coliform bacteria have multiplied around the injection well here. However this has been shown to occur elsewhere (eg. Vecchioli 1970 for wastewater injection into a granular aquifer), and moreover, bacteria other than coliforms have the potential to cause clogging.

Consequently, sample collection during a pumping phase which followed the fourth season considered a greater diversity of microorganisms. This showed that Pseudomonas spp. counts were not sufficiently high (<10^4 cfu per 100 mL) as to warrant concern from a clogging perspective. Heterotrophic iron bacteria were found to be present in much higher numbers (up to ~10^6 cfu per mL), but these levels were not greater than those found in either the injectant or the ambient groundwater.

The data suggests that microbial growth is probably not a significant concern with respect to the clogging potential at this site with the low nutrient content of stormwater. However, potential for regrowth has been demonstrated at this site when nutrient conditions are favourable which would have implications for aquifer clogging if the nutrient status of the stormwater were to change dramatically.

4.3 *Carbonate reactions*

In general, there is more bicarbonate in the groundwater around the injection well than expected by conservative mixing between low salinity stormwater and the higher salinity receiving groundwater. Dissolution tends to be greatest during and immediately following injection, and is accentuated by CO_2 production, predominantly from oxidation of organic matter. When the groundwater returns to an anaerobic condition, dissolution rates steadily decline. . On average, the stochiometric estimate of bicarbonate produced by reactions was ~0.7 mMol L^{-1}. Therefore the estimated quantity of calcite minerals dissolved from the aquifer in the second season is in the order of 2,300 kg. Clearly this is a significant flux in terms of clogging, and recalling that this is equivalent to over twice the amount of sediment injected into or recovered from the aquifer in that same season. Given that the calcium carbonate content of the aquifer ranges from 10 to 70%, this quantity is small with respect to the total present in the aquifer. Approximate calculations suggest <0.01% of the aquifer mass was lost in the injection zone.
Dissolving the calcite, which is present as a cementing agent for minerals such as quartz, has the potential to impact on ASR in a number of ways. The porosity, and hence the permeability of the

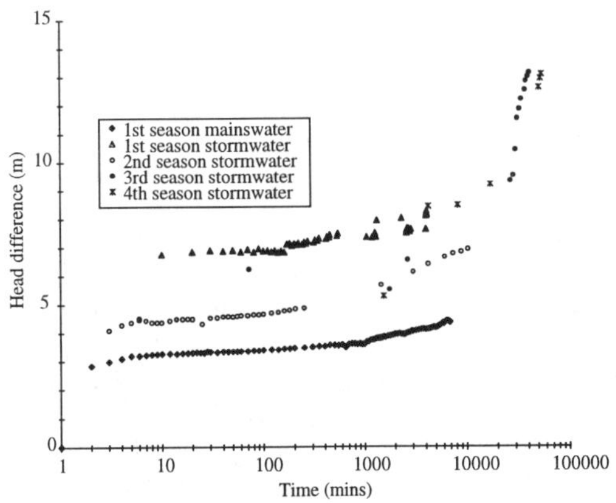

Figure 4. Semi-log plot of head difference between the 25 m and 65 m observation wells for periods up to 50,000 minutes. *Note that for in the first season mainswater and stormwater are treated independently.*

aquifer would probably increase. This has been shown to occur in the laboratory through column experiments with materials from the same aquifer system at a nearby ASR experimental site (Rinck-Pfeiffer et al. 1998). Further, dissolution can mobilise sand grains which can then be extracted during airlifting. In addition, these same sediments may also penetrate further into the aquifer by radial flow. This has been observed at the 65 m observation well during the first three seasons especially, where suspended solids peaked at over 700 mg L^{-1} during periods of injection, and whose composition was consistent with aquifer sediments. Under such circumstances there is the potential for collapse of the injection well.

4.4 *Piezometric response*

The water level differences as a function of the injection time in each season are shown by Figure 4 (mainswater and stormwater have been distinguished in the first season). It is quite apparent that the response for each season differs. The initial head difference value reflects the different injection rates and temperature in each season. The range of head differences for the first 10,000 minutes (~7 days) are lowest for mainswater, and highest for the third and fourth season's stormwater injection. This appears to be related to the concentration of suspended solids in the stormwater, which vary from season to season and also within seasons (Table 3). Average concentrations in the stormwater range from 29 mg L^{-1} in the second season to 169 mg L^{-1} in the fourth.
The continual rise in the piezometric head differences must be attributed to clogging of the aquifer between observation wells. This is the case even for mainswater, where suspended sediment concentrations are probably negligible (inferred but not measured). Normalised to a temperature of 20°C and injection rate of 20 Ls^{-1}, rates of clogging measured in the first 50,000 minutes (~35 days) range from 63 m yr^{-1} in the fourth season to 154 m yr^{-1} in the first. These rates appear to be dependent not only on the suspended solid concentration, but also on the temperature, rate and duration of injection.
Recalling from section 4.2 that it is likely that the rate of unclogging through calcite dissolution exceeds the rate of clogging by injected particles (Table 2). The spatial distribution of calcite dissolution is expected to influence the effectiveness of unclogging.
Only the third and fourth seasons extended beyond 50,000 minutes, but were discontinuous due to a brief redevelopment event at 37 days and 41 days respectively. After these events rates of clogging

Table 3. Clogging and unclogging parameters for each season

Injection season	Suspended solids[I] mg L^{-1}	Mass flux[II] kg	Flow rate Ls^{-1}	Volume injected ML	Specific capacity[III] m^3d^{-1} / m
first (mains)	~0 (-)	~0	15.5	6.5	178
first	138 (97)	722	19	19.0	153
second	29 (8)	948	15	32.7	143
third	66 (65)	4280	18	64.9	148
fourth	169 (110)	22400	17	132.8	158

[I] average value with standard deviation in brackets for stormwater
[II] product of volume and concentration
[III] injection well after 24 hours of injection

were stabilised but not reduced due to the high suspended sediment concentrations in the stormwater. Specific capacities 24 hours after mid-injection redevelopment in the third and fourth seasons were ~95 m^3d^{-1} / m, or about 60% of the initial conditions. The specific capacity of the injection well at the commencement of each injection season is similar (Table 3), indicating that redevelopment techniques are effective in restoring injection rates.

5 CONCLUSIONS

The principal conclusions of this study are:

1. Some clogging by particulates occur when wetland-treated stormwater is injected into this aquifer.
2. Rates of clogging are variable and depend on suspended solid concentrations, temperature, injection rate and duration of injection.
3. Calcite dissolution increases the porosity and permeability of the carbonate aquifer, offsetting the clogging effects of injected sediment.
4. Redevelopment of the injection well was required after around 40 days, and appeared to be effective in restoring the hydraulic characteristics of the aquifer.
5. Further analysis is required to assess the longer-term stability of the injection well and transmissivity of the aquifer. This may involve modelling (Osei-Bonsu 1996; Pérez-Paricio & Carrera 1998) and extended field efforts.

6 ACKNOWLEDGEMENTS

We are appreciative of the support of our collaborators, Primary Industries and Resources, South Australia, and in particular the technical staff of the "Groundwater Resources" section who make the field work and logistics possible. Financial support for the project has been provided by CSIRO, the Urban Water Research Association of Australia, and the Hickinbotham Group. This paper benefited through discussion with Dr Santo Ragusa on microbiological aspects.

REFERENCES

Dillon, P.J. & P. Pavelic. 1998. Environmental guidelines for aquifer storage and recovery: Australian experience. *TISAR'98: Proc. 3rd Int. Symp. on Artificial Recharge of Groundwater*, 21-25 September, 1998, Amsterdam (these proceedings).

Gerges, N.Z., X.P. Sibenaler & D. Armstrong. 1995. Experience in injecting stormwater into aquifers to enhance irrigation water supplies in South Australia. In:"Artificial Recharge of Groundwater II", [Eds. A.I. Johnson & R.D.G. Pyne], *Proc. 2nd Int. Symp. on Artificial Recharge of Groundwater*, 17-22 July, 1994, Orlando, Florida, p.436-445.

Olsthoorn, T.N. 1982. Clogging of recharge wells, main subjects, KIWA-Communications 72, 150pp.

Osei-Bonsu, K. 1996. Clogging by sediments in injected fluid flowing radially in a confined aquifer. Flinders University of South Australia, School of Earth Sciences, PhD Thesis.

Pavelic, P. & P.J. Dillon. 1997. Review of international experience in injecting natural and reclaimed waters into aquifers for storage and reuse. *Centre for Groundwater Studies Report* No. 74.

Pérez-Paricio, A. & J. Carrera. 1998. A conceptual and numerical model to characterize the clogging of artificial groundwater recharge systems. *TISAR'98: Proc. 3rd Int. Symp. on Artificial Recharge of Groundwater*, 21-25 September, 1998, Amsterdam (these proceedings).

Rattray, K.J., P.J. Dillon, A.L. Herczeg & P. Pavelic. 1996. Geochemical process in aquifers receiving injected stormwater. *Centre for Groundwater Studies Report*. No. 65.

Rinck-Pfeiffer, S.M., S.R. Ragusa, T. Vandevelde & P.J. Dillon. 1998. Water quality changes and hydraulic properties through laboratory columns. *TISAR'98: Proc. 3rd Int. Symp. on Artificial Recharge of Groundwater*, 21-25 September, 1998, Amsterdam (these proceedings).

Vecchioli, J. (1970) A note on bacterial growth around a recharge well at Bay Park, Long Island, New York, *Water Resources Research* 6(5):1415-1419.

Artificial Recharge of Groundwater, Peters et al. (eds) © 1998 Balkema, Rotterdam, ISBN 90 5809 017 5

Effects of natural channel bed filtration prior to deep well injection

S.W.van Duijvenbode & T.N.Olsthoorn
Amsterdam Water Supply, Vogelenzang, Netherlands

ABSTRACT: This paper presents the results of a deep well recharge experiment with four injection wells in the Amsterdam Dune Area. This experiment is a pilot study as part of a possible artificial recharge system comprising seventy injection wells, with a total capacity of 13 million m^3/year. Whether this experiment with four injection wells will be expanded in the future depends on the costs, possible alternatives and the success of the present test. The pretreated river water from the Rhine used in the deep well recharge system is transported by an artificial water supply channel system in the dunes. The four injection wells are fed by a 300 meter long drainage system, which acts as a natural filter embedded in the bottom of the water supply channel. Since the start of this recharge experiment in June 1994, more than three million m^3 water have been injected. The recharge capacity of the system is maintained by conditioning the bottom filter and the injection wells.

1 INTRODUCTION

The Amsterdam Water Supply is considering to build an artificial recharge system comprising 70 injection wells, with a total capacity of 13 million m^3/year (Olsthoorn 1993). This system is completely embedded in the existing infrastructure of the Amsterdam Dune Water Catchment Area (fig. 1).

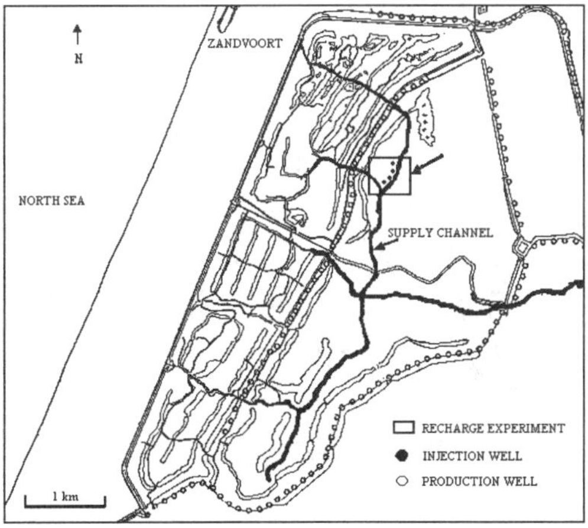

Figure 1. The Amsterdam Dune Water Catchment Area and the location of the recharge experiment.

A pilot-project with four injection wells and a recharge capacity of 0.7 million m^3/year started in June 1994. The four wells are fed by a 300 meter long drainage system, which is embedded in the bottom of the water supply channel (fig. 2).

At the intake point, 55 km from the recharge area, the river water from the Rhine is pretreated by coagulation and rapid sand filtration. The pretreated water is piped to the dune area, where it is further transported by an artificial open supply channel system in the dunes. The use of an existing open water transportation system is advantageous from the ecological point of view. However, some physical water quality parameters degrade because of biological activity and catchment of fine particles.

Before injection, the water is polished by a slow sand filter, constructed in the bottom of the supply channel (fig. 2). This slow sand filtration makes the water suitable for injection. The fine rounded aeolic dune sand (150 μm) in the filter, 0.8 m thick, is separated from the underlying gravel pack by nicolon fabric. To prevent an inflow of iron-rich groundwater, a non-permeable plastic sheet was placed under the gravel pack of the drainage system. The water collected by the drainage system is injected in a confined aquifer on 30 to 60 m deep. The disign infiltration rate at the borehole wall is 0.20 m/h.

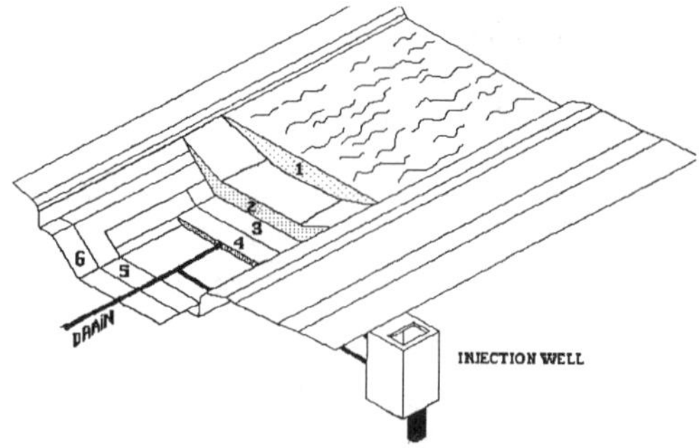

Figure 2. Cross-section of the bottomfilter in the supply channel. 1: pretreated river water, 2: fine aeolic dune sand, 3: nicolon fabric, 4: gravel pack, 5: non-permeable sheet, 6: old bottom of the water supply channel.

The continuous infiltration of water with a moderate physical quality leads to a gradual increase of the hydraulic resistance in the filter (Van Duijvenbode 1996, 1997a). Accumulation of (in)organic matter takes place in the surface layer (<15 cm) of the filter. A low oxygen saturation and an increased oxygen consumption during summer periods cause anaerobic conditions in especially the central part of the filter (fig. 9). Reduction of ferric iron ($Fe(OH)_3$), present on the surface of the filter sand grains, leads to the mobilisation of ferrous iron (Fe^{2+})(fig. 10). The available dissolved organic carbon (CH_2O, 2.0 mg/l) in the infiltrate acts as an electron donor in this process (Zender & Stumm 1988). After passage of the filter, different types of water (Eh-status) are mixed in the drain. At low oxygen concentrations ($O_2 < 1.0$ mg/l), chemical oxidation of ferrous iron is prevented and specific chemotrophic bacteria may develop (Seppänen 1988, Hanert 1981). The growth and erosion of the chemotrophic bacteria (particullarly iron bacteria) lead to biological clogging in the injection well. This biological clogging is caused by the bacteria material itself and the production of an organic slime or biofilm (Hanert 1981, Pedersen & Hallbeck 1985). Chemical oxidation and precipitation of ferrous iron in the drain will take place under subaerobic conditions. The accumulation of biomass and iron flocks in the gravel pack and well screen slots might be responsible for the increase of the resistance in the injection well during summer periods.

2 OBJECTIVES

This pilot-project is aimed at answering the following questions: a) the durability and operating-safety of the artificial recharge system and b) how to retain or avoid a possible loss of recharge capacity? Therefore, we need to know which factors are involved in the loss of recharge capacity and which techniques are effective for the conditioning of the filter and the injection wells. To answer these questions, the recharge system is hydrologically, physically, biochemically and visually investigated. Futhermore, some techniques for cleaning the filter and injection wells have been investigated.

3 RESULTS

3.1 *Hydrological monitoring of the filter*

The hydraulic resistance in the filter increases gradually at a rate of 0.25 cm/m^3/h a year (fig. 3). To maintain the filter capacity, the surface layer of the filter must be cleaned, because increase in resistance concentrates in this area In order to do this under wet conditions, a mobile cleaning robot was developed (fig. 4)(Storm van Leeuwen, 1996).

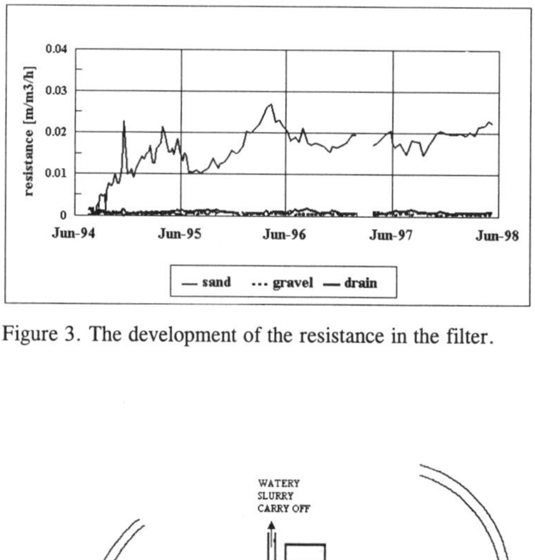

Figure 3. The development of the resistance in the filter.

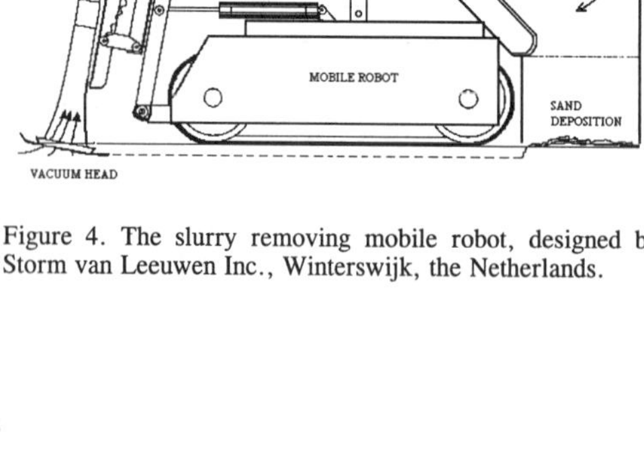

Figure 4. The slurry removing mobile robot, designed by Storm van Leeuwen Inc., Winterswijk, the Netherlands.

The mobile robot removes the material of the surface layer of the filter (< 10 cm) with great precision (Storm van Leeuwen, 1997). The watery sand-slurry fraction is led to the separator on the back of the robot. After separation, the slurry is transported to the waterside and the sand replaced to the filter. In March 1997, this mobile slurry removing system came into use (Van Duijvenbode 1997b). This reduced the risistance of the filter a bit. The resistance has remained the same till begin 1998.

3.2 *Hydrological monitoring of the injection wells*

The hydraulic resistance in the injection wells is separated into the resistance in gravel pack and well screen slots and that on the borehole wall. Figure 5 shows the development of the resistance in injection well 2. The development of the hydraulic resistance in the other three injection wells is comparable with injection well 2.

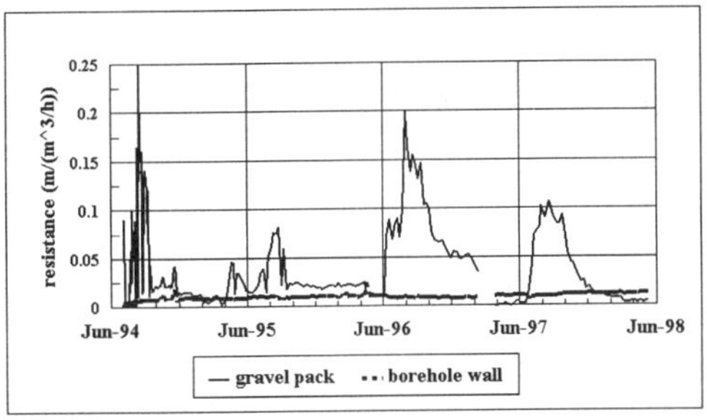

Figure 5. The development of the hydraulic resistance in injection well 2.

The development of hydraulic resistance on the borehole wall is gradual and limited to about 0.40 $cm/m^3/h$ per one million m^3 of infiltrated water (Van Duijvenbode 1998). The hydraulic resistance in gravel pack and well screen slots is more variable and temperature-dependent. In July and August, the resistance reaches its highest level (10 - 20 $cm/m^3/h$). In the period thereafter, the resistance disappears almost completely. Still, some gradual increase of resistance remains, due to chemical clogging in the injection well. This clogging can be removed by air surging as were done in March 1997 (Van Duijvenbode 1997c).

3.3 *Physical monitoring*

The Membrane Filter Index (MFI) is indicative for the physical quality of the water and its ability to be injected. When the MFI of the infiltrate is beneath 3.0 s/l^2, there is little risk of clogging.

Figure 6 shows the MFI values of the pretreated river water before and after detention in the open water supply channel system and the infiltrate used for injection.

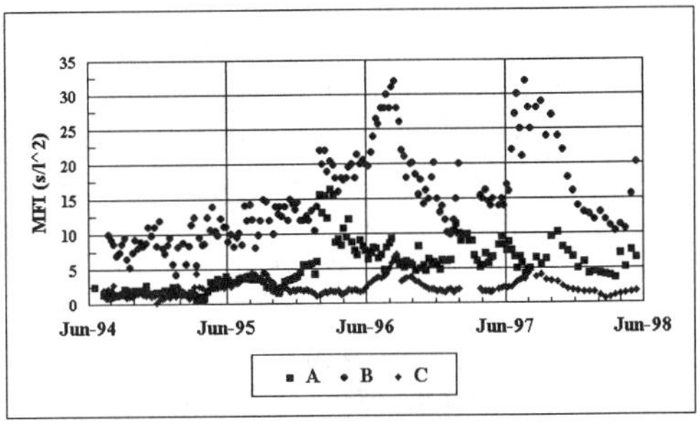

Figure 6. The Membrane Filter Index. Pretreated riverwater before (A) and after (B) detention in the open water supply channel, (C) the infiltrate of injection well 3.

During the detention of the water in the open supply channel, the physical quality of the water declines by a factor 3-5. The filterpassage, however, recovers the physical quality of the water by a factor 8-10. Most of the time, except in July and August, the MFI value of the infiltrate is beneath 3.0 s/l^2.

3.4 *Biochemical monitoring*

To study the effect of the detention of the water in the open supply channel and the filter on the chemical composition of the water, several parameters, like temperature, oxygen, dissolved organic carbon and ferrous iron are monitored (Van Duijvenbode 1997a/b). Temperature and light-intensity influence the oxygen concentrations in the supply channel (figs. 7, 8). In July and August, the oxygen concentration of the infiltrate becomes critical.

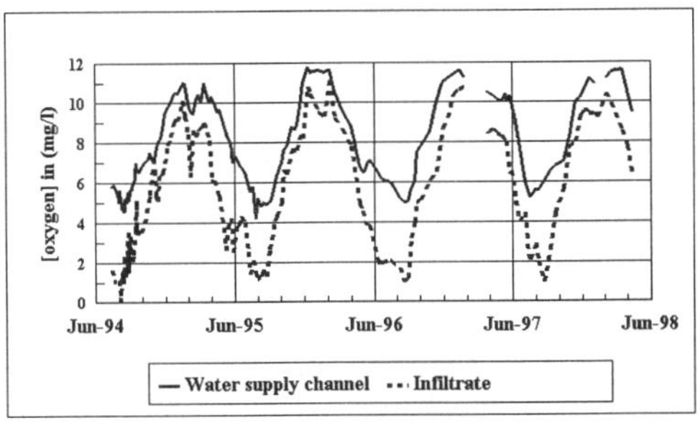

Figure 7. Seasonal fluctuation of the [oxygen] in the open water supply channel and the infiltrate of injection well 3.

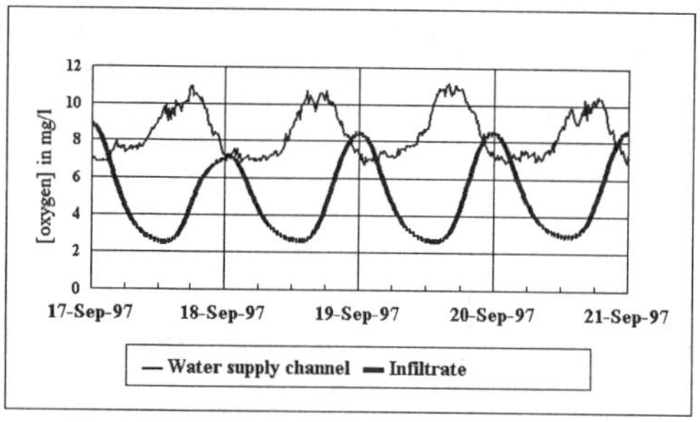

Figure 8. Day/night fluctuations of the [oxygen] in the open water supply channel and the infiltrate of injection well 3.

The oxygen consumption during filterpassage depends on a) the temperature of the infiltrate and b) the degree of accumulation of (in)organic material in the surface layer of the filter. Especially in the central part of the filter, where the accumulation of (in)organic material is high, the infiltrate becomes anaerobic during summer periods (fig. 9). An extra flow of dissolved ferrous iron (Fe^{2+}) can be demonstrated (fig. 10).

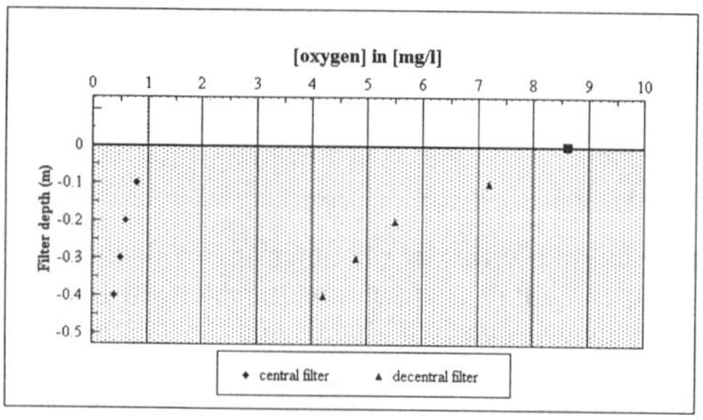

Figure 9: [Oxygen] profile of infiltrate during filterpassage (July 1996), ■ = [oxygen] in the water supply channel.

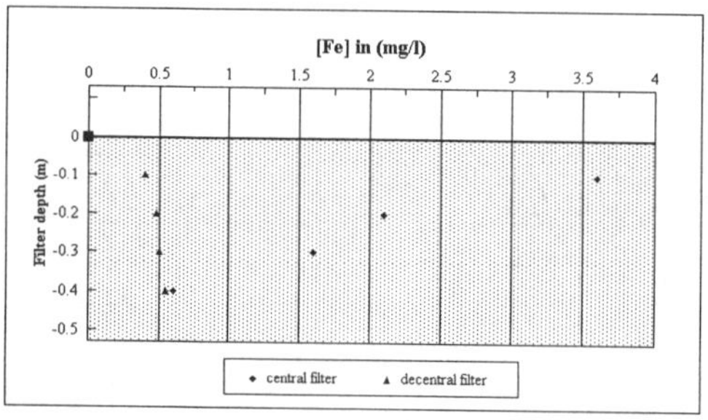

Figure 10: [Fe^{2+}] profile of infiltrate during filterpassage (July 1996), ■ = [Fe^{2+}] in the water supply channel.

3.5 *Visual monitoring*

The drain and injection wells have been inspected several times by a video-camera system (fig. 11). The growth of iron bacteria and precipitation of iron during summer periods is not uniformly distributed in the drain (Van Duijvenbode 1997c).

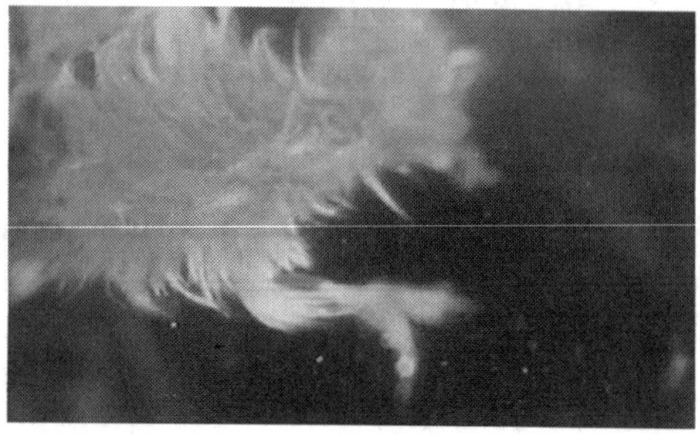

Figure 11. The growth of iron bacteria in the drain during summer 1994 (object enlargement: 3x).

The degree of accumulation of (in)organic material in the overlying filter seems to stimulate the development and growth of potential clogging material in the drain and injection well.

3.6 *Capacity monitoring*

Figure 12 reflects the total amount of recharged water with the four injection wells between June 1994 and May 1998. Almost four years after the start of this artificial recharge experiment with four injection wells, more than 3.25 million m^3 water have been injected.

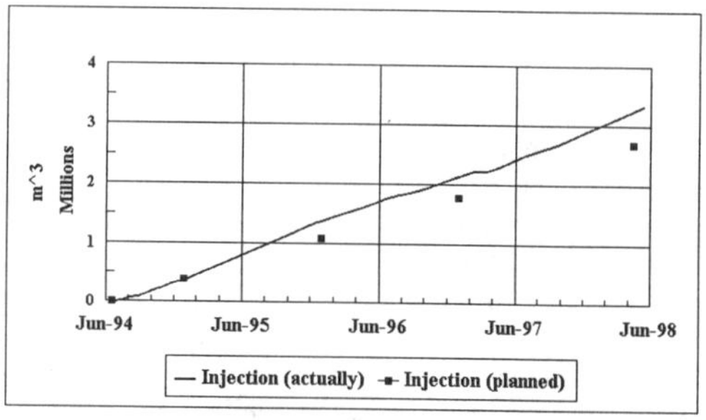

Figure 12. The total netto recharge capacity of the four injection wells.

At this moment, the average injection rate is 24 m^3/h per injection well (planned 20 m^3/h). The recharge capacity of the system can still be maintained by conditioning the filter and the injection wells. The maximum recharge capacity is only partially restricted by the resistance in the filter and injection wells, when the temperature of the water is respectively beneath 7°C and above 17°C (Van Duijvenbode 1997a, 1998).

4 CONCLUSION

- The recharge capacity of the four injection wells can be retained by cleaning the filter by a specially engineered mobile slurry removing system and by removing the biochemical clogging in the injection wells by air surging.
- A maximum recharge capacity can be realised at an infiltrate temperature between 7 and 17°C.
- The physical quality of the pretreated riverwater during its detention in the open water supply channel system declines with a factor 3-5.
- The slow sand filtration fulfills objectives as it improves the physical water quality with a factor 8-10 and makes the water suitable for injection.
- Accumulation of (in)organic matter in the upper layer of the filter seems to stimulate the development and growth of potential clogging material in the drain and injection well during summer periods.

REFERENCES

Hanert, H.H. (1981). The genus Gallionella. *Springer-Verlag,* Berlin, FRG, 509-515.

Olsthoorn, T.N. & P.T.W.J. Kamps (1993). Substitution of 13 mln m^3/year surface-infiltration by deep well recharge (In Dutch). *Amsterdam Water Supply report.*

Pedersen, K. & E.L. Hallbeck (1985). Rapid biofilm development in deep groundwater by Gallionella ferruginea. *Vatten*, 41: 263-265.

Seppänen, H. (1988). Groundwater, a living ecosysteem. *Water Sci.Tech.*, 20:95-100.

Storm van Leeuwen (1996). Design of slurry removing mobile robot (In Dutch).

Storm van Leeuwen (1997). Cleaning the filter of the recharge system in the Amstedam Dune Water Catchment Area (In Dutch).

Van Duijvenbode, S.W. (1996). Interim report of the deep well recharge-experiment (in the Amstedam Dune Water Catchment Area (In Dutch). *Amsterdam Water Supply report.*

Van Duijvenbode, S.W. (1997a). Deep well recharge in the Amstedam Dune Water Catchment Area (In Dutch). *Amsterdam Water Supply report.*

Van Duijvenbode, S.W. (1997b). Cleaning the injection wells and the filter of the recharge system (In Dutch). *Amsterdam Water Supply report.*

Van Duijvenbode, S.W. (1997c). Camera-inspection of the drain and injection wells of the recharge system (In Dutch). *Amsterdam Water Supply report.*

Van Duijvenbode, S.W. (1998). Deep well recharge in the Amstedam Dune Water Catchment Area, a real option (In Dutch). *Amsterdam Water Supply report.*

Zender, A.J.B. & W. Stumm (1988). Geochemistry and biogeochemistry of anaerobic habitats. *Biology of anaerobic micro-organisms*, 1-38. John Wiley and Sons, New York.

Artificial Recharge of Groundwater, Peters et al. (eds) © *1998 Balkema, Rotterdam, ISBN 90 5809 017 5*

Microbial growth and clogging in sand column experiments simulating artificial recharge of groundwater

H.-J. Albrechtsen, R. Boe-Hansen, M. Henze & P. S. Mikkelsen
Department of Environmental Science and Engineering, Technical University of Denmark, Lyngby, Denmark

ABSTRACT: Clogging in sand columns infiltrated with untreated lake water was studied in laboratory experiments. During the clogging a substantial oxygen consumption in the sand, and reduction of the dissolved organic carbon (DOC) and assimilable organic carbon (AOC) over the column was observed. The co-occurence of bacteria with the organic particles demonstrated a high bacterial activity in the clogging layer.

1 INTRODUCTION

Artificial recharge of groundwater may be used for several purposes: storage, thermo-stabilization, and quality improvement, especially microbial quality. Infiltration may also be used for more complete water treatment, including infiltration of raw water or less treated water (e.g. Peters 1996). Clogging may occur during infiltration as a result of deposition of particles and microbial growth in the infiltration material.

Even small concentrations of dissolved organic carbon available for microorganisms may result in clogging due to microbial growth. This has been observed for pre-treated water (Hijnen & van der Kooij 1992) and in artificial systems under denitrifying conditions (Holm et al. 1997). Such activity may alter the redox conditions deeper in the sand bed which may affect the attenuation of organic carbon and thereby the resulting water quality.

This study focused on microbial growth during clogging and the aim was to gain information on the effects of specific parameters by carrying out a phenomenological study.

2 MATERIALS AND METHODS

2.1 *Experimental set-up*

Infiltration water used in the experiments was collected at the Arrenæs pilot infiltration plant, Zealand, Denmark (for further details see Brandt, 1998). The water originated from the Lake Arresø and was not pre-treated in any way. It was collected in the late summer and autumn 1997 where the experiments were carried out.

The experimental set-up was designed to mimic open reservoir infiltration although it was not the intention to perform a precise experimental simulation of a specific field site. The set-up consisted of a reservoir with the infiltration water, a peristaltic pump and a sand column with down-flow (Fig. 1). The column was constructed in stainless steel (diameter: 10 cm height: 25 cm) and had a 13 mm (inner diameter) PEX tube attached.

The infiltration water was stored in a stainless steel reservoir and it was kept at 5°C and slowly stirred to prevent changes during storage.

Quartz sand (0.0-0.4 mm, (no. 00), d_{60}/d_{10}: 2.0, Sdr. Vissing Sand A/S, Brædstrup, Denmark) was packed in the stainless steel column. To remove any organic matter from the sand, it was soaked in 10% HNO_3 for 24 hours, washed in Milli-Q water and heated to 220°C for 8 hours. To ensure a homogeneous distribution of microorganisms down through the column, the sand was preinoculated by soaking in infiltration water for 24 hours before the columns were packed. The sand column was at all times saturated, since the outlet tubing was higher than the upper surface of the column. A 5 cm high water phase was always present on top on the sand column to facilitate a homogeneous infiltration over the sand surface. Sampling ports were installed in the steel column, allowing for collection of water samples from the inlet, the outlet and down through the column.

The water was pumped from the reservoir to the tube on top of the sand column with a maximum flow corresponding to an infiltration rate of 13 cm/h.

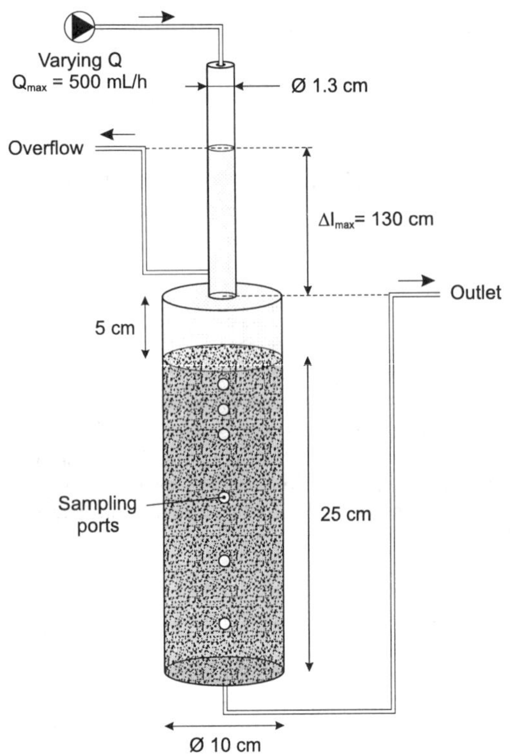

Figure 1. The experimental set-up.

In the first phase of an experimental run the infiltration rate was controlled by the inlet pump. When the column started to clog, the hydraulic conductivity and thus also the headloss of the column changed resulting in an increasing water table in the tube. An overflow was placed 1.3 m + 5 cm above the sand surface. After this level was reached, the headloss was constant while the infiltration rate decreased. By monitoring the headloss and infiltration rate, the hydraulic conductivity of the sand column could be determined according to Darcy's law.

The experiments were carried out in the dark at 15°C, similar to the average annual temperature of lake water in Denmark. When each experiment was terminated, the column was opened and the sediment down through the column was investigated.

2.2 *Analyses*

Water or sediment samples were analysed for redox sensitive parameters (oxygen, nitrate, and methane) and organic matter content (dissolved, suspended and sediment associated). The total number of bacteria was counted by the Acridine Orange Direct Count (AODC) method (Albrechtsen & Winding 1992). Heterotrophic plate count (HPC) was performed by the use of R_2A agar (Reasoner and Geldreich 1985), incubated at 7 days at 20°C. The particle size distribution was measured by Coulter counting. Sediment samples from different depths down through the column were shaken with 10 mL of water per 10 mg of sediment and the number and size distribution of particles was measured by Coulter counting.

To characterise water with respect to the risk of clogging during infiltration, water samples were investigated with special attention to parameters which could influence the microbial turnover, especially organic matter (dissolved and particulate). The amount and character of the organic matter was quantified in terms of dissolved organic carbon (DOC) and assimilable organic carbon (AOC, modified after van der Kooij et al. 1982). OUR (oxygen uptake rate) (Kristensen et al. 1992) was used to asses the respiration (turnover) of the microorganisms (including algae). The OUR method is based on a series of batch experiments in which the oxygen consumption due to respiration is measured on-line.

All results from analysis of sediment samples are expressed per gram of dry weight.

3 RESULTS AND DISCUSSION

The experiments showed that the experimental set-up worked satisfactory, and that the water quality in terms of concentration of organic matter seemed stable during the storage (up to three weeks). The total number of bacteria (AODC) also remained fairly constant, although a weak increase in the heterotrophic plate count was observed.

The columns were designed to simulate open-basin infiltration and the columns were successfully operated at conditions close to the open-basin infiltration site at the Arresø pilot infiltration plant.

Table 1. Characterisation of water from Lake Arresø, used for infiltration in the column experiments. [1]: total alkalinity

Parameter	Units	Values
pH		7.5-7.7
TAL[1]	meq/L	0.45-0.50
NO_3^-	mg-N/L	0.10
O_2	mg-O_2/L	6-10
Particles	L^{-1}	8.0-8.6 x10^5
DOC	mg-C/L	16-24
AOC	µg Ac-C/L	>300
HPC (R_2A)	CFU/mL	0.23-36 x10^5
AODC	cells/mL	3.7-12x10^6

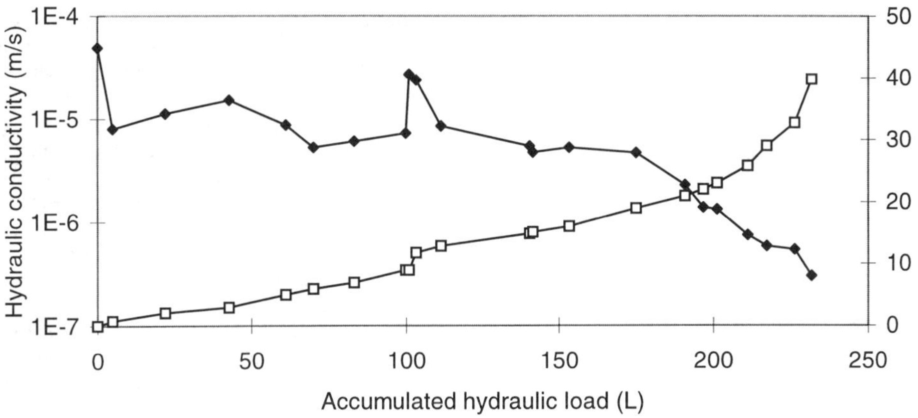

Figure 2. The change in the hydraulic conductivity (◆) as a function of the total amount of water infiltrated into the column with an area of 79 cm^2 during 43 days. The time (□) used for the infiltration is shown at the left axis.

3.1 *Hydraulic conductivity*

The flow rate decreased during the experiment where the hydraulic conductivity decreased from 5×10^{-5} m/s to 3×10^{-7} m/s (Fig. 2). The hydraulic conductivity rapidly decreased from 5×10^{-5} m/s to 8×10^{-6} m/s during the first day, when approximately 5 L had been infiltrated on the column. This decrease was mainly caused by the development of a thin (approx. 1 mm thick) clogging layer. The change in hydraulic conductivity may thus be due to a significant clogging.

The hydraulic conductivity decreased steeply to about 3×10^{-7} m/s in the final phase of the experiment after 18 days when approximately 175 L had been infiltrated through the column. This steep derease in hydraulic conductivity may be due to the lower infiltration rate allowing an accelerating packing of the upper clogging layer including fine particles. At the low flow rate the fine particles may clog the small pores whereas they at higher flow rates may have been dispersed deeper into the sand bed.

The experiments took place in the late summer and fall, and therefore filtration of algae may have contributed substantially to the clogging observed.

3.2 *Organic carbon*

Water samples from Lake Arresø, infiltrated in the column experiments, had a DOC content in the range 16-24 mg-C/L with a mean content of 20 mg-C/L (Table 1). For comparison, DOC in the outlet was in the range 16-22 mg-C/L (Table 2) with a mean content of 18 mg-C/L. Consequently an average removal of 2 mg-C/L of dissolved organic carbon was observed. This removal was caused by accumulation and microbial turnover in the column.

Particles were also removed from the water phase since comparisons between samples from the inlet and the outlet of the column furthermore showed 80-90% removal of particles.

Table 2. The concentrations of dissolved oxygen (DO), dissolved organic carbon (DOC) and total number of bacteria enumerated by AODC measured in the inlet and in the outlet during the experiment. [1]: not measured

Time Days	DO mg O_2/L		DOC mg TOC/L		Bacteria x 10^6 Cells/mL	
	In	Out	In	Out	In	Out
1	-[1]	-	19.1	16.5	12.4	4.5
2	10.0	6.6	-	16.1	7.6	3.8
9	7.0	3.7	17.8	-	5.5	4.4
15	8.0	4.3	22.7	18.3	6.6	5.8
23	7.1	0.2	23.6	21.6	4.5	6.6
33	5.6	0	21.7	19.0	4.5	6.6
43	6.3	0	20.5	18.1	4.7	6.4

The content of organic carbon in the sediment (Fig. 3) clearly showed an accumulation of organic matter (2.8-3.3 mg-C/g dry weight) in the top of the column, but also that the organic matter had penetrated approximately 6 cm into the column (0.5-2.3 mg/g dry weight. The number of particles in the sediment was also clearly higher in the layer from the top to a depth of 6-7 cm and the particles were larger in the

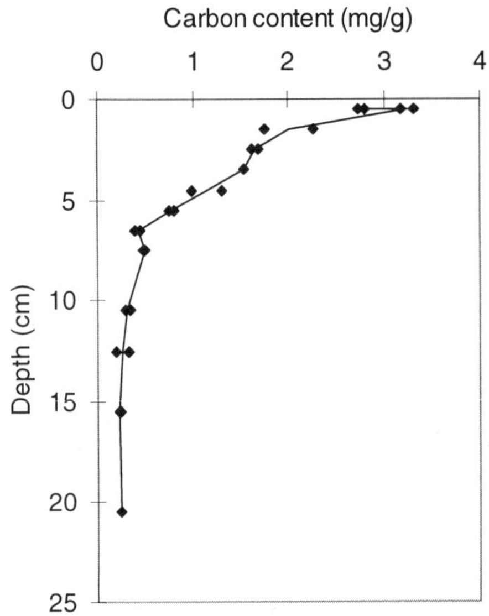

Figure 3. Content of organic carbon down through the sand column at the end of the experiment.

upper layers. In the upper layers 8-15% of the particles were larger than 5μ whereas only 5-9% of the particles deeper in the column were larger than 5μ. 20-30% were smaller than 2μ in the upper layer, whereas 23-36% were smaller than 2μ deeper in the column.

The peak in the hydraulic conductivity after infiltration of approximately 110 L was due to disturbance of the column because of sampling from the sampling ports own through the column.

These results show that the sediment column efficiently retained organic particles, functioning as a filter.

The content of AOC was high in the infiltration water (>300 μg-Ac-C/L, Table 1) which showed a substantial content of substrate available for microbial growth in the infiltrating water. The AOC in water in the outlet of the column was much smaller (<60 μg-Ac-C/L) than in the inlet. Since it is unlikely that dissolved organic matter to a high degree should be retained in the column due to filtration this indicated a substantial microbial activity and microbial degradation of the organic matter in the column.

3.3 *Redox conditions*

The column experiments showed that oxygen was depleted very rapidly under saturated conditions due to biological processes, although it was not possible to determine to which extent the oxygen consumption was due to algae respiration or to bacterial turnover. The initial oxygen uptake rate (OUR) of the Arresø lake water was 0.1 mg O_2/L/h, which was close to the detection limit of this method.

During the first 21 days of the column experiment, oxygen was present in the outlet (Table 2). A profile of the oxygen concentration down through the column at day 28 showed that oxygen was depleted at the first sampling point in the column at a depth of 3 cm in the sand column, probably due to respiration of the turn over of organic material in the clogging layer.

Nitrate was totally removed as well, but it should be noticed that the nitrate level of inlet water was very low, therefore denitrifying conditions cannot be documented. There was a small production of Fe(II)

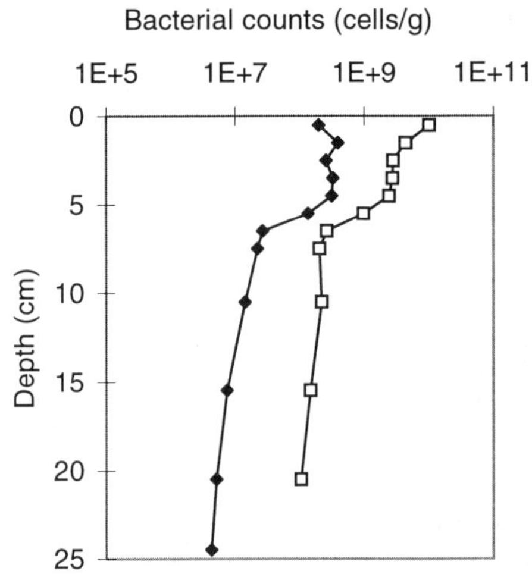

Figure 4. Number of bacteria, enumerated as heterotrophic plate count at R_2A (CFU) (◆) and total number of bacterial (AODC) (□) down through the sand column at the end of the experiment.

but no traces of methane or no smell of H_2S or visual signs of sulphide precipitations.

This strongly indicated a substantial microbial activity in the clogging layer and that oxygen was the electron acceptor responsible for the major part of the biological turnover in the column experiments performed.

3.4 *Microbiology*

The heterotrophic plate count varied in infiltrating water (0.23 - 36 x 10^5 CFU/mL) (Table 1) as well as in the outlet (0.16 - 39 x10^5 CFU/mL). During the experiment the total number of bacteria (AODC) in the outlet was relatively stable at 3.9 - 7.1 x10^6

cells/mL, and during the last 20 days the number was higher in the outlet than in the inlet (Table 2).

The total number of suspended bacteria (AODC) in samples collected from the sampling ports down through the column remained constant down through the column over time.

The number of suspended bacteria and the variation in these numbers reflected a balance between filtration, growth and detachment of the bacteria. Therefore the present data for suspended bacteria only provided little information on the clogging process compared with data for the sediment.

Analysis of sediment samples from the column showed that the major part of the biomass was situated in the upper 5 cm of the column. The total number of bacteria (AODC) reached values as high as 1×10^{10} cells/g wet weight and the heterotrophic plate counts reached values of 4×10^8 CFU/mL (Fig. 4).

4 CONCLUDING REMARKS

The substantial oxygen consumption in the sediment may be due to respiration from algae or bacteria. However, the reduction of dissolved organic carbon and AOC over the column and the co-occurence of bacteria with the organic particles caught in the sand column by filtering processes clearly demonstrated a high bacterial activity in the clogging layer.

This study is part of an ongoing project and the importance of the microbial processes for the clogging compared with the filtering are currently under investigation. These studies also include infiltration of filtered lake water, and filtered lake water with different size fractions of particles. Finally, sterilised columns infiltrated with sterilised infiltration water will be used as control.

5 ACKNOWLEDGEMENT

This research is part of the 'Artificial recharge of groundwater Project' funded by EU under the EU *Environment and Climate* Program.

REFERENCES

Albrechtsen, H.-J. & A. Winding 1992. Microbial biomass and activity in subsurface sediment from Vejen, Denmark. *Microb.Ecol.* 23:303-317.

Brandt, G. 1998. Water quality and storage aspects - Arrenæs artificial recharge trial plant. This issue.

Hijnen, W.A.M. & D. van der Kooij 1992. The effect of low concentrations of assimilable organic carbon (AOC) in water on biological clogging of sand beds. *Wat.Res.* 26:963-972.

Holm, J., P. Engesgaard, K.H. Jensen, M. Henze, & H.-J. Albrechtsen 1997. Effects of biomass growth on the hydrodynamic properties of groundwater aquifers. Abstract. *EOS Transactions*, 78, (46 - Suppl.), F232.

Kristensen, G.H., P.E. Jørgensen & M. Henze·1992. Characterisation of functional micro-organism groups and substrate in activated sludge and wastewater by AUR, NUR, and OUR. *Wat.Sci.Techol.* 25:43-57.

Peters, J. H. 1996. Are there any blueprints for artificial recharge? In: Artificial Recharge of Groundwater. Proceeding of an International Symposium, Helsinki, Finland. June 3-5, 1996. A.-L. Kivimäki & T.Suokko (eds.) Nordic Hydrological programme NHP Report No. 38. 257-271.

Reasoner, D.J. & E.E. Geldreich 1985. A new medium for the enumeration and subculture of bacteria from potable water. *Appl.Environ.Microbiol.* 49:1-7.

van der Kooij, D., A. Visser & W.A.M. Hijnen 1982. Determining the concentration of easily assimilable organic carbon in drinking water. *J.Am.Water Works Assoc.* 74:540-545.

1.4 Recharge of treated waste water

Artificial Recharge of Groundwater, Peters et al. (eds) © 1998 Balkema, Rotterdam, ISBN 90 5809 017 5

The San Gabriel Valley Recycled Water Demonstration Project

Adam S. Hutchinson
CH2M Hill, Santa Ana, Calif., USA

Greg Woodside
CH2M Hill, Santa Ana, Calif., USA (Presently: Orange County Water District, Fountain Valley, Calif., USA)

ABSTRACT: Recycled water represents a critical resource for satisfying growing water demands in arid Southern California. This paper presents an evaluation of the San Gabriel Valley Recycled Water Demonstration Project, located in Los Angeles County, California. Recharge of tertiary-treated recycled water into the San Gabriel groundwater basin is proposed in a 3 kilometer section of the San Gabriel River. We present an evaluation of the project based on proposed regulatory criteria, including maximum infiltration rates and percentage of recycled water at production wells. Field testing of infiltration rates was conducted with 8 test basins, and indicated that infiltration rates ranged from 1.5 to 15 meters per day. A multi-layer finite-element groundwater model was utilized using the CFEST code to simulate the migration of recycled water in groundwater and estimate the amount of recycled water at nearby production wells.

1 INTRODUCTION

Recycled water represents one of the last "new" significant sources of untapped water supply in Southern California. Groundwater pumping in many groundwater basins exceeds natural recharge. Imported water from Northern California and the Colorado River is used to make up the shortfall. Water imported from outside Southern California is purchased and recharged into groundwater basins, or directly delivered to consumers to satisfy water demands that cannot be met with local groundwater supplies. The future availability of imported water is less certain, and the cost of imported water has and will likely continue to increase. The combination of unreliable future supply and rising costs has increased interest in evaluating the use of recycled water for groundwater recharge.

The San Gabriel Valley Recycled Water Demonstration Project evaluates the recharge of recycled water in the San Gabriel Basin. The San Gabriel Basin is an alluvial groundwater basin covering approximately 440 square kilometers in Los Angeles County, California (Fig. 1). Drinking water for over one million people is provided by wells in the basin.

The Demonstration Project proposes to recharge up to 33,000 cubic meters per day of tertiary-treated recycled water in a 3 kilometer stretch of the San Gabriel River. Treated recycled water from the San Jose Creek Water Reclamation Plant would be transported through a 12-kilometer long pipeline to the recharge area in the unlined San Gabriel River (Fig. 2). Treatment at the Water Reclamation Plant includes oxidation, multi-media filtration, and disinfection

The remainder of this paper discusses the following issues related to evaluation of this proposed project: 1) evaluation of infiltration rates, 2) regulatory criteria, 3) evaluation of project in light of the regulatory criteria.

2 EVALUATION OF INFILTRATION RATES

Infiltration rates in the proposed recharge area were evaluated to assess compliance with regulatory criteria, and determine if the desired amount of recharge (up to 33,000 cubic meters per day) could be recharged in the recharge area. The recharge area consists of unconsolidated alluvial sediments, and the dominant grain size is coarse sand, with fine-grained gravel to boulders present in some areas. Finer grained sediments such as fine sand to silt are less abundant, but do occur in localized areas. The hydraulic conductivity of the aquifer materials is estimated to average about 100 meters per day, but may be as high as 500 meters per day in localized areas.

Infiltration rates were tested with eight test basins. Test basins were located along the 3 kilometer stretch of the San Gabriel River, and were

Figure 1. - Project Location
San Gabriel Valley Recycled Water Demonstration Project

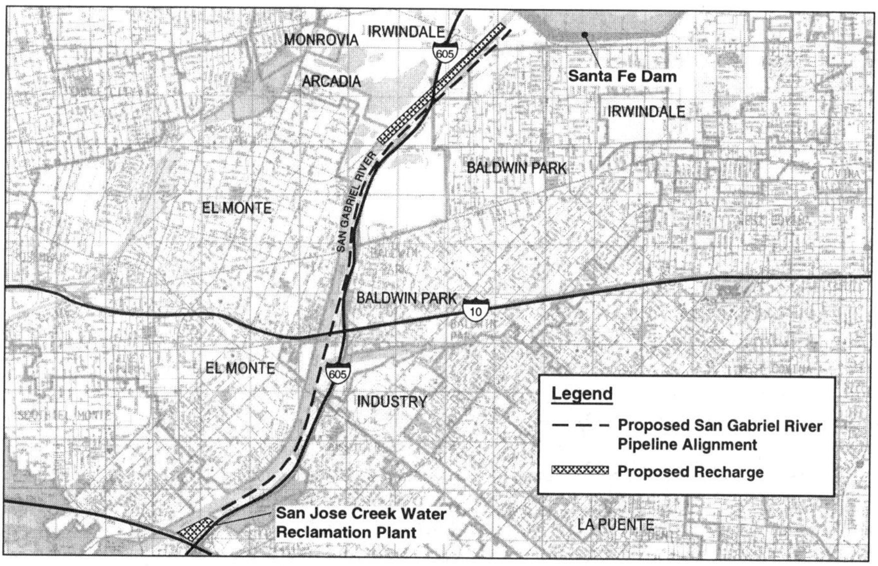

Figure 2. - General Location of Proposed Recharge Area and Pipeline Alignment
San Gabriel Valley Recycled Water Demonstration Project

0 0.8 1.6 kilometers
Approximate Scale

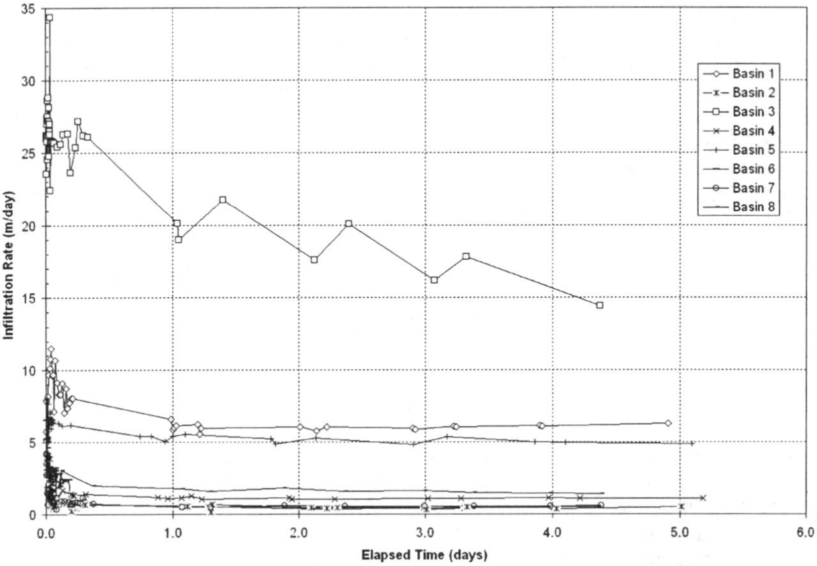

Figure 3. - Results of Infiltration Rate Testing
San Gabriel Valley Recycled Water Demonstration Project

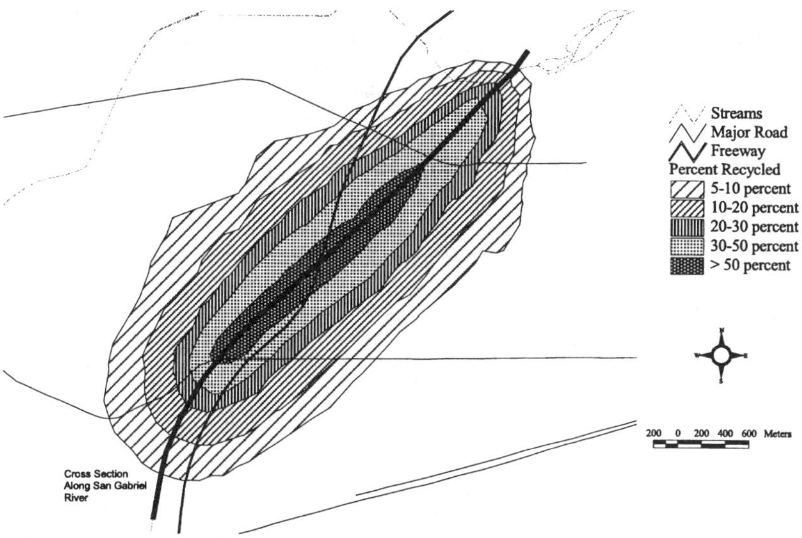

Figure 4. - Percent Recycled Water Top of Water Table
San Gabriel Valley Recycled Water Demonstration Project

constructed as 3-meter by 3-meter wooden boxes. The height of the side of each box was 0.3 meters. Constant head infiltration tests were conducted by piping water into each test basin, and measuring the flow rate into the test basin with a flow meter. The basins required about 10 to 30 minutes to fill with water to a depth of about 0.25 meters. Constant head conditions were maintained after this initial filling period. Infiltration rates under constant head conditions varied from about 1.5 meters per day to 15 meters per day (Fig. 3).

3 REGULATORY CRITERIA

Groundwater recharge of recycled water in California is subject to regulation by the State of California. Proposed criteria for surface recharge of recycled water include:

1. The travel time of recycled water to potable production wells must be greater than 6 months (i.e., recycled water produced at a potable water well must have been in groundwater for longer than 6 months).
2. The travel distance to potable production wells must be greater than 150 meters.
3. The percentage of recycled water at potable production wells must not be greater than 20 percent. The maximum percentage criterion is evaluated based on a 5-year moving average.

Regulatory agencies also require an assessment of project impacts on inorganic groundwater quality, as measured by chloride, sulfate, total dissolved solids, and nitrate.

Proposed regulatory criteria also include a limit on the maximum infiltration rate. This limit is 11 meters per day when the minimum depth to groundwater is 6 meters. Based on the infiltration testing summarized in Figure 3, the maximum infiltration rate criterion is satisfied at 7 of the 8 test site in the recharge area. The lone site that does not meet the criterion is believed to be unrepresentative of infiltration rates across the majority of the recharge area.

4 EVALUATION OF PROJECT

A multi-layer finite-element model was developed to evaluate the project's conformance to the proposed regulatory criteria and impacts on inorganic water quality. The CFEST computer code (Gupta et al. 1987) was utilized to develop the 3-dimensional model.

Vertical layering is incorporated into the model to allow for resolution of vertical concentration differences and vertically-distributed groundwater production and recharge. Because of the relatively high horizontal and vertical hydraulic conductivity in the recharge area, vertical head differences are typically less than 2 meters. In portions of the basin removed from the recharge area, vertical head differences are larger and range up to 10 meters or more. Horizontal to vertical hydraulic conductivity anisotropy is accounted for by specifying vertical hydraulic conductivity to be ten percent of horizontal hydraulic conductivity.

Reported groundwater production and measured recharge at spreading grounds and in the San Gabriel River are accounted for. The model also includes estimated recharge from deep percolation of precipitation.

Groundwater flow is simulated for an 18-year period at over 5000 nodes using a quarter-year (91 day) time step. Recharge during the 18-year period is based on historical recharge rates for the period from 1977 to 1995. The 18-year period is repeated twice to achieve a longer (36-year) simulation period.

Recycled water is recharged in the model when basin groundwater levels are sufficiently low to permit recharge. Existing basin management guidelines prohibit recharge of supplemental water when groundwater levels are too high. The net effect of the guidelines that dictate when recycled water can be recharged is that recycled water is recharged about 65 percent of the time in the model simulation. Up to 12 million cubic meters is recharged on an annual basis in some years. The average recharge rate is 10 million cubic meters per year.

4.1 *Solute transport modeling*

The particle tracking portion of the CFEST code is used to evaluate the minimum 6-month travel time criterion (CH2M HILL 1996). The solute transport portion of the CFEST code is utilized to simulate the percentage of recycled water in groundwater.

The amount of recycled water in groundwater can be represented by the percentage of recycled water relative to water from other sources. Before surface recharge, recycled water mixes with local water in the river, if local water is present. Mixing of recycled water with local water partially dilutes the recycled water before infiltration occurs. After surface recharge, recycled water mixes with groundwater from other sources (such as local water recharge upgradient), and flows away from the recharge area.

As the recycled water moves with groundwater, the recycled water and groundwater mix together. As this mixing occurs, the percentage of recycled water relative to water from other sources decreases. This mixing process results in dilution of recycled water, and is numerically simu-

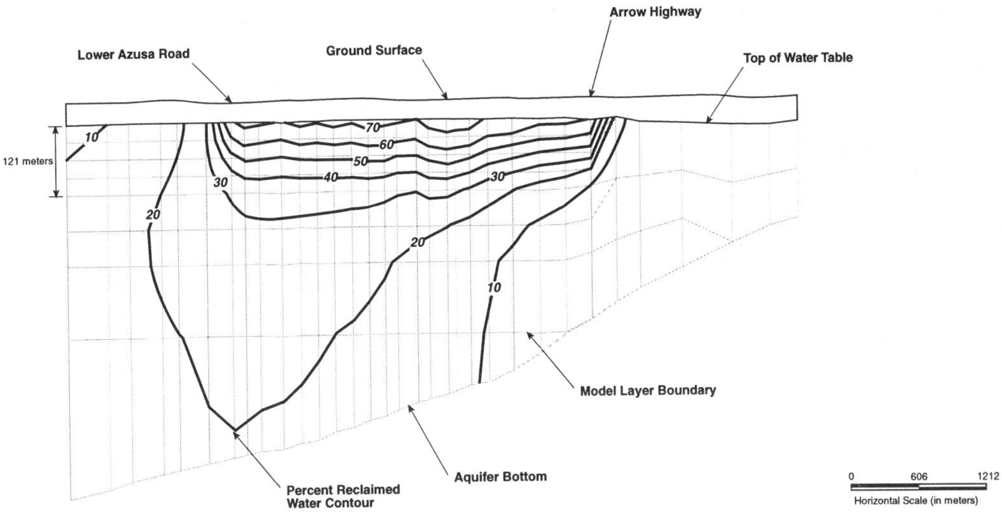

Figure 5. - Recycled Water Distribution in Cross Section Along San Gabriel River After 35 Years

lated according to the advection-dispersion equation. Longitudinal and transverse dispersivity values of 91 meters and 9 meters are used.

Figure 4 shows the percentage of recycled water in groundwater at the top of the groundwater table after 10 years of recharge. In the area directly adjacent to the river, the percentage of recycled water is greater than 50 percent. The estimated amount of recycled water at the top of the water table is 20 percent up to 500 meters from the recharge area.

Figure 5 shows the variation with depth of the estimated percentage of recycled water in a cross-section along the length of the recharge area. Percentages of recycled water are highest directly beneath and adjacent to the river. The percentage of recycled water decreases with depth, but concentrations up to 20 percent are estimated at depths up to 300 meters or more below the top of the groundwater table.

Sensitivity analyses with both lower and higher dispersivity values were conducted (CH2M HILL 1996). The sensitivity analysis indicated that the estimated number of wells impacted by the project were unchanged when the dispersivity was increased, but that the estimated number of wells impacted increased when the dispersivity was decreased. With a lower dispersivity, less mixing occurs between the recycled water and groundwater, resulting in larger amounts of recycled water at some wells.

4.2 *Water quality impact evaluation*

Changes in inorganic water quality parameters such as chloride and sulfate are estimated using the percentage of recycled water in groundwater. The concentration increase for a constituent X in groundwater is calculated as follows:

$$\Delta X = \% \text{ recycled water} * (X_{recycled} - X_{groundwater})/100$$

where $X_{recycled}$ = Concentration of constituent X in recycled water source and $X_{groundwater}$ = Concentration of constituent X in groundwater.

This approach assumes that the constituent, such as chloride or sulfate, has a relatively constant initial background value in groundwater. Evaluation of background water quality data indicates that this assumption is reasonable (CH2M HILL 1996).

Using this approach, the estimated chloride increase after 35 years of recharge is estimated by the model. Chloride increases in the range of 9 to 18 mg/L occur at distances up to about 2000 meters from the recharge area. Sulfate and nitrate increases are also estimated using this methodology. Nitrate increases are estimated with and without denitrification.

5 CONCLUSIONS

1. Field testing of infiltration rates was conducted using a new methodology that has been accepted by the regulatory agencies.
2. A multi-layer finite-element model of recycled water migration in groundwater was developed with the CFEST code.
3. The recycled water model was applied to estimate the impact of the recycled water recharge on nearby

production wells. The model was also used to estimate potential increases in total dissolved solids, chloride, sulfate, and nitrate.

4. No active public water supply wells were within the 150 meter exclusion area.

5. No active public water supply wells produced recycled water with a travel time of less than 6 months.

6. Two active public water supply wells were estimated to produce greater than 20 percent recycled water.

7. During project operation, the model will be updated and used to evaluate regulatory compliance.

REFERENCES

CH2M HILL. *Groundwater Impact Assessment Report, San Gabriel River Recycled Water Demonstration Project.* September 1966.

Gupta, S. K., C. R. Cole, C. T. Kincaid, & A. M. Monti. *Coupled Fluid, Energy, and Solute Transport Model.* Office of Nuclear Waste Isolation, Battelle Memorial Institute. Columbus, Ohio. October 1987.

Artificial Recharge of Groundwater, Peters et al. (eds) © *1998 Balkema, Rotterdam, ISBN 90 5809 017 5*

Wastewater Aquifer Storage and Recovery (ASR) – Towards sustainable reuse in South Australia

C.B.Bosher & T.O.Simms
United Water International Pty Limited, Adelaide, S.A., Australia

B.Kracman
South Australian Government Project Delivery Task Force, S.A., Australia

ABSTRACT: Sustainable use of water resources is an issue which faces many regions throughout the world, particularly in South Australia, the driest state in the driest continent in the world. This paper describes a series of activities, including aquifer storage and recovery, which combine to form the basis of an holistic, sustainable water reuse initiative in Adelaide, SA. The irrigation needs of the US$30m/year horticultural industry in the Virginia Triangle to the north of the city of Adelaide have been matched with the requirements of an Environmental Improvement Programme at the Bolivar Waste Water Treatment Plant. A 120ML/day Dissolved Air Flotation/Filtration Plant will supply sufficient reclaimed water to the irrigators to meet their increased irrigation requirements, whilst alleviating pressures of over exploitation and salinisation of groundwater supplies and pollution of the current receiving water in the nearby Gulf St. Vincent. Aquifer storage will be practiced during periods of reduced demand in the winter with stored water being recovered the following summer, thus optimising the benefits of this unique and innovative effluent reuse scheme.

1 INTRODUCTION

As the conventional use of water resources approaches the limits of sustainability in areas of South Australia, the need to address reclamation and appropriate reuse is becoming critical. At the same time, significant long term environmental impacts have been observed (e.g. mangrove forest and sea grass degradation) along the South Australian coastline particularly adjacent to the state capital, Adelaide. A major contribution to this degradation is eutrophication, a consequence of discharging secondary treated sewage effluent and urban storm water to the marine and estuarine environments.

This situation is particularly evident to the North of the City where the Bolivar Wastewater Treatment Plant (WWTP) currently discharges up to 50,000 ML/year of nutrient rich effluent to the shallow inshore waters of Gulf St. Vincent. At the same time the horticultural industry in the nearby Virginia Triangle is lacking sufficient irrigation water to fully meet its needs. At an estimated usage of 16,000 ML/year of pumped ground water, the current horticultural drain on confined aquifers beneath the Virginia Triangle has lowered groundwater pressures allowing lateral ingress and vertical seepage of saline water. Soil quality and horticultural production are consequently threatened by increases in salinity. Insufficient water is an industry constraint as the industry has the potential to double the current annual farm gate value of US$30m (SA Centre for Economic Studies 1995).

An holistic and sustainable solution has been developed which meets the irrigation requirements of the horticultural community whilst satisfying environmental compliance obligations. A number of capital works projects are being implemented to deliver this solution, one of which is the Virginia Pipeline Scheme (VPS).

This scheme is a significant national irrigation project which will reuse reclaimed water from the upgraded Bolivar plant, thereby providing irrigation water for over 250 growers in the horticultural community.

Aquifer Storage and Recovery (ASR) is a further key element to balancing seasonal supply and demand for water and enhancing economic return to the state of South Australia. Accordingly, a 3 year, US$2m, research project is currently underway to determine the technical feasibility, environmental sustainability and economic viability of ASR using treated wastewater. Once complete, the outcomes of this project will form the basis of a model enabling transfer of this reuse technology worldwide.

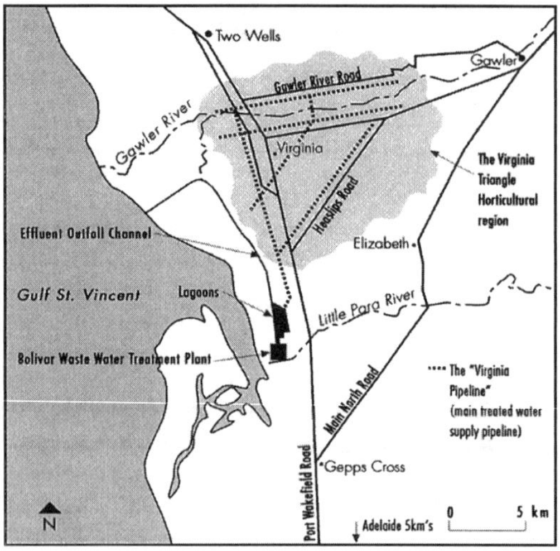

Figure 1. Location of Bolivar WWTP and the Virginia Triangle

The success of a reuse scheme of this magnitude and nature is dependent on the delivery of a number of constituent elements including:
- brokering commercially viable irrigation agreements
- provision of adequately treated wastewater
- optimised supply vs. demand of irrigation water.

2 BROKERING IRRIGATION AGREEMENTS

The role of the South Australian Government's Project Delivery Task Force has been central in brokering commercially viable agreements between the key players. These are the irrigators, their selected private sector pipeline developer Euratech Ltd and SA Water Corporation, the government agency which owns the Bolivar WWTP. This asset is operated on behalf of SA Water by United Water International Pty. Ltd.

The Virginia Pipeline Scheme agreements were negotiated on the basis of a build, own, operate, transfer (BOOT) contract. SA Water will be the bulk water supplier and ultimate transferee of the asset, Euratech will be the BOOT Proprietor and the Virginia growers the water customers. From the outset of negotiations it was clear that the overall project could deliver major financial and environmental benefits to all parties. However, the project was promoted in an era of public and private sector frugality, lessened risk appetite and at a time when all Australian governments were moving toward greater cost recovery on water infrastructure projects.

The limits to what the horticulture industry would pay for water were based on historic price rather than on economic analysis of current production costs. There were also limits on the level, if any, to which the government would subsidise new water projects, particularly when the economic arguments indicated that industry had the capacity to absorb the full cost. Additionally there were differing points of view as to whom, and to what level, the project benefits would accrue.

An acceptable transaction was negotiated despite the fact that the project's financial viability in the early years will be sensitive to a number of factors. This required rigor in configuring the commercial elements of the transaction on the basis of two key principles. These principles were:
- Innovation to minimise risks in order that the parties could accept lower rewards in the collective interest.
- Apportioning risks to those parties best able to manage them.

Because of the attraction of reduced risk, a variable length BOOT concept was embraced up until the final negotiating session, whereupon Euratech decided to seek to move to a fixed length BOOT position. Because of the work being done on ASR, Euratech saw the opportunity to use this technology, if proven, to reduce risk, increase water sales and hence revenue. Concessions were traded and the BOOT period fixed at 20 years.

In addition Euratech is wholly owned by a pipe manufacturer. This allowed him to put forward an attractive commercial concept based on his pipes and fittings being used in the project at a pre-agreed and competitive price. This assured him of a base return on products supplied to the project and he was able to accept a lower than usual return on equity.

Risk was also accepted by the irrigators in embracing the concept of "take or pay" water sales contracts. Under this arrangement the irrigators pay for the annual contracted amount of water regardless of whether they use it or not. Hence some risk passes to those who are the best qualified to assess their water needs and take the necessary on farm management measures.

During the course of negotiations, a set of contract documents were developed which included :
- Customer contracts between Euratech and each irrigator.
- Water Supply, Pipeline Maintenance and Transfer Agreement between Euratech and SA Water.
- Subscription and Shareholder Loan Agreement between Euratech and SA Water. This was the key financial document.
- Design and Construction Management Agreement between Euratech, his Consulting Engineer and SA Water.

These agreements were negotiated on the understanding that the required treated effluent quality for irrigation use would be achieved through implementation of an Environmental Improvement

Programme (EIP) upgrade at Bolivar WWTP. In parrallel with this, ASR studies were to be undertaken in the North Adelaide Plains.

3 DAF/F TREATMENT, RECLAIMED WATER QUALITY AND PUBLIC HEALTH ISSUES

SA Water is required to undertake an EIP at the Bolivar WWTP. This EIP is being implemented under the management of United Water, who operate and maintain Adelaide's water and wastewater treatment plants, water distribution and wastewater collection networks on behalf of SA Water. A concept design has been completed and construction is underway of a 120ML/day Dissolved Air Flotation and Filtration (DAF/F) plant to treat the effluent from the stabilisation lagoons at the Bolivar WWTP. This upgrade will meet the requirements of the EIP for Bolivar, which aims to reduce nitrogen discharge to sea by 80% and maximise reuse.

Figure 2. Bolivar WWTP

Growers proposing to use the reclaimed wastewater requested a quality permitting the spraying of all crops, including salad vegetables with minimal restrictions regarding application. When supply exceeds demand this is the quality of water which will be used to recharge the aquifer for subsequent reuse. In May 1995 the South Australian Health Commission (SAHC) approved the reclaimed water quality. This is known as Bolivar reuse quality and is intermediate between classes A and B of the Guidelines for the Use of Reclaimed Water in South Australia. Bolivar reuse quality is defined by the following parameters (KME, 1996) :

- suspended solids (SS) $\leq$ 20mg/L (mean)
- faecal coliform (90 % ile) $\leq$ 100 orgs./100mL
- pathogens $\prec$ 1/50L (objective nil)
- unrestricted spray irrigation permitted.

The main reason behind the SAHC relaxing the Bolivar Reuse Quality guidelines to a level between Classes A and B is due to the tertiary treatment effected by generally 30 days and not less than 16 days detention in the three stage stabilisation lagoons following the secondary biological treatment. The large pondage area (total 350 hectares) tends to even out trade waste peaks, breaks down many complex chemicals and causes, in general, several log removals of pathogens. (United Water International, 1997).

A DAF/F pilot plant pictured below was set up at Bolivar WWTP in December 1996 and used for 3 months to verify the treated effluent quality achieved. The plant was operated under a range of conditions in which key operational parameters were systematically varied to assess performance limits.

Figure 3. DAF/F pilot plant rig at Bolivar WWTP

The trials clearly demonstrated that the DAF/F process is able to produce the required water quality. By the end of the trial period, a robust set of operating criteria had been established, within which this treated water quality was met, even when the plant was not being operated in an optimal manner (Bosher and Mobius, 1997).

Over the trial period average post-filtration suspended solids achieved was 6 mg/L; a mean removal efficiency of 95%. Faecal coliform removal without disinfection was variable but averaged 86%; in many samples meeting the required reuse quality without the need for post-filtration disinfection. Pathogen levels were reduced to below detection levels in all but one sample.

A detailed comparison of chlorination and UV disinfection was undertaken as part of the concept design. In addition, a limited number of bench scale chlorination tests were conducted on treated waste water samples to assess further improvement in final reclaimed water quality. With disinfection, no faecal coliforms were detected in any of the samples analysed.

On the basis of cost and because of its residual effects, chlorination of the DAF/F treated effluent will be adopted. This will assist in controlling biological slime growth in the distribution pipelines.

To give confidence in meeting pathogen levels a thorough understanding of the filtration step was critical. Five specific filter media combinations were trialed in the external filter columns shown below.

Figure 4. Filtration column trials at Bolivar WWTP

The optimum arrangement was found to be a dual-media sand/anthracite configuration, capable of achieving filter run times up to 16 hours. A small number of filtration runs are shown in figure 5, illustrating the characteristic head loss development observed for this media combination.

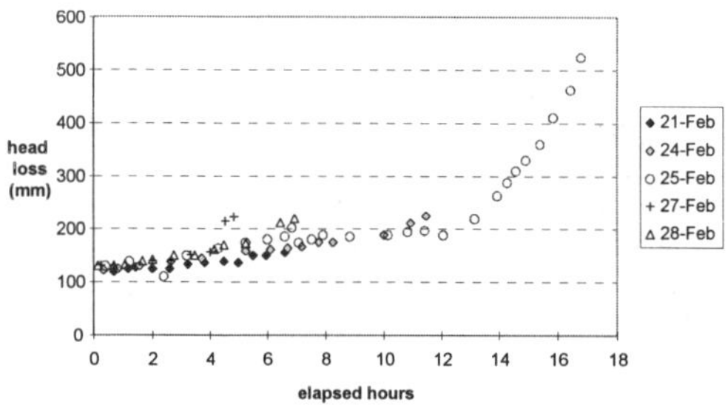

Figure 5. Typical head loss development for dual media filter

Subsequent bench scale treatment has been used to achieve this reclaimed water quality as part of column tests on core samples to assess the technical viability of the ASR aspect of the reuse scheme. (Rinck-Pfeiffer et al. 1998).

4 AQUIFER STORAGE AND RECOVERY

As discussed above, a key element of the Virginia Pipeline reuse scheme is the implementation of ASR. Under the current proposal it is estimated that 20% of the annual production of reclaimed water will be injected into the lower of two confined aquifers (T2) in the North Adelaide Plain during winter. See Figure 6.

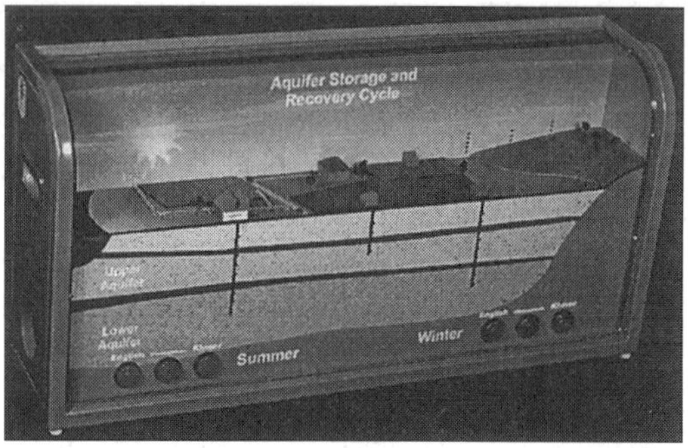

Figure 6. ASR cycle demonstration model constructed for communication with the irrigators and general community.

This will be subsequently recovered for irrigation during periods of high demand in summer. Already there has been considerable success in injecting stormwater into this aquifer (Dillon *et al.* 1997) and examples are presented in Gerges and Howles (1998) and Pavelic *et al.* (1998). These have co-evolved with guidelines for ASR with reclaimed water as described by Dillon and Pavelic (1998).

There are a number of extensive aquifers which lie beneath the North Adelaide Plains from which groundwater has been abstracted for the past 50 years for irrigation purposes. Prior to the 1950's the piezometric surface was approximately 10 metres above ground level. Groundwater levels have now dropped such that levels at the end of the summer pumping season are 65 metres below ground level.

Injection into the brackish confined sandy limestone aquifer T2, at approximately 100 metres below ground level, will substantially increase groundwater hydraulic pressure depending on conjunctive use practices by irrigators, to pre-1950's levels. In addition, higher pressures will ameliorate the ingress of saline water.

However, treated wastewater injection can have major impacts on the hydraulic properties, bio-geochemistry and microbial ecology of the aquifer.

These may include bore clogging, interaction of chemical species with aquifer matrix material, fate of pathogens and other microbiota, suspended solids and mixing of water within the aquifer.

With this in mind, a 3 year research project is underway to determine the technical feasibility, environmental sustainability and economic viability of ASR using the DAF/F treated wastewater. The project is expected to demonstrate whether any potential health and environmental risks associated with the practice can be controlled effectively within a strict quality regulation and monitoring regime.

A consortium has been formed to manage and undertake this project comprising state government and research agencies and an industry partner with international experience to ensure world's best practice is adopted. Members are :

- SA Water Corporation
- SA Government Project Delivery Task Force
- Primary Industries and Resources, SA
- Commonwealth Scientific and Industrial Research Organisation (through Centre for Groundwater Studies)
- United Water International Pty Ltd.

The project includes 2 PhD studies in biochemical clogging and biochemistry and microbiological regrowth and transport.

The 3 stage project has the following activities and objectives:

Stage 1 - Stage 1 (complete) centred around laboratory studies of the effects of continuous injection of tertiary treated wastewater into columns packed with aquifer material on water quality and biological clogging. This is the subject of a further paper presented at this symposium.(Rinck-Pfeiffer *et al*). Field work, shown below in Figure 7, involved borehole drilling to retrieve core samples from the T2 aquifer. Additional field tests included geophysical logging, draw-down pump tests and groundwater sampling.

An economic study and risk analysis was also carried out confirming the viability of the project and allowing progress to Stage 2.

Central to the success of the project is the need to demonstrate early in the project to the SA Environmental Protection Agency, SA Department of Environmental, Heritage and Aboriginal Affairs (DEHAA), SA Health Commission and the general public that there will be no adverse effects on the potential beneficial uses of groundwater nor on public health. To this end, the consortium has established an ongoing Communications Plan to allay concerns by explaining the project aims and results to the irrigators and wider community. Key activities areas include :

- Local community consultation
- Information sheets for the media and the community
- Briefing sessions for Members of Parliament, Chief Executives and Local Governments
- Public forums
- An ASR demonstration interactive model in 3 languages (see figure 6), and educational videos.

Figure 7. Borehole drilling at Bolivar WWTP

Stage 2 - Field trials using injection and observation wells will be carried out to provide data to confirm laboratory technical findings, refine the financial model, reduce uncertainties and better define and quantify risks. A range of new field observation techniques will be used, including cross-hole tomography, in-situ oxygen sensors, freon tracing and pathogen survival monitoring cells. The relationships between water quality, aquifer retention time, extraction volumes, extraction / injection well locations, will be evaluated. Models of the process will be developed and tested against the field data.

Stage 3 - From field trials and models developed, a fully operational ASR scheme will be designed. This will include operation and maintenance procedures together with generic guidelines for whole system design at other locations.

This unique project is generating significant new knowledge in the practice of ASR using irrigation quality reclaimed water. The consortium agreement includes mechanisms for capturing intellectual property generated and an agreed framework and vehicle for commercialisation of the process.

5 CONCLUSIONS

1. The Virginia Pipeline Scheme is a significant national irrigation project which will reuse reclaimed water from the Bolivar WWTP thereby providing

irrigation water for over 250 growers in the horticultural community. Aquifer storage and recovery is a key element to balancing seasonal supply and demand for water, and reducing effluent discharge to a stressed marine environment.

2. Successful completion of the DAF/F pilot plant trials was an important step towards the 1998 implementation of the full scale plant of this sustainable reuse project.

3. Central to the achievements to date has been the role of the South Australian Government's Project Delivery Task Force in brokering commercially viable agreements between the key players i.e. the irrigators, their selected private sector developer Euratech Ltd., and SA Water.

4. A 3 year research project is currently underway to determine the technical feasibility, environmental sustainability and economic viability of ASR using the DAF/F treated water. The project will demonstrate that any potential health risks associated with the practice can be controlled effectively within a strict quality regulation and monitoring regime.

5. A consortium has been formed to manage and undertake this ASR research project comprising state government and research agencies and industry. It draws on the international experience of the industry partner to complement South Australian expertise.

REFERENCES

Bosher, C. & W. Mobius 1997. *Proposed Dissolved Air Flotation Filtration Plant atthe Bolivar Wastewater Treatment Plant - Pilot Plant Trials Report.* United Water International Pty Ltd. report to the South Australian Water Corporation.

Dillon, P., P. Pavelic, X. Sibanaler, N. Gerges, R. & Clark 1997. *Storing stormwater run off in aquifers*. Australia Water and Wastewater Association J. Water, 24(4)7-11.

Dillon, P., & P. Pavelic 1998. *Environmental guidelines for aquifer storage and recovery: Australian experience.* Centre for Groundwater Studies & CSIRO Land and Water PMB2, Glen Osmond, SA 5064, Australia.

Gerges, N.Z. & S.R. Howles 1998. ASR, Hydraulic and Salinity response in unconfined / confined aquifers. *ibid.*

KME (1996) *Virginia Pipeline Scheme Treated Water Study,* Final Report, July 1996.

Pavelic, P., P. Dillon, K.E. Barry, A.L. Herczeg, K.J. Rattray, P. Hekmeijer, & Gerges, N.Z. 1998. Well clogging effects determined from mass balance and hydraulic response at a stormwater ASR site. *ibid.*

Rinck-Pfeiffer, S.M. & S.R. Ragusa 1998. *Column Experiments to evaluate clogging & bio-geochemical reactions in the vicinity of an effluent injection well.* Third Int. Symp. on Artificial Recharge, Amsterdam, Holland, 21-25 September 1998.

SA Centre for Economic Studies 1995. *Audit of Horticulture on the Northern Adelaide Plains.* Report prepared for The City of Munno Para, July 1995.

United Water International Pty Ltd *Proposed Dissolved Air Flotation Filtration Plant at the Bolivar Wastewater Treatment Plant,* Concept Design Report, Contract UW 172/96, April 1997.

Artificial Recharge of Groundwater, Peters et al. (eds) © 1998 Balkema, Rotterdam, ISBN 90 5809 017 5

Column experiments to evaluate clogging and biogeochemical reactions in the vicinity of an effluent injection well

S.M.Rinck-Pfeiffer
Compagnie Générale des Eaux, Anjou Recherche, Paris, France & The Flinders University of South Australia, S.A., Australia

S.R.Ragusa
Centre for Groundwater Studies and CSIRO Land and Water, S.A., Australia

T.Vandevelde
Compagnie Générale des Eaux, Paris, France

ABSTRACT: Laboratory columns were used to study biogeochemical processes resulting from the continuous injection of treated wastewater into an aquifer matrix. The results showed that despite levels of suspended solids (SS) between 3-4 mg/L in the influent, flow rates (hydraulic conductivities) were maintained at 20% to 50% of initial flow through three identical columns for a period of 23 days. An initial drop in the hydraulic conductivity (K) through the columns was evident early in the experiment, and initially decreased from 3 m/day to 1.4 m/day in the first 5 days of the experimental run. This was thought to be due to clogging by suspended solids as SS were filtered out on the surface of the aquifer material at the inlet end of the columns initially. The physical bioclogging was relieved midway through the experiment due to calcite dissolution and resulting formation of fissures within the columns allowing increased flow. Nitrification of nitrite and initially ammonia to nitrate and denitrification of nitrate to nitrogen gas in the columns was evident from the results obtained and provided an insight into the type of oxidation-reduction reactions could be expected during ASR using treated wastewater.

1 INTRODUCTION

Aquifer Storage and Recovery (ASR) is a process whereby surface water is injected into a groundwater system and stored for subsequent reuse. A three year ASR project is underway in the Northern Adelaide Plains, 30 km from Adelaide, capital of South Australia. The aim is to store DAF/F (Dissolved Air Flotation/Filtration) treated wastewater in a sandy-limestone aquifer for subsequent recovery and non-potable reuse (irrigation). The project is summarised in detail by Bosher & Kracman (1998) ibid.. Research and development involving detailed hydrogeological and geochemical investigations, minimisation of clogging and pathogen attenuation is currently under way to assess the feasibility, environmental sustainability, and economic viability of ASR at that site. A consortium has been formed to undertake this project including federal and state government agencies, academia and industry.

The injection of treated wastewater underground can have major impacts on the biogeochemistry of the aquifer. Such impacts may for example include changes in groundwater quality and clogging of porous material. “Clogging” or “plugging” during recharge usually results in an increased head buildup in the injection well, thereby changing the hydraulic characteristics of the well and severely restricting further injection of the wastewater. Previous researchers (Taylor & Jaffe 1990; Shaw et al. 1985, Vandeviviere & Baveye 1992, Osei-Bonsu, 1996) have documented a number of causes leading to clogging. These include bacterial production of polysaccharides and biomass, entrapment of suspended solids, entrained air, gas binding and dispersion of clay within the formation. The degree of pre-treatment prior to recharge is an important factor in minimising the extent of clogging and degradation of the surrounding groundwater.

The main water quality parameters of concern when injecting wastewater for ASR purposes, are the suspended solids (SS) content, the level of total dissolved solids as well as the concentration of nutrients that could lead to biological growth and thereby clogging in the vicinity of the injection well or further within the aquifer system.

Suspended solids present in the recharge water are filtered out from the water and accumulate on the aquifer-injection well interface. This filtration and accumulation phenomenon entails a reduction in permeability of the aquifer thus decreasing the hydraulic flow and injection efficiencies. Previous experimental results have shown that recharge water used for ASR should have levels of SS < 2mg/L to avoid major clogging problems (Okubo & Matsumoto 1983), although levels in excess of 25 mg/L have not caused clogging in the same aquifer system at a nearby stormwater ASR site (Pavelic et al. 1998 ibid.). Injectant water quality needs to be controlled and reuse quality guidelines were set by the South Australian Health Commission (SAHC) for irrigation with the water to be injected in this particular ASR project. The guidelines include specifications for suspended solids, faecal coliforms and pathogen levels.

The main purpose of this paper is to investigate the effect of continuous injection of post-secondary treated wastewater into columns packed with aquifer material, on water quality and biological clogging. The laboratory columns were used as a tool to assess the biochemical compatibility of the recharge water with the aquifer material and the native groundwater before field trials proceed.

2 MATERIALS AND METHODS

2.1 *Packing of laboratory columns*

A core of a sandy limestone aquifer (aquifer T2) was taken from a depth of 117 metres in the Northern Adelaide Plains, South Australia. The recovered core was packed in PVC tubes, flushed with nitrogen gas on-site and stored at 4 degrees. The core was partly sectioned and analysed for mineralogy and hydraulic properties (data not shown). The X-ray diffraction pattern of the core material showed that it was predominantly quartz and calcite.

A subsample of the core material was taken in an anaerobic chamber flushed with argon and packed into columns. The edges of the core material were removed and the inner piece of the core was submerged in groundwater collected at the field site. The material was crushed, sieved through a 1mm sieve, and packed into three columns.

Breakthrough curves were obtained using a bromide tracer in native groundwater prior to the inoculation of the columns with additional groundwater and then wastewater, to determine whether the columns were packed properly and that no preferential flow occurred. The columns were shaded from light by using aluminium foil, and the room temperature was kept at 20 ±2°C (the temperature of the groundwater at the depth of the core), for the entire length of the experiment. The columns were initially dosed with groundwater from the field site before passage of the treated wastewater, this allowed the stabilisation and the inoculation of the columns.

2.2 *Quality of post-secondary treated sewage*

The secondary treated wastewater was obtained from the Bolivar treatment plant in South Australia. The existing treatment at Bolivar consists of a secondary treatment followed by approximately 30 days retention in open lagoons. The wastewater was obtained from a channel collecting effluent from 2 lagoon streams and was then post treated using a mini flocculation/filtration pilot plant to simulate the water quality used for injection at the future ASR site. The water obtained after postsecondary treatment generally had a low content of suspended solids, nitrogen, organic matter and phosphorus. The main parameters are listed in Table 1.

Table 1: Average water quality of treated wastewater and native groundwater.

Parameter	Wastewater (treated) mg/L	Native groundwater mg/L
Biological Oxygen Demand (BOD)	2.0-3.0	------*
Total Kjehldal Nitrogen (TKN)	2.5-3.5	0.21
Phosphorus (as P)	0.2-0.3	0.039
Ammonium-N	0.2-0.3	0.056
Nitrate + Nitrite (as N)	6-10	0.008
Suspended Solids (SS)	3-4	60

* *Value not measured*

2.3 *Experimental Apparatus*

Treated wastewater was pumped through three columns (diameter 2.5cm, length 16 cm) packed with aquifer core material (section 2.1)at a rate of 5 L/day/column (Darcian velocity of 10^{-2} m/d) using a peristaltic pump. The development of clogging and the resulting decrease in hydraulic conductivity was monitored using pressure sensors located along the column and connected to a data logger. The pressure sensors acted as a means of measuring pressure heads at different points along the columns and at the inlet and outlet of the columns.

The velocity of wastewater (influent) pumped through the columns corresponded to that expected at the future ASR site in the vicinity of the well.

Column influent water quality combined with daily flow and pressure measurements and other observations such as bacterial biomass and polysaccharide production, were correlated with changes in hydraulic conductivity and water quality.

2.4 *Analytical Methods*

Temperature, pH, dissolved oxygen, conductance (EC), turbidity, flow rates, pressure changes along the columns, and COD were measured daily in the laboratory. Nutrients, major ions, metals and some microbiological analyses were performed by the Australian Water Quality Centre, Bolivar, South Australia.

3 RESULTS and DISCUSSION

The wastewater used for this column experiment varied slightly in quality over the experimental period. As the residence time of the water in the column is short, the effluent data shown in the figures below is represented as Ceffluent/Cinfluent (Ce/Ci) to account for these changes.

3.1 *Changes in Hydraulic Conductivity*

The change in hydraulic conductivity in different parts of the columns was calculated to display where most of the clogging had occured. The hydraulic conductivity (K) results shown in Fig. 1, correspond to the average K in meters/day of three columns running concurrently for a period of 23 days. The results obtained suggest that there was an initial drop in K over the entire length (0-16 cm) of each column, from 3 m/day to 1 m/day in the first 7 days of the experimental run. Hydraulic conductivities were lowest at the inlet end of the columns (0-3 cm), between 0.2 to 0.8 m/day, compared to the remaining part of the columns (3-16 cm) where conductivities remained between 1.4 to 2 m/day after an initial decrease from 3m/day.

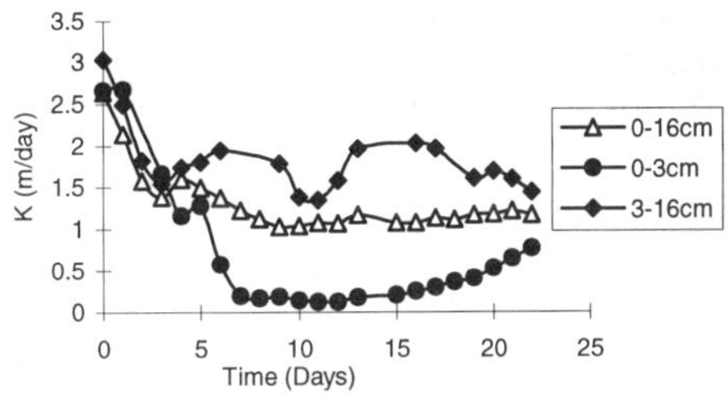

Figure 1. Decrease in hydraulic conductivity (K) after passage of treated wastewater for a period of 23 days.

The early drop in K can be explained by an initial filtration of suspended material on the surface of the aquifer matrix at the inlet end of the column. Together with bacterial polysaccharides found in higher concentrations at the interface of the columns than along the columns (results not shown). The sudden increase after day 13 and the levelling off of K is probably due to calcite dissolution. This process is clearly evident from the increase in calcium levels in the effluent from day 8 onwards (Fig. 2), and is due to the high percentage of carbonate minerals (49%) present in the aquifer material contained in the columns. A concurrent increase in pH, alkalinity (plotted) and a slight increase in EC was also observed and are additional indicators of calcite dissolution.

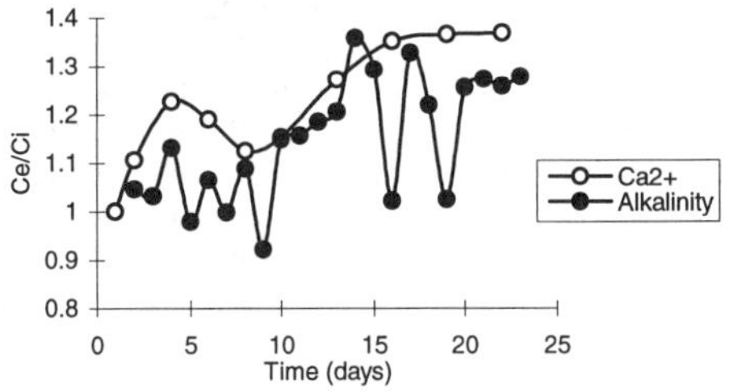

Figure 2. Dissolution of calcium and changes in alkalinity in the columns (average of 3 columns shown) after passage of treated wastewater for 23 days.

Fissures were observed at the inlet end of the columns. The fissures widened with the increasing dissolution of the calcite, and may possibly explain the increase in hydraulic conductivity rate on day 14 (Fig. 1). These results parallel those observed in the field where stormwater has been injected into this aquifer (Pavelic et al. 1998 ibid).

3.2 *Ammonium and Nitrite Removal*

Ammonium and nitrite were biologically transformed to nitrate in the first stage of the experiment, (days 4-8) by a process known as nitrification. This process can clearly be demonstrated from the results shown in Figure 3, where ammonium and nitrite significantly decreased coinciding with a significant increase in nitrate. Chemoautotrophic nitrifying bacteria have probably adhered to the aquifer material in the columns, during the early stages of the experiment when the DO levels were still high. Ammonium was consumed in the first 5 days of the experiment (Fig. 3). This could be attributed to the nitrification of ammonium to nitrite and nitrate by nitrifying bacteria and/or the assimilation of ammonium by bacteria present in the columns using ammonium as an nitrogen source.

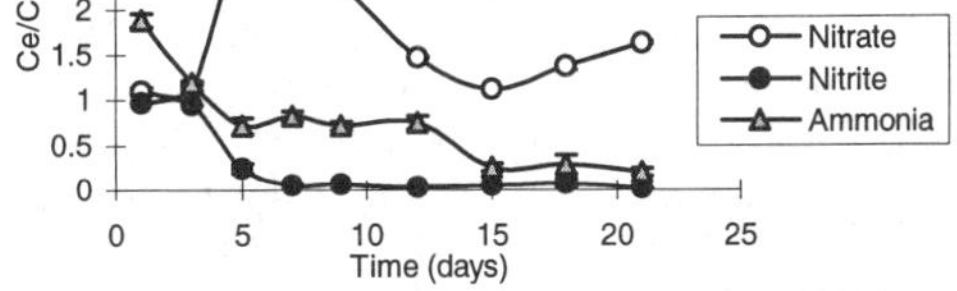

Figure 3. Changes in nitrate, nitrite and ammonium levels in the columns over the 23 day experimental run. The data plotted represents an average result for 3 replicate columns.

Towards the latter stage of the experiment, on day 9, when the DO levels in column effluent were found to be relatively low, at about 1-2 mg/L (Fig. 4), nitrification was being limited by the low levels of DO present within the columns. Nitrate started to decrease after day 8, indicating denitrification conditions developed, at least lower in the columns, where nitrate was converted to nitrogen gas (N_2). This coincided with conditions of low DO within the columns.

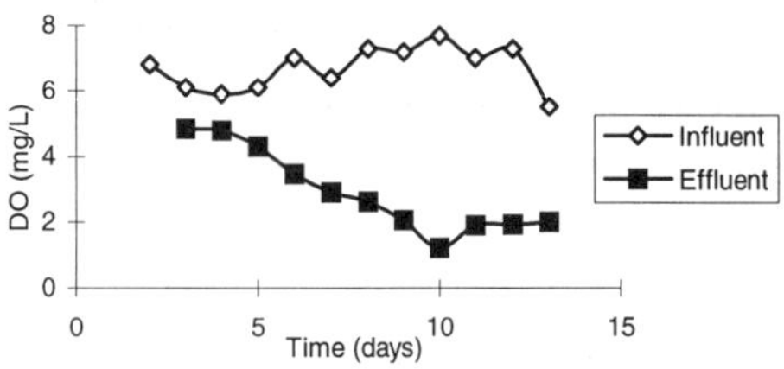

Figure 4: Change of dissolved oxygen (DO) levels in the three columns. Average of 3 columns shown for effluent data.

The DO of the three columns changed with time in the same manner (Fig. 4). The DO in effluent from all three columns decreased rapidly for the first 10 days, and remained almost constant at a level of 1-2 mg/L for the remaining 13 days of the experiment. However, both aerobic and anaerobic processes occured in the columns during the length of the experiment. It is apparent that anaerobic conditions were present at the effluent end of the columns, but as DO was still present in the influent (Fig. 4), then presumably aerobic reactions (nitrification) took place within the column near the influent end. No specific conclusion can be made at this stage as to the distribution of DO along the columns.

3.3 *Turbidity of the injected wastewater*

The post-secondary treated influent (sewage) contained a low level of suspended solids which was not totally removed by the post treatment process. As organic suspended material may have a big impact on the clogging of the aquifer material in the columns, monitoring of SS levels in the influent and effluents was performed on a daily basis. The turbidity of the influent was in the range of 0.8-1.3 NTU's corresponding to 3-4 mg/L SS. The effluent turbidities for all three columns ranged from 0.15-0.3 NTU's, corresponding to 1-2 mg/L SS. This clearly indicates filtration of suspended solids on the aquifer matrix. Suspended solids have adhered to the grain surface at the inlet of the columns, forming deposits in the pores thereby reducing the hydraulic conductivity and effective porosity. The restriction in hydraulic conductivity due to this filtration process was evident at the early stage of the experiment when the hydraulic conductivity rates decreased (Fig. 1), and the pressures at the inlet of the three columns increased (data not shown). As calcite dissolution proceeded, fissures were formed allowing preferential flow of the influent, thereby increasing the hydraulic conductivity slightly and stabilising the pressures along the columns.

4 CONCLUSIONS

Small-scale laboratory column experiments were used to predict biogeochemical transformations resulting from injecting post-secondary treated wastewater into an aquifer system. The most important conclusions resulting from this work are:

1. Clogging in the columns in the early stage of the experiment was due to filtration of suspended solids and polysaccharide production at the inlet of the columns. There was some recovery of hydraulic conductivities after about 10 days, possibly due to calcite dissolution processes affecting the flow rates by creating fissures through the columns. This shows that in the long term, prolonged calcite dissolution will significantly alter the hydraulic properties of the aquifer system used for ASR purposes, due to the formation of subsurface channels (secondary porosity).
2. Nitrification occurred during the early stages of the experiment, where nitrite was completely removed from the injected water by being oxidised to nitrate. When oxygen became limiting halfway through the experimental run, denitrification occurred.

ACKNOWLEDGMENTS

This study is a collaborative research exercise between the Centre for Groundwater Studies, CSIRO Land & Water, Primary Industries & Resources, United Water International Pty Ltd, SA Water and SA Government Project Delivery Task Force. Contributions from members of each of these organisations were invaluable in conducting this research. I am grateful to Dr. Peter Dillon, Dr. John Hutson and Dr. Howard Fallowfield for their supervision. Chris Bosher from United Water Pty Ltd, is also acknowledged for his help in the design and set up of a small post-secondary treatment pilot plant at the Bolivar Wastewater Treatment Plant. We are also deeply grateful to the Association Nationale de la Recherche Technique (ANRT), Paris, France, for funding this work through the grant of a CIFRE bursary. My thanks are extended to the staff at the

Mechanical Engineering Department at the Flinders University, for the construction of the equipment, and the Australian Water Quality and Water Treatment (AWQC) for assistance with chemical and biological analyses.

REFERENCES

Bosher, C. & B. Kracman 1998. Wastewater aquifer storage and recovery (ASR) towards sustainable water reuse in South Australia. *TISAR'98: Proc. 3rd Int. Symp. On Artificial Recharge of Groundwater*, 21-25 September, 1998, Amsterdam (these proceedings).

Pavelic, P., P.J. Dillon, K.E. Barry, A.L. Herczeg, K.J. Rattray, P. Hekmeier & N.Z. Gerges 1998. Well clogging effects determined from mass balances and hydraulic response at a stormwater ASR site. *TISAR'98: Proc. 3rd Int. Symp. On Artificial Recharge of Groundwater*, 21-25 September, 1998, Amsterdam (these proceedings).

Okubo, T. & J. Matsumoto 1983. Biological clogging of sand and changes of organic constituents during artificial recharge. *Water Research* 17(7):813-821.

Osei-Bonsu, K. 1996. Clogging by sediments in injected fluid flowing radially in a confined aquifer. Flinders University of South Australia, School of Earth Sciences, PhD Thesis.

Shaw, J.C., Bramhill, B., Wardlaw, N.C. & Costerton 1985. Bacterial fouling in a model core system. *Appl. Environ. Microbiol.* 49 (3): 693-701.

Taylor, S.W.& P.R. Jaffe 1990. Biofilm growth and the related changes in the physical properties porous medium. *Water Resour. Res.* 26 (9): 2153-2159.

Vandeviviere, P. and P. Baveye 1992. Saturated hydraulic conductivity reduction caused by aerobic bacteria in sand columns. *Soil Sci. Soc. Am.* J. 56 (1): 1-13.

Artificial Recharge of Groundwater, Peters et al. (eds) © 1998 Balkema, Rotterdam, ISBN 90 5809 017 5

Wastewater treatment and groundwater recharge for reuse in agriculture: Dan Region reclamation project, Shafdan

N. Icekson-Tal & R. Blanc
Mekorot Water Company Limited, Ramle, Israel

ABSTRACT: The Dan Region Reclamation Project is the largest wastewater reclamation project in Israel. The wastewater from the Greater Tel Aviv metropolitan area (220 sq. km., 1.5 million inhabitants) is biologically treated, most of it (93%) in an activated sludge treatment plant, and the rest (7%) in recirculated facultative oxidation ponds. On average 303,000 m^3 are treated every day. The effluent is then further treated through recharge into the aquifer and subsequent recovery using a special soil-aquifer treatment (SAT) developed at the Dan Region Project. The SAT adds to purification of the effluents by processes of slow sand filtration, chemical precipitation, adsorption, ion exchange, biological degradation, nitrification and denitrification. In addition to the water quality improvement, SAT constitutes a seasonal and multiannual water storage system. Water recovered after soil aquifer treatment is of extremely high quality, and can be used for unlimited agricultural irrigation. In 1997, a total of some 103 million m^3 were recharged and recovered in the Dan Region Reclamation Project, and since 1977 851 million m^3 have been reclaimed through SAT.

1 INTRODUCTION

The Dan Region Reclamation Project (known by its Hebrew acronym Shafdan) is the largest wastewater treatment and reclamation project in Israel. Wastewater treatment serves to reduce environmental pollution caused to the Mediterranean and local rivers; reclamation serves to economically utilize the only significant source of water not yet exploited to its fullest in Israel. With its growing population and economy, and its severely limited water resources, Israel must find innovative ways to increase its water potential. With groundwater and surface water being utilized to capacity, it is envisioned that by 2010 reused wastewater will constitute one fifth of the total supply, and one-third of agricultural demand (Shelef, *1990)*. In the Dan Region Project, the wastewater from the Greater Tel Aviv metropolitan area (220 sq. km., 1.5 million inhabitants, 1.9 million p.e. including industrial input) is biologically treated, most of it (93%) in an activated sludge treatment plant, and the rest (7%) in a 200 hectare system of recirculated oxidation ponds. On average 303,000 m^3 are treated every day. This constitutes between one-quarter and one-third of the entire country's household wastewater discharge. The treated effluent is subsequently recharged into the aquifer through a system of recharge basins, and is subsequently pumped up and conveyed to the southern part of the country where it is used for irrigation. Figure 1 (following page) shows the general layout of the project, and figure 2 is the schematic flowchart.

2 THE WASTEWATER TREATMENT PLANT

The influent wastewater collected from the Dan region consists of ~90% household sewage and ~10% industrial sewage (hydraulically, though the industrial organic input is considered to be 400,000 p.e). Typical wastewater parameters are an average BOD of 422 mg/l (COD of 1,020 mg/l), suspended solids 380 mg/l, and total nitrogen of over 63 mg/l.
Biological treatment of the wastewater is achieved in two parallel treatment systems. Stage 1 of the project, in existence since the late sixties, consists of ~200 hectares of recirculated facultative oxidation ponds and polishing ponds. 15,000-20,000 m^3, roughly 7% of the influent raw sewage, are treated in Stage 1 per day. Retention time in this system is

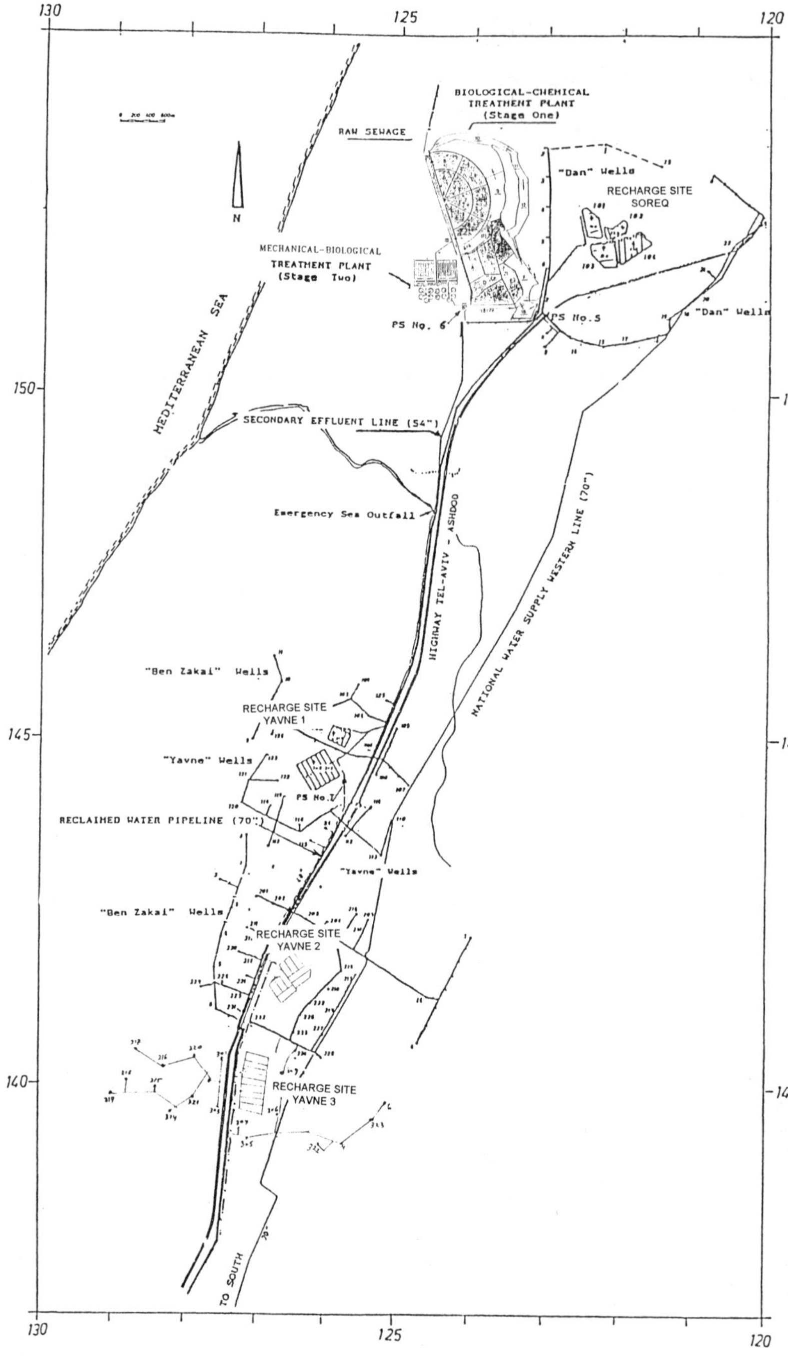

Figure 1. General layout of the Dan Region wastewater treatment and reclamation project

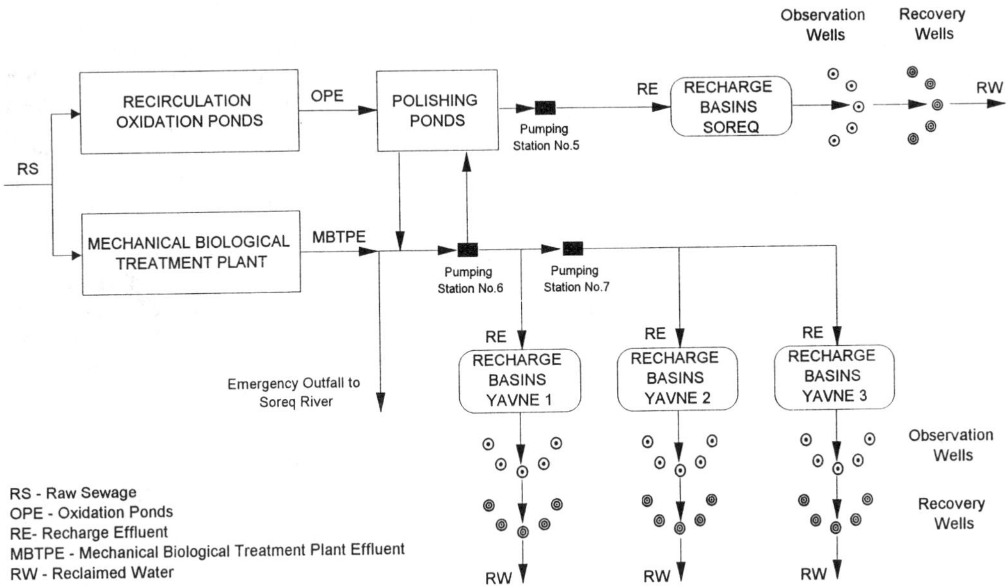

Figure 2. Schematic diagram of the Dan Region wastewater treatment and reclamation project

Table 1. Main parameters of biological treatment, annual averages 1997 (Soffer et al., 1997)

Parameter	Raw Sewage	Polishing Pond effluent	Removal (%)	Activated sludge effluent	Removal (%)
Suspended Solids (105°), mg/l	380	32	91.6	8	97.9
BOD, mg/l	422	19	95.5	6	98.6
COD, mg/l	1,020	105	89.7	53	94.8
COD(filtered), mg/l	339	60	82.3	42	87.6
Ammonia(filtered) - N, mg/l	39.9	9.4	76.4	7.3	81.7
Total N, mg/l	63.3	16.5	73.9	10.8	82.9
Phosphorus, mg/l	14.4	5.6	61.1	2.7	81.2
Total Bacteria, log conc.	7.17	5.80	19.1	5.46	1.7
Coliforms, log. Conc.	8.22	5.45	33.7	5.34	2.9
Faecal Coliforms, log conc.	7.27	5.12	29.6	4.29	3.0
Chloride, mg/l	297	318	--	289	--
Chromium, µg/l	135	7	94.8	7	94.8
Boron, mg/l	0.62	0.60	--	0.55	11.3

60-90 days. The oxidation pond effluents are of lower quality than the activated sludge effluents, typical values being suspended solids concentrations of 32 mg/l, BOD of 19 mg/l, and ammonia levels of 9.4 mg/l. The main parameters of the polishing pond effluent and activated sludge effluent quality, in annual averages, are listed in Table 1. Because of the limited efficiency of treatment in this stage, the polishing pond effluents are mixed at a ratio of approximately 1:2 (ponds : AS effluents) with activated sludge effluents in order to improve effluent quality before recharge in the SAT.

Most of the wastewater is treated in stage 2 of the project, a completely-mixed activated sludge plant with single-stage simultaneous nitrification-denitrification. 93% of the sewage, on average 282,000 m^3 a day, is treated in this part of the plant. During the winter months, storm events can cause flows of up to 437,000 m^3 (over 50% above average flow), though these events are infrequent. Treatment consists of pre-treatment (bar-screens, grit removal and scum removal), an anaerobic selector (for filament control and Enhanced Biological Phosphrus Removal), aeration with horizontal Mammoth rotors (36 per reactor) in four 55,000 m^3 aeration tanks, and secondary settling in twelve 52 m. diameter clarifiers. There is no primary settling. The aeration tanks consist of alternating aerobic and anoxic zones, so that nitrification-denitrification take place simultaneously, together with carbonaceous BOD removal. Following the activated sludge treatment suspended solids are reduced by 97% to 8 mg/l, BOD is reduced to 9 mg/l (97%), and ammonia is reduced by 80 % to 6 mg/l.

3 THE SOIL-AQUIFER TREATMENT (SAT)

The biologically-treated effluent is further improved through soil-aquifer treatment (SAT). Soil aquifer treatment consists of intermittent flooding and drying of recharge basins with the treated effluents, so that each of the basins is flooded for one day and then left to dry for 2-4 days. Effluents from stage 1 (a 1:2 mixture of polishing pond effluents and the activated sludge plant effluent) are spread in the Soreq recharge basins, consisting of ~25 ha. divided into 18 sub-basins for operational flexibility. Effluents from stage 2 (the activated sludge plant) are spread in the Yavne 1, Yavne 2 and recently-added Yavne 3 spreading fields, consisting of ~48 ha divided into 28 sub-basins. Hydraulic loads vary between 80-180 m/year, depending on the infiltration capacity of the basins. In 1997 some 18 million m^3 were recharged in the Soreq recharge basins (6.6 mcm from the polishing ponds, 11.1 mcm activated sludge effluent), and 85.3 million m^3 were recharged in the Yavne recharge basins. In all, in 1997 some 103.1 million m^3 were recharged, and about 117 million m^3 were reclaimed, the difference between recharge and recovery being made up of native aquifer water and rainfall replenishment. Since 1977, when full operation of the SAT system began, some 851 million m^3 of treated wastewater were recharged into the aquifer.

The SAT system used at the Dan Region Plant was developed on-site (Idelovitch and Michail, 1984). The system is located over an underground zone dedicated entirely to effluent treatment and storage, where no drinking-water wells are located. This constitutes only a small part of the regional aquifer. The surface soil of the recharge basins consists primarily of uniform, fine dune sand, interspersed by sandstone aggregates and thin loamy sand horizons. The site is located above the central part of the Coastal Aquifer (Pleistocene), about 4 km. east of the Mediterranean Sea. This aquifer is composed mainly of calcareaous sandstone, and is divided into four subaquifers which are seperated by silt and clay layers. The uppermost aquifer is phreatic and extends about 1.5 to 2 km. inland from the coast, terminating west of the oxidation ponds site (see figure 1). The other three aquifers extend 4-6 km. inland from the coast, where they unify into one aquifer. The "Saqiye" clay formation delineates the lower boundry of the Coastal Aquifer.

The treated effluent percolates vertically through 15-43 meters of the unsaturated zone and, once reaching the groundwater, spreading radially (horizontally) through the saturated zone outward from the spreading basins to a series of recovery wells surrounding the recharge area. Some 100 recovery wells, located radially 300-1500 m. from the recharge basins, pump the recharged water from a depth of 100-120 meters. The water produced is pumped through a single, 87 km. long pipeline (the Third Pipeline) with six floating reservoirs along it, to the southern, arid part of the country where it is used for unlimited agricultural irrigation.

Except for water production, the production wells serve to confine the movement of the recharged effluent within the saturated zone, thereby preventing mixing with the native aquifer water. A ring of observation wells, located 30-800 m. from the recharge basins in between the recharge basins and the recovery wells, serve to monitor the recharged water quality and flow direction. Effluent movement through the saturated zone can be monitored through concentrations of certain conservative elements, such as chlorides and boron, which are not reduced to any significant extent through the SAT (or, in fact, through the biological treatment).

SAT serves to improve the effluent quality to near-drinking-water level, through treatment in both the unsaturated and the saturated zones. Residence time between infiltration and recovery is ~100-300 days. Treatment includes physical processes of filtration, occurring in the top layer of soil; physiochemical processes such as adsorption, ion exchange and precipitation, occurring throughout

Table 2. Main water quality parameters of SAT, compared with activated sludge plant effluents (Stage 2) used for recharge, annual averages 1997 (Soffer et al. 1997).

Parameter	Before SAT	After SAT (OW 017/s)*	Removal, %
Suspended Solids (105°), mg/l	8	0	100.0
BOD, mg/l	6	<5	>98.0
COD, mg/l	53	6	88.7
COD(filtered), mg/l	42	6	85.7
Ammonia(filtered) - N, mg/l	7.3	0.07	99.0
Total N, mg/l	10.8	3.19	70.5
Phosphorus, mg/l	2.7	0.02	99.3
Total Bacteria, no/ml	290,300	4,430	98.5
Coliforms, MPN/100 ml	220,300	0	100.0
Fecal Coliforms, log conc.	19,500	0	100.0
Chloride, mg/l	289	266	--
Chromium, µg/l	7	<3	>57.0
Boron, mg/l	0.55	0.54	1.8

*Observation Well 017 shallow

the soil profile; and biological processes such as breakdown of carbonaceous organic matter and full nitrification and partial denitrification, which occur throughout the unsaturated and saturated zones (Kanarek and Michail, 1994). Filtration accounts for the removal of much of the particulate material and turbidity, and to a certain extent removal of microorganisms (natural dieoff accounts for additional microbial improvement). Chemical processes are responsible for removal of phosphorus and some trace elements (Cr, Cd, Cu, Mo, Ni and Se), while biological processes improve COD/BOD removal, and are responsible for complete nitrification and partial dentirification. In all, quality of the reclaimed water is suitable for unrestricted agricultural use (including vegetables to be eaten raw) , industrial uses, and non-potable municipal use (such as lawn irrigation). At present, however, the water is used only for agriculture. Table 2 presents the quality of reclaimed water in a representative observation well, following recharge of activated sludge plant effluents.

4 CONCLUSION

The Dan Region Reclamation Plant is an integrated solution for urban wastewater treatment and water reclamation for agricultural use. Biological treatment in lagoons and activated sludge serve to reduce most of the wastewater organic load, thereby producing an effluent which can be used for groundwater recharge through SAT. SAT serves to further improve the water quality, as well as serve as a large-capacity, seasonal and multiannual storage system for water in a country where most of the irrigation is done in the dry summer months. The combination of environmental, health and water-resource benefits is unique in Israel, and can serve as a model for other large cities in water-scarce regions of the world.

REFERENCES

Idelovitch, E. & M. Michail. 1984 JWPCF 56:936

Kanarek, A. & M. Michail. 1994 Contamination and Control of Groundwater in Aquifer Recharge, in Groundwater Contamination and Control, ed. U. Zoller, Marcel Dekker, Inc., Publishers, New York

Shelef, G. 1990 The role of wastewater in water resource management in Israel, Wat. Sci.Tech. 23, (10-12), pp. 2081-2089

Soffer, D., N. Icekson-Tal, M. Michail, D. Sherer & R. Blanc. 1997 Groundwater Recharge with Municipal Effluent, Recharge Basins Soreq, Yavne 1, Yavne 2, and Yavne 3, Dan Region Reclamation Project, year 1996 Annual Report. Mekorot Water Co. Ltd. Central district, Dan Region Unit, Tel Aviv, Israel.

1.5 Pollutants and bankfiltration

Artificial Recharge of Groundwater, Peters et al. (eds) © 1998 Balkema, Rotterdam, ISBN 90 5809 017 5

Removal of organic matter during bank filtration

A-L. Kivimäki, K. Lahti & T. Hatva
Finnish Environment Institute, Helsinki, Finland

S. M. Tuominen
Helsinki University of Technology, Finland

I. T. Miettinen
National Public Health Institute, Kuopio, Finland

ABSTRACT: The quality and quantity of organic matter in raw water affects the purification processes in artificial recharge of groundwater, and thus final drinking water quality. The removal of organic matter during induced recharge of artificial groundwater was studied at two Finnish bank filtration plants (Kangasala and Nokia). Water samples from lake, observation tubes and pumping wells were taken monthly over two years. The amount of organic carbon (AOC/TOC), molecular size distribution of organic carbon and microbial activity were studied. The reduction of TOC and AOC in the Lake Vesijärvi (Kangasala) bank filtration zone was 29% and 72%, respectively. In the Lake Vihnusjärvi (Nokia) bank filtration zone the reduction in TOC content was 55%, and in AOC content 40%. The highest molecular weight fractions of organic matter were removed at the beginning of infiltration. At the end of filtration the lowest molecular weight fractions were predominant. In both study areas, the reduction in organic carbon occured mostly in the littoral region. Later, during the underground passage, no remarkable decrease in AOC content occurred, whereas the content of TOC still decreased. Results suggest that in saturated, reducing conditions only minor microbial degradation will occur. The decrease in TOC probably resulted from sorption of humic substances into soil particles.

1 INTRODUCTION

In Finland the proportion of bank filtrated artificial groundwater in the municipal drinking water production is around 9 %. A large number of the waterworks use bank filtered water quite unintentionally, or they may even try to prevent the bank filtration, because of the poor quality of surface water (Kivimäki 1995). However, on account of the insufficient natural groundwater resources in some parts of Finland, both direct and indirect methods of artificial recharge will be more widely used in the water supply in future.

Since pre-treatment of lake water cannot be used in bank filtration, the quality of raw water is of concern. High content of organic material in the soils of Finnish forested areas and peatlands generates a high runoff of organic matter to the lakes, thus resulting in a high concentration of total organic carbon (TOC) and organic acidity in lake waters. The results of the Nordic Lake Survey in 1995 indicate that the median TOC in Finnish lakes is 7.6 mg l^{-1} (Henriksen et al. 1997). The median COD_{Mn} in lake water used as raw water at Finnish artificial groundwater plants is 7.5 mg l^{-1} (Hatva 1996). At bank infiltration plants COD_{Mn} values ranging from 20 to 31 mg l^{-1} may occur in raw water sources. (Kivimäki 1995). Data on concentrations of assimilable organic carbon (AOC) in Finnish drinking waters is quite sparse. According to Miettinen et al. (1997) AOC is 390±130 μg l^{-1} and 170±60 μg l^{-1} in drinking waters produced from surface water and groundwater, respectively.

Dissolved organic carbon in lakes consists of a mixture of high molecular weight, refractory humic substances, and organic compounds of low molecular weight, which are more labile (Kaplan & Newbold 1993). In Finland, TOC mainly consists of high molecular weight components (Vartiainen et al. 1987). Humus itself is not harmful for human health, but humic substances have indirect effects on drinking water production. Organic matter may result in bacterial regrowth in the water distribution network. Disinfection of the water with chlorine may cause the formation of mutagenic or carcinogenic compounds. Furthermore, humic substances in groundwater are able to interact with less soluble organic pollutants and transport them in the aquifer (Thurman 1985).

Microbes living in sediments and soil surface layers in the littoral region of lakes decompose organic compounds. Bacterial uptake and growth efficiency are dependent upon the source and fraction of DOC (Kaplan & Newbold 1993). Therefore, knowledge on the characteristics of organic matter as well as factors regulating microbial activity are still needed in order to optimize water purification.

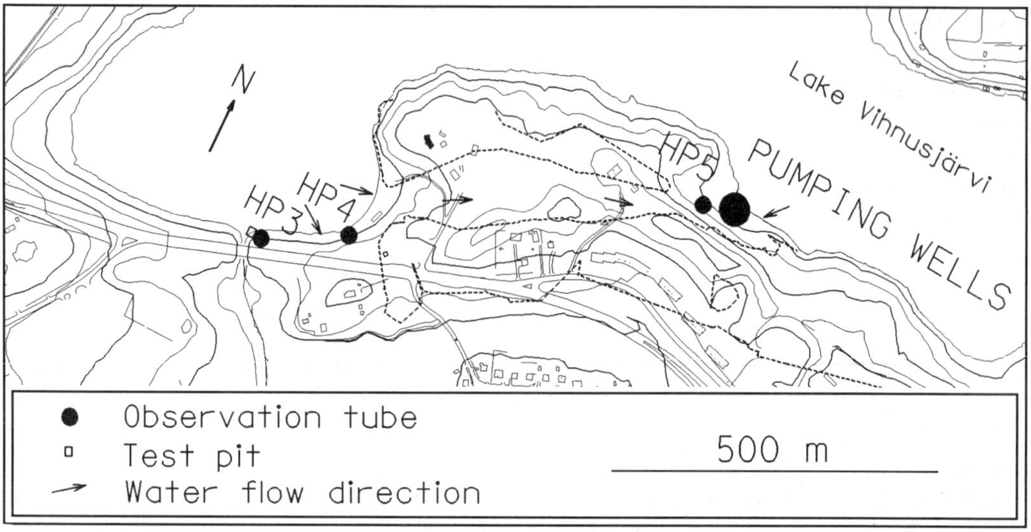

Figure 1. Nokia waterworks by Lake Vihnusjärvi

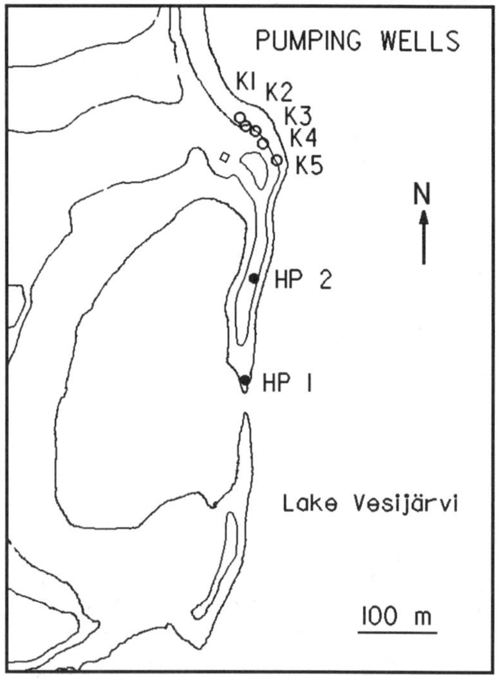

Figure 2. Kangasala waterworks by Lake Vesijärvi.

We studied the removal of organic matter during induced recharge of artificial groundwater at two bank filtration plants. Biological lability of organic matter was studied by means of parameters indicating the amount of organic carbon, molecular size of organic carbon and microbial activity. Seasonal changes in factors affecting biodegradation such as temperature,oxygen content and amount of nutrients were also examined.

2 MATERIALS AND METHODS

2.1 *Study areas and sampling*

Removal of organic matter during induced recharge of artificial groundwater was investigated at two Finnish bank filtration plants using different lake water quality. In both cases, the waterworks were situated in an esker area, glacifluvial formation composed mainly of sand and gravel layers (Figs 1 and 2). The natural yield of the Nokia aquifer has a maximum of 1000 m^3 d^{-1}, and the amount of artificial groundwater recharged by pit infiltration is around 750 m^3 d^{-1}. Since the amount of water pumped at groundwater intake is at maximum 5000 m^3 d^{-1}, the majority of the water is bank filtered. In Nokia the permeability of the shore line is quite low because of the topmost clayey and silty layer with thickness of around 3.5 m and hydraulic conductivity $K=8x10^{-7}$ m s^{-1}. This layer retards the effective entering of lake water into the aquifer, except in the vicinity of the observation tube HP3, where the canal cuts the esker. Most of the bank filtration takes place in the shore line 500-800 metres west from the groundwater intake area, but to some extent also in the vicinity of the pumping wells. The bank filtered water flows northeast towards the pumping wells along the narrow and deep bedrock depression filled with gravel and sand. On the basis of groundwater modeling, the residence time from the main bank filtration zone to the groundwater intake area is three to four months.

In Kangasala esker the estimated yield of the aquifer is as low as 150 m^3 d^{-1}, and the amount of water produced is on average 3500 m^3 d^{-1}. Thus nearly all of the pumped water is bank filtered, and the entry of lake water is concentrated near the coarse-grained gravelly shoreline. The estimated residence time in the longitudinal direction of esker from the observation tube HP1 to the pumping wells is approximately one, to one and a half months.

We took water samples from lake, observation tubes and pumping wells on a monthly basis throughout two successive sampling periods. Groundwater samples were taken with the Grunfos MP1 submersible pump at two levels of observation tubes, with screens covering the entire saturated zone. Before sampling, the tubes were purged for 30 minutes at a rate of 8 - 10 litres per minute. The first sampling period lasted from September 1995 to October 1996, and the second from March 1997 to February 1998. Temperature and oxygen concentration were measured monthly in situ from observation tubes at several levels at intervals of 0.5 - 1.2 m with the YSI 57 DO Meter and YSI 3000 T-L-C Meter. In order to analyse microbial biomass and activity in soil layers, we took soil samples from the bank filtration zone with auger drilling in August 1995, 1996 and 1997.

2.2 *Water quality analyses*

Chemical oxygen demand (COD_{Mn}, oxidation with permanganate) was determined according to the Finnish Standard (SFS 3036, 1981).

The amount of total organic carbon (TOC) was measured by the combustion-infrared method according to ISO 8245 (1987). The combustion temperature was 900 °C, and the instrument used was Dohrman DC 90.

AOC (assimilable organic carbon) was analyzed by a modified method originally developed by van der Kooij and co-workers (1982). In order to ensure that organic carbon was the only factor limiting the bacterial growth, nutrients (4.55g $(NH_4)_2SO_4$, 0.2g KH_2PO_4, 0.1g $MgSO_4{\cdot}7H_2O$, 0.1g $CaCl_2{\cdot}2H_2O$, 0.2g NaCl) were added into the samples (Miettinen et al. 1996). AOC was analyzed using *Pseudomonas fluorescens* strain P17 biotype 7.2 (ATCC 49642) and *Spirillum NOX* (ATCC 49643) test strains.

The molecular weight fractions of organic carbon (UV absorbing humus fractions) were measured by high pressure size exclusion chromatography (HPSEC) using TSK-gel G3000SW columns. Analyses were performed with 0.01 M sodium acetate solution at a flow rate of 0.7 ml min^{-1} (Vartiainen et al. 1987).

The number of heterotrophic bacteria was determined by cultivation on R2A medium (Difco 1826) at 20±2 °C for 7 days. The quantity of dissolved oxygen was determined by means of the titrimetric method (SFS 3040, 1990).

2.3 *Analyses of microbial activity in soil samples*

The adenosine triphosphate content (ATP) of soil samples taken from Nokia esker was analyzed according to Vanhala & Ahtiainen (1994) with the exception that 10 g instead of 4 g fresh-weight sieved soil and sediment samples were used.

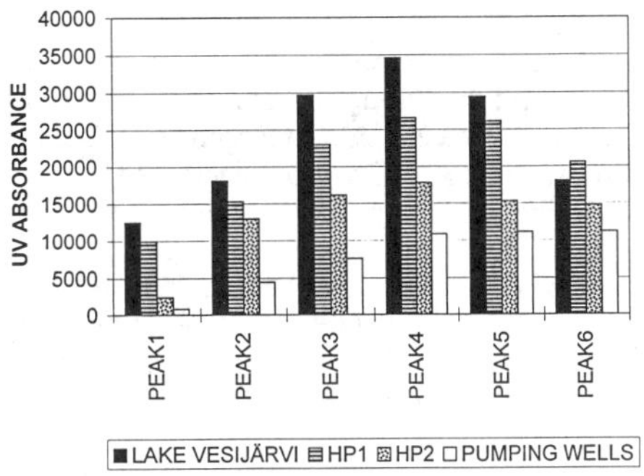

Figure 3. The median values (n=8) of UV absorbing humus fractions in Kangasala study area (lake water, HP1, HP2, pumping wells). Peak 1 indicates the highest molecular weight fraction and peak 6 the lowest weight fraction.

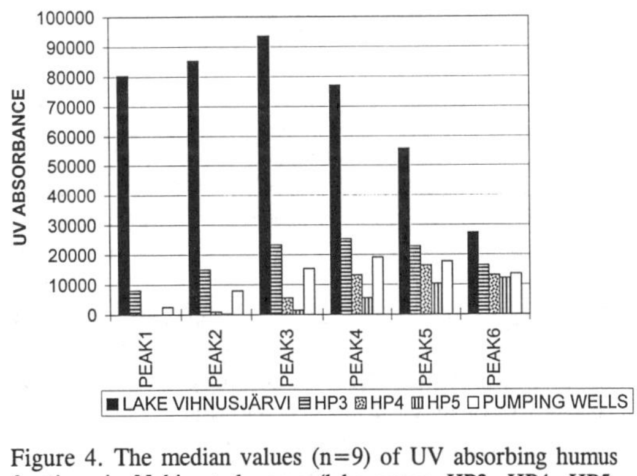

Figure 4. The median values (n=9) of UV absorbing humus fractions in Nokia study area (lake water, HP3, HP4, HP5, pumping wells).

3 RESULTS

3.1 *Quality of lake waters*

The quantity and quality of organic carbon was different in Lake Vesijärvi in Kangasala and Lake Vihnusjärvi in Nokia. TOC ranged in Lake Vesijärvi from 4.3 to 5.3 mg l^{-1}, and COD_{Mn} from 3.4 to 4.3 mg l^{-1}. In Lake Vihnusjärvi the range in TOC was from 8.0 to 12.1 mg l^{-1}, and in COD_{Mn} from 5.4 to 13.4 mg l^{-1}.

The distribution of molecular size fractions of organic matter indicated that in Lake Vesijärvi the organic matter consisted equally of different molecular weight fractions, whereas in Lake Vihnusjärvi the high molecular weight fractions dominated (Figs 3 and 4).

The amount of AOC was higher in Lake Vesijärvi (median 86 μg l^{-1}) than in Lake Vihnusjärvi (median 57 μg l^{-1}), especially during the first period (median values 166 μg l^{-1} and 85 μg l^{-1}, respectively). During the second sampling period AOC decreased in both lakes. A distinct maximum in AOC content was discovered in both lakes in September 1997.

The median colony counts of heterotrophic bacteria in Lake Vesijärvi were 580 CFU ml^{-1} (the first period) and 895 CFU ml^{-1} (the second period), whereas in Lake Vihnusjärvi the median values during the first and the second sampling periods were 2725 CFU ml^{-1} and 4900 CFU ml^{-1}, respectively. In both lakes and sampling periods, peaks in microbial counts occurred in May and September. In addition to the seasonal variation, the number of heterotrophic bacteria in both lakes was slightly higher during the second sampling period. At the beginning of September 1997 a slight peak in chlorophyll-*a* was also discovered in Lake Vihnusjärvi, owing to the occurrence of cyanobacteria.

3.2 *Removal of organic matter*

Both TOC and AOC decreased right at the beginning of underground passage in both waterworks. Reductions of TOC and AOC in Lake Vesijärvi bank filtration zone were 29% and 72%, respectively, whereas in Lake Vihnusjärvi bank filtration zone the reduction in TOC content was 55% and in AOC content 40% (Table 1). The seasonal variations of TOC and AOC in bank filtered water followed the trends in lake water, indicating peak values in AOC content in early autumn, especially in September 1997. There were no remarkable changes in AOC content from bank filtration zone towards the pumping wells.

In general, six different molecular weight fractions of organic carbon (UV absorbing humus peaks) were discovered from the lake water samples. The highest molecular weight fractions of humus were already removed at the beginning of filtration. The decrease in high and medium molecular weight fractions continued along the underground passage. At the end of filtration the lowest molecular weight fractions were predominant (Figs 3 and 4).

The changes in number of heterotophic bacteria in bank filtered water corresponded to seasonal variation in content of AOC in lake waters. The maximum peaks in number of heterotrophic bacteria were discovered in early summer and early autumn. In the observation tubes close to the bank shore the median values in the Vesijärvi area were 445 CFU ml^{-1} (the first sampling period) and 655 CFU ml^{-1} (the second sampling period), whereas in the Vihnusjärvi bank filtration zone the median values were 300 CFU ml^{-1} and 370 CFU ml^{-1}.

Table 1. The mean values and standard deviations of TOC and AOC in Kangasala and Nokia study areas. Number of observations in parentheses.

Sampling point	TOC (mg l^{-1})	AOC (μg C l^{-1})
KANGASALA STUDY AREA		
Lake Vesijärvi	4.8 ± 0.3 (20)	121 ± 126 (11)
HP1	3.4 ± 0.5 (20)	34 ± 24 (7)
HP2	2.4 ± 0.6 (20)	31 ± 20 (5)
Pumping wells	1.5 ± 0.3 (20)	30 ± 24 (11)
NOKIA STUDY AREA		
Lake Vihnusjärvi	10.0 ± 1.0 (23)	62 ± 30 (17)
HP3	4.5 ± 0.7 (23)	37 ± 23 (16)
HP4	2.2 ± 0.8 (23)	25 ± 23 (16)
HP5	1.7 ± 0.2 (20)	33 ± 16 (10)
Pumping wells	3.1 ± 0.3 (20)	45 ± 31 (16)

Oxygen concentration was extremely low in bank filtered water during summer and early autumn. The seasonal variation in oxygen concentration was more clearly observed in the Vesijärvi bank filtration zone than in the Lake Vihnusjärvi area. There the oxygen content in bank filtered water (observation tube HP1) was highest from December to April (10 - 12 mg l^{-1}), and lowest from July to October (below 1 mg l^{-1}). The temperature ranged from < 1 °C to 20 °C. The maximum and minimum values of temperature were measured in July-October and January-April, respectively.

3.3. *Microbial activity in soil layers*

Biological activity analyzed as ATP content was highest in the first 0.3 m of the humic surface layer of the esker area, whereas only minor or non-existent activity was detected in deeper soil layers. In soil samples taken very close to the shoreline of Lake Vihnusjärvi from depths of 2.5 m and 4.5 m, only minor ATP contents were detected (0.01 - 0.02 $\mu g\ g^{-1}$). In deeper layers of the bank filtration zone, a few metres from the shoreline, the ATP content was below the detection limit.

4 DISCUSSION

The content of TOC was different in Lake Vesijärvi (median 4.8 mg l^{-1}) and Lake Vihnusjärvi (median 10.0 mg l^{-1}). The content of AOC had a wider seasonal variation than TOC, ranging in Lake Vesijärvi from 32 to 469 $\mu g\ l^{-1}$, and in Lake Vihnusjärvi from 24 to 108 $\mu g\ l^{-1}$. The share of AOC in total organic carbon seemed to be greater in Lake Vesijärvi (1.8 %) than in Lake Vihnusjärvi (1.1%). This might be due to the fact that organic matter in Lake Vihnusjärvi consisted mainly of high molecular weight compounds, which obviously are refractory, or the AOC method was not able to measure these compounds. In Lake Vihnusjärvi the number of heterotrophic bacteria was clearly higher than in Lake Vesijärvi although the AOC concentrations were usually lower.

The reduction in TOC content was higher in the Lake Vihnusjärvi bank filtration zone (55%) than in the Lake Vesijärvi bank filtration zone (29%). The material in the bank shore of Lake Vesijärvi is coarse-grained, thus allowing more effective bank filtration and shorter residence times. The highest molecular weight fractions of humic substances were removed first, during a quite short underground passage especially in the Lake Vihnusjärvi bank filtration zone. The reduction of the highest molecular weight fractions was 16-21% in Lake Vesijärvi area, and 82-90% in the Lake Vihnusjärvi area. Thus, the removal of humus was more efficient when water from Lake Vihnusjärvi was filtered. This might be due to the clay material in Lake Vihnusjärvi bank shore, and adsorption of high molecular weight compounds into alumina surface, as suggested by Thurman (1985). Also Miettinen and co-workers (1994) found that organic matter was retained in the soil layers, even if it might be remobilized as a result of seasonal changes in water quality.

The higher number of heterotrophic bacteria in filtered water in the Lake Vesijärvi bank filtration zone, compared to the Lake Vihnusjärvi site indicated more effective biodegradation, resulting in 72% reduction of AOC content. In both study areas, after removal of organic matter in the littoral region, there was no remarkable decrease in AOC content, whereas the content of TOC still decreased. This suggests that in saturated reducing conditions only minor, if any, microbial degradation occurred. Also the extremely low ATP contents in the Vihnusjärvi bank filtration zone indicated poor microbial activity in sandy aquifer material. The decrease in TOC towards the pumping wells probably resulted from sorption of refractory humic substances into soil particles. The role of anaerobic degradation of organic carbon still needs to be studied.

As for the pumping wells, the median TOC and AOC contents in Kangasala waterworks were 1.5 mg l^{-1} and 16.0 $\mu g\ l^{-1}$; and in Nokia waterworks 3.1 mg l^{-1} and 35.5 $\mu g\ l^{-1}$. According to the Finnish drinking water directive (Anon. 1994) the TOC content should not exceed 2.0 mg l^{-1}. Thus, with regard to TOC value, the water in Nokia waterworks did not constantly fulfil the quality standards. The reason for this is probably that a minor amount of lake water seeps into the aquifer even in the vicinity of the pumping wells.

Reduction in the content of total organic carbon was not the only goal of bank filtration, but also reduction in the content of microbially labile organic compounds. Guidelines for the highest acceptable AOC concentration in drinking water ranges from 10 $\mu g\ l^{-1}$(van der Kooij 1992) to 50 $\mu g\ l^{-1}$ (LeChevallier et al. 1987). The content of AOC at both waterworks periodically exceeded the upper limit value, mainly during the summer months. As a consequence, the number of heterotrophic bacteria also occasionally exceeded the limit value of 100 CFU ml^{-1} (Anon. 1994). Miettinen and co-workers (1996) have earlier presented slightly higher AOC concentrations of 110±60 $\mu g\ l^{-1}$ in the pumping wells of bank filtration plant, using humus-rich lake water as raw water.

The decomposition of organic matter in sediment layers resulted in extremely low concentrations of oxygen in bank filtered water during summer and early autumn. In the Lake Vesijärvi bank filtration zone there was a wide seasonal variation in oxygen content (0 to 12 mg l^{-1}). Towards the groundwater intake area the variation diminished, yet in the pumping wells the variation in oxygen content was from 1.3 mg l^{-1} to 6.9 mg l^{-1} (median 3.5 mg l^{-1}). In the Nokia waterworks by Lake Vihnusjärvi the higher amount of organic matter and clayey layers in aquifer material resulted in constantly low oxygen content, the range in the pumping wells being from 0.1 mg l^{-1} to 2.8 mg l^{-1} (median 1.7 mg l^{-1}). Especially in the Nokia waterworks the reducing conditions in soil caused high contents of iron in the pumping wells (median 4.9 mg l^{-1}). Therefore, iron needs to be removed using post treatment methods.

ACKNOWLEDGEMENTS

This research was supported by the EC Environment and Climate Research Programme (contract: ENV4-CT95-0071) and the Ministry of Agriculture and Forestry. The authors thank Janet Sharpe for revising the English.

REFERENCES

Anonymous. 1994. Guidelines for quality requirements and surveillance of drinking water. Directive 74/94. Finnish Ministry of Social Affairs and Health.

Hatva, T. 1996. Artificial Groundwater Recharge in Finland. In Kivimäki, A-L. & Suokko, T. (eds.). 1996. Artificial Recharge of Groundwater. Proceedings of an International Symposium, Helsinki, Finland, June 3 - 5, 1996. NHP Report No 38, 3-12.

Henriksen, A., B.L. Skjelvåle, J. Mannio, A. Wilander, J.P. Jensen, T. Moiseenko, R. Harriman, T.S. Traaen, E. Fjeld, J. Vuorenmaa, P. Kortelainen & M. Forsius 1997. Results of national lake surveys 1995 in Finland, Norway, Sweden, Denmark, Russian Kola, Russian Karelia, Scotland and Wales. NIVA Report SNO 3645-97. Norwegian Institute for Water Research, Oslo. 43 p.

ISO 8245, International Organization for Standardization. 1987. Water quality - Guidelines for the determination of total organic carbon (TOC), ISO 8245:1987.

Kaplan, L.A. & J.D. Newbold 1993. Biogeochemistry of dissolved organic carbon entering streams. In Ford, T.E.(ed.). 1993. Aquatic Microbiology - An ecological approach. Blackwell Scientific Publications. Boston. ISBN 0-86542-225-7, 139-165.

Kivimäki, A-L. 1995. Rantaimeytys tekopohjaveden muodostamismenetelmänä. Vesi- ja ympäristö hallituksen monistesarja nro 573. Helsinki. 51 p. (Artificial groundwater recharge with bank filtration. Publications of the National Board of Waters and the Environment. In Finnish).

Kooij, D. van der, W.A.M. Hijnen & A. Visser 1982. Determinating the concentration of easily assimilable organic carbon in drinking water. *J. AWWA* 74, 540-545.

Kooij, D. van der. 1992. Assimilable organic carbon as an indicator of bacterial regrowth. *J. Amer. Water Assoc.* 84(2), 57-66.

LeChevallier, M.W., W. Schultz & R.G. Lee 1991. Bacterial nutrients in drinking water. *Appl. Environ. Microb.* 57, 857-862.

Miettinen, I.T., P.J. Martikainen & T. Vartiainen 1994. Humus transformation at the bank filtration plant. *Wat.Sci.Technol.* 30, 179-187.

Miettinen, I.T., T. Vartiainen & P.J. Martikainen 1996. Bacterial enzyme activities in ground water during bank filtration of lake water. *Wat. Res.* Vol.30, No 10, 2495-2501.

Miettinen, I., T. Vartiainen & P.J. Martikainen 1997. Microbial growth and assimilable organic carbon in Finnish drinking waters. *Wat.Sci.Tech.* Vol.35, No.11-12, 301-306.

SFS 3036, Finnish Standards Association. 1981. Determination of chemical oxygen demand (COD_{Mn}-value or $KMnO_4$ number) in water oxidation with permanganate, 2nd Edition. Finnish Standards Association (SFS), Helsinki.

SFS 3040, Finnish Standards Association. 1990. Determination of dissolved oxygen content in water. Titrimetric method, 2nd edition. Finnish Standards Association (SFS), Helsinki.

Thurman, E.M. 1985. Humic Substances in Groundwater. In Aiken, G.R., McKnight, D.M., Wershaw, R.L. & MacCarthy, P. (eds). Humic Substances in Soil, Sediment and Water. Jon Wiley & Sons, Inc. USA,. 87-103.

Vanhala, P.T. & J.H. Ahtiainen 1994. Soil respiration, ATP content and Photobacterium toxicity tests as indicators of metal pollution in soil. *Environmental Toxicology and Water Quality*. Vol.9, 115-121.

Vartiainen, T., A. Liimatainen & P. Kauranen 1987. The use of TSK-columns in determination of the quality and quantity of humus in different raw waters and drinking waters with high humus content. *Sci. Total. Environ.* 62, 75-84.

Artificial Recharge of Groundwater, Peters et al. (eds) © 1998 Balkema, Rotterdam, ISBN 90 5809 017 5

Long-term sustainability in artificial groundwater recharge

Cristina Frycklund
Royal Institute of Technology, Division of Land and Water Resources, Stockholm, Sweden

ABSTRACT: A gradual decrease in infiltration rates has been seen over a period of decades in infiltration ponds in Sweden. This indicates that long-term clogging is taking place below infiltration ponds. In addition some surface waters used for artificial groundwater recharge contain low-degradable anthropogenic pollutants. For a sustainable drinking water production, in line with Agenda 21, these compounds, undesired in the subsurface and groundwater need to be removed in the initial stage of infiltration, *i.e.* in the filtersand. To attain knowledge about how to reach these goals, further research into the mechanisms involved in the retention and transport of various compounds through porous media is crucial.

1 BACKGROUND

The history of artificial groundwater recharge is more than a hundred years long. Much of the current knowledge of the chemical and physical processes behind the desired quality changes is empirical knowledge, gained through trial and error. The technique has by and large developed through situations where questions and problems arising during the practical application has been solved by trial. In e.g. Germany and the Netherlands, however, research on this subject has been carried out during the last 25-45 years. Nevertheless, basic research leading to an understanding of the mechanisms is still lacking in a number of areas for optimal design of artificial recharge schemes. This is especially true for Boreal conditions.

Artificial groundwater recharge is used for various conditions and purposes. The quality of the infiltration water can be anything from that of the lowest point in the catchment area for a river running through densely populated and industrial areas, to a pristine mountain creek in a sparsely populated area (tab. 1). When the purpose of recharge is to produce drinking water, the demands on the quality changes during the passage through the subsurface vary correspondingly. In some applications, though, the purpose is more related to quantity and storage than to quality. But even in such situations it is important to have control over the water quality changes so that the aquifer is not "consumed" or made unfit for future use by pollution or reduction of porosity due to accumulation of fines.

The objectives of this paper is *i.* to elucidate some risks associated with artificial groundwater recharge in order to avoid future problems which would require long-term remediation of aquifers.

2 LONG-TERM CLOGGING

In Uppsala, 80 km north of Stockholm artificial recharge has been practiced since the early 1950´s. A river cutting through agricultural landscape supplies the infiltration water and an imposing esker, allowing a residence time of 6-8 months, constitutes the aquifer. After 35-40 years of untroubled operation, the infiltrating capacity gradually decreased in a way not affected by the regular maintenance of filtersand. Examination of the porous media down to a depth of 3 meters revealed silty layers highly enriched in iron and manganese in the upper part of the natural geologic material, under the infiltration ponds (VBB Viak 1992). The silty layers were found at different levels, 0-1 m below the bottom of the filtersand, hence not distinctly concentrated to the interface between the filtersand and the natural geologic material.

A nearby artificial recharge scheme has recently begun to show the same signs of decreasing infiltration capacity, after satisfactory operation during 20 years (S. Ahlgren, Uppsala Municipality, pers. comm.) Also at other artificial recharge schemes in Sweden, where a turbid water is infiltrated and a clear water abstracted in the production wells, there is concern about the long-term effect and risks for clogging.

Table 1. Selected parameters for infiltration water quality at some artificial recharge schemes in Europe and guide-levels and limits for European drinking water according to Council Directive 1980.

Parameter	Unit	Södertälje, SE	Uppsala, SE [a]	Huhtiniemi SF [b]	Scheveningen Rhine, pre-treated, NL [c]	Hengsen, Ruhr, D	Magdeburg Ohre, D [d]	EU guide-level in drinking-water [e]
COD-Mn	O_2 mg/l	1-6	9-18	20	9.8	2.6[§]	9.0[*]	2
Alkalinity HCO_3	mg/l	41-49	140-240	15	162	98	n.s.	>30 [#]
Turbidity	NTU	1-2	1-13	1	n.s.	n.s.	n.s	1 [£]
Fe	mg/l	0.05-0.1	0.06-0.7	0.1	0.04	0.2	1.6[**]	<0.05
El. cond.	mS/m	16-18	43	6	80	n.s.	62	40
E. coli.	CFU/ 100 ml	<1	10-1600	3	n.s.	10^4 - 10^5 [f]	n.s.	--
Temp.	°C	2-14	0-20[g]	2-17¤	13	n.s.	11[f]	12

Single values = average values
* average for 6 years
** average for 3 years
*** average for 4 years
n.s. = not stated
μS/ cm = 10 mS/m

§ Dissolved Org Carbon
¤ groundwater temp
refers to softened water
£ unit: mg/l SiO_2

Sources:
a. Stockholm Vatten et al. 1997
b. Kivimäki 1992
c. Stuyfzand 1993, p.133
d. Kaatz 1997
e. Council Directive 1980
f. Preuss & Nehrkorn 1996
g. Morosini 1989

2.1 *Suggested course of events*

The potential and actual clogging below infiltration ponds are best viewed from a geological historical perspective. The ridgeformed eskers now used for artificial recharge of groundwater were once created by rivers of glacial meltwater, carrying minerogenic matter of various dimensions. Frequent changes in flow velocity resulted in a layering of sediments with different texture (Fig. 1). Since the sediments were deposited by flowing water, the layers often appear slanting, with an inclination from the direction of flow.

During periods of low flow velocity, small silt particles settled on top of coarser material. Today, water percolating vertically through the filtersand is diverted along the low conductive layers, resulting in a perched water table. Furthermore, during unsaturated flow, sandy layers on top of gravelly layers with stones and pebbles, conduct the water laterally. The larger the horizontal component of the flow vector is, the better are the chances for suspended matter to settle out from the water, thus causing a secondary build-up of the low conductive layers of primary origin. The lateral flow may also lead to an extension of these layers along the direction of flow.

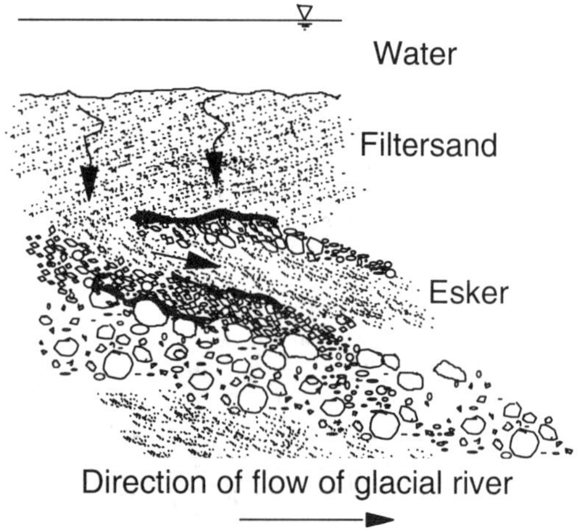

Figure1. Profile through infiltration pond in glacifluvial material. Silty layers, of primary and secondary origin, are seen on top of coarser material.

2.2 *Observations*

Tendencies of perched water table were observed just above the interface between the filtersand and the natural geologic material (Frycklund et al. 1994).

Here, the deepest part of the filtersand had the highest water content, despite a homogeneous texture within the filtersand. In the interface between filtersand and the underlying natural material, a layer of silty material has been found in several Swedish infiltration ponds (Frycklund & Jacks 1993, Frycklund et al. 1994, Frycklund & Jacks 1997). These layers were highly enriched in iron and most often also in manganese. The reason for the metal accumulation has not been clarified. It may be attributed to presence of iron and manganese bacteria which here find a more convenient habitat than in the regularly disturbed filtersand environment. The silty layers most likely protect the deeper levels of the aquifer, by catching fine material supplied with the infiltration water.

2.3 *The literature*

The literature dealing with the transport and retention of particles and colloids in porous media is quite voluminous. Handbooks (*e.g.* Hiemenz & Rajagopalan 1997) and literature reviews (*e.g.* McDowell-Boyer et al. 1986, Ryan & Elimelech 1996) as well as reports of process oriented studies (*e.g.* Sherard et al. 1984, McDowell-Boyer 1992, Harvey at al. 1993) are available. Despite this apparingly thorough research carried out, the results are often difficult to generalize and make practical use of (Harvey et al. 1993). In addition, the present modeling tools often have little practical value for predicting clogging development in natural systems, at least more than qualitatively (Hiemenz & Rajagopalan 1997). The complex semi-natural systems of artificial groundwater recharge would hence require site-specific studies if deep clogging is to be mitigated.

2.4 *Present studies*

One such site-specific study, aiming at getting insight in the retention and transport of colloids in filtersand, is currently being evaluated. Infiltration water was continuously led through one meter long filtersand columns. The study showed that the retention of colloids in the column gradually increased (Fig. 2) (Frycklund, in prep.) After an initial phase of one month, the concentration of colloids in effluent water was 10-20% of that in infiltration water. One of the columns was run with saturated flow conditions and the other with unsaturated ditto. Unsaturated flow generally resulted in a less turbid effluent water than did saturated flow. These differences were, however, small. Further analysis of the column filtersand will be made, focusing on localization of the sites for removal of colloidal and suspended clay.

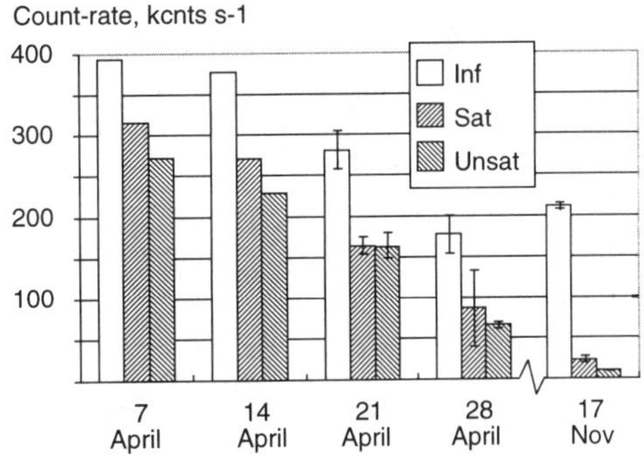

Figure 2. Concentration of colloids in infiltration water and effluent water from two filtersand columns, one with saturated and the other with unsaturated flow. Concentration derived from light scattering measurements and indicated as kilocounts per second. Standard deviation indicated with vertical lines.

This study is part of a larger research project, funded by the European Commission and currently running on its final year ("Artificial Recharge of Groundwater"). Other parts of that project will hopefully throw more light on the processes connected to artificial recharge of groundwater and clogging.

3 NATURAL ORGANIC MATTER AND ANTHROPOGENIC POLLUTANTS

In a study of the retention of natural organic matter (NOM) during pond infiltration, it was found that large amounts of NOM was introduced to the geologic material below the filtersand (Frycklund 1995). In this study at best 50% of the organic matter in the infiltration water passed through the filtersand at 1.15 m depth. During colder periods the retention was lower and up to 80% of initial concentrations of organic carbon would pass through the filtersand. Although there are indications of a continued decomposition of the organic matter at depth (Frycklund et al. 1994), little is known about the long-term effect of this introduction of organic matter in the subsurface.

Some pesticides which are found in infiltration waters, such as lindan, mecoprop and proturone (Stuyfzand 1997; Zullei-Seibert et al. 1997) are retained in the subsurface to varying degrees during artificial groundwater recharge. The consequences of the presence of these compounds in the subsurface are, to the best of our knowledge, not known.
Recently, attention has been given to the increasing

presence of brominated flame resistance agents in the environment. In view of their ability to mimic hormones, spreading of these compounds in natural environments, including the potential subsurface introduction by artificial groundwater recharge, is of great concern.

4 CONCLUSIONS

Infiltration of surface water into aquifers has several implications important from a long-term perspective. Minerogenic material, NOM and anthropogenic pollutants, such as pesticides and various low-degradable compounds, are introduced to and retained in the aquifer. This may reduce the hydraulic conductivity. It may also affect the type and extent of microbiota in the subsurface, the oxygen status and in turn the redox conditions.

All the above may reduce or eliminate the opportunities for future generations to utilize these aquifers for their drinking water supply. Remediation of an aquifer once ruined seem incalculable. Precautionary actions are feasible.

For a sustainable drinking water production, in line with Agenda 21, artificial groundwater recharge through ponds should aim at retaining all those compounds undesired in the subsurface and the groundwater already in the filtersand. Methods which optimize early elimination of such compounds under controlled forms are desirable. Until enough knowledge is gained of how to achieve this ideal process, research must continue in the strive to find such methods. Furthermore, choices of infiltration water supply must be carefully made, considering the discussed aspects and applying appropriate pretreatment.

5 ACKNOWLEDGEMENT

Some of the studies presented in this paper have been financially supported by the EC Environment and Climate Research Programme (Contract No ENV4-CT95-0071, Artificial Recharge of Groundwater).

REFERENCES

Council Directive of 15 July 1980 relating to the quality of water intended for human consumption (80/778/EEC). *Official Journal of the European Communities No* L 229/11-29.

Frycklund, C. 1998. Colloids in infiltration water and their retention in filtersand. In: Artificial Recharge of Groundwater for public water supply. Potential and limitations in Boreal conditions. *Doctoral thesis*. Div. of Land and Water Resources, Royal Inst. of Technology, Stockholm.

Frycklund, C. 1995. Total organic carbon retention by filtersand in infiltration pond for artificial groundwater recharge. *Aqua Fennica* 25:5-14.

Frycklund, C. & G. Jacks 1993. Konstgjord infiltration vid Emmaboda Vattenverk. (Artificial Groundwater Recharge at Emmaboda Waterworks) *Vatten* 49:49-55.

Frycklund, C. & G. Jacks 1997. Iron and artificial recharge of groundwater. *Boreal Env. Res.* 2:171-181.

Frycklund, C., G. Jacks, P-O Johansson & K. Lekander 1994. Konstgjord grundvattenbildning - Processtudier vid inducerad infiltration och bassänginfiltration. (Artificial groundwater recharge - processes associated with induced recharge and basin recharge.) *VA-Forsk Report* No 1994-08. 55 pp. (In Swedish with English summary.)

Frycklund, C., B. Sundlöf & G. Jacks. Limestone filters in artificial recharge of groundwater. Manuscript submitted to *Env. Technology*.

Harvey, R.W., N.E. Kinner, D. MacDonald, D.W. Metge & A. Bunn 1993. Role of physical heterogeneity in the interpretation of small-scale laboratory and field observations of bacteria, microbial sized microsphere, and bromide transport through aquifer sediments. *Water res. res.* 29:8, 2713-2721.

Hiemenz, P. C. & R. Rajagopalan 1997. *Principles of colloid and surface chemistry*. Third edition. Marcel Dekker, New York. 650 pp.

Kaatz, K.-H. 1997. 34 Jahre Grundwasseranreicherung in der Colbitz-Letzlinger Heide. 2. Deutsch-Niederländischer Workshop. Künstliche Grundwasseranreicherung. Schwerte, Germany, Oct 28-29 1997. Deutscher Verein des Gas- und Wasserfaches e.V., Bonn, *DVGW-Schriftenreihe Wasser* Nr 90: 199-218.

Kivimäki, A-L. 1992. Tekopohjavesilaitokset Suomessa. *Vesi- ja Ympäristöhallinon Julkaisuja* sarja A No 98 (The artificial groundwater plants in Finland. *Publication of the Water and Environment Administration*, series A98. In Finnish).

McDowell-Boyer, L.M. 1992. Chemical mobilization of micron-sized particles in saturated porous media under steady flow conditions. *Env. Sci. and Tech.* 26,3:586-593.

McDowell-Boyer, L.M., J.R. Hunt & N. Sitar 1986. Particle transport through porous media. *Water res. res.* 22(13): 1901-1921.

Morosini, M. 1989. The artificial recharge of Tunåsen, Uppsala. A hydrochemical considera-

tion. *Licentiate thesis*, No 139, Quaternary Division, Uppsala University.
Preuss, G. & A. Nehrkorn 1996. Succession of microbial communities during bank filtration and artificial groundwater recharge. Proceedings of the International Symposium on Artificial Recharge of Groundwater, *Nordic Hydrological Programme Report* No 38: 215-221.
Ryan, J.N. & M. Elimelech 1996. Colloid mobilization and transport in groundwater. *Colloids and Surfaces A*. Vol 107, 1-56
Sherard, J.L., L.P. Dunnigan & J.R. Talbot 1984. Basic properties of sand and gravel filters. *J. Geotech. Eng.* 110(6) 684-700.
Stockholm Vatten, Norrvatten, J&W, Kungliga Tekniska Högskolan, VBB Viak. 1996. Storstockholms framtida vattenförsörjning. Undersökning av möjligheterna till konstgjord grundvattenbildning. Etapp 1. (The future water supply of the Stockholm region - possibilities for artificial recharge. Phase I. In Swedish). *Internal report Stockholm Vatten*. 188 pp.
Stuyfzand, P.J. 1993. Hydrochemistry and hydrology of the coastal dune area of the Western Netherlands. *Doctoral thesis*, Vrije Universiteit van Amsterdam.
VBB Viak AB, 1992. Sammanställning och utvärdering av utförda undersökningar vid Tunåsens infiltrationsplats i Uppsala. (Report over investigations at the infiltration site in Uppsala) *Report, VBB Viak AB* 1992-11-25.
Zullei-Seibert, N., C. Skark, D. Schwarz & U. Gatzemann1997. Auswirkungen von Pflanzenschutzmitteln unterschiedlicher Herkunft auf die Grundwasseranreicherung. 2. Deutsch-Niederländischer Workshop. Künstliche Grundwasseranreicherung. Schwerte, Germany, Oct 28-29 1997. Deutscher Verein des Gas- und Wasserfaches e.V., Bonn, *DVGW-Schriftenreihe Wasser* Nr 90: 101-114.

Artificial Recharge of Groundwater, Peters et al. (eds) © 1998 Balkema, Rotterdam, ISBN 90 5809 017 5

Fate of pollutants during artificial recharge and bank filtration in the Netherlands

Pieter J. Stuyfzand
Kiwa Research and Consultancy, Nieuwegein, Netherlands

ABSTRACT: The results are presented from a quality survey on 3 sites with Artificial Recharge (AR) and 2 with Bank infiltration (BI), using rows of monitoring wells. The main differences between AR and BI systems relate to: the selective intake and pretreatment (AR only), the ability to manipulate the spreading pond (AR only), the transport length and detention time in the aquifer (BI >> AR), interaction with anoxic Holocene aquitards (BI > AR), and complexicity of the whole system (AR > BI). Bank filtrate exhibited 4 times more different organic micropollutants than dune filtrate. The redox environment proved here, with calcite saturated waters, to be the chemical 'master-variable'. For many micropollutants the specific 'redoxe barrier' could be established, where these pollutants degrade or precipitate. The current concentrations of many pollutants in AR and BI systems are in general lower than during surveys in the 80s.

1 INTRODUCTION

Aquifer passage is considered an essential advantage of Artificial Recharge (AR) and Bank Infiltration (BI) systems over surface storage reservoirs combined with direct purification techniques. This is mainly because of the elimination of pathogens, more natural removal of chemical pollutants, attenuation of quality fluctuations and storage in a medium protected from atmospheric fall-out (Asano [ed] 1985, Stuyfzand 1989ab).

In this contribution some results are presented from an extensive quality survey on 3 sites with AR and 2 with BI, using rows of monitoring wells (Stuyfzand & Lüers 1996). This study was initiated in order to predict the future concentration levels of pollutants in the raw water of AR and BI systems. Uncertainty arose as drinking water prepared from surface waters was observed to contain many organic micropollutants (OMPs), even though most OMPs were <0.1 μg/l (Van Genderen et al. 1994). This contrasted with a very significant improvement of both the quality of the river Rhine and the pretreatment for AR systems during the past 20 years.

Here we focus on: a) a comparison of BI and AR systems; b) the redox environment, which dictates the behaviour of many pollutants in our calcite-saturated AR and BI systems; c) a comparison of concentrations measured in the early 80s and mid 90s; and d) future developments.

2 SITES AND METHODS

The 5 drinking water plants using resp. AR (3) and BI (2), which are studied here, are indicated in Figure 1. Water samples were collected there twice in the period 1994-1995, using rows of 4-5 observation wells (approximately along a flow path), in between infiltrating surface water and groundwater recovery. The 21 monitoring wells showed a wide range in subsoil travel distances and redox environments.

The samples were analysed for main constituents (using conventional techniques), 20 trace elements (TEs) and many organic micropollutants (OMPs). The TEs were analysed using AAS-techniques with element specific matrix modifiers and background corrections (Van der Jagt & Stuyfzand 1996). B, Be, Cd, Hg, Pb, Sb, Se, Sn, Te, Tl and V were preconcentrated prior to analysis, using microwave evaporation.

Chlorinated hydrocarbons, aromatics and selected pesticides were analysed using standard group specific methods. For screening purposes GC-MS was applied after adsorption to XAD-resin at pH7 and pH2 (Noordsij et al. 1985).

Translation of the observed concentration changes downgradient in the aquifer in 1994-1995, into general conclusions on the behaviour of pollutants, required the following:

1) a reconstruction of the pollution history of the surface water (prior to infiltration); 2) linking of

the various concentration changes downgradient ('response type curves') to the appropriate impulse curve (historical input signal); 3) comparison of the recent data from the observation wells with previous results; 4) literature research on physical and chemical characteristics of various pollutants; and 5) simulation of the behaviour of pollutants using the finite element transport code 'INFOMI 3.1' (Dijkhuis & Stuyfzand 1997), which is 1-dimensional for organic and inorganic pollutants, first in a basin and subsequently in a watersaturated aquifer.

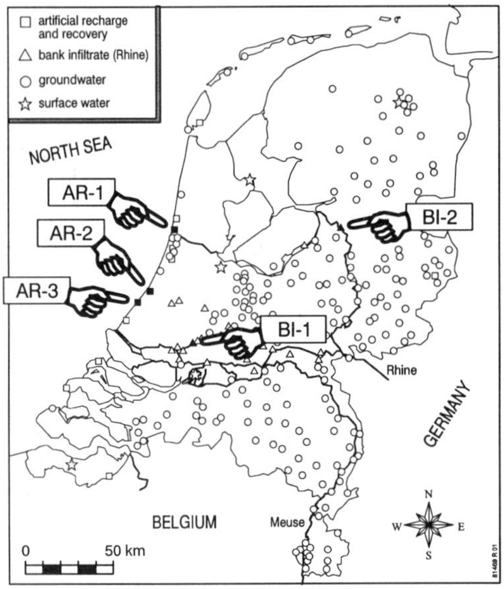

Figure 1. Location of the 5 study sites amidst all pumping stations for drinking water supply in the Netherlands anno 1994, with subdivision according to their source. AR-1 = Wijk aan Zee, using pretreated lake IJssel water; AR-2 = site Katwijk using pretreated river Meuse water; AR-3 = site Scheveningen, as 2; BI-1 = site Opperduit, river Rhine water; BI-2 = site Zwolle, river Rhine water. AR = Artificial Recharge; BI = Bank Infiltration.

3 AR- AND BI-SYSTEMS IN THE NETHERLANDS

3.1 *General aspects*

The general characteristics of the current AR- and BI-systems in the Netherlands are summarized in Table 1. The main differences between both systems relate to: the selective intake and pretreatment (AR only), the ability to manipulate the spreading pond (AR only), the transport length and detention time in the aquifer (BI >> AR), interaction with anoxic Holocene aquitards (BI > AR), and complexity of the whole system (AR > BI).

The aquifer passage, including groundwater flow and the spatial distribution of redox environments, is schematized for both systems in Figure 2.

The aquifers are generally composed of uncemented, fine to medium grained sands containing ca. 0.2% organic matter.

3.2 *Differences in water quality*

Bank filtrate exhibited 4 times more different OMPs (Figure 3) and a 2-3 times higher lumped concentration of OMPs than dune filtrate.

In addition the bank filtrates show higher concentrations of group 3C and 4C substances and lower concentrations of group 1C and 2C constituents (Table 2).

Table 1. General characteristics of the current bank infiltration systems and artificial recharge systems (using basins) for drinking water preparation, in the Netherlands.

	Bank infiltration	Artificial recharge
WHOLE SYSTEM (anno 1994)		
Start of surface water infiltration	1300 AD[@]	1955 AD
Number of sites in the Netherlands	27	11
Size (range in Mm^3/y)	1-11	1-57
Possibility of 2 intake points	-	+
Relation river discharge - recharge rate	+	-
Selective intake	-	++
Pretreatment (near the intake)	-	+++[A]
Transport-chlorination	-	++
Post-treatment	+++[B]	++[C]
THE INFILTRATING WATER COURSE		
Drying up of the river/basin	-	+
Clogging	-	++
Sludge removal from basin/river	-	++/+
Quality changes in recharge basins	-	+
AQUIFER PASSAGE		
Passage of Holocene aquitards	20-100%	5-30%
Permeability aquifer (K_h, in m/d)	50	12
Redox environment[†]	An + DAn	SOx + An
Modal travel time (d)	1000	70
Modal distance travelled (m)	500	70
Modal flow velocity groundwater (m/d)	0.5	1
Variations in detention time (y)	2-100	0.1-30
RECOVERY		
Admixing authochtonous groundwater	5-80%	5-20%
Clogging type of wells[†]	An	Ox

[@] = spontaneous; - = not the case; + = of minor importance c.q. low; ++ = of great importance c.q. high; +++ = of paramount importance c.q. very high. [A] = coagulation, pH-correction, rapid sand filtration; [B] = with many exceptions: aeration, rapid sand filtration, aeration, filtration over activated carbon; [C] = with many exceptions: hardness correction, aeration, rapid sand filtration, filtration over activated carbon; [†]: An = anoxic; DAn = deep anoxic; Ox = oxic; SOx = (sub)oxic.

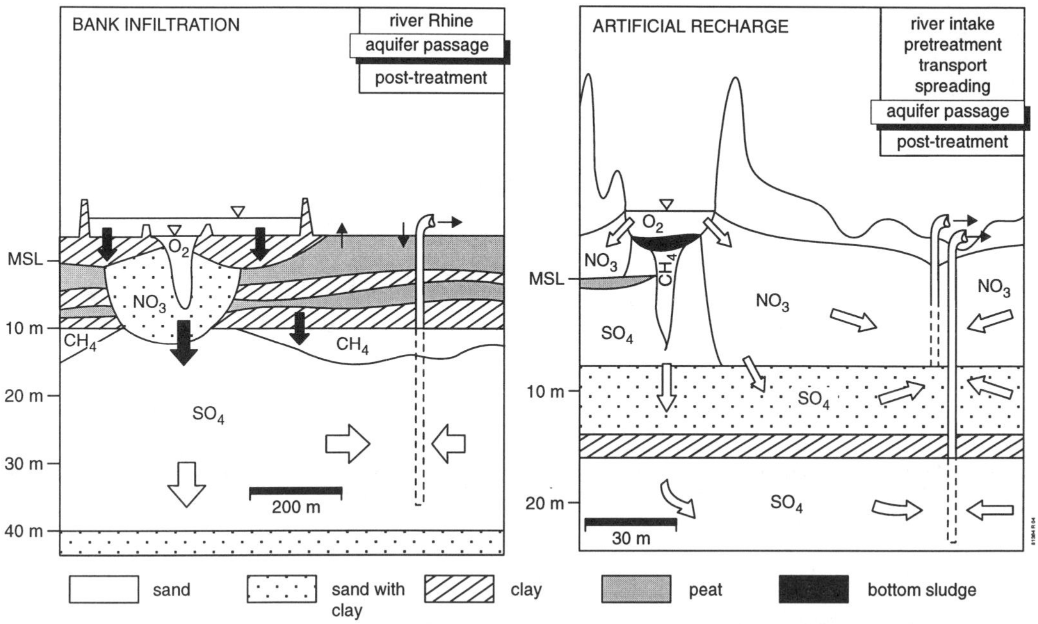

Figure 2. A schematized cross section over a representative Rhine bank infiltration system and an artificial recharge system, showing the position of aquitards, groundwater flow and the following redox environments: O_2 = oxic (equilibrium with atmosphere); NO_3 = suboxic (oxygen depleted (10-100%), nitrate > 1 mg NO_3/l, sulphate stable); SO_4 = anoxic (no oxygen, no nitrate, sulphate stable); CH_4 = deep anoxic (no oxygen, no nitrate, sulphate depleted (10-100%) and/or methane present (>0.5 mg/l)).

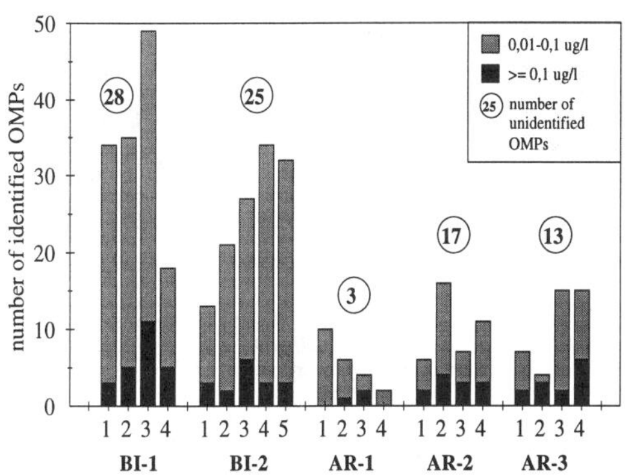

Figure 3. Rhine bank filtrate contains more different OMPs than dune filtrate. Shown is the total amount of OMPs detected with GC/MS in pH7-XAD-isolates in November 1994 and June 1995, with distinction between 2 concentration levels. 1-5 = observation wells in order of increasing distance to banks of river or basin.

4 THE REDOX ENVIRONMENT: A MASTER VARIABLE FOR AR AND BI

The redox environment (see also Figure 2) proved in this study, with calcite saturated waters, to be the chemical 'master-variable'. For many micropollutants the specific 'redox barrier' could be established, the redox environment where these pollutants degrade or precipitate (Table 2). For instance the deep anoxic environment (no oxygen, no nitrate, ferro-ions and H_2S in solution) poses a reliable redox barrier in aquifers to amongst others: Se and atrazine (Figure 4), chlorate, AMPA, diuron and many heavy metals (which (co)precipitate as sulphide-minerals). Anoxic and deep anoxic aquifers with a high content of pyrite and organic matter constitute very stable redox barriers, contrary to sludges at the bottom of rivers and basins.

5 HISTORICAL DEVELOPMENTS

5.1 *In the surface waters*

The general trends in water quality of the river Rhine near both BI-sites, and of the pretreated surface water for AR-sites 2 and 3, are shown in Figure 5. Where data are lacking, on TEs prior to the early 70s and on OMPs prior to the early 80s or 90s, concentration levels have been estimated on the basis of sediment core analyses (Salomons & De Groot 1978, Beurskens et al. 1993) and pesti-

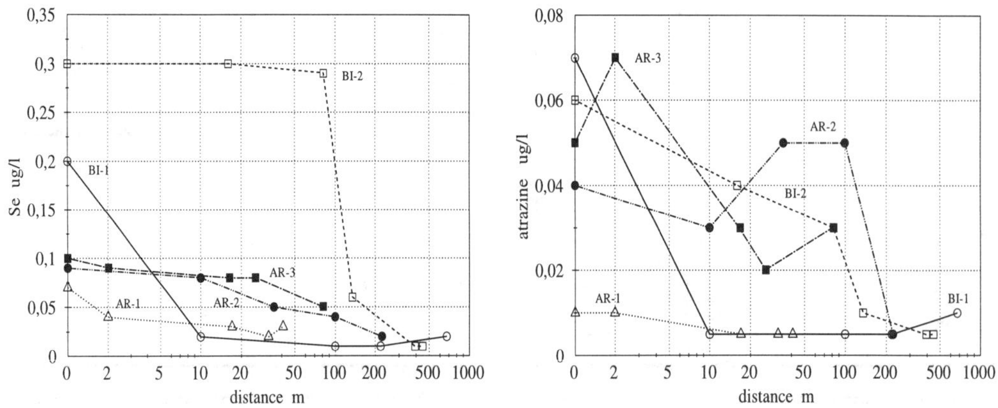

Figure 4. Concentration evolution downgradient in the aquifers for selenium and atrazine at the 5 study sites, in 1994-1995. Most AR-wells are (sub)oxic, each BI-1-well is (deep) anoxic. At the BI-2 site the (sub)oxic front advanced to ca. 130 m into an anoxic aquifer.

Table 2. Subdivision of constituents of surface water prior to infiltration (either BI or AR), in 12 groups on the basis of their behaviour during aquifer passage and as a function of the redox environment. R = Retardation factor.

Behaviour type (1-4) Redox environment (A-D)	1. STRONG AND LONG TERM REMOVAL (>70%) (T½<30 d, and/or R≥100)	2. MEDIUM OR TEMPORARY REMOVAL (20-70%) (T½=30-120 d and/or R=10-100)	3. HARDLY ANY REMOVAL (<20%) (T½>120 d and/or R<10)	4. INCREASE
A. (SUB)OXIC##	Group 1A: NH_4, BTEX, THM (excl. chloroform), NTA?, DOP-sorbofuranose, MCA, lactones, PCF?, palmitic acid, MCPP, MCPA	Group 2A: O_2, PO_4, As, Cd, Co, Cu, Ni, Sb, Se, Te, V, Zn, chloroform, simazine, dinoseb, TCB, tri(iso)butylphosphate, tetraglyme?, EDTA?	Group 3A: O_2, NO_3, 1,2-dichloroethane, tri, tetra, per, trichloroethane, AMPA, atrazin, diuron, metolachlor, t-butylphenol, triphenylfosfinoxide, 2,4-DCA, N-methylbenzamide?, TCFA	Group 4A: SO_4@, As@
B. ANOXIC##	Group 1B: O_2, NO_3, chlorate, TCFA, THM, tetra, PCF?, trichloroethane, AMPA, atrazine, simazine, diuron, dinoseb, palmitic acid	Group 2B: As, Cd, Co, Cu, Ni, Sb, Se, Te, V, Zn tetraglyme?, TCB, per	Group 3B: SO_4, 1,2-dichloroethylene, tri, MCPA, MCPP, t-butylcresol, tri(iso)butyl-phosphate, lactones, triphenylfosfin-oxide, DOP-sorbofuranose, diacetone-glucose, EDTA?, 1,2-dichloroethane	Group 4B: Fe, Mn, NH_4, PO_4 As
C. DEEP ANOXIC##	Group 1C: O_2, NO_3, SO_4, chlorate, Cd, Co, Cu, Ni, Sb, Se, Te, Zn, THM, dinoseb, t-butylcresol, per, trichloroethane, tetra, AMPA, atrazine, simazine, diuron, triphenylfosfinoxide, TCFA, tri(chloropropyl)phosphate	Group 2C: SO_4 diphenylsulfon 1,2-dichloroethane 1,2-dichloroethylene	Group 3C: palmitic acid, DOP-sorbofuranose, EDTA?, diacetoneglucose, MCPA, MCPP, benzene, xylene, tri(iso)-butylphosphate, dimethylaniline, lactones	Group 4C: H_2S, CH_4, Fe, Mn, NH_4, PO_4, DOC, I
D. ALL REDOX ENVIRONMENTS (A-C)	Group 1D: susp. solids, lanthanides, actinides, Al, Be, Cr, Cs, Hg, Pb, chlorite, most PAH, PCB and OCP, bromoform, DCA, nitrobenzene, nitro-toluene, chloronitrobenzene, pathogenes	Group 2D: K, B, Ba, F, Li, ^{90}Sr diglyme	Group 3D: Cl, HCO_3, Na, K, Ca, Mg, Br, F, ^3H, ^2H, ^{18}O, bentazone, DEET?, CAE, N-butylbenzenesulphonamide, trichloro-ethylphosphate, jodocyclohexane, 1,3,3-trimethyloxindole, diglyme	Group 4D: Ca, HCO_3, SiO_2

BTEX = benzene, toluene, ethylbenzene and xylene; CAE = chloroalkylethers; DCA = dichloroaniline; DCB = dichlorobenzene; DOP-sorbofuranose = di-orthopropylidene-1-sorbofuranose; MCA = (mono)chloroaniline; OCP = organochlorine pesticides; PCF = pentachlorophenol; TCB = trichlorobenzene; TCFA = tetrachloro-orthophtalic acid; THM = trihalomethanes; @ = only if pyrite oxidizes; ##: Criteria for redox environments: (Sub)oxic: if $O_2 \geq 1$ or $NO_3 \geq 1$ mg/l; Anoxic: if $O_2 < 0{,}5$ and $NO_3 < 0{,}5$ mg/l, $SO_4 \geq 0{,}9\ (SO_4)_O$ and $CH_4 < 0{,}5$ mg/l; Deep anoxic: if $O_2 < 0{,}5$ and $NO_3 < 0{,}5$ mg/l, $SO_4 < 0{,}9\ (SO_4)_O$ or $CH_4 > 0{,}5$ mg/l. $(SO_4)_O$ = original sulphate concentration (to be calculated through linear regression with Cl^-)

cides product information (year of introduction and production-stop). The overall picture is a kind of pollution wave starting in the 50s, culminating in the early 70s and terminating in the late 80s. This curve holds especially for the easy to remove pollutants, notably those bound to suspended solids (like PAHs, PCBs, heavy metals) and those which are easily oxidizable (like NH_4), and for early black list substances like organochlorinepesticides (OCPs). The curve for pretreated surface water is, in addition, influenced by a switch from Rhine to Meuse intake in 1976, and a declining chlorination for transport (necessary to kill mussel larvae).
More recent pollution waves include those of diacetoneglucose (low since 1988), bentazone (low since 1989), atrazine and AMPA.

5.2 *In the bank and dune filtrates*

The current concentrations of many pollutants in Rhine bank filtrate (Figure 6) and coastal dune filtrates (Figure 7) are in general lower than during surveys in the 80s (Noordsij et al. 1985, Stuyfzand 1989ab).
These positive developments are due to spectacular quality improvements in the river Rhine since about 1975, and, in case of artificial recharge, also to a reduced chlorination for transport, an improved pretreatment, and (on the 3 AR-sites) a switch to a better source of surface water. The improvements include in most cases a decrease for: soluble salts like NaCl and $CaSO_4$; DOC, NH_4 and PO_4 (Figure 6); and the more hydrophylic organics like bentazone, mecoprop (MCPP), volatile halogenated aliphatics (Figure 7), 1,3,3-trimethyloxindole, chloroanilines, chlorobenzenes and chloroalkylethers (Figure 6). There were no improvements for nitrate and most heavy metals, hydrophobic OMPs remained below detection.

6 FUTURE DEVELOPMENTS

Some well fields pumping Rhine bank filtrate, are likely to receive higher concentrations of many constituents of group 1C in Table 2 (like nitrate, copper, nickel, AMPA, atrazine and diuron). On the other hand these well fields will pump bank filtrate showing lower concentrations of constituents belonging to group 1A and 4C in Table 2 (like benzene, calcium, ammonia, iron, arsenic and diphenylsulfon). This relates to well fields where the riverbed is cutting through the Holocene superficial aquitard and where regular floods prevent the river bed to clog by sludge. The reason is that in this situation the raised oxygen and nitrate concentrations in the river Rhine and reduced load of oxydizable solutes (like NH_4^+ and DOC) will stimulate the (sub)oxic zone to expand downgradient. Further analytical progress and a higher resolution of mass spectra in cleaner groundwater will permit the detection of many 'new' OMPs. Our study revealed already 43 OMPs which had not been detected in bank and dune filtrate to date.

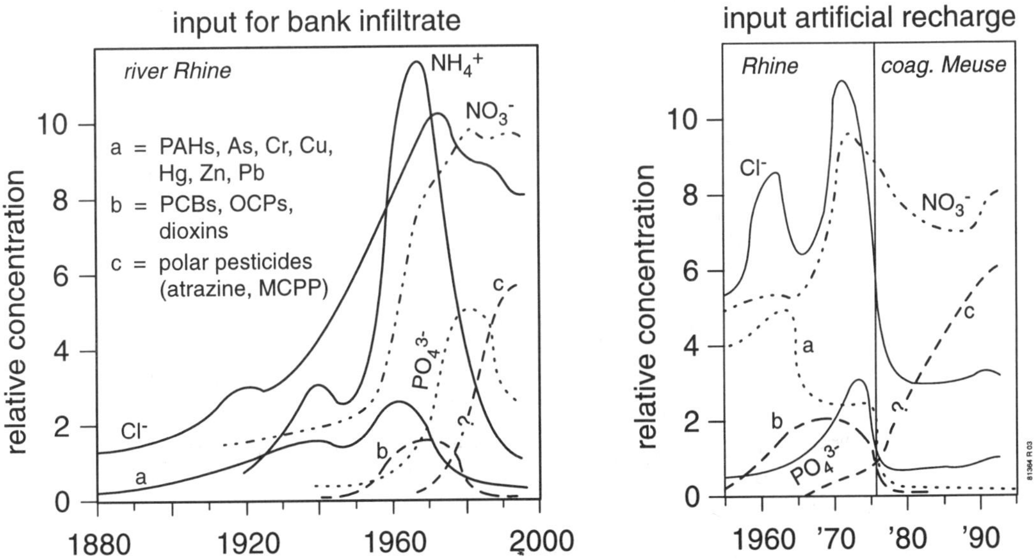

Figure 5. Schematized quality development of the river Rhine for both BI-sites, and pretreated surface water (Rhine, since 1976 Meuse) for AR-sites 2 and 3. Based on many sources including sediment cores.

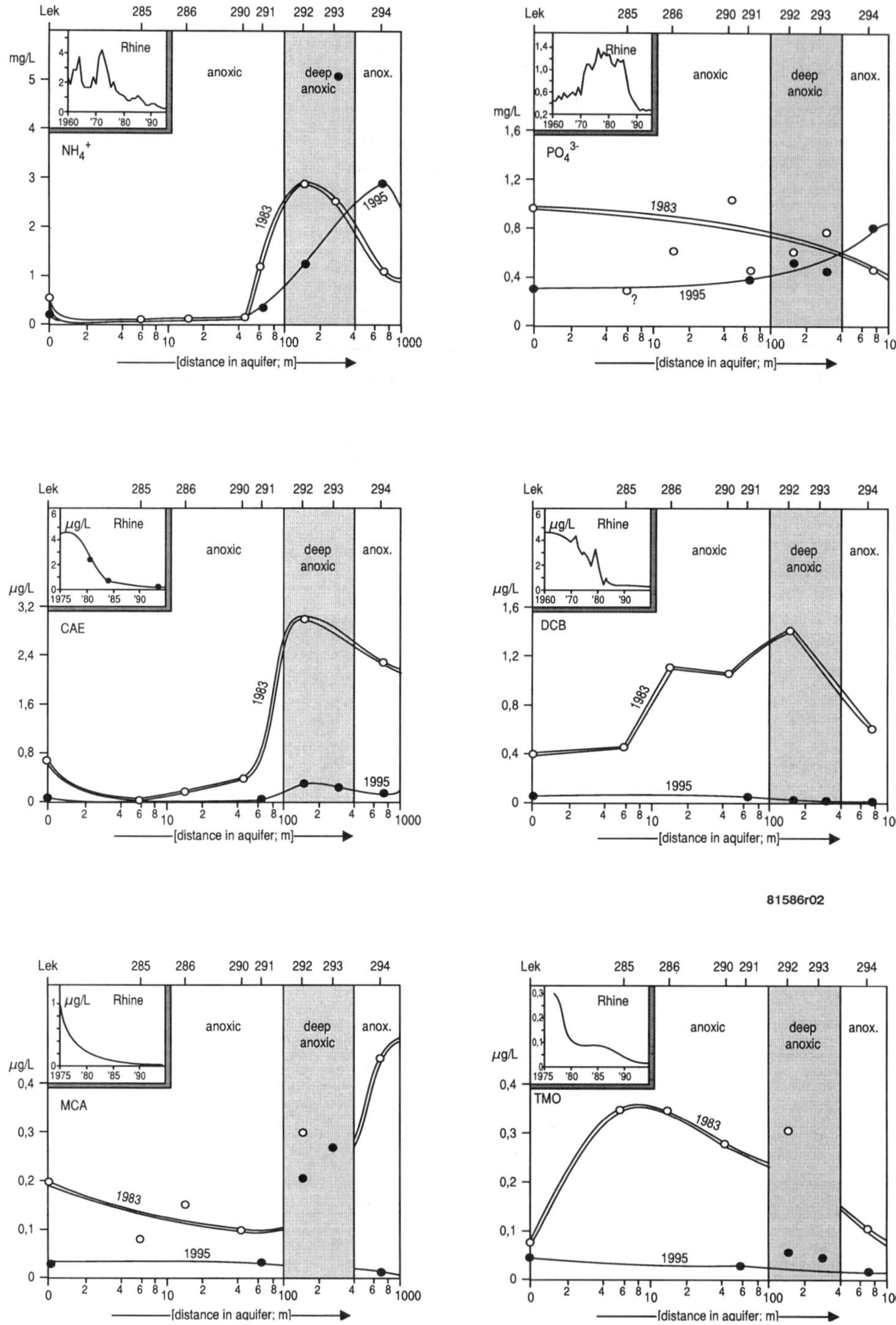

Figure 6. Quality evolution downgradient in the aquifer fed by river Rhine water on site BI-1, in 1983 and 1995. The inset shows concentration trends in the Rhine. CAE = sum of chloroalkylethers; DCB = sum of all dichlorobenzenes; MCA = ortho-chloroaniline; TMO - 1.3.3-trimethyloxindole.

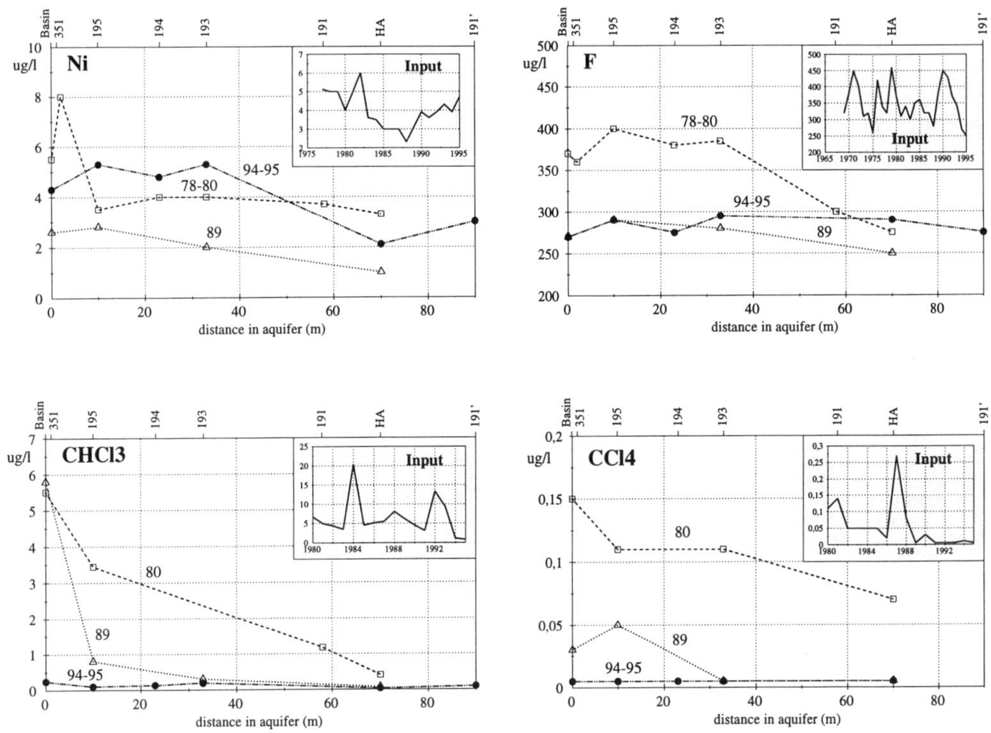

Figure 7. Quality evolution downgradient in the aquifer recharged with pretreated river Meuse water on site AR-3, in 1980, 1989 and 1995. The inset shows concentration trends in the recharge water.

REFERENCES

Beurskens, J.E.M., G.A.J. Mol, H.L. Barreveld, B. Munster, H.J. van & Winkels 1993. Geochronology of priority pollutants in a sedimentation area of the Rhine river. Envir. Toxicol. & Chem. 12, 1549-1566.

Dijkhuis, L.G. & P.J. Stuyfzand 1997. INFOMI 3.1: een 1-D transportmodel voor microverontreinigingen in een waterloop en in de bodem na infiltratie. Kiwa-rapport SWI 96.219.

Jagt, H. van der & P.J. Stuyfzand 1996. Methods for trace element analysis in surface water, atomic spectrometry in particular. Fresenius J. Anal. Chem. 354, 32-40.

Noordsij, A.,L.M. Puijker, M.A. van der & Gaag 1985. The quality of drinking water prepared from bank-filtered water in the Netherlands. Sci. Total Environ. 47, 273-292.

Salomons, W. & A.J. de Groot 1978. Pollution history of trace metals in sediments, as affected by the Rhine river. In 'Environmental biogeochemistry and geomicrobiology' (ed. by Krumbein, W.E), Ann Arbor Sci. Publ. 1, 149-162.

Stuyfzand, P.J. 1989a. Hydrology and water quality aspects of Rhine bank ground water in The Netherlands. J. Hydrol. 106, 341-363.

Stuyfzand, P.J. 1989b. Quality changes of river Rhine and Meuse water upon basin recharge in the Netherlands' coastal dunes : 30 years of experience. In : 'Artificial recharge of Groundwater' (ed. by A.I. Johnson & D.J. Finlayson), Proc. Int. Symp. Anaheim USA, 1988, Am. Soc. Civil Eng., N.Y., 233-245.

Stuyfzand, P.J. & F. Lüers 1996. Gedrag van milieugevaarlijke stoffen bij oeverinfiltratie en kunstmatige infiltratie; effecten van bodempassage gemeten langs stroombanen. Kiwa-Meded. 125, 272p.

Van Genderen, J., Th.H.M. Noij & J.A. Van Leerdam 1994. Inventarisatie en toxicologische evaluatie van organische microverontreinigingen. RIWA-rapport dec. 1994.

1.6 Microbiological aspects

Artificial Recharge of Groundwater, Peters et al. (eds) © 1998 Balkema, Rotterdam, ISBN 90 5809 017 5

DNA-analysis to study the microbial diversity in recharged groundwater

B. Eschweiler, B. Kilb, B. Kuhlmann, G. Preuß & E. Ziemann
Institute for Water Research, Schwerte, Germany

ABSTRACT: A molecularbiological approach was used to analyse the bacterial populations in surface- and groundwater. The bacterial DNA was isolated and purified by adapting and modifying standard protocols with attention to the special characteristics of the analysed samples. PCR (Polymerase chain reaction) with universal primers was used to amplify a variable region of the 16S rRNA gene. Based on their sequence differences similar sized PCR products were separated by DGGE (Denaturing gradient gel electrophoresis). DGGE-patterns from ground- and surface water as well as lake sediment were compared with regard to common features and differences in species composition. Samples from groundwater with different hydrogeochemical conditions revealed great differences in DGGE-patterns indicating a characteristic species composition in these habitats. Related samples showed less or no differences in DGGE patterns. Temporal variation in species composition was demonstrated in surface water over a period of 6 month.

1 INTRODUCTION

Microbial activity is one of the most important processes during bank filtration and artificial recharge of groundwater. The degradation of organic matter and the reduction of electron acceptors like oxygen, nitrate and sulphate during the infiltration and underground passage are caused by microbial processes. In addition, transport, adsorption and mobilisation processes are influenced by microbial activities. The succession of redox zones and the quality of groundwater are influenced by the structure and physiological state of the microbial communities in the aquifer. The high diversity of bacterial species that can be observed in these habitats is of major importance for the degradation of contaminants and groundwater quality.

Bacterial diversity in environmental samples is usually determined by a characterisation of isolated strains. A problem for the analysis and characterisation of microbial communities is the inability to culture most of the bacteria species present in the sample. Therefore isolated bacteria may account for only a minor portion of the total bacterial diversity originally present in the sample. This problem is particularly severe in oligotrophic habitats like groundwater, where approx. only 0.1 to 1% of the bacterial species are culturable (Amann et al. 1995).

A new approach in microbial ecology is based on the analysis of the bacterial genetic information (DNA) without prior cultivation. This direct molecular approach is culture independent and has greatly enhanced the ability to assess bacterial diversity in ecosystems such as groundwater and subsurface environments. It is also important to note that organisms identified by direct DNA analysis are often demonstrated to be absent from the cultivable fraction of the population (Suzuki et al. 1997, Santegoeds et al. 1996). Therefore the molecular approach might allow an increasingly comprehensive and more precise picture of the microbial populations of different environmental habitats to be revealed.

Today the most frequently used target for the molecular approach is the bacterial 16S rDNA containing genetic information essential for the bacterial metabolism. By using PCR (Polymerase Chain Reaction) amplification (Saiki et al. 1985), gene fragments from variable regions of the 16S rDNA can be selectively amplified. By selecting the specific reaction conditions during PCR, the monitoring of single species, functional groups or part of the total diversity is made possible. Subsequent analysis of the PCR products by DGGE (Denaturing gradient gel electrophoresis) (Fischer & Lerman 1983, Muyzer et al. 1993) reveals the number of species and characteristic diversity patterns of the analysed sample.

In the present study, samples from recharged groundwater, bank filtrate and lake sediment were analysed by PCR and subsequent DGGE. PCR conditions were selected to amplify 16S rRNA gene fragments from total diversity. Similarities and differencies between DGGE-patterns were compared.

2 MATERIAL AND METHODS

2.1 *Overall strategy and approach*

The general protocol in Figure 1 shows the major steps during extraction and analysis of bacterial DNA from liquid and solid samples.

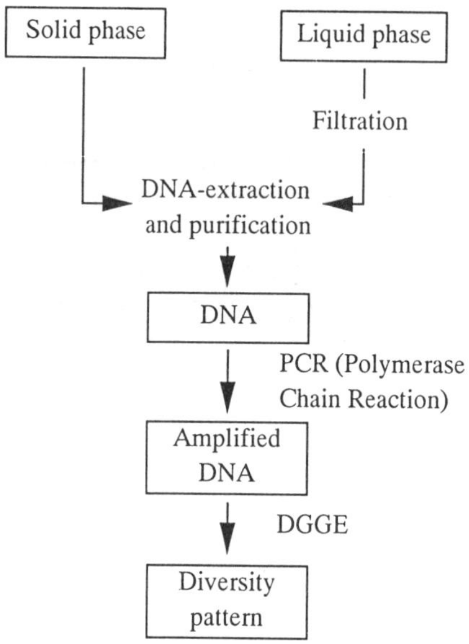

Figure 1. Flowchart of the protocol used to analyse bacterial diversity.

2.2 *Site description*

The study site is located near the river Ruhr in NW-Germany. In this area the Dortmund Water Works produces drinking water by artificial recharge of groundwater. This field site comprises an area of 0.5 km^2 and is completely surrounded by river and lake water. After pre-filtration through fine-gravel, the river water is infiltrated in the aquifer by slow sand filters and after underground passage abstracted from the aquifer for drinking water supply. Different zones of bank filtration and artificial recharge can be distinguished by the different geochemical and hydraulic situations as described earlier (Schöttler & Schulte-Ebbert 1995). A well equipped monitoring network of several observation wells exists throughout the whole area allowing sampling from surface and groundwater with different hydrochemical and microbiological properties.

2.3 *Sampling*

Surface- and groundwater samples were collected from the river Ruhr or different wells on the study site. Samples from groundwater were collected with a submergeable pump and pumping over night (pump rate: 1 - 1.5 litre min^{-1}). Temperature, pH, E_h and oxygen were measured through an in-line flow cell and additional samples were taken to determine the physical and chemical characteristics.

All samples were filled in sterile 1 liter wheaton bottles, stored on ice and processed immediately upon arrival in the laboratory.

2.4 *Extraction of DNA*

Surfacewater (200 ml) and groundwater samples (750 ml) were filtered through 25 mm diameter, 0.2 µm pore size Durapore filters (Millipore Corp.) and the filters were cut in small pieces and placed in 2 ml microcentrifuge tubes. For DNA - extraction from solid phases 200 - 400 mg sample material were lysed directly as described below with the addition of a small amount (spatula tip) polyvinylpolypyrrolidone (PVPP) to remove phenolic impurities. DNA was extracted by a modified protocol of Tsai and Olson (1991). To the filter (solid sample) 480 µl lysozyme solution (50 mM Tris-HCl [pH 8.0], 1 mM EDTA, 100 mM NaCl, 15 mg of lysozyme ml^{-1}) was added and the mixture incubated at 37°C in a shaking water bath for 90 min., followed by 120 µl SDS solution (50 mM Tris-HCl [pH 8.0], 1 mM EDTA, 100 mM NaCl, 10% sodium dodecyl sulfate [SDS]) and an additional incubation at 37°C for 20 min. The sample was frozen in dry ice and thawed in a 70°C water bath, and this freeze-thawing cycle was repeated four times.

Lysis efficiency was determined by using bacterial reference strains (*E. coli*, DSM 498; *Bacillus subtilis*, DSM 347; *Pseudomonas putida*, DSM 548) and enrichment cultures from groundwater in P-medium (peptone from meat 1.0 g, glucose 0.1 g, K_2HPO_4 0.1 g, $FeSO_4 \times 7H_2O$, tap water to 1000 ml, pH 7.2) Cells were were counted before and after lysis by acridine orange staining (Hobbie et al. 1977) in combination with epifluorescent microscopy and determination of colony forming units (cfu) on P-medium (21°C incubation).

Excess RNA was removed by addition of 3 µl RNase (28.8 mg ml^{-1}, Sigma Corp.) and incubation at 37°C for 40 min. After a quick spin the DNA containing solution was transferred to a fresh tube. Bacterial genomic DNA was extracted and purified from cell lysate with a QIAamp spin column (QIAGEN Inc.) according to the manufacturer's instruction, followed by ethanol precipitation (Maniatis et al. 1982). The resulting DNA pellet was resuspended in 20 µl TE buffer (10 mM Tris-HCl [pH 7.5], 1 mM EDTA).

2.5 *Primer design and PCR*

A fragment from the 16S rDNA was amplified by PCR. This fragment contains variable regions, facilitating the subsequent discrimination of species by DGGE and conserved regions as primer starting points. These conserved regions are identical among all different eubacterial species. Two universal bacterial 16S rDNA primers were used to amplify a 527 bp fragment from total genomic DNA: 27fGC (5′ - **CGCCCGCCGCGCCCCGCGCCCGTCCC GCCGCCCCCGCCCC**AGAGTTTGATCA/CTG GCTCAG-3′) and 517r (5′-ATTACCGCGGCTGCT GG-3′). Primer position are given according to the *E. coli* numbering (Brosius et al. 1981). The 27fGC-primer contain a GC-clamp (bold) to increase the thermal stability of the PCR-product during DGGE. PCR was performed in a total volume of 100 µl containing 1 X PCR buffer (10 mM Tris-HCl [pH 8.3], 50 mM KCl, 1.5 mM $MgCl_2$, 0.001% gelantine), 200 µM each dNTP, 0.5 µM each primer, 0.3 $mgml^{-1}$ bovine serum albumin (BSA), and 2.5 U AmpliTaq DNA polymerase (Perkin-Elmer Corp.), 10 µl DNA solution. The temperature cycle for the PCR was 60 sec of denaturation at 94°C, 60 sec of annealing (see below), and 90 sec of primer extension at 72°C. During an initial touchdown cycle, the annealing temperature was lowered from 68°C to 58°C in intervals of 1°C per cycle. The additional annealing cycles were done at 58°C. This touch-down-procedure reduces the formation of spurious by-products during the amplification process (Don et al. 1991, Muyzer et al. 1993). Positive controls containing purified DNA (2 ng) from *E. coli* were included in all sets of amplifications along with negative controls (no DNA added).

2.6 *Detection of amplified products*

PCR-amplified DNA was detected by gel electrophoresis. Aliquots of amplified samples (10 µl) were separated by gel electrophoresis on 1.2% agarose gels, stained with ethidium bromide, visualized with a UV transilluminator, and photographed.

2.7 *DGGE - Denaturing Gradient Gel Electrophoresis*

DGGE was performed with the Bio-Rad D GENE System. The polyacrylamide gels (7.5%) contained a denaturing gradient from 40 to 70% (100% denaturant: 7 M urea and 40% formamide) and were run in 0.5 x TAE buffer (40 mM Tris base [pH 7.4], 20 mM sodium acetate, 1 mM EDTA). The optimum denaturant gradient was determined by performing perpendicular DGGE according to the instruction manual (Bio-Rad Lab.). Electrophoresis was performed at constant voltage (70 V) and temperature (56°C) for 16 h. After electrophoresis the gels were stained with silver nitrate according to Heukeshoven and Dernick (1985), dried and photographed.

3 RESULTS

3.1 *Extraction and purification of DNA*

The efficiency of the lysis protocol was determined with bacterial reference strains and enrichment cultures from oxic and anoxic groundwater. The final protocol lysed most of the bacteria present (> 99.9%) guaranteeing access to most of the bacterial diversity.

A major prerequisite for the successful amplification and analysis of bacterial DNA by PCR was the adaptation and modification of standard protocols to extract and purify nucleic acids from environmental samples. DNA fragmentation due to excess shear forces as well as the copurification of inhibitors had to be avoided.

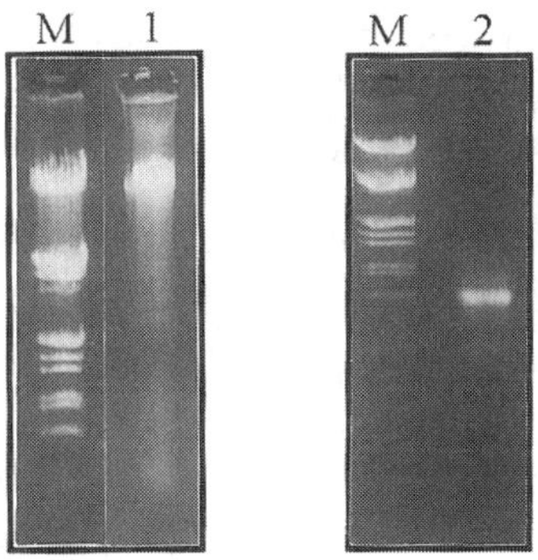

Figure 2. Isolated DNA from groundwater enrichment culture (lane 1) and PCR-amplified 16S rDNA from groundwater (lane 2). M: size standard (phage lambda DNA x Hind III x Eco RI)

Isolated DNA from groundwater enrichment culture is shown in Figure 2 (lane 1). The isolated DNA is mostly high molecular indicating a low mechanical stress on the DNA during extraction and purification. DNA isolated directly from groundwater is not shown due to the low bacterial numbers (i.e. DNA content below detection limit) in these samples. Figure 2 shows also the successful amplification of DNA from groundwater (lane 2).

3.2 *DGGE - Denaturing Gradient Gel Electrophoresis*

After the successful extraction and amplification of DNA from various environmental samples the bacterial diversity originally present was now reflected by the different PCR-products. DGGE was used to separate these PCR-products and obtain patterns characteristics for different hydrogeochemical situations related to artificial recharge and bank filtration. DGGE gels are analysed based cn the number and overall arrangement of DNA banding patterns. Comparing two samples, DNA bands migrating at identical positions in the gel indicate bacterial species present in both samples. Different positions reveal species not identical in the two samples.

DGGE-patterns from groundwater with different hydrogeochemical characteristics are shown in Figure 3.

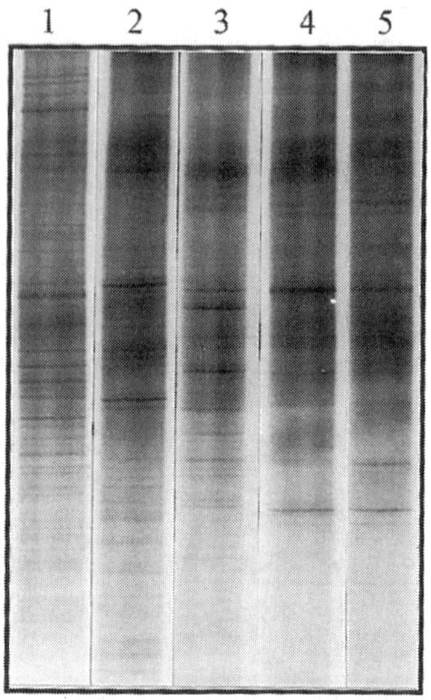

Figure 3. DGGE-patterns from groundwater with different hydrogeochemical characteristics. Lane 1: Anoxic, Lane 2: Changing redox conditions, Lane 3: Oxic, Lane 4: Oxic 2.31 m, Lane 5: oxic 1.41m

Patterns from anoxic groundwater (lane 1) are different from those obtained with oxic groundwater (lane 3) or groundwater from a well with changing redox-conditions (lane 2). On the other hand samples from one well with oxic groundwater but different depth (lane 4 and 5) showed identical patterns indicating a similar or identical bacterial population. Sampling site 3 is only 50 m apart from site 4 and 5 and interestingly these samples show DGGE patterns similar but not identical to each other.

In order to demonstrate temporal changes in microbial populations DGGE-patterns from surface water (river Ruhr) were recorded over a period of 6 months. The changing patterns in Figure 4 exhibits that certain species are found over a limited period of time while others occur in all samples.

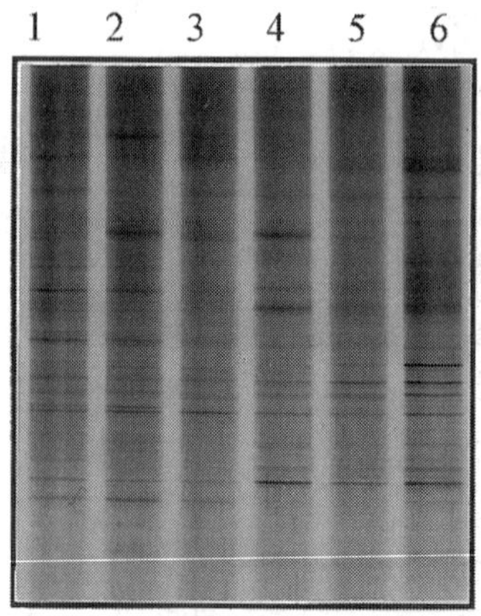

Figure 4. DGGE-patterns from surface water (river Ruhr). Lane 1 - 6: samples from March - August 1997.

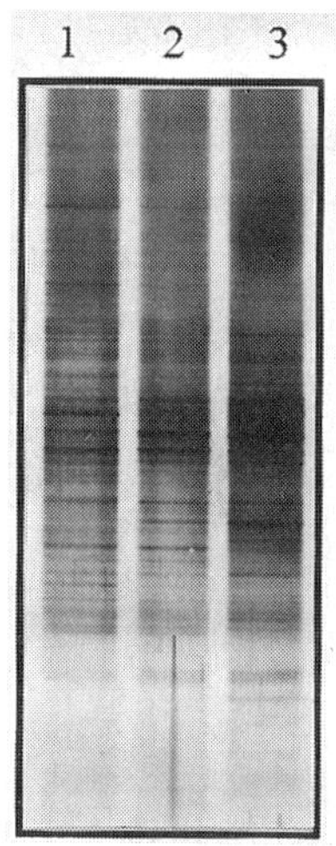

Figure 5. DGGE-patterns from lake-sediment (lake Hengsen). Lane 1: 0 cm; Lane 2: 6 cm; Lane 3: 24 cm.

DGGE-patterns from lake sediment are shown in Figure 5. Sediment samples were taken with a corer (diameter 7 cm) and analysed at different depths as indicated in the figure. Diverse patterns are obvious in all three samples indicating a species-rich bacterial population. Therefore these highly diverse samples are difficult to analyse and compare (see below). Still it is apparent that some DNA-bands are present at only one or two depths whereas others appear in all three samples. Additionally, DGGE-patterns from sediment samples usually contain more DNA-bands then groundwater samples.

4 DISCUSSION

A major problem when working with environmental samples like surface-, groundwater and lake sediment is the presence of certain impurities that are known to inhibit the PCR. The Taq polymerase enzyme used to amplify the DNA is sensitive to compounds such as iron (De Leon et al. 1991) and humic acids (Picard et al. 1992, Tebbe & Vahjen 1993). Other unknown substances that are present in the sample may also act as inhibitors to PCR. Therefore the copurification of these inhibitors had to be avoided and several precautions during the DNA purification and PCR were necessary to yield pure DNA and PCR products. The addition of PVPP (Holben et al. 1988) during DNA extraction was found to be necessary for samples with a high humic acid content. Other inhibitors were bound to BSA during PCR (Kreader 1996) and increased the PCR efficiency remarkably. A third option was dilution of the extracted DNA 1:10 - 1:1000. By implementation of these precautions during extraction and PCR the finally established protocol is suitable to separate inhibitory compounds from ground- and surface water. Furthermore the protocol can be used for environmental samples like soil and lake-sediment, usually regarded as very difficult for DNA-analysis involving PCR.

Besides its major advantages, DNA analysis in combination with PCR may introduce its own biases. Because of selective priming, certain sequences may be slightly discriminated compared with others. This would result in great differences after multiple cycles of amplification leading to a shift in species composition apart from the natural situation. On the other hand results based on sequence information revealed the ability of the PCR approach to identify species that would have never been detected with other techniques (Suzuki et al. 1997, Santegoeds et al. 1996).

When analysing highly diverse samples with universal primers, one problem inevitably arises: to some extent, the PCR products reflect the diversity originally present in the sample thus creating a very complex banding patterns during DGGE separation. This problem is of less importance when using more specific primer combinations (single species, functional groups). In the present study - by applying the universal approach - the DGGE-patterns were sometimes difficult to read and interpret. Samples taken from totally different hydrogeochemical conditions were easier to analyse because of their prominent differences (Fig. 3). In contrast, the bacterial diversity analysed in related samples, e.g. similar hydrogeochemical conditions or different depth from one well, showed great similarities in DGGE-patterns and the potential differences - if there were any - were difficult to analyse. To address this problem we are currently testing digitalized gel-imaging and analysis for the evaluation of complex DGGE-patterns.

5 FUTURE PERSPECTIVES

In the present study environmental samples such as ground- and surface water were analysed by applying PCR amplification and subsequent DGGE. Highly diverse samples can be compared with regard to characteristic DGGE patterns. Based on the separation by DGGE one future task could be the identification of single species by DNA sequencing or hybridisation with DNA probes of prominent DNA-bands.

6 ACKNOWLEDGEMENTS

This project is supported by the EC Environment and Climate Research Programme (Contract: ENV4-CT95-0071) and the DVGW (Deutscher Verein des Gas- und Wasserfaches e.V.)

REFERENCES

Amann, R.I., W. Ludwig & K.-H. Schleifer 1995. Phylogenetic Identification and In Situ Detection of Individual Microbial Cells without Cultivation. *Microbiol. Rev.* 59:143-169.

Brosius, J., T.J. Dull, D.D. Sleeter & H.F. Noller 1981. Gene Organisation and Primary Structure of a Ribosomal RNA Operon from *Escherichia coli*. *J. Mol. Biol.* 148:107-127.

De Leon, R., C. Shieh, R.S. Baric & D. Sobsey 1991. Detection of enteroviruses and hepatitis A virus in environmental samples by gene probes and polymerase chain reaction, p. 833-853. *Proc. Water Quality Tech. Conf.*, San Diego, Calif.

Don, R.H., P.T. Cox, B.J. Wainwright, K. Baker & J.S. Mattick 1991. „Touchdown“ PCR to circumvent spurious priming during gene amplification. *Nucleic Acids Res.* 19:4008.

Fischer, S.G. & L.S. Lerman 1983. DNA fragments differing by single substitutions are separated in denaturing gradient gels: correspondence with melting theory. *Proc. Natl. Acad. Sci.* USA 80:1579-1583.

Heukeshoven, J. & R. Dernick 1985. Simplified method for silver staining of proteins in polyacrylamide gels and the mechanisms of silver staining. *Electrophoresis* 6:103-112.

Hobbie, J.E., R.J. Daley & S. Jasper 1977. Use of Nucleopore Filters for Counting Bacteria by Fluorescence Microscopy. *Appl. Environ. Microbiol.* 33:1225-1228.

Holben, W.E., J.K. Jansson, B.K. Chelm & J.M. Tiedje 1988. DNA Probe Method for the Detection of Specific Microorganisms in the Soil Bacterial Community. *Appl. Environ. Microbiol.* 54:703-711.

Kreader, C.A. 1996. Relief of Amplification Inhibition in PCR with Bovine Serum Albumin or T4 Gene 32 Protein. *Appl. Environ. Microbiol.* 62:1102-1106.

Maniatis, T., E.F. Fritsch & J. Sambrook 1982. *Molecular cloning: a laboratory manual.* Cold Spring Harbor Laboratory, Cold Spring Harbor, N.Y.

Muyzer, G., E.C. De Waal & A.G. Uitterlinden 1993. Profiling complex Microbial Populations by Denaturing Gradient Gel Electrophoresis Analysis of Polymerase Chain Reaction-Amplified Genes Coding for 16S rRNA. *Appl. Environ. Microbiol.* 59:695-700.

Picard, C., C. Ponsonnet, E. Paget, X. Nesme & P. Simonet 1992. Detection and Enumeration of Bacteria in Soil by Direct DNA Extraction and Polymerase Chain Reaction. *Appl. Environ. Microbiol.* 58:2717-2722.

Saiki, R.K., S. Scharf, F. Faloona, K.B. Mullis, G.T. Horn & H.A. Ehrlich 1985. Enzymatic amplification of ß-globin genomic sequences and restriction site analysis for diagnosis of sickle cell anemia. *Science* 230:1350-1354.

Santegoeds, C.M., S.C. Nold & D.M. Ward 1996. Denaturing Gradient Gel Electrophoresis Used to Monitor the Enrichment Culture of Aerobic Chemoorganotrophic Bacteria from a Hot Spring Cyanobacterial Mat. *Appl. Environ. Microbiol.* 62:3922-3928.

Schöttler, U. & U. Schulte-Ebbert (eds.) 1995. *Schadstoffe im Grundwasser - Band III*, Verhalten von Schadstoffen im Untergrund bei der Infiltration von Oberflächenwasser am Beispiel des Untersuchungsgebietes „Insel Hengsen" im Ruhrtal bei Schwerte. Deutsche Forschungsgemeinschaft, Bonn 1995.

Suzuki, M.T., M.S. Rappé, Z.W. Haimberger, H. Winfield, N. Adair, J. Ströbel & S.J. Giovannoni 1997. Bacterial Diversity among Small-Subunit rRNA Gene Clones and Cellular Isolates from the same Seawater Sample. *Appl. Environ. Microbiol.* 63:983-989.

Tebbe C.C. & W. Vahjen 1993. Interference of Humic Acids and DNA Extracted Directly from Soil in Detection and Transformation of Recombinant DNA from Bacteria and Yeast. *Appl. Environ. Microbiol.* 59:2657-2665.

Tsai, Y.-L. & B.H. Olson 1991. Rapid Method for Direct Extraction of DNA from Soil and Sediments. *Appl. Environ. Microbiol.* 57:1070-1074.

Artificial Recharge of Groundwater, Peters et al. (eds) © 1998 Balkema, Rotterdam, ISBN 90 5809 017 5

Temporal and spatial distribution patterns of Crustaceans inhabiting a slow sand filter

P. Rumm & H. Schmidt
Arbeitsgruppe Zoomorphologie, FB Biologie, Carl von Ossietzky Universität Oldenburg, Germany
W. Hollwedel
Varel, Germany

ABSTRACT: A one year study of the meiofauna of a slow sand filter revealed 100 taxa. Most species belong to the Crustaceans (48 species out of 3 groups - copepods, ostracods and cladocerans). A statistical analysis was applied to find temporal and spatial distribution patterns. It showed the 48 species to occur in typical assemblages within distinct areas of the slow sand filter.

1 INTRODUCTION

Slow sand filtration plays an increasingly role in the supply of drinking water. Even though the biological, chemical and physical processes within slow sand filters have already been thoroughly investigated, they are not yet completely understood (Brinks & Parks 1996, Weber-Shirk & Dick 1997a,b). Bacterial processes are important for the mineralization of infiltrated dissolved and particular organic matter (Weber-Shirk & Dick 1997a). Microorganisms also are a basic nutritional source for detritus feeders and their predators.

From an ecological point of view, slow sand filters have a biotrophic (food) web similar to that in transition zones between surface water and natural groundwater, the "surface water/groundwater-ecotone" (Gibert et al. 1997). Natural aquatic systems have a biological layer, the benthos (Danielopol 1980). This biologically active zone (the upper filter layer) is also present in slow sand filters. It is the stratum where most of the biotic and abiotic processes take place. It is characterized by high densities of bacteria, protozoa and meiofauna, high quantities of organic matter (due to sedimentation processes and algal growth), and hydrological connections between surface water and groundwater.

The role of the meiofauna in slow sand filters is supposed to be similar to the one in groundwaters. The permanent locomotory and feeding activities of this fauna keep the interstices free from organic material that might clog them (Husmann 1976, Schminke & Glatzel 1988, Westphal 1996).

Investigations of fauna in slow sand filters were carried out by Noll (1974) and Husmann (1974). Noll (1974) studied the faunal and algal colonization of permanently and intermittently used slow sand filters. Husmann (1958, 1974) investigated the interstitial fauna of the sand layer. Deeper areas such as the gravel zone beneath the sand layer have not yet been investigated, although both parts are hydrologically connected.

There is a lack of data on the distribution and abundance of the fauna, not only in the sand layer, but also in the gravel zone beneath it. In a joint project with Gelsenwasser AG (Gelsenkirchen, Germany) the investigation of a slow sand filter in the upper Ruhr valley aimed at the following questions:

1. Which and how many species inhabit the sand layer and the deeper zones of the slow sand filter?
2. Are there temporal (i.e. seasonal) patterns in the occurrence of the specific filter fauna?
3. Do typical species assemblages exist within the different zones: biologically active zone, sand layer, gravel zone and collecting well?

2 STUDY SITE

2.1 *Geography*

The slow sand filter is part of the waterworks Echthausen. It is located at the River Ruhr (7°50'34" E, 51°30'57" N). Figure 1 shows the investigation area located at the River Ruhr.

2.2 *The slow sand filter*

The slow sand filter No. 8 (waterworks Echthausen) is designed for artificial recharge of groundwater. It is 20 m wide, 350 m long and has a total filtering area of 7000 m². It is covered with a 50 cm layer of coarse and fine sand. The sand layer rests on natural gravel of the Ruhr valley. The source water is taken from the River Ruhr. Before its infiltration, the

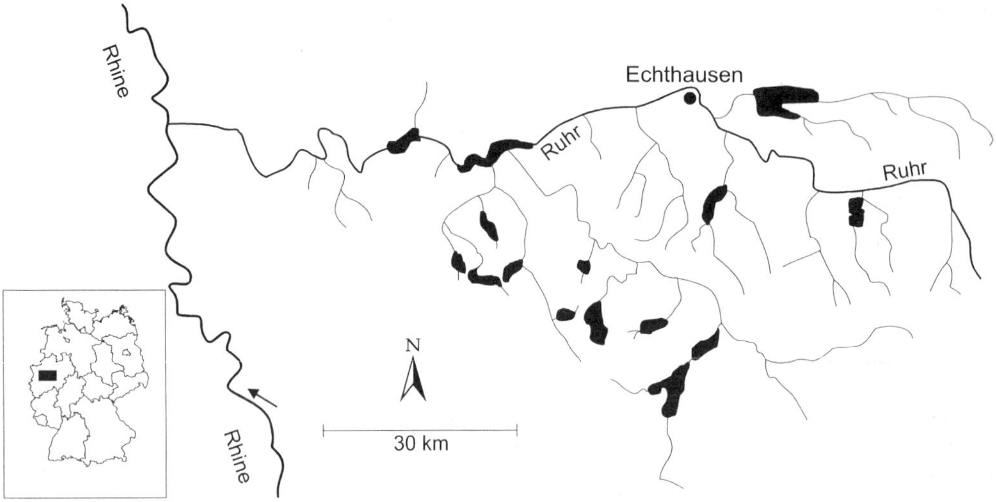

Figure 1. Area of the River Ruhr and its tributaries. • indicates the place where the sample were taken.

source water is pre-treated by passing a sedimentation bassin, where floating particles are removed. If necessary, the pre-treatment processes are supported by optionally adding activated carbon or/and aluminium chloride ($AlCl_3$).

3 METHODS

Five permanent observation wells were driven into the slow sand filter, each one reaching to a different depth. The first well is 40 cm deep, which is within the sand layer, the following are 120 cm, 190 cm, 310 cm 470 cm deep, which are all within the gravel zone. From each well a monthly sample of 100 l was taken over one year. Each sample was filtered through a plankton net (mesh width 50 µm). Samples of 50 cm³ were taken from the biologically active zone. Further samples were taken from the collecting well at the end of the infiltration gallery with a 2 m long plankton net (Westphal 1996, Rumm et al. 1997). The average volume of filtered water per sample taken out of the collecting well was 1000000 l.

All specimens were sorted and identified under a dissection microscope. A description of the method in detail is given in DVWK-Schriften 120 (1997, p.80).

The faunistic data form the base for the calculation of the biological index (BI), the presence and dominance. A cluster analysis (using percentage simillarity index) was applied to determine typical species assemblages.

4 RESULTS

4.1 *Cluster analysis and species-site-table*

The investigated slow sand filter is colonized by 48 species of crustaceans of the following 3 groups: Copepoda (18 species), Ostracoda (11 species), and Cladocera (19 species). The copepods are represented by the Harpacticoida (10 species) and Cyclopoida (8 species). Concerning the habitat of the analysed fauna there are 39 species belonging to surface waters (epigean fauna) and 9 species which are confined to the stygobionts (hypogean or groundwater fauna). These stygobionts fulfill their complete life cycle exclusively in groundwater habitats.

In total we have found 100 taxa to live in the slow sand filter. The cluster analysis shows 6 species clusters (1-6) and 4 station clusters (I-IV) (Fig. 2).

Station cluster I – the biologically active zone (A). Here the stygoxenes (animals without any affinity to groundwater) which constitute species cluster 6 can be found.

Station cluster II – the sand layer (B; 0-40 cm depth) housing the species of species cluster 5.

Station cluster III – station F (470 cm) and G (collecting well). These stations closely resemble the groundwater. Mainly stygobionts belonging to species clusters 2 and 4 can be found.

Station cluster IV – unites the stations C (120 cm), D (190 cm) and E (310 cm). Cyclopoida (Copepoda) are the dominating faunal component. They constitute species cluster 3. The ostracod *Pseudocandona zschokkei* forms species cluster 1.

Indices are calculated for 24 out of 48 species of Ostracoda, Cyclopoida and Harpacticoida, each having a dominance of more than 1%.

cw = collecting well; baz = biologically active zone

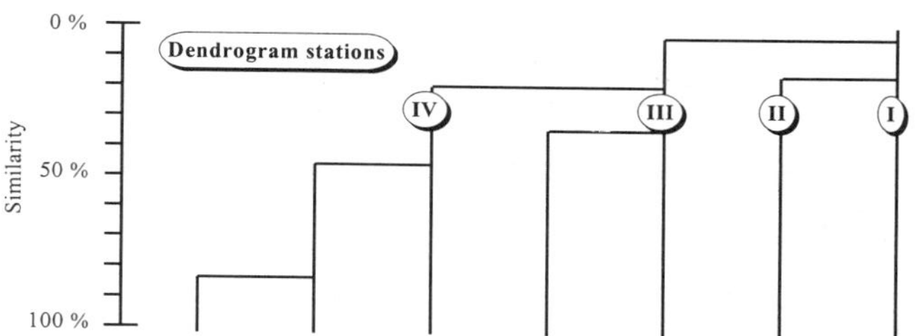

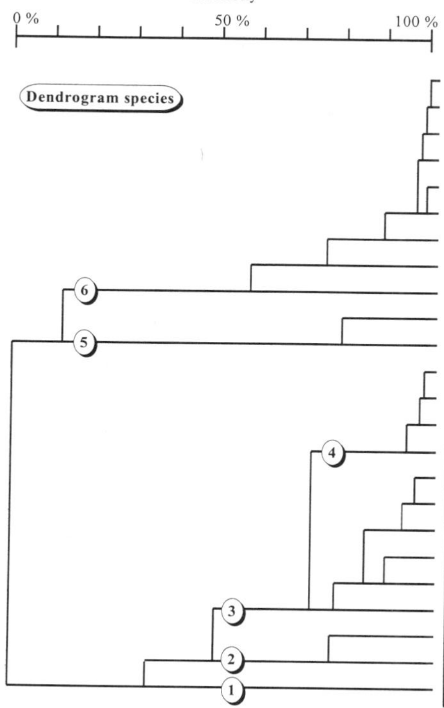

ID	species / station	Cluster	D 190cm	E 310cm	C 120cm	F 470cm	G cw	B 40cm	A baz
1	Alona guttata	6	.	.	.	.	.	.	**131**
2	Candona neglecta		.	.	.	.	.	.	**76**
3	Cypridopsis vidua		7	.	.	.	.	.	**821**
4	Bryocamptus pygmaeus		.	.	4	.	.	20	**1.004**
5	Alona affinis		2	.	2	.	5	1	**627**
6	Cypria ophtalmica		4	1	1	2	23	.	**1.167**
7	Chydorus sphaericus		1	.	1	.	13	1	**131**
8	Bryocamptus minutus		.	.	6	3	2	55	**196**
9	Paracyclops fimbriatus						67		**90**
10	Phyllognathopus viguieri	5	51	206	209	55	21	**3.674**	643
11	Epactophanes richardi		21	156	136	16	56	**655**	144
12	Alona rectangula	4	7	1	6	3	**6.231**	.	59
13	Bryocamptus echinatus		2	1	.	.	**416**	.	6
14	Mixtacandona laisi laisi		1	.	.	.	**218**	1	.
15	Graeteriella unisetigera		2		47		**1.548**	38	
16	Diacyclops bisetosus	3	160	203	34	104	**2.728**	57	
17	Diacyclops languidoides		58	91	8	91	**1.428**	22	2
18	Acanthocyclops robustus		10	15		3	**213**		
19	Diacyclops bicuspidatus		23	26	3	14	**200**	2	5
20	Diacyclops languidus		110	46	14	18	**805**	14	
21	Eucyclops serrulatus		9	5		2	**300**	1	90
22	Chappuisius inopinus	2	.	11	1	**73**	**82**	.	.
23	Chappuisius singeri		.	3	.	**40**	**108**	1	.
24	Pseudocandona zschokkei	1	**1.720**	**1.822**	**261**	6	**1.822**	.	27

Figure 2. Species-site-table with the dendrograms of cluster analysis

4.2 *Presence, Dominance and Biological Index (BI)*

The BI shows the 3 species *Diacyclops bisetosus, Phyllognathopus viguieri* and *Pseudocandona zschokkei* to take the first 3 positions. They are characteristic for 3 different habitats. In table 1 the values for the analysed data are given. The 24 most dominant species are numbered (ID in table 1).
- *Diacyclops bisetosus* for the collecting well.
- *Phyllognathopus viguieri* for the sand layer
- *Pseudocandona zschokkei* for the gravel zone

4.3 *Temporal occurrence*

Biologically active zone: The harpacticoid copepod *Bryocamptus pygmaeus* and the cladoceran *Alona affinis* appear from late summer until winter. The ostracod *Cypria ophtalmica* increases in abundance from spring to summer and is absent in winter.

Sand and gravel zone: *Pseudocandona zschokkei* (Ostracoda) can regularly be found but varies in abundance. *Diacyclops languidoides* and *Diacyclops bisetosus* (Copepoda, Cyclopoida) show both, variable occurrence and high individual numbers. *Phyllognathopus viguieri* and *Epactophanes richardi* (both harpacticiod copepods) can often be found but show strong variability in abundance. *Chappuisius inopinus* and *Graeteriella unisetigera* (both stygobiont copepods) were only found sporadically in summer with high abundance.

Collecting well: The stygobiont copepods *Chappuisius inopinus* and *Graeteriella unisetigera* occur with high abundance from spring to summer. *Alona rectangula* (Cladocera) and *Bryocamptus echinatus* (Copepoda) are mainly present from summer to autumn.

Some species show regular seasonal patterns in occurrence and abundance and in specific zones (*Pseudocandona zschokkei, Cypria ophtalmica, Phyllognathopus viguieri*). Other species show clear temporal patterns with maxima in summer and minima (or complete absence) in winter (*Bryocamptus pygmaeus, Alona affinis*). The species of the third group show no regular seasonal pattern and are only of sporadic occurrence (*Epactophanes richardi* and *Cypridopsis vidua*).

Table 1. Species list in taxonomic order

ID	species list	dominance (%)	presence (%)	rank BI	BI
	Cyclopoida				
16	*Diacyclops bisetosus*	10	86	1	71
17	*Diacyclops languidoides**	5	100	4	53
20	*Diacyclops languidus*	3	86	6	44
15	*Graeteriella unisetigera**	5	57	7	29
19	*Diacyclops bicuspidatus*	1	100	10	19
18	*Acanthocyclops robustus*	1	57	15	10
21	*Eucyclops serrulatus*	1	86	17	9
9	*Paracyclops fimbriatus*	1	29	47	-
	Harpacticoida				
10	*Phyllognathopus viguieri**	15	100	2	69
11	*Epactophanes richardi*	4	100	5	51
4	*Bryocamptus pygmaeus*	3	43	9	19
8	*Bryocamptus minutus*	1	71	11	19
22	*Chappuisius inopinus**	1	57	12	14
	*Parastenocaris cf. glacialis**	0,1	71	16	9
23	*Chappuisius singeri**	1	57	19	9
13	*Bryocamptus echinatus*	1	57	20	6
	Canthocamptus staphylinus	0,1	14	28	-
	Nitocra hibernica	0,1	14	33	-
	Ostracoda				
24	*Pseudocandona zschokkei**	18	86	3	57
6	*Cypria ophtalmica*	4	86	13	14
3	*Cypridopsis vidua*	3	29	14	11
14	*Mixtacandona laisi laisi**	1	43	23	3
	Candona candida	0,1	29	35	-
	Herpetocypris reptans	0,1	14	37	-
2	*Candona neglecta*	0,2	14	38	-
	Chlamydotheca spec.	-	14	39	-
	Heterocypris incongruens	-	14	40	-
	*Pseudocandona triquetra**	0,2	14	46	-
	Isocypris beauchampi	0,1	14	48	-
	Cladocera (Anomopoda)				
12	*Alona rectangula*	20	86	8	20
5	*Alona affinis*	2	71	18	9
1	*Alona guttata*	1	14	21	4
6	*Chydorus sphaericus*	1	71	22	3
	Camptocercus rectirostris	-	14	24	-
	Alonella nana	-	14	25	-
	Pleuroxus denticulatus	-	14	26	-
	Simocephalus vetulus	-	14	27	-
	Graptoleberis testudinaria	0,1	14	29	-
	Pleuroxus aduncus	0,1	29	30	-
	Acroperus harpae	0,1	14	31	-
	Drepanothrix dentata	-	14	32	-
	Alona costata	-	29	34	-
	Bosmina longirostris	-	29	36	-
	Chydorus piger	-	14	41	-
	Ceriodaphnia quadrangula	-	14	42	-
	Macrothrix hirsuticornis	-	14	43	-
	Eurycercus lamellatus	-	14	44	-
	Alona quadrangularis	-	14	45	-

* = stygobionts;
ID numbers correspond with ID from figure 1;
BI = biological index

5 CONCLUSIONS

1. Slow sand filters are inhabited by a great number of different meiofaunal organisms. They occur both in high abundance and high species numbers.

2. The statistical methods used (cluster analysis, calculation of presence and biological index) reveal a distinct spatial distribution of the meiofauna, but the permanently filtered water together with its organic components makes the slow sand filter a highly dynamic biological system.

3. The meiofauna can be found at all depths of the slow sand filter. The high abundances indicate good living conditions for the epigean fauna as well as for the groundwater animals, the stygobionts.

4. The high abundance of the meiofauna constitutes an important component of the slow sand filter system. Metabolic and locomotory activities (bioturbation) of the fauna support water cleaning processes and prevent the interstices from clogging.

6 ACKNOWLEDGEMENTS

This research was conducted under a grant of Gelsenwasser. Special thanks are due to Prof. Dr. H.K. Schminke, Dr. B. Pätsch, Dr. B. Westphal and Dipl.-Biol. Christine Schönberg who gave useful comments on the manuscript. The Wolfgang-Schulenberg-Stipendienprogramm (University of Oldenburg) and Gelsenwasser supported the participation on TISAR 98.

REFERENCES

Brinks, D.R. & S. Parks. 1996. Update on slow sand/advanced biological filtration research. In Graham, N. & R. Collins (eds), *Advances in slow sand and alternative biological filtration*. Chichester: J. Wiley & Sons.

Danielopol, D.L. 1980. The role of the limnologist in groundwater studies. *Int. Rev. ges. Hydrobiol.* 65: 777-791.

DVWK-Fachausschuß "Grundwasserbiologie". 1997. *Parameter und Methoden der biologischen Charakterisierung des Untergrundes - Feststoffe und Wasser*. Heft 120. Bonn: DVWK e.V.

Gibert, J., J. Mathieu & F. Fournier 1997. *Groundwater/surface water ecotones: biological and hydrological interactions and managament options*. Cambridge. University Press, U.K.

Husmann, S. 1958. Untersuchungen über die Sandlückenfauna der bremischen Langsamfilter. *Abhandl. Braunschw. Wiss. Ges.* Bd. 10: 94 -116.

-- 1974. Die ökologische Bedeutung der Mehrzellenfauna bei der natürlichen und künstlichen Sandfiltration. *Wiss. Ber. Unters. Plan. Stadtwerke Wiesbaden AG*. Nr 2: 173-183.

-- 1976. Langsamsandfilter als Lebensräume von Grundwassertieren mit Befunden aus Wiesbaden-Schierstein. *Forschung und Entwicklung in der Wasserwerkspraxis Stadtwerke Wiesbaden AG*. Nr. 3: 83-92.

Noll, M. 1974. Die Besiedlung von Langsamsandfiltern und ihre Beeinflussung durch Pestizide. *Veröffentlichungen des Institut für Wasserforschung GmbH Dortmund und der Dortmunder Stadtwerke AG*. Nr.19: 1-179.

Rumm, P., H. Schmidt & H.K. Schminke 1997. Organismenaustrag aus Langsamsandfiltern. *gwf-Wasser/Abwasser*. 138(7): 355-361.

Schminke, H.K. & Th. Glatzel 1988. Besonderheiten und ökologische Rolle der Grundwassertiere. *Z. dt. geol. Ges.* 139: 383 - 392.

Weber-Shirk, M.L. & R.I. Dick 1997a. Biological mechanisms in slow sand filters. *J.A.W.W.A.* 89(2): 72-83.

Weber-Shirk, M.L. & R.I. Dick 1997b. Physical-chemical mechanisms in slow sand filters. *J.A.W.W.A.* 89(1): 87-100.

Westphal, B. 1996. Planktonalgen und Metazoen in Trinkwasserversorgungsanlagen. *gwf-Wasser/Abwasser*. 137(5): 271-275.

Artificial Recharge of Groundwater, Peters et al. (eds) © 1998 Balkema, Rotterdam, ISBN 90 5809 017 5

Fate of pathogens and consequences for the design of artificial recharge systems

J.H. Peters, M.J.C. van Baar, P.J. Nobel & A.J. Vogelaar
Kiwa Research and Consultancy, Nieuwegein, Netherlands

J.F. Schijven
RIVM, National Institute of Public Health and the Environment, Bilthoven & Delft University of Technology, Civil Engineering and Geosciences, Netherlands

W. Hoogenboezem, J. Bergsma & A. Stakelbeek
PWN Waterworks of North-Holland, Bloemendaal, Netherlands

ABSTRACT: A field experiment has been conducted at the Castricum artificial recharge system in the Netherlands. Two types of bacteriophages (MS2 and PRD1) were dosed as model organisms. The water's flow pattern was first calculated and then verified in 6 monitoring wells. Then sodium chloride was added for 7 days as a tracer. Following this, bacteriophages were added to the artificial recharge water for 11 days. The effectiveness of the soil in removing these micro-organisms was worked out from the levels at which they were present in the monitoring wells. For the test set-up described and for flow through fine dune sand, the experiment indicates an elimination rate for the phages studied of 3 log units in the first 2.5 m; 6 log units for 10 days' passage through the aquifer (15 m) and 8 log units for 25 days (30 m). Fate and behaviour of human viruses might be explained and understood from these results, so that an artificial recharge system design required for virological reliability can be determined.

1 INTRODUCTION

Water companies want to minimize the side-effects of artificial recharge (AR) on the environment and therefore want to limit the space taken up by existing artificial recharge projects (Peters 1996). Any redesign, renovation or new design for an artificial recharge system automatically raises the question of what the minimum residence time has to be. This is different for each of the companies operating artificial recharge systems in the Netherlands and is often based on the work of Knorr (1937, 1951) and Knoll (1982). It should also be pointed out that pre-treatment of the artificial recharge water removes several log units of bacteria and viruses (Nobel & Van der Kooij 1997). The question arises as to whether a shorter passage through the soil would suffice. If the residence time could be shorter, there would be possibilities for intensification and for reducing the amount of space taken up by artificial recharge systems.

2 PURPOSE

The purpose of the field experiment is to study the behaviour of bacteriophages in their passage through the soil and the processes that affect that behaviour. The aim is to find out how AR systems should be designed and arranged in order to guarantee that the water is free of pathogens.

3 WHY BACTERIOPHAGES?

We concentrate on 'virus-type-organisms' because of their small size, low inactivation rates, high infectivity and low adsorption in sands. On the basis of a maximum infection rate of 1 per 10,000 people per year as a result of drinking-water consumption, we can calculate that human pathogenic viruses must be absent in 4500 m^3. Given the numbers of enteroviruses in surface water, it is necessary to achieve a removal rate of 5 up to 8 log units. This does not yet take into account peak loads, for instance as a result of a sewage overflow after heavy rainfall (Schijven et al. 1996). On the other hand, a concentration of all type of micro-organisms already reduces a number of log units in case of pre-treatment prior to artificial recharge.
Given the extremely low permissible concentration in drinking water, measurements cannot directly demonstrate whether removal of viruses is sufficient. Using bacteriophages as model organisms offers a solution to this problem. They can be added in high numbers, thereby making it possible to measure the rate of removal. It is unthinkable to work with human viruses in a recharge system that is in operation. However, adding bacteriophages does not present health risks. Moreover, it is relatively inexpensive to produce phages, take samples and make analyses. The model phages chosen are MS2 and PRD1. MS2 can be seen as a relatively conservative model organism for virus transport,

owing to its low binding (Bales et al. 1993, Powelson et al. 1993, Jin et al. 1997). PRD1 adsorbs at a higher rate in many cases but can, even at 20°C, survive for a long time (Yahya et al. 1993).

4 GENERAL DESCRIPTION OF THE CASTRICUM ARTIFICIAL RECHARGE SYSTEM

The Castricum artificial recharge system was constructed in the mid nineteen fifties, in old dune reclamation areas to the west of the village of Castricum. The system was last expanded in 1972. Technical renovations were carried out between 1990 and 1993 together with nature conservation and development (Peters et al. 1992). The system now covers around 150 hectares and has a capacity of 25 million m^3/year, see figure 1. The infiltration rate is 0.3 $m^3/m^2/d$ on average and up to 0.5 $m^3/m^2/d$ locally.

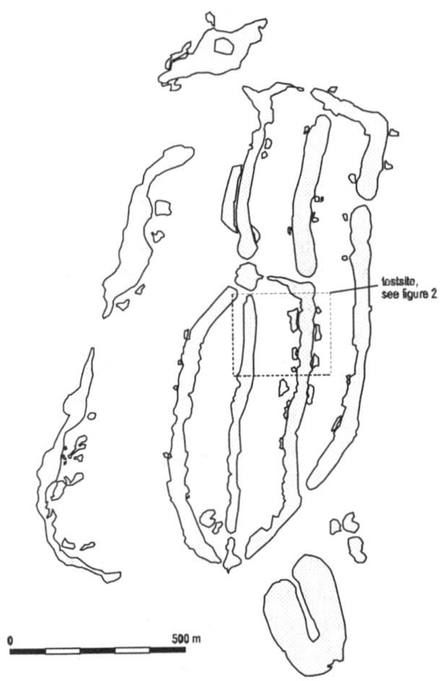

Figure 1. Castricum AR system showing size and dimensions of basins. Location of test site is indicated

The pre-treatment is carried out elsewhere, near the surface water's point of intake, and consists of coagulation, sedimentation, rapid sand filtration and partially active carbon filtration. The area open water is approximately 22 hectares. The water is retrieved using around 600 shallow wells approximately 9 m in depth, distributed over 12 separate suction lines. The abstraction can be adjusted for each suction line. Up to 8 m^3/hour can be abstracted per well. The water level in the infiltration basins is a fairly constant 2.85 m+NAP (NAP is approximately mean sea level). The infiltration basins are an average of 0.8 m deep. The ground level is approximately 3 m+NAP. The phreatic aquifer is around 10 m thick and consists of fine, aeolian dune sands. Of this sand 12% passes through a sieve with openings of 125 μm and 92% through openings of 300 μm. Hydraulic conductivity is about 12 m/d (at a water temperature of 5°C, the average temperature during the tracer test). Effective porosity with respect to flow is approximately 0.38. Below this top layer is around 15 m of silt-containing sand with a lower hydraulic conductivity.

5 WHAT WAS DONE PREVIOUSLY?

Prior to the experiment described, an experiment was conducted in the winter of 1995/1996 at the nearby Kieftenvlak artificial recharge system. In this experiment, the removal of naturally occurring viruses and bacteriophages during dune recharge was determined. Here too, water is recharged via basins. However, the background levels of enteroviruses and F-specific RNA bacteriophages in this pre-treated water turned out to be low. This meant that the results of this experiment, in which monitoring wells were used at distances of 2 and 4 metres from the bank of a basin, could not be used to obtain reliable estimates of the removal of viruses and phages. The results will be published elsewhere, but formed the reason for conducting the follow-up experiment with dosing bacteriophages, which is described here. Large numbers of bacteriophages were added to a recharge basin, so that they could be detected at greater distances from the bank in measurable concentrations.

6 SET-UP OF FIELD EXPERIMENT

A set-up and site were chosen with a known flow pattern and especially one that can be kept constant. The test site is shown in figure 2.

Throughout the test period, the abstraction rate was 330 m^3/hour from a series of 55 wells, i.e. 6 m^3/hour/well. The distance between the extraction wells at the research site is 10 m. The distance from basin V to extraction well PCQ414 is 63 m. Minimum travel time of the water is approximately 40 days. In October 1996, on a flow path almost perpendicular to the bank of basin V, 7 monitoring wells were installed with 0.25 m long filters, at those places where prior calculations had indicated a residence time of 1 day to a few weeks.

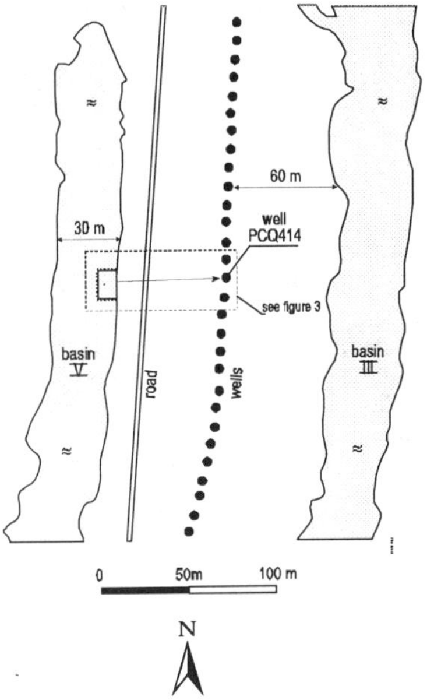

Figure 2. Detail of testsite in Castricum recharge system

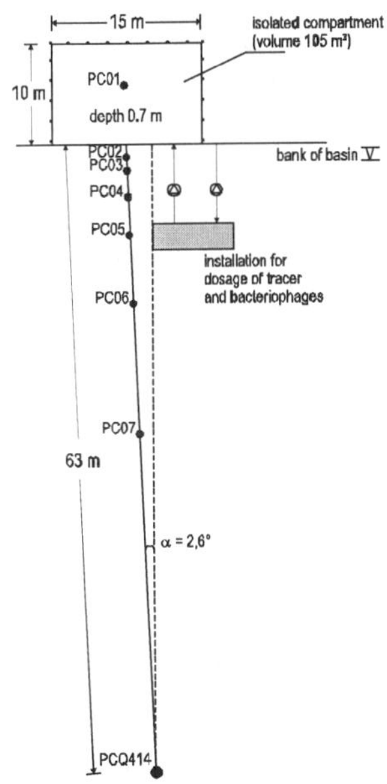

Figure 3. Test site in detail. Wells PCO1/7 were put on a flow path deviating a little from a line perpendicular to bank as was clear from measured ground water levels

The ground water flow model MODFLOW was calibrated on the basis of the ground water level measured in the monitoring wells. Following this, a 150 m^2 compartment was created using PVC sheeting along wooden poles in basin V, see figure 3. The water level in the compartment was and remained the same as the level outside of it (2.85 m+NAP). Monitoring wells and the flow that occurred during the test are shown in cross-section in figure 4.

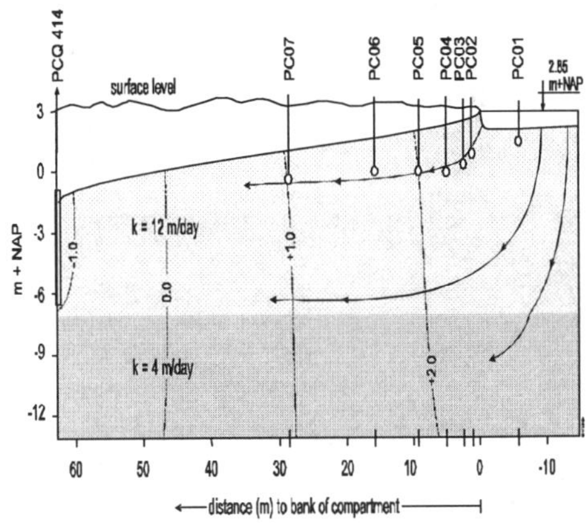

Figure 4. Monitoring wells and flow pattern in vertical cross-section

7 TRACER TEST

Sodium chloride was initially added to the compartment as a tracer. The aim was to verify the occurring flow, which had been calculated prior to the experiment. The chloride content of the artificial recharge water is approximately 150 mg/l. The Electrical Conductivity (EC) is approximately 900 μS/cm. From 2 December 1996, for exactly 7 days, the chloride concentration and the EC of the water recharging the compartment were increased to around 750 mg/l and 2700 μS/cm respectively. A total of around 600 kg of salt was added as NaCl. To measure the breakthrough of the salt front, the chloride concentration was regularly determined on the basis of samples taken from the wells. During 35 days the EC was regularly measured at 10 minute intervals in 6 monitoring wells by means of fixed EC-sensors with a data logger (Vogelaar et al. 1997). Several litres of water per hour were abstracted from all the monitoring wells to prevent stagnant water above the screens distorting the measurements. Figure 5 shows the results of the sensor readings, converted to chloride concentration. The calibration ratio used was Cl(mg/l) = 0.329EC(μS/cm) - 148. As expected, the breakthrough curves show that NaCl behaves as a conservative tracer. The same maxima were achieved in all the monitoring wells as in the compartment. The dispersion could also be calculated from the breakthrough curves. The test proved to be an especially suitable method for determining the actual residence time in the various monitoring wells, see table 1.

Table 1. Distance of monitoring wells to recharge basin V and travel times as measured in the tracer test

Monitoring Well	Horizontal distance (m) to basin	Real distance along flow path	Travel time (d) as measured in tracer test
PCO1	-5.8	-	-
PCO2	1.4	2.4	1.7
PCO3	2.7	3.7	2.5
PCO4	5.3	6.4	4.3
PCO5	9.1	10.2	7.0
PCO6	16.0	17.1	11.9
PCO7	29.0	30.1	26.2

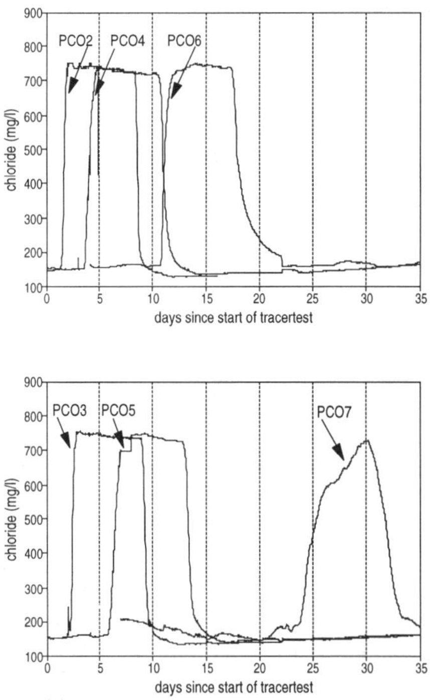

Figure 5. Breakthrough curves for chloride (measured as EC) in the 6 monitoring wells PCO2/7

8 EXPERIMENT WITH BACTERIOPHAGES

Starting on 27 January 1997, bacteriophages MS2 and PRD1 were added to the recharge water in the compartment. Dosage lasted for exactly 11 days. The experiment was carried out in winter time because it is especially in this season's low water temperatures that little inactivation of the micro-organisms can be expected. MS2 and PRD1 were added in the compartment, at concentrations of 10^8 pfp/l, and 10^7 pfp/l respectively. During a 120-day sampling and measuring campaign, both types of bacteriophages were counted in approximately 250 water samples. Results of the occurrence of MS2 in the compartment, in PCO2 and in PCO4 are illustrated in figure 6.

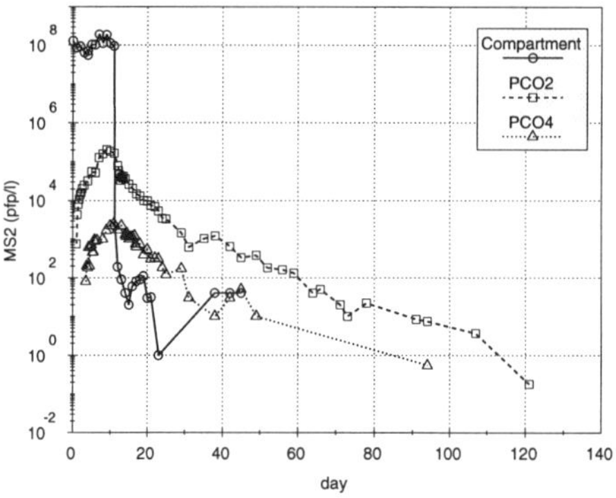

Figure 6. MS2 (pfp/l) in compartment and monitoring wells PCO2 & PCO4 during 120 days starting January 27, 1997

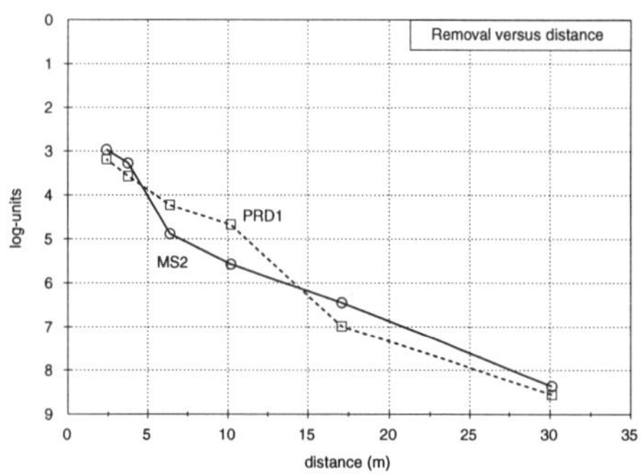

Figure 7. Removal (log units) of MS2 and PRD1 versus distance (m)

Figure 7 shows the removal rate as a function of the distance along a flow path to the compartment. Figure 8 shows the removal rate with respect to the residence time for the water (as determined using the tracer test). The vertical movement of the water means that the distance along a flow path deviates somewhat from the horizontal distance of the monitoring wells up to the compartment, see table 1.

Figures 7 and 8 show removal in log units, assuming that a 'steady state' maximum has been reached in the monitoring wells. The logarithm of the ratio between the maximum concentration in the compartment and in the various monitoring wells is shown. For the test set-up described and for flow through fine dune sand, the experiment indicates an elimination rate for the phages studied of more than 6 log units for 10 day's passage through the soil, or 15 m (8 log units for 25 days and 30 m). It should be pointed out that a removal rate of 3 log units occurs during the flow to the first well PCO2, at 1.4 m from the basin. The removal rate is lower in the rest of the aquifer.

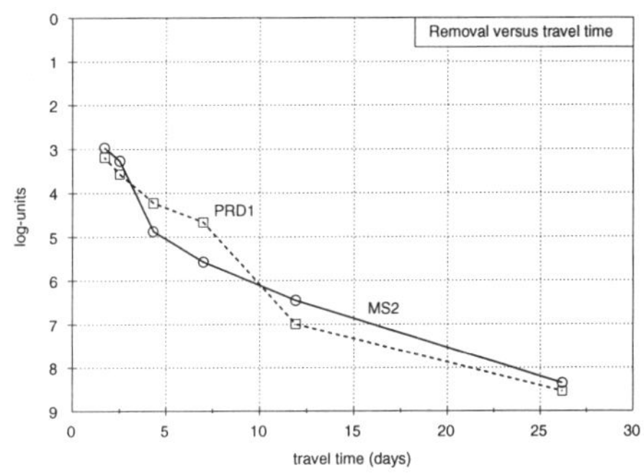

Figure 8. Removal (log units) of MS2 and PRD1 versus travel time (days)

9 DISCUSSION AND EVALUATION

When setting up an artificial recharge system, it is not just the microbiological aspects that have to be considered. There are also clear relationships to the mixing of water from various parts of the artificial recharge system, to micro-contaminants that are taken up from or left behind in the aquifer, to the removal of short-life isotopes (radioactivity), and to the behaviour of other constituents associated with suspended matter. This project is just concerned with the aspect of microbiological reliability.
The experiment described is one of a series intended to determine the processes that form the barrier to pathogens (is it time, distance, fineness of the sand, redox, temperature or a combination of these factors?). There is also the question of whether the required minimum residence time in days (or distance in metres) is determined by the coarseness of the sediment and whether the problem with well recharge is the same as that with surface spreading. After all, in the case of surface spreading, the water passes a biologically active interface between water and soil that is known to be very efficient at removing micro-organisms, as is shown by research into slow sand filtration. The measurements described here indicate a removal rate of 3 log units between the basin and the first monitoring well at 1.4 m from the bank. Moreover, in the case of surface spreading the flow is often through fine sand (aeolian deposits) and a sequence of redox environments. In the case of well recharge, there may be recharge and recovery in the same, possibly coarse, layer.
More research is needed to interpret the results in terms of the behaviour of human viruses and the set-up of artificial recharge systems required for virological reliability. Other work is in preparation (Schijven et al. in prep). It is an in-depth study of the modelling, simulation and explanation of the behaviour (binding, desorption, inactivation) of micro-organisms.

10 ACKNOWLEDGEMENTS

The description relates to a joint study by Kiwa Research and Consultancy, the National Institute of Public Health and the Environment, and PWN Waterworks of North-Holland. The Kiwa part of the study is being conducted as part of the research programme of the Dutch water companies and receives financial support from the European Union, within the scope of the 'Environment and Climate Research Programme' (contract ENV4-CT95-0071). The part conducted by the National Institute of Public Health and the Environment is commissioned by the Ministry of Housing, Spatial Planning and Environment. Thanks are extended to the many workers who helped make the experiment a success: the people who built the test arrangements, carried out the experiments, took the samples, made the analyses and processed the data.

REFERENCES

Bales, R.C., S. Li, K.M. Maquire, M.T. Yahya & C.P. Gerba 1993. MS2 and poliovirus in porous media: hydrophobic effects and chemical perturbations. *Water Resources Research*, 29, pp 957-963.

Jin, Y, M.V. Yates, S.S. Thompson & W.A. Jury 1997. Sorption of viruses during flow through saturated sand columns. *Environmental Science Technology*, 31, pp 548-555.

Knoll, K.H. 1982. Persistenz anthropogener Mikroorganismen in Porengrundwasserleitern. *Forum Städte Hygiene*. Jahrgang 33, p 270-275.

Knorr, M. 1937. Die Schutzzonenfrage in der Trink-wasserhygiene. *Das Gas- und Wasserfach*, 80 Jahrgang, p 330-334 & 350-355.

Knorr, M. 1951. Zur hygienischen Beurteilung der Ergänzung und des Schutzes grosser Grundwasservorkommen. *Wasser*, 92 Jahrgang, p 104-110 & 151-155.

Nobel, P.J. & D. van der Kooij 1997. Verwijdering van pathogene micro-organismen uit water door middel van bodempassage, enkele achtergronden en uitgangspunten. Kiwa-report SWI 97.141, in Dutch.

Peters, J.H., Q.L. Slings & A. Stakelbeek 1992. Open infiltratie Nieuwe Stijl. Tijdschrift *H_2O*, volume 25, no 19, p 532-537, in Dutch.

Peters, J.H. 1996. Are there any blueprints for artificial recharge? *Proceedings of an International Symposium on Artificial Recharge of ground water*, Helsinki, Nordic Hydrological Programma NHP report no 38.

Powelson, D.K., C.P. Gerba & M.T. Yahya 1993. Virus transport and removal in wastewater during artificial recharge. *Water Resources Research*, 27, pp 583-590.

Schijven, J.F., G.J. Medema, A.C.M. de Nijs & J.G. Elzenga 1996. Emissie en verspreiding van Cryptosporidium, Giardia en enterovirussen via huishoudelijk afvalwater. RIVM-report 289202014.

Schijven, J.F., et al. (in prep). Modelling of transport of bacteriophages MS2 and PRD1 during artificial recharge.

Vogelaar, A.J., M.J.C. van Baar, J.H. Peters & J. Bergsma 1997. Gedrag van pathogene micro-organismen bij bodempassage, verslag van de NaCl-tracerproef in december 1996 te Castricum, Kiwa report SWI 97.115, in Dutch.

Yahya, M.T., L. Galsomies, C.P. Gerba & R.C. Bales 1993. Survival of bacteriophages MS2 and PRD1 in ground water. *Water Science Technology*, 27, pp 409-412.

1.7 Hydrological aspects

Artificial Recharge of Groundwater, Peters et al. (eds) © 1998 Balkema, Rotterdam, ISBN 90 5809 017 5

Predicting infiltration and mounding, and managing problem soils

H. Bouwer
US Water Conservation Laboratory, ARS-USDA, Phoenix, Ariz., USA

ABSTRACT: Artificial recharge of groundwater with fresh water and sewage effluent can be expected to increase worldwide. Techniques are presented to predict infiltration rates of soils for recharge system design, and to predict where groundwater must be pumped and to what depth to prevent undue groundwater rises in the recharge area. Less permeable soils can also be used for recharge if carefully managed.

1 INTRODUCTION

Artificial recharge of groundwater with fresh water or sewage effluent must be increasingly used to meet the water demands of increasing populations in the face of uncertain global climate changes which may produce more weather extremes like excessive rain, droughts, tornados, and unusual temperatures. More long-term storage will be necessary to store water in times of water surplus for use in times of water shortage. More water also must be reused. Underground storage by artificial recharge is receiving more and more interest because the traditional way of storing water behind dams is not always possible and certainly not always as "glamorous" as it used to be when "taming the river" was the thing to do. Dams are being increasingly criticized for ecological, environmental, socio-cultural, public health, evaporation, health, economic, safety, and other reasons, including poor sustainability prospects because all dams eventually silt up (Pearce 1992; Knoppers & Van Hulst 1995; and Devine 1995). Thus, integrated water management must increasingly include artificial recharge for conjunctive use of surface water and groundwater, and water reuse (Bouwer 1997). Of all the liquid fresh water in the world, already less than 2% occurs as surface water in streams and lakes and more than 98% as groundwater (Bouwer 1978) and there is lots of room for more of the latter! When sewage effluent is used for artificial recharge of groundwater, additional benefits are obtained due to the physical, chemical, and biological processes that go on in the underground environment and that give the effluent "soil-aquifer treatment" (SAT) or "geopurification". Also, the aquifer offers seasonal or longer storage benefits to absorb differences between supply and demand of effluent water, and it breaks the pipe-to-pipe connection of water reuse which is very important where there are cultural or religious objections against water reuse. Even in the United States, direct potable water reuse is not considered acceptable and indirect reuse where the treated effluent first goes through surface water or groundwater is recommended (McEwen & Richardson 1996). The groundwater route is preferred because, in contrast to the surface water route, there are no evaporation losses, the water is not vulnerable to secondary contamination by animals or humans, and there is no algae growth. Besides, recharge gives underground treatment (SAT) and storage benefits.

2 SURFACE SYSTEMS

The preferred recharge systems for surface water and sewage effluent are shallow basins or other surface infiltration systems. The systems can be constructed in streambeds or flood plains (in-channel systems) using T-dikes or dams to spread the water in the channel, or they can be constructed off-channel with a system of shallow basins. Each basin should be hydraulically independent so that it can be flooded, dried, and cleaned according to its best schedule. If sewage effluent is used, it usually has had primary and secondary treatment and some disinfection before recharge. Sometimes nitrification-denitrification and sand filtration (tertiary treatment) are also given. In some parts of the world, primary treatment may be all that is available (Rice & Bouwer, 1984). Permeable soils like fine to coarse sands or even sandy gravels are preferred for artificial recharge of high quality water (fresh or reclaimed), especially for large flows which would otherwise require a lot of land. Such soils could infiltrate 100 to 300 m per year, including drying times for infiltration recovery and cleaning. However, if SAT benefits need to be maximized, finer soils like loamy sands or sandy loams may be preferable. Accumulated infiltration may then be about 50 to 100m per year. Groundwater tables below infiltration systems should be relatively deep, like 2 or 3m or more.

With the increasing interest in artificial recharge, suitable permeable soils will not always be available and finer textured soils like sandy loams and even light loams may have to be used. Such soils may only infiltrate 20 to 50 m/yr. Considering that evaporation losses from a year-round system may range from 0.5 m/yr in cool temperate climates to 2 m/yr in warm dry climates, the systems are still very efficient with evaporation losses often less than 5%. However, they will require a lot of land, especially for large flows, so that careful soil investigations are necessary (including surveys, backhoe pits, borings, and infiltration tests with cylinder infiltrometers and test basins) to estimate land requirements and project feasibility. A quick, simple test for single cylinder infiltrometers has recently been developed (Bouwer et al. 1998). In some cases, water needs to be stored underground for long periods (decades) to meet future demands. Estimates may then also have to be made of how much water can be stored or "banked" underground, and where groundwater must eventually be pumped and to what level to prevent waterlogging of the recharge area (Bouwer et al. 1998).

3 CYLINDER INFILTROMETERS

After promising soils have been identified by studying soil maps and doing on-site soil inspections, "wet" infiltration tests should be performed. These were typically done with metal, cylinder infiltrometers about 0.3m in diameter. However, use of such small infiltrometers can seriously overestimate large-area infiltration rates because of lateral flow (divergence) around the cylinder due to capillary suction in the soil (Bouwer 1986; Bouwer et al. 1998). Double ring or "buffered" infiltrometers are not much better because the divergence also causes over-estimation of infiltration in the center portion of the cylinder. The obvious approach then has been to use larger infiltration test areas like 3 x 3m bermed areas, where divergence or "edge" effects are less significant. These tests are laborious and they can also require large volumes of water because it may take more than a day to reach or approach "final" infiltration rates. A better approach would be to use conventional single cylinders with significant water depth to speed up the infiltration process so that tests can be completed in a relatively short time (a few hours, for example), and to correct the results for water depth in cylinder, limited depth of wetting, and lateral flow in the soil below the cylinder to calculate the hydraulic conductivity K of the wetted zone, which then should be about equal to the long-term infiltration rate of a large, basin-size area.

The infiltrometers used for this procedure are single steel cylinders 0.6m in diameter and 0.3m high with beveled edge (Fig. 1) that are driven or pushed about 2cm into the soil. The cylinder is carefully filled with the proper water to the top, and clock time is recorded. Water is allowed to drop about 5 to 10cm, the drop is measured with a ruler, clock time is recorded, and the cylinder is filled again to the top. This is repeated for about 6 hours or when the accumulated infiltration has reached about 50cm, whichever comes first. The last drop y_n of the water level is measured and clock time is recorded to obtain the time increment Δt_n for y_n. A shovel is used to dig outside the cylinder to determine the distance x of lateral wetting or divergence (Fig. 1). The infiltration rate i_n inside the cylinder during the last water level drop is calculated as $y_n/\Delta t_n$. The corresponding downward flow rate or flux i_w in the wetted area below a cylinder of radius r is then calculated as

$$i_w = \frac{i_n \pi r^2}{\pi (r + x)^2} \tag{1}$$

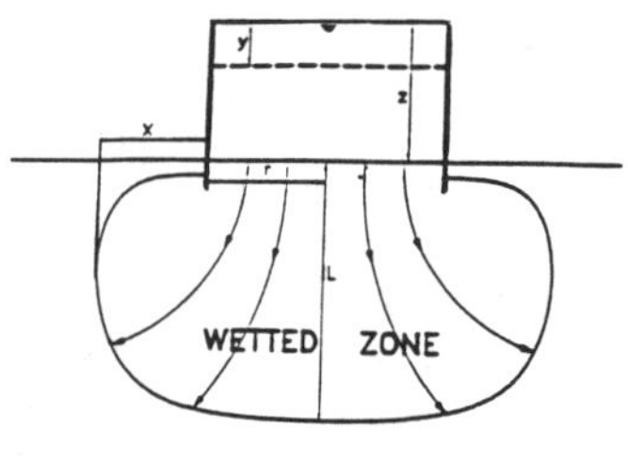

Figure 1. Geometry and symbols for single-ring infiltrometer.

The depth L of the wet front at the end of the test is calculated from the accumulated drop y_t of the water level in the cylinder as

$$L = \frac{y_t \pi r^2}{n \pi (r + x)^2} \tag{2}$$

where n is the fillable porosity of the soil. The value of n is estimated from soil texture and initial water content. For example, n may be about 0.3 for dry uniform soils, 0.2 for moderately moist soils, and 0.1 for relatively wet soils. Well-graded soils would have lower values of n than uniform soils. The value of L can also be determined by digging down with a shovel right after the test and seeing how deep the soil has been wetted. This works best if the soil initially is fairly dry and there is good contrast between wet and dry soil, and there are no rocks. Applying Darcy's equation to the downward flow in the wetted zone then yields

$$i_w = \frac{K (z + L - h_{we})}{L} \tag{3}$$

where z is the average depth of water in the cylinder during the last water level drop. The term h_{we} is the water-entry value of the soil and it is used to estimate the suction at the wet front as it moves downward. The water-entry value is about half the air-entry value (due to hysteresis) and may be estimated as follows (in cm of water; see also Leij et al. 1997, and references therein):

coarse sands	-5
medium sands	-10
fine sands	-15
loamy sands - sandy loams	-25
loams	-35
structured clays (aggregated)	-35
dispersed clays	-100

Since K is now the only unknown in Equation 3, it can be solved for K as

$$K = i_w L / (z + L - h_{we}) \tag{4}$$

This calculated value of K can be used as an estimate of long-term infiltration rates in large and shallow inundated areas, without clogging of the surface and without restricting layers deeper down. Because of entrapped air, K of the wetted zone is less than K at saturation, for example, about 0.5 of K at saturation for sandy soils, and about 0.25 of K at saturation for finer soils. If the K-values calculated with Equation 4 look promising for an infiltration system, the next step is to put in some test basins of about 0.2 ha each for long-term flooding to evaluate clogging effects and potential for infiltration reduction by restricting layers deeper down. In field tests, good agreement has been obtained between predicted infiltration rates (K in Equation 4) and those of larger basins. For example, six infiltrometers installed in a field west of Phoenix, Arizona gave an average K of 40cm/day. Two test basins of 0.3 ha each in the same field yielded final infiltration rates of 30 and 35cm/day (Bouwer et al. 1998). If the infiltrometer tests give infiltration rates that are too low for surface infiltration systems to be feasible, alternate systems such as excavated basins, recharge trenches, recharge shafts or vadose zone wells, or aquifer wells can be considered.

The above procedure is by no means exact. However, in view of spatial variability (vertical as well as horizontal) of soil properties, exact procedures and measuring water level drops very accurately with vernier equipped hook gages are not necessary. The main idea is to account somehow for divergence and limited depth of wetting, rather than applying a flat reduction percentage to convert cylinder infiltration rates to long-term large area infiltration rates, as has sometimes been done. Because of spatial variability, cylinder infiltration tests should be carried out at various locations within a given site. Finally, the resulting infiltration rates should never be expressed in more than two significant figures!

4 GROUNDWATER MOUNDS

Numerous papers have been published on the rise of groundwater mounds on the aquifer in response to infiltration from a recharge system, and some also on the fall of the mound after infiltration has stopped (Glover 1964; Hantush 1967; Marino 1975 a,b, Warner et al. 1989).

Of most interest to operators and managers, often, is the long-term effect of recharge on groundwater; i.e. where will the groundwater mound be 10, 20 or 50 years from now, how much water can be stored or "banked" underground, will the whole recharge area become waterlogged, will infiltration rates decrease, will the groundwater rise into landfills or basements or buildings, or become too high below cemeteries, and how and where must the water be recovered from the aquifer to prevent undue groundwater rises in the recharge area? Computer models can be used to simulate regional recharge inputs and pumped well outputs for the aquifer. However, a quick idea about ultimate or quasi-equilibrium mound heights can be obtained from a steady state analysis where the mound is considered to be in equilibrium with a constant water table at some depth and at some distance from the infiltration system. The constant "far-away" water table is then obtained by groundwater pumping, discharge into surface water like rivers or lakes, or some other "control".

Usually, recharge systems consist of a number of basins or other infiltration facilities. Steady state equations were developed for two general geometries of the entire recharge area (Bouwer et al. 1998): 1. the basins form a long strip with a length of at least 5 times the width so that it performs about the same as an infinitely long strip (Glover 1964), and 2. the basins are in a round or square or irregular area that can be handled as an equivalent circular area. For the long strip (Fig. 2) the groundwater flow away from the strip was taken as linear horizontal flow (Dupuit-Forchheimer flow). Below the infiltration area, the lateral flow was assumed to increase linearly with distance from the center. The lateral flow was then assumed to be constant between the edge of the recharge system at distance W/2 from the center and the constant control water table at distance L_n from the edge (Fig. 2). This yielded the equation

$$H_c - H_n = \frac{iw}{2T}\left(\frac{W}{4} + L_n\right) \tag{5}$$

for the ultimate rise of the groundwater mound below the center of the recharge strip when there is equilibrium between recharge and pumping from the aquifer (Bouwer et al. 1998). The symbols in this equation are: i is the average infiltration rate in recharge area (total recharge rate divided by total area) and T is the transmissivity of the aquifer for the average aquifer thickness. The other symbols are as shown in Figure 2.

For a round or square recharge area, the groundwater flow will be radially away from the area. The equilibrium height of the mound below the center of the recharge system above the constant groundwater table at distance R_n from the center of the recharge system then can be calculated with radial flow theory (Bouwer et al. 1998) as

$$H_c - H_n = \frac{iR^2}{4T}\left(1 + 2\ell n\frac{R_n}{R}\right) \tag{6}$$

where R is the radius or equivalent radius of the recharge area, R_n is the distance from the center of the recharge area to the water-table control area, and the other symbols are as defined in Figure 2.

Perched groundwater mounds can form on layers of reduced permeability that restrict the downward flow in the vadose zone. For large infiltration systems and restricting layers of large horizontal extent, the flow will be vertically downward and the equilibrium height of the perched mound can be calculated as (Bouwer et al. 1998)

$$L_p = L_r \frac{\frac{i}{K_r} - 1}{1 - \frac{i}{K_s}} \tag{7}$$

where L_p = height of perching mound above restricting layer

L_r = thickness of restricting layer

i = infiltration rate

K_r = hydraulic conductivity of restricting layer

K_s = hydraulic conductivity of soil above restricting layer

Often, i will be much smaller than K_s because surface soils are finer textured than deeper soils, or there is a clogging layer on the surface soil that reduces infiltration. Also, i often will be much larger than K_r. For these conditions, Equation 7 can be simplified to $L_p = i\,L_r/K_r$.

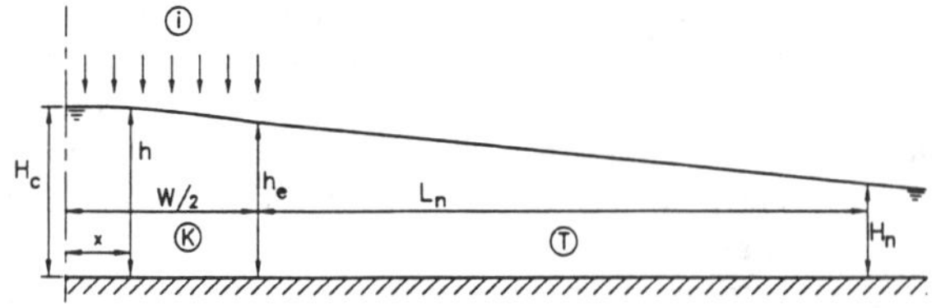

Figure 2. Geometry and symbols for groundwater mound below long infiltration area (strip).

Heights of perched groundwater mounds below small infiltration areas and/or above non-continuous perching layers (lenses etc.) will be less than L_p calculated with Equation 7 because there will then also be lateral flow in the mound.

5 CLOGGING LAYERS

Clogging layers consist of organic and inorganic materials that were suspended in the water and accumulated on the bottom of recharge basins. Also, algae and biofilms that can grow on the bottom and restrict infiltration. Clogging layers develop even if the water for recharge is very clear and contains essentially no nutrients (N and P) or organic carbon, particularly on relatively fine soils. The thickness of the clogging layer may range from less than 1 mm to more than several dm. Since their permeability is a lot less than that of the soil material on which they form, clogging layers reduce infiltration rates. As a matter of fact, the entire management of infiltration basins for groundwater recharge is often dominated by controlling the clogging layer. If it builds up too much and reduces infiltration rates too much, basins are dried so that the clogging layer cracks and curls up. When the basin is flooded again, infiltration rates will then be high again, only to go down when clogging develops again. Periodically, the clogging layer must be removed by scraping or "shaving" the bottom. Tilling the soil to work the clogging layer into the ground gives temporary relief, but in the long run the whole tillage layer could become a clogging layer and must be removed in its entirety to restore infiltration rates. Clogging layers are com-pressible by water depth, especially when they are mostly organic. Shallow basins are then preferred because the clogging layer then tends to be loose and more permeable (Bouwer & Rice 1989). Because of compression, increasing the water depth in recharge basins has actually produced decreases in infiltration rates!

Finer soils are more vulnerable to clogging than coarser soils, because their pores are smaller and easier to plug. They can also exhibit fine particle movement where clay and fine silt particles move down a small distance (maybe less than 1 mm) by the infiltrating water to form a mini-clay layer that restricts further downward flow. This "wash-out wash-in" phenomenon and other crust forming processes are well documented in the soils literature (Sumner & Stewart 1992) and should be understood for optimum basin management. More research also is needed, for example, on the effect of rolling or otherwise compacting and smoothing the soil after cleaning and disking or harrowing the soil at the end of a dryup. Mild compaction and smoothing may reduce fine-particle movement so as to not start a wash-out wash-in process that already restricts infiltration at the start of a new flooding period.

Other management aspects of recharge on "challenging" soils include avoiding using water that has a tendency to disperse the clay particles due to the combination of a low salt content and a high sodium adsorption ratio, avoiding exposure of the soil to rain impact that could disperse the clay and move fine soil particles to form crusts, use of gypsum or other soil amendments to keep clay particles in a coagulated or aggregated condition (Bouwer 1978), and breaking up "hard-setting" soils as can occur for sandy loams and similar mixtures of sand and clay. After prolonged wetting, drying these soils can then form extremely hard surfaces that make disking or harrowing essentially impossible. Such tillage should then be done while the soil is still slightly moist. Basins in soils containing clay and fine silt must be designed and managed to avoid all possible soil erosion, because once the fine particles become suspended in the water, they take a long time to settle out so that muddy water can spread all over the basin. The last particles to settle out are also the finest, so that the entire wetted area of the basin then gets covered with a thin layer of fines that reduces infiltration rates. Thus, banks or berms should be vegetated or covered with soil cement or other material to avoid any erosion due to rainfall runoff or wave action. Inlet structures should be designed with low inflow rates and/or energy dissipaters to keep overland flow rates when filling the basin below erosive values.

While clogging layers reduce infiltration rates and recharge capacities, they also function as active biofilters that can significantly improve the quality of the water as it passes through the layer, especially if this water is of low quality as in recharge with treated sewage effluent or contaminated surface water. First of all, the clogging layer is a physical filter than could retain small suspended particles, including the larger microorganisms like cryptosporidium. Because of its high organic carbon content and microbiological activity, the clogging layer also is an effective biological filter. Since oxygen levels in the clogging layer may decrease from aerobic at the top to anaerobic at the bottom, there is a whole range of oxygen conditions whereby different organic compounds can be degraded, including disinfection byproducts and other more refractory compounds through secondary utilization and cometabolism. Nitrate can be denitrified. Clogging layers can also be effective in constructed wetlands for treatment of sewage effluent. The clogging layer then not only reduces the infiltration and flow of water to underlying groundwater, it also improves the quality of that water. Thus, "leaky" wetlands may combine effluent treatment and groundwater recharge benefits.

The beneficial effects of drying and removing clogging layers on infiltration rates are thus at the expense of a decrease in water quality improvement. Thirty years ago this was already observed at the Flushing Meadows Project in Phoenix, Arizona, where fecal coliform levels in groundwater about 3m below effluent infiltration basins always showed definite peaks of several 100 per 100 ml about 10 days after a new flooding period was started and the newly infiltrated water had reached the intake of the well which was at a depth of about 10m (Bouwer et al. 1974). As the clogging layer developed with continued flooding, fecal coliform levels declined to essentially zero in a few weeks. Antagonistic effects from increased bacterial activity in the soil itself after flooding was resumed may also have played a role. Other biological and chemical constituents may have had similar breakthroughs to the aquifer after the clogging layer was removed or dried up. The challenge thus is to manage clogging layers so that their treatment functions are maintained while their clogging effects are controlled to acceptable levels.

6 REFERENCES

Bouwer, H. 1978. Groundwater Hydrology. Mc Graw-Hill, Inc., New York, New York, 480 pp.

Bouwer, H. 1986. Intake rate: Cylinder infiltrometer. Methods of Soil Analysis, Part 1, Physical and Mineralogical Methods, Agronomy Monograph No. 9 (2nd Edition) Am. Soc. Agron. - Soil Sci. Soc. Am., Madison, Wisconsin, 825-855.

Bouwer, H. 1997. Role of groundwater recharge and water reuse in integrated water management. Arabian J. for Science and Eng. 22(1C):123-131.

Bouwer, H., & R.C. Rice. 1989. Effect of water depth in groundwater recharge basins on infiltration rate. J. Irrig. and Drain. Eng. 115(4): 556-568.

Bouwer, H., J.C. Lance, & M.S. Riggs. 1974. High-rate land treatment II: Water quality and economic aspects of the Flushing Meadows Project. J. Water Poll. Contr. Fed. 46(5):844-858.

Bouwer, H., J.T. Back & J.M. Oliver. 1998. Predicting infiltration and ground water mounding for artificial recharge. Submitted to *Ground Water*.

Devine, Robert A. 1995. The trouble with dams. *Atlantic Monthly*, August 1995, p. 67-73.

Glover, R.E. 1964. Ground water movement, US Bureau of Reclamation, Engin. Monogr. 31,67 pp.

Hantush, M.S. 1967. Growth and decay of ground water mounds in response to uniform percolation. Water Resour. Res. 3:227-234.

Knoppers, R. & W. van Hulst. 1995 De Keerzijde van de Dam. Jan van Arkel, Utrecht. The Netherlands, 142 pp.

Leij, F.J., W.B. Russell, & S.M. Lesch. 1997. Closed-form expressions for water retention and conductivity data. Groundwater 35:848-858.

Marino, M.A. 1975a. Artificial ground water recharge, I. Circular recharging area, J. Hydrol. 25:201-208,

Marino, M.A. 1975b. Artificial ground water recharge, II. Rectangular recharging area, J. Hydrol. 26:29-37.

McEwen, B. & T. Richardson. 1996. Indirect potable reuse: committee report. Proc. 1996 Water Reuse Conference, Am. Water Works Assoc. and Water Environment Fed., San Diego, California, Feb. 1996, p. 486-503.

Pearce, F. 1992. The Dammed. The Bodley Head, London, 276 pp.

Rice, R.C. & H. Bouwer. 1984. Soil-aquifer treatment using primary effluent. *J. Water Pollut. Control. Fed.* 56(1):848.

Sumner, N.E. & B.A. Stewart (eds.). 1992. *Soil Crusting: Chemical and Physical Processes*, Lewis Publishers, Boca Raton, Florida, 372 pp.

Warner, J.W., D. Molden, M. Chehata, & D.K. Sunada, 1989. Mathematical analysis of artificial recharge from basins. Water Resources Bull. 25:401-411.

Artificial Recharge of Groundwater, Peters et al. (eds) © 1998 Balkema, Rotterdam, ISBN 90 5809 017 5

Tracking the movement of recharge water after infiltration

Y. Fujita, J. Zhou, E. Orwin & M. Reinhard
Department of Civil and Environmental Engineering, Stanford University, Calif., USA
M. L. Davisson & G. B. Hudson
Isotope Sciences Division, Lawrence Livermore National Laboratory, Calif., USA

ABSTRACT: An approach was developed and tested to track the subsurface movement of water infiltrated from a recharge basin. Conservative isotope tracers ($^{18}O/^{16}O$ and $^{3}H/^{3}He$) and selected chemical water quality parameters were monitored to differentiate between waters of different sources. A preliminary evaluation of the method was conducted at a field site in Orange County, California, USA. The isotopic results were used as conservative tracers to validate conservative and non-conservative transport of the water quality constituents. In particular sulfate, nitrate and ethylenediaminetetraacetic acid (EDTA) concentrations resulted in estimated travel times to wells, however, only sulfate appeared to behave conservatively in this system.

1 INTRODUCTION

One of the major challenges facing artificial recharge practitioners is the tracking of water once it has been introduced into the subsurface. Monitoring the movement of subsurface water and changes in its quality is particularly important when the water used for recharge is of wastewater origin. Tracing water quality in the subsurface is necessary in the State of California because of proposed regulations that stipulate minimum residence times and travel distances for managed recharge with reclaimed municipal wastewater (Asano 1993). However, tracking this water quality with time is difficult because of complex subsurface flowpaths and mixing with water already present in the aquifer (i.e. local groundwater). In Orange County, California the major source of recharge water is the Santa Ana River (SAR), which receives a high load of tertiary-treated municipal effluents. The Orange County Water District (OCWD) controls river flows to maximize infiltration, and also utilizes several off-river basins to supplement the riverbed recharge capacity.

Stanford University has extensive experience in characterizing the quality of Orange County's water resources (McCarty & Reinhard 1980, Reinhard et al. 1994), and recently developed methods using water quality data to track the fate of infiltrated water constituents in the Orange County groundwater basin. The approach involves monitoring selected chemical water quality parameters that differentiate water sources. In particular, xenobiotic trace organic contaminants such as ethylenediaminetetraacetic acid (EDTA), nitrilotriacetic acid (NTA), and alkylphenol ethoxycarboxylates (APECs) serve as unique indicators of wastewater origin (Fujita et al. 1996). Some of these compounds may be of potential health concern. These compounds were previously identified in wastewater effluents and receiving waters (Ahel et al. 1987, Alder et al. 1990, Xue et al. 1995), and were also detected in treated effluents and groundwater in Orange County (Fujita et al. 1996). Other parameters include dissolved organic carbon (DOC), sulfate, and nitrate, which are routinely measured by water supply agencies. While these latter measurements are not necessarily indicative of wastewater-derived organic pollution, they could be used to demonstrate regulatory compliance with the proposed subsurface residence times.

In addition, isotopic data of $^{18}O/^{16}O$ ratios and ^{3}H-^{3}He ages were generated by Lawrence Livermore National Laboratory (LLNL). These data provide ultra-conservative tracers of molecular water and a basis to quantify fate and transport of water quality parameters. An experiment to test the approach was conducted in Orange County in late 1996, utilizing an off-river infiltration basin. Several monitoring and production wells are available in the immediate vicinity of the basin. The SAR water that is ordinarily diverted to the basin was replaced for 50 days with imported Colorado River water (COR). The COR water has a different $^{18}O/^{16}O$ than SAR water and has a distinctly different water quality character. The goal of the experiment was to determine if it was possible to follow the chemical "signature" of the COR water after infiltration.

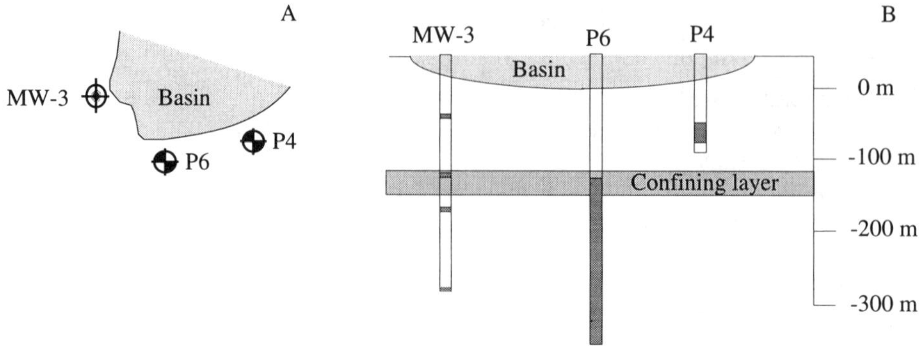

Figure 1. Schematic of study area; not to scale. A. Relative location of wells. B. Vertical profile of well screening depths.

2 METHODS

Samples from the recharge basin, SAR, two production wells (P4 and P6), and four levels of a multi-level monitoring well (MW-3) were analyzed for the study. A schematic of the study site is shown in Figure 1. All three of the wells are located within approximately 50 meters of the basin periphery. The screening depths are listed in Table 1. There is a hydrogeologic confining layer approximately 120 m below the basin. Samples for water quality measurements were collected monthly by OCWD personnel, and shipped overnight on ice to Stanford. Samples were collected for analysis of ^{18}O and ^{3}H in glass, air-tight sealed bottles. The sampling for noble gases required a copper tube pinch-clamp assembly that sealed a water sample in a copper tube free of any gas bubbles. The sampling frequency for the isotopic analyses varied depending on the calculated ^{3}H-^{3}He ages (Table 1).

2.1 *Methods for Chemical Parameters*

Upon arrival at Stanford, all samples were immediately filtered through 0.2 μm nylon filters, and stored at 4 °C until analysis to inhibit microbial activity. Concentrations of DOC and the specific organic compounds were measured using methods described previously (Fujita & Reinhard 1997, Fujita et al. 1996). Nitrate and sulfate concentrations were measured by ion chromatography at Stanford. Nitrate measurements for the well water and the recharge basin samples collected in October 1996 were performed at OCWD.

2.2 *Methods for Isotopic Analyses*

The ^{18}O was measured with a standardized CO_2 equilibration method (Epstein & Mayeda 1953), and the purified CO_2 was measured on a VG Prism isotope ratio mass spectrometer at LLNL. Results are

Table 1. Screening depths, estimated ages based on $^{3}H/^{3}He$ measurements, and $\delta^{18}O$ values for the wells prior to the experiment.

Well	Screening Depth (m)	Estimated Age (years)	$\delta^{18}O$
P4	60-87	< 1	-7.53
P6	131-360	18	-8.10
MW-3/1	61-67	< 1	-7.33
MW-3/2	137-143	3	-7.58
MW-3/3	177-183	5	-8.03
MW-3/4	273-279	25	-7.73

reported as per mil deviation from a standard in δ notation where

$$\delta^{18}O = \left[\frac{^{18}O/^{16}O}{\left(^{18}O/^{16}O\right)_{std}} - 1 \right] * 1000 \;. \qquad (1)$$

The $(^{18}O/^{16}O)_{std}$ is the isotopic ratio of the U. S. National Bureau of Standards' "Standard Mean Ocean Water". Tritium was analyzed by the helium-accumulation method (Surano et al., 1992). Noble gases were cryogenically extracted from water collected in copper tubes and the isotopic abundance of helium was measured on a VG-5400 noble gas mass spectrometer at LLNL. Ages are calculated using methods similar to Schlosser et al. (1988).

3 RESULTS

3.1 *Initial Conditions*

Previous isotopic work (Davisson et al. 1996) showed that the $\delta^{18}O$ value of COR water (-11.1 ±0.2 per mil) in Orange County was distinctly lower than local groundwater (-7.5 ±1.0 per mil), providing an excellent conservative water tracer. The ^{3}H-

Table 2. Water quality data for Colorado River (COR), Santa Ana River (SAR), and for well P4 and well P6 prior to the experiment.

Parameter	COR [a]	SAR [b]	P4 [c]	P6 [c]
DOC (mg-C/L)	2.2 ± 0.2	5.5 ± 2.1	1.9	0.7
Sulfate (mg/L)	262.6 ± 20.5	111.6 ± 27.1	102.0	141.0
Nitrate (mg/L)	0.5 ± 0.3	21.8 ± 8.8	13.9	17.2
EDTA (μg/L)	0.1 ± 0.2	7.0 ± 3.8	1.1	*n. d.*
NTA (μg/L)	*n. d.*	0.19 ± 0.18	*n. d.*	*n. d.*
APEC (μg/L) [d]	0.2 ± 0.2	1.7 ± 1.3	*n. d.*	0.2

[a] Average of six samples ± one standard deviation, for October 1996 to March 1997.
[b] Average of thirty samples ± one standard deviation, for October 1994 to April 1997.
[c] Average of two samples, August and September 1996.
[d] Total alkylphenol ethoxycarboxylates, with propylated base ions of *m/z* 235, 249 and 275 (Ding et al.).
n.d. Not detected, at detection limit of 0.1 μg/L.

^{3}He ages previously measured indicated ages <1 year for MW-3/1 and P4 (Table 1). Ages of the remaining sample sites ranged from 3 to 25 years and generally correlated with depth. Table 2 contains a summary of the previous water quality data for the COR, SAR, P4, and P6 water sources. Well MW-3 was not completed until shortly before the initiation of the experiment, and therefore pre-experiment data for this well were not available.

The COR and SAR water exhibited significant differences in chemical composition. Sulfate concentrations in the COR water were approximately twice as high as in the SAR water. Nitrate concentrations were more than an order of magnitude less in the COR water than in the SAR water. EDTA concentrations were around the detection limit (0.1 μg/L) in the COR water, but on the order of 5 μg/L in the SAR water. DOC, NTA and APEC concentrations were generally higher in the SAR water than in the COR water, but varied significantly, such that the lower values in the SAR water approached the high values in the COR water. This variability rendered these parameters less suitable as specific source indicators.

Both of the production well waters prior to the experiment exhibited lower DOC concentrations than the surface waters. DOC removal is known to occur during groundwater recharge by infiltration, and is attributed primarily to biological activity (Quanrud et al. 1996). Sulfate and nitrate concentrations at P4 and P6 were similar to those observed in the SAR water. EDTA was detected in both of the pre-experiment P4 water samples, but not in the P6 water. Similar to the surface waters, NTA was just at or below the detection limit in the groundwater samples. APEC was detected in just one of the P6 samples, but not in the P4 samples.

3.2 *Experimental Results*

The COR water was diverted into the recharge basin starting October 1, 1996, and was present in the basin for 50 days. A total of approximately 7.4 million m^3 of COR water was percolated. Immediately prior to the introduction of the COR water, all of the SAR water was removed from the basin, and no additional SAR water was added until the COR water was fully percolated. The $\delta^{18}O$ measurements (Fig. 2) indicated that values for the groundwater at P4 and MW-3/1 decreased in ≤30 days to the isotopic value of the recharged COR water. These isotopic values persisted for approximately 30 days before returning to initial values. MW-3/2 after approximately 6 months decreased to a $\delta^{18}O$ value suggesting a 50% mixture of COR water with local groundwater. The remaining samples showed no significant isotopic variation, suggesting a negligible amount of COR water recharged the sampling sites during the experiment. This latter result is consistent with the older ^{3}H-^{3}He ages for these locations.

Figures 3 - 5 show the results of the analyses for sulfate, nitrate, and EDTA, respectively, at the various sampling sites. Within 1 month after the introduction of the COR water, water quality analyses indicated a change in water quality at P4 and MW-3/1. In P4 an increase in sulfate above the initial 102 mg/L was observed after 2 weeks, and by day 45 sulfate levels had risen to >250 mg/L, similar to the recharged COR water. Between approximately 75 and 105 days, the sulfate concentrations in P4 and MW-3/1 returned to the initial concentration. This change in sulfate was similar to the $\delta^{18}O$ results, suggesting little or no dilution with native groundwater, and a conservative behavior. Nitrate concentrations in P4 and MW-3/1 also clearly showed the arrival of the COR water, decreasing after 2 weeks and reaching levels similar to the COR water within 45 days. Nitrate returned to initial concentra-

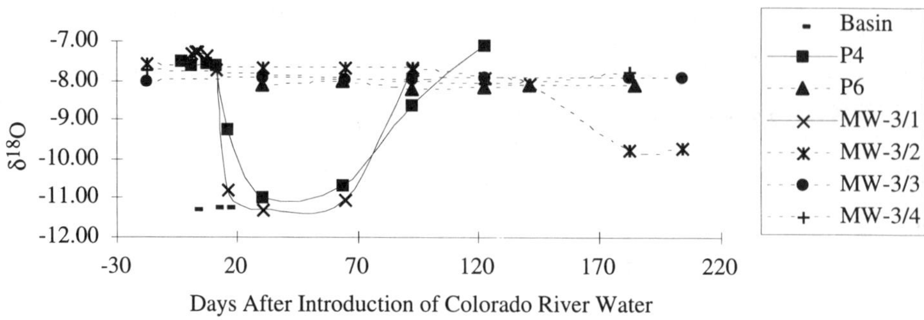

Figure 2. $\delta^{18}O$ values at sampling sites during experiment.

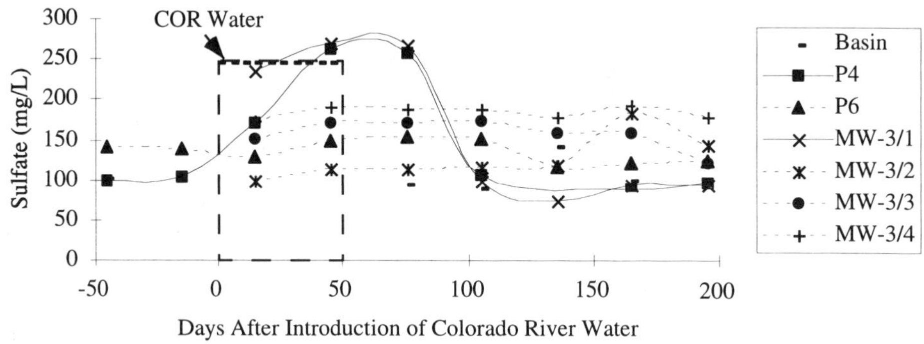

Figure 3. Sulfate concentrations at sampling sites during experiment.

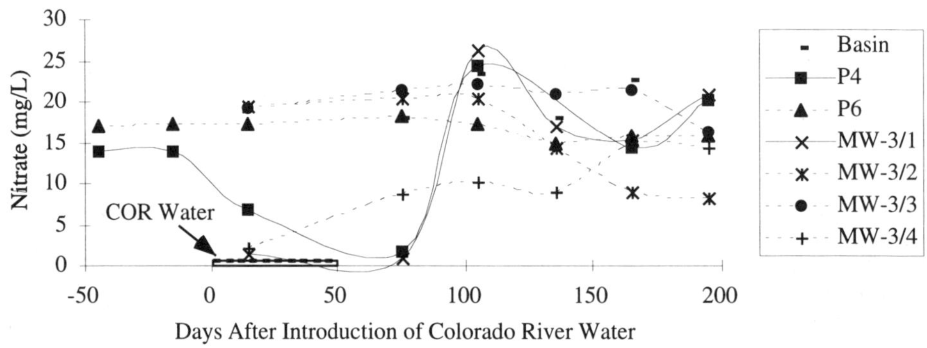

Figure 4. Nitrate concentrations at sampling sites during experiment.

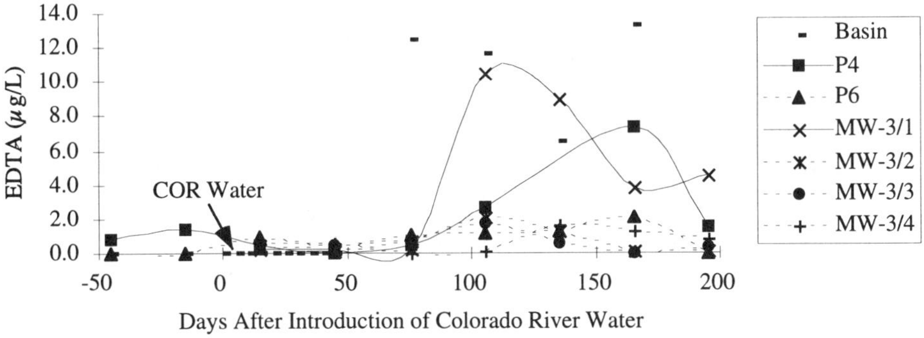

Figure 5. EDTA concentrations at sampling sites during experiment.

tions by 105 days. Finally, EDTA concentrations in P4 decreased to <0.5 µg/L during the period when COR water was recharging this well, and returned to approximately initial concentrations by 105 days. Concentrations in P4 and MW-3/1 both continue to increase, but then did not co-vary afterwards.

MW-3/2 showed an increase in sulfate and decrease in nitrate and EDTA approximately 165 days after the start of the experiment. However, the sulfate concentrations reached only approximately 180 mg/L, rather than the 250+ mg/L levels observed at MW-3/1 and P4. This suggested that significant dilution occurred at MW-3/2. Using the 50% mixture implied by the $\delta^{18}O$ results (Fig. 2), an expected sulfate concentration would be 187 mg/L. The actual sulfate concentration measured at 165 days was 183 mg/L, indicating further the conservative behavior of sulfate in this system. Similar calculations for nitrate result in an estimated concentration of 11.2 mg/L, as compared to the measured value of 8.9 mg/L. With EDTA, the discrepancy was even greater, with a calculated value of 3.55 µg/L and a measured value of 0.1 µg/L. For both nitrate and EDTA, large standard deviations for the measurements over time in the COR and SAR waters (Table 2) may affect the results, but generally it appears that these parameters, particularly EDTA, are not conservative tracers.

As expected from the $\delta^{18}O$ results, Well P6 and the third and fourth levels of MW-3 did not show any of the changes in water quality indicative of the COR water. Except for some variability in nitrate concentrations at MW-3/4, the sulfate, nitrate, and EDTA concentrations at these sampling sites did not change significantly.

4 CONCLUSIONS

The results of this initial experiment suggest that chemical water quality parameters can be used to monitor the movement of recharged water in aquifers. For the two source waters tested, sulfate, nitrate and EDTA were the most distinctive source-specific indicators. However, only sulfate appeared to be unambiguously conservative in this system, as confirmed by the isotopic measurements. Nitrate measurements were more variable within each recharge component, but may also be subject to denitrification after recharge, as has been observed in other systems (Wilson et al. 1995). The behavior of EDTA was of particular interest, because EDTA concentrations were lower in the groundwater than would be expected if it was a conservative tracer. In addition, EDTA concentrations differed significantly between wells P4 and MW-3/1, even though $\delta^{18}O$ values co-varied. This finding suggests that some removal of EDTA possibly occurred during subsurface transport. Mechanisms of removal are unknown. While microbial degradation of EDTA has been shown to occur in some laboratory tests (Nörtemann 1992, Lauff et al. 1990, Tiedje 1977), other investigators have reported little or no degradation (Hinck et al. 1997, Bolton et al. 1993). EDTA is generally very resistant to biological sewage treatment, and previous research has indicated that it is not significantly removed in aerobic infiltration recharge zones (Egli et al. 1990). Removal of EDTA by sorption onto sediments is a possible removal mechanism (Nowack & Sigg 1996). Further research investigating subsurface behavior of EDTA is recommended, as are additional tests under various recharge conditions to verify the chemical indicator approach described here.

5 ACKNOWLEDGMENTS

The Orange County Water District provided technical support and funding for this work. In addition, work was performed under the auspices of the U.S. Department of Energy by the Lawrence Livermore National Laboratory under contract W-7405-Eng-48.

6 REFERENCES

Ahel, M., T. Conrad & W. Giger. 1987. Persistent Organic Chemicals in Sewage Effluents. 3. Determinations of Nonylphenoxy Carboxylic Acids by High-Resolution Gas Chromatography/Mass Spectrometry and High-Performance Liquid Chromatography. *Environ. Sci. Technol.* 21 (7):697-703.

Alder, A. C., H. Siegrist, W. Gujer & W. Giger. 1990. Behaviour of NTA and EDTA in Biological Wastewater Treatment. *Water Res.* 24 (6):733-742.

Asano, T. 1993. Proposed California Regulations for Groundwater Recharge with Reclaimed Municipal Wastewater. Water Sci. Technol. 27: 157-164.

Bolton, J., H., S. W. Li, D. J. Workman & D. C. Girvin. 1993. Biodegradation of Synthetic Chelates in Subsurface Sediments from the Southeast Coastal Plain. *J. Environ. Quality* 22:125-132.

Davisson, M. L., G. B. Hudson, R. L. Herndon, S. Niemeyer & J. Beiriger. 1996. Report on the feasibility of using isotopes to source and age-date groundwater in the Orange County Water District's Forebay region, Orange County, California. Lawrence Livermore National Laboratory, UCRL-ID-123953, 31 pp.

Ding, W.-H., J. Wu, M. Semadeni & M. Reinhard. Chemical Wastewater Indicators in Groundwater Recharged with Reclaimed Wastewater: Santa Ana River, California. *In preparation.*

Egli, T., M. Bally & T. Uetz. 1990. Microbial Degradation of Chelating Agents Used in Detergents with Special Reference to Nitrilotriacetic Acid (NTA). *Biodegradation* 2:121-132.

Epstein, S. & T. Mayeda. 1953. Variation of the 18O Content of Waters from Natural Sources. *Geochim.Cosmochim. Acta* 4:213-224.

Fujita, Y., W.-H. Ding & M. Reinhard. 1996. Identification of Wastewater Dissolved Organic Carbon (DOC) Characteristics in Reclaimed Wastewater and Recharged Groundwater. Water Environ. Res. 68: 867-876.

Fujita, Y. & M. Reinhard. 1997. Identification of Metabolites from the Biological Transformation of the Nonionic Surfactant Residue Octylphenoxyacetic acid and its Brominated Analog. *Environ. Sci. Technol.* 31 (5): 1518-1524.

Fujita, Y., W.-H. Ding, J. Zhou, E. Orwin, T. Rivlin & M. Reinhard. 1997. Draft Final Report: Organic Contaminant Behavior During Groundwater Recharge with Reclaimed Wastewater and Santa Ana River Water: A Laboratory and Field Evaluation. Submitted to Orange County Water District, October 1997.

Hinck, M. L., J. Ferguson & J. Puhaakka. 1997. Resistance of EDTA and DTPA to Aerobic Biodegradation. *Water Sci. Technol.* 35 (2/3):25-31.

Lauff, J. J., D. B. Steele, L. A. Coogan & J. M. Breitfeller. 1990. Degradation of the Ferric Chelate of EDTA by a Pure Culture of an *Agrobacterium* sp. *Appl. Environ. Microbiol.* 56 (11):3346-3353.

McCarty, P. L. & M. Reinhard. 1980. Trace Organics Removal by Advanced Wastewater Treatment. *J. Water Pollut. Control Fed.* 52 (7):1907-1922.

Nörtemann, B. 1992. Total Degradation of EDTA by Mixed Cultures and a Bacterial Isolate. *Appl. Environ. Microbiol.* 58 (2):671-676.

Nowack, B. & L. Sigg. 1996. Adsorption of EDTA and Metal-EDTA Complexes onto Goethite. *J.Colloid Interface Sci.* 177:106-121.

Quanrud, D. M., R. G. Arnold, L. G. Wilson & M. H. Conklin. 1996. Effect of Soil Type on Water Quality Improvement During Soil Aquifer Treatment. *Water Sci. Technol.* 33 (10-11):419-431.

Reinhard, M., W.-H. Ding & Y. Fujita. 1994. Organic Carbon Characterization of Advanced Treated Wastewater at Water Factory 21, Orange County. Department of Civil Engineering, Stanford University.

Schlosser, P., M. Stute, H. Dörr, C. Sonntag & K. O. Münnich. 1988. Tritium/^{3}He dating of shallow groundwater, *Earth, Planet. Sci. Lett.*, 89, 353-362.

Surano, K. A., G. B. Hudson, R. A. Failor, J. M. Sims, R. C. Holland, S. C. MacLean & J. C. Garrison. 1992. Helium-3 Mass Spectrometry for Low-Level Tritium Analysis of Environmental Samples. *J. Radioanal. Nuclear Chem.-Art.* 161 (2):443-453.

Tiedje, J. M. 1977. Influence of Environmental Parameters on EDTA Biodegradation in Soils and Sediments. *J. Environ. Qual.* 6 (1):21-26.

Wilson, L. G., G. L. Amy, C. P. Gerba, H. Gordon, B. Johnson & J. Miller. 1995. Water Quality Changes During Soil Aquifer Treatment of Tertiary Effluent. *Water Environ. Res.* 67 (3):371-376.

Xue, H., L. Sigg & F. G. Kari. 1995. Speciation of EDTA in Natural Waters: Exchange Kinetics of Fe-EDTA in River Water. *Environ. Sci. Technol.* 29 (1):59-68.

Artificial Recharge of Groundwater, Peters et al. (eds) © 1998 Balkema, Rotterdam, ISBN 90 5809 017 5

Optimisation of artificial groundwater recharge system with infiltration galleries

C.K.Vidanaarachchi, Y.Zhou & J.C.Nonner
International Institute for Infrastructural, Hydraulic and Environmental Engineering (IHE), Delft, Netherlands

S.E.Meijer
Water Company of Gelderland, Velp, Netherlands

ABSTRACT: The Water Company of Gelderland has considered to implement a system of artificial recharge with infiltration galleries to compensate the groundwater extraction to preserve the natural environment. A 3-D generic groundwater simulation model with an option for particle tracking analysis coupled with an optimisation procedure was used to design and optimise the infiltration-pumping system. Two systems of artificial recharge, the closed system and the parallel system, have been designed and analysed. The water level in and the size and length of galleries were varied to determine the required infiltration rate. The number, location and rate of pumping wells were optimally determined with the coupled simulation-optimisation model. The results shown that both systems can achieve the target infiltration and pumping rate while satisfying the two design criteria for drawdown and travel time.

1 INTRODUCTION

The Water Company of Gelderland is responsible for the supply of drinking water in the province Gelderland, The Netherlands. The source for the water supply is groundwater. One of the pumping stations is located in Vorden which is surrounded by natural forests and farm lands. Allowable extraction rate for the pumping station at Vorden is three million cubic metres per year. But the company extracts less water because of negative impacts of groundwater level decrease on the environment.

To meet the ever increasing demand, it is not possible to increase extraction from the natural groundwater resource because over- extraction leads to the lowering of the groundwater level, which causes two main problems:
(1) influences on natural vegetation, marsh lands, springs, and brooks etc; and
(2) reduction of agricultural production.

One of the solutions is to compensate the extraction by means of artificial recharge (AR). Artificial recharge of groundwater using subsurface galleries was successfully implemented by RWW in the Ruhr valley near Essen-Kettwig of Germany (Bundermann 1993). The pre-treated Ruhr water is led to the galleries with a total length of 1250 m. The total infiltration capacity is 2000 m^3/hour. The system has been operated for 6 years without the decrease of infiltration capacity. The infiltration gallery system is suitable for the area with thin semi-permeable confining layer on the top of the main aquifer. The system has advantages of easy implementation, good protection from pollution, minimum or no evaporation loss, easy maintenance, and preservation of natural environment (landscape, natural groundwater heads, vegetation etc.). Therefore, the Water Company of Gelderland has considered to implement such a system in the Vorden area.

The main objective of this research is to design a system of infiltration galleries to maximise the extraction while satisfying the following two design criteria: (1) the changes to the natural groundwater level should be insignificant; and (2) the minimum underground passage of the infiltrated water should be more than 60 days before it is extracted.

The first criterion controls the impact on the natural environment and on the agricultural production of the area, while meeting the second criterion produces water free from pathogenic microorganisms.

A 3-D generic groundwater simulation model with

an option for particle tracking analysis coupled with an optimisation procedure was used to design and optimise the infiltration pumping system. The methodology has been developed and tested for optimal design of artificial recharge systems with open water pond (Jonoski et al. 1997). Two systems of artificial recharge galleries were design and optimised in this study.

2 DEVELOPMENT OF A SIMULATION MODEL

2.1 *Conceptual model*

Conceptual model has been developed using geological and hydrogeological data of the Vorden area. The main aquifer system in the area of Gelderland are two sandy layers of Pleistocene age: the eolian fine sand on the top, and the fluvioglacial and fluviatile coarse sand below. The ground surface has a slope of 0.06% from south east towards northwest and the tertiary base has a slope of 0.18% in the same direction. The section of the conceptual model is presented in Figure 2.1.

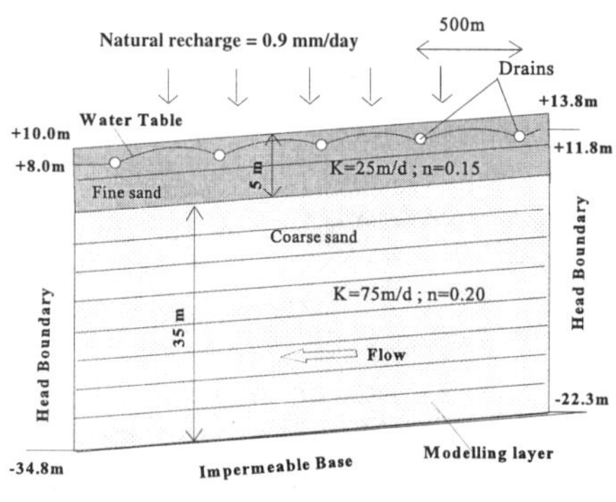

Figure 2.1. Conceptual model

2.2 *Modelling of natural groundwater flow*

For the setup of a flow model, a representative area of 6750m x 6750m has been considered, where Vorden pumping station is located in the centre. The distance from centre of the model to the boundary has been chosen such that no influence would occur on the boundary due to the operation of the artificial recharge system. The total area covered by the model is 45 km^2. Model boundaries are chosen as hydraulic boundaries: two streamlines in the north and the south were selected as no flow boundaries, and two equipotential lines in the east and the west were defined as constant head boundaries.

The model area was discretised by 67 gridlines in the x direction and 67 in the y direction, and the aquifer was divided into 10 modelling layers. The model has 44890 cells in total. In the central part of the model, where the infiltration galleries and the pumping wells are located, the model has a grid size of 20m x 20m and grid size gradually increase to 300m x 300m at the boundary. The actual hydrogeology of the area does not require fine vertical discretisation, but with the introduction of the 10 modelling layers, the position of the artificial recharge galleries and screens of the pumping wells can be precisely located and modelled. Hydrogeological parameters and the hydrological stresses (natural recharge and drainage) were taken as average values in the Vorden area. For the setup of this generic groundwater flow model the model code MODFLOW (McDonalad & Harabaugh 1988) was used in combination with the Processing Modflow, PMWIN (Chiang & Kinzelbach 1993). The package MODPATH (Pollock 1994) was used for the particle tracking calculations.

Natural transient model was built for the purpose of analysis of the response of the designed system under transient conditions. The time-variant specified-head package (Leake & Prudic 1991) was used to simulate eastern and western head boundaries under transient condition.

3 DESIGN AND ANALYSIS OF AR SYSTEMS

Design of AR system is a trial and error approach with the simulation model. During the design, the length of the infiltration galleries, position of the wells, and pumping rate of the wells were varied to obtain a solution satisfying two design criteria.

The drawdown criterion for the design was specified as ±2.5 cm at a distance of 500m from galleries. Using simulation model, drawdowns were calculated and checked to see whether they complied with the criterion. If the drawdown criterion was not satisfied, a satisfactory solution might be obtained by changing water level in the gallery. Once the drawdown criterion was satisfied, the travel time was calculated and checked with 60 days constraint. When the system satisfied both of these criteria, particle travel time distribution curve was constructed as an indicator of the consistency of extracted water quality.

With the aforementioned procedure, two systems of artificial infiltration galleries were identified and analysed. They are:

- the closed system of galleries; and
- the parallel system of galleries.

3.1 *Design of the closed system of galleries*

In the closed system, galleries formed a square shape and penetrated through the top fine sand layer to the coarse sand aquifer as a ditch of 2 m width (Fig. 3.1). Sides of the galleries were lined with concrete so that infiltration took place from the bottom of galleries. Nine pumping wells were located inside the system, each with a capacity of 1000 m^3/day, yielding the total target extraction rate of 9000 m^3/day. Well screens were placed in 8th and 9th modelling layers (approximately 24 m to 32 m below the surface). Schematic cross-section of the system is shown in Figure 3.2.

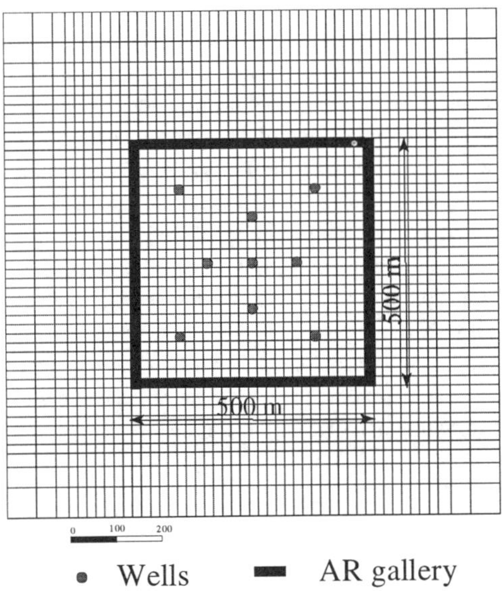

Figure 3.1. Layout of the closed system of gallery and wells.

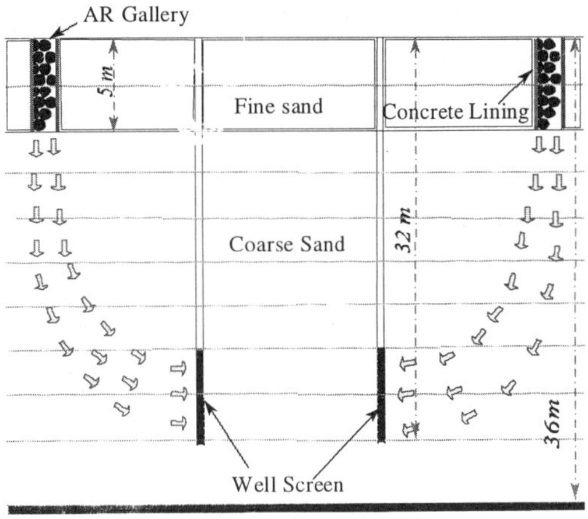

Figure 3.2. Schematic cross section of the closed gallery system along the direction of flow.

Simulation of galleries and pumping wells was carried out in 2 steps.

In the first step, galleries were introduced in the natural flow model and simulated by a very high hydraulic conductivity (1000 m/day). Pumping wells were not introduced yet. This simulation created the natural distribution of groundwater heads which were used as reference levels to calculate drawdowns caused by artificial recharge and pumping.

In the second step, galleries were simulated as an external source with General Head Boundary (GHB) package of MODFLOW. Water levels in the GHB cells were specified as the same natural heads simulated in the first step. Pumping wells were introduced to extract infiltrated water from galleries. The recharge from galleries was assigned to the top face of the 3rd layer so that infiltration takes place only from the bottom of the gallery. This simulates a system where sides of galleries are concrete lined.

The comparison of the water balance calculations for the natural flow and the artificial recharge and pumping (Table 3.1) shows that the extraction is supplied by the artificial recharge from the closed system of galleries and by the reduction of natural drainage. The particle travel time distribution (Fig. 3.3) shows that about 85% of the extracted water is of the age ranging from 60 days to 500 days. The mixing of water with different ages ensures the extracted water with consistent quality.

Table 3.1 Water balance of the closed system

Components	Units	In	Out
Boundaries	m^3/day	6855	10522
Drainage	m^3/day	-	27248
Natural recharge	m^3/day	31781	-
Artificial recharge	m^3/day	8134	-
Wells	m^3/day	-	9000
Total	m^3/day	46770	46770

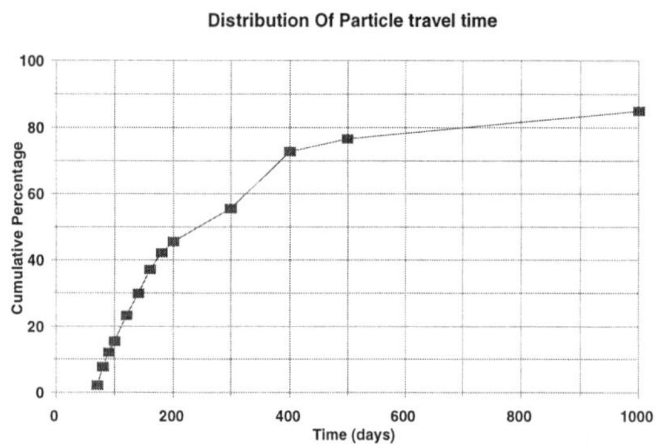

Figure 3.3 Distribution of particle travel time of the close system of galleries.

3.2 *Design of the parallel system of galleries*

In the parallel system, two separate line galleries, each of 500 m in length, were placed in parallel to each other at 500 m apart, perpendicular to the natural groundwater flow direction. Nine pumping wells were placed between two line galleries, each with a pumping rate of 1000 m^3/day.

Simulation of the parallel system of galleries and pumping wells was carried out with the same steps for the closed system. Water level in the two parallel galleries were varied step by step until the drawdown criterion was satisfied. When water level in the two line galleries was increased to 0.25 m above the natural groundwater level in the aquifer, the total recharge from the galleries was more or less the same as the total extraction rate (Table 3.2). Drawdown and travel time criteria were also satisfied. The particle travel time distribution is similar to the closed system of galleries.

Table 3.2 Water balance for the parallel system

Components	Units	In	Out
Boundaries	m^3/day	6778	10611
Drainage	m^3/day	-	28045
Natural recharge	m^3/day	31781	-
Artificial recharge	m^3/day	9097	-
Wells	m^3/day	-	9000
Total	m^3/day	47656	47656

4. OPTIMISATION OF AR SYSTEMS

4.1 *Formulation of the optimisation problem*

The major design aspects of the AR systems are the length and water level of the galleries, the number, location and rate of pumping wells. The previously described systems were designed purely on a trial and error approach. As indicated by Jonoski et al. (1997), coupling of a simulation model with an optimisation procedure is necessary to find the best design solution. During the optimisation, the length of the gallery and the water level in the gallery were kept at the same values as before. The variables during optimisation were then reduced to the number of wells, the position of the wells, and the extraction rate of the wells.

The groundwater management software package MODMAN (Greenwald 1994) was used to formulate the optimisation problem. It consists of an objective function and a set of constraints. The objective was to maximise the total extraction rate, i.e.

$$Maximise\ Z = \sum_{j=1}^{m} q_j \tag{1}$$

where q_j is pumping rate of each pumping well; and m is the total number of pumping wells.

The maximisation of (1) was subjected to the following constraints:

$$-2.5 < s_i < 2.5 \quad (2)$$
$$T_k \geq 60 \ days \quad (3)$$
$$q_j > 0 \quad (4)$$

where s_i is drawdown at control location i, which is predefined at 500 m distance from the gallery in layer one; n is the total number of control locations; T_k is travel time of the fastest particle to each pumping well.

The constraint on drawdown (2) is effectuated on the pumping rates through an additional equation:

$$s_i = \sum_{j=1}^{n} \beta_{ij} \, q_j \quad (5)$$

where β_{ij} is the response coefficient in terms of drawdown at control location i by a unit pumping rate at pumping well j. A matrix consisting of all β_{ij} is called a response matrix which was generated by the flow model based on the principle of linear superposition. Once the response matrix is known, drawdowns caused by any combination of pumping wells are easily calculated with (5).

The equations (1) to (5) forms a mathematical programming and was solved by an optimisation code LINDO (Schrage 1991). However, this optimisation procedure could be used only to solve the problem with drawdown constraints. Travel time constraints cannot be incorporated as a linear problem since travel time has a nonlinear response to the pumping rates (Jonoski et al.1997). To overcome the difficulty of incorporating the minimum travel time criterion, the following procedure was adopted.

Minimum travel time depends on the pumping rate and the distance between a well and the artificial recharge gallery. Therefore, in the first instance an optimal solution was found only with drawdown constraints. Then, the travel time was calculated for all selected pumping wells using the optimum (maximum) pumping rates. If the travel time is shorter than 60 days, pumping rate was reduced gradually to see the variation in travel time. If travel time cannot be controlled only by the pumping rate then some wells extracting water with short travel times were excluded from the potential well locations. This elimination procedure ultimately found a good design system satisfying all the design criteria.

4.2 *Optimisation of the closed system*

20 potential pumping wells were placed inside the closed system of galleries. 184 drawdown control locations were specified outside the galleries with a distance of approximately 500 m. Results of this optimal solution shown that some wells are extracting water with travel time less than 60 days and total pumping rate is 15291 m^3/day which is more than target rate.

In the next trial, wells that were found with low particle travel times were eliminated from the potential well locations. This resulted in a new optimal solution with the travel time to all the wells above 60 days. 10 pumping wells were selected to be active from the potential wells, yielding the maximum extraction rate of 11227 m^3/day. The water balance calculation shown that nearly 97% of the extraction was supplied by the artificial recharge and the remaining was from the reduction of the natural drainage.

4.3 *Optimisation of the parallel system*

For the optimisation of the parallel system the same drawdown locations were used as in the closed system. 25 potential pumping wells were placed between two line galleries. The optimal solution selected 10 pumping wells to be active. The maximum extraction rate was 11376 m^3/day. Minimum travel time to all pumping wells was more than 60 days except one well where minimum travel time was 58 days but very close to the criterion. Again, nearly 97% of the extraction was from the artificial recharge and 3% was from the reduction of the drainage.

5 COMPARISON OF THE TWO SYSTEMS

In order to compare the performance of the two systems, sensitivities of the two systems on hydraulic conductivity and porosity were analysed and the responses of the two systems to the transient conditions were evaluated. The results of the comparison are presented in Table 5.1.

The area required for the implementation of the two systems is very small. Both systems are most suitable for the areas where a main aquifer is covered by a thin impermeable or less permeable layer. The total length of the galleries in the closed system is the twice as that in the parallel system. However, in order to produce the same amount of artificial recharge, water level in the galleries of the parallel system has to be increased 0.25 m above the natural level.

Sensitivity analysis indicated that the hydraulic conductivity has high influence on the flow rate, artificial recharge and drawdowns of both systems. With increasing hydraulic conductivity, flow through boundaries and artificial recharge from the galleries are increased, but drawdowns become negative due to the increase of the artificial recharge. The hydraulic conductivity has small influence on the travel time in the closed system, but large influence in the parallel system. The travel time becomes shorter with higher hydraulic conductivity. Porosity influences only travel time. Travel time is shorter with smaller porosity.

Table 5.1 Comparison of the two systems

Aspects	The closed system	The parallel system
Coverage area	500 m x 500 m	500 m x 500 m
Length of the gallery	2000 m	1000 m
Depth of the gallery	5 m	5 m
Water level in the galley	Natural level	0.25 m above the natural level
Influence of hydraulic conductivity	high on flow rate; marginal on travel time	high on flow rate; large on travel time
Influence of porosity	no on flow rate; high on travel time	no on flow rate; high on travel time
Response to transient conditions	small violation on drawdowns; no on travel time	small violation on drawdowns; no on travel time

The operation of both systems has little influence on the seasonal variations of the natural groundwater flow since the system is designed such that it creates a local quasi-steady flow system which is superimposed on the natural regional flow system.

6 REFERENCES

Bundermann, G. 1993. Infiltraion von aufbereitetem flubwasseser uber sickerschlitzgraben. *Deutch-Niederlandischer workshop, Kunstliche grundwasseranreicherung. Castricum, Holland.*

Chiang, W.H & W. Kinzelbach 1993. *Processing Modflow, pre and post processors for simulation of flow and contaminants in groundwater systems with MODFLOW, MODPATH, MT3D*, version 3.0, Hamburg, Heidelberg.

Greenwald, R.M. 1994 *MODMAN, MODflow MANagement: An optimisation module for MODFLOW*, Version 3.0, Geotrans Inc.sterling, Virginia.

Jonoski, A. Y. Zhou, J. Nonner & S. Meijer 1997. Model-aided design and optimisation of artificial recharge pumping systems. *Hydrological Sciences Journal*: 42(6): 937-953.

Leak, L.K & D.E. Prudic 1991. *Documentation of a computer programme to simulate aquifer system compaction using the modular finite-difference groundwater flow model.* U.S. Geological Survey.

McDonald, M.G & A.W. Harbaugh 1988. *A modular three-dimensional finite-difference groundwater flow model (MODFLOW)*, USGS, Scientific Software Group, Washington D.C.

Pollock, D.W 1994. *User's guide for MODPATH / MODPATH plot*, version 3: A particle tracking post-processing package for MODFLOW, the USGS finite difference groundwater flow model, USGS, Reston, Virginia.

Schrage, L. 1991. *User's manual for linear, integer and quadratic programming with LINDO*, Release 5.0. The scientific press, San Francisco.

Artificial Recharge of Groundwater, Peters et al. (eds) © *1998 Balkema, Rotterdam, ISBN 90 5809 017 5*

Ground water management system in Vienna – An evaluation after three years of operation

J.E. Dreher & A. Gunatilaka
Verbundplan, Vienna, Austria

ABSTRACT: After the construction of Freudenau Danube hydro power plant, in order to maintain the ground water regime in the 2nd and 20th districts of Vienna (30 km² area) unaltered a ground water management system has been developed. Sealing walls along the river bank have been constructed and the ground water flow is regulated with the help of withdrawing and recharging wells placed on either side of the walls. The direct infiltration of contaminants from the Danube is prevented through continuous control of the water quality and regular monitoring in an elaborate net of ground water wells. The major changes in the ground water management area were associated with problems related to clogging processes in the river bed as well as with a general improvement of water quality. This paper presents an evaluation of the management system after three years of operation of the well systems.

1 INTRODUCTION

The fully automated ground water management system in the 2nd and 20th districts of the city of Vienna is in operation since three years. This system has been installed during the construction of the Freudenau hydro power plant on the Danube to prevent undesirable changes in the ground water quality and quantity. As a precaution a 13 km long, 12 to 40 m deep double walled sealing system was constructed along the right river bank. The ground water management system was designed to adjust ground water levels during high water periods and also to counteract the receding ground water table, a problem connected to river regulation (for details see Dreher & Gunatilaka 1996). It was designed to improve the overall ground water situation as well as to prevent flooding of existing houses (nearly 4000) in the densely settled areas. The proposed management system has also resulted in the revitalisation of the riparian woods in the region, which are threatened of drying up due to falling ground water level.

As the aquifer is used intermittently to feed the city water works, the maintenance and also improvement of ground water quality is an utmost necessity. Hence compensatory measures have been developed (e.g. re-oxygenation of the water before infiltration) and also an intensive water quality monitoring programme has been undertaken since 1989 (Gunatilaka & Dreher 1996).

One of the anticipated problems related to artificial recharge and withdrawal of ground water during the proposed management in the 2nd and 20th Vienna districts is connected with clogging processes of the river bed. The authors use the ground water model described below for the evaluation of the reduction of infiltration rates and leakage through the sealing walls as well as for bankfiltrate.

Changes in ground water quality are analysed statistically using the data collected from the monitoring network.

2 MANAGEMENT SYSTEM

The ground water management system is composed of following elements: a ground water model, recharging and abstracting wells, regulating wells and controlling wells. All these components of the ground water management system are connected to a control centre at the Freundenau power plant through a data transmission system.

The ground water level is controlled with the aid of 21 pairs of wells, an abstracting and an infiltrating well situated on each side of the sealing walls (Figure 1) constructed in a stretch of about 13 kilometres. Another four wells are installed for the supply of surface waters in the reparian woods.

The abstraction and the infiltration wells are of the same type, large bore wells of similar construction which have been specially designed to achieve higher pumping capacity. This wells have an overall bore hole diameter of 3 m and 800 mm filter pipe ID. The pumping capacities of the wells vary between 20 and 180 l s^{-1} depending on the local aquifer properties.

2.1 *Site description*

Enclosed between Danube and Danube Canal, the aquifer of the 2nd and 20th districts of Vienna is strongly influenced by the water level fluctuations of both running water systems. The aquifer shows high transmissivity near the Danube bank but has very heterogeneous permeability that vary between 10^{-4} and 0.04 m s^{-1} which has resulted as a consequence from the geomorphological activity of the former Danube. The shallow aquifer is lined at the bottom with tertiary impervious clay layers of variable thickness ranging from a few centimetres to some meters wich are located at a depth of 8 to 20 m (Dreher 1991). In the management area the amplitude of the ground water level fluctuations range between 3 and 4 meters near the Danube Canal and Danube respectively. Near the Danube banks in the whole 13 km stretch the corresponding infiltration rates can be of more than 5 m^3 s^{-1} per meter water level difference. The exfiltration from ground water into the Danube (upwelling) is rather less intensive. The ground water streams from the Danube to the Danube Canal giving rise to an average infiltration flow of 0.2 m^3 s^{-1}.

Ground water quality in this area shows local anthropogenic influences. High conductivity levels observed points out to both point and diffused sources. The most common contaminant that contribute to high conductivity is chloride that infiltrates into ground water mainly through street salting during the winter. Another possible source is leakage from abandoned factories and depots. Some of the observed contamination can be traced back to old waste dumps. Dissolved organic content in ground water is comparatively high, most probably as a consequence of elevated THC and VOC levels. Local contamination of aliphatic hydrocarbons, volatile and halogenated hydrocarbons are sometimes reported. During the construction of the power plant a number of hazardous waste dumps were removed and the area was subjected to a thorough clean up.

2.2 *Model approach*

For the simulation of the ground water flow for the whole project area a two dimensional ground water model was developed (Dreher 1991). This model for transient ground water flow is based on a representative data set that was derived from hydrographical data observed over a long period. With the aid of the model the development of an optimal management concept as well as the design of the well system (abstraction and recharging wells) was possible. The location of the well system (Figure1) as well as the abstraction and the infiltration rates (Table 1) necessary for the ground water management have been optimised through model simulation runs.

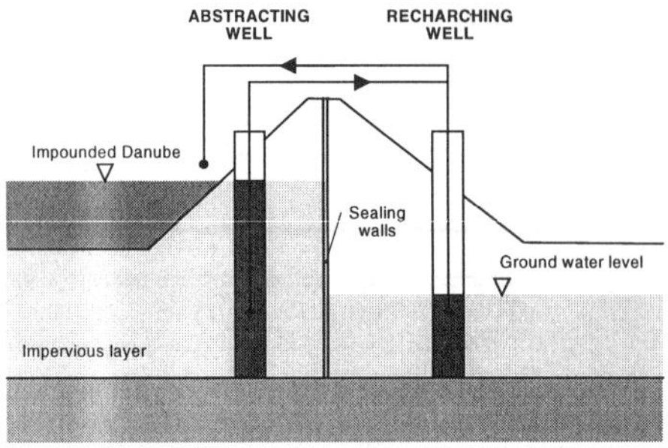

Figure 1. Schematic presentation of a well system installed in the ground water management area.

The management of the complete well system uses this ground water model, whereby the amount of infiltration water at the Danube banks is calculated from the observed flow of the impounded Danube. The pumping periods, direction of flow and rates are computed using observed and simulated behaviour of the groundwater based on calibrated aquifer properties. The quantity of water infiltrating across the river banks is calculated from the observed flow of the impounded Danube. This model is further employed for the controlling of the ground water table in the management area.

Table 1. Pumping rate statistics for infiltration and extraction. During management operations, depending on the ground water level, water is extracted from or infiltrated into the aquifer. The maximum extractable or rechargeable volume from 21 well pairs is given as the total.

Pumping procedure	Pumping rate (l s^{-1}) Min	Mean	Max	Total
Infiltration	15	47,9	80	1197
Extraction	15	20,1	25	422

2.3 *Water quality control*

The direct infiltration of contaminants from the Danube is prevented through continuous control of the water quality and regular monitoring in an elaborate net of ground water wells (190 ground water wells for both water quality and quantity measurements and 20 surface water stations). A total of six on-line stations are used for the continuous monitoring of water quality (chemical and biological) of which three are equipped for ground water and the rest for direct river water control (Gunatilaka & Dreher 1996). In case of an accident or the presence of a pollution plume the continuous analysis and installed alarm systems will prevent pumping of a contaminated bank filtrate into the aquifer.

3 GROUND WATER MANAGEMENT AFTER THREE YEARS OF OPERATION

3.1 *Management strategies*

The optimisation of the management strategies was obtained through the ground water flow model. The minimum recharging time step was found to be of one day while the infiltration rates were kept constant. After evaluation of the recorded data of the ground water management using the ground water model a minimum recharging time step of six hours was found to give better results. This gives the possibility of more automatic control of the recharging process and at the same time also to avoid over pumping during low water periods, where minimum ground water level has to be maintained.

3.2 *Problems related to bankfiltration*

Two re-calibration runs of the model were necessary during the three years of operation. This was carried out by integrating observed water level data, pumping rates and precipitation records into the model. Through transient model simulations (time step of 6 hours) it was possible to estimate clogging of the river bed and the reduction of pumping capacities of the abstracting wells during the period of partial filling of the reservoir (beginning with Feb. 1996).

The result of the model calibration mentioned above shows the distribution of the infiltration rates beneath (tertiary sandy layers) and through the installed sealing walls along the right river bank. The first step was to introduce a starting leakage factor into the model as a boundary condition at the beginning of the phase of partial filling of the reservoir. Figure 2 shows the initial exchange flow between Danube and the aquifer behind the sealing walls (the large infiltration rates at km 1931,35 follow from a defective sealing wall stretch that has been subsequently replaced).

The next step was to describe the rate of decay of the infiltration rates through clogging processes and, experimentally fit the model to the observed data by changing the decay constant.

Figure 3 shows the development of the clogging processes on the river bed from the beginning of the phase of partial filling of the reservoir of the Freudenau power plant for the duration of one year. The sealing up process is depicted as the decay of the total infiltration rates per meter of water head difference against the time (in months). These results were calculated with the aid of the above mentioned ground water model.

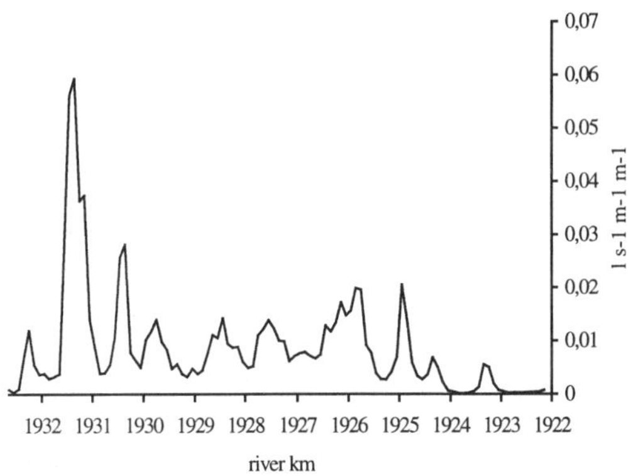

Figure 2. Specific infiltration rate at the right bank of the Danube at the beginning of the partial filling of the reservoir (initial boundary conditions). The ground water management extends from river km 1922 – 1932,5.

The best fit of the model runs to the observed ground water level data was achieved with a zero order decay rate, assuming the decreasing permeability of the corresponding river bank stretch. The results obtained corrected for effect of water temperature on viscosity are given in Figure 3, which shows an increase of permeability during the summer period. The average river water temperature during winter and summer are 5°C and 20°C respectively.

From the results of the model simulation it can also be stated that the reduction on permeability also depends on the depth of the water profile in the reservoir which varies from 2,8 to 8,3 metres in relation to the mean water level. Here an additional reduction of the leakage factor of 2% per metre head difference may be assumed.

Even if the infiltration decay at the river bank is of more than 50% in one year (Figure 3) the same cannot be assumed for the reduction of the specific yield of the abstraction wells. It also has to be emphasised that the recharging is a discontinuous process.

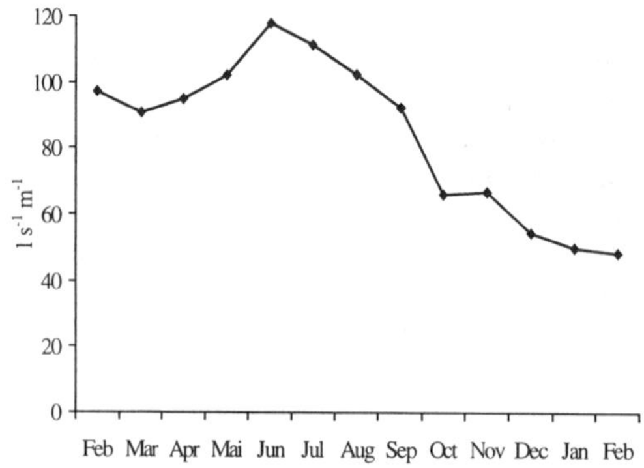

Figure 3. Total specific infiltration rate along the right river bank during one year after the partial filling of the hydro power reservoir. The gradual reduction of the infiltration rate is a consequence of the intensifying of the clogging processes in the river bed.

During the two years after partial filling of the Freudenau reservoir one can observe a decreasing of only 20% to 30% in the specific yield of the wells. The explanation may be the existence of different flow pathways under the river bed and the bank. An on going study at the Technical University of Vienna should throw more light on these aspects.

3.3 *Ground water quality*

A total of 120 monitoring wells with a long water quality monitoring record are located in the management area of ca. 30 km^2. During the period from 1994 – 1997, they were sampled regularly (monthly or biweekly) and analysed for major constituents (standard methods), trace metals (AAS and ICP) and organic micro pollutants (GC, GC-MS, HPLC). The main results as a statistical summary (median values) are listed in Table 2.

During bank filtration, the river water undergoes a number of changes, where microbial processes are playing a major role. It appears that the dissolved organic content in the Danube water is sufficient to support these degradation processes. These processes are temperature and substrate dependent and may consume an appreciable amount of oxygen, leading to oxygen depletion. The reduced conditions can trigger denitrification processes resulting in comparatively higher ammonium concentrations. Matthess (1983) and Carbiener & Tremolieres (1990), Gunatilaka et al. (1994) reported that river waters in general during the passage through the river banks are subjected to a strong clean up as some of the DOC, NH_4^+ and NO_3^- are retained or liberated as gaseous products during microbial transformations.

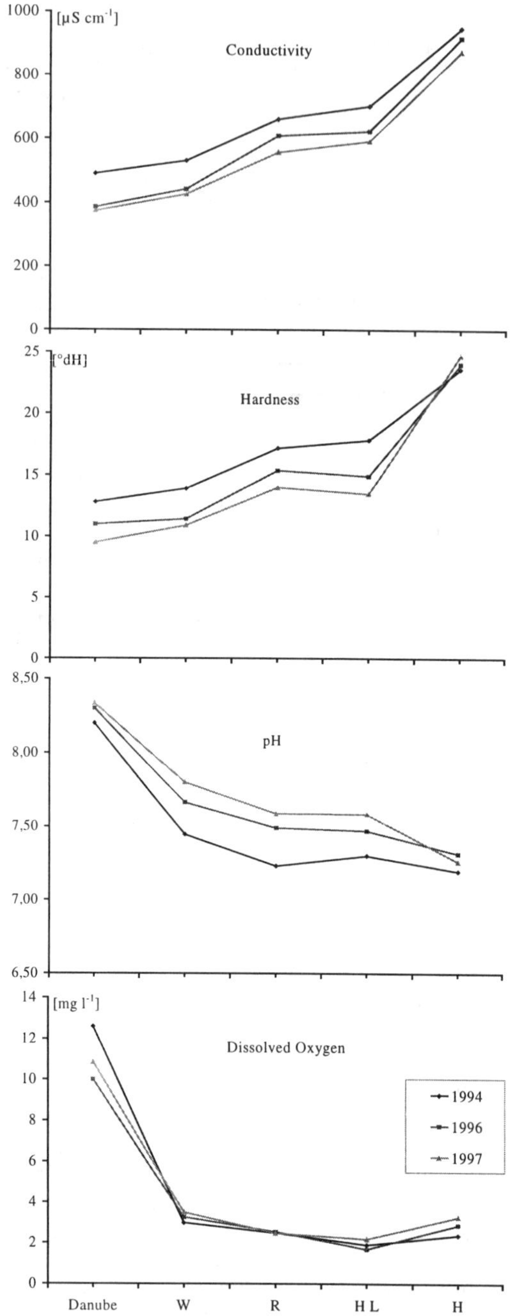

Figure 4. Conductivity; Hardness, pH and Dissolved Oxygen – distribution in the ground water management area (2nd and 20th Districts of Vienna), between Danube and Danube Canal. Data represented here are median values for the respective years for Danube river, W = abstraction wells, R = recharging wells, HL = Wells 50 m away from recharging wells and H = remaining wells from the hinterland (downgrading wells)

Table 2. Concentration of major constituents and some selected pollutants (median values) in the Danube (river km 1932) and bankfiltrate(river km 1928 - 1932)

Constituents	unit	Danube 1994 (n=12)	Danube 1997 (n=15)	Bankfiltrate 1994 (n=24)	Bankfiltrate 1997 (n=12)
Ca^{2+}	mg/l	62	64	110	55
Mg^{2+}	mg/l	12	12	24,5	12
Na^{+}	mg/l	8,6	7,65	22,84	8,8
K^{+}	mg/l	2,3	2,35	4,91	2,55
HCO_3^-	mg/l	195	198	330	189
SO_4^{2-}	mg/l	28	25,5	78	26,2
Cl^-	mg/l	15	13,5	42,5	14,2
NH_4-N	mg/l	0,09	0,105	0,012	<0,01
NO_3-N	mg/l	2,35	2	6,99	1,96
PO_4-P	mg/l	0,045	0,027	0,024	0,078
F^-	mg/l	0,175	0,14	<0,18	<0,1
Fe	mg/l	0,2	0,3	<0,02	<0,01
Mn	mg/l	0,015	0,02	<0,02	<0,01
DOC	mg/l	3	4,3	1,26	1,1
$KMNO_4$ cons.	mg/l	12	12	2,8	2,3
Al	µg/l	0,001	0,001	<0,02	<0,01
As	µg/l		0,001	0,0011	0,0035
Hg	µg/l	<0,0002	<0.0002	<0,0009	<0.0002
B	µg/l		0,08	<0,08	0,19
Cd	µg/l	0,0002	<0,0001	<0,0002	<0,0002
Pb	µg/l	0,0007	<0,0001	<0,0014	<0,0001
Cr	µg/l	0,0006	<0,001	<0,03	<0,001
Cu	µg/l	0,002	0,002	<0,0014	0,002
Ni	µg/l	0,001	0,001	<0,0011	<0,001
Zn	µg/l	0,004	0,0055	<0,02	<0,01
Alachlor	µg/l	<0,1	<0,1	<0,05	<0,05
Aldrin	µg/l	<0,03	<0,03	<0,01	<0,01
Atrazin	µg/l	<0,1	<0,1	<0,09	<0,05
Chlordan (isomers)	µg/l	<0,1	<0,1	<0,09	<0,02
Dieldrin	µg/l	<0,03	<0,03	<0,018	<0,01
Heptachlor	µg/l	<0,1	<0,1	<0,09	<0,01
Heptachlorepoxide	µg/l	<0,1	<0,1	<0,09	<0,02
Hexachlorobenzene	µg/l	<0,01	<0,01	<0,013	<0,01
Total PAHs (5 ind.)	µg/l	<0,028	<0,0285	<0,10	<0,3
PCBs (6 individual)	µg/l	<0,01	<0,01	<0,10	<0,01
Total bacteria	col./ml	2950	2050	134	66
Total coliforms	in 100ml	2200	2900	0	0
Faecal coliforms	in 100ml	13	445	0	0
Faecal streptococci	in 100ml	220	60	0	0

A comparison between the Danube, extraction wells (W), recharging wells (R) and downgrading wells (HL and H) show that the chemical characters of the ground water changes gradually during the underground passage (Table 2 & Figure 4). Through implementation of management procedures, the system is continuously flushed to maintain predefined water levels in the aquifer (regulation with the ground water model), which in turn lead to a chain of interlinked chemical changes, through an increase in the oxygen levels, redox-potentials and the pH. The results also show that some of the pollutants and pathogens are completely eliminated during the aquifer passage.

After three years of artificial recharge activity, the concentration of the major constituents in the bank filtrate show a clear decrease and the changes are summarised as follows:

– calcium, magnesium and bicarbonate concentrations have nearly halved leading to decrease in hardness,

– sulphate and chloride concentrations have dropped to one third that of the Danube water,

– the above changes (major cations and anions) have contributed to a gradual drop in the electrical conductivity (Figure 4),

– ammonium concentrations show a clear reduction during the passage (Table 2); the average ammonium concentrations recorded in three on-line ground water quality monitoring stations located in the management area, ranges between 3-5 µg/l,

– compared to Danube water, iron and manganese are found only as traces,

– DOC concentration in the bank filtrate drops nearly to one third and permanganate consumption drops four fold respectively in comparison to that of the Danube, showing that they are either degradated or mineralised. Detailed analysis of the DOC fractions showed that a large fraction is refractory, the major constituent being humic substances,

– most of the trace metals are retained during the passage but there is a slight enrichment of geogenic components such as arsenic,

– pesticide and herbicide concentrations as well as other micro organic pollutants (VOCs and PCBs) are rather low in the Danube water and the results show that there is no recurrence in the bank filtrate,

– the microbiological quality of the bank filtrate improves tremendously with twenty fold reduction of the total bacterial counts and total elimination of coliforms and streptococci.

Sub-oxic conditions prevailing in the aquifer are responsible for the break down of nitrates; microbiological processes and adsorption to clay fractions would have been mainly responsible for elimination of ammonium. Microbial processes also help in breaking down of some of the organic micro pollutants. They could be either degradated or absorbed (see Matthes 1983, Hem 1970).

Absorption of metals during the underground passage is common and hence low metal concentrations in the bank filtrate. However, slow desorption of for e.g. PO_4^{3-}, As, Cd and Ni may take place during or long travel passage (long retention time); in the management area, the retention time, depending on permeability varies between 3 – 6 years.

4 CONCLUSION

1. During three years of operation of the described ground water management system, optimisation of the management guidelines have been undertaken. This was achieved through improvements in the step time for the regulation of ground water recharge.

2. The management procedures show positive developments through maintenance of pollution prevention in the aquifer. The results confirm that there is no infiltration of inorganic or organic pollutants.
3. Furthermore there is no contamination through pathogens showing improvement of the microbiological quality of the ground water.
4. The induced flushing of the aquifer has lead to slight improvement of the oxygen climate (sub-oxic) and also reduction of ionic strength and hardness.
5. The development of clogging processes in the river bed are inevitable and there is a rapid reduction of infiltration rates and a slower decline of the specific yield from the abstraction wells.

REFERENCES

Carbiener, R. & M. Tremoliérs 1990. The Rhine rift valley ground water - river interactions: evolution of their suceptibility to pollution. *Regl. Rivers* 5: 375-389.

Dreher, J. 1991. Ground water management in the city of Vienna, Case Study. *IAHS Publications No. 202:229*. Vienna.

Dreher, J.& A. Gunatilaka 1996. Ground water management in the city of Vienna. Modelling, Testing & Monitoring for Hydro Power Plants - II. *Int. J. Hydro Power & Dams* 2: 545 - 554.

Gunatilaka, A., Dreher, J. & A. Richter 1994. Dissolved organic carbon dynamics in a groundwater environment. *Verh. Internat. Verein. Limnol.* 25: 1420 -1425.

Gunatilaka A. & J.E. Dreher 1996. Use of early warning systems as a tool for surface and ground water quality monitoring. *Proc. IAWQ and IWSA Symp. on Metropolitan Areas and Rivers. Rome, May 1996. TSI - River quality surveying and monitoring methods* - 2: 200-211.

Hem, J.D. 1970. Study and interpretation of the chemical characteristics of natural water. 2nd Ed. *US Geol. Surv. Water Supply Pub.*, 1473:363p., Washington, DC.

Matthess, G. 1983. *Properties of ground water, Wiley Interscience*, NY. 389p.

1.8 Well recharge case studies

Artificial Recharge of Groundwater, Peters et al. (eds) © 1998 Balkema, Rotterdam, ISBN 90 5809 017 5

Creation of potable water reserve in Kuwait through artificial recharge

A. Mukhopadhyay & J. Al-Sulaimi
Kuwait Institute for Scientific Research, Kuwait

A. A. Al-Sumait
Ministry of Electricity and Water, Kuwait

ABSTRACT: Kuwait, an arid country, depends almost entirely on desalination plants for its freshwater requirements. A large storage capacity is required to ensure water availability during a long-drawn emergency, when the desalination plants may be out of commission. Storage of water in the aquifers is a viable option for this purpose. Field trials have demonstrated that the storage and subsequent recovery of fresh water from the two major aquifers of Kuwait are technically feasible. An area has been selected for recharging the Dammam Limestone aquifer for creation of a potable water reserve. Injection at a rate of 650 to 1300 m^3/d through 35 to 40 wells has been planned at this site, over a period of 5 to 10 years. Recovery at a rate of 45 Ml/d at a salinity level of 1500 mg/l or less, over a period of one year, is envisaged during an emergency period.

1 INTRODUCTION

Kuwait is an arid country with an average rainfall of 110 mm. Except for some small accumulations of fresh water in the near-surface aquifer in the northern part of the country, ground water is mostly brackish to saline. Most of the fresh water consumed in Kuwait is produced by the desalination of seawater. A surface storage capacity of 9.856 Mm^3 of fresh water is maintained for emergency use. This storage capacity is enough to meet the water demand for a period of about 12 days, at the average fresh water demand at 1996 level (838,000 m^3/d) (MEW 1997). At a reduced per capita consumption level of 30 l/d, the stored water can, however, last for around seven months with a population level of 1.5 million. The existing storage capacity is not adequate, because:

1. In the event of major desalination plant breakdowns or of major civil disturbances resulting in substantial destruction in desalination capacity, a lead time of 12 days may not be long enough for bringing back the plants on stream. In the case of total destruction, even seven months will not be sufficient to recommission the plants.
2. The surface storage facilities themselves are easy targets for sabotage, and a safer storage space is needed for water security of the country.

With the present fresh water storage levels and the general uncertainty associated with the functioning of desalination plants during an emergency, continuous supply of fresh water cannot be assured for all times. One of the obvious solutions for this problem is the provision of large aquifer storage of fresh water to meet the internal demands for a period of about one year, albeit on a restricted scale.

2 WATER DEMAND REQUIREMENTS

From the latest consumption figures (1996), it is assumed that the average per capita demand of fresh water in Kuwait is 471 l/d (MEW 1997). Under emergency conditions, this figure could be reduced to 30 l/d (Agnew & Anderson 1992). Use of fresh water under these conditions will be restricted to drinking and cooking. The rest of the water demand (toilets, bathroom, and laundry) can be met from brackish water supply. Assuming a population level of 1.5 million people, minimum requirement of fresh water under emergency conditions works out to be 45,000 m^3/d. Assuming that the emergency will last for a year, total volume of water to be supplied under emergency conditions is around 16.4 Mm^3.

3 AQUIFERS IN KUWAIT

There are two major aquifers in Kuwait: Mio-Pliocene clastic deposits of the Kuwait Group, and unconformably underlying Middle Eocene Dammam Limestone Formation. The former is unconfined to semi-confined, and is composed of sandy horizons with variable permeability due to the presence of clay and calcretic/gypcritic layers. The Dammam Formation is confined and consists of dolomitic limestone. Ground water in both the aquifers is generally brackish (2,500-10,000 mg/l), changing to saline (10,000 to >150,000 mg/l) towards the northeast. The underflow from Saudi Arabia is the main source for recharge. Rainfall precipitation is a minor source of local recharge in the depressions of north Kuwait, where fresh water lenses of limited extent occur in the upper part of the Kuwait Group, specially at Rawdhatain, Umm Al-Aish, and Abdally areas. The aquifers are heavily pumped, especially in the southern and southwestern parts of the State, resulting in large declines in the potentiometric heads.

4 AVAILABILITY OF WATER FOR RECHARGE

The present installed capacity of the desalination plants in Kuwait is about 1.064 Mm^3/d (MEW 1997). If the capacity of the desalination plant under construction at Az-Zour is taken into consideration, the installed capacity reaches a little over 1.173 Mm^3/d. Assuming an operating efficiency of 80%, the production capacity will be about 938,000 m^3/d when the Az-Zour plant comes on stream. If the production capacity at high temperature (110°C) is considered, the total installed capacity will reach 1.304 Mm^3/d, and the available capacity will be at a level of 1.043 Mm^3/d (including the Az-Zour plant). The daily average consumption of fresh water in 1996 was almost 838,000 m^3/d (MEW 1997). Assuming a 5% yearly increase, the daily average consumption of fresh water in Kuwait will reach a level of 1.018 Mm^3/d in the year 2000. Thus, a surplus capacity of about 25,000 m^3/d is expected to be available in the year 2000, and can be utilized for artificial recharge. It is anticipated that further additions to the installed capacity will be made to meet the rising demand, and also to replace the ageing units, and surplus capacity of a similar order or more will continue to be available over the years, which can be utilized for creating a strategic reserve of fresh water underground through artificial recharge.

5 ARTIFICIAL RECHARGE STUDIES IN KUWAIT

Artificial recharge of the Kuwait Group aquifer was first tried by Parsons in 1964 (Senay 1977), using a recharge pit in the Rawdhatain depressions for the purpose of collecting surface run-off during occasional rainstorms. The study concluded that the average infiltration capacity of the recharge pit was 9.1 m/d, and 4-8 mg/l chlorine in solution was required to counteract the clogging effect of bacterial and algal growth. Except on two occasions (once in 1970, and once in 1977), no significant recharge through the recharge pit was, however, recorded.

Recharge experiments through injection wells were undertaken by the Ministry of Electricity and Water (MEW) at the Rawdhatain Water Field during 1972-73 and in 1977 to study the clogging effects of injection and quality changes during the recovery phase (Senay 1977). In the first test, clogging was observed within 100 min of recharge, but this was cleared within a few minutes of backpumping. In the second recharge test, no clogging effect was noticed, and it was estimated that, during recovery, about 10% of the water was recovered without undergoing any quality change, and a total of about 52% was recovered with a quality of less than 1000 mg/l total dissolved solids (Senay 1977).

During 1989-92, MEW retained the Kuwait Institute for Scientific Research (KISR) to undertake artificial recharge and recovery experiments through three existing water production wells (Mukhopadhyay et al. 1994). The highlights of these experiments are presented in Table 1

These injection-withdrawal experiments demonstrated that it was technically possible to inject water in the aquifers of Kuwait, and to recover a part of this water, mixed with native ground water in variable proportions and meeting the quality standard adopted by the water management authorities in Kuwait. The clogging problem, however, makes injection easier in the carbonate Dammam Formation compared with that in the clastic Kuwait Group. It was, therefore, recommended that artificial recharge on pilot scale should start with the Dammam Formation.

6 SELECTION OF SITE FOR ARTIFICIAL RECHARGE

Based on the field experiments, a set of criteria was developed for selecting a site for artificial recharge (Mukhopadhyay et al. 1994). These were as follows:
1. Transmissivity of the aquifer should be moderate (150 - 400 m^2/d).
2. Salinity of the ground water should not exceed 5,000 mg/l.
3. The lateral and vertical hydraulic gradient should be as low as possible.

Also, the site should be as near as possible to the injection water source and the demand center. To reduce the cost of infrastructure, it should be as close as possible to the road, power grid, water pipeline, etc.

Based on these criteria, a 10 km x 10 km site was selected for artificial recharge of the Dammam Formation aquifer with desalinated water to create a reserve of potable water for emergency use (Fig. 1)(Viswanathan & Mukhopadhyay 1992).

7 COMPATIBILITY STUDY

A laboratory investigation was carried out to investigate the degree of compatibility of the desalinated water with the Dammam Formation aquifer at the selected site (Mukhopadhyay et al. 1997). A well was drilled to a depth of 271 m within the selected site.

Table 1. Highlights of the injection - recovery tests conducted by the Kuwait Institute for Scientific Research.

Recharge details	Injection wells		
	SU-10	C-105	SU-135A*
Aquifer injected	Dammam Formation	Dammam Formation	Kuwait Group
Aquifer transmissivity (m^2/d)	50	≥ 4000	250
Formation water quality (TDS) (mg/l)	3,500	2,700	4,000
Injected water quality (TDS) (mg/l)	310 - 410	2,700**	4,000***
Average injection rate (m^3/d)	590	5,900	655 - 820
Total number of injection days (d)	30	30	0.9
Total injection volume (m^3)	16,416	171,700	642
Volume of water recovered at TDS ≤ 1500 mg/l (m^3)	2,727	25,730	409
Recovery efficiency****	15 - 20 %	10 - 15 %	65 - 75 %
Redevelopment procedure followed	Back pumping	Back pumping	Acid cleaning with surging

* Second injection test.
** Mixed with fluorescein as tracer.
*** Mixed with tritium as tracer.
****Recovery efficiency was calculated using the expression:
Recovery efficiency = 100 x (Volume of water recovered with TDS ≤ 1500 mg/l)/(Volume of water injected)

The major part of the Dammam Formation was conventionally cored in the well, and the samples of the ground water were collected. Core flow tests of selected core plugs were carried out using synthetic Dammam Formation water and desalinated water. The changes in hydraulic conductivity during the flow tests were monitored. Only a minor reduction in hydraulic conductivity was noted after the flow tests with desalinated water. The variation of the total suspended solids in the outflow water suggested that the movement of fines within the pore spaces was the most likely cause of this reduction. Geochemical simulation indicated that very little scaling was expected when the desalinated water would be injected in the aquifer. It was concluded that the injection of desalinated water in the Dammam Formation aquifer at the selected recharge site should not cause any major compatibility problem.

8 WELL FIELD DESIGN

8.1 *Volume of injection water*

The demand for water under emergency conditions is expected to be 45Ml/d. Assuming supply requirements of one year, the total volume of water to be recovered is estimated to be 16,400 Ml. Assuming a conservative recovery efficiency of 20% for water with total dissolved solids content of 1500 mg/l or less, the total volume of water to be injected is 82,000 Ml.

8.2 *Injection rate*

Rates of injection mainly depend on the aquifer transmissivity, well hydraulics, clogging rate, and the availability of injection water. From the past experience of injection experiments, to avoid excessive head buildups in the wells, the injection rates should be restricted in the range 650-1300 m^3/d. At these rates of

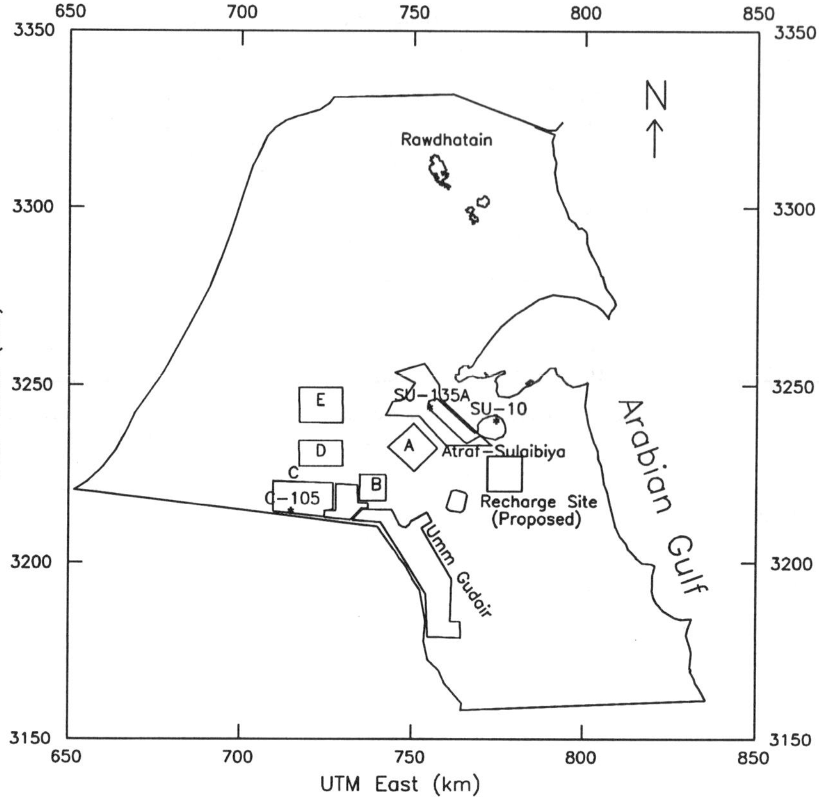

Figure 1. Location of water fields, experimental recharge wells, and selected site for recharge of desalinated water in Kuwait.

injection, clogging effect and head buildups are manageable.

8.3 *Number of wells and their arrangement*

With an injection rate of 1300 m^3/d and the same rate of recovery, a total of 35 wells will be required to inject and produce a total of 45 Ml/d. A number of possible arrangement of wells was investigated using a well-flow model to select a preferred arrangement of wells that would give a minimum head buildup and, at the same time, a compact arrangement of wells that will also minimise the cost of pipeline for distributing injection and recovered water. After several trials, a hexagonal arrangement of 37 wells with an average interwell distance of 1500 m was found to be an optimum one (Fig. 2)(Viswanathan & Mukhopadhyay 1992). With an injection rate of 1300 m^3/d, five years will be required to inject a total volume of 82,000 Ml, and the injection time has to be increasedto 10 years if the injection rate is maintained at a lower level of 650 m^3/d per well.

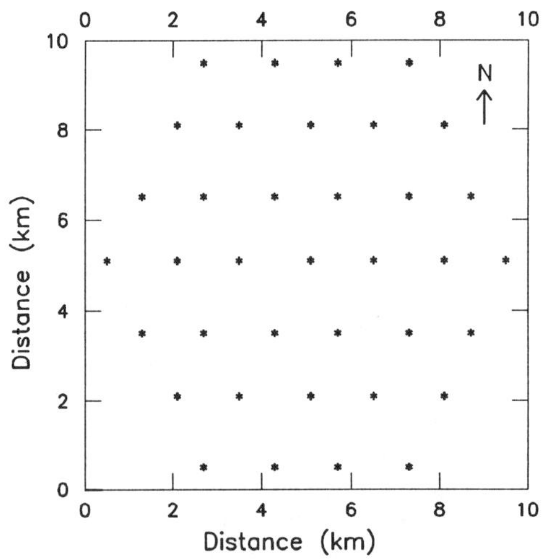

Figure 2. Tentative arrangement of recharge wells.

Assuming a well-loss of 10 m and confined aquifer condition, the maximum head buildup for the higher injection rate was estimated to be 96 m after five years of injection, and 59 m after 10 years of injection for the lower rate. Assuming an average depth of the Dammam Formation potentiometric surface at the recharge site at 20 m below the ground surface, the maximum rise of the potentiometric head at the wells at the end of the injection period was estimated to be about 75 m and 40 m above the ground surface, respectively. This will require especial wellhead and well completion design that should be able to withstand injection under pressure.

Neglecting the effects of dispersion and assumimg uniform porosity distribution, the radius of the injected water around each well at the end of the injection period (10 years for an injection rate of 650 m^3/d, and five years for an injection rate of 1300 m^3/d) was estimated to be about 310 m for an average porosity of 10% and 435 m for an average porosity of 5%, assuming an injection face of 80 m. The expression:

$$r^2 = Q/3.14h\phi_e \quad (1)$$

was used in this estimate, where:

r = radius of injected water around the well
Q = total volume of water injected in the well
h = effective injection height in the well
ϕ_e = effective porosity

Effect of dispersion on the radius of injected water may be estimated using the approximate expression of Bear (1972) as follows:

$$r' = r + 2.3(D.r)^{0.5} \quad (2)$$

where:

r' = radius of injection with dispersion
D = dispersion coefficent

Assuming an average dispersion coefficent (D) of 20 m for limestone, the radius of injection (with dispersion effect) computes to about 490 m for an average porosity of 10% and 650 m for an average porosity of 5%. If secondary porosity (fracture, solution vug) predominates, it is, however, difficult to estimate the radius of injection zone, as injected water will flow preferentially along the fracures and/or solution vugs.

8.4 *Pipeline requirements*

Maximum volume of water to be injected and recovered during an emergency is planned to be 45 Ml/d or 0.52 m^3/s. With an assumed flow velocity of 2 m/s, the required pipe diameter for this flow rate is 0.58 m or approximately 24". Approximate distance of the selected recharge site from the nearest desalinated water distribution system is 20 km. Total head loss in the pipe (mild steel) of this length is about 150 m. The average difference in elevation between the nearest source of desalinated water and the recharge site is 60 m. Assuming a maximum wellhead pressure of 75 m and losses in the valves, bends, etc., to be 10 m, the total head required for injection is about 300 m. Booster pumps will be required to provide this injection head. A fusion bonded epoxy coating for the mild steel pipeline is recommended to overcome corrosion problems. Even if the injection is carried out at half the rate as noted above (i.e., 22.5 Ml/d) for a ten-year period, the size of the pipeline will remain the same, as the production rate during an emergency will still be maintained at around 45 Ml/d. The maximum flow velocity during injection will be reduced to 1 m/s under this condition, and total injection head (taking into consideration losses in pipeline, valves, and bends; elevation difference; and maximum wellhead pressure) required to be delivered by the booster pumps at this reduced injection rate will be about 200 m.

8.5 *Pretreatment requirements*

The laboratory study on compatibility has indicated that no major geochemical incompatibility problem is to be expected from the injection of desalinated water in the Dammam Formation aquifer at the selected recharge site. Presence of 2-4 mg/l of free chlorine in the desalinated water supplied to Kuwait City should preclude biofouling. Clogging from suspended solids and entrapped air may, however, progressively reduce injection capacity. Removal of suspended solids using online pressure filters is planned. To reduce the clogging from air entrapment, the following measures are suggested: i) the use of online air release valves; ii) injection under pressure; iii) maintenance of full pipe condition by reducing the in-hole injection pipe diameter, and the use of a throttling valve at the bottom of the injection pipe; iv) placing of the injection pipe shoe below the water level in the well; and v) maintaining the temperature of the injection water equal or above the prevailing aquifer temperature (30-35°C) by burying the conveying pipeline 2-3 m below the ground surface and/or heating the injection water with the help of electrical heating coils, at least during the winter.

9 PROJECT IMPLEMENTATION PLAN

A two-stage implementation of the project is suggested. In the initial phase, seven central injection wells in the pre-defined hexagonal pattern will be drilled. An adequate number of observation wells should also be drilled to monitor both the hydraulic effects of injection and the transport pattern of injection water. Geophysical well logging, flow-metering, and if possible, conventional coring are to be conducted in these wells. The recharge should continue for a year or more, alternating with recovery periods if possible. Detailed monitoring of potentiometry and water quality in the injection and the observation wells will be carried out throughout this experimental phase. Statistical and numerical analysis of the data will be carried out, and more detailed information on aquifer characteristics (transmissivity, storage coefficient, leakage coefficient, dynamic porosity, dispersivity, distribution of permeable zones, inter-aquifer flow through borehole) will be derived. The final design of the wells and the field, pretreatment requirements, injection and recovery schedule, and redevelopment procedures to be followed will be decided on the basis of this analysis and experience gained during the operation of the initial phase. A numerical modelling of various recharge-recovery options at the selected recharge site is currently under study to obtain some preliminary idea about an optimum strategy for artificial recharge of the Dammam Formation at the selected site that will maximise the recovery efficiency under the prevailing hydrogeological condition.

10 CONCLUSIONS

1. Kuwait is an arid country that depends almost completely on the desalination plants for its fresh water requirements. It, therefore, needs a large storage system to meet water demands during emergencies, when desalination plants may be out of commission due to various reasons (natural disasters, industrial accidents, war, etc.). A minimum storage capacity of 16,400 Ml of water is required to meet the very basic per capita requirement of 30 l/d for a population of 1.5 million people, in case such an emergency lasts for a year.

2. The most economic option for storage of water is in the aquifers. Building surface storage facilities is very costly and also takes a lot of space. Surface storages are also vulnerable to destruction and sabotage. Kuwait exploits useable brackish ground water from two aquifers: clastic Kuwait Group and carbonate Dammam Formation. Field trials have demonstrated that storage of fresh water in these brackish aquifers is a feasible proposition.

3. A site for recharge in the Dammam Formation has been selected, and a preliminary design of the aquifer storage facility has been drawn. A total of 37 injection wells, arranged in a hexagonal pattern with an interwell distance of 1500 m, have been proposed with an injection rate of 650 to 1300 m^3/d per well. With a conservative estimate of recovery efficiency of 20% for water with a total dissolved solids content of 1500 mg/l or lower, the total volume of injected water will be 82,000 Ml, over five to ten year period. Recent laboratory investigations have suggested that no major geochemical compatibility problem is expected due to the introduction of desalinated water in the Dammam Formation at the selected site. Steps to filter out suspended solids and eliminate formation of air bubbles in the injection stream are to be taken, however, to avoid clogging of the injection phase.

ACKNOWLEDGEMENT

The permissions from the managements of the Kuwait Institute for Scientific Research and Ministry of Electricity and Water for presentation of the paper (Publication No. KISR5212) are gratefully acknowledged.

REFERENCES

Agnew, C. & E. Anderson. 1992. *Water resources in the arid realm*. New York: Routledge.

Bear, J. 1972. *Dynamics of fluid in porous media*. New York: Elsevier Publishing.

MEW. 1997. *Statistical Year Book: Water*, Ministry of Electricity and Water, Kuwait.

Mukhopadhyay, A., E. Al-Awadi, M. N. Al-Senafy, & P. C. Smith. 1997. Compatibility of desalinated water with the Dammam Formation aquifer at a pilot recharge site in Kuwait. Final Report submitted to Kuwait Foundation for the Advancement of Sciences, Kuwait. Kuwait Institute for Scientific Research, Report No. KISR5021.

Mukhopadhyay, A., J. Al-Sulaimi, & A. Al-Sumait. 1994. Lessons from the latest artificial recharge experiments in Kuwait. In A. I. Johnson & R. D. G. Pyne (eds.), *Proc. 2nd Int. Symp. Artificial Recharge of Ground Water, Orlando, 17-22 July. 1995*:790-799. New York: American Society of Civil Engineers.

Senay, Y. 1977. Groundwater resources and artificial recharge in Rawdhatain water field. Open File Report. Ministry of Electricity and Water, Kuwait.

Viswanathan, M. N. & A. Mukhopadhyay. 1992. Assessment of artificial ground water recharge in Kuwait. Final Report. Vol. IV: Selection and design of pilot projects. Submitted to Ministry of Electricity and Water, Kuwait. Kuwait Institute for Scientific Research Report No. KISR 4125.

Artificial Recharge of Groundwater, Peters et al. (eds) © 1998 Balkema, Rotterdam, ISBN 90 5809 017 5

Deep well recharge in a polder area near the river Rhine

H.Timmer
Watermaatschappij Zuid-Holland Oost, Gouda, Netherlands

P.J.Stuyfzand
Kiwa Research and Consultancy, Nieuwegein, Netherlands

ABSTRACT: By using field tests, Watermaatschappij Zuid-Holland Oost (WZHO) investigated the possibility of recharging pretreated water from the Rhine, in deep sand layers in the adjacent polder area. The study focused on well clogging and the effects of aquifer passage on the quality of the recharged water. The purpose was to obtain information for the design of a deep well recharge system in practice, and for the sizing of pre- and posttreatment. This paper describes the tests and results from July 1996 to March 1998. The conclusion is that deep well recharge of pretreated Rhine water in the polder area is very promising indeed. Pretreatment of water from the Rhine with rapid sand filtration, coagulation and filtration over activated carbon is needed however, to obtain water with a negligible clogging potential and satisfying EU-regulations. Posttreatment could be as simple as aeration followed by a single rapid sand filtration.

1 INTRODUCTION

Watermaatschappij Zuid-Holland Oost is one of the three water supply companies in the province of Zuid-Holland, with its own method of production of drinking water. The Rotterdam region obtains drinking water from directly treated water from the river Meuse. The Hague and the surrounding area are supplied from the dunes, where water from the river Meuse is recharged by ponds and recovered. WZHO is the groundwater pillar in the province, producing drinking water from groundwater and river Rhine after bank filtration. The idea behind this provincial trio of sources is to spread the risks: if a serious disaster occurs on one of the rivers, it still remains possible to supply drinking water.

WZHO is committed to groundwater. In reply to increasing demands for drinking water, the answer was therefore always to abstract more groundwater. In the last ten years, however, this reflex action has led to increasing resistance from environmentalists, who link groundwater abstraction to degradation of the area's ecological value. Because of this development, WZHO has begun searching for other production methods, for example, deep well recharge. This particular method combines the positive aspects of surface water and groundwater: negligible environmental impacts, good water quality, thanks to a lengthy aquifer passage, and a high guarantee of supply in the event of a disaster.

Deep well recharge for supplying drinking water has already been used for several years on a test scale in the dune area of the Netherlands. The application of this method was new, however, for the polder area in the west of the country. WZHO started its first field trials of deep well recharge in 1994. Since 1996, the tests have focused on the recharge and recovery of pretreated water from the Rhine near Langerak. Depending on the results, WZHO will start to use deep well recharge at the beginning of the next century, to meet the increasing need for drinking water.

In 1994, Kiwa identified the possible sources for deep well recharge. These were evaluated for water quality, guarantee of supply and cost. Based on the results, it was decided to conduct tests, from mid-1996 until spring of 1998, with pretreated Rhine water from the company NV Watertransportmaatschappij Rijn-Kennemerland (WRK) in Nieuwegein. The pretreatment employed by WRK consists of

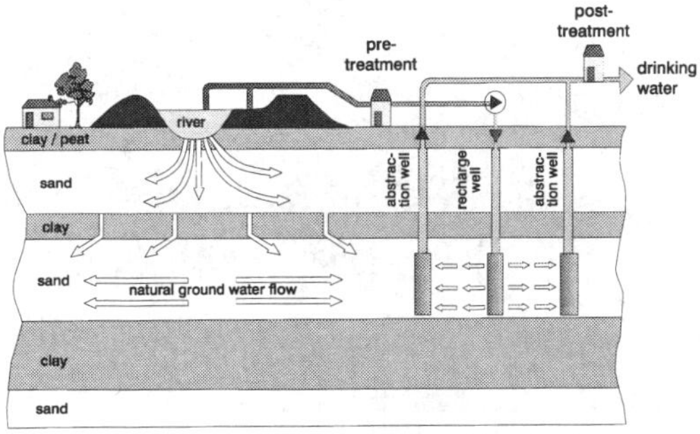

Figure 1. Drinking water preparation using deep well recharge

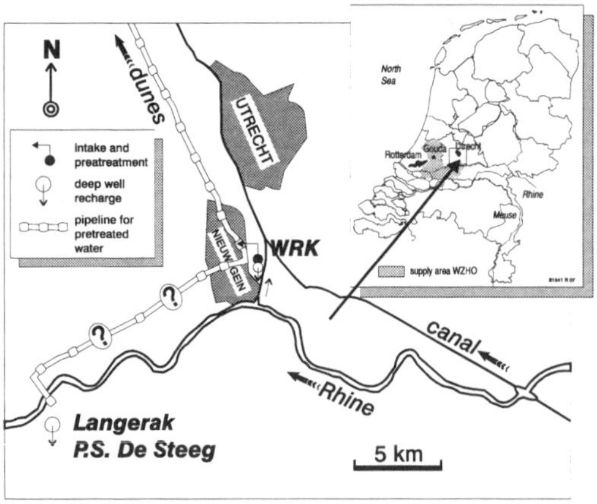

Figure 2. WZHO supply area and site of test locations

coagulation, sedimentation and rapid sandfiltration.

As the future location for a test-size deep well recharge system, WZHO selected Langerak (fig.2). The reasons were, on the one hand, hydrogeological, the presence of a high-quality aquifer, and, on the other, infrastructural, there being a direct connection to the main artery of the drinking water supply system at the existing treatment plant De Steeg (fig. 3).

2 EXPERIMENTAL DESIGN

The tests carried out can be divided into two main groups: treatment tests and recharge tests.

2.1 *Treatment tests*

The purpose of these tests was to determine the supplementary (pre)treatment steps needed to make the pretreated Rhine water from WRK suitable for deep well recharging in Langerak. The criteria for the effluent were:

- The water had to satisfy the standards contained in Dutch legislation covering recharge water, intended to protect the quality of the soil.
- The water had to have a low clogging potential. Indicators of this include:
- MFI (membrane filtration index): less than 2 to 3 s/l^2, and
- AOC (assimilable organic carbon) content: less than 10 μg/l.

These standards are based on the results of many Dutch tests, in particular, those conducted in the dune area (Van der Kooij 1992). It is worth noting that, even without treatment, water from WRK meets the AOC standard. Apart from one or two conclusions, this paper will not go further into the design or

Figure 3. Aerial view of Langerak site (sky pictures)

results of the treatment tests.

2.2 *Recharge tests*

The recharge tests were intended to provide insight into:

- well clogging (causes and cures), and
- changes in water quality during aquifer passage.

They were needed to produce information for the design of a full-size deep well recharge system and for the sizing of a posttreatment plant for water recovered from the soil.

2.3 *Locations and design of recharge tests*

Since there was no transportation pipeline between Nieuwegein (site of WRK-plant) and Langerak (Fig. 2), it was not possible to recharge water from WRK in Langerak. A pipeline will only come after it has been proved that deep well recharging of water from WRK is both possible and desirable. For the recharge tests, this meant they were conducted at the WRK site in Nieuwegein with the future water, as well as at future site Langerak, however with the locally produced drinking water.

It is possible to translate the results of the Nieuwegein tests to the future situation in Langerak, because the experimental design aimed at an identical recharge situation for both locations. The design took into account recharging in comparable hydrogeochemical environments (the anoxic Harderwijk formation), and similarities in quality of the recharge water, drilling method and finishing of the recharge well, recharge speed, etc.

To ensure constant compliance with the Dutch standards for soil protection, an activated carbon filter was installed ahead of the recharge well in Nieuwegein as a precautionary measure. This was needed in Nieuwegein where polluted river Rhine is pretreated.

In Langerak, high quality drinking water, prepared from local groundwater by aeration and rapid sand filtration, was used as recharge water. To make the oxidation capacity of this drinking water comparable with the water from WRK; NaNO3 was dosed for the entire test period. NaCl was added as a tracer for one month only.

To obtain supplementary information on clogging and chemical reactions during aquifer passage, parallel-column tests on soil samples from both Langerak and Nieuwegein were conducted.

2.4 *Simulation with parallel columns*

Apart from the deep well experiments, recharge tests were also performed on a small scale using a set of parallel columns (Schippers et al. 1989). Each column was filled with a layer of filtering material

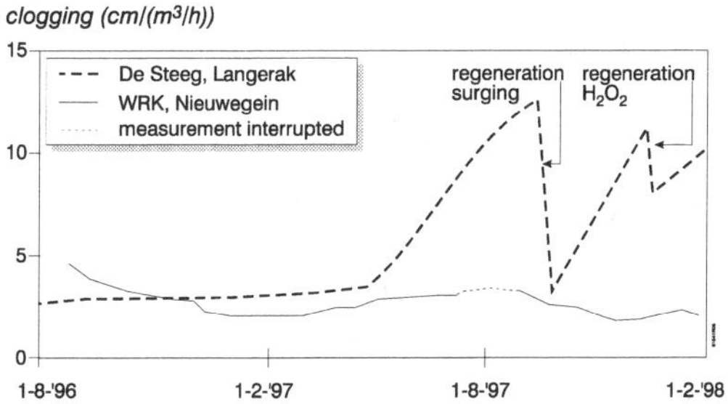

Figure 4. Clogging development in recharge wells at De Steeg and WRK

and a layer of aquifer material obtained with a core catcher during auger hole drilling. By allowing recharge water to flow through the column, processes occurring around a recharge well are simulated aboveground. In addition, it was a simple matter to flush an aquifer sample from Langerak with water from WRK.

Depending on the processes being monitored (well clogging or quality changes); parallel columns were either hydrological or chemical. By rapidly flushing the columns, data on clogging development and quality changes were obtained for extrapolation to longer time scales.

3 RESULTS OF RECHARGE TESTS

3.1 *Clogging of recharge wells*

The recharge well at the WRK location did not clog. Apparently, activated carbon filtration following the existing WRK pretreatment is enough to prevent clogging. This is supported by the results of the pretreatment tests. Supplementary slow or rapid sand filtration produces hardly any improvement in quality.

The recharge well at Langerak clogged, however, specifically in the filtering material (gravel pack). Figure 4 shows the clogging development at both locations. Strange enough, the difference in clogging behaviour cannot be explained by the conventional indicators, MFI and AOC content. Further study into the nature and origin of the material reveals that the clogging of the recharge well at Langerak was most likely caused by flocculation of residual iron particles, $Fe(OH)_3$, in the drinking water from the local treatment plant. Although the concentration of dissolved iron in the drinking water is only 0.04 mg/l, this is still enough to cause a clogging. Particle size measurements show that the size of particles responsible for clogging lies between approximately 15 and 40 μm.

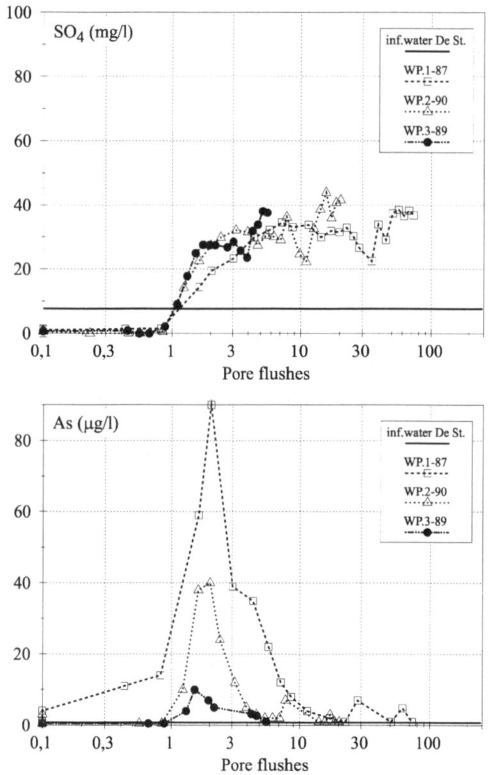

Figure 5. Changes in sulphate and arsenic (total) during aquifer passage of drinking water at Langerak

3.2 *Quality changes during aquifer passage*

The quality changes during aquifer passage have been described and quantified by Stuyfzand & Timmer (1998), and were further modelled by Brun et al. (1998) and Saaltink et al. (1998) using different models.

Oxygen- and nitrate-rich recharge water gradually displaced the original deep anoxic groundwater. Through redox reactions, with oxygen and nitrate organic material and pyrite are oxidised which produces especially, hydrogen carbonate, acid and sulphate. Moreover, cation exchange causes the addition of iron, manganese and ammonium to the recharge water.

Because of calcite supersaturation of the recharged drinking water at Langerak, no calcite dissolved here not withstanding acid production due to pyrite oxidation. At Nieuwegein, on the other hand, calcite did dissolve, the increase in calcium concentration lying between 2 and 25 mg/l.

During aquifer passage, the amounts of the trace elements arsenic and nickel temporarily increased at the two locations. Both elements arise from oxidising pyrite, but are subsequently re-adsorbed by the aquifer matrix. This is shown for arsenic in figure 5.

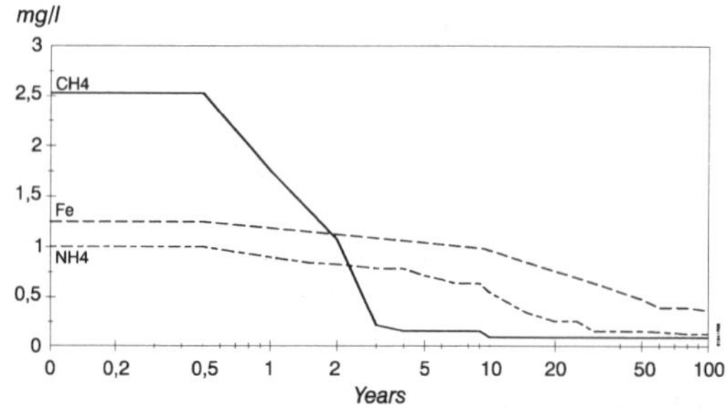

Figure 6. Changes in some parameters of the recovered river Rhine water after deep well recharge at Langerak as predicted by the model 'EASY-LEACHER' (Stuyfzand, 1998)

3.3 *Predicting the water quality after deep well recharge*

With the aid of a groundwater model, WZHO has calculated the hydrological constraints of a future deep well recharge and recovery system. The aim is an average residence-time range of 1 year and a minimum residence time of 3 months. The first parameter is for smoothing out temperature fluctuations of the surface water in the subsoil; the second is for bacteriological reliability.

Based on the above, the EASY-LEACHER model of Kiwa has been used to predict the quality of the recovered water after deep well recharge. Figure 6 shows the most significant changes. Employing these predictions, it is possible to size a pre- and posttreatment plant.

4 CONCLUSIONS

1 The field tests show that supplementary pretreatment of water from WRK with activated carbon is enough to satisfy the standards for soil protection and prevent well clogging.
2 Based on the measurements at the two locations, it is possible to quantify the soil reactions if water from WRK is recharged at Langerak. The most significant reactions are oxidation of pyrite and organic material, cation exchange, dissolution of lime, and a temporary slight increase in arsenic and nickel content.
3 The tests provide sufficient information for the design and sizing of a posttreatment process for the water recovered after deep well recharge. Based on the water quality predictions, aeration and a single rapid sand filtration step should be adequate.
4 Deep well recharge in the polder area with water from the Rhine has very good perspectives.

ACKNOWLEDGEMENTS

The EC Environment, Climate Research Programme (Contract), supported the research regarding water quality changes.

REFERENCES

Brun, A., F.D. Cristensen, J.S. Cristiansen, P.J. Stuyfzand & H. Timmer 1998. Waterquality modelling at Langerak deep-well recharge site. (this volume).

Saaltink, M.W., C.Ayora, P.J. Stuyfzand & H. Timmer 1998. Modelling the effects of deep artificial recharge on groundwater quality. (this volume).

Kooij, D. van der 1992. Assimilable organic carbon as an indicator of bacterial regrowth. J. *American Water Works Association* (84) 57-65.

Schippers, J.C., J. Verdouw en G.J. Zweere 1989. Het voorspellen van de verstoppingssnelheid van infiltratieputten. Kiwa, Nieuwegein. Mededeling 106 (in Dutch).

Stuyfzand, P.J. 1998. Simple models for reactive transport of pollutants and main constituents during artificial recharge and bankfiltration: infomi 3.1 and easy-leacher. (this volume).

Stuyfzand P.J.& H. Timmer 1998. Injecting oxic surface water into an anoxic aquifer along the river Rhine; Water-rock interactions including redox dynamics. Kiwa-report, Nieuwegein in prep.; EU-contract (ENV4-CT95-0071) and WZHO

Timmer, H., J.P.H. Bierling & J.D. Verdel 1997. Proefproject diepinfiltratie; 2e interim-rapportage over infiltratieproef De Steeg, infiltratieproef WRK en kolomproeven. WZHO, Gouda (in Dutch)

Artificial Recharge of Groundwater, Peters et al. (eds) © 1998 Balkema, Rotterdam, ISBN 90 5809 017 5

Simulation of the influence of recharged cold groundwater on temperature and level of surrounding groundwater

J.Akiyama, Y.Katsuragi, N.Goto & H.Abiko
Japan Ground Water Development Company Limited, Yamagata, Japan

T.Yokoyama
Department of Mechanical System Engineering, Yamagata University, Japan

ABSTRACT: More than 20% of Japanese live in the snowy area, therefore snow removal and snow control on roads have been essential since early times in Japan. In 1980, snow melting system without sprinkling ground water was developed in order to prevent land subsidence and to maintain ground water as a precious resource. Such snow melting systems are popular in the snowy area of Japan in these days because of its convenience, low running costs, and superiority in preservation of the environment. However, many pumping wells and many recharge wells are designed to be required in the limited area in some cases, and it is required to study or predict the influence of recharged cold ground water on temperature and level of the other pumping wells nearby. Simulation by numerical analysis supported by actual cold water recharge test in field, provides appropriate locations of each well with minimum mutual interference.

1 GENERAL

Ground water, in Japan, used to be utilized for many years for both drinking and agriculture because of its good properties. Consumption of ground water, however, for agriculture was decreased remarkably and on the other hand, consumption for industry was increased because of industrialization during the 1980's. And consumption of ground water for snow melting systems was started during 1960's and has been increased because of the modern transportation in the snowy area of Japan. Snow removal and snow control on roads have been essential because more than 20% of Japanese live in the snowy area, which occupies about 60% of the country of Japan.

Snow melting system with sprinkling ground water was developed in 1961, and the very heavy snowfall in 1963 made it popular in the snowy area of Japan. Such systems worked well for a long time, but it is not so convenient for pedestrians and shops beside the roads because of splashing by vehicles passing by. And it wastes much ground water without recovery, and it sometimes causes troubles such as lack of ground water, saltwater intrusion and land subsidence.

In 1980, snow melting system without sprinkling ground water was developed in order to prevent land subsidence and to maintain ground water as a precious resource. This system utilizes only heat energy of ground water for melting snow and recharges used ground water into the aquifer again without sprinkling. It has an economical advantage because it does not need any fuel nor electricity as a heat source for melting snow. Such snow melting systems without sprinkling ground water are popular for clearing snow on roads or sidewalks in the snowy area of Japan in these days because of its convenience, low running costs and superiority in preservation of the environment.

It has been sometimes observed that several pumping wells and recharge wells were located in a limited area, but the influence of recharged cold ground water on temperature and level of surrounding ground water has not been discussed seriously.

It is, however, required to study and predict the influence of recharged cold ground water in order to decide the locations of pumping wells and recharge wells with minimum interference in such cases as readjustments of town lots where many pumping wells and many recharge wells will be provided in a limited area. The simulation by numerical analysis, supported by the field cold water recharge test consists of a pumping well, a recharge well, an observation well and a cooling facility, provides appropriate location of each well.

This paper introduces the outline of snow melting systems without sprinkling ground water, the results of the field cold water recharge tests and the typical results of the simulation by numerical analysis.

2 SNOW MELTING WITHOUT SPRINKLING GROUND WATER

2.1 *Snow melting system without sprinkling ground water*

This system consists of two wells and a heat radiation area as shown in Figure 1. One well, called "pumping well", is for pumping ground water and the other, called "recharge well", is for infiltrating used ground water into the aquifer again. Heat radiation pipes of small diameter are embedded underneath the surface of roads, sidewalks, ramps, and parking lots, etc., where snow is to be controlled. Ground water pumped from the pumping well is led to the embedded radiation pipes and radiates its heat energy to warm the pavements above and melts the snow there on. And then it is infiltrated to the aquifer again through the recharge well.

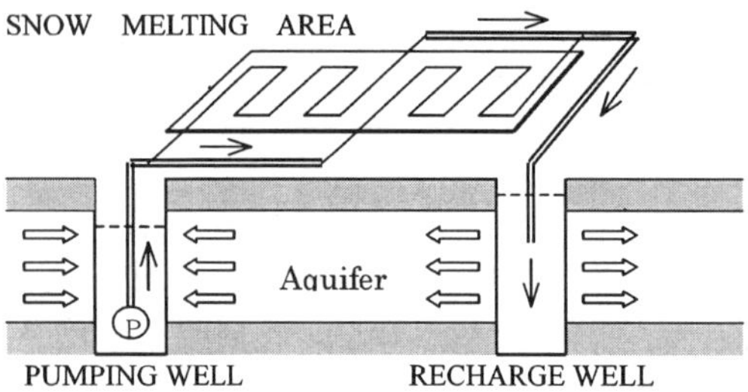

Figure 1. System components.

2.2 *Advantages of the system*

1. It is convenient for pedestrians because no water sprinkled nor splashed.
2. It will not cause troubles such as lack of ground water and land subsidence because it does not wastes ground water.
3. It proves to be energy-saving and has low running costs because it utilizes thermal energy of ground water which has not been used.

Thus, around 700 installations of this system cover more than 480,000 m2, as of December 1997.

3 OUTLINE OF THE PLAN

3.1 *Location and outline of the plan*

The plan of readjustment of town lots, which we examined in this paper, is the west side of Yamagata Station in Yamagata city, Japan. 400m in east-west direction and 800m in north-south direction area is planned to be readjusted for higher quality roads, parks and public facilities.

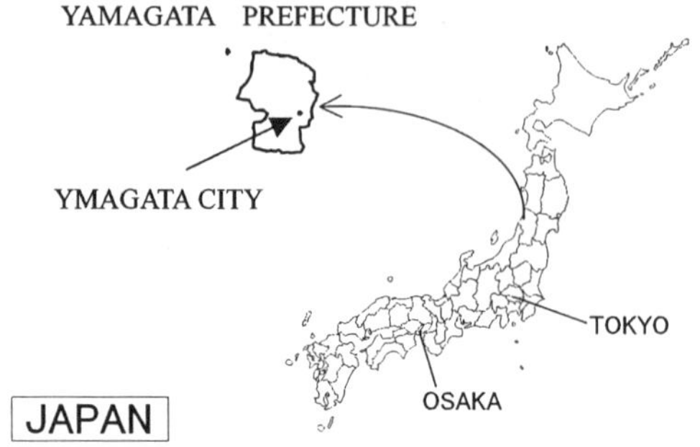

Figure 2. Location Map.

3.2 *Area of snow melting system*

Snow melting system without sprinkling ground water is planned to be installed on the main roads, all the sidewalks and square in front of the station. And it will cover more than 40,000 m2.

3.3 *Geological conditions of the site*

The site is located at the south-east of the Yamagata Basin and at the edge of the alluvial fan of the Mamigasaki River flowing from the Zao Mountains. Geological map is shown in Figure 3.

The top formation in made of alluvial deposits mainly by gravel with some clay layer. Base formations of Yamagata Basin are very thick, expected 200 to 300 m, and made of Quaternary alluvial bed such as gravel, sand, clay and silt.

Ground water has been rich in Yamagata city. However, land subsidence caused by over pumping for agriculture and industry was observed around 20 years before. In 1985, guideline for restriction of pumping ground water was established and also ground water after snow melting is recommended to be recharged.

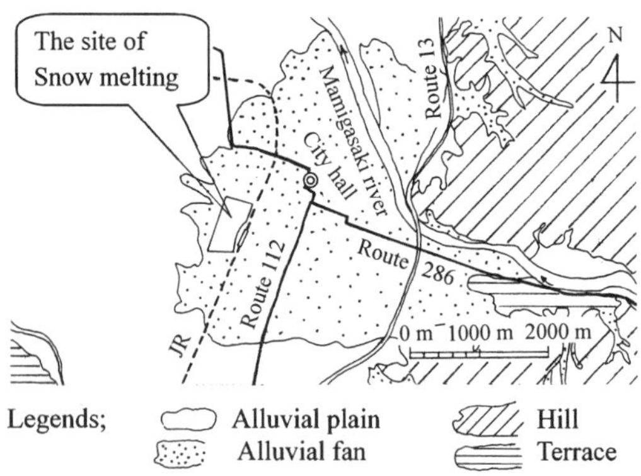

Figure 3. Geological Map

4 WELL PLAN FOR DESIGN OF SNOW MELTING

Ground water of which temperature is 14 to 15 ℃ suitable for melting snow is obtained from 100 to 150 m in depth from ground level in Yamagata city. And 0.4 l/min·m2 of such ground water is designed to be required for melting snow there. Therefore, it comes 16 m3/min by 40,000 m2 and it requires 27 pumping wells with approximately 0.6 m3/min pumping capacity each. Considering classification of roads, sidewalks and squares, 29 pumping wells and also 29 recharge wells are required to be provided in the site. Such congestion of many wells needs to answer the following questions:

1. Pumping of many wells simultaneously causes lowering of the ground water level around the site or not.
2. Recharging of much cold ground water causes lowering of the ground water temperature around the site or not.

5 STEPS TO EVALUATE

Steps to evaluate the positions of wells for designing of the snow melting system on the site is described here under.

1. All the pumping wells and the recharge wells are positioned tentatively, and the changes of ground water level and temperature during system operation are simulated by numerical analysis.
2. Positions of the wells are changed by try and error to minimize mutual interference of cold water recharge by numerical analysis to determine appropriate positions of all the wells.
3. Parameters for numerical analysis are obtained by actual field tests of cold water recharge with observation of water level and temperature at the experimental pumping, recharge and observation wells.

6 THEORETICAL ANALYSIS

The porous media is classified into two phases. The first is the fluid phase, the other is the solid phase. Both phases exist at the same location like a front- and back page. Each phase is continuous and connected together through the total surface "*ns*" of particles with surface heat conductance α. The governing equation of heat transfer in fluid phase is;

$$c\rho_f \frac{\partial \theta_f}{\partial t} = D\left(\frac{\partial^2 \theta_f}{\partial x^2} + \frac{\partial^2 \theta_f}{\partial y^2} + \frac{\partial^2 \theta_f}{\partial z^2}\right) - \frac{c\rho_f}{\varepsilon}\left(u\frac{\partial \theta_f}{\partial x} + v\frac{\partial \theta_f}{\partial y} + w\frac{\partial \theta_f}{\partial z}\right) - \frac{ns\,\alpha}{\varepsilon}\left(\theta_f - \theta_p\right) \quad (6.1)$$

$$u = -K_h \frac{\partial h_0}{\partial x}, \quad v = -K_h \frac{\partial h_0}{\partial y},$$

$$w = -K_v\left(\frac{\partial h_0}{\partial z} - 1 + \frac{\rho}{\rho_0}\right) \quad (6.2)$$

$$h = \frac{p}{g\rho} + z, \quad h_0 = \frac{p}{g\rho_0} + z \quad (6.3)$$

where $c\rho$ = volumetric heat, θ = temperature, t = time, D = thermal dispersivity, ε = porosity, (u,v,w) = the darcy velocity in x, y and z direction, K = temperature-dependent hydraulic conductivities, p = pressure, h = head potential ; subscript f = fluid, p = particle, h = horizontal, v = vertical, 0 = standard .

ρ, ρ_0 are densities at any temperature and at the standard temperature of 15 ℃, therefore "h_0" is somewhat different from "h", but its variation is just considered in equation (6.2). Namely Boussinesque approximation is stated. The continuity equation of mass is given as;

$$S_b \frac{\partial h}{\partial t} = \frac{\partial u}{\partial x} + \frac{\partial v}{\partial y} + \frac{\partial w}{\partial z}, \quad \textit{confined aquifer} \quad (6.4)$$

$$\varepsilon \frac{\partial h}{\partial t} = \frac{\partial (hu)}{\partial x} + \frac{\partial (hv)}{\partial y} + w, \quad \textit{free aquifer} \quad (6.5)$$

where *Sb* is storage coefficient.

Negligible difference in radial temperature field was resulted under coarse sand (Yokoyama et al. 1996). At the particle-phase the rate of uniform temperature is governed by the equation;

$$c\rho_p \frac{\partial \theta_p}{\partial t} = r_c \lambda_p \left(\frac{\partial^2 \theta_p}{\partial x^2} + \frac{\partial^2 \theta_p}{\partial y^2} + \frac{\partial^2 \theta_p}{\partial z^2}\right) - \frac{ns\,\alpha}{1-\varepsilon}(\theta_p - \theta_f) \quad (6.6)$$

where θp = denotes temperature particles, rc = contact ratio, λ = thermal conductivity.

7 FIELD TESTS FOR NUMERICAL ANALYSIS

The experimental wells, of which specifications are shown in Table 1, are provided in east-west line at the west side of the site. Screens for the pumping well are installed lower than –50 m to obtain ground water with its temperature around 15 ℃.

Table 1. Specification of the experimental wells.

Well Name	Depth	Caliber	Depth of screen	Length
Pumping	150m	250 ㎜	62.0m-150.0m	60.5m
Recharge	100m	250 ㎜	28.5m- 94.5m	40.5m
Observation	150m	100 ㎜	29.0m-150.0m	82.5m

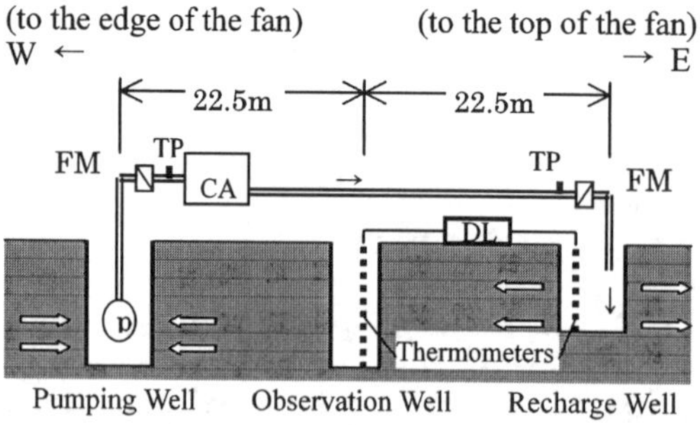

FM = Flow meter , TP = Thermometer in pipes , CA = Cooling apparatus , DL= Data logger , p = Pump

Figure 4. Outline of device for field tests.

And screens for the recharge well are installed lower than –30 m, as not to interfere shallow wells nearby. Screens for the observation well are installed to cover all the aquifers concerned.

Temperatures of ground water at each level in the observation well are measured during cold water recharge test. Ground water pumped from the pumping well is cooled down by the cooling apparatus and infiltrated through the recharge well in the same conditions of snow melting system operation as shown in Figure 4.

8 RESULTS OF FIELD TESTS AND MODELING OF FORMATIONS

8.1 *Results of pump test*

It is found by drilling works of the experimental wells that a free aquifer and 4 confined aquifers exist at the site. Results of pumping and recharge tests for each well are shown in Table 2. It is confirmed that pumping quantity shown in Table 2 is within the governmental restriction of pumping ground water in Yamagata city and such quantity can be recharged thoroughly through the recharge well of this specifications.

Table 2. Results of pumping and recharge test for each experimental well.

Well name	Pumping	Observation	Recharge
Date	1997.1.10	1997.12.20	1997.1.7
NL (GL-m)	16.49	14.81	14.21
Qp (m^3/min)	0.60	0.15	0.54
PL (GL –m)	17.19	16.32	15.83
Tw (℃)	15.5	15.7	14.8
Ow-s (m)	0.06	--	0.21
Qr (m^3/min)	--	--	0.60
RL (GL-m)	--	--	13.55*

*1997.1.20, NL is GL-14.87m before the test.

NL = Natural water level , Qp = Pumping quantity ,PL = Pumping water level , Tw = pumping water temperature , Ow-s = drawdown on observation well ,Qr= Recharge quantity , RL = Recharge water level.

8.2 *Modeling of formations*

Geological numerical analysis model, shown in Figure 5, approximate roughly to the formations at the site based on the data from the experimental wells and other existing wells nearby. Formations are divided by 20 m cell in vertical direction. No.8 and No.9 are free cells and divided by ground water level. All aquifers are made of gravel. An impermeability layer between free ground water and aquifer 1 is not considered to the model because its thickness is only a few meters.

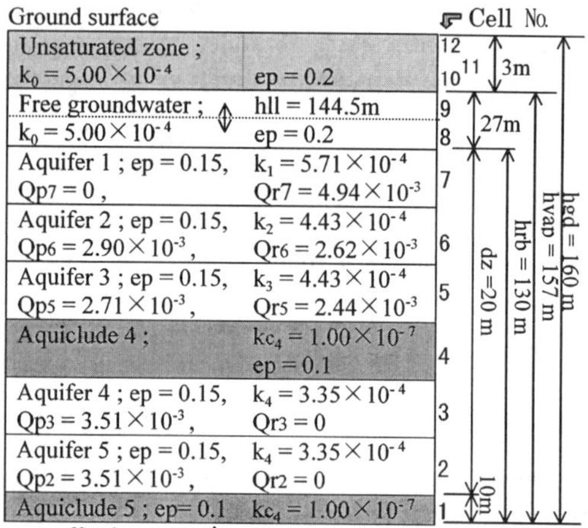

ep = effective porosity, k = permeability coefficient (m/s), Qp = pumping quantity per one cell (m3/s·cell), Qr = recharge quantity per one cell (m3/s·cell), dz = a unit of devided cells,hrb = the lowest high of fluctuated water level, hvap = thickness of unsaturated zone, hgd = high from the lowest plane, hll = high of water level in the first stage on (x =0 , y =0).

Figure 5. Geological numerical analysis model in vertical cross section.

8.3 *Comparison between field tests and numerical analysis results*

Cold water recharge test with 0.6 m3/min. pumping and recharge volume is performed with 7 days continuous operation followed by 10 days stop and 8 hours/day operation with 3 days. And the level and the temperature of ground water were measured during 14 days after above cold water recharge operation. Average temperature of cold water was recorded as 10.2 ℃, and this is the same as the designed ground water temperature after snow melting. Figure 6 shows records of ground water level measurement.

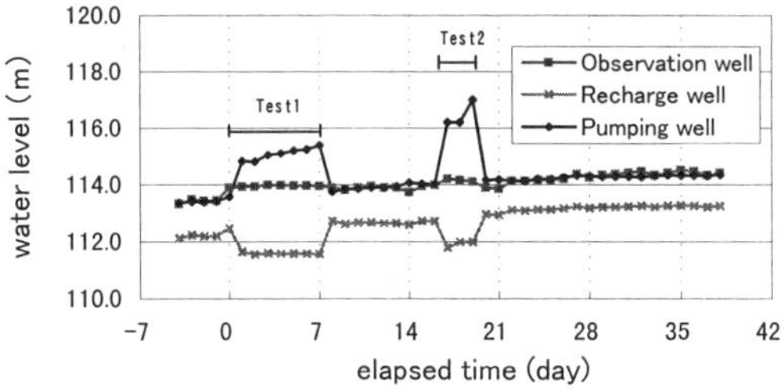

Figure 6. Records of the field test on ground water level measurement.

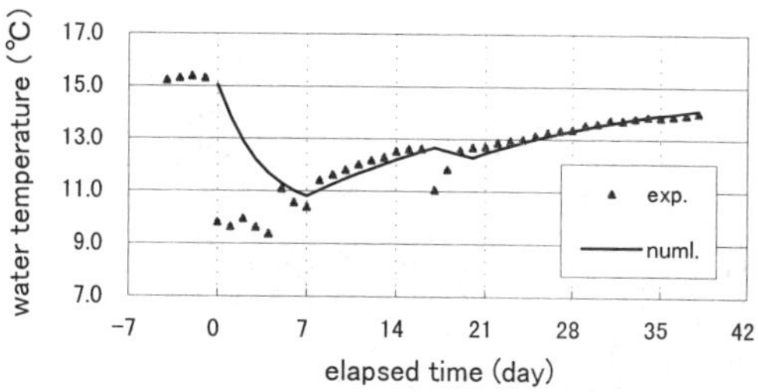

a. Comparison at Recharge well

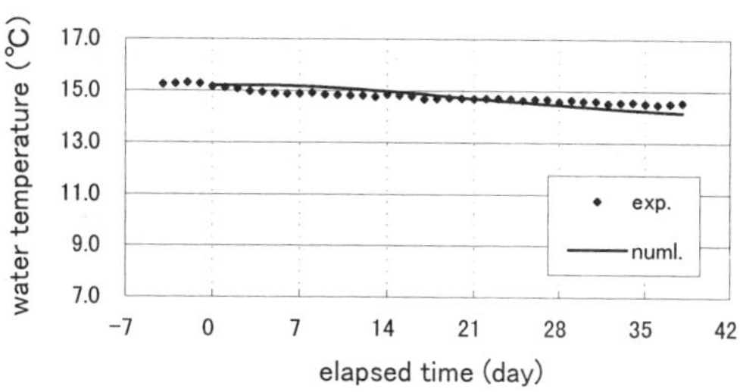

b. Comparison at Observation well

Figure 7.a,b. Comparison between the experiment and the numerical in Aquifer 2.

Other physical properties and recharge conditions are shown in Table 3. Ground water flow direction and inclination at this area is determined to N 65 W with a gradient of 6/1000.

Changes of ground water temperature in aquifer 2 are measured and simulated. Those in the recharge well and in the observation well are shown in Figure 7 a, b. Both seem to be similar with small differences caused from the conditions that a performance of the cooling tower is affected by the ambient temperature. Similar phenomena are observed in aquifer 1 and aquifer 3. Aquifer 4, however, is not affected by cold water recharge operation.

Therefore, it is confident that model for numerical analysis is appropriate.

Table 3. Physical properties.

D	10 W/mK	θ_0	15.2℃	d_{A1}	0.003 m
Rc	0.1	θ_i	10.2℃	d_{A2}	0.005 m
α	10 w/m² K	duz	0.01 M	d_{A3}	0.005 m
λ_p	0.0596 W/mK	dfg	0.01 m	d_{A4}	0.007 m

θ_0 = water temperature of the first stage, θ_i = recharge water temperature, d = particle diameter of aquifer.; subscript uz = unsaturated zone, fg = free groundwater, A1-A4 = Aquifer1-Aquifer4.

9 PREDICTION FOR WHOLE THE SITE

9.1 *Well positions and horizontal cell arrangement*

Area for snow melting system will be divided in sections of 1500 m2 which are calculated by the unit water requirement of 0.4 l/min·m2 and well capacity of 0.6 m3/min. The pumping well and the recharge well for each section are positioned in a distance of 50 m which is proved to be enough to prevent affect from cold water recharge by the field test. Positions of the wells and cell arrangements for analysis are shown in Figure 8. Analysis is performed on the conditions that a well is supposed to be positioned in the center of the closest cell and model of geoiogical formations is the same as Figure 5 and "hll" is 143.5 m.

9.2 *Simulated ground water level and temperature*

Ground water level and temperature is simulated by numerical analysis after 5 days continuous cold water recharge operation because it hardly snows more than 5 days continuously in Yamagata city. Conditions of analysis are shown in Table 4.

Expected ground water level for aquifer 2 and 4 is shown in Figure 9 a, b. In aquifer 4, where cold water is not recharged, about 0.8 m lowering is expected around well No.20 but it is negligible in practical. In aquifer2, where ground water is pumped and recharged, about 0.4 m raising or about 0.1 m lowering of ground water level is expected partially, but it can be said that no-change is expected in practical. Same phenomena are observed in aquifer 3 also. In aquifer 1, where ground water is only recharged, raising of ground water of 0.5 m in some area is calculated but it is negligible in practical.

Table 4. Conditions of analysis.

Well No.	Almost	6,7,18,19	25,26
Quantity*	0.60m³/min	0.32m³/min	0.54m³/min
Recharge tp	←	10 ℃	→
Initial tp	←	15 ℃	→
Atmospheric tp	←	1.8℃	→

* = Pumping or recharge temperature ; tp = temperature.

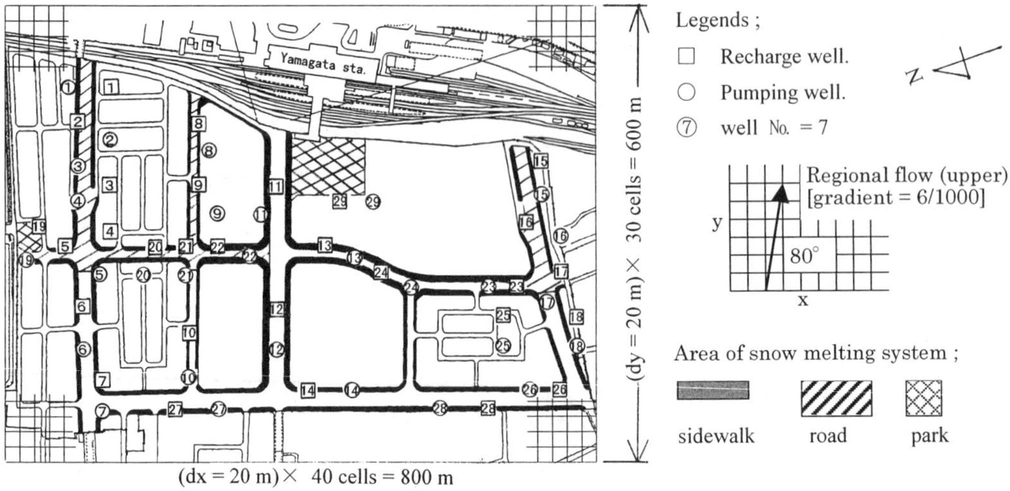

Figure 8 Positions of the wells and cell arrangements for analysis.

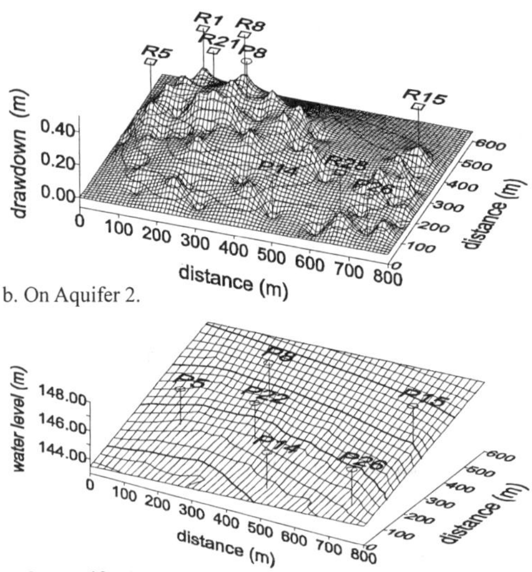

b. On Aquifer 2.

a. On Aquifer 4.

Figure 9. a, b. Expected ground water level (5 days after).

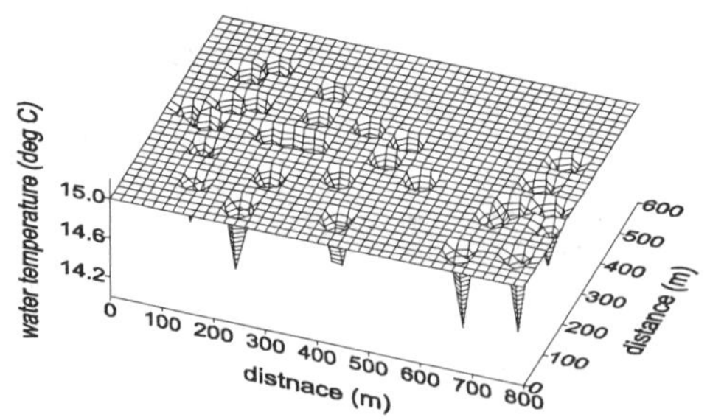

Figure 10. Expected ground water temperature on Aquifer 2 (5 days after).

Expected ground water temperature is also simulated for 5 days during operation and also 7days after operation and those for aquifer 2 are shown in Figure 10. The area where ground water temperature is affected by cold water recharge is limited up to 30 m. And such phenomena are disappeared after 7 days period except only small area close to the recharge well. And it is found that such affected area does not expand nor move.

10 CONCLUSION

It is confirmed that the simulation by numerical analysis for the influence of recharged cold ground water on level and temperature of surrounding ground water is effective. In case many wells for snow melting system are planned to be installed in a limited area, all the pumping wells and the recharge wells are positioned appropriately to minimize mutual interference by this simulation.

This simulation is confirmed and traced by only a short period of operation, therefore the effectiveness of this simulation for the seasonal operation based on the actual snow melting system operation will be confirmed in the coming winter season.

REFERENCES

Yokoyama T. et al. 1996 Treatment of Transient Heat Transfer among Fluids and Particles in Free Aquifers, *JSME int. Journal* in submission, Japan.

Artificial Recharge of Groundwater, Peters et al. (eds) © *1998 Balkema, Rotterdam, ISBN 90 5809 017 5*

ASR feasibility testing of the semi-confined Cretaceous Chalk aquifer of South Essex, England

M.C.Cook
Anglian Water Services, Cambridge, UK

S. Moncaster
Golder Associates (UK) Limited, UK

ABSTRACT: Anglian Water Services supply 1263 Ml of potable water per day to 3.9 million customers spread over an area of 27500 sq. kms in eastern England which includes up to a 100 Ml/d export to neighbouring water companies. The company are currently examining the feasibility of using aquifer storage and recovery (ASR) to supplement available groundwater resources. This paper presents the results of a pilot scale ASR test which was conducted at an Anglian Water site to the north of Colchester in the county of Essex, England. The test was based upon the injection and recovery of 2800 m^3 of treated water. The target aquifer was the semi-confined Chalk. The results of the test confirm that treated water can be injected into the Chalk at high rates but that the rate of re-abstraction will be limited by the large draw downs which were experienced. The test also confirms that diffusion exchange processes within the Chalk and interaction between the injected water and the overlying Tertiary strata, are both likely to impact on the injected/stored water quality. To further investigate the potential for developing an ASR scheme at the site, larger scale testing is being conducted.

1 INTRODUCTION

Anglian Water Services is the main subsidiary of Anglian Water Plc and is responsible for the abstraction, treatment and distribution of water supplies as well as the collection, treatment and disposal of waste water. Anglian Water Services serves 3.9 million water supply customers and 5.1 million waste water customers in an area of 27500 sq. kms (Fig. 1). A total of 1263 Ml/day is treated and put into supply of which 50% is groundwater and 50% is from surface reservoirs or direct intakes.

The other subsidiary, Anglian Water International, deals with non regulated operations, principally outside the UK.

An additional 2 million customers are served under concessions to water and waste water utilities.

Groundwater is abstracted from aquifer systems which range from confined to unconfined and from fissure flow to integranular flow (Fig. 2).

The average rainfall in the Anglian Water region is 596mm per annum. Rainfall since April 1995 has been 85% of the long term average. This has resulted in low river flows affecting reservoir refill potential and groundwater levels at or below minimum recorded levels.

Aquifer Storage and Recovery (ASR) is being evaluated as a water resources management tool to store surplus winter supplies in order to meet short term peak demands in the summer.

Figure 1. Anglian Water Region.

Anglian Water has completed a programme of investigation aimed at evaluating the potential of ASR in the supply area and more specifically in the confined/semi-confined Chalk of Essex.

1. N. Lincolnshire Chalk
2. S. Lincolnshire Chalk
3. N. Lincolnshire Limestone
4. C. Lincolnshire Limestone
5. S. Lincolnshire Limestone
6. Spilsby Sandstone
7. Sherwood Sandstone
8. Lower Greensand
9. Minor Aquifers
10. Sandringham Sands
11. Outcrop Chalk
12. Confined Chalk
13. Crag

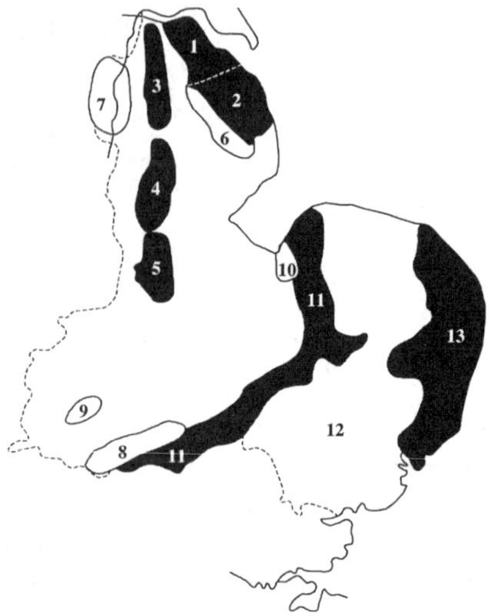

Figure 2. Aquifer Systems within the Anglian Water Region

2 ASR PROGRAMME AND SITE SELECTION

The Anglian Water ASR investigation programme comprises the following phases:

Phase I - Scoping Study
Phase II - Site Selection
Phase III- ASR Pilot Testing Programme
Phase IV - ASR Scheme Development

The major aquifers in the Anglian Water supply area have been considered in terms of their potential for ASR under a scoping study (Phase I). The scoping study considered the following evaluation criteria for each of the major aquifer units:

- hydrogeological conditions
- water quality in the aquifer and recharge water
- potential supply of additional water
- potential demands on an ASR scheme

The study concluded that the Lower Greensand aquifer of Bedfordshire and the Chalk aquifer of Suffolk and Essex had potential for ASR. Such a development would prove beneficial in terms of recharging parts of the Chalk aquifer where groundwater abstraction had resulted in piezometric depressions and to meet short term peak demands during drought periods.

A decision modelling approach was used to rank five sites in the Essex Chalk in terms of suitability for ASR pilot investigations (Phase III). The key criteria used for decision modelling are

- Hydrogeology (hydrochemical/hydraulic)
- Infrastructure (recharge/discharge)
- Potential for future development

Two preferred sites were later discounted due to difficulties obtaining a continuous supply of treated water for ASR injection purposes. The site selected for the Phase III investigation is located at Great Horkesley water treatment works. The facility supplies 280 l/s (max.) of treated groundwater to the town of Colchester.

3 PHASE III - TESTING OBJECTIVES AND TEST DESIGN

The near surface geology at the Great Horkesley test site is summarised in Figure 3;

The low permeability London Clay supports a perched water table in the Pleistocene sands and gravels and confines groundwaters in the Lower London Tertiary deposits and the Chalk.

Groundwater is abstracted from the Chalk aquifer 3 - 4 km to the north of the site and is used for public water supply. The abstraction is sustained, in part, by leakage from the overlying formations of the Lower London Tertiary Group (Woolwich and Reading Beds), particularly in areas where the London Clay cover has been eroded. Such conditions do not extend to the site and as such, it is believed that there is only minor recharge of the Chalk from the overlying strata. The position of the site on a groundwater divide further reduces the potential for active groundwater circulation.

The quality of the Chalk groundwater at the site is

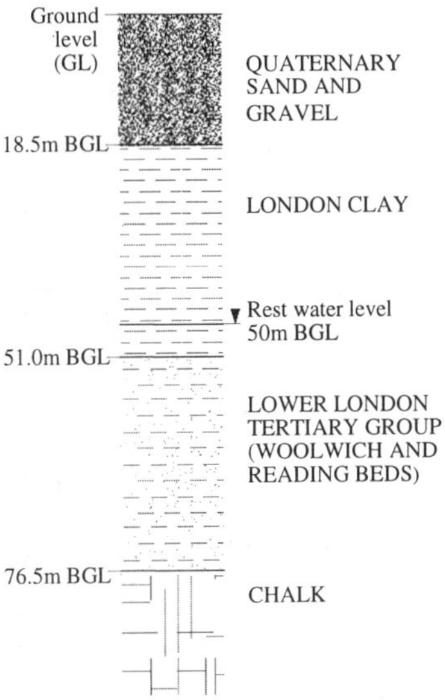

Figure 3. Site Geology

dominated by sodium, bicarbonate and chloride. In addition, however, significant concentrations of calcium, magnesium, potassium, chloride and sulphate are also present. These characteristics are believed to have evolved as a consequence of the mixing of modern recharge waters with older, in-situ, saline groundwaters. Sodium concentrations are also believed to have been further increased as a consequence of ion exchange beneath the London Clay cover (Lloyd et al, 1981). The hydrochemistry thus appears to confirm details of the conceptual hydraulic model.

The objective of the Phase III works was to determine whether or not a large scale ASR trial at Great Horkesley would be successful. Specific issues investigated included the following:

- whether or not it would be possible to inject and recover the treated water at sufficiently high rates;
- whether or not the quality of the injected water would be maintained in storage.

Phase II modelling indicated that the aquifer hydraulics were likely to be suitable for a large scale trial. The Phase II investigation also indicated, however, that the following processes may impact on the injected and/or stored water quality;

- mixing with marginal quality in-situ groundwater in the Chalk fractures;
- diffusion exchange with marginal quality in-situ groundwater in the Chalk matrix;
- pyrite oxidation within the Tertiary deposits;
- ion exchange within the Tertiary deposits;
- mineral dissolution and precipitation reactions, both within the Chalk and the Tertiary deposits.

In order to make a preliminary assessment of these issues, a small scale cycle test was designed. This comprised injection of 2.8×10^3 m^3 of treated water into the Chalk, a short period of storage and then abstraction of a volume of 7.8×10^3 m^3. Owing to restrictions on the availability of treated water, injection was restricted to four approximately 10 hour intervals commencing at 08:00 on consecutive days. Injection and recovery were at rates of approximately 20 l/s.

Borehole and aquifer performance were assessed by interpretation of the water level variations which were observed in the ASR pilot borehole, an adjacent Chalk observation borehole and a borehole completed into the Tertiary deposits. The Tertiary and Chalk observation boreholes were located within 8 m and 250 m of the pilot borehole respectively.

The potential for a hydrochemical impact was assessed by comparing the recovered water quality to the in-situ water quality (in both the Chalk and the Tertiary deposits) and the mineral composition of the Lower London Tertiaries.

4 PHASE III - ENGINEERING

The ASR Chalk pilot and Tertiary observation boreholes were both constructed specifically for the purposes of this investigation.

The ASR pilot borehole was drilled to 120 m below ground level (BGL) in order to fully penetrate the effective thickness of the Chalk aquifer. It was completed using 450 mm mild steel casing which was grouted 10 m into the top of the Chalk (Fig. 4). A direct circulation fluid flush technique was used in construction and the borehole was developed on completion by a combination of acidisation and air-lifting.

The Tertiary observation borehole was completed with 100 mm uPVC casing; plain to 55 m BGL and slotted with a 250 μm geotextile wrap screen between 55 and 70 m BGL. This borehole was drilled into clay rich strata immediately above the Chalk. A direct circulation fluid flush technique, using a polymer additive, was used in construction and the borehole was developed by the circulation of a hypochlorite solution and air-lifting. The section of the borehole between 55 and 70 m BGL was cored.

The ASR pilot borehole is completed with 75 mm diameter injection main, 100 mm diameter logging/water level monitoring tube, 75 mm diameter rising main and submersible pump (60 hp). The above equipment is suspended on a sealed flange plate.

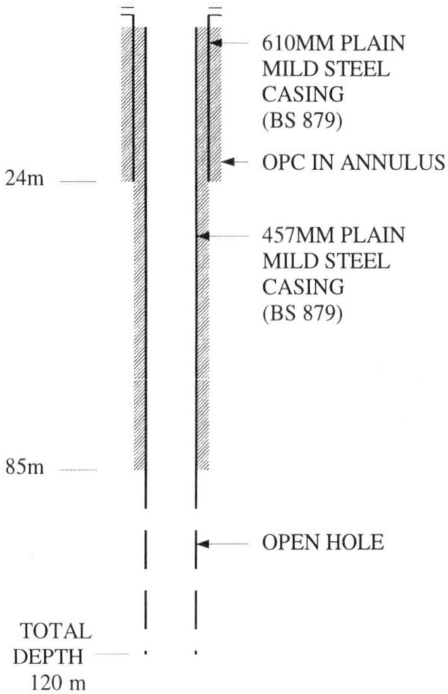

Figure 4. ASR Pilot Borehole Construction

5 PHASE III - RESULTS

Figures 5 and 6 indicate that injection produced a water level rise of 15 m in the pilot borehole, while five days post injection reabstraction induced a draw-down of 23 m. Injection produced little or no impact in the Tertiary borehole while post injection reabstraction induced a 3 m draw-down. Water levels in the Chalk observation borehole rose by 2 m during each phase of injection and fell by 5 m during the subsequent reabstraction.

Analysis of hydraulic response data from a pre ASR programme indicated that the Chalk in the vicinity of the site has a transmissivity of 100 m^2/d and storativity of 10^{-4}. The analysis shows no evidence of any leakage effects. This indicates that at the scale of the test, any hydraulic interaction between the Chalk and the Tertiary deposits was probably restricted to a small area in close proximity to the pilot borehole.

The results of the hydrochemical sampling and analysis are presented in Table 1. These confirm that the in-situ Chalk groundwater is of a sodium-bicarbonate-chloride type with significant concentrations of calcium, sulphate, magnesium and potassium. Sodium and potassium concentrations exceed those which are permitted in UK drinking waters (Water Supply (Water Quality) Regulations 1989). The injected water, which is derived from the Chalk well-field to the north of the site, is of a calcium-bicarbonate type with significant, concentrations of sodium, chloride and sulphate. With the exception of calcium, the concentrations of all of the major ions are lower in the injected water than in the in-situ water.

The groundwater stored in the Tertiary deposits is of a sodium-chloride-sulphate type with relatively low concentrations of calcium and bicarbonate and traces of both nitrate and ammoniacal nitrogen. The concentrations of sodium and chloride are similar to those which are observed in the Chalk. There is, however, a marked excess of sodium relative to chloride.

From Figure 7 it can be seen that the quality of the abstracted water varies as a function of the duration of the recovery. With the exception of calcium, concentrations of the major ions show an increase over that period which is equivalent to the recovery of the volume injected. By contrast, concentrations of calcium decline. The pattern of concentration variation is marked in all cases by initially rapid rates of change. These gradually diminish with time such that during the period in which "excess" groundwater was recovered, concentrations of all of the major ions stabilised. With the exception of calcium, bicarbonate and sulphate, the stabilised concentrations are slightly less than those which are observed in the in-situ Chalk groundwater. Calcium and bicarbonate stabilised at concentrations which are significantly lower; sulphate at concentrations which are higher.

The mineral assemblage which was present in the sampled core of the Tertiary deposits is given in Table 2 below.

Table 2. Lower London Tertiary Group - Mineralogy

Mineral	Concentration (%)
Quartz	71.7 - 89.1
Calcite	0 - 6.2
Kaolinite	1.6 - 6.0
Plagioclase	2.1 - 6.7
Dolomite	1.7 - 4.7
Illite/Mica	1.2 - 2.2
Chlorite	0 - 2.2
Alkali feldspar	0 - 2.55
Anhydrite	0 - 1.5
Pyrite	0 - 2.5

The mineralogy of the sampled material is thus dominated by quartz, as would be expected in predominantly sandy strata. However, significant concentrations of carbonate minerals, clays, feldspars and anhydrite/pyrite are also shown to be present.

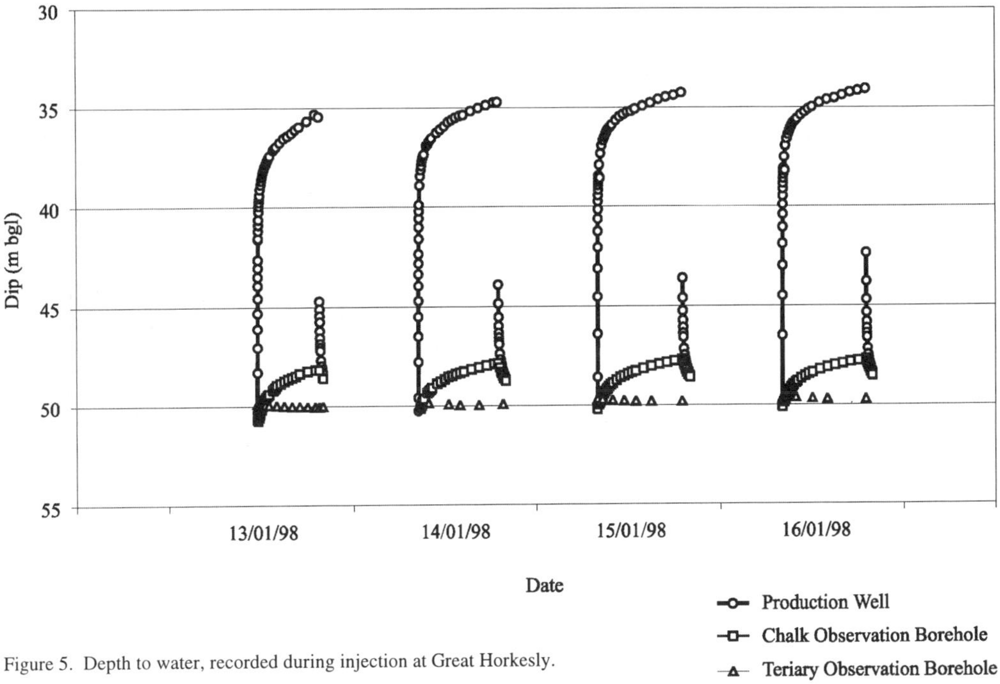

Figure 5. Depth to water, recorded during injection at Great Horkesly.

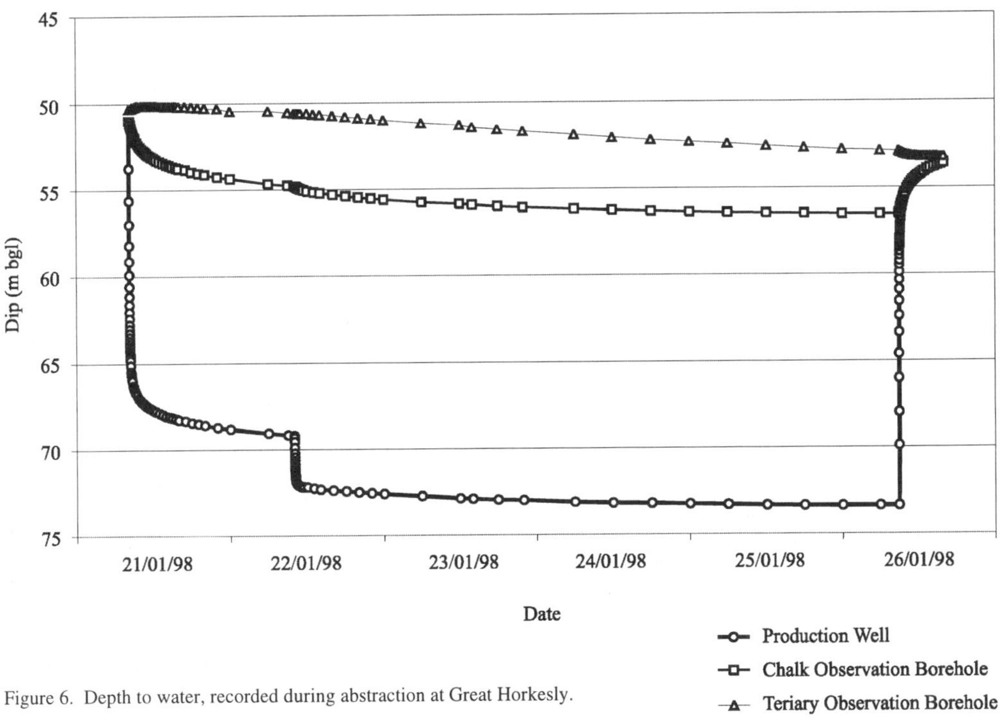

Figure 6. Depth to water, recorded during abstraction at Great Horkesly.

Table 1. Hydrochemical Data

Sample	Sampling Date and/or Time	(Solute Concentration mg/l)							
		Ca^{2+}	Mg^{2+}	Na^{+}	K^{+}	Cl^{-}	HCO_3^{-}	SO_4^{2-}	NO_3^{-}
In-situ Groundwater	18/11/97	70	40	177	23	236	381	82	0
Injected Water	composite sample	96	20	62	10	104	351	47	3
Tertiary Groundwater	21/01/98	31	13	183	43	202	60	194	5
Abstracted Water									
Sample A-1	21/01/98 08:45	82	29	114	17	171	291	72	0
Sample A-2	21/01/98 09:15	76	29	117	18	175	292	69	0
Sample A-3	21/01/98 09:45	75	29	117	18	176	288	70	0
Sample A-4	21/01/98 10:15	73	30	122	18	181	289	73	0
Sample A-5	21/01/98 11:00	75	30	125	19	185	289	72	0
Sample A-6	21/01/98 12:00	73	31	126	19	184	292	73	0
Sample A-7	21/01/98 14:00	70	31	128	19	191	291	77	0
Sample A-8	21/01/98 16:00	70	32	134	19	192	299	76	0
Sample A-9	21/01/98 18:00	70	32	136	20	199	297	79	0
Sample A-10	21/01/98 19:50	70	33	138	20	197	297	78	0
Sample A-11	22/01/98 08:30	66	34	143	20	212	300	78	0
Sample A-12	22/01/98 14:05	67	35	147	21	208	308	84	0
Sample A-13	22/01/98 19:45	66	34	147	20	214	308	83	0
Sample A-14	23/01/98 08:45	62	36	150	20	223	312	88	0
Sample A-15	23/01/98 18:30	64	35	151	20	226	310	87	0
Sample A-16	24/01/98 12:30	63	35	151	20	224	317	85	0
Sample A-17	25/01/98 12:30	60	35	155	22	224	311	86	0
Sample A-18	26/01/98 08:00	61	36	157	21	222	314	82	0
Sample A-19	26/01/98 08:05	61	36	158	21	222	313	81	0
Sample A-20	26/01/98 08:10	61	35	155	20	224	309	79	0

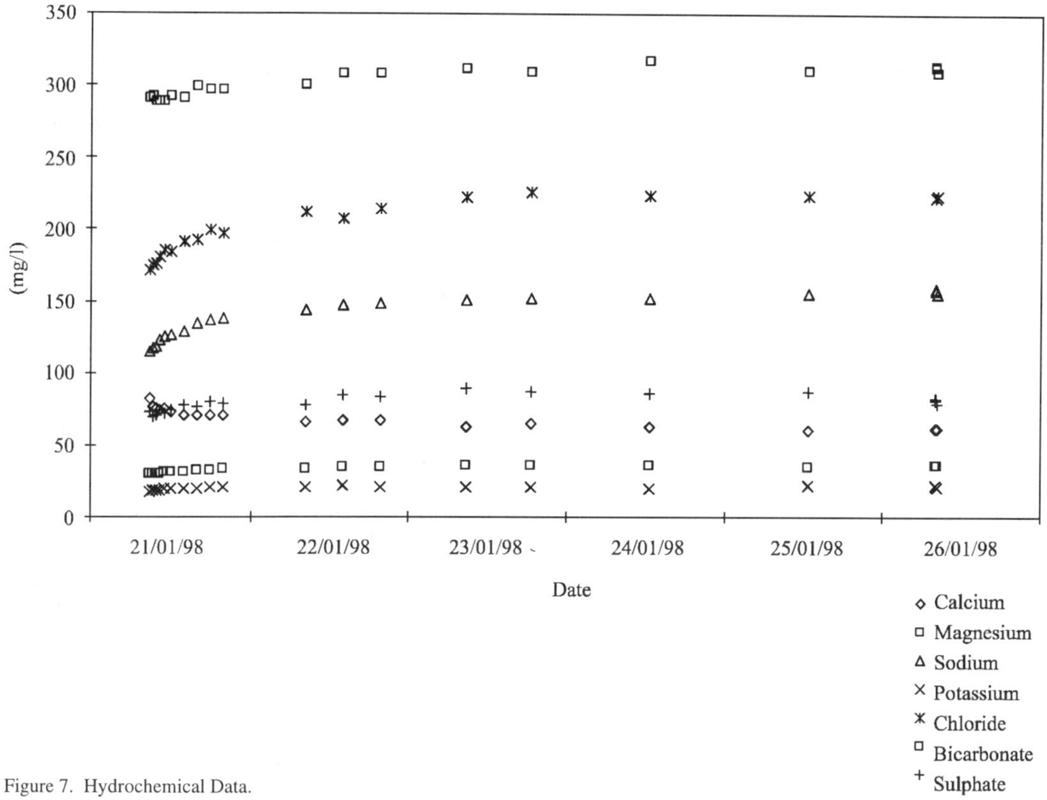

Figure 7. Hydrochemical Data.

6 PHASE III - DISCUSSION

Given the large depth to groundwater at the site, it can be seen that high rates of treated water injection can be accommodated with little or no risk of excessive head build up in the borehole.

By contrast, the 23 m drawdown can be seen to have lowered water levels in the Chalk to the level of the junction with the base of the Tertiary deposits. Given both the adverse geochemical impacts (pyrite oxidation and ion exchange) which are associated with de-watering the Tertiary deposits (Kinneburgh et al, 1994) and the poor borehole performance which would arise from de-watering the aquifer, it appears that recovery from individual boreholes may be restricted to rates which are less than 20 l/s.

From Table 1 it can be seen that throughout the recovery phase the recovered water quality was poorer than that which was injected. Specifically, concentrations of potassium remained in excess of those permitted in UK drinking waters while concentrations of sodium increased from below to above the permitted limit (12 mg/l and 150 mg/l respectively from the (Water Supply (Water Quality) Regulations 1989). At the scale of the test, therefore, it is apparent that the injected water quality cannot be maintained in storage. Possible explanations are given below;

6.1 *Advection-dispersion mixing*

This process acts at the margins of the injected water mass and results in mixing between the injected and in-situ waters. Advection-dispersion mixing impacts are well documented and, where the quality of the in-situ water is marginal, the process typically results in contamination of the water which is recovered during the first few test and/or operational cycles.

Given that advection-dispersion impacts are restricted to the margins of the injected water mass, the injected water which is stored closer to the borehole should maintain its quality. Under such circumstances, the quality of the water which is initially recovered will approximate that which is injected. Examination of Table 1 shows this clearly was not the case and suggests that processes other than advection-dispersion were responsible for the poor quality of the water which was recovered.

Possible alternative and/or contributory processes include off-site migration of the injected water and diffusion exchange.

6.2 *Off-site migration*

Problems associated with off site migration typically arise in circumstances where hydraulic gradients are steep and/or aquifer transmissivity is high. Neither conditions are found at the test site and as such the possibility that off-site migration is responsible for the poor quality of the recovered water is considered to be remote.

6.3 *Diffusion exchange*

The water which was injected into the Chalk will have flowed away from the borehole through a network of interconnected fractures. Owing to low permeability, no significant flows are likely to have been induced in the Chalk rock (or matrix). Groundwater held within the matrix is thus likely to have remained in-place during the test and in areas close to the borehole, have been in contact with the injected water. As a consequence of molecular diffusion processes, solutes will have been exchanged between the two types of water across fracture surfaces. This process may have resulted in the quality of the injected water evolving towards that of the in-situ groundwater.

Evidence to support a diffusion impact includes the poor quality of all of the water which was recovered. The lack of any reasonable quality water suggests that the process responsible operated throughout the region in which the injected water was stored and was not constrained to the margin of the injected water mass. In addition, the initially high rates of change of concentration differ from those which would be expected from an advection-dispersion impact alone.

6.4 *Interaction with Tertiary Strata*

The recovery of sulphate at concentrations which are in excess of those present in either the injected or in-situ Chalk groundwater suggests that the test produced some impact in the Tertiary strata.

The simplest explanation is that the drawdown in the Chalk during abstraction induced leakage from the Tertiary deposits in the vicinity of the borehole. Given that the Tertiary groundwaters contain relatively high concentrations of sulphate, this may have resulted in the elevated sulphate concentrations which were measured. However, migration of the oxygen saturated injected water into the base of the Tertiaries could also have resulted in pyrite oxidation and the release of sulphate; a process which may then have been exacerbated by localised de-watering and the entry of air into the formation. Similarly, the dissolution of any anhydrite which is present at the base of the Tertiary deposits would also release sulphate.

Whichever process is responsible, the recovery of excessive concentrations of sulphate confirms some form of hydro-geochemical interaction between the Chalk and the overlying tertiary deposits. Given that the UK Water Supply (Water Quality) Regulations 1989 specify a maximum sulphate concentration of 250 mg/l, it is apparent that any interaction which generates additional sulphate may prevent successful development of an ASR scheme at the site.

7 PHASE III - LONG DURATION TESTING

In view of the above, the priorities for future investigations can be seen to include (a) identifying the processes which are responsible for the generally poor quality of the water which was recovered and (b) identifying the processes which resulted in the sulphate impact. To enable these issues to be investigated a larger scale test has been designed. This comprises injection of a volume of 20×10^3 m^3 of treated water, a period of storage and then abstraction of a volume in excess of that injected. The aim is to replicate the responses observed during the Phase III test and to accentuate those which are scale dependent. The test is currently under way and field work will be completed by early April 1998.

8 SUMMARY AND CONCLUSION

Anglian Water Services are currently investigating the potential for using ASR to provide additional groundwater resources that can be used to meet short term peak demands.

The investigation identified the semi- confined Chalk of Essex and Suffolk as an aquifer with the potential for ASR development. Subsequent work located a suitable test site at Great Horkesley, near Colchester. Chalk groundwater at the Great Horkesley site is confined by a thick cover of Tertiary deposits and is of marginal quality.

The work which has been completed at Great Horkesley includes the construction of a pilot ASR borehole and a Tertiary observation borehole. A small scale cycle test has been completed; this was based on the injection and recovery of 2800 m^3 of treated water. The test confirmed that large volumes of treated water could be injected into the Chalk but that recovery would probably be limited by comparatively low borehole yields and the associated hydraulic and hydrochemical impacts. The test also demonstrated that maintaining the quality of the injected water could be problematical. Processes which are thought to have impacted on the recovered water quality include advection-dispersion and diffusion exchange within the Chalk and a combination of processes associated with the Tertiary deposits. These include leakage and mixing, pyrite oxidation and/or anhydrite dissolution.

To investigate further the potential for developing the site, a larger scale test has been designed. This is currently being implemented. The field works will be completed by April 1998, test data interpretation shortly thereafter.

REFERENCES

D.G. Kinneburgh, I.N. Gale, P.L. Smedley, W.G. Darling, J.M. West, R.T. Kimblin, A. Parker, J.E. Rae, P.J. Aldous, and M.J. O'Shea, 1994. "The effects of historic abstraction of groundwater from the London Basin aquifers on groundwater quality" *Applied Geochemistry*. 9, pp 175-195.

J.W. Lloyd, D. Harker, and A. Baxendale, 1981. "Recharge mechanisms and groundwater flow in the Chalk and drift deposits of southern East Anglia" *Q.J. eng. Geol. London*. 14, pp 87-96.

1.9 Pollutants and basin recharge

Artificial Recharge of Groundwater, Peters et al. (eds) © 1998 Balkema, Rotterdam, ISBN 90 5809 017 5

Microparticle facilitated transport of contaminants during artificial groundwater recharge

T. Hofmann & U. Schöttler
Institute for Water Research, Schwerte, Germany

ABSTRACT: Many substances show a strong tendency to sorb readily to the solid phase of an aquifer. Commonly they are regarded to be immobile or retarded. However, colloidal and suspended microparticles may be mobile also in the subsurface environment. Thus, microparticles should be taken into account to predict contaminant transport. A particle load at the "Hengsen-Island" test site (Dortmund, Germany) of less than 20 µg/l inorganic microparticles was observed, compared to 1.2 – 3.6 mg/l of macromolecular organic carbon. These values can increase by several orders of magnitude at the bankfiltration site during hydraulic mobilisation. No or only little mobilisation was observed at the slow sand filter site, where hydraulic changes are frequent due to the operating water works. There was a very good correlation between data from particle analysis and microbiology.

1 INTRODUCTION

Artificial groundwater recharge is a widespread technique to treat surface water for drinking water supply and to increase the quantity of natural groundwater reserves. In general, the recharge of groundwater reservoirs can be enhanced by injection wells, infiltration ponds, slow sand filters or bank-filtration. Pre-treated river water is often used for artificial recharge of groundwater although it does have a higher content of colloidal and suspended particles than groundwater.

During the infiltration and pre-treatment process colloidal and suspended particles and many contaminants are retained or reduced. This is due to several chemical, physical, and biological processes along the infiltration path of the water. The elimination of substances is a function of the hydrochemical and hydrobiological boundary conditions in the infiltration system. Apart from real elimination many contaminants are held back by filtration.

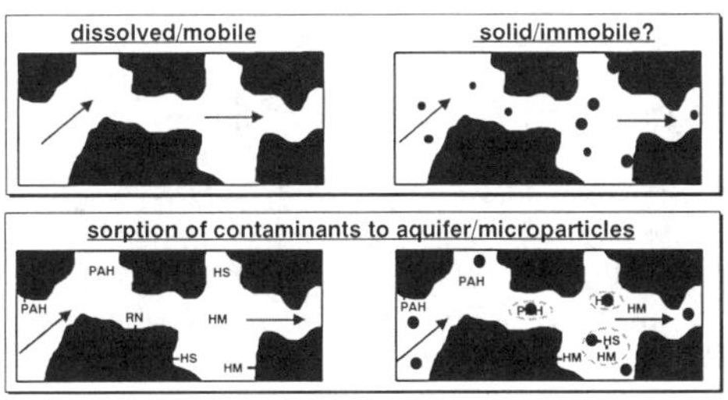

Figure 1. Transport mechanisms of contaminants in subsurface environments; RN = Radionucleides; HM = Heavy Metals; HS = Humic Substances

The ability to predict contaminant transport during all stages of artificial groundwater recharge is one of the main objectives to provide a sustainable management of recharge plants. Therefore major pathways and transport mechanisms must be defined. Most approaches to describe and predict transport phenomena treat groundwater as a two phase system: a mobile aqueous phase and an immobile solid phase. Contaminants partition between this two systems and those which have a strong tendency to sorb readily to the solid phase are considered to be virtually immobile or retarded (Fig. 1).

However, colloidal and suspended particles also may be mobile during artificial groundwater recharge. The composition of these microparticles is expected to be chemically similar to that of the surface of the aquifer material. Thus, they can also sorb organic and inorganic contaminants (McCarthy & Zachara 1989). Predictions on the transport of contaminants may fail if microparticle facilitated transport is not taken into consideration.

In Table 1 results from previous research projects at the Institute for Water Research concerning microparticle transport are listed (Hofmann & Schulte-Ebbert 1997, Schöttler & Schulte-Ebbert 1995).

Table 1. Dissolved and microparticle bound transport

Group	Substance	Transport Dissolved	Transport Colloidal
Microorganisms	Enterobacteria/Virus	-	XXX
Heavy metals	Fe, Mn	XX	XX
	Cd, Ni, Pb, Zn	XX	XX
Radionuclides	U, Th, Co, Cs	X	XXX
PAH	Fluoranthene	XX	XX
	Benzo(a)pyrene	X	XXX
	Indeno(1,2,3-cd)pyrene	X	XXX
Triazine	Atrazine	XXX	X
	Desethylatrazine	XXX	X
	Simazine	XXX	X
Biogenic substances	Toluene	X	XX
	m-Xylene	X	XX
	Geosmin	X	XX
	Pentadecene-1	X	XX
	Methylsulfides	X	XX

- = no influence XX = significant influence
X = little influence XXX = dominating influence

Many nonpolar organic and biogenic substances, e.g. polycyclic aromatic hydrocarbons, geosmine, toluene and m-xylene have a strong tendency to adsorb to the solid phase. Inorganic trace substances, like heavy metals and radionuclides can show the same tendency (Backhus & Geschwend 1990, McCarthy & Zachara 1989, Stumm 1987) or build up particles themselves.

This article describes the major mechanisms of microparticle origin and transport and presents the results from a monitoring campaign of an artificial groundwater recharge plant. This recharge plant, "Hengsen-Island", is operated by the Dortmunder Energie und Wasser GmbH and is located within the Ruhr valley close to Dortmund/Germany.

2 ORIGIN AND TRANSPORT OF PARTICLES

2.1 *Formation and origin of particles*

In this paper we refer to microparticles (or particles) as potential mobile solids including macromolecular (or "dissolved") components of organic carbon. They can cover a broad size range from a few nm up to 10-20 µm. Thus, common filtration techniques in groundwater quality monitoring, i.e. filtration through 0.45 µm filters, may exaggerate the concentration of dissolved elements and underestimate the total element content. Microparticles can be colloidal stable or suspended. The colloidal stability and transport behaviour of a microparticle is a complex function of the surface charge, density, surface chemistry, water chemistry and flow velocity. These conditions change along the flow path in the aquifer. Thus, for practical reasons colloids and suspended particles are not differentiated.

Many different particles can be expected in artificial recharge systems. River water shows for example strong temporal changes in particle concentration, composition and size distribution, depending on hydrological fluctuations (Ledin 1996). Apart from the transport of particles from the pre-treated water and from the vadose zone, the aquifer itself can consist to a large part of small particles < 20 µm.

These particles were deposited during sedimentation of the aquifer or formed by post-depositional weathering of thermodynamically unstable primary minerals. The composition of these particles depends on the sedimentation conditions, the chemistry of the pore water solution and the mineral content of the original geologic material. Dominating inorganic particles are aluminosilicates, oxides and hydroxides of aluminium, iron, manganese as well as calcium and magnesium carbonates and silica (Jenkins & Lion 1993, Schöttler & Schulte-Ebbert 1995).

A second group of particles in the aquifer is formed by precipitation or mobilisation due to a geochemical gradient along the flow path. Particles can form or be mobilised due to changes of parameters like pH, Eh, electrical conductivity or the content of dissolved gases. Changes in the hydrochemical environment are frequently observed in artificial recharge systems. Bankfiltration of river water leads at the test site "Hengsen-Island" to anoxic conditions in the aquifer by microbial reduction of O_2, NO_3^-, NO_2^-, MnO_2, Fe(III)-oxide, SO_4^{2-} and CO_2 (Schöttler & Schulte-Ebbert 1995). Iron and Manganese coatings of the aquifer material are dissolved during this process and liberate formerly fixed particles and elements in these coatings. Nevertheless, the total content of iron and manganese, released by dissolution of these cements, is limited by the solubility of carbonates and sulfides. Geochemical modeling with PHREEQC predicts at "Hengsen-Island" the precipitation of iron and manganese carbonates and sulfides, which may lead to the formation of new particles because supersaturation is reached. These particles can change their crystal structure during transport in the groundwater and modify their physical and chemical properties. For example, the transition of ferrihydrite to goethite leads to larger crystals and hence to different sorption and transport properties (Cornell & Schwertmann 1996).

Biogenic substances are from a third group of microparticles. Organic macromolecules and humic substances can be released from the aquifer material or brought in by the infiltrated surface water. Bacteria attached to the matrix may become unattached or can be infiltrated. They may release extracellular matter or cell fragments. Another rare but very important and conceivably dangerous group are spores and viruses. Most of these organic substances and micro-organisms have a strong tendency to adsorb to

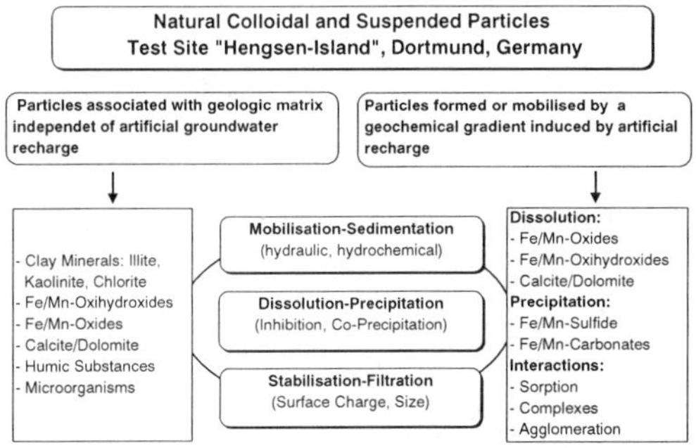

Figure 2. Interactions between microparticles detected at the "Hengsen-Island" test site

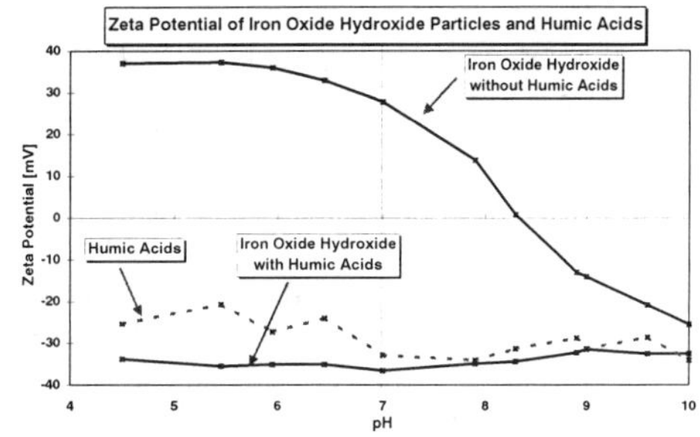

Figure 3. Charge reversal of ferrihydrite due to the adsorption of humic substances

the solid phase and therefore also to inorganic particles or build up aggregates with them. This effect

can give rise to serious problems during the disinfection of drinking water.

Figure 2 shows the interactions between particles which were detected with SEM/EDS, XRD and common groundwater analytical techniques at the test site "Hengsen-Island".

2.2 *Transport and filtration of particles*

Suspended or colloidal particles can be mobile in the aquifer if they are not susceptible to filtration and stable with respect to agglomeration and dissolution. Filtration effects can be either straining filtration or physico-chemical collection. While the maximum size of mobile particles is limited by straining filtration and pore velocity - for example on the monitored test site "Insel Hengsen“ 5-10 µm - concentration and size distribution of particles in the nanometer size range are more controlled by physico-chemical filtration. A minimum filtration rate can be observed at a size range close to 1 µm, where straining filtration is less effective in most pore aquifers and physico-chemical filtration does not yet reach its maximum (Bedbur 1993).

The collection of particles on the aquifer material strongly depends on the surface charge. An opposite surface charge of particle and aquifer matrix enhances collection. Differently charged particles favour coagulation and consequently sedimentation. This effect is even more complex regarding clay minerals, because the edges and faces may show opposite charge. A charge reversal of particles can occur due to adsorption processes of polyelectrolytic substances. The adsorption of humic substances to positive charged particles like ferrihydrite can cause a charge reversal and facilitate transport (Fig. 3).

3 SAMPLING AND ANALYTICAL PROTOCOL

3.1 *Site description*

The test site is located within the Ruhr valley close to Dortmund in NW-Germany. The area "Hengsen-Island" is completely surrounded by the Ruhr river and a storage lake. After prefiltration through fine gravel filters the river water is recharged in the north-eastern part of the test site by slow sand filtration (Fig. 4). The oxic infiltrated water is captured by a drain tile after an average residence time of 3 days in the aquifer. Due to fluctuations in the demand of the water works this part of the test site is influenced by very strong hydraulic changes.

The south-western part of the test site is only influenced by bankfiltration and not by the slow sand filters or the drain tile. Schöttler & Schulte-Ebbert (1995) observed a complete reduction of O_2, NO_3^-, NO_2^-, MnO_2, Fe(III)-oxide and SO_4^{2-}, but only minor hydraulic changes.

3.2 *Groundwater sampling*

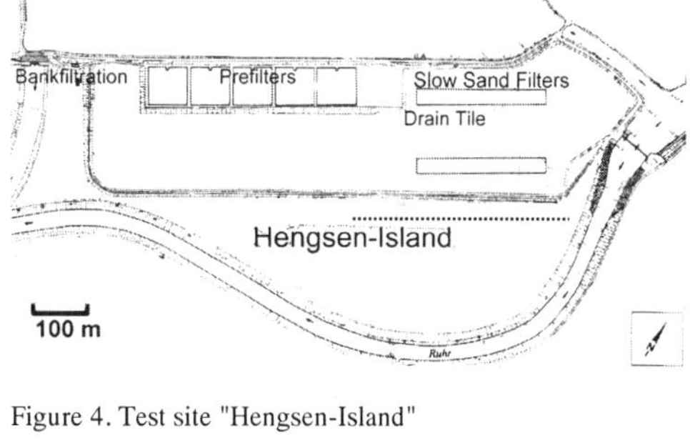

Figure 4. Test site "Hengsen-Island"

In general, common sampling procedures are linked to changes in the hydrochemical boundary conditions, e.g. changes in Eh, pH, temperature, exposure to daylight, and the content of dissolved gases, mainly oxygen and carbondioxide. In anoxic environments with high contents of dissolved iron even the smallest amount of oxygen will immediately (within a few seconds) induce the precipitation of iron oxyhydroxide particles and thus destroy any information about the natural conditions. Small changes in the hydraulic conditions due to pumping during groundwater sampling will mobilise particles. This effect is extremely strong in the well itself. To provide representative sampling hydraulic effects must be minimised, and a sedimentation effect of whirled up particles in the well itself must be prevented. Furthermore, any contact of the sample with the atmosphere (for anoxic samples) must be excluded.

Groundwater samples were taken with a submersible Grundfos MP1 pump. We use rigid PVC tubing from the pump to the sampling vessel in order to minimise gas diffusion through the sampling material (10 times lower for rigid PVC than for plasticised PVC tubing). Samples are taken in argon preflushed, flow through gas sampling vessels in order to prevent any contact with the atmosphere. Following this protocol it is possible to store samples several days without any alteration in the particle concentration and size distribution. Before sampling observation wells are purged for a minimum time of 24 hours up to 72 hours to prevent any contamination by whirled up particles in the well itself. The pumping rate was adjusted to a constant value between 1 - 1.5 l/min in an aquifer with a hydraulic conductivity between $1E^{-03}$ - $1E^{-01}$ m/s. A drawdown of less than 1 cm can be observed. Samples were taken after the particle concentration reached a constant value. We call this protocol "long-time-sampling" to achieve information about the natural particle content in the groundwater.

In order to mobilise the particle depot in the aquifer and to study the time and pumping rate dependency of this depot a "shock-sampling" protocol was applied. This means that after taking a "long-time" sample the pumping rate is abruptly increased to 15 – 17 l/min and the decrease in particle concentration is monitored after a drastic initial increase.

3.3 *Particle and groundwater analysis*

Particle concentration and size distribution in a size range between 0.01 – 2 μm were measured with PCS (photon correlation spectroscopy). The measurements were performed with a Malvern ZetaSizer 5000 equipped with a 2 W Spectra-Physics argon-laser and a flow through cell. This technique uses the principle of temporal fluctuations of the scattered light intensity, which is based on the Brownian motion of particles (Berne & Pecora 1976). The average scattered light intensity can also be used to determine the particle concentration if suitable reference particles are available. There was a good correlation between scattered light intensity of natural reference particles extracted from "Hengsen-Island" and particle concentration down to 20 μg/l (Fig. 5).

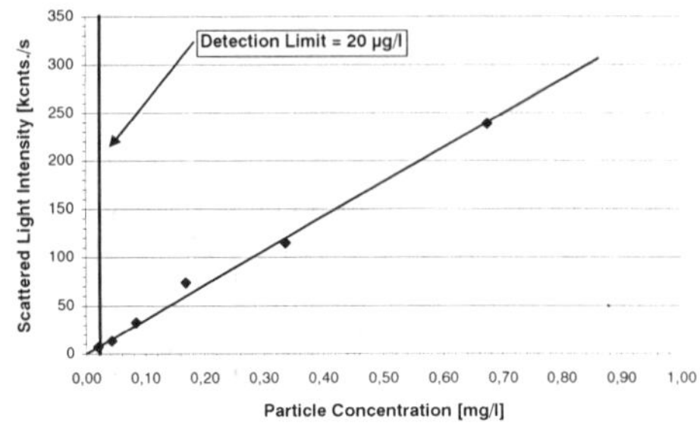

Figure 5. Correlation between scattered light intensity and particle concentration of reference colloids

Particle concentration and size range above 0.5 μm were measured in-line with a Galai CIS-1 by Schäfer, Mainz University (Schäfer et al. 1997). This detection system measures particles with the laser extinction technique based on the time-of-transition principle (Aharonson et al. 1986). At "Hengsen-Island" the detection limit for this method was about 1000 particles/ml or 1 μg/l (assuming a density of 2 g/cm^3 for particles).

In addition all water samples were analysed for major anions/cations, heavy metals and colony forming units (CFU). Temperature, pH, Eh, electric conductivity, and oxygen were measured through an in-line flow cell.

4 RESULTS

4.1 *Natural particle load at "Hengsen-Island"*

The crucial point in determining the natural particle load is to minimise the influence of the sampling itself. In order to get information about the particle transport without hydraulic influence due to sampling the observation wells were purged until no further changes occurred. In general, sampling was performed after 24 - 72 h of purging, depending on the sensitivity of the system.

Total particle concentration in all observation wells was below or close to the detection limit of 20 μg/l (PCS). The concentration of particles larger than 0.5 μm was below or close to 1 – 15 μg/l (CIS). The macromolecular organic carbon measured as DOC varied between 3.6 mg/l at the bankfiltration site of "Hengsen-Island" and 1.2 mg/l in the slow sand filter area. Colony forming units showed a

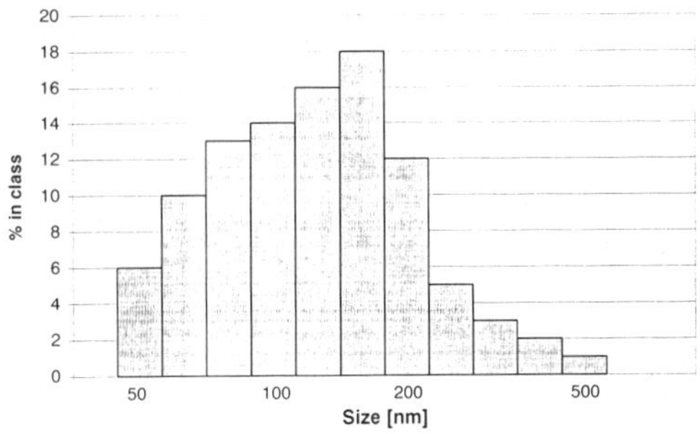

Figure 6. Particle size distribution measured with PCS, "long-time-sampling"

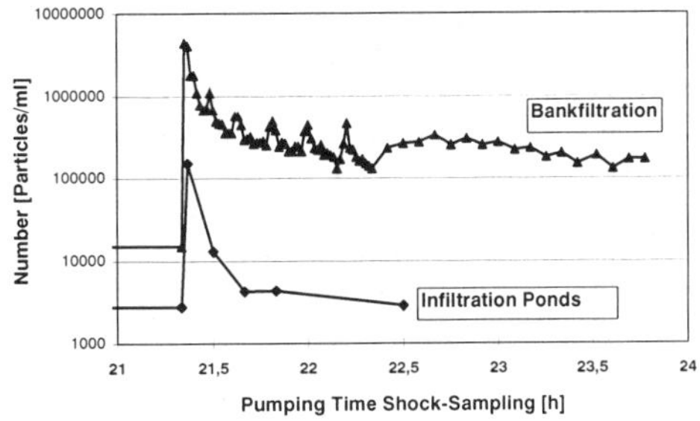

Figure 8. Particle concentration measured online with CIS

sharp decrease after the first hours of purging and reached especially in anoxic observation wells values as low as 1 - 20 CFU/ml. The particle size measured with PCS showed a maximum at 170 nm. Only few particles were larger than 500 nm. The size distribution of a representative sample is shown in Figure 6. Corresponding results were obtained using scanning electron microscopy on particles collected on polycarbonate membrane filters with a pore size of 50 nm.

4.2 *Mobilisation of particles*

Hydraulic changes are frequent in an operating artificial recharge plant. In order to simulate these effects and to study the particle depot in the aquifer the previous described "shock-sampling" protocol was applied. Directly after increasing the pumping rate the particle concentration increased by several orders of magnitude. At the bankfiltration site the particle concentration was between 18.5 – 196 mg/l during "shock-sampling" (measured gravimetrically after collection on 200 nm polycarbonate membrane filters). At the slow sand filter site, were hydraulic changes are frequent due to the operating water works, less than 2.0 mg/l of particles can be mobilised. The concentration of colloidal stable particles determined with PCS varied between 2 mg/l and the upper limit of the PCS configuration (20 mg/l) at the bankfiltration site and 0.02 – 0.82 mg/l at the slow sand filter site. Size range increased significant to an average diameter of 1.7 μm (Fig. 7). Organic carbon was not significantly altered by shock sampling procedure.

The increase in the particle concentration was accompanied by higher heavy metal concentrations. Mn, Fe, Zn and SiO_2 increased by 6 – 89 % as compared to samples collected by following the "long-time-sampling" protocol. Ni, Cr, Pb, Cu increased by 100 - 300 % at the bankfiltration site. There is no or only a very insignificant increase in heavy metals at the slow sand filter site.

In-flow CIS measurements clarify these effects. In Figure 8 the different behaviour of an observation well in the slow sand filter and the bankfiltration part of the test site is shown. After reaching stable conditions at 21 h there is a sharp increase in the particle concentration due to the "shock-sampling" procedure. The observation well at the bankfiltration site showed a slow and constant decrease subse-

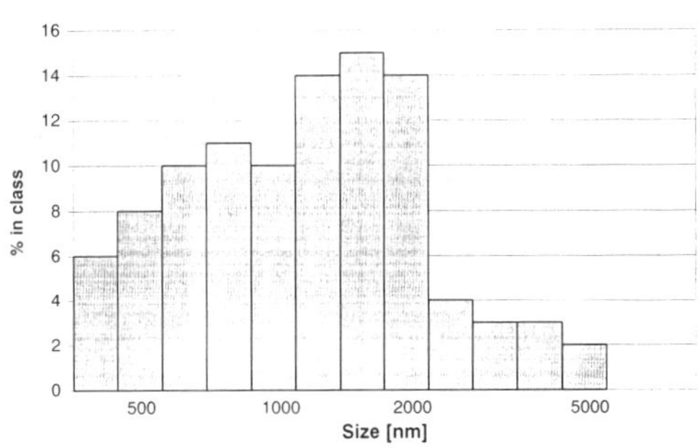

Figure 7. Particle size distribution measured with PCS, "shock-sampling"

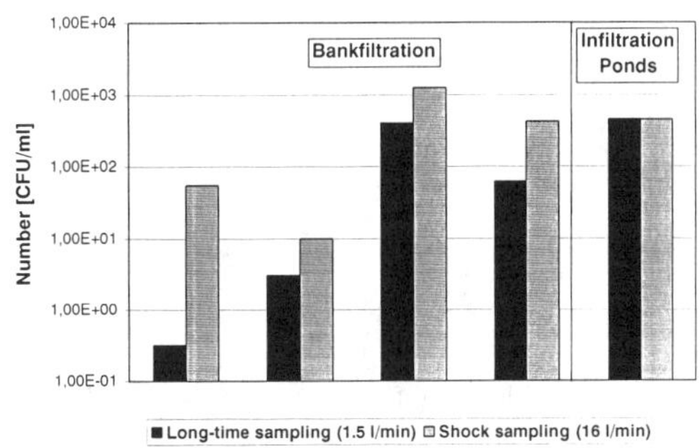

Figure 9. Colony forming units during "long-time" and "shock-sampling"

quently. In contrast, the observation well at the slow sand filter site showed a lower increase in particle concentration during shock sampling and a decrease to the previous value within only 30 min. Obviously the influence of 50 years of water works operations are visible: at the bankfiltration site, where only small hydraulic fluctuations occur due to constant bank filtration, a huge particle depot is mobilised. At the slow sand filter site, where hydraulic changes are frequent due to the operation of the water works, the reservoir of mobilisable particles has been largely depleted.

If these results are compared with the microbiological data there is also a very good correlation. The colony forming units increase significantly at the bankfiltration site during "shock-sampling" (4 samples). On contrary, their is no increase at the slow sand filter site (Fig. 9).

5 CONCLUSIONS

1. The natural inorganic particle load at the test site "Hengsen Island" is lower than 20 µg/l if adequate sampling methods, i.e. no oxidation of anoxic samples and 24 – 72 h purging is applied. The macromolecular organic carbon concentration varies between 1.2 mg/l at the slow sand filter site and 3.6 mg/l at the bankfiltration site due to different infiltration processes.

2. The size range distribution measured with PCS shows a maximum at 170 nm during "long-time-sampling". Colloidal particles during "shock-sampling" show an average size of 1.7 µm. This data shows a good correlation with results from scanning electron microscopy.

3. There is a significant mobilisation of particles during the simulation of hydraulic changes at the bankfiltration site, which is normally not influenced by the operating water works. There is no mobilisation of particles or colony forming units at the slow sand filter site, where hydraulic changes are frequent.

4. A very good correlation between inorganic particle mobilisation and the mobilisation of microorganisms measured as colony forming units was demonstrated.

6 ACKNOWLEDGEMENTS

This project is supported by the EC Environment and Climate Research Programme (ENV4-CT95-0071) and the DVGW (Deutscher Verein des Gas- und Wasserfaches e.V.).

REFERENCES

Aharonson et al. 1986. Galai CIS-1. A novel approach to aerosol particle size analysis. *J. Aerosol Sci.*, 17/3: 530-536.

Backhus D. A. & P. M. Gschwend 1990. Fluorescent polycyclic aromatic hydrocarbons as probes for studying the impact of colloids on pollutant transport in groundwater. *Environ. Sci. Technol.* 24: 1214-1223.

Bedbur E. 1993. *Laboruntersuchungen zum Einfluß sedimentologischer und hydraulischer Parameter auf die Filterwirkung gleichförmiger Sande.* Dissertation, Christian-Albrechts-Universität, Kiel.

Berne B. & R. Pecora 1976. *Dynamic Light Scattering with Applications to Chemistry, Biology and Physics.* Wiley-Interscience, New York.

Cornell R. M. & U. Schwertmann 1996. *The Iron Oxides. Structure, Properties, Reactions, Occurrence and Uses.* VCH, Weinheim.

Hofmann T. & U. Schulte-Ebbert 1997. Mikropartikel und künstliche Grundwasseranreicherung. In: 2. Deutsch-Niederländischer Workshop Künstliche Grundwasseranreicherung, *DVGW-Schriftenreihe Wasser* 30: 115-130.

Jenkins M. B. & L. W. Lion 1993. Mobile bacteria and transport of polynuclear aromatic hydrocarbons in porous media. *Appl. Environ. Microbiol.* 59: 3306-3313.

Ledin et al. 1996. Temporal changes in concentration, composition and size distribution of the colloidal phase in Lake Risten, Sweden. *The Sci. of the Total Environ.*, 188: 87-99.

McCarthy J. F. & J. M. Zachara 1989. Subsurface transport of contaminants. *Environ. Sci. Technol.* 23: 496-502.

Schäfer T. et al. 1997. Untersuchungen zum partikulären Fe-Mn-Oxid-Hydroxid-Transport und Einfluß der Hydraulik eines Aquifers. In: Coldewey W.G., Löhnert E.T. (eds.). *Grundwasser im Ruhrgebiet. Probleme, Aufgaben, Lösungen*, Bochum : 249-254, Sven von Loga.

Schöttler U. & U. Schulte-Ebbert 1995. *Schadstoffe im Grundwasser* - Band III. Verhalten von Schadstoffen bei der Infiltration von Oberflächenwasser am Beispiel des Untersuchungsgebietes "Insel Hengsen" im Ruhrtal bei Schwerte. Deutsche Forschungsgemeinschaft, Bonn.

Stumm W. 1987. *Aquatic Surface Chemistry.* John Wiley, New York.

Artificial Recharge of Groundwater, Peters et al. (eds) © 1998 Balkema, Rotterdam, ISBN 90 5809 017 5

Fate of cyanobacterial hepatotoxins in artificial recharge of groundwater and in bank filtration

K. Lahti, J. Vaitomaa & A-L. Kivimäki
Finnish Environment Institute, Helsinki, Finland

K. Sivonen
Department of Applied Chemistry and Microbiology, University of Helsinki, Finland

ABSTRACT: Cyanobacteria form mass occurrences in raw water sources of waterworks and may represent a health hazard to water consumers by production of potent hepatotoxins, microcystins. The aim of this study was to assess the fate of microcystins in artificial recharge of groundwater. Laboratory experiments with hepatotoxic *Microcystis* and *Anabaena* species showed that cyanobacterial cells were efficiently removed by filtration through soil and sediment columns. However, in soil columns the removal of microcystins was less efficient than the removal of biomass. In sediment columns the biodegradation of microcystins was rapid. In field studies at two bank filtration sites in Finland traces of microcystins, usually below 0.1 $\mu g\ l^{-1}$, and single cells or filaments of cyanobacteria were occasionally detected in observation tubes and pumping wells of the waterworks. The isolated bacteria capable of degrading microcystins have all been aerobic and therefore the anaerobic conditions common in bank filtration zones probably hinder degradation of microcystins.

1 INTRODUCTION

Cyanobacteria (blue-green algae) produce mass occurrences in several lakes serving as raw water sources of waterworks. In addition to problems with odour and clogging, some cyanobacterial strains may produce potent toxins (Carmichael 1992). Cyclic peptide hepatotoxins, microcystins, may represent health hazards to water consumers if present in drinking water (Falconer 1994). Microcystins are potent liver toxins which inhibit serine/threonine protein phosphatases 1 and 2A and act as tumour promoters (Honkanen et al. 1990, Nishiwaki-Matsushima et al. 1992). Activated carbon filtration or ozonation is needed to supplement conventional surface water treatment processes in order to remove dissolved microcystins from drinking water (Keijola et al. 1988, Fawell et al. 1993, Drikas 1994). Studies on microcystin removal in slow sand filtration or in basin filtration are sparse and the results have varied (Keijola et al. 1988, Annadotter 1993, Lahti et al. 1996).

The aim of this study was to assess the fate of microcystins in artificial recharge of groundwater, especially in bank filtration. Removal of toxins was investigated both at field sites and in the laboratory. The role of adsorption and biodegradation in toxin removal was studied with soil and sediment columns.

2 MATERIALS AND METHODS

2.1 *Column experiments*

Two laboratory experiments with soil and sediment columns were performed during summer 1996. In the first experiment *Microcystis sp.* strain 98, producing microcystin-LR and demethylmicrocystin- LR and -RR (Sivonen et al. 1995), was added to oligotrophic water from lake Päijänne and filtered through the columns. In the second experiment filamentous *Anabaena sp.* strain 90, producing microcystin-LR, -RR and demethylmicrocystin-LR (Sivonen et al. 1992) and geosmin-producing, non-toxic *Oscillatoria sp.* strain 18G (Sivonen 1990), were used. The cyanobacterial strains for the experiments were grown for 14 days in Z8-broth (Kotai 1972), *Anabaena* strain 90 in Z8-broth without nitrogen, at room temperature under continuos illumination. Light intensity on the surface of the culture vessels was 20 $\mu mol\ m^{-2}\ s^{-1}$. Fresh raw water was produced every other day during the experimental period of 9 to 14 days by suspending concentrated fresh cyanobacterial cells in lake water. The average cyanobacterial biomass in raw water with *Microcystis* 98 was 12 $mg\ l^{-1}$ (wet weight) and with *Anabaena* 90 and *Oscillatoria* 18G the total average cyanobacterial biomass was 21 $mg\ l^{-1}$.

Four plastic columns (diameter 0.1 m, height

0.25 m) were filled with soil profiles taken from an esker used for artificial recharge of groundwater in central Finland. Two of the columns were filled with humic surface layer and poorly sorted silty material with sand and gravel taken from the surface to a depth of 0.25 m (hydraulic conductivity from $9x10^{-8}$ to $2x10^{-5}$ m s^{-1}). The other two columns were filled with soil from a depth of 0.25 m to 0.5 m, consisting of sand and gravel (hydraulic conductivity from $9x10^{-4}$ to $2x10^{-3}$ m s^{-1}). Water with toxin-producing cyanobacterial cultures was pumped first onto the duplicate columns with soil from 0 m to 0.25 m and the percolated water was further led through the other two columns containing soil from 0.25 m to 0.5 m.

Duplicate sediment columns (acryl, diameter 0.1 m, height 1.1 m) were filled with sandy gravel dug from the littoral of Lake Vihnusjärvi near the bank filtration zone. The material was sieved through a 0.016 m net. Hydraulic conductivity of the sediment columns was $2x10^{-3}$ m s^{-1}.

The laboratory experiments on microcystin removal were performed in the dark at 14 ± 2 °C. Raw water with toxic cyanobacteria was pumped at a rate of 5 ml min^{-1}, corresponding to a hydraulic surface load of 0.038 m h^{-1}. The sediment columns were maintained saturated with a 0.1 m layer of water above the sediment surfaces.

The detention of water in soil and sediment columns was determined by NaCl tracer test during the one week stabilization period before addition of toxic cyanobacterial strains into the lake water. The mean detention time in sediment columns was 6.2 h and in soil columns 5.25 h.

2.2 *Field studies*

The occurrence of microcystins in raw water sources, observation tubes and pumping wells at two bank filtration plants, Forssa and Jokioinen, was monitored during the summer and autumn of 1997. Both of the plants are situated in southern Finland. Jokioinen waterworks is situated on a narrow esker area consisting mainly of sand and gravel (hydraulic conductivity 10^{-1} - 10^{-5} m s^{-1}). In the vicinity of the lakeshore the topmost layers (thickness of 4-6 m) consist of fine-grained sand mixed with silty material (hydraulic conductivity 10^{-4} - 10^{-9} m s^{-1}). Most of the water extracted from pumping wells (60 - 80 %) is bank filtered. Lake Rehtijärvi is the raw water source for Jokioinen. Its mean depth is 9.2 m, surface area 40 ha and 70 % of the total drainage area is cultivated land. The lake is eutrophic and anoxic conditions are common in the hypolimnion. The groundwater table in the pumping area is 2 to 3 m below the surface level of lake Rehtijärvi.

Forssa waterworks is also situated in the esker area, the thickness of the saturated zone ranging from 15 to 60 metres. The thin uppermost layer consists of clayey and silty material, the main grain size in esker material being sand, in deeper layers mixed with gravel. The amount of water extracted from pumping wells is around 2000 m^3 d^{-1}, of which 70 % is bank filtered. The raw water source of Forssa is lake Kaukjärvi, which is a eutrophic lake with a mean depth of 3.4 m and a surface area of 2 km^2.

Weekly composite samples from the surface to 2 m depth were taken from the lakes and, when microcystins were detected, additional samples from observation tubes and pumping wells were also taken (Fig. 1 and 2).

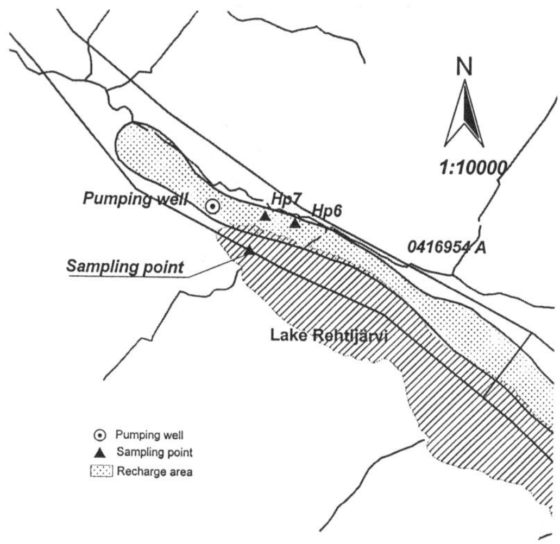

Figure 1. Observation tubes (Hp6, Hp7), pumping well and sampling point at the raw water source of Jokioinen waterworks in Lake Rehtijärvi.

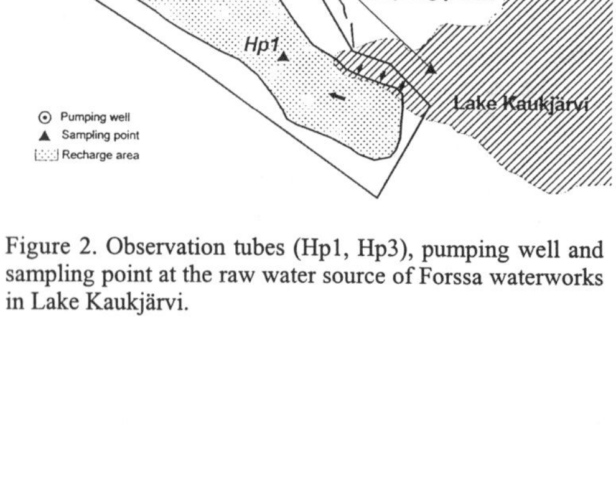

Figure 2. Observation tubes (Hp1, Hp3), pumping well and sampling point at the raw water source of Forssa waterworks in Lake Kaukjärvi.

2.3 *Water quality and soil analyses*

Both in laboratory and field studies the microcystin concentrations in water were determined by immunoassay (Envirogard[R] Microcystins Plate Kit). After repeated freezing at -20 °C and thawing, the samples were filtered through 0.8/0.2 µm membrane syringe filters (Acrodisc, Gelman). Samples of the observation tubes and intake wells were concentrated either by evaporation or in some cases also by filtration through Bond Elut C_{18} cartridges (Varian USA) according to Harada et al. (1988). The detection limit for total microcystin was 0.01 µg l^{-1}. Samples (1 to 2 litres) were also fractionated and prepared according to Tsuji et al. (1994) for the HPLC analysis. The HPLC equipment and procedures were described by Lahti et al. (1996).

The semiquantitative analysis of cyanobacterial occurrence and species distribution of cyanobacteria in lake waters, observation tubes and intake wells were determined from Lugol-preserved samples under an inverted microscope at magnification 125 and 250x. In column experiments the numbers of cyanobacterial cells or filaments were counted from raw and filtered water at magnification 400x. Chlorophyll - a was analysed from lake waters in field experiments according to Finnish standard method SFS 5772 (1993).

The role of adsorption and degradation in microcystin removal was calculated on the basis of toxin amounts in raw and percolated water and the amount of microcystin in soil and sediment material at the end of the experiments. Samples for microcystin analysis from soil and sediment material were taken from different depths of the columns. Subsamples of 2 g were mixed with 8 ml of distilled water, repeatedly frozen at -20 °C and thawed. After ultrasonication for 15 min in a Branson 5200 bath the samples were centrifuged for 15 min at 300 x g and the supernatants were filtered through 0.45 µm syringe filters (Acrodisc) for the microcystin immunoassay (Envirogard[R] Microcystins Plate Kit).

3 RESULTS AND DISCUSSION

3.1 *Column experiments*

Cyanobacterial cells were retained efficiently in both soil and sediment columns (Figs. 3 and 4). The removal of filamentous cyanobacteria was more efficient than that of *Microcystis* 98, which occurred as single cells and not as colonies as in natural lake water. At the end of the experiments, after 9 or 14 days, the cyanobacterial biomass in percolated water was reduced 98.4 - 99.9 % from that of pumped raw water. Experiments with three of the four soil columns were stopped after 9 days due to clogging in silty layers.

The removal of microcystin toxins in the soil columns was much lower than removal of cyanobacterial biomass in both experiments (Figs. 5 and 6). Decrease in microcystin concentration in percolated water was detected after 9 days of filtration in the soil column which was not clogged (Fig. 6). However, in sediment columns the concentration of microcystin started to decrease already after a few days in percolated water and the removal was 98 - 99.9 % by the end of the experiment (Figs. 5 and 6).

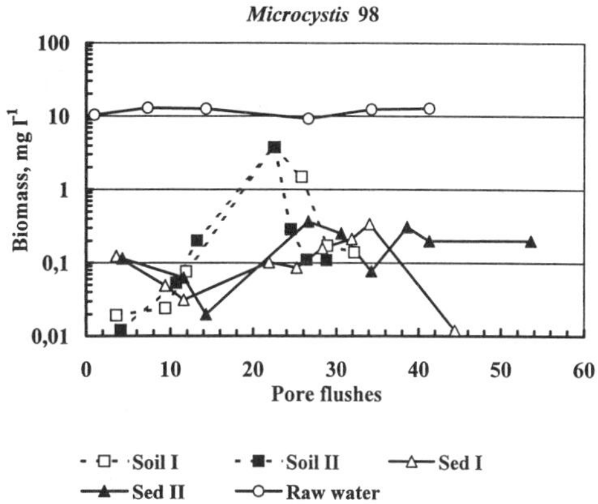

Figure 3. Cyanobacterial biomass in raw water and in water percolated through soil and sediment columns - experiment with *Microcystis* 98.

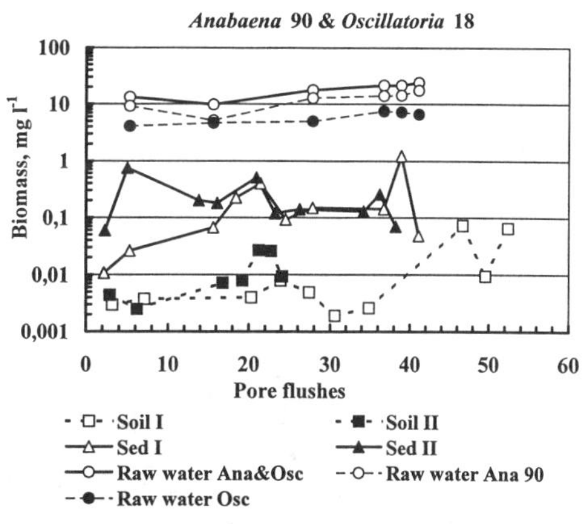

Figure 4. Cyanobacterial biomass in raw water and in water percolated through soil and sediment columns - experiment with *Anabaena* 90 and *Oscillatoria* 18G.

The calculations based on the amount of microcystin in raw water and percolated water, as well as on the toxin content of soil and sediment material at the end of the experiments, showed that biodegradation of microcystins played a major role in toxin removal (Table 1). Adsorption of toxins to particles accounted for 2 to 25 % of the toxin removal. Biodegradation was more efficient in sediment columns, probably due to the presence of natural bacteria capable of microcystin degradation. These results are in accordance with the studies of Rapala et al.(1994) showing differences in the onset of biodegradation with different sediments. Bacteria capable of degrading microcystins were more common in sediments taken from lakes with a recent history of cyanobacterial blooms than from humic or oligotrophic lakes. The esker soil, even with an active humic surface layer, appears not to contain high numbers of bacteria capable of microcystin degradation. The lag period of 9 days is rather similar to the lag phases in microcystin degradation reported from experiments with activated sludge or lake water after copper treatment (Jones & Orr 1994, Lam et al. 1995).

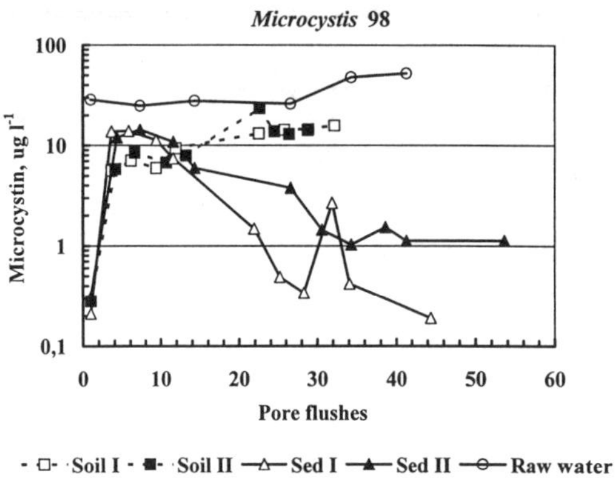

Figure 5. Microcystin concentration in raw water and in water percolated through soil and sediment columns - experiment with *Microcystis* 98.

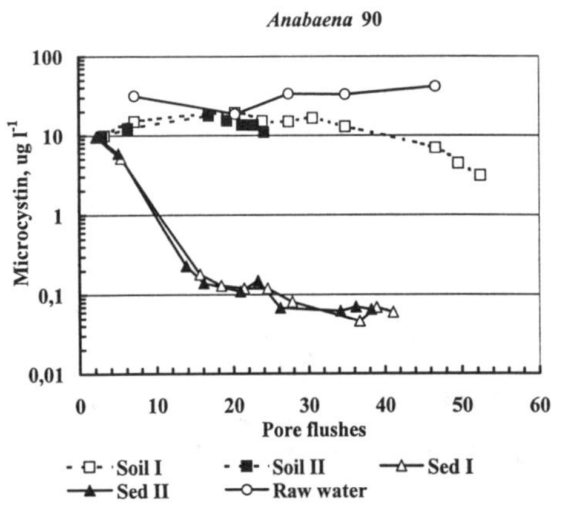

Figure 6. Microcystin concentration in raw water and in water percolated through soil and sediment columns - experiment with *Anabaena* 90 and *Oscillatoria* 18G.

3.2 *Field experiments*

Concentration of microcystins in the upper 2 metres of lake Rehtijärvi ranged from less than 0.1 to 1.9 μg l^{-1} during the period from July to October 1997. Dominance of *Planktothrix agardhii* corresponded with microcystin occurrence (Fig. 7). Microcystins and cyanobacterial cells were occasionally detected in samples taken from the observation tubes and the pumping well. The distance from the lake to the intake well was 0.1 km. However, the concentrations of microcystins were usually below 0.1 μg l^{-1} (0.03 - 0.32 μg l^{-1}) determined both with immunoassay and HPLC. A few cyanobacteria belonging to Oscillatoriales, Chroococcales and Nostocales were detected in every sample taken from the pumping well.

The species *P. agardhii* usually occurs deeper in the water than bloom-forming species of the genera *Microcystis*, *Anabaena* or *Aphanizomenon* and thus the microcystin concentrations in lake Rehtijärvi have probably been higher near the thermocline than at the surface. This, together with the short

Table 1. Mass balance of microcystins in column experiments with *Microcystis* 98 and *Anabaena* 90 (Percentage in parenthesis).

	Sediment columns		Soil columns	
	I	II	I	II
Microcystis 98				
Input, μg	3558	3558	2039	1728
Outflow, μg	535 (15)	520 (15)	643 (31)	541(31)
Adsorbed, μg	291 (8)	910 (25)	159 (8)	163(10)
Degraded, μg	2730(77)	2126(60)	1236(61)	1024(59)
Anabaena 90				
Input, μg	3085	3085	3085	1550
Outflow, μg	217 (7)	237 (8)	1270(41)	708(46)
Adsorbed, μg	727(24)	732 (24)	73 (2)	155(10)
Degraded, μg	2140(69)	2115(68)	1742(57)	688(44)

infiltration distance, would also explain the occurrence of microcystins in bank filtered water.

In lake Kaukjärvi, microcystins were detected by immunoassay from the end of July to late September (Fig. 8). Microcystin occurrence was not correlated with any dominant cyanobacterial species. However, *Anabaena flos-aquae* and *A. crassa* formed blooms at the surface of the lake and water samples from these blooms contained 44 and 57 μg l^{-1} microcystins, respectively. In composite lake water samples, August 11 and September 15, analysed with HPLC no microcystins were detected, although with the immunoassay method the total microcystin concentrations were 0.4 and 0.9 μg l^{1}, respectively.

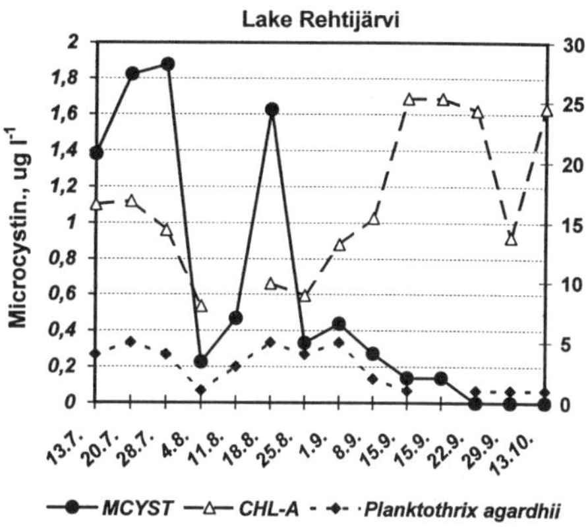

Figure 7. Microcystin and chlorophyll-a concentrations and abundance of *Planktothrix agardhii* (scale 1-5) in Lake Rehtijärvi during summer 1997.

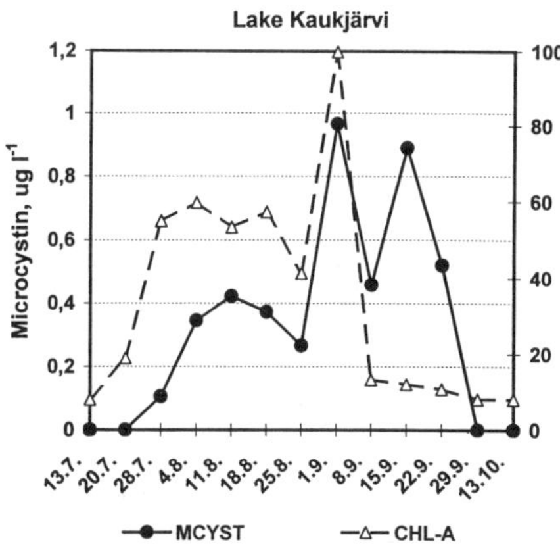

Figure 8. Microcystin and chlorophyll-a concentration in Lake Kaukjärvi, Forssa, during summer 1997.

Microcystins were also detected in August in observation tubes and in the pumping well at concentrations of 0.02 - 0.16 μg l^{-1} with the immunoassay but not with HPLC. Filaments belonging to Oscillatoriales were occasionally detected in observation tubes, but less seldom than at Jokionen.

The discrepancy between the results of the two methods used for microcystin analysis may be due to unspecific binding of compounds present in water to antibodies produced against microcystin-LR, or to failures in sensitivity of the HPLC method used. *Artemia* bioassay showed some toxicity of a bloom sample containing *Anabaena*. This sample was taken in August from the surface of lake Kaukjärvi (data not shown). Further studies on toxin identification are in progress.

Removal of microcystins from lake water in bank filtration plants was usually over 90 %. Although the microcystin concentrations detected in observation tubes and pumping wells were low, they confirm the persistent nature of these compounds and the limited ability of microbes to degrade them. The isolated bacteria capable of degrading microcystin have all been aerobic (Jones et al. 1994, Lahti 1997) and therefore the anaerobic conditions common in bank filtration zones probably hinder degradation of dissolved microcystins entering the aquifer.

ACKNOWLEDGEMENTS

This research was supported by the EC Environment and Climate Research Programme (contract: ENV4-CT95-0071) and the Ministry of Agriculture and Forestry. The authors thank Michael Bailey for revising the English and Juhani Gustafsson for skillful assistance in drawing the figures.

REFERENCES

Annadotter, H. 1993. Cyanobacterial toxins in drink ing water - a study of two Swedish water plants and a literary survey. *VAV-Forsk rapport* nr 1993-03. Stockholm. p. 120.

Carmichael, W.W. 1992. Cyanobacteria secondary metabolites - the cyanotoxins. A review. *J.Appl.Bact.*72:445-459.

Drikas, M. 1994. Removal of cyanobacterial toxins by water treatment processes. In: *Cyanobacteria- A global perspective.* Australian Centre for Water Quality, Adelaide. pp. 30-44.

Falconer, I.R. 1994. Health problems from exposure to cyanobacteria and proposed safety guidelines for drinking and recreational water. In: Codd G.A., Jefferies T.M., Keevil C.W. & Potter E.

(eds.), *Detection methods for cyanobacterial toxins*. Royal Society of Chemistry, Cambridge. pp. 3-10.

Fawell, J.K., J. Hart, H.A. James & W. Parr 1993. Blue-green algae and their toxins - analysis, toxicity, treatment and environmental control. *Water Supply*11(3/4):109-121.

Harada, K.-i., M. Suzuki, A.M. Dahlem, V.R. Beasley, W.W. Carmichael & K.L.Jr. Rinehart 1988. Improved method for purification of toxic peptides produced by cyanobacteria. *Toxicon* 26(5):433-439.

Honkanen, R.E., J. Zwiller, R.E Moore, S.L. Daily, B.S. Khatra, M. Dukelow & A.L. Boynton 1990. Characterization of microcystin-LR, a potent inhibitor of type 1 and type 2A protein phosphatases. *J. Biol. Chem.* 265(32):19401-19404.

Jones, G.J., D.G. Bourne, R.L. Blakeley & H. Doelle 1994. Degradation of the cyanobacterial hepatotoxin microcystin by aquatic bacteria. *Nat. Toxins* 2:228-235.

Jones, G.J. & P.T Orr 1994. Release and degradation of microcystin following algicide treatment of a *Microcystis aeruginosa* bloom in a recreational lake, as determined by HPLC and protein phosphatase inhibition assay. *Wat. Res.* 28(4):871-876.

Keijola, A.M., K. Himberg, A.L. Esala, K. Sivonen & L. Hiisvirta 1988. Removal of cyanobacterial toxins in water treatment processes: laboratory and pilot-scale experiments. *Toxic. Assess.* 3:643-656.

Kotai, J. 1972. Instructions for preparation of modi fied nutrient solution Z8 for algae. *Norwegian Institute for Water Research*. p. 5 B-11/69. Blindern, Oslo.

Lahti, K. 1997. Cyanobacterial hepatotoxins and drinking water supplies - aspects of monitoring and potential health risks. *Monogr. Bor. Environ. Res.* No. 4. Tammer-Paino Oy, Tampere. 40 p.

Lahti, K., J. Kilponen, A.-L. Kivimäki, K. Erkomaa & K. Sivonen 1996. Removal of cyanobacteria and their hepatotoxins in soil and sediment columns. Proc. Intern. Symp. on *Artificial Recharge of Groundwater* (eds. A.-L. Kivimäki & T. Suokko). Nordic Hydrol. Prog. Series, NHP Report No 38. 187-195.

Lam, A. K.-Y., P.M. Fedorak & E.E. Prepas 1995. Biotransformation of the cyanobacterial hepatotoxin microcystin-LR, as determined by HPLC and protein phosphatase bioassay. *Environ. Sci. Technol.* 29:242-246.

Nishiwaki-Matsushima, R., T. Ohta, S. Nishiwaki, M. Suganuma, K. Kohyama, T. Ishikawa, W.W. Carmichael & H. Fujiki 1992. Liver tumor promotion by the cyanobacterial cyclic peptide toxin microcystin-LR. *J.Cancer Res. Clin. Oncol.* 118:420-424.

Rapala, J., K. Lahti, K. Sivonen & S.I. Niemelä 1994. Biodegradability and adsorption on lake sediments of cyanobacterial hepatotoxins and anatoxin-a. *Lett. Appl. Microbiol.* 19:423-428.

SFS 5772, 1993. Determination of chlorophyll-a in water. Extraction with ethanol. Spectrophotometric method. *Finnish Association for Standardisation*. p. 3 (in Finnish).

Sivonen, K. 1990. *Toxic cyanobacteria in Finnish freshwaters and the Baltic Sea.* Reports from Department of Microbiology, University of Helsinki 39/1990. 87 pp.

Sivonen, K., M. Namikoshi, W.R. Evans, W.W. Carmichael, F. Sun, L. Rouhiainen, R. Luukkainen & K.L. Rinehart 1992. Isolation and characterization of a variety of microcystins from seven strains of the cyanobacterial genus *Anabaena*. *Appl. Environ. Microbiol.*58:2495-2500.

Sivonen, K., M. Namikoshi, R. Luukkainen, M. Färdig, L. Rouhiainen, W.R. Evans, W.W. Carmichael, K.L. Rinehart & S.I. Niemelä 1995. Variation of cyanobacterial hepatotoxins in Finland. In: Munavar M. & M. Luotola(eds.), *The contaminants in the Nordic ecosystem: Dynamics, processes & fate*. Ecovision World Monograf Series. Academic Publishing, Amsterdam, the Netherlands. pp. 163-169.

Tsuji, K., S. Naito, F. Kondo, M.F. Watanabe, S. Suzuki, H. Nakazawa, M. Suzuki, T. Shimada & K.-I. Harada 1994. A clean-up method for analysis of trace amounts of microcystins in lake water. *Toxicon* 32:1251-1259.

Artificial Recharge of Groundwater, Peters et al. (eds) © 1998 Balkema, Rotterdam, ISBN 90 5809 017 5

Arrenæs artificial recharge trial plant, Denmark – Hydrological and chemical aspects

G. Brandt
Copenhagen Water Supply, Denmark

ABSTRACT: The groundwater model and the trial results indicate, that it is possible to efficiently infiltrate lake water with a very high algae content. Especially the fact, that in the period when the water has the highest algae content (in summer) it is possible to use water without pretreatment (sedimentation) for surface spreading on the grass covered areas. The water quality of the "new" groundwater is generally good although the nitrate content exceeds the highest guideline value. The nitrate content is expected to drop with time, as the nitrate probably derives from former agricultural activities at the site. In Denmark, artificial recharge is not a widely used supplement to groundwater and surface water abstraction in water supply, as is the case in many other countries. This paper presents the results of a monitoring programme of a trial artificial recharge plant operating since 1995. The groundwater model shows that a high plant efficiency can be obtained when groundwater abstraction is more or less equal to the amount of water that is recharged artificially.

1 INTRODUCTION

In Denmark drinking water supply is almost exclusively based on groundwater abstraction. The danes therefore have little experience in the operation of artificial recharge plants. In order to introduce artificial recharge as a supplement to the traditional water distribution and treatment, the Copenhagen Water Supply in coordination with Frederiksborg County has established a trial plant in the northeastern part of the island Zealand.

The trial plant was completed in November 1994 and has been functioning since April 1995. Water is pumped from Lake Arresø to the artificial recharge site, where particulate organic material is sedimented in two sedimentation basins. The water is then led to the infiltration basins and troughs as well as grass covered areas with surface spreading (Passow 1996).

The trial plant area has recieved 584,693 m^3 lakewater in the period 1995 to 1997, of which 304,211 m^3 in 1997. In 1997, a total of 90,743 m^3 water was abstracted from the groundwater zone.

The water table rises to a maximum of around 30 cm. Ground water levels are continuously monitored by electronic level sounders.

2 WATER QUALITY

A "dynamic" monitoring programme with an action/reaction pattern is carried out as an integral part of the trial plant itself (Jensen & Jørgensen 1994). If for example pesticides are found in concentrations above a defined threshold level in the lakewater, percolate samples from the unsaturated zone will be analysed. If these samples also have a pesticide content above the threshold level, then groundwater samples are subsequently analysed.

2.1 *Lake water*

The lake water has a chloride content of about 65 mg/l. The pesticide metabolite 2,6-diclorbenzamide was found in 8 out of 9 lakewater samples, and in 3 samples the pesticide content was higher than 0.15 µg/l (Haagensen 1997). The algae toxin Microcystin has been found in some of the lakewater samples. An "alarmsystem" has therefore been introduced for algae toxins.

2.2 *Percolate (unsaturated zone)*

In-situ sampling of pore water in the unsaturated zone is carried out in several bores at different levels. 2,6-dichlorbenzamide is the substance among the tested pesticides and their metabolites that occurs in highest

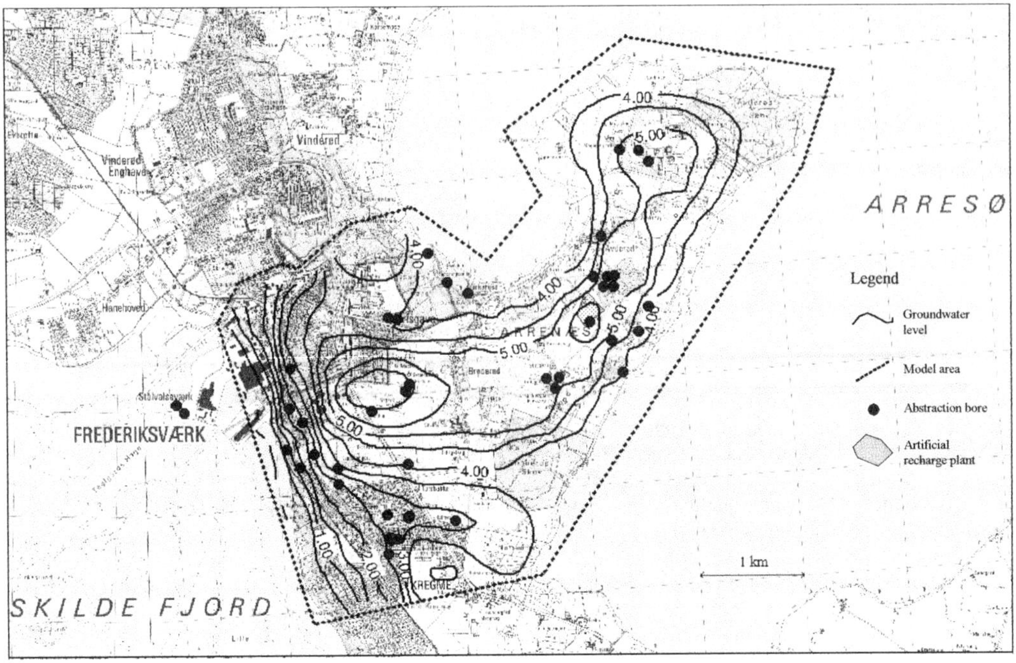

Figure 1. Observed groundwater level August 1992

concentrations in the percolate samples (up to 0.24 µg/l). The "mother" product dichlobenil is only found in high concentrations in the percolate from one bore, where a rapid decay of the substance is seen between 1 and 5 meters depth. The use of commercial products containing dichlobenil was forbidden in Denmark at the beginning of 1997.

The highest concentrations of dinoseb have been detected in percolate samples from 5 meters below ground level.

The potassium permanganate consumption results imply degradation of organic matter with depth.

2.3 *Groundwater*

The background chloride level in the native groundwater zone is 36 mg/l (before artificial recharge). In

Table 1 Comparison between native groundwater, lake water and renovated groundwater quality

	Native Groundwater	Lake Water	Renovated Groundwater
pH	7.5	9.0	7.3
Cl^- (mg/l)	31	65	45
NO_3^- (mg/l)	56	2	19
SO_4^- (mg/l)	43	50	75
AOC (µg/l)	50	No values	40
Atrazin (µg/l)	-	-	-
Deisopropylatrazin (µg/l)	-	-	0.046
Desethylatrazin (µg/l)	-	0.029	0.030
Dichlobenil (µg/l)	-	0.95	-
2.6-dichlorbenzamid (µg/l)	-	0.15 (highest value)	-

several of the monitoring wells groundwater has a chloride content between 36 and 55 mg/l, which shows that a mixing of infiltrated lakewater and groundwater has taken place.

The sulphate content in the groundwater exceeds the advisory guideline level (50 mg/l) but not the threshold level. The reason for this is to be investigated in more detail.

The nitrate levels in the groundwater periodically exceed the highest guideline level (50 mg/l). This is probably due to the washing out of fertilizer from former agricultural activities from the upper soil layers. This will stop at some point, as the trial plant area no longer recieves any fertilizer. The nitrate content in the groundwater zone was at the same level in 1995 (before artificial recharge of lake water).

The development in AOC-levels (assimilated orga nic carbon) in the groundwater zone is generally under 40 μg/l.

The pesticide atrazin has been detected in concentrations of up to 0.03 μg/l, while its metabolites desethylatrazin and deisopropylatrazin have been found in higher concentrations (up to 0.046 μg/l). Dinoseb has been detected three times in one bore with values between 0.11 and 0.51 μg/l (more than five times higher than the guideline value). In Denmark atrazin use was forbidden in 1994 and dinoseb use was forbidden in 1990.

The zinc content in the groundwater has decreased from 40-97 μg/l in 1995 to 18-20 μg/l in 1996.

A lowering of the nitrate and phosphate levels can be expected in the future, as the percolating lakewater has lower concentrations of these substances than the "old" native groundwater under the trial plant site (Brandt 1996).

3 GROUNDWATER MODELLING

The purpose of groundwater modelling is to establish a dynamic tool to evaluate the development of the aquifer recieving artificial recharge of lakewater at the site (Petersen 1997). The SHE-model (3D-model) evaluates and focuses on catchment areas, distribution and transport times of the artificially recharged water, fluctuations in groundwater level and the influence on contaminated sites in the vicinity of the site.

The model is based on existing data on geology, hydrogeology, meteorology, land use and other drinking water abstraction units in the area (Brandt 1997). The Arrenæs peninsula is a moraine landscape formed during the last ice age. The resulting meltwater sands are underlain by limestone.

The Prequaternary limestone surface is relatively even and lies 30 to 37 metres below sea level. A till layer lies directly above the limestone and is between 5 and 15 metres thick. The meltwater sand above the till layer has a thickness of ca. 40 metres at the trial site.

There are two groundwater aquifers in the area. The groundwater level in the limestone aquifer is 5 metres above sea level in the eastern part and 1 meter above sea level in the western part near Roskilde Fjord. At the trial site, the groundwater level in the meltwater sand is 5 metres above sea level (fig. 1). To the north and south the groundwater level falls to ca. 3.5 metres above sea level at the lake shore. The groundwater flow is from the central part of the peninsula to Lake Arresø and Roskilde Fjord.

The regional, local and detailed models have been calibrated with an observed stationary groundwater level from 1992, as well as single groundwater level soundings in the vicinity of the site. The net size for the three different models is respectively 100, 50 and 25 metres.

Vertically the model is built up of 5 layers of varying thickness. In layer 1, the upper layer, where unsaturated, vertical flow occurs (fig. 2), different soil types have been included based on soil classification maps. In the uppermost layers comprising the meltwater sands, clay lenses have been included based on what is known as far as distribution and thickness

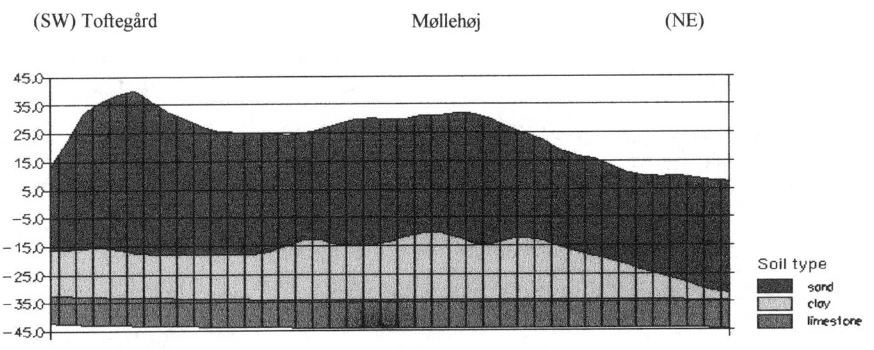

Figure 2. Geological model. Cross-section through the artificial recharge plant.

are concerned.

Based on simulations with a recharge of 60 and 180 m^3/hour travel times, groundwater level development and the distribution of artificially recharged lake water have been estimated.

The travel time through the unsaturated zone from ground level to the groundwater table has been calculatated to be around 25 days (60 m^3/hour). With less infiltration, transport times grow whereas with larger infiltration rates transport times get smaller. This is due to the increase in hydraulic conductivity, and therefore water's travel time, with rising water content.

4 STORAGE ASPECTS

The hydrological model has been used to illustrate the hydrological conditions for four different operation scenarios (see table 1). For each scenario the following was calculated: Variations in groundwater level with time, level changes in relationship to the natural groundwater level, distribution of the artificially recharged water, plant efficiency, travel times from artificial recharge to abstraction well as well as the influence on a contaminated site in the vicinity.

The modelling results show that the influence on the groundwater level is less than 1 metre at a distance of 300 metres from the site - even in the scenario with the heaviest hydraulic loading of the site. This is due to unconfined aquifer conditions in the area. Model calculations show that groundwater flow is more or less equal in all directions.

With a constant infiltration of 1 million m^3 of lake-water a year and no groundwater abstraction, a new "stationary" groundwater table is established after 2 years of artificial recharge.

5 GROUNDWATER ABSTRACTION

In operation scenario 3 and 4, where groundwater is abstracted by three wells, the drawdown (cone of depression) at abstraction well 3 is reduced by a factor 2 compared to operation scenario 1 and 2. The catchment area maps show that the catchment area is largest in October where there is a water surplus (Petersen 1997).

The model shows, that the largest part of the artifici-ally infiltrated water flows from the site to the active abstraction wells. A small part of the infiltrated water wil flow past the abstraction wells. The model indicates that the water which is artificially recharged in the western recharge units is not recovered by any of the abstraction wells.

An efficiency calculation for every operation sce-nario has been carried out (how much "new" water is actually abstracted). Comparison of the efficiency of the abstraction wells in the four operation scenarios shows well 1 right next to the site abstracts nearly 100% artificially recharged water. Water abstracted from abstraction well 4 consists of 75% artificially recharged water.

It is expected, that the production of "new" ground-water will be increased by a higher loading of the site in 1997/1998 than was the case in 1996.

6 DISCUSSION

The Copenhagen Water Supply hopes that artificial recharge of groundwater will become an integral part of the future water supply system.

The trial results so far seem to be positive as far as operational procedures, water quality and abstraction are concerned.

A tracer test is planned in 1998 (Brun 1997). The purpose of the tracer test is to see how fast infiltration occurs under the infiltration basins in a vertical direction at a given hydraulic loading. The existing groundwater model will be used as a basis for the tracer test. The test should also be able provide insight into preferential (finger) flow under the basin floor.

In 1997, infiltration basin east 1 was prepared for the tracer test. Four multilevel bores with a total of twenty suction cells (miniscreens) were completed.

Table 2. Artificial recharge and groundwater abstraction in the four operation scenarios in m^3 per year.

	Yearly Recharge	Well 1 Abstraction	Well 3 Abstraction	Well X Abstraction
Scenario 1	480,000	250,000	250,000	0
Scenario 2	960,000	500,000	500,000	0
Scenario 3	480,000	135,000	135,000	155,000
Scenario 4	960,000	330,000	330,000	270,000

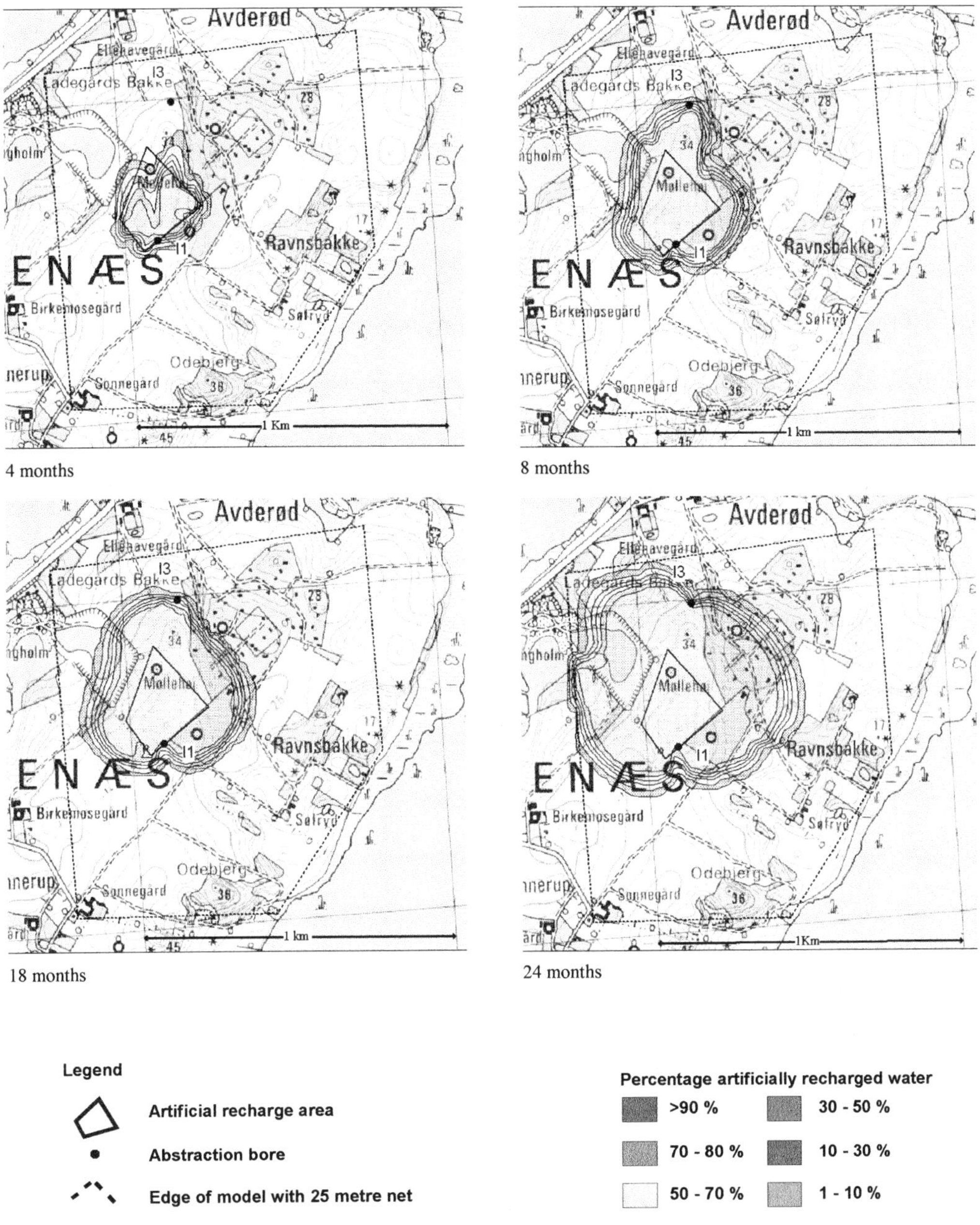

Figure 3. Distribution of artificially recharged lake water. Scenario 2.

Sixteen more observation sites will be established in the near future. A part of the basin (7x7 m) will be closed off from the rest of the basin. About 8 kg of tracer material (NaBr) dissolved in 20 litres of water will be applied to the basin by sprinkling in order to achieve a constant bromide concentration of ca. 500 mg/l.

7 CONCLUSION

The knowledge that the trial plant has given/will give on the decomposition and mobility of various pollutants is of interest to other areas of water supply.

REFERENCES

Brandt, G. 1996. Water Quality Development at the Arrenæs Artificial Recharge Plant, Denmark. In A. Kivimäki and T. Suokko (eds), *Artificial recharge of groundwater -Proceedings of an International Symposium, Helsinki, Finland, June 3-5, 1996:* 155-160. Nordic hydrological programme, NHP Report No 38.

Brandt, G. 1997. Arrenæsprojektet - Grundvands model, operationsplan, foreløbelige resultater. IDAmiljø møde om kunstig infiltration, 1997 (in danish). The Arrenæs project - Groundwater model, operation plan, trial results).

Brun, A. 1997. Oplæg vedrørende tracerforsøg ved Arrenæs infiltrationsanlæg, Bassin 1, øst. Vandkvalitetsinstituttet (in danish). (Project on tracer test at the Arrenæs artificial recharge plant).

Haagensen, K. 1997. Arrenæs-projektet. Opbygning af anlægget og moniteringsprogram. IDAmiljø møde om kunstig infiltration, 1997 (in danish). (The Arrenæs project - Trial plant buildup and monitoring programme).

Jensen, B. & C. Jørgensen, 1994. Opstilling af moniteringsprogram for forsøgsanlæg til kunstig infiltration af vand fra Arresø. Vandkvalitets-instituttet (in danish). (Monitoring programme for the Arrenæs trial plant).

Passow, J. 1996. The Arrenæs Trial Recharge Plant - Design and Operation. In A. Kivimäki and T. Suokko (eds), *Artificial recharge of groundwater - Proceedings of an International Symposium, Helsinki, Finland, June 3-5, 1996:.* 99-104.Nordic hydrological programme, NHP Report No 38.

Petersen, N. 1997. Grundvandsmodel for Arrenæs. RAMBØLL (in danish).(Groundwater model for Arrenæs).

Artificial Recharge of Groundwater, Peters et al. (eds) © 1998 Balkema, Rotterdam, ISBN 90 5809 017 5

Fate and behaviour of atrazin in the Meijendel artificial recharge system

H.G.de Jonge & J.A.Stäb
Dune Water Company of South Holland, The Hague, Netherlands

ABSTRACT: An overview is given of the presence of atrazin in the recharge system of the Dune Water Company of South Holland (DZH). The infiltrated load per annum of atrazin was at its highest in 1990 (10 kg) and decreased to 1-2 kg in 1997. In the conventional surface infiltration system, which is only slightly anaerobic, approximately 50% of the atrazin is removed during dune passage, while in the highly anaerobic well-recharge system of Waalsdorp, over 90% of the atrazin is removed from the infiltrated water. In order to improve taste, odour and colour and to further reduce the remaining concentrations of pesticides, activated carbon is added as part of the posttreatment for the production of drinking water.
An important question remains as to whether atrazin break-down products are still present in drinking water but cannot be detected by the techniques currently in use, or whether break-down products adsorb on soil particles.

1 INTRODUCTION

The Dune Water Company of South Holland (DZH) uses water from the river Meuse as the main source for the production of drinking water. In the eighties and nineties, DZH was confronted with high concentrations of triazines (atrazin and simazin) in the water abstracted from the river Meuse

In this paper we present the way in which DZH has coped with this phenomenon. Firstly, the occurrence of triazines in the rivers Meuse and Rhine will be described, along with their properties and behaviour, including those of some of their break-down products. Then the water supply system of DZH will be described, together with the fate and behaviour of atrazin in the system. Finally, future developments will be discussed.

2 ATRAZIN IN THE RIVERS MEUSE AND RHINE

The first results available on a regular basis concerning atrazin concentrations in the rivers Meuse and Rhine are from the year 1988. In figure 1, atrazin concentrations are given that were measured in the river Rhine at Lobith and Andijk, and in the river Meuse at Eijsden and Keizersveer. The figure clearly shows that over the last 10 years, atrazin concentrations in the Meuse have been significantly higher than in the Rhine. The concentrations are highest in

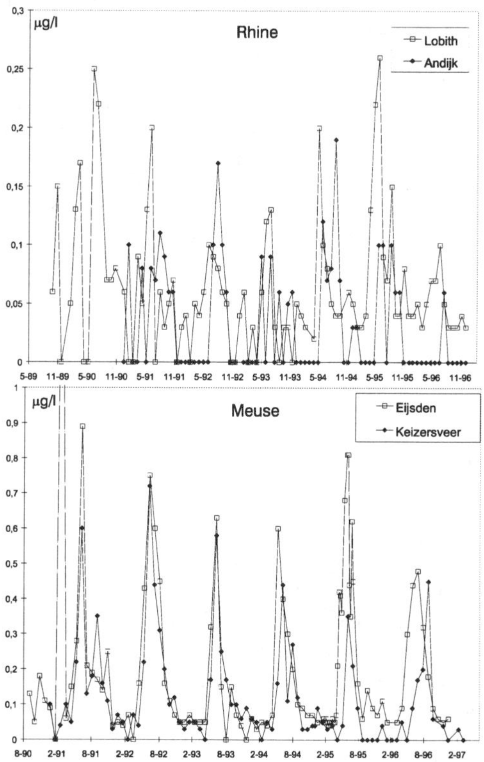

Figure 1. Atrazin concentrations (µg/l) in the rivers Rhine and Meuse

summer, when the pesticide is used during the cultivation period of corn. Application is generally only once a year in June (van Rijn et al. 1995). This results in a very clear seasonal variation in the atrazin concentrations in the rivers. The higher concentrations of atrazin in the river Meuse than in the river Rhine are in the first place due to the fact that the use of atrazin has been prohibited in Germany for many years, while in the Netherlands, its application is still legal. Secondly, in the Dutch part of the Meuse basin a lot of corn is grown, which increases the atrazin concentrations in the water of the Meuse.

Furthermore, figure 1 shows that atrazin levels have not decreased much in recent years. This corresponds to the fact that the use of atrazin on corn has not decreased significantly over the last few years.

3 PROPERTIES AND BEHAVIOUR OF ATRAZINE, DIURON AND GLYPHOSATE

Atrazin, diuron and glyphosate are herbicides that are used in high quantities in the Netherlands for weed control. Farmers generally use herbicides to control weeds before the emergence of the crop; another application is to kill the entire grassland before grass is sown again, to obtain good-quality meadows. Local councils use herbicides to keep roads and pavements clear of weeds.

In figure 2, chemical structures of these three compounds are shown. In Table 1, an overview of their use in the Netherlands is given (adapted from van Rijn et al. 1995). Over the last five years, there has been a move away from atrazin, through diuron, to glyphosate. In general, it can be said that the last-named exhibits lower toxicity and accumulation, and is less persistent (in the environment).

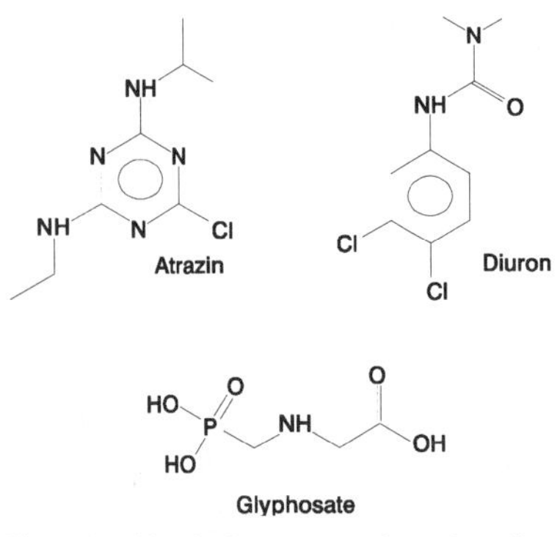

Figure 2. Chemical structures of atrazine, diuron and glyphosate

Table 1. Use in the Netherlands, and properties of atrazin, diuron and glyphosate

atrazin		
use in the Netherlands		
corn	1 x April to July	0.8-1.5 kg/ha
asperagus	1 x May to July	1-2 kg/ha
waste land	1 x whole year	2.5-15 kg/ha
toxicity		
birds	LC 50	5 g/kg feed
fish	LC 50	6.3 mg/l
crustacea	LC 50	5.7 mg/l
insects	LC 50	0.7 mg/l
algae	NOEC	0.01 mg/l
humans	ADI	0.7 mg/kg b.wt.
(break-down) half life in fresh water 4 months		
break-down products		
desisopropyl atrazine		
desethyl atrazine		
desisopropyl hydroxy atrazin		
2-hydroxy-4,6-diamino-1,3,5,-triazin		
2-chloro-4,6-diamino-1,3,5,-triazin		
2-hydroxy atrazin		
diuron		
use in the Netherlands		
asparagus	1 x April to may	0.4-1.6 kg/ha
streets, roads	1 x Mar to Sep	1-16 kg/ha
apples, pears	1 x Mar to April	0.8-2.8 kg/ha
toxicity		
birds	low toxicity	
fish	LC 50	5.6 mg/l
crustacea	LC 50	1.4 mg/l
insects	low toxicity	
algae	unknown	
humans	ADI	2 mg/kg b.wt.
(break-down) half life in fresh water 1 m-2 y		
break-down products		
3-(3,4-dichlorophenyl)-1-methylurea		
3,4-dichlorophenylurea		
3,4-dichloroanilin		
glyphosate		
use in the Netherlands		
potatoes	1 x Aug to Oct	0.7-2 kg/ha
grass killing	1 x Jan to Sep	0.5-1 kg/ha
onions, corn	1 x Feb to June	0.4-2.1 kg/ha
toxicity		
birds	low toxicity	
fish	low toxicity	
crustacea	low toxicity	
insects	low toxicity	
algae	NOEC	1 mg/l
humans	ADI	300 mg/kg b.wt.
(break-down) half life in fresh water 3 d-1 m		
break-down products		
aminomethyl phosphonic acid (AMPA)		

A low persistence (or fast break-down in the environment) implies, in principle, that the compound is less hazardous than more persistent compounds. However, the role of break-down products is not fully recognized in many cases. In figure 3, the possible break-down products of atrazin are given. It is known that some break-down products, specifically desethyl atrazin and desisopropyl atrazin, often occur in fresh water in concentrations comparable to the parent compound (van Rijn et al. 1995).

For many other break-down products, reliable analytical determination methods do not exist or are fairly expensive. This means that reliable figures on the concentrations of these compounds are scarce. It is likely, however, that significant levels of break-down products are present owing to laboratory experiments.

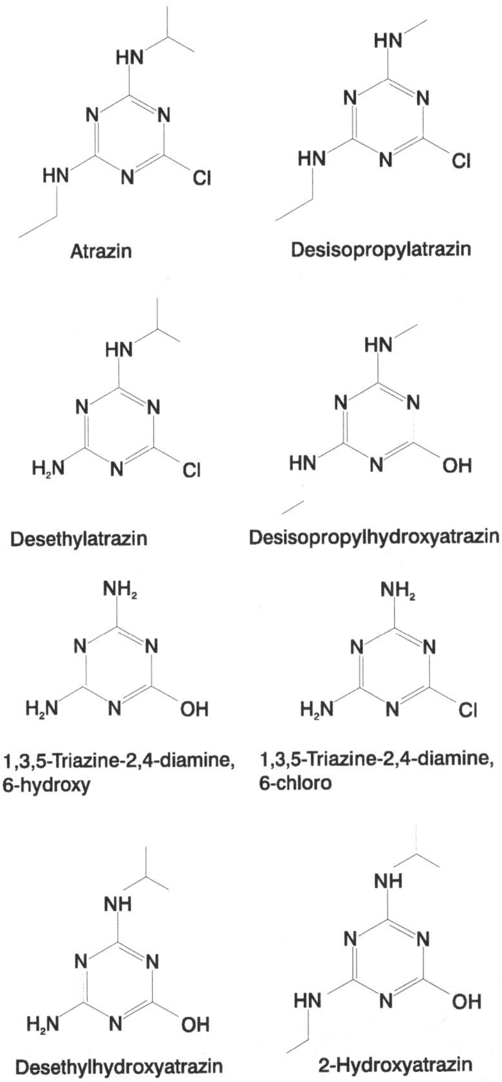

Figure 3. Possible break-down products of atrazin

4 THE WATER SUPPLY SYSTEM OF DZH

The source for the production of drinking water at DZH is found in the Afgedamde Maas, a tributary of the river Meuse with only a very slight flow and high suitability for use as a sedimentation and supply basin (Fig. 4). The water is kept here for about two months, which is long enough to allow the river water to undergo a fairly extensive natural purification. Iron sulphate is added to the water to reduce the level of phosphates. This considerably improves the quality of the water. In periods with much precipitation, water from the adjacent polders is pumped into the Afgedamde Maas. This is some-times more than 50 % of the abstracted water. Water from the Afgedamde Maas has been abstracted since 1976 and is pumped to Bergambacht. In Bergambacht, the water is pretreated, allowing sediment to be removed by rapid sand filtration units. The water is then ready to be recharged in Meijendel, Berkheide and Solleveld, the dune areas of DZH (Fig. 4)

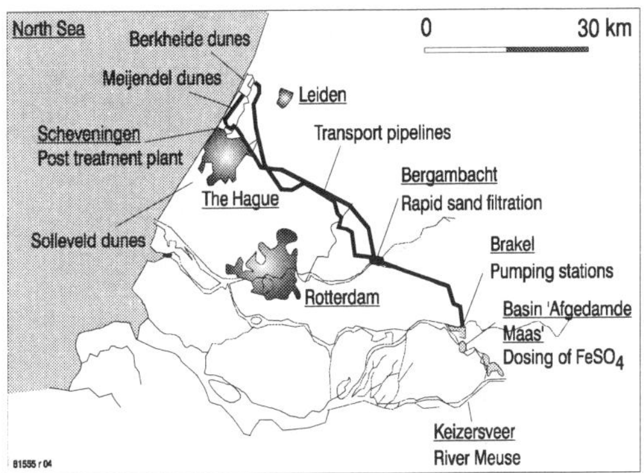

Figure 4. Transport to the recharge system in the dunes of Meijendel, Berkheide and Solleveld

In Meijendel, the water is spread by using some 25 recharge basins. During dune passage, the water becomes bacteriologically and virologically safe, and is given an even quality and constant temperature. after a retention of between two and three months, most of the water is recovered. The water that is not recovered seeps slowly into the ground, where it recharges the fresh groundwater reservoir. The recovery is achieved by means of drainage pipes and shallow wells (Fig. 5)

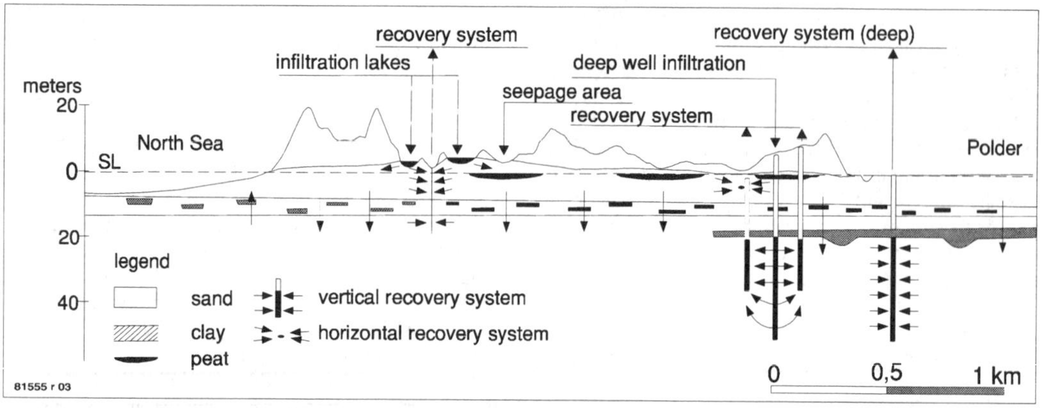

Figure 5. Recharge system of DZH

From the dune area, the water is pumped to Scheveningen for further treatment. Here the post-treatment consists of the addition of activated carbon (granular form), water-softening, rapid sand filtration and slow sand filtration. After the water has passed the slow sand filtration unit, it is ready for consumption. No further disinfection, such as chlorination is necessary.

Dune infiltration by means of recharge basins forms the most important part of the water supply system of DZH. Since 1990, however, an alternative infiltration Meuse has been used. In this infiltration technique, river water is infiltrated in a deeper aquifer in the dune area, by means of wells (Fig. 6). The river water is extensively pretreated. The pretreatment of the water that is already treated in Bergambacht with rapid sand filtration consists of coagulation, flotation and another rapid sand filtration.

5 FATE AND BEHAVIOUR OF ATRAZIN IN THE WATER SUPPLY SYSTEM OF DZH

Results of atrazin determinations are available since 1988. From 1988 until now, the fresh-water intake of the dune waterworks has been around 1 million m^3 per week. The atrazin load has been calculated at six different locations in the production process, and is plotted in figure 7. The locations are:

- The river Meuse at Keizersveer (Meuse)
- The intake of river water at Brakel (Brakel)
- The (pretreated) river water before infiltration (Influent)
- The effluent of the recharge system in the dune area (Effluent)
- The effluent of the rapid sand filters (RSF)
- The drinking water at pumping station Scheveningen (DW)

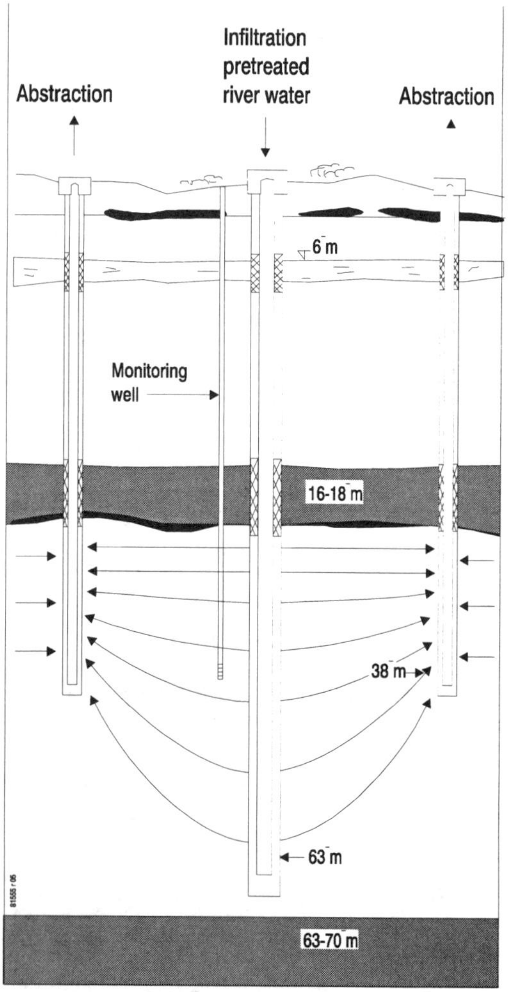

Figure 6. Waalsdorp deep well infiltration system

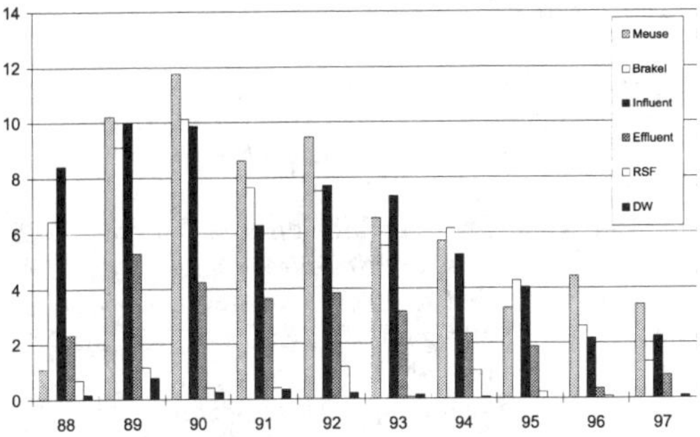

Figure 7. Atrazin load (kg) at six different locations

For the calculations, the atrazin concentrations below the detection limit of (generally) 0.05 μg/l were set to zero. For comparison purposes, a hypothetical load was calculated for the location Keizersveer by multiplying the intake volume of DZH by the atrazin concentration at Keizersveer. In figure 7, it can be seen that the atrazin load was highest around 1990. Generally, the load at the first three locations was more or less the same. There was no reduction in the Afgedamde Maas basin. After this, about half of the atrazin load is removed by the recharge system in the Meijendel dunes. The water abstracted from the recharge system is a mix of water passing through both (partially) aerobic and anaerobic environments. The partial removal of atrazin is in agreement with the literature, which states that atrazin in a (partially) aerobic environment (with nitrate and a low concentration of iron) is not, or only slightly, broken down in a period of up to 100 days (Stuyfzand 1997)

In an anaerobic environment (no nitrate and considerable iron), atrazin is almost completely broken down within 60 days. This is illustrated by figure 8, which shows the water produced by the Waalsdorp deep well infiltration and recharge system.

The atrazin still present in the water abstracted from the dune area, is removed during the posttreatment process, in particular, by the dosing with activated carbon. Although from the start, in 1955, this dosing was primarily intended for the improvement of taste, odour and colour, in the eighties the emphasis shifted to the removal of pesticides. A lot of research has gone into determining the most efficient type and amount of dosage. Currently, activated carbon derived from coal is used and dosed in an amount of 3 mg/l.

Thanks to the above-mentioned processes, the drinking water produced has been distributed in compliance with the current standard for the atrazin content of drinking water in the Netherlands, that is, below 0.1 μg/l.

6 FUTURE DEVELOPMENTS

For new production systems, such as the well-recharge systems already in use and the projected ones, DZH has decided in an agreement with the provincial government of South Holland to install an additional pretreatment step that uses granular activated carbon.

Apart from implementing extra treatment, DZH is involved in a project together with Rijkswaterstaat, a national body responsible for protecting and maintaining the quality standards of surface waters in the Netherlands and other partners. This project aims at improving the quality of the surface water in the Afgedamde Maas.

In recent years, there has been a shift away from the use of non-polar pesticides to polar ones, the aim being to prevent pesticide accumulation in the food chain. An undesirable side-effect of this policy is the fact that these pesticides, because they do not adsorb easily owing to their polarity, are difficult to remove from the water phase.

Examples of these types of pesticides are atrazin, diuron and glyphosate. Diuron and the main breakdown product of glyphosate, aminomethyl phosphonic acid (AMPA), are also found in surface water throughout the year. Figure 9 shows the concentration of AMPA and diuron, together with that of atrazin, in the water of the river Meuse at Keizersveer and in the water at the intake in Brakel. The figure

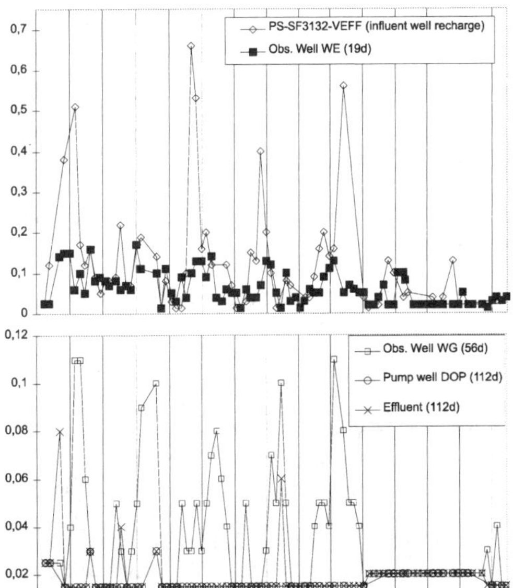

Figure 8. Presence of atrazin (mg/l) in the Waalsdorp well-recharge system, at influent stage and after. Retention times of 19, 56 and 112 days

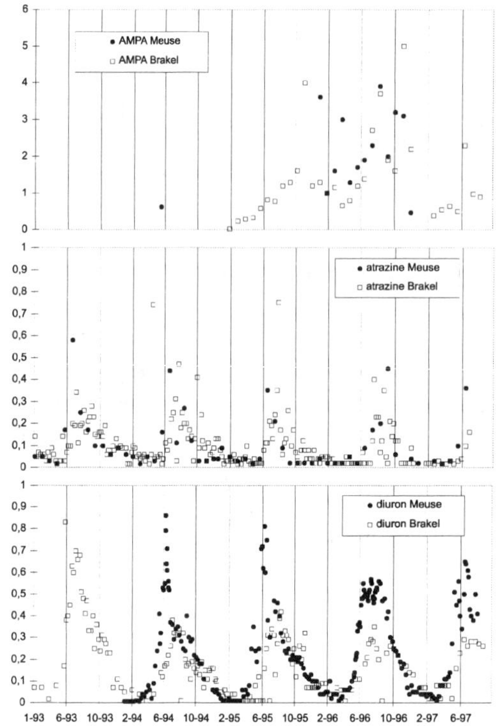

Figure 9. AMPA, atrazin and diuron (μg/l) in the river Meuse and at the intake at Brakel

shows that although the concentrations of atrazin are decreasing, diuron, and more recently AMPA, is present in much higher quantities. In the discussions concerning the redrafting of the Water Act, much attention is being paid to the question of whether and how to apply standards for the break-down products of pesticides. Most probably the same standard will be applied for toxicologically relevant break-down product.

In the field of pesticides, inventory studies still show compounds for which the environmental fate is not clear. The admission of new (polar) pesticides will add to this list, and the emphasis on the presence of (again polar) break-down products poses new analytical challenges. Aside from the presence of pesticides, surface water screenings have shown that many other organic micropollutants are still occurring in surface water (Stäb et al. 1998). Inventory studies by RIWA (RIWA 1997) list surfactants, whitening agents, odour compounds, biocides and bulk chemicals as relevant.

REFERENCES

RIWA. 1997. Chemical compound studies, a series of reports prepared under supervision of the "Stofstudies" working group of RIWA. Amsterdam: RIWA

Rijn, J.P. van, N.M. van Straalen & J. Willems. 1995. *Handboek Bestrijdingsmiddelen gebruik en milieu-effecten* (in Dutch). Amsterdam: VU.

Stäb, J.A., S. Vellinga, E. Brussee & R. Kerkhoff. 1998. Screening of Target Compounds by means of Gas Chromatography Mass Spectrometry: Dune Water Company South Holland. Laboratory report no. 202005

Stuyfzand, P.J., 1998. Fate of pollutants during artificial recharge and bank filtration in the Netherlands . This volume.

1.10 Other types of recharge

Artificial Recharge of Groundwater, Peters et al. (eds) © 1998 Balkema, Rotterdam, ISBN 90 5809 017 5

Artificial recharge schemes in water resources development in Oman

Nasir Mohammad Al Battashi & Syed Rashid Ali
Ministry of Water Resources, Sultanate of Oman

ABSTRACT: The Sultanate of Oman, located at the south-eastern corner of the Arabian peninsula is classified within the arid, semi desert climatic region, with a mean annual rainfall of 100 mm / year and high evaporation rate. The Sultanate depends on limited groundwater resources to secure its water requirements. The artificial recharge schemes in Oman are based on the concept of recharge dams which intercept rare torrential floods from intermittent wadis and release them in a controlled way. Since 1985 to date, 17 recharge dams of capacity between 0.13 to 11.5 million m^3 have been constructed. This paper discusses the artificial recharge from two selected dams, one in the coastal drainage and the other in the desert drainage system. During their five year operation, 20 to 22 flood events have occurred. The dams have contributed annual recharge of varying magnitude, resulting in an increase in storage and improvement in groundwater quality.

1 GENERAL DESCRIPTION.

1.1 *Climate*

The Sultanate of Oman is located at the south eastern corner of the Arabian Peninsula It lies between latitudes 16-27 N which is under the influence of the equatorial convergence zone. The Sultanate is classified within the arid, semi desert climatic region. The rough estimated volume of annual rainfall over the Sultanate may be in the range of $10\text{-}12 \times 10^9\ m^3$. However the reliable amount does not exceed 10 percent which precipitate over the mountain ranges and the coastal plains. Annual rainfall varies from 100 mm at the coastal plain to almost 300 mm over the highest peaks and less than 50 mm in the desert plain. The Sultanate depends upon groundwater resources to secure its water requirement for domestic, agricultural, and industrial purposes. No perennial surface water sources exist except for a few springs emerging from springlines or base flow from wadi courses.

1.2 *Surface runoff*

Surface runoff is generated during intense rainstorm or due to successive rainy days, part of which infiltrates the alluvial deposits that extend from foothills to the coast or desert. However, during a normal event, a significant part of the floods still reaches the sea or desert. In-spite of the regional and local differences, wadi regimes in different parts of the country have much in common. The floods produce hydrograph with a very short time of rise, high peaks and quick recessions. The average frequency of floods ranges from one in several years in the arid part to 4-5 per year in Northern Oman. The estimated magnitude of flood water lost to the sea or desert during a wet year can be as high as 260 million m^3. A number of exceptional floods occurred in Oman during the period 1995 -1997. During this period the recharge dams retained 268 and spilled 364 million m^3 of flood water. For this period the flood losses to the sea or desert were estimated to be over 1.8 billion m^3 compared to 260 million m^3 as the average long term losses.

1.3 *Groundwater*

Groundwater occurs in many different types of aquifer in Oman. Sizes of aquifer system vary from small narrow wadi fills, to extensive platform aquifer systems of large aerial extent. The alluvial aquifers, are widely scattered in the country and have limited storage potential. vulnerable to seasonal depletion of resources in dry years. The regional Quaternary aquifers represent the most important system that occurs in the coastal and interior plains attaining a maximum thickness of 600 m near the coast. Both are favorably located to source of recharge. Over-exploitation in the coastal zone causes increase in water salinity. The hard rock aquifers of the Hajar Super group in the mountainous region also store and transmit certain quantities of water. They yield considerable quantities to spring, wadis and aflaj. Groundwater occurs in the deep aquifers of the Fars and Hadramawt Groups in the Karstic limestone of Taqa formation and in the confined Umm-Er Radhuma formation of regional extent in Dhofar area. The statistics show roughly that the renewable resources meet only the 70 percent of the present demand. Agricultural demand is about 93 percent of the total. Measures adopted to reduce the deficit include tapping new unexploited aquifers, improvement of potential through artificial recharge, controlled use of ˙ groundwater through regulations. Artificial recharge by recharge dams has been used over the last ten years as a method of enhancing the groundwater in Oman. Because of the rapidly increasing water demand the primary objective of these dams is now to offset the potential water deficits and to minimize landward migration of fresh-salt water interface in coastal areas. The dams also provide measures of flood protection. A monitoring network exists at each recharge dam site, with data collected since 1985.

1.4 *Recharge dam studies and construction*

The Sultanate depends on limited groundwater resources to secure its water requirements. The artificial recharge schemes in Oman are based on the concept of recharge dams which intercept rare torrential floods from intermittent wadis and release flows in a controlled way in order to increase groundwater recharge and thus reduce losses to sea or desert. Since the water resources information is never or seldom available to the extent needed for planning and design of recharge or other schemes, the recharge schemes in Oman have been planned and designed on the basis of limited available information on the hydrological and hydrogeological systems. The studies on recharge schemes have involved a good deal of interpretive and correlation techniques for the assessment of hydrological quantities such as, volume of runoff, peak discharges, probable maximum flood during the preparatory and feasibility stage. The hydrogeological studies include water balances making use of appropriate models for quantification of recharge with and without a recharge dam. In general, the studies faced scarcity of data. The assessment on technical feasibility and economic viability of recharge schemes was achieved by developing a set of articulate and integrated scientific approaches. Between 1985 and to date, 17 recharge dams of capacity between 0.13 and 11.5 million m^3 have been constructed in Oman. The hydraulic structures are built across the main drainage and their crest height and length varies between 5 to 21 m and 135 to 9000 m respectively. Recharge dams in Oman have a total capacity of 56.3 million m^3 and to date, the total flood retained is 437 million m^3. To demonstrate the role of recharge dams in water resources development, the following sections describe the execution and performance of two selected recharge dams located in a desert and coastal environment. These are namely Wadi Al Fulayj Sur recharge dam at the coast in the Sharqiyah region and Wadi Al Kabir dam in the interior region.

2. WADI AL KABIR RECHARGE DAM

2.1 *History and background.*

Wadi Al Fulayj is situated in the Sharqiyah Region of Northern Oman. Sur is the main regional town. Irrigated agriculture covers 655 hectares, is supplied from alluvial aquifers with unsustainable yields, under natural recharge condition. A decrease in groundwater levels causes fluctuating changes in yields and groundwater quality. To improve the water potential, feasibility studies for a recharge dam were conducted in 1989 and dam construction was completed in December 1991.

2.2 *Location and design.*

The dam is located about 10 km upstream from the sea coast and 3 km upstream from Sur gardens in the main wadi (Figure 1). The hydraulic structure comprises an earth-fill dam with a concrete spillway built across the valley and a downstream stilling basin made of rock fill gabions. The dam is 7.5 m high with a storage capacity of 0.78 million m^3. The spillway, with provision of two controlled outlets (diameter 0.8 meter) is 430 m long and designed to discharge a maximum flood of 6600 m^3/s.

2.3 *Rainfall and hydrology*

Wadi Al-Fulayj drains an area of over 750 km^2 into the Gulf of Oman (Figure 1). The larger part of the catchment is covered by rocks as cliffs or low hills near the coast, the high peaks rising to an elevation of 2327 meters above sea level in the west and southwest. In general the catchment area consists of steep to moderate mountainous terrain at the head of the catchment and a gentle sloping alluvial plain near the coast. The rainfall over the catchment is low and erratic. Most of the rainfall occurs during the winter (primarily in February and March) due to depressions moving southward from the Mediterranean Sea, tracking roughly parallel with the mountain range in Oman. These events are generally larger in extent, longer in duration, but of lower intensity than summer convective storms.

Summer rainfall is more sporadic and more localized and depends on the location of Inter-Tropical Convergence Zone over the Arabian Peninsula. The past record shows 1983, 1986, and 1990 as wet years and 1984, 1985 and 1991 as dry years. Average annual rainfall varies from 125 mm /year in the alluvial plain to 200 mm/ year at altitude.

2.4 *Surface flows*

The main wadi channel is nonperennial, about 40 km long above the dam and 48 km up to the Gulf of Oman. Short duration flows occur during rain storms. Flow is gauged at the dam site and a few kilometer upstream at Rafsha. The mean annual discharge at the Rafsha gauge station is 5.2×10^6 m^3 for the period 1986 to 1996 with a maximum of 17×10^6 m^3 recorded in February 1990. Annual flow is about 15 percent of the annual rainfall for the above period. OTT chart recorders are installed to record inflow and outflows at the dam. For the period of operation till April 97, a total of 24 flood events, have occurred at the dam. The dam has spilled 15 times. The reservoir size is smaller than a 1 in 5 return flood. This however does not reflect the dam as under-designed, when, other deciding factors like storage capacity of the aquifer, engineering solution, capital cost for a larger structure and potential threat to the safety of downstream areas in case of dam failure, are considered.

2.5 *Geology and hydrogeology*

The rocks exposed in the basin are of sedimentary origin of Tertiary Age. Geological formations exposed over the larger part of the catchment includes Usawa and Abat formations (Early-Middle Eocene) represented by bioclastic marl, marly limestone, calcarenitic beds, with subordinate sandstone, siltstone and yellow glauconitic marl. Elsewhere, the rocks of Tahawa Seeb formations include polymict breccia olstoliths marl, bioclastic limestone and conglomerate. Commonly the rocks dip at a gentle angle and vertical faults occur in the southwestern part of the catchment. The unconsolidated deposits (recent to sub-recent) up to 80 m thick, are found within the wadi . Downstream of the dam, the alluvial fill widens towards the coast, flanked by marly outcrops. The unconsolidated deposits downstream of the dam are the principal aquifer, with water of potable quality down to 70 m, depth. Generally the production wells are up to 40 m deep with aquifer thickness of up to 30 m. In the target area the water level ranges between 7 to12 m that drop to 18 m below land surface in dry periods. The hydraulic gradient varies from 0.0025 to 0.0015. The aquifers have transmissivity of up to 7000 m^2 / day and groundwater is unconfined.

3 WADI AL KABIR RECHARGE DAM

3.1 *History and background*

Wadi Al Kabir catchment is a hydrologic basin of the desert inland drainage system in the Dhahirah region in Oman (Figure1). In a reconnaissance study for recharge schemes carried out in 1985 in Oman, Wadi Al Kabir recharge scheme was identified and ranked 9th amongst 58 on priority list. The study identified an overall long term positive balance for the Al-Kabir catchment. This could revert to deficit since modern techniques of water extraction allowed large quantities to be developed which in turn effect the yields of traditional aflaj and wells for sustained use. The project area experienced decline in water level and aflaj yields in individual years. In 1989 a detailed study for technical and economic feasibility of a recharge dam was carried out and a detailed design prepared.

3.2 *Location and design*

The Wadi Al Kabir Recharge Dam is built on the main wadi, about 7 km from Araqi town. The structure comprises an earth-fill embankment with both sides protected by rock rip-rap, and a spillway and stilling basin made of gabions. The hydraulic structure is designed with one controlled culvert outlet that leads the flow in the main wadi. The Wadi Al Kabir recharge dam with a crest height 8.9 m, length, 2664 m and a storage capacity of 0.5 million m^3 was commissioned in December 1990.

3.3 *Catchment rainfall and hydrology.*

Wadi Al Kabir catchment, an area of 757 km^2, drains to the inland desert system in Oman. The catchment contains a network of 6 rain gauge stations, in operation since 1991. The catchment received 56.6 , 140.8, and 70.1 mm of rainfall during the hydrological years 1990/91, 1991/92, and 1992/ 93 respectively. The catchment mean annual rainfall for the period of dam operation is 116 mm, with minimum of 55.15 mm in 1990-91 and maximum of 180.77 mm in 1994-95. The mean winter and summer rainfall for the years of dam operation is 41.8 and 68.35 mm respectively. Until 1994, the catchment experienced dry conditions

3.4 *Runoff and surface flows*

Wadi Al Kabir is nonperennial. Short duration flows occur during the storm period only. A reconnaissance study derived 12.2 million m^3 as simulated mean annual runoff, at the recharge site

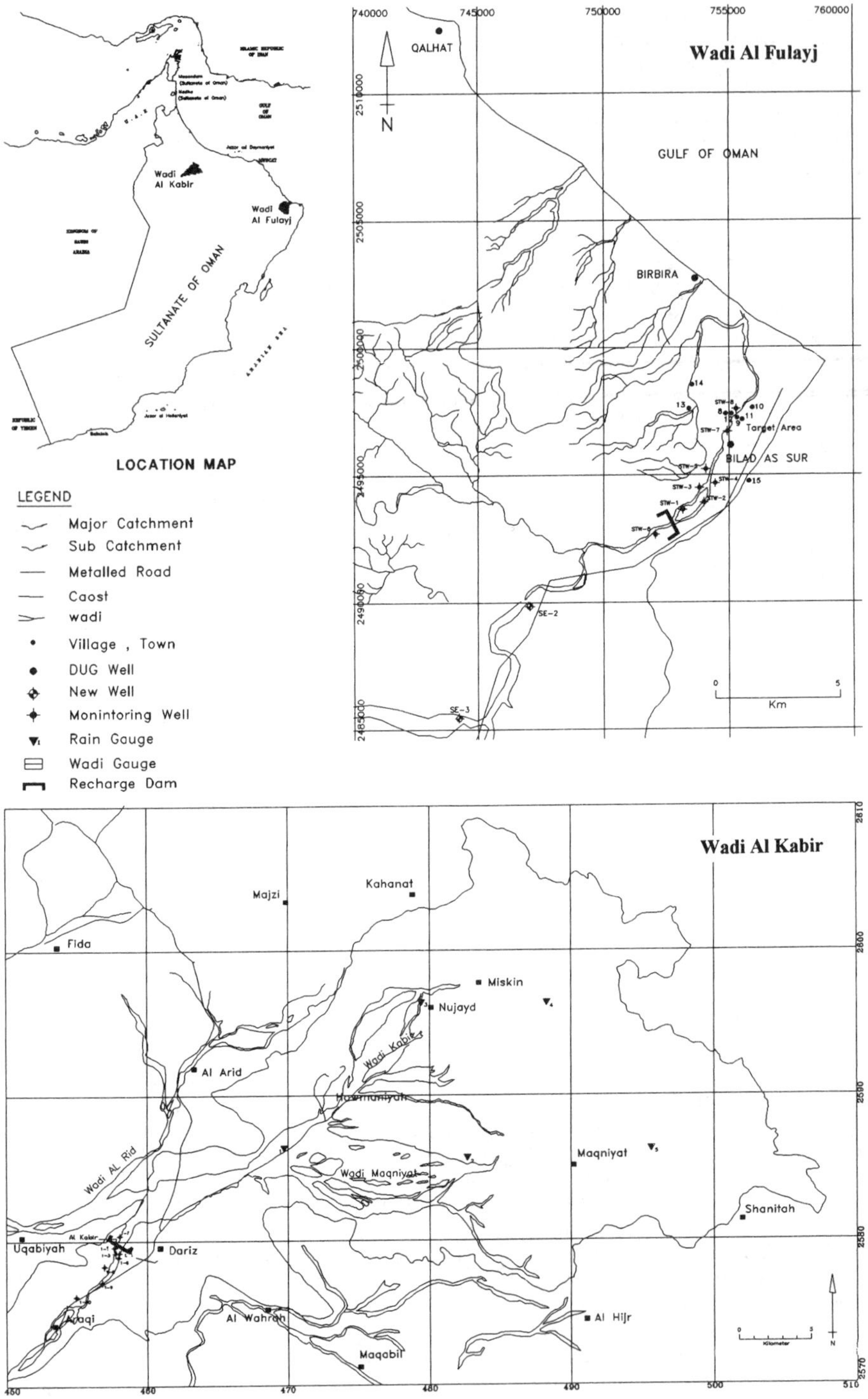

Fig. 1 Locations of Wadi Al Kabir and Wadi Al Fulayj recharge dams

using rainfall-runoff model HYRUN. Floods at the dam are monitored by an automatic Ott R-20 strip chart type recorder. Up to the end of the hydrologic year 1996-97 some 26 floods have reached the dam. The initial three years, of operation 1991 to 1993, were dry with a limited number of events, smaller than the reservoir capacity. The summer rainfall with only two events, in the year 1994-95 created a sizable flood with a total volume of 7.4 million m^3. The flood during the hydrologic year 1995 -96 totaled 14.32 million m^3 of which single event in December 1996 accounted for 13.9 million m^3.

3.5 *Geology and hydrogeology*

The eastern part of the catchment comprises ophiolite rocks which are mainly harzburgite and dunite. These rocks are extensively fractured and weathered in the gabbroic zone. The rocks in the central part and near the recharge dam are of sedimentary origin, consisting of oolitic, litholoclastic, silicified to marly limestone, red and, white cherts. Hard rocks in general are not believed to be the potential aquifers. The alluvium consist of unconsolidated / semi-consolidated clastic deposits, occupying the areas in-between the wadi channels and plain extending downstream to the dam. The alluvial deposits near the dam have a limited thickness that hardly exceeds 40 m and consists of inter -layers of cemented to non-cemented gravel, that form the aquifer. The depth to water ranges from 13.5 to 17 m below land surface, reducing to 17 to 22 m in dry season. The aquifer's transmissivity values obtained from past studies ranged from 200 to 700 m^2 / day. However transmissivity in channel deposits can exceed 7000 m^2/ day. This variation is due to difference in degree of cementation and the thickness of the aquifer at individual sites. The storage coefficient values range from 0.01 to 0.02. The average thickness of the aquifer is 18 m. The hydraulic gradient is 0.007.

4. ARTIFICAL RECHARGE.

Artificial recharge is estimated by applying the following.

a) The change in storage, represented by a rise in water level during each flood event.

b) Seventy five percent of volume retained as inflows to dams is considered to be infiltrated.

c) Evaluation of water balances over the years of dam operation.

Artificial recharge is computed from an average rise in water level for each event as the change in storage over the area under the influence of the dam. The change in storage or recharge for an event is given by:

$$aV = ah \times A \times S \qquad (1)$$

where aV = change in storage in m^3 , ah = rise in water level in m, S = dimensionless aquifer storage coefficient and A = the area in m.^{2} The groundwater head at the dam also results in an increase in the sub-surface flow component of recharge in the alluvial section of the wadi, calculated with the help of Darcy's law

4.1 Wadi Al Fulayj recharge dam

During 6 years of operation since 1992, the dam retained a total volume of 21.4 million m^3 of water from 27 flood events. There are 12 observation wells monitored monthly for water level and some for water quality. Frequency of observation increases to daily during flood periods. Depending upon the flood size, the average rise in water level varies from 4 to 7 m in the target area, which is about 6 km^2. Using known parameters, the annual artificial recharge for the years of dam operation is calculated, as 2.5 million m^3. This is in agreement with the 2.6 million m^3 annual recharge assumed to be 75 percent of the 21.4 million m^3 flood water retained in the dam in 6 years of operation. Evaluation on annual water balance for the postdam periods shows an equilibrium state between net recharge component which includes 3.0 million m^3 subsurface flow near the dam axis, 1.5 and 2.5 million m^3 as addition to storage (natural and artificial recharge) against an estimated annual use of 7.5 million m^{3}in the area.

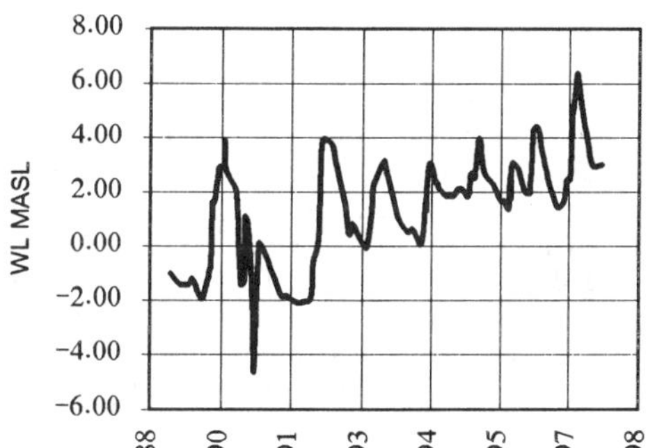

Figure 2.Hydrograph of well STW-3, Al Fulayj recharge dam

The hydrograph of well STW-3 shows a sustained rise in storage after the commissioning of the dam. Water quality improvement occurs in most of the dug wells . The electrical conductivity in wells close to the coast recovers from 10,000 to below 6000 micro Siemens/cm. The recharge from the dam is therefore effective to control the inland migration of the salt - fresh water interface.

4.2 *Wadi Al Kabir recharge dam*

During 6 years of operation since 1992, the dam retained a total of 8.4 million m^3 of water from 36 flood events. The first three years were dry with small floods. The single flood in Dec.95 amounted to over 13 million m^3 which spilled over the dam. In the year 1997 the dam retained 2.2 million m^3 of water. The observation wells monitored show rises in water level, ranging from 2 to 5 m. The recharge, as increment in storage, is computed to be 0.66 million m^3 per year during the dry period and 1.6 million m^3 during wet years.

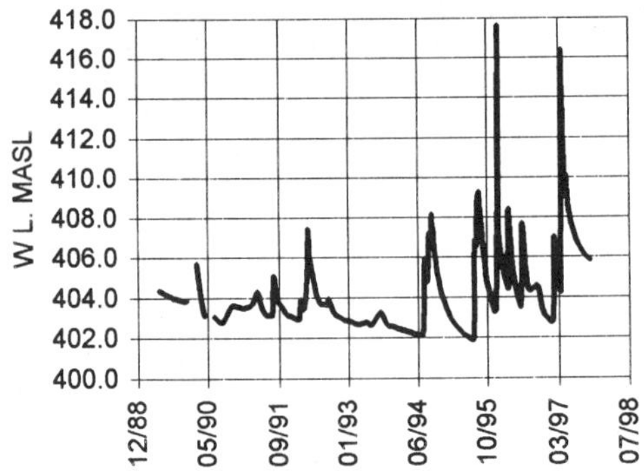

Figure 3.Hydrograph of well I-1, Wadi Al Kabir recharge dam

This assessment is in agreement with the approximation that 75 percent of the floods retained at the dam, during the six years of operation goes as recharge to groundwater. The annual water demand in the area is of the order of 9 million m^3. Recharge to groundwater as subsurface flow is computed to vary annually between 2. 68 to 6.10 million m^3 during dry and wet years respectively. The water deficit attained during dry period is overcome to some extent by artificial recharge from the dam. .

5 CONCLUSIONS

Conservation of surface runoff that is lost to the sea or desert, is considered important to improve the groundwater potential and to offset the adverse effects of groundwater mining which have occurred, due to over-exploitation in many parts of Oman. The regional Quaternary aquifer systems have favorable locations for artificial recharge where dams are constructed. During the period of operation, the 17 recharge dams have stored 437 million m^3 of water. The analyses show that recharge dams have been effective in many basins in reducing water deficits and improving water quality. The operation and monitoring of dams provides an understanding of the principal processes governing recharge in the wadi systems which is very valuable in planning future recharge schemes in Oman .

REFERENCES.

Atkins, W.S. April 1989. Groundwater Recharge Scheme for Ibri - Araqi Area Feasibility Studies Engineering Assessment Report. Ministry of Agriculture and Fisheries Oman

Atkins, W.S. May 1989. Groundwater Recharge Scheme for Ibri - Araqi Area Feasibility Studies. Data Analysis Report .Ministry of Agriculture and Fisheries Oman

Hydroconsult, 1985. Catchment Water Conservation and Recharge Scheme for Irrigation Reconnaissance Study. Ministry of Agriculture and Fisheries Oman.

DTCD. UN 1982, Groundwater in the Eastern Mediterranean and Western Asia. Natural Resources / Water Series No.9

Euroconsult, TNO, IHE, 1996 Future of Recharge Dams and Similar Schemes in Oman. Main Report.

Hydroconsult, 1986. Preliminary Study for Recharge Schemes Volume 1 Ibri Araqi. Final Report. Ministry of Agriculture and Fisheries Oman.

Mott MacDonald International, 1994e Detailed Hydrotechnical Recharge Studies in Dhahira Region Wadi Kabir Wadi Lusayl.

Artificial Recharge of Groundwater, Peters et al. (eds) © 1998 Balkema, Rotterdam, ISBN 90 5809 017 5

Artificial recharge of a lake excavated for gravel extraction

M.H.A.Juhász-Holterman
Waterleiding Maatschappij Limburg, Maastricht, Netherlands

J.H.Peters
Kiwa Research and Consultancy, Nieuwegein, Netherlands

J.J.G.M.Geerts
DHV Water, Amersfoort, Netherlands

ABSTRACT: For the production of drinking water, Waterleiding Maatschappij Limburg (WML) will be using more surface water in order to save groundwater. Natural processes have been chosen to give the surface water the character of groundwater by means of recharge. In Heel, water from the river Maas is used, the water being pumped to a reservoir with a volume of 25 million m^3. The reservoir 'De Lange Vlieter' is the result of gravel extraction from an area 1 km distant from the river Maas. In recent years, different studies have been made of the feasibility and design of the project. As from 2002, 20 million m^3 of bank-filtered water will be recovered annually. The reservoir itself will be continuously supplemented by river water. The bank-filtered water will be recovered divided over three well-fields, with in total approximately 30 wells. The design of the well-fields will be such that water in the subsoil will have a residence time of at least 30 days, resulting in pumped water being free of pathogenic micro-organisms. It is judged that disinfection of the pumped water is therefore not necessary. For the design of the posttreatment process for this water, a rapid decrease in its iron concentration is expected. The manganese concentration will also drop, although less rapidly.

1 INTRODUCTION

In the province of Limburg, in the south-east of the Netherlands, a good 81 million m^3 of water are distributed by Waterleiding Maatschappij Limburg (WML) every year. Until recently, this water was produced mainly from groundwater. The expectation is that the demand for drinking water will increase by 30% over the next few decades.

Mainly through water management measures, the once-high water table of the Netherlands has declined. This means a precarious existence for many unique wetland areas. Hence, the government has decided that the use of groundwater must drop. To save groundwater WML is producing more and more drinking water from surface water.

On the other hand, WML wants to employ natural processes as much as possible so that the water entering the treatment stage has the character of groundwater. For this reason aquifer recharge, among other processes, is used in Heel. In recent years, various studies have been conducted that demonstrate the feasibility and the design of the project. In 2002, Waterproductiebedrijf Heel (WPH) can start operating at an annual capacity of 20 million m^3. To reach this goal WML is putting in an all-out effort.

Following a short description of the complete project, this article will cover the aspects of the study relevant to the recharge process. These include the special reservoir for the water to be recharged, the location of the wells with respect to the travel time, travel time range, water quality characteristics during aquifer passage, and clogging of the banks of the reservoir.

2 THE PROJECT

WPH uses the river Maas as a source (Fig. 1). This rain-fed river has unpredictable water quality and discharge, so that direct use of its water for producing drinking water would entail significant costs. By making use of the local geological situation, this problem can be resolved.

In the region during the pleistocene period, the Maas deposited a 10-to-30-metre thick gravel aquifer, plus coarse sand and stones, as a wide strip along the then river bed. The gravel aquifer, which is almost entirely filled with groundwater, has been used for many gravel excavations. This has led to the creation of several deep lakes with open connections to the groundwater. One of these former excavated lakes is WPH's present reservoir. Today, this reservoir, with its 120-ha surface area and 35-metre depth, is separated from the River Maas by barriers. The unique feature of this lake is its relatively high level combined with the stability of the water table.

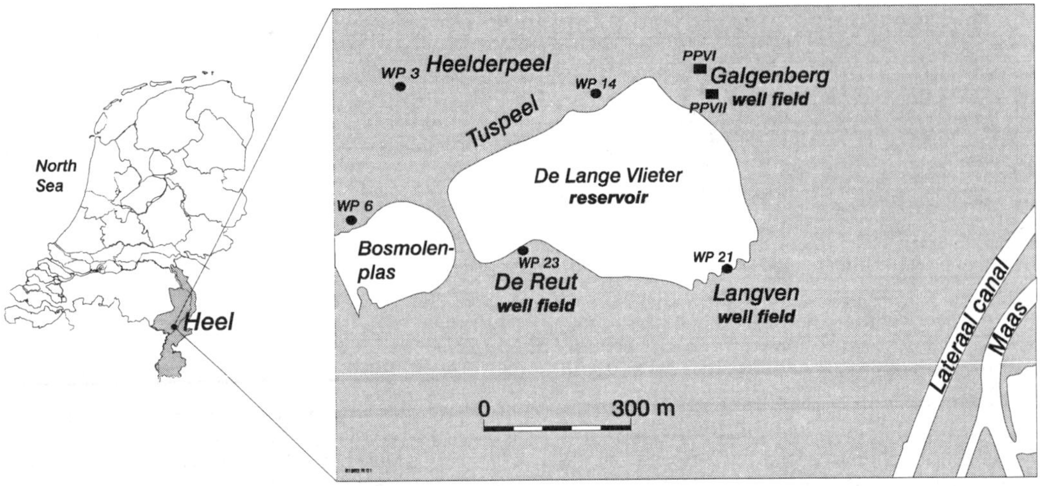

Figure 1. Layout of WPH

Groundwater leaks into the reservoir on the northwest side and flows out again on the other side. The lake is drained by the Maas and the Lateraalkanaal parallel to the river, both having a lower level than the reservoir. Hence, there is already natural recharge of the reservoir water.

When the construction work is finished in 2001, water from the Maas will be pumped to the reservoir, where it will first arrive in a separate section used as an analysis basin. After a residence time of at least one week, the water will flow into the main reservoir, where its average stay will be more than one year. The water will recharge through the bank of the reservoir, and, after aquifer passage, 30 abstraction wells will pump it to the surface. The water will then be treated by aeration, rapid sand filtration and activated-carbon filtration, followed by a UV safety disinfection (Fig. 2).

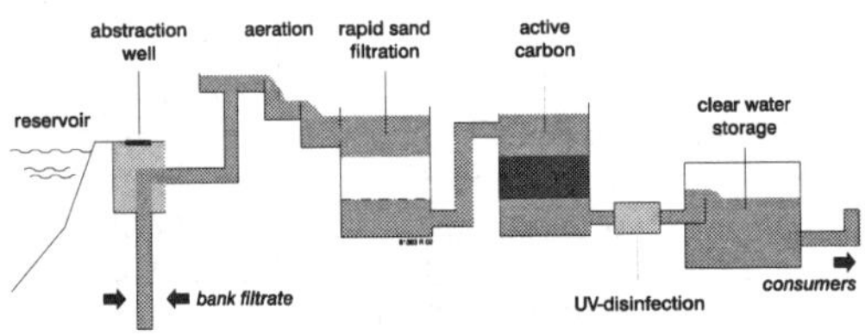

Figure 2. From Maas water to drinking water

3 WHY BANK FILTRATION ?

The reservoir is a deep gravel extraction lake, containing at this moment mainly groundwater. Originally, the intention was to use this water directly, and to replenish the volume abstracted with river water. Although natural processes and equalization greatly improve the quality of the water entering the reservoir, aquifer passage has nevertheless been selected, mainly because this removes the need for disinfection. However, before this could be finally decided, many studies were carried out.

3.1 *Soil structure and hydrogeological modelling*

First, a regional model (Diepenbeek 1994) was converted to one more suitable for also modelling the bank in detail (DHV 1996). To do this properly, apart from the geological structure of the subsoil it was also important to know the gravel production history. During gravel extraction, bed sediment was not only recovered, but also partly deposited in the reservoir and along its bank. This artificial construction of the reservoir bed and bank is highly significant for the location and potential of the recharge process.

The subsoil has the following natural structure: starting at ground level, there is a 7-metre layer of surface sand with embedded layers of loam; then 18 metres of gravel with stones and layers of sand; next comes 20 metres of coarse gravelly sand; and underneath this, 80 metres of fine, sediment-rich sands. Deeper than 125 metres below ground level are layers of clay. The water table is approximately 5 metres below surface level.

The gravel and some of the sand underneath it have been removed from the site of the present reservoir. The fine, sandy top soil could not be used and was therefore deposited along the southern part of the bank to a height of a few metres above the water level. The lower-lying areas thus created are known as Langven and De Reut (Fig. 1).

At Galgenberg, there was a choice of the gravel or the sand aquifer for the location of the recovery filters. At Langven and De Reut, by contrast, the gravel aquifer and part of the sand aquifer have been excavated, so that wells can only be situated in the sand aquifer.

3.2 *Hydrogeological model and recharge*

We now deal with the importance of the hydrogeological model study for the feasibility of recovery by means of wells. As already mentioned, the reservoir has an open connection with the groundwater. The travelling time of reservoir water to the wells must be long enough to achieve the intended disinfection.

Calculations have been made to identify possible locations for wells, so that there will be no damage to the local wetlands. The analysis has shown that it would be best to carry out half the recovery at Galgenberg, using wells in the gravel. The other half of the recovery will be divided over a large number of wells in the sand aquifers at De Reut and Langven. Various well-field configurations have also been analysed to see if they satisfy the required minimum travelling time of 30 days. The final locations for the wells and the individual flow rates are such that the degradation of the natural wetlands is minor, yet the minimum travelling time per well is achieved. This does require the water level in the reservoir to be raised.

Pump tests have, of course, been carried out in all three well-fields. Since these tests produce only very local data, particularly concerning soil and bank resistance, a reservoir test has also been carried out. This test shows that mainly the top metres of the bank are highly permeable.

Moreover, since the reservoir bed is practically impermeable, the calculated entrance rate through its bank is very high. There is a good chance that the bank, now highly permeable, will become clogged, causing further degradation of the surrounding nature reserves. To avoid this problem, a design for a silt remover is being produced that can clean the bank as and when needed. A layer of sediment on the bank might also have advantages. Impurities are trapped and the micro-biological removal of pathogens between bank and wells can be increased. If the decline of the groundwater is greater than predicted, the level in the reservoir can still be raised a few more decimetres.

3.3 *Changes in water quality between the Maas and the wells*

To predict the quality of the water to be pumped, apart from the travelling time, travelling time range, origin and subsoil, a considerable number of chemical and biological processes must also be considered.

Naturally, the quality of the source must be thoroughly known, just as the present quality of water in the reservoir and subsoil, the characteristics of the reservoir, and the geochemical characteristics of the soil. Only then it is possible to predict what happens to the water during its stay in the reservoir and its subsequent aquifer passage.

3.4 *Quality of the Maas and the reservoir*

The quality of the water from the source, the Maas, is known. Among other measures, it is continuously monitored 40 km upstream in Eijsden. In the water from the Maas, there are both pathogenic micro-organisms and micro-contaminants. In the reservoir, these mix, break down, evaporate and settle. To accurately simulate these processes, the reservoir model BEKWAAM has been developed (Benoist et al. 1997).

Some substances, such as certain herbicides, occur in high concentrations in the Maas. The concentrations in the reservoir can be kept within acceptable limits by temporarily stopping intake of water from the Maas. Even though certain substances are broken down or diluted during aquifer passage, their concentrations in the recharge water must be low enough for them not to threaten the quality of the soil or the groundwater (Peters et al. 1997). Accordingly, the government has specified 73 parameters for the recharge water in the reservoir.

Pathogenic micro-organisms are always present in the surface water of the Maas, and their removal in the reservoir is bound to be imperfect (DHV Water 1996a). The decimal elimination capacities (DEC) of some very well-known pathogenic micro-organisms, such as giardia, cryptosporidia and enteroviruses, are shown in table 1. The table also shows what removal capacities are required. Taking into account the series

Table 1. Predicted and required DEC

Treatment stage	Decimal Elimination Capacity		
	Enteroviruses	Giardia	Cryptosporidia
Reservoir[1]	> 2.5	> 2.1	> 1.4
Aquifer passage[2]	> 4	> 4	> 4
Posttreatment:			
aeration/rapid filtration	1	2	2
activated carbon	0	0	0
UV-disinfection	2	0.5	0
Total	>9.5	>8.6	>7.4
Required	7.7	4.9	4.5

[1] residence time: 1 year

[2] residence time: 30 days minimum, soil composition: gravel

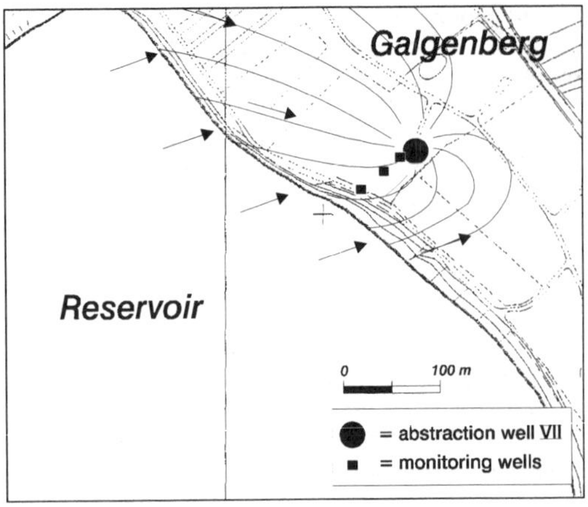

Figure 3. Site of test wells and flow lines

of linked barriers, the removal capacity is sufficient. The aquifer passage plays an important role in this, however. To be certain that, in particular, the removal within the coarse-gravel aquifer will be adequate, field trials have been conducted at a test site alongside the reservoir.

The concentrations or values of other factors relevant to treatment are also greatly influenced by artificial recharge. Ones that come to mind are iron, manganese, ammonia, temperature and hardness. Given that the raw material is surface water, it seems strange initially that iron and manganese occur in the wells. However, it should be remembered that the reservoir was excavated in an originally anaerobic aquifer, in which the presence of considerable amounts of dissolved iron and manganese is quite normal. For the predictions of the chemical parameters, one of the test sites used was the same as for the pathogen study. This field study is described below in more detail.

3.5 *Field study and predicted changes due to aquifer passage*

By the eastern part of the reservoir's bank, in the Galgenberg well-field, a line comprising observation wells and a pumping well (PP VII) has been constructed (Fig. 3 and 4). A large number of physical measurements have been collected from these wells, such as pump test data, flow measurements, temperature readings, and standard water level measurements. In addition, a precise description of the soil composition has been obtained from soil samples (Laeven &

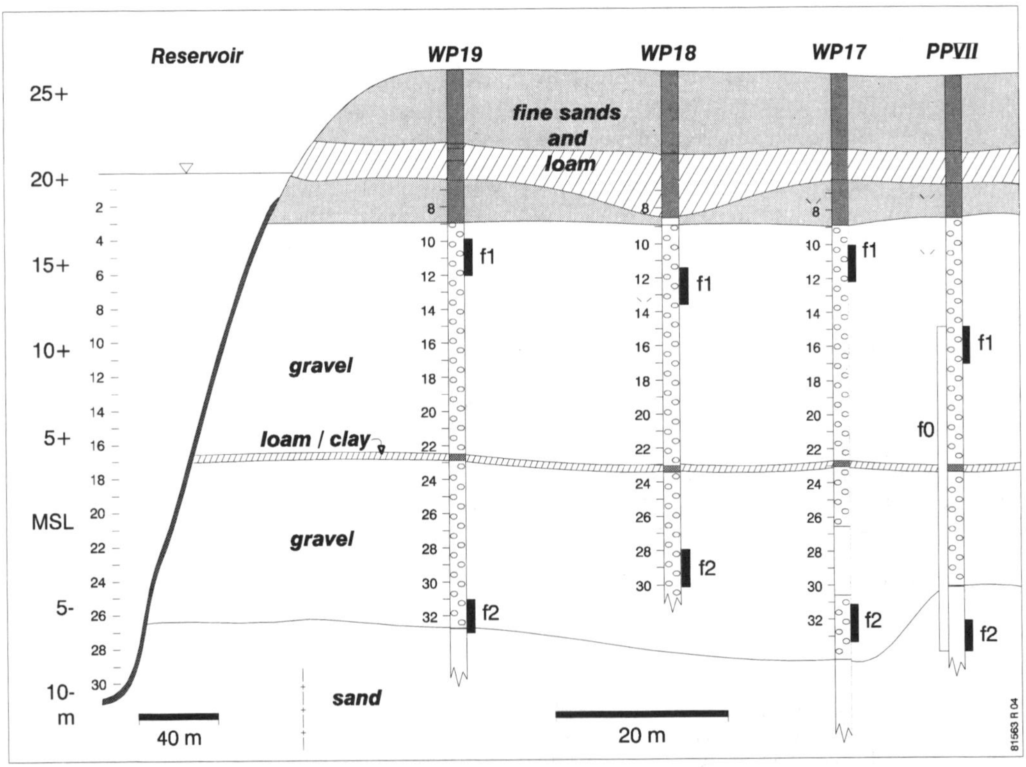

Figure 4. Cross-section of the measurement line. MSL is Mean Sea Level (m)

Stuyfzand 1997).

The travelling time of the water along the measurement line was calculated and measured after the test well with a capacity of 140 m^3/h began operating. Here, too, the regional hydrogeology plays an important role. The flow lines from the reservoir to the test well do not in fact take the shortest route, but leave the reservoir in a more northerly direction. The flow is thus heavily influenced by the natural flow of the groundwater (Fig. 3).

On the basis of the measured and calculated travelling times, a further study has begun that should provide data on changes to the reservoir water that occur in the coarse-gravel aquifer and the sand aquifer underneath. This data, supplemented by data from the other well locations, would need to be translated for the situation where total recovery is 20 million m^3 per annum. At the observation wells 16 to 19 and the pumping well, a large number of parameters were measured, biological as well as chemical.

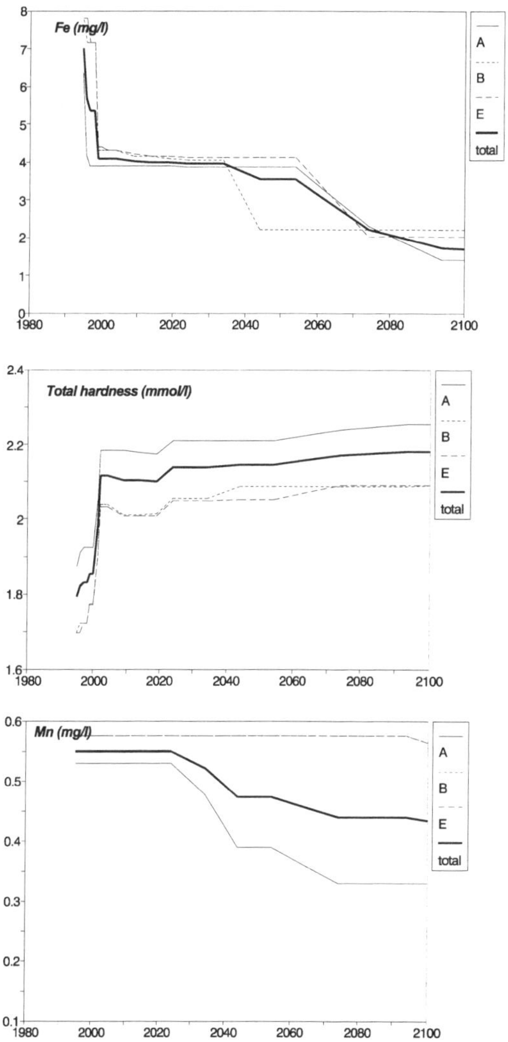

Figure 5. Predictions of the factors Fe, Mn and hardness for the well-fields A (Galgenberg), B (Langven) and E (De Reut) (Laeven & Stuyfzand, 1997)

From the physical preliminary study and the hydrogeological model study, it was established that, if water were abstracted simultaneously from the test well and from a pumping well some distance away, the travelling time from the reservoir to observation well 19 would be approximately 30 days. By examining the filter for pathogens, it could be shown that the removal capacity of micro-organisms in the subsoil was at least 3.5 log-units (Theunissen et al. 1996). It was not possible to determine the precise removal effect, because micro-organisms were at that time not present in the reservoir water in quantities high enough. In any case, at the time of the study, no water from the Maas was being let in yet.

The change in the chemical groundwater quality resulting from reservoir water being recharged was also measured and related to the travel time and travel time range for the flow line (Laeven & Stuyfzand 1997). The results of this have been extrapolated to find the future water quality of the pumped water in the three well-fields. These results are shown in figure 5.

Figure 5 indicates that the iron concentration in the pumped water drops quickly due to replacement of the original groundwater by reservoir water. The manganese concentration will drop less quickly. Small amounts of iron and manganese will always be present in the recovered water because it will never be possible to avoid pumping a small amount (approximately 15%) of natural groundwater.

In the reservoir, natural softening occurs. However, because of the aquifer passage, the hardness of the water will increase again.

REFERENCES

Benoist, A.P., A.G. Brinkman, J.P.M.J.A. van Diepenbeek & J.M.J. Waals, 1997. BEKWAAM, A model fit for reservoir design and management, Proceedings IAWQ-IWSA Conference Reservoir Management and Water Supply, Prague.

DHV Water, 1996. Milieu-effectrapportage, deelrapport geohydrologie, Amersfoort: DHV Water (in Dutch).

DHV Water, 1996a. Desinfectiestrategie Panheel. PD 96-37, Amersfoort: DHV Water (in Dutch).

Diepenbeek, P.M.J.A. van, 1994, Geohydrologisch basismodel Spaarbekken Panheel. Maastricht WML (in Dutch).

Juhasz, M.H.A., C.G.E.M. van Beek &

C.J.G. Janssen, 1997. Heel, putverstopping/ontwerp putten Fase 1. Maastricht/Nieuwegein. NV Waterleiding Maatschappij Limburg/ Kiwa (in Dutch).
Laeven, M.P & P.J. Stuyfzand, 1997. Voorspelling grondstofkwaliteit Waterproductiebedrijf Heel, eindrapport 1996. Nieuwegein: Kiwa. Reportnr. KOA 96.155 (in Dutch).
Peters, J.H. et al, 1997. Kwaliteit infiltrerend bekkenwater waterproductiebedrijf Heel. Nieuwegein: Kiwa. Reportnr. KOA 97.232 (in Dutch).
Theunissen, J.H.H., R. van der Heide, M. Bahar, H.A.M. de Bruin, P.F.M. Teunis & G.J. Medema, 1996. Virusverwijdering door oeverfiltratie bij proefproject Panheel. Bilthoven: RIVM. Reportnr. 289202011 (in Dutch).

Artificial Recharge of Groundwater, Peters et al. (eds) © *1998 Balkema, Rotterdam, ISBN 90 5809 017 5*

Sprinkling infiltration in Finland: Effects on forest soil, percolation water and vegetation

H-S. Helmisaari, V. Kitunen, A-J. Lindroos, I. Lumme, S. Monni, P. Nöjd, L. Paavolainen, E. Pesonen, M. Salemaa & A. Smolander
Vantaa Research Centre, Finnish Forest Research Institute, Finland

J. Derome
Rovaniemi Research Station, Finnish Forest Research Institute, Finland

ABSTRACT: In spring 1996 the Finnish Forest Research Institute started a three-year study on the effects of sprinkling infiltration on forest soil, percolation water and vegetation. The aim of the research project was to determine the effects of sprinkling infiltration and surface runoff on soil nutrient concentrations, soil organic matter, the microbiological characteristics of the different soil layers, the nutrient status and growth of the tree stand and understorey vegetation, and the quality of percolation water. The results obtained during the first two years indicate changes in soil hydrology, an increase in the pH and base cation concentration of the soil, and changes in nitrogen cycling and the species abundance of the understorey vegetation. Nitrification increased strongly in the soil on the infiltration plots but, due to the large amount of infiltrated water, the nitrate concentrations in the groundwater were very low (mean value 0.2 mg/l).

1 INTRODUCTION

Finland is one of the few countries using surface recharge/sprinkling infiltration in the artificial recharging of groundwater. This method has been in use at some waterworks since the 1970's. During the past few years, a number of new sprinkling infiltration plants have been established on forested eskers, and several are currently being planned. Many cities in southern and central Finland are planning in the near future to meet at least part of their water requirements by utilising artificially infiltrated groundwater. By the year 2010 the amount of groundwater and artificial groundwater as a percentage of the total water consumption in Finland is estimated to reach 70-75 %, of which artificially recharged groundwater will account for about 15-20 % (Hatva 1996).

Sprinkling infiltration is considered to be a rather natural infiltration technique, since the raw surface water is sprinkled directly onto the forest soil from a network of pipes, and therefore does not cause as much direct disturbance to the vegetation and soil surface as e.g. basin recharge. Water purification is assumed to take place via the same processes as for precipitation; microbiological, biological and chemical processes, such as adsorption and ion exchange.

1.1 *Soil changes*

The amount of water annually infiltrated greatly exceeds the annual precipitation, and this may lead to changes in the hydrological and biochemical processes in the forest soil. The chemical composition of the infiltration water also differs from that of precipitation. The surface water may contain considerable amounts of mineral nutrients. The pH of the surface water is always much higher than the typical values for precipitation and percolation water in Finland. For example, the median pH of lake water in the Finnish lake survey carried out in 1987 was 6.3 (Mannio and Vuorenmaa 1995), and the pH of percolation water collected below the organic layer has been reported to vary between 4.3-5.0 in a number of forest soils in Finland (Lindroos et al. 1995, 1998, Helmisaari and Mälkönen 1989, Soveri and Ahlberg 1990).

If the pH of forest soil increases as a result of infiltration, this may cause changes in the biochemical cycling of nitrogen especially. The median pH of the forest soil in southern Finland is about 4.0 (Tamminen and Starr 1990). Nitrification (the production of nitrite and nitrate from ammonium by microbes) is usually rare in Finnish forest soils due to the low pH and relatively low ammonium availability (Martikainen 1996, Paavolainen and Smolander 1998, Smolander et al. 1998). A rise in soil pH, e.g. due to liming (Smolander et al. 1995), can increase nitrification, and the leaching of nitrate can pose a risk for groundwater quality.

If there is increased production of nitrate in the soil, and the soil is saturated with water, N_2O production may also increase. N_2O is formed during denitrification from nitrate but it can also be formed as a by-product in nitrification. Even though N_2O is a strong greenhouse gas, the onset of denitrification can be considered to have a beneficial effect on groundwater quality because it decreases the amount of leachable nitrate in the soil.

1.2 *Vegetation changes*

The understorey vegetation may be changed by infiltration as a result of changes in the microclimate and environmental conditions. Soil moisture, fertility, acidity and temperature all affect plant vitality and growth. Infiltration may change the species composition; species which favour moist, fertile soil may increase in abundance, and species of dry, relatively infertile sites may decrease. Waterlogging may cause a lack of oxygen for plant roots (Blom and Voesenek 1996), especially on flat terrain. Trees may also benefit from infiltration, especially during dry summers.

1.3 *Aims of the research*

The quality and extent of environmental impacts resulting from sprinkling infiltration depend on the quality and amount of infiltrated water, the soil and vegetation in the infiltration area, and the timing and duration of infiltration. There is no long-term experience about the environmental impacts of sprinkling infiltration.

In spring 1996, the Finnish Forest Research Institute started a three-year study on the effects of sprinkling infiltration on forest soil, percolation water and vegetation. The project is being sponsored by the water utilities of the cities of Hämeenlinna, Jyväskylä, Mikkeli, Tampere-Valkeakoski area, Turku area and Tuusula area, the Häme Environmental Institute and the Ministry of Agriculture and Forestry.

The aim of the research project is to study the effects of sprinkling infiltration and surface runoff on soil nutrient concentrations, soil organic matter, the microbiological characteristics of the soil, the nutrient status and growth of the tree stand and understorey vegetation, and the quality of percolation water.

2 METHODS

2.1 *Study area*

The project was carried out during 1996 - 1998 on study plots established on the Ahvenisto esker in Hämeenlinna, southern Finland. The artificial recharge of groundwater was started in 1976 using infiltration basins. The quality of artificially recharged groundwater has been good, apart from the high iron concentrations.

Sprinkling infiltration was started in December 1995 in order to improve the oxidizing conditions and water purification efficiency in the infiltration area. Infiltration water is obtained from a near-by lake.

The sprinkling irrigation area used in this study on the Ahvenisto esker is situated on a relatively steep and stony, eastern slope. The soil consists of till and glaciofluvial sediments. The forest site type is the relatively fertile Oxalis-Maianthemum type, according to the classification system by Cajander (1949).

2.2 *Study plots, irrigation method and amounts*

In May 1996, five study plots (size 625 m^2) were established in the area, one for wintertime infiltration, one for continuous summertime infiltration and one for periodical summertime infiltration, and two control plots. Wintertime infiltration was carried out during November-May, continuous summertime infiltration during June-October, and periodical summertime infiltration from the middle of June to the middle of July, and from the middle of August to the middle of September.

The amounts of irrigation water applied were more than 2000 times annual precipitation in the area (Table 1).

Table 1. The irrigation amounts, m^3/m^2.

Plot	1996	1997
Wintertime	1778	1726
Continuous summertime	1118	1258
Periodical summertime	492	634

2.3 *Research methods*

Soil samples were taken from the organic layer three times during the growing season using a soil auger. Sixty samples per plot were bulked to give three composite samples per plot. These samples were used for the pH measurements (soil/water slurry) and the microbiological studies. Soil samples were also taken from the organic and mineral soil layers (0-10 cm mineral soil) in June 1996. 24 samples per plot were bulked to give three composite samples per plot. The samples were taken about one meter down the slope below each irrigation pipe. The total concentrations of P, K, Ca, Mg, Al, Fe, Cu, Mn, Zn and Na were determined on the organic layer samples by dry ashing (550 °C) followed by extraction with concentrated HCL and analysis by ICP/AES. The extractable K, Ca and Mg concentrations were determined by extraction with ammonium acetate (pH 4.65) followed by analysis by ICP/AES. Ammonium and nitrate nitrogen were extracted with 1 M KCL and the extract analysed by flow injection analysis (FIA).

Net nitrogen mineralization and net nitrification were determined in laboratory incubation experiments using soil samples taken three times during the growing season. The production of ammonium and nitrate nitrogen was measured in the incubations (Smolander et al. 1995). The moisture content of the soil sample was maintained at 60 % and the incubation performed at +14 °C for 42 days. Ammonium and nitrate nitrogen were analysed as described above. The amount of nitrogen dioxide released during denitrification was measured in the field with gas sampling chambers installed on the soil surface (Martikainen et al. 1993).

Water samples were collected continuously during the growing season by means of plate lysimeters (6 replicates in each plot) below the 5-10 cm thick organic layer, and by means of suction cup lysimeters (6 replicates in each plot) of depths 40 and 100 cm below the ground surface. Monthly groundwater samples were taken from the observation pipes all the year round. The pH was measured, and the concentrations of P, K, Ca, Mg, Al, Fe, Cu, Mn, Zn and Na were determined by ICP-AES. The concentrations of total nitrogen, NH_4-N and NO_3-N were determined by FIA. The concentrations of organic carbon were determined on a TOC analyzer.

The species abundance of the understorey was determined by estimating the percentage cover on 15 one m^2 sub-plots on each plot. The area of eroded soil was estimated. Changes in the diameter growth of 19 trees were measured using automatic girth bands once every hour. The biomasses of the fine roots (< 2 mm diameter) of grasses, shrubs and trees were determined on 15 soil samples per plot taken three times during the growing season (Helmisaari 1995).

The nutrient status of the trees was determined on needle samples taken each year in early June from 10 trees per plot, two branches per tree. Nitrogen was determined on a CHN analyser (LECO) and the other nutrients by ICP/AES.

3 RESULTS AND DISCUSSION

3.1 *Changes in the soil and percolation water*

The very large amount of infiltration water altered the fluxes of water and nutrients in the soil. These changes will eventually have an effect on the quality of the groundwater. The results obtained during the first two years indicated changes in soil hydrology, an increase in the pH and base cation concentration of the soil and changes in nitrogen cycling and the species abundance of the understorey vegetation.

The high pH (around 7.3) and high Ca, K and Mg concentrations in the infiltration water affected the acidity (Fig. 1) and nutrient content of the organic layer, as well as of the percolation water. The pH of the organic layer increased rapidly to 6.7 from the original value of 4.7. The mean pH of the infiltrated percolation water was 6.9 and the mean pH of the

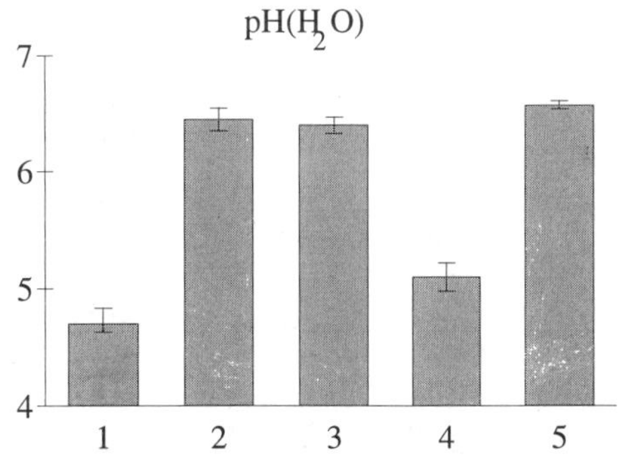

Figure 1. The pH of the organic layer on plots 1-5 (plots 1 and 4 = control, 2 = continuous summertime infiltration, 3 = periodical summertime infiltration, 5 = wintertime infiltration). Means with SD of three sampling dates during the growing season in 1996.

groundwater was 7.0. After cessation of the infiltration, the hydrogen ions in the rainwater will gradually decrease the soil pH as a result of cation exchange. However, this may take a very long time because the pH of the soil had not decreased within two years after cessation of the infiltration.

The resulting elevated pH of the forest soil has affected the ecosystem in several ways, the most significant of which were the changes in nitrogen cycling. Nitrification increased strongly in the soil on the infiltration plots, and continued even after infiltration had ceased (Fig. 2). The initiation of net nitrification was probably due to an increase in the pH and the availability of ammonium. In laboratory studies with acid forest soils, in a pH gradient from 4.2 to 6.2, higher pH values always resulted in the higher production of nitrate (Paavolainen and Smolander 1998). Nitrification is normally a relatively rare occurence in forest soil in Finland.

Production of N_2O, which is a greenhouse gas, was also enhanced in the soil (Lindroos et al. 1998), but gradually ceased after cessation of the infiltration. The origin of N_2O in the studied site is not known, but nitrification and denitrification are the most important processes in the biological N_2O production. However, N_2O production decreases the amount of nitrate likely to be leached from the soil.

Nitrate was leached from the organic layer in the percolation water. The large amount of infiltration water diluted the nitrate concentrations in the percolation water, but high concentrations were measured after infiltration was stopped (Table 2).

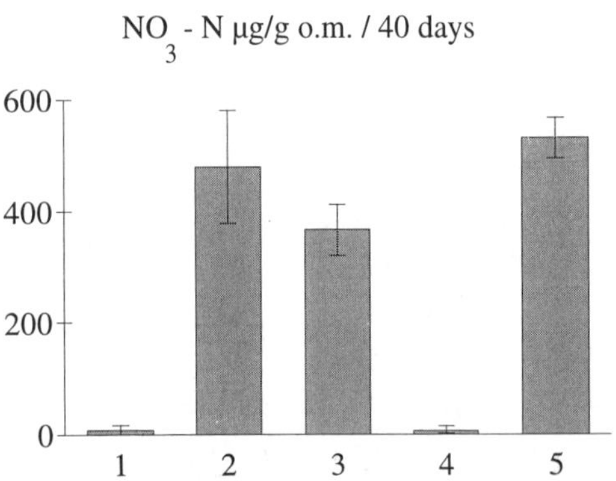

Figure 2. Net nitrification in the organic layer. Laboratory incubation (40 days) at constant moisture content (60 % of the water holding capacity) and temperature (14°C). Means with SD of three samplings during the growing season in 1996.

Table 2. NO_3-N (mg/l) in the surface water, percolation water and groundwater (mean values in 1996 and 1997, 9-54 observations). The percolation water samples represent wintertime infiltration and infiltration break during the summer in plot 5.

	Wintertime infiltration	Summertime no infiltration
Surface water	0.2	-
Percolation water	0.2	10.6
Groundwater	0.2	0.3

However, the nitrate concentrations in the groundwater remained low (mean value 0.2 mg/l). The organic carbon concentrations in groundwater were also low (approximately 5.5 mg/l). The soil retained most of the organic carbon released by infiltration.

The iron concentrations in the observation pipes close to the infiltration area fluctuated around 0.2 mg/l, which is the upper limit for household water. It is too early to say whether infiltration had a reducing effect on the groundwater iron concentrations. Due to the large amounts of infiltration water, the base cation concentrations in percolation water and groundwater reflected the concentrations in the infiltrated surface water. The quality of the groundwater was generally very good.

3.2 *Changes in the vegetation*

The chemical and hydrological changes in the soil changed the species composition of the understorey vegetation. Part of the produced nitrate was taken up by grasses and herbs that increased in biomass (Figs. 3 a-b). Mosses and dwarf shrubs suffered from the increased moisture and nutrient availability. In August 1997, the percentage of eroded soil was highest on the plot continuously infiltrated throughout the year (Fig. 3 c). Wintertime infiltration increased soil erosion but the soil was able to partly recover during the breaks in summertime infiltration.

There were no clear differences in the diameter growth of trees between the infiltration and control plots. Some Norway spruce trees grew especially well on the winter infiltration plot. The increased nitrogen availability may have contributed to the increased growth. The increased nitrogen availability was also reflected in the higher needle nitrogen concentrations. The availability of nitrogen generally restricts tree growth on mineral soils in Finland.

Infiltration did not increase fine root mortality. On the contrary, Norway spruce fine roots grew better in the soil of the summertime infiltration plot than in the soil of the control plot.

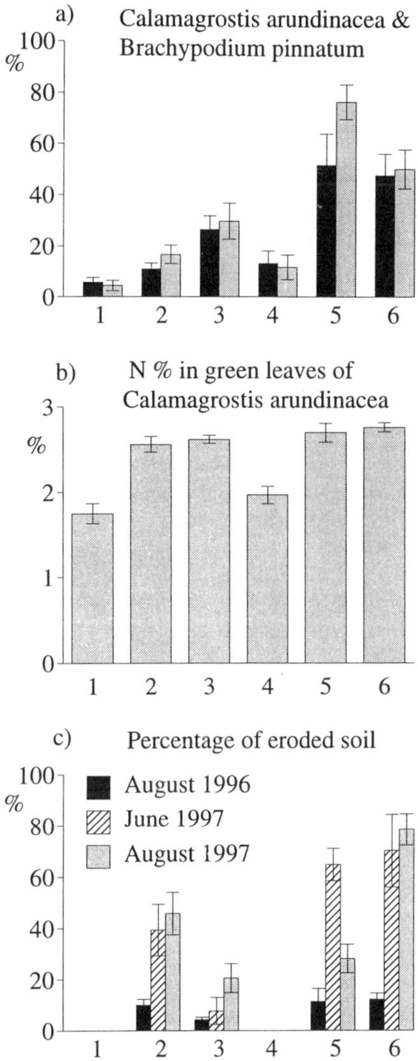

Figure 3 a) The mean (with s.e.m) percentage cover of two grass species in 1996 and 1997 estimated from 10-15 (size one square meter) sub-plots on the infiltration and control plots (numbering 1-5 see Fig. 1., plot 6=continuous infiltration throughout the year), b) nitrogen concentrations in *Calamagrostis arundinacea* in August 1997 and c) the percentage of eroded soil on the same sub-plots.

4 CONCLUSIONS

Sprinkling infiltration may become an important method for artificial recharge. Knowledge of the changes in soil processes during and after infiltration is necessary for minimizing the possible negative environmental effects of this method.

Based on these results, the increased nitrification and leaching of nitrate seems not to be a risk for the groundwater quality during infiltration. However, after the cessation of infiltration, when the infiltrated water no longer dilutes the nitrate, the concentrations in the percolation water increase and this may pose a threat to groundwater. However, this depends on the size of the infiltration area in relation to the size of the whole aquifer. Further studies are needed to find out how long time the increase in nitrification may last after cessation of infiltration.

REFERENCES

Blom, C.W. P.M. & L.A.C.J. Voesenek 1996. Flooding:the survival strategies of plants. *Trends in Ecology and Evolution* 11(7):290-295.

Cajander, A.K. 1949. Forest types and their significances. *Acta Forestalia Fennica* 56(5):1-71.

Hatva, T. 1996. Artificial groundwater recharge in Finland. In: A-L. Kivimäki, & T. Suokko (eds.), Artificial recharge of groundwater. Proceedings of an international symposium, Helsinki, Finland, June 3-5, 1996. *NHP Report* 38:3-12.

Helmisaari, H-S. 1995. Nutrient cycling in Scots pine stands in eastern Finland. *Plant and Soil* 168-169:327-336.

Helmisaari, H-S. & E. Mälkönen 1989. Acidity and nutrient content of throughfall and soil leachate in three Pinus sylvestris stands. *Scand. J. For. Res.* 4:13-28.

Lindroos, A-J., J. Derome & K. Niska 1995. The relationship between dissolved organic matter and percolation water chemistry in Northern Finland. *Water, Air and Soil Pollution* 79: 191-200.

Lindroos, A-J., J. Derome & K. Niska 1998. The effect of emissions from Cu-Ni smelters at Nikel, NW Russia, on the quality of bulk deposition, stand throughfall and percolation water in four Scots pine stands in Northern Norway and Finland. *Chemosphere* 36(4-5):1101-1106.

Lindroos, A-J., L. Paavolainen, A. Smolander, J. Derome and H-S. Helmisaari 1998. Changes in

nitrogen transformations in forest soil as a result of sprinkling infiltration. Submitted manuscript.
Mannio, J. and J. Vuorenmaa 1995. Regional monitoring of lake acidification in Finland. *Water, Air and Soil Pollution* 85:571-576.
Martikainen, P.J., M. Lehtonen, K. Lång, W. De Boer and A. Ferm 1993. Nitrification and nitrous oxide production potentials in aerobic soil samples from the soil profile of a Finnish coniferous site receiving high ammonium deposition. *FEMS Microbiol. Ecol.* 13:113-122.
Martikainen, P. 1996. Microbial processes in boreal forest soils as affected by forest management practises and atmospheric stress. *Soil Biochemistry* 9:195-232.
Paavolainen, L. and A. Smolander 1998. Nitrification and denitrification in soil from a clear-cut Norway spruce (*Picea abies*) stand. *Soil Biology and Biochemistry* (in press).
Smolander, A., V. Kitunen, O. Priha and E. Mälkönen 1995. Nitrogen transformations in limed and nitrogen fertilized soil in Norway spruce stands. *Plant and Soil* 172:107-115.
Smolander, A., O. Priha, L. Paavolainen, J. Steer and E. Mälkönen 1998. Nitrogen and carbon transformations before and after clear-cutting in repeatedly N-fertilized and limed forest soil. *Soil Biology and Biochemistry* (in press).
Soveri, J. and T. Ahlberg 1990. Effects of air pollutants on chemical characteristics of soil water and groundwater. In P. Kauppi, P. Anttila & K. Kenttämies (eds.), *Acidification in Finland.* Springer-Verlag, Berlin, pp. 865-881.
Tamminen, P & M. Starr 1990. A survey of forest soil properties related to soil acidification in Southern Finland. In P. Kauppi, P Anttila & K. Kenttämies (eds.), *Acidification in Finland.* Springer-Verlag, Berlin, pp. 235-251.

Artificial Recharge of Groundwater, Peters et al. (eds) © *1998 Balkema, Rotterdam, ISBN 90 5809 017 5*

An innovative drinking water concept based on double aquifer passage

H.J. Roelofs
Waterworks of Central Netherlands, Utrecht, Netherlands

ABSTRACT: Groundwater abstraction has to be reduced at many places in the Netherlands to stop the loss of wetlands. Waterworks of central Netherlands (WMN) has therefore developed the DOPASS concept for some of its future drinking water production. This innovative concept is based on a double aquifer passage, the use of surface water and natural processes. Bank filtrate will be pumped from shallow wells in the floodplains of the river Rhine and then transported to the treatment plant. The treated water will be used for artificial recharge in a thick phreatic aquifer at an existing groundwater extraction site and will be extracted with several months detention together with the local groundwater. The first aquifer passage removes the suspended material from the river water and levels off quality differences. The main functions of the second aquifer passage are disinfection, a further levelling off of seasonal and quality differences and buffering for disaster situations.

1 INTRODUCTION

WMN currently supplies more than 80 million cubic metres of drinking water to 1.2 million residents as well as companies and organizations in its supply area. For many years, WMN has only used groundwater as the raw water for its drinking water production. In the Province of Utrecht, groundwater is mainly abstracted from a more than 150 metre thick course-sand aquifer. The advantage of groundwater based drinking water production is a high basic quality of raw water and only limited posttreatment is required, which means there is little environmental impact and that the process is relatively cheap. The reason for the high basic quality is the water's long residence time in the deep subsoil, during which the various virological and bacteriological contaminants are broken down by natural purification processes.

The downside of this is that groundwater abstractions lower the groundwater level. This can slowly but surely undermine the basis of a healthy and varied natural landscape. In the Netherlands, this results in a loss of wetlands. It is therefore not surprising that one of the four priorities set for the drinking water industry in the Second National Environmental Policy Plan (Ministry . . . 1993) is to reduce the loss of wetlands. To reverse the trend, it will be necessary to restore the water table.

The Province of Utrecht has outlined a policy for creating new opportunities for the natural environment in its Water Management Plan (Province Utrecht 1992). The falling water table in the Utrecht region is attributable to a number of factors. Besides abstractions for drinking water, industry and agriculture, the empoldering of the province of Flevoland and urbanization play a major role. In concrete terms, WMN has agreed with the provincial authorities that the present permitted yearcapacity for groundwater abstractions will be reduced by 9 million cubic metres in the year 2000. At the same time, WMN will also have to produce more drinking water than it does at present, in order to satisfy an estimated increase in annual demand of 6 million cubic metres. Consequently, the task WMN faces is to find a new source of raw water for the production of 15 million cubic metres of drinking water per year. Using surface water is the only alternative.

2 INNOVATIVE CONCEPT

The new project will have to be worked out so that surface water is used in a way that causes the least environmental impact. From the environmental point of view, there are two sides to using surface water for drinking water production. On the one hand, surface water is available in almost unlimited quantities, certainly in the large rivers in the Netherlands, and its extraction has no impact on the groundwater level. On the other hand, the 'most environment-friendly' method is to use groundwater for drinking water production. The excellent basic quality of this raw water can only be equalled using much more intensive treatment processes.

In view of this, an innovative concept was sought in which the advantages of groundwater and surface water could be combined. This was found in a concept that employs a double aquifer passage. The *first* aquifer passage takes place upon extracting the surface water: instead of abstracting the water directly, it is abstracted indirectly, at a short distance from the river. The *second* aquifer passage takes place at the end of the drinking water production process: before the water is distributed, it undergoes deep artificial recharging and is re-extracted with several months detention.

This concept, known as DOPASS (double passage), is shown in figure 1.

The surface water is pumped up in wells in the floodplains at a short distance from the river. This 'bank filtrate' has already undergone one aquifer passage at the time it is pumped up. The bank filtrate originates for about 90% in the surface water. The aquifer passage acts as a filter; natural processes remove various contaminants from the water. After extraction, the bank filtrate undergoes pretreatment, after which it undergoes a second aquifer passage in the range of hills known as the Utrechtse Heuvelrug. The virological and bacteriological contaminants are further broken down during this passage.

Overall there are several advantages to using this concept:

- Reliable drinking water supplies are assured by the system's components for re-extraction and post-treatment. Operational reliability is assured by the components: bank filtrate extraction - transport - pretreatment - transport - well recharge. The operational management of transport and pretreatment takes place throughout the year and is more or less independent of drinking water demand. Therefore, the plant can have a low peak factor (110%).
- The treatment operations are minimized by the aquifer passage before bank filtrate extraction and the aquifer passage following pretreatment, in the stage between well recharge and recovery. Furthermore the technology employed for the pretreatment process can be simple because there is time to make risk assessments in the event of serious river water contamination, in a calamity situation for example. It is at least 6 weeks before the river water is pumped up into the extraction plant.
- Drinking water extraction in the floodplains fits in well with other functions, such as nature restora-tion projects and floodplain widening by means of flow systems for increasing protection against flooding.

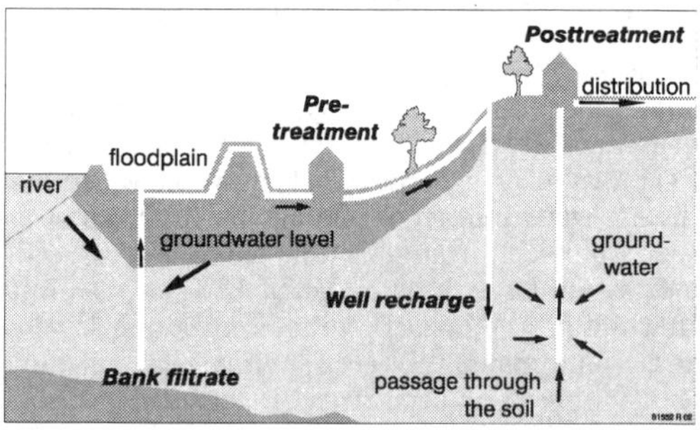

Figure 1. Diagram of the DOPASS concept

The DOPASS concept is an innovative, sustainable concept. The double aquifer passage means that nature's ability to purify the water is used to the full, without any environmental impact in the form of harmful consequences or irreversible changes in the subsoil.

3 THE WHOLE IS MORE VALUABLE THAN THE SUM OF ITS PARTS

This concept is based on previous experiences with a small-scale trial project carried out by WMN in the late nineteen seventies and early nineteen eighties. The trial project and the project being implemented consist of the following components: bank filtrate extraction, influent transport, pretreatment, effluent transport, well recharge and recovery, post-treatment and distribution. The draft hydraulic design of the first phase of the DOPASS project is shown in figure 2.

The components are described below from the point of view of achieving the synergetic advantages the concept has to offer. The various components are in fact inseparably linked, both figuratively as well as literally by the interconnecting pipelines. Each component contributes to the entire system but it is only the combination that makes the whole more valuable than the sum of its parts.

3.1 *Bank filtrate*

Because the water is pumped up nearby the bank of a river, the river bed acts as a natural filter. The natural self-cleansing effect of the river means there is little likelihood of the bed silting up. The water quality changes owing to the aquifer passage between the river and the extraction wells and any fluctuations in the water's composition are levelled off. The levelling off occurs because of dispersal in the residence time as a result of the irregular inflow of water to the extraction wells and because the water mixes with the groundwater in the aquifer. This is shown in figure 3.

Water quality changes during bank filtration owing to:

- biological processes in the soil, such as the oxidation of organic matter and dissolved carbon, deni-

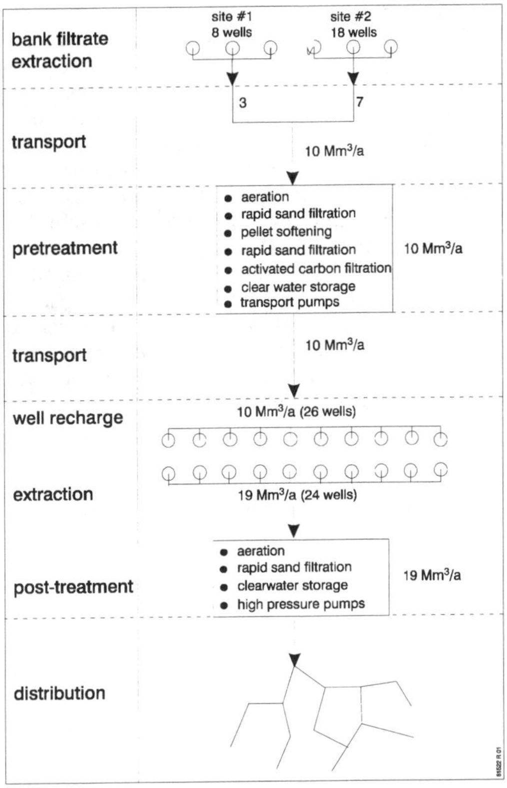

Figure 2. Draft hydraulic design of phase 1 of DOPASS project

trification and a reduction of ferric and manganese compounds;
- sludge formation on the river bed in which the metals and organic microcontaminants adsorbed by suspended matter are left behind.

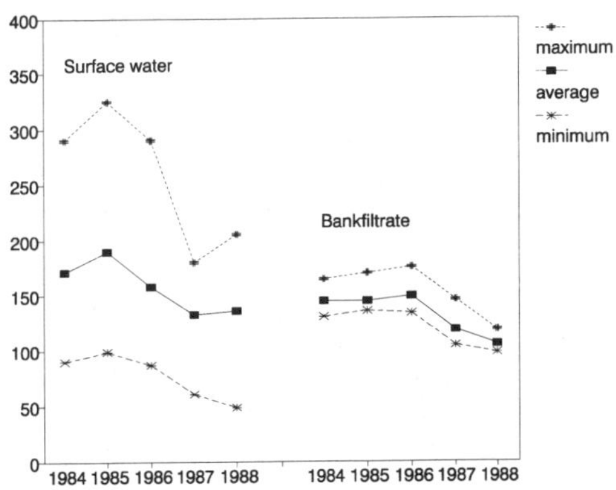

Figure 3. Levelling off chloride concentrations during bank filtration in Remmerden (period 1984-1988) (Vogelaar, 1996)

The actual composition of the extracted bank filtrate is not known because the river water takes on the characteristics of groundwater during the aquifer passage and the quality of the water's composition in the initial period of water extraction is not constant. A forecast of the expected quality has been made on the basis of water quality inspections (see table 1). Pretreatment continues to be necessary before the water can be used for artificial recharge.

Table 1. Quality of composition of surface water and bank filtrate (Laeven & Reijnen, 1997)

Parameter	Unit	Surface water	Bank filtrate
Acidity	-	8.1	7.4 - 7.5
Chloride	mg/l	108	132 - 137
Total hardness	mmol/l	2.15	2.4 - 2.5
Iron	mg/l	1.1	0.3 - 2.6
Manganese	mg/l	0.07	0.7 - 1.5
Methane	mg/l	-	13 - 65
TIC[1]	mmol/l	2.7	3.2 - 3.3
Sulfate	mg/l	54	72 - 75
NPOC[2]	mg C/l	3.7	1.0 - 3.0
Pesticides		-	river water quality

[1] TIC= total inorganic carbon, [2] NPOC= not purgable organic carbon

3.2 *Pretreatment*

The treatment process has to be capable of producing drinking water that meets the present and future requirements of the Dutch Water Supply Decree. The DOPASS concept consists of two treatment stages. The first stage is the central pretreatment of the bank filtrate. No disinfection takes place in this stage as this is carried out naturally after well recharge, during the aquifer passage. The second stage is the post-treatment of the re-extracted water.

The first precondition for the pretreatment is producing water that meets the requirements for well recharge. Organic matter and organic microcontaminants are also removed. The second precondition is that biologically reliable water must be produced, so that no chlorination is needed prior to transport. The third precondition is that the water's hardness is corrected. Water softening is carried out to meet the requirement for the water's copper-dissolving capacity. The extracted bank filtrate first undergoes conventional treatment in the form of intensive aeration and rapid sand filtration. Hardness is reduced by pellet-softening. This is followed by rapid sand filtration and double activated-carbon filtration intended to remove the apolar and slightly polar organic microcontaminants.

Trials are underway to determine the possibilities

for optionally including oxidative techniques using biological activated-carbon filtration in the treatment process when polar microcontaminants are present. The aquifer passage means there is time to carry out a risk assessment in the event of the river becoming seriously contaminated. In that case it would be possible to employ additional oxidative techniques in the treatment process.

Pretreatment is part of the total process of bank filtrate abstraction, pretreatment and artificial recharging. The aforementioned system components will be controlled centrally after pretreatment. The nature of the DOPASS concept means that drinking water supplies from the Soestduinen pumping station will not be jeopardized in the event of pretreatment not taking place. Operational reliability is guaranteed by the modular design of the process components, which enables the separate modules to be taken out of service for maintenance.

3.3 *Transport pipeline*

The starting point of the DOPASS concept is to ensure a continuous supply of water on demand. Transport occurs at an even rate throughout the year, on the basis of daily, round-the-clock bank filtrate extraction. The transport system must be operationally reliable to enable the process to continue with minimum disruption. This is also important for operational management and the pretreatment capacity. If the transport system has to be regularly taken out of service, for maintenance for instance, the number of production hours will be lower. This will necessitate an increase in the capacity of bank filtration, pretreatment and transport, to guarantee the annual production figure of 10 million cubic metres.

The transport system in the floodplains is therefore a multi-pipe system, which enables at least 2/3 of annual capacity to be maintained in the event of maintenance work. The maintenance frequency is high for the transport system from the extraction plant to the pretreatment stage: the anticipated iron sludge formation will make cleaning necessary at least twice a year.

The maintenance frequency is extremely low in the effluent pipeline after the pretreatment section: it is judged that only one pipeline is necessary for this.

3.4 *Soestduinen pumping station: well recharge, recovery, post-treatment and distribution*

The existing pumping station at Soestduinen is at present permitted to abstract 9 million cubic metres of groundwater per year. This water undergoes simple treatment before being distributed. The pumping station is being redesigned for the well recharge, recovery, post-treatment and distribution operations. Wherever possible, the existing infrastructure will be used in the new design.

The first precondition for well recharge and recovery is that it must be possible to completely re-extract all the recharged water in due course. The second precondition is that the recharged water must be hygienically safe after recharging. Matching the design of the well fields used for well recharge to those used for re-extraction, and withdrawing more water than is artificially recharged will ensure that both preconditions are satisfied.

Achieving bacteriological and virological reliability will entail locating the well fields of both systems some distance from each other, so that the recharged water remains in the subsoil for a given period. Table 2 shows the requirements to achieve complete bacteriological and virological reliability using surface water. It also shows the purifying effect achieved during the first and second aquifer passage and during pretreatment using the DOPASS concept. It follows from this that the bacteriological and virological reliability in the DOPASS concept is amply achieved.

Table 2. Effect of the components of the DOPASS concept on achieving bacteriological and virological reliability (in log-units) (Laeven & Medema, 1998).

component	*bacterias*	*Viruses*	*protozoa*[1]
bank filtr. extr. [2]	1	1	1
pre-treatment	1 - 2	0 - 1	1 - 2
well rech. / extr. [3]	10 - 30	10 - 24	10 - 30
total effect	12 - 33	11 - 26	12 - 33
required effect [4]	5	8	6 - 7

1 *Cryptosporidium and Giardia;*
2 worst case (short cut during flooding of banks);
3 minimum residence time: 90 days;
4 based on the maximum concentration in surface water and required concentrations in drinking water.

Using the pumping station's existing infrastructure will enable more water to be abstracted than is artificially recharged. In total, the artificial recharge capacity plus the currently permitted volume may be abstracted. In the new situation there will be a permit for the well recharge and recovery of 10 million cubic metres per year, plus the existing groundwater extraction permit for 9 million cubic metres per year. Calculations show that a volume of at least 125% of the artificial recharge volume will have to be extracted to achieve complete re-extraction of the artificially recharged water. The re-extracted water will be mixed with local, natural groundwater. This will reduce the chloride content and hardness to acceptable levels.

The post-treatment stage has an annual capacity of 19 million cubic metres and consists of a compact aeration and rapid sand filtration stage. This design has been chosen because the abstracted water is expected to exceed the permitted levels for iron, manganese and ammonium. The peak factor for the post-treatment plant is around 155% and is determined on the basis of reliable supplies.

3.5 *Conclusion*

The DOPASS concept is an innovative, sound and sustainable concept in which double aquifer passage enables the maximum benefit to be derived from nature's ability to purify water.

REFERENCES

Jutte, F.A. & H.J. Roelofs 1995. Winning van oeverfiltraat in combinatie met diepinfiltratie. *H2O* 4(28): 115-117 (in Dutch).

Laeven, M.P. & G.K. Reijnen 1997. Bijstelling prognose waterkwaliteit OEDI-project deel 1: oeverfiltraat en voorzuivering. Kiwa, Nieuwegein. Rapport KOA 97.137 (in Dutch).

Ministry of Housing, Planning and Environment & Ministry of Economical Affairs & Ministry of Agriculture 1993. Second National Policy Plan. SDU, Den Haag (in Dutch).

N.V. Waterleidingbedrijf Midden-Nederland 1996. MER Oeverfiltraatwinning en Diepinfiltratie. WMN, Utrecht (in Dutch).

Province Utrecht 1992. Water Management Plan Province Utrecht (in Dutch).

Vogelaar, A.J. 1996. Winning, zuivering en diepinfiltratie van oeverfiltraat. Een overzicht van 10 jaar ervaring in Remmerden en Leersum. Kiwa, Nieuwegein. Rapport KOA 96.055 (in Dutch).

1.11 Storage and recovery efficiency

Artificial Recharge of Groundwater, Peters et al. (eds) © 1998 Balkema, Rotterdam, ISBN 90 5809 017 5

Aquifer storage recovery: Recent developments in the United States

R. David G. Pyne
CH2M Hill, Gainesville, Fla., USA

ABSTRACT: Aquifer storage recovery is a technology for storing large water volumes underground through wells during times when it is available and recovering this water from the same wells when needed to meet peak, emergency, long-term storage or other needs. Twenty-seven ASR systems are operational in twelve states within the United States, with about 50 more in various stages of development. Recovery capacities range from 2 Ml/d to over 400 Ml/d, storing fresh water in fresh, brackish and other non-potable aquifers at depths up to 900 m and at construction costs typically less than half those of other water supply alternatives. Six of the operating sites now have over 10 years of operating experience, providing a basis for evaluation of their long-term performance. With the success of these potable water ASR systems, many new sites are being developed for storage of drinking water and also high quality water from other sources, including surface water, groundwater from different aquifers, and reclaimed wastewater.

1 BACKGROUND

Aquifer storage recovery (ASR) wells have been used for seasonal water storage in the United States since the first system became operational at Wildwood, New Jersey in 1969. All 27 operational ASR systems are storing treated drinking water, mostly in confined artesian aquifers. Many of these aquifers contain native water that is brackish, or contains high nitrate, fluoride, iron, manganese or other constituents at concentrations too high for potable use without treatment. The stored water displaces the native water around each ASR well, forming a bubble of fresh water within the aquifer. Typically the same volume of water stored can be recovered, once a buffer zone has been created around each well. The same well is used for both recharge and recovery, and the pump in the well is also used to periodically backflush the well during recharge to remove accumulated suspended solids, thereby maintaining recharge capacity. The recovered water does not require retreatment prior to distribution, although in the United States it receives additional disinfection. At a few sites, pH adjustment is also required.

The principal factor that is driving the rapid implementation of ASR technology is the relatively low unit cost for meeting peak demands. Typically this is less than half the cost of other conventional alternatives, and in some cases the savings exceeds 90%, particularly where ASR enables reduction in size, or elimination of the need for, expansion of water treatment and/or surface reservoir facilities.

Since a majority of ASR sites use storage zones that contain native water unsuitable for direct potable use, it is important to establish the recovery efficiency of each ASR system, defined as the percentage of water stored that can be recovered while still meeting applicable water quality standards in the distribution system. This is normally established during testing, through several cycles of recharge and recovery designed to address recovery efficiency as well as other performance characteristics of interest at each site. These other characteristics may include plugging, backflushing frequency, disinfection byproduct reduction, hydraulic response, geochemical and water quality changes during storage, and other issues. Typically these criteria are ranked in order of importance, so that the test program is designed to answer the most important questions with the highest level of confidence.

In addition to an improved understanding of recovery efficiency in brackish aquifers, there have been several recent advances in ASR technology relating to corrosion control, cascading control, geochemistry, long-term storage, and other issues. A significant development has been the reduction in disinfection byproducts that has been observed at many sites during ASR storage. Both recovery efficiency and DBP reduction may be particularly pertinent at such time as the level of interest in ASR technology may develop at any new potential site to the point where conceptual development and licensing of facilities is ready to proceed.

2 RECOVERY EFFICIENCY

Where recovery efficiency is the most important characteristic to be addressed, it is usual for the cycle test program to incorporate an initial small volume cycle to demonstrate hydraulic and water quality response. This might include less than one week of recharge, no storage period, and recovery almost to background concentration in the aquifer, as measured by a suitable conservative constituent such as chloride or total dissolved solids (TDS) concentration. Then three or four cycles would be conducted, each identical in design, recharging a fixed, larger volume of water and then recovering it immediately at the end of recharge until a predetermined concentration is reached for the conservative constituent of greatest interest. Water that is left underground during each cycle then forms part of the buffer zone around the well, continuing to build with successive cycles until the ultimate recovery efficiency is achieved. In most cases, it is possible to achieve 100% recovery efficiency after a few cycles at the same volume, once the buffer zone is formed.

Volumes stored in each test cycle should be sufficient to theoretically displace a cylinder of water around the well that has a radius approximating at least one thickness of the storage zone. For example, a confined sand aquifer with a thickness of 40m and an estimated effective porosity of 25% would require test cycles of at least 50 Ml volume. This is an approximate guideline. Where this is not attainable due to large aquifer thickness or limited available water for recharge, results may need to be interpreted with the understanding that local mixing in the vicinity of the borehole or screen may indicate lower recovery efficiency than could be achieved in long-term operations at higher volumes.

Density stratification does not appear to be a significant mechanism for constraining recovery efficiency in storage zones with TDS concentrations below about 5000 mg/l. At much higher TDS concentrations, the stored water will tend to float on top of the native groundwater, particularly during extended storage periods. At one ASR test site in Marathon, Florida, the storage zone was a tightly confined sand aquifer at a depth of about 130 m. Hydraulic characteristics of this zone include transmissivity - 212 m^2/d; storativity - 3.2 x 10^{-4}, and thickness 13m. The zone contains seawater with a TDS concentration of 37,000 mg/l TDS. Density stratification was the primary mechanism for limiting recovery efficiency to a maximum of about 70%. As the storage period increased, recovery efficiency dropped linearly. At 120 days storage, recovery efficiency was about 45%. Of particular interest at this site, however, was the demonstration that trickle recharge at a rate of about 2.5 lps during the storage period was sufficient to maintain the recoverable storage volume of about 60 Ml. Since the purpose of this ASR site was to provide emergency storage during the hurricane season, this was a low cost, viable solution to their emergency water storage needs. Testing results showed that 60 Ml of drinking water could be stored prior to the hurricane season (June to November), and maintained during the hurricane season with a trickle flow of 2.5 lps. Assuming no emergency, then this water would be recovered at a rate of about 3 Ml/d per well to help meet peak system demands during the following winter season in December and January. Figure 1 shows recovery efficiency versus storage time for the Marathon ASR site in the Florida Keys. Storage time is measured between the mid-points of recharge and recovery. Cycles 1 through 10 were conducted with no trickle recharge flow during the storage period whereas Cycle 11 incorporated the trickle flow for about 80 days.

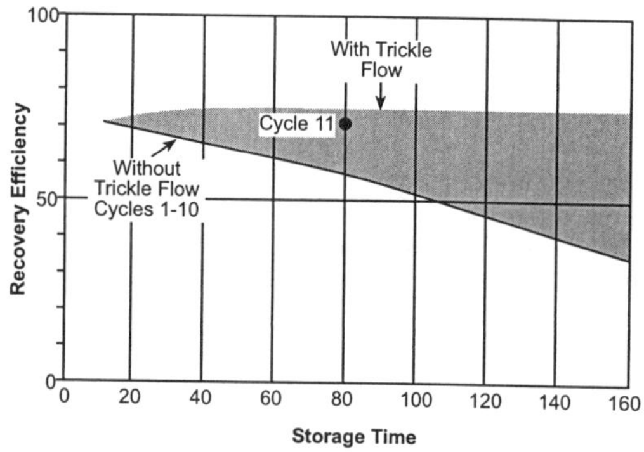

Figure 1. Marathon Aquifer Storage Recovery Cycle Testing Results, 1990 to 1995

The Boynton Beach, Florida, ASR site has been in operation since 1993, storing treated drinking water in a sandy, limestone, confined aquifer at a depth of 240 to 270 m. Transmissivity of this aquifer is 870 m^2/d; The storage zone contains brackish water with a TDS concentration of 3900 mg/l, making it the most brackish ASR site to date with some long term operating history. Typical storage volumes for individual cycles are about 150 Ml while recharge and recovery rates are up to about 5.4 Ml/d in the single ASR well. Much larger volumes could be stored in this aquifer, however the City has determined that this volume is adequate to meet their peaking needs. As shown in Figure 2, initial recovery efficiency was under 30%, however efficiency climbed to about

90% after 11 cycles during a period of four years. Two subsequent cycles have not demonstrated any further increase in recovery efficiency, for reasons that are still being evaluated. It is expected that recovery efficiency will achieve 100% at this site in the near future. The total volume of water in the buffer zone is about 750 Ml. Stored during times when system water demand was below peak levels, the value of this water is best represented by its marginal cost, including only power and chemicals. This value is about US$45,000, or about 7% of the total capital investment in the engineering, construction and testing of the ASR system. This is a relatively small investment in a system capable of storing large volumes of water and meeting peak demands with up to about 8 Ml/d from their ASR well.

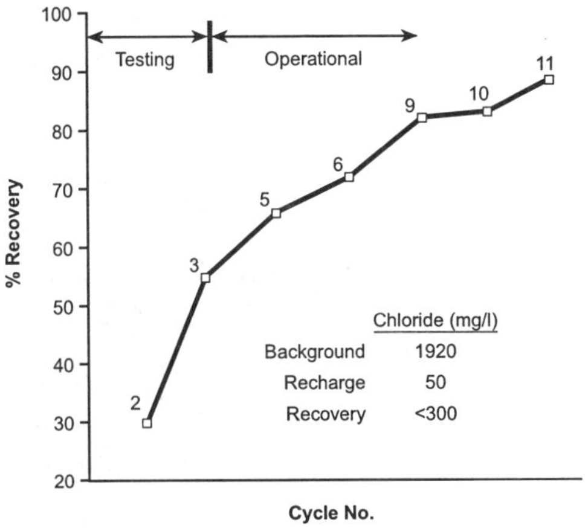

Figure 2. Boynton Beach Aquifer Storage Recovery Cycle Testing Results, 1993-1997

Based upon the operational experience from Boynton Beach, Marathon and several other ASR sites, the concept of a Target Storage Volume (TSV) has evolved. This is the volume required in aquifer storage so that projected water demands during recovery can be met while meeting flow, volume and water quality goals with an acceptable level of reliability. In a fresh water storage zone, the TSV may be the same as the projected water demand during recovery. In a storage zone containing poor quality water, the TSV will be larger than the projected water demand, providing a buffer zone around the well. Once the TSV has been achieved, all subsequent water stored can be recovered, thereby achieving 100% recovery efficiency. Where potential losses exist, such as lateral flow of the storage bubble during long-term storage, or density stratification in saline aquifers, less than 100% recovery efficiency may be achievable.

Both the Boynton Beach and Marathon ASR sites obtain their drinking water for recharge from water distribution systems supplied from wellfields. Although these wellfields experience some small seasonal variability in yield, the principal variability is in system demand for drinking water, whether due to seasonal or emergency needs. Consequently ASR storage volumes are determined based upon meeting variations in demand. At the majority of ASR sites, however, the ASR storage volume is also based upon seasonal and long-term variability in supply and water quality. Typical of these sites is the Peace River ASR system in southwest Florida.

The Peace River is a highly variable source, both in terms of flow and quality. Due to regulatory and environmental constraints, periods of 2 to 3 months each year with no diversions from the river are fairly common, while a duration of 6 months is the longest projected period of no diversion based upon over 60 years of streamflow record. Furthermore, river water quality is difficult to treat about 14% of the time due to high algae concentrations. In addition to this seasonal variation in supply, seasonal variation in demand creates a need for supplemental storage. The ASR system at Peace River has been developed since 1985 and now includes 9 wells with a total recovery yield of about 28 Ml/d. The storage zone for 8 of these wells is a confined, limestone artesian aquifer at a depth of 180 to 280 m containing water with a total dissolved solids concentration of about 800 mg/l. Typical well yields are about 4 Ml/d. Transmissivity of this aquifer averages about 560 m^2/d, while storativity is 1×10^{-4} and leakance is 8.5×10^{-3}/day. The 9th well is in an overlying similar aquifer at a depth of 125 to 155 m. A 10th ASR well has been constructed in a highly productive but more saline aquifer (TDS = 3700 mg/l) at a depth of 400 to 470m, however testing has not yet been completed on this well. Other facilities at the site include a 45 Ml/d water treatment plant and a 2.3 mcm raw water surface reservoir. The treatment plant is being expanded to a capacity of 68 Ml/d and 14 additional ASR wells will be constructed.

Recently an analysis was completed for the first 10 years of operation at the Peace River ASR site. Based upon an ASR simulation model developed in 1985 and updated in 1988, expected water quality response during recovery, in terms of TDS, was compared to actual water quality during recovery. The period of record ended with two consecutive years in which recovery volumes matched or approximated those expected during the design ("once in 50 year") drought. Each year had 5 to 6 months' duration of no withdrawal from the river. At the beginning of the first design withdrawal event, which was caused not by low flows but by poor raw water quality, the volume of water in storage was 6.4 mcm compared to a target storage volume of 10.2 mcm. At the end of this period, recharge resumed for a few months but was only able to increase total storage to 5 mcm. At the end of the second drought, which was an extreme low river flow event, the volume of water in storage had declined to 3.2 mcm. Previous modeling, based upon early testing and operating performance, had indicated that the buffer zone required at this site is about 5 mcm. Withdrawals that reduce the volume in storage below 5 mcm would be expected to cause TDS levels exceeding drinking water standards of 500 mg/l in the water going to the distribution system. Actual system performance during this extreme sequence of events was better than expected. Individual wells demonstrated a wide range of performance characteristics, ranging from minimal to substantial mixing between the stored water and the surrounding brackish water. However average TDS for all wells during recovery exceeded the drinking standards only two times, with a maximum concentration of 545 mg/l. All other daily samples were below 500 mg/l. The ASR system performed better than expected considering the volume of water in storage, which was less than the target storage volume, and the quality of the recharge water, which varied between 160 and 480 mg/l TDS.

Figures 3 and 4 illustrate the typical mixing relationships for drinking water stored in aquifers containing brackish water or other conservative constituents unsuitable for potable use without treatment. As shown in Figure 3, successive cycles at the same volume tend to flush out the storage zone around the ASR well, steadily forming a buffer zone that serves essentially as the "walls of a tank." The shape of the water quality response curve for each cycle shows increased mixing as recovery approaches 100% of the stored water volume. Figure 4 shows the initial ASR cycle water quality response curve for several different sites. The variety of curve shapes reflects different aquifer types, thicknesses, well construction, and cycle testing volumes. For each of the sites shown in Figure 4, successive cycles demonstrated improvement in recovered water quality as a result of the formation of a buffer zone around the well. As confirmed by the rapidly increasing number of operational ASR sites in brackish aquifers, Figure 4 suggests that careful engineering design and operation can usually achieve desired ASR goals in terms of recovered water quality, even in aquifers containing poor water quality.

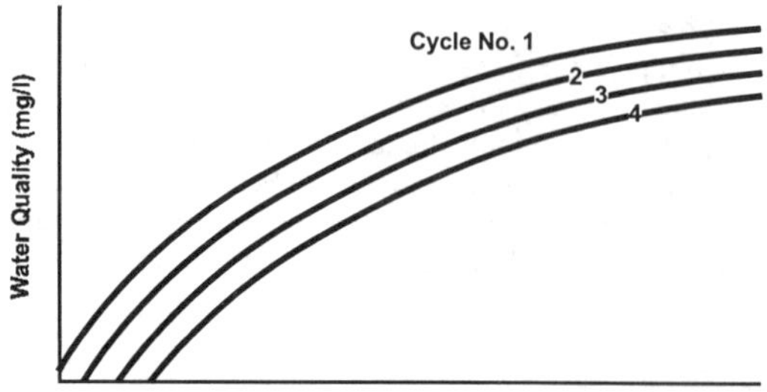

Figure 3. Water Quality Improvement in Successive ASR Cycles of Equal Volume

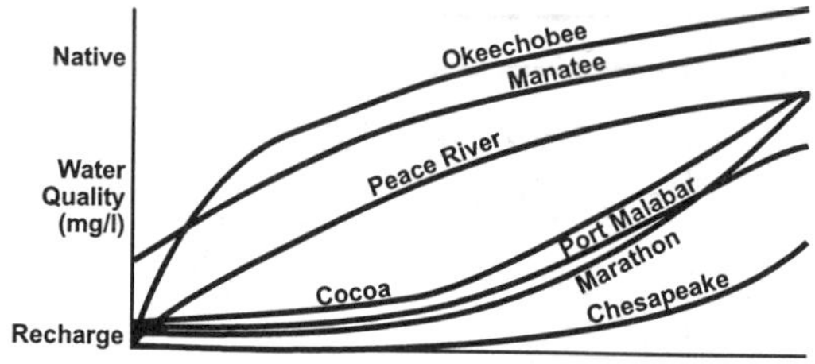

Figure 4. Water Quality During Initial ASR Cycle Recovery

3 DISINFECTION BYPRODUCT REDUCTION

During recent years, increasing attention has focused on the fate of disinfection byproducts (DBPs) during storage in ASR wells. The American Water Works Association Research Foundation has published research on this subject (1996), based upon detailed field investigations at four ASR sites in the United States and one ASR site in the United Kingdom. Among the conclusions from that investigation were the following:

1. THMs (trihalomethanes) and HAAs (haloacetic acids) are removed from chlorinated drinking water during aquifer storage over a period of several weeks.
2. HAA removal precedes THM removal
3. The more highly brominated species tend to be eliminated earliest.
4. In most cases, THM removal does not appear to occur until anoxic conditions develop, and it frequently follows the onset of denitrification. HAA removal occurs under aerobic conditions. A biological mechanism is suggested, including DBP removal under both anoxic and aerobic conditions.
5. THM and HAA precursor concentrations (formation potentials) decreased at most of the sites investigated; THM precursor concentrations exhibited no clear pattern.

Results from these five sites, and others since then, have shown that HAA concentrations are removed in days while THM concentrations are removed in weeks of storage.

Of some significance is 1997 data from testing at the Rome Avenue Park ASR site for the City of Tampa, Florida. This site is located in a residential area in the distribution system. The storage zone is a semi-confined limestone artesian aquifer containing fresh water (TDS = 234 mg/l). The storage interval is between 95 and 125 m below ground level. Recovery efficiency is not a significant factor at this site since the storage zone is fresh and fluoride added to the drinking water at the water treatment plant to meet drinking standards prior to recharge remains in the water recovered from the ASR well and going to the distribution system. However the drinking water is disinfected and contains 55 ug/l THM and 50 ug/l HAA. Testing showed that after ASR storage for 90 days, and redisinfection of the recovered water prior to redistribution, THM concentrations were only 35 ug/l and HAA concentrations were only 3 ug/l. Fluoride measurements showed that mixing and dilution did not account for the observed reduction in DBP concentrations.

During ASR storage at the Tampa site, reduction in DBP formation potential clearly occurred. This is significant since many water utility systems are now incorporating ASR into their plans for meeting increasing peak and emergency demands, and are also moving forward with parallel plans to add ozonation and/or membrane treatment to meet pending reductions in regulatory standards for disinfection byproducts. For many of these utilities, meeting the HAA standard is more critical, and this may only be needed for a few days or weeks per year when water quality from the surface water source is high in DBP formation potential. The opportunity exists for many of these water utilities to help meet their peak demands with ASR wells, and also to provide sufficient ASR capacity (storage volume, recovery rate, number and location of wells) so that during times when DBPs may exceed the new standards, ASR recovery can be implemented, blending low DBP recovered water with high DBP water from the water treatment plant and thereby meeting the standards.

In Europe, there exists greater sensitivity to disinfection practices and DBP concentrations than currently in the United States. This is an important issue for ASR wells since experience suggests that maintaining sterile conditions in the well, well screen and gravel pack during ASR operations helps to control well plugging due to bacterial activity. As discussed above, bacteria are naturally present and active in the subsurface. Disinfection is believed to control bacterial activity so that it only occurs away from the well where any resulting reductions in hydraulic conductivity are less likely to cause a reduction in well yield or recharge rate. This is more likely to be important in sand and clayey sand aquifers than in more coarse-grained or consolidated aquifers. While the absence of a disinfectant residual in an ASR well does not necessarily lead to bacterial plugging, the presence of a residual in the well will certainly help to ensure that significant plugging does not occur due to bacterial causes.

At such time as the level of interest in ASR technology may develop at any potential site to the point where conceptual design of an initial ASR system is required, this DBP issue will undoubtedly arise. One approach is to treat the water with membrane treatment or other processes prior to recharge through wells, without disinfection, to minimize bacterial activity in the ASR well. An alternate approach is to add disinfectant to the water going to the ASR well to control downhole bacterial activity. The research and experience described above may be helpful in demonstrating the fate of the resultant DBPs during ASR storage, providing an indication that such DBPs will probably be reduced significantly in the recovered water. Site specific testing will still be required to achieve the level of confidence needed on this issue.

4 CONCLUSIONS

Aquifer storage recovery is proving to be a viable, cost-effective technology for storing large volumes of water to meet peak, long-term and emergency water needs and to achieve many other objectives. ASR is being applied throughout the United States, and also in Canada, England, Australia, Israel and other countries.

Most ASR sites are storing drinking water in confined aquifers containing water quality that is brackish or contains constituents such as nitrates, fluorides, iron, manganese, hydrogen sulfide, all unsuitable for drinking purposes except following treatment. Mixing between the drinking water and the native water in the aquifer can be controlled in most situations by the proper design and operation of ASR wells, so that recovered water quality is acceptable. Operation includes development of a Buffer Zone surrounding the ASR well to contain the stored water, and development of a Target Storage Volume for each well so that recovered water will meet flow, volume and water quality criteria with acceptable reliability.

Wellhead disinfection is frequently practiced during recharge of ASR wells, to control downhole plugging from bacterial activity in the well screen and gravel pack. It is reasonable to anticipate that the resulting disinfection byproducts and their formation potential will be significantly reduced in the recovered water. Testing will be needed to confirm this at any specific ASR site.

REFERENCES

Pyne, R., G. David, Singer, C. Philip & Miller, T. Cass 1996. *Aquifer Storage Recovery of Treated Drinking Water,* American Water Works Association Research Foundation, Denver, Colorado, 176p.

Pyne, R. & G. David 1995. *Groundwater Recharge and Wells: A Guide to Aquifer Storage Recovery.* Lewis Publishers, CRC Press, 376 p.

Artificial Recharge of Groundwater, Peters et al. (eds) © 1998 Balkema, Rotterdam, ISBN 90 5809 017 5

The use of modelling to predict the behaviour of ASR systems

M.J. Streetly
Entec UK Limited, Shrewsbury, UK

ABSTRACT: Development of an ASR scheme involves injecting and recovering water over a number of cycles. A finite difference solute transport model was used to simulate the operation of an ASR scheme in a borehole in a radially symmetrical, single porosity aquifer with no prevailing hydraulic gradient or density differences between injected and background water. Under these conditions, the dispersivity of the aquifer had the greatest effect on recovery efficiency whilst diffusion and permeability had no effect. The effect of other parameters could be summarised by their effect on the radius on the bubble of injected water in the ground (r_b); parameter combinations that increased r_b led to higher recovery efficiencies. The practice of increasing the volume injected on successive cycles will therefore lead to an over-estimate of the improvements in recovery efficiency over subsequent cycles.

1 THE ROLE OF CYCLE TESTING IN THE DEVELOPMENT OF ASR SCHEMES

The use of boreholes to artificially recharge, store and recover large volumes of potable water (Aquifer Storage and Recovery - ASR) has become increasingly popular over the last decade. ASR schemes are usually developed in a series of phases (Pyne 1994).

Investigation Phase (drilling, pumping tests and initial injection/abstraction cycle tests): The aim of this phase is to identify and quantify where possible the issues that are likely to affect the cost/feasibility of a full scale ASR scheme. The issues identified will affect the design of the initial cycle testing. For instance, where mixing with brackish native water is an issue it is usual to carry out a number of cycles to measure improvement in the quality of recovered water over successive cycles. Where geochemical reactions are an issue it is usual to start with relatively small cycles with long rest periods in order to give reactions time to proceed.

Aquifer Conditioning/Buffer Zone Development Phase: The investigation phase should provide the information needed to optimise the design of the aquifer conditioning phase in which a buffer zone of mixed water is developed around the injected water to protect it from the native water. In highly dispersive aquifers with poor background quality a significant number of cycles may be required during this phase. This may add to the costs, and more importantly the time, needed to develop a fully operational ASR scheme. Generally there is a net injection of water during this phase, the added water being considered as a permanent investment in the buffer zone.

Operational Phase: The aim of the aquifer conditioning phase is usually that, by the end, cycles can be performed in which the same volume of usable water is abstracted as was injected. (Schemes can also be designed as net injection or net abstraction schemes if storage is not the only operational benefit required).

Whilst there are many published cases of ASR schemes in which this format has been followed, there is relatively little information on the criteria used to design the cycle testing programme. The aim of this paper is to present the results of a modelling exercise which illustrates some of the factors that control the recovery efficiency during cycle testing. This provides some insights into the factors that need to be considered in designing a cycle testing programme.

2 METHODOLOGY

The SWIFT III code (Reeves 1986) was used to simulate the effects of injection/abstraction cycles on a single borehole in a homogenous, single porosity, confined aquifer. SWIFT III is a finite difference code capable of simulating not only the

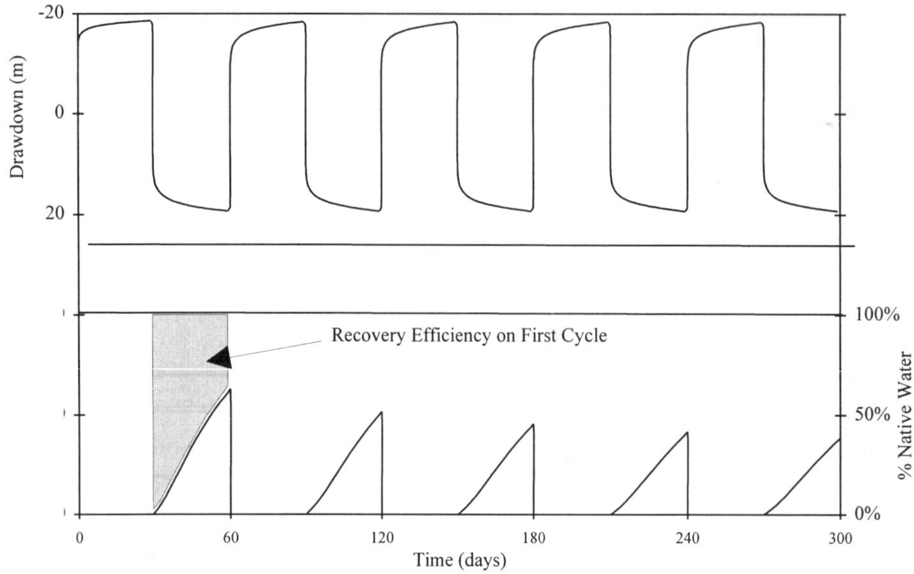

Figure 1. Standard Sequence of Cycles Modelled

hydraulic aspects of the testing but also solute transport and fluid density in both single and dual porosity aquifers. For the present study the model was configured in a radial format with 200 cells spaced at logarithmic intervals from the well (0.15 m) to the outer boundary (20 000 m). No density differences between injected and reabstracted water were applied.

A standard set of aquifer parameters was selected (See Table 1) together with a standard test format of 5 cycles of 30 days injection followed by 30 days abstraction (See Figure 1). Sensitivity runs were then developed to assess the effect of different aquifer parameters and test configurations on recovery efficiency. In each series of runs one parameter was varied by a factor of 0.25, 0.5, 2.0, 4.0.

Table 1: Parameters used in Model Runs

Parameter	Run	Value Min.	Std.	Max.	Unit
Transmissivity	a	50	200	800	m^2/d
Aquifer Thickness	b	12.5	50	200	m
Porosity	c	0.013	0.05	0.2	-
Dispersivity	d	2.5	10	40	m
Diffusion Coefficient	e	0.86	3.5	14	10^{-5} m^2/d
Pumping rate	f	500	2000	8000	m^3/d
Cycle length	g	8	30	120	d
Rest period	h	-	0	30	d

3 RESULTS

The results of these sensitivity runs are given in Table 2 for the recovery efficiency on the first and fifth cycles.

The "recovery efficiency" of each cycle was defined as the volume of injected water recovered. Note that this is different from the standard definition of recovery efficiency as the volume of usable water recovered. The latter is a useful criterion for operational schemes whereas the former provides more information on the actual mixing processes occurring in the ground.

As would be expected, dispersivity has a very significant effect on recovery efficiency (Fig 2 and 3).

Permeability and diffusion have no effect on recovery efficiency. The latter result demonstrates that, in a single porosity aquifer, diffusion is completely masked by the effects of mechanical dispersion. Diffusion is however very significant in dual porosity aquifers.

It should be remembered that, in reality these parameters do not vary entirely independently; high dispersivity may often be associated with high transmissivity for instance.

Apart from dispersivity, the parameters that affected recovery efficiency are all related to the size of the bubble of water injected (pumping rate, aquifer thickness, length of cycle and porosity); a larger radius of bubble (r_b) is associated with higher recovery efficiencies. Where;

$$r_b^2 = \frac{\text{(Volume injected)}}{(\pi \text{ x aquifer thickness x porosity})}$$

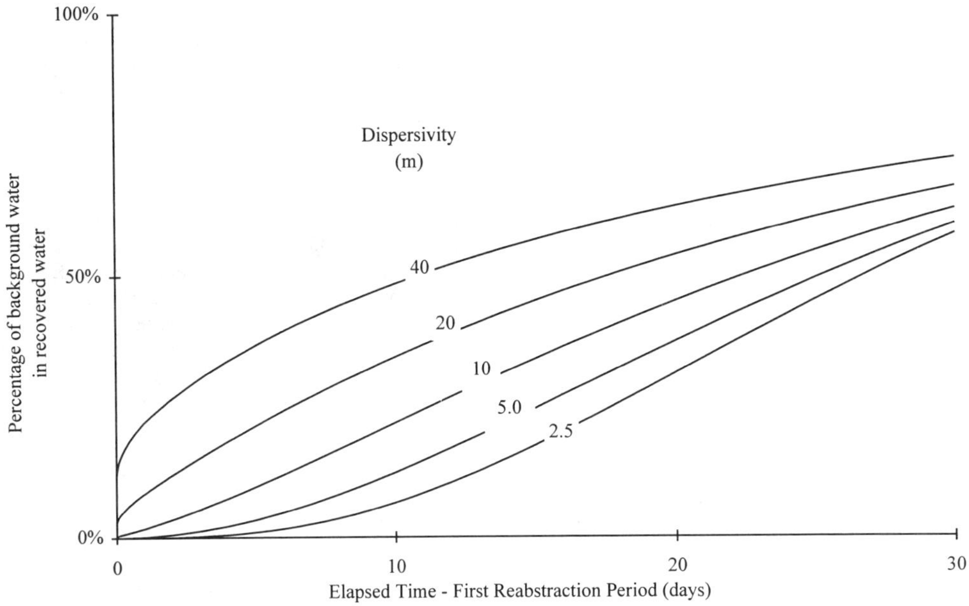

Figure 2. Effect of Dispersivity on Water Quality on First Reabstraction Cycle

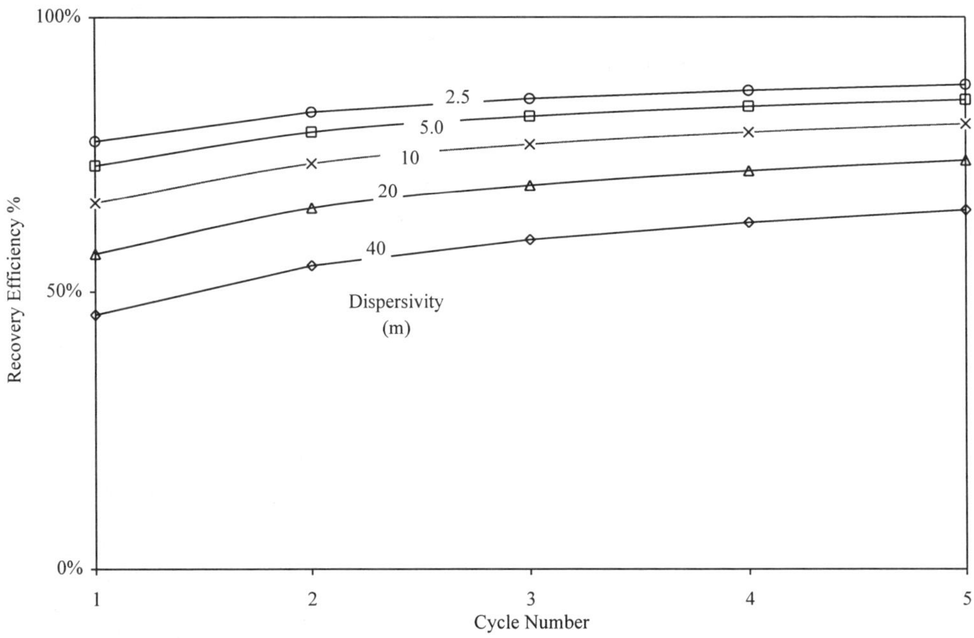

Figure 3. Effect of Dispersivity on Recovery Efficiency

Table 2: Recovery Efficiency on First and Fifth Cycles

Run	Cycle	Recovery Efficiency when Parameter is Multiplied by:				
		0.25	0.5	1.0	2.0	4.0
a	1	66%	66%	66%	66%	66%
a	5	80%	80%	80%	80%	80%
b	1	73%	70%	66%	62%	57%
b	5	85%	83%	80%	77%	74%
c	1	73%	70%	66%	62%	57%
c	5	85%	83%	80%	77%	74%
d	1	77%	73%	66%	57%	46%
d	5	88%	85%	80%	74%	65%
e	1	66%	66%	66%	66%	66%
e	5	80%	80%	80%	80%	80%
f	1	57%	62%	66%	70%	73%
f	5	74%	77%	80%	83%	85%
g	1	54%	61%	66%	71%	74%
g	5	72%	77%	80%	83%	86%
h	1	66%	66%	66%	66%	66%
h	5	80%	80%	80%	80%	80%

Plotting recovery efficiency from the runs in which these parameters were varied against r_b shows a good correlation (Fig 4), although the effect of cycle length is marginally greater than the effect of rate, porosity and aquifer thickness which all give identical results.

An important implication of these results is that, as the volume injected clearly affects the recovery efficiency, injecting a larger volume will increase the recovery efficiency. The common practise of increasing the duration of cycles during the investigative phase may thus lead to over-estimation of improvements in recovery efficiency on subsequent cycles unless the volume factor is corrected for.

This conclusion is supported by the analytical solutions developed by Gelhar and Collins. (Gelhar & Collins 1971).

These model runs were all for an aquifer with no prevailing hydraulic gradient. It is important to bear in mind that, if there is a gradient, the aquifer parameters may take on a new or different significance. Thus, whilst when there is no gradient low porosities will be associated with higher recovery efficiencies, they will lead to larger lateral displacement when hydraulic gradients are present. Likewise, permeability has no effect on recovery efficiency when there is no gradient but is highly significant when there is a hydraulic gradient across the site.

The length of rest period between injection and recovery also had no effect on recovery efficiency. This emphasises the fact that in a single porosity aquifer any reductions in recovery efficiency associated with length of storage will be due to displacement by the hydraulic gradient, density effects or geochemical reactions not dispersion or diffusion. The implication is that rest periods during testing should only be used to address the former aspects and that monitoring should be designed accordingly.

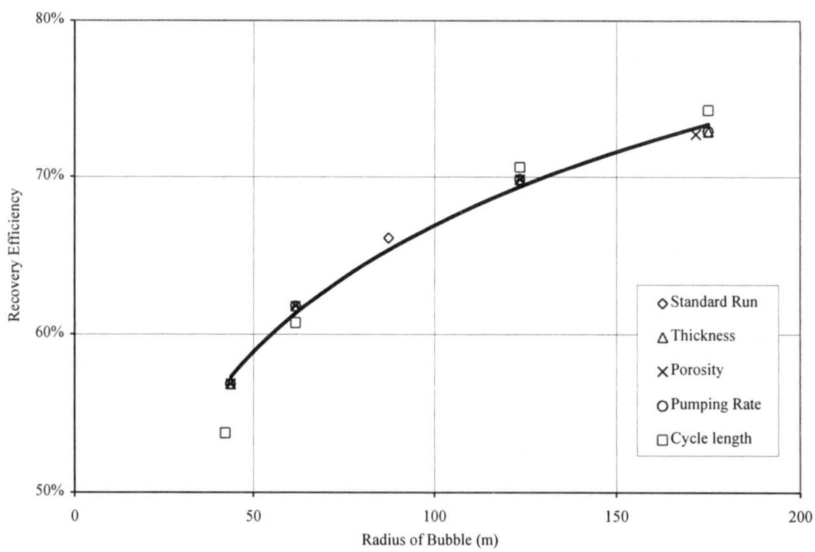

Figure 4. Effect of 'Bubble 'radius on Recovery Efficiciency on First Cycle

4 CONCLUSIONS

In geochemically stable, single porosity aquifers with no prevailing hydraulic gradient, dispersivity is the most important parameter controlling recovery efficiency.

The effect of most other parameters can be summarised by consideration of their effect on the radius of the bubble of injected water (r_b). Parameter sets that lead to a large r_b (large injection rates, long cycles, low porosity and thin aquifers) will have higher recovery efficiencies than those that lead to smaller values of r_b. Diffusion and permeability have little effect on recovery efficiency in these conditions.

However, where there is a hydraulic gradient across the site, some parameters have different effects; high permeability and low porosity lead to increased lateral displacement).

The duration of the rest period between injection and recovery has no effect when there is no prevailing hydraulic gradient. Thus rest periods should only be included in a test programme when the effect of hydraulic gradient, density differences or geochemical reactions are considered to be significant, or in dual porosity aquifers in which diffusion will be an important process.

As a larger volume injected itself leads to higher recovery efficiency, cycles during the investigation phase should involve injecting the same volume wherever possible. The common practice of injecting larger volumes on successive cycles will lead to over-estimation of the improvements in recovery efficiency to be gained on successive cycles.

5 REFERENCES

Gelhar, L.W. & M.A.Collins 1971. General Analysis of Longitudinal Dispersion in Nonuniform Flow. *Wat. Res. Res*. Vol. 7 No. 6: 1511-1521

Reeves, M., D.S. Ward, N.D. Johns & R.M.Cranwell 1986. *Theory and implementation of SWIFT II The Sandia Waste-Isolation Flow and Transport Model for Fractured Media, Release 4.84*, NUREG/CR-3328, SAND83-1159, Sandia National Laboratory, Alberquerque, New Mexico.

Pyne, R. D. G. 1994. *Groundwater recharge and wells: a guide to aquifer storage and recovery.* CRC Press, Florida

Artificial Recharge of Groundwater, Peters et al. (eds) © *1998 Balkema, Rotterdam, ISBN 90 5809 017 5*

ASR, hydraulic and salinity response in unconfined/confined aquifers

N.Z.Gerges, S.R.Howles & K.J.Dennis
Primary Industries and Resources, Adelaide, S.A., Australia

ABSTRACT: The hydraulic and salinity response to aquifer storage and recovery of a highly saline, unconfined, permeable limestone aquifer, and a deep confined low permeability limestone aquifer are discussed. A town water supply is being developed in a highly saline unconfined highly transmissive limestone aquifer, in a location where toxic algae make river water non potable during summer. The injection of 890 ML over five cycles developed a 'sacrificial lens' with a radius beyond 145 m. Trial recovery through the 1997/98 summer indicated the scheme could deliver 20 ML of UV sterilised, filtered potable water. Clogging has not been shown to be a problem to date. Water will need to be injected to sustain the 'sacrificial lens' during the year, and prior to summer to ensure sufficient storage. An irrigation scheme has been developed in a saline deep confined low transmissivity limestone aquifer. A volume of 75 ML of storm water was pressure injected at a rate of 8 L/s with an injection head of -70 m. During repeated cycles of injection, followed by backflushing, the specific capacity increased indicating dissolution close to the well. Recovery indicated 40% with storm water salinity, 90% less than 1,000 mg/L, 100% less than 1,300 mg/L. These examples in carbonate aquifers indicate the difference in injecting and recovering water from a confined low transmissivity aquifer where the injected 'bubble' remains localised with little decay, and from an unconfined high transmissivity aquifer where the injected lens decays rapidly.

1 INTRODUCTION

South Australia currently leads the nation in experimenting and implementing a policy of reuse of storm water/treated effluent, particularly for irrigation. This involves aquifer storage and recovery (ASR), a management tool which enables conjunctive surface water and groundwater resource management as an alternative to traditional allocation systems. The objectives of ASR in South Australia are:

1. The development of a key water resource management tool which will assist in maintaining groundwater systems for current and future development.
2. Reducing the reliance of urban users on imported water with high infrastructure costs related to storage and distribution.
3. Reducing the use of mains water treated to drinking water standards for irrigation.
4. Developing economically viable alternatives where traditional water supply is impractical.

Research and development in South Australia has been unique as it has focussed on the injection, storage and recovery of river and storm water into shallow unconfined and deep confined aquifers. The hydraulic and salinity response to ASR of a highly saline unconfined high transmissivity limestone aquifer, and a saline deep confined low transmissivity limestone aquifer are discussed. These examples are the most interesting of the sites under investigation close to Adelaide, capital of South Australia.

Terminology adopted in this paper is that 'depth to water' is signed positive, and 'injection head' above initial depth to water is signed negative, as is 'head above ground'. Both discharge and injection rates are signed positive. A well should behave in the same manner under both discharge and injection with a compatible water.

2 ASR IN A HIGHLY SALINE, UNCONFINED, HIGH TRANSMISSIVITY AQUIFER

2.1 *Initial investigation*

Clayton, a small town at the lower end of the River Murray system in South Australia, draws its water supply from the river which in summer months can suffer pollution from toxic algal blooms. An ASR scheme was envisaged in which river water could be injected into an underlying aquifer during the winter months, when toxic algal counts are low, and this water pumped during the summer during an algal outbreak. Clayton requires 20 ML during the summer months, and has an annual demand of between 40 and 70 ML.

A 115 m deep injection/exploration well was completed (15/3/95) with 92 m of open hole within an unconfined fractured and karstic Tertiary limestone. A final depth to water 14.5 m, a salinity of 37,445 mg/L and an airlift yield of 25 L/s were recorded. An observation well was drilled at a distance of 145 m.

Step drawdown test (1/5/95 steps: 4.5, 10.1, 15.2, 20.2 L/s) resulted in each of the steps attaining steady state indicating the well was not sufficiently stressed to test it properly. The specific capacity for step-1 was 32 L/s/m of drawdown. The well was plugged back (28/7/95) to 50 m to provide more control over injection. Further discharge testing was not undertaken due to budget constraints.

The initial trial injection (31/7/95 to 9/8/95, 15 L/s) resulted in the injection of 10 ML, and the development of -2.5 m of injection head. Injection testing (22/8/95 to 28/9/95, 15 L/s) resulted in the injection of 49 ML, and the development of -2.0 m of injection head. The trial injection indicated the well was readily able to accept the injection of lake water, but was less efficient than when the well was discharging. This was believed to be due to density differences and dissolved gasses. Suspended solids were not considered to be a problem as the clogging effect was noticed immediately.

The injected water was left in residence for 287 days (28/9/95 to 11/7/96) during which conductivity profiles indicated the slow decay of the injected lens. It was believed that if a sufficiently large volume of water was injected it would contain potable water at its core. As a result, four observation wells were drilled (30/1/96 radii: 5 m, 10 m, 25 m) to determine the extent and vertical profile of the lens. Drilling revealed the existence of a clay layer at 29 m. It was unclear whether this was a regional feature indicating the possibility of a (?)confined aquifer underlying the clay. Due to budget constraints injection into the well continued with no further modification.

2.2 *Stage 2 investigation*

A decision was made to inject up to 500 ML of lake water at a rate of 40 L/s to form what may be termed a 'sacrificial lens' to act as a buffer and transition zone for the decay process to the native groundwater. Within this lens the potable water supply lens would be injected. A telemetry system was installed to monitor the injection rate and head distribution. Injection (11/7/96 to 31/10/97, 40 L/s) resulted in the injection of 300 ML and the development of -4 m of injection head. No boundaries were intersected during the test. On 31/10/96 conductivity profiles indicated that the 25 m observation well contained 500 mg/L water and injection was stopped. The salinity of the water at the 145 m well was 8,000 mg/L.

The injected water was left in residence for 90 days (31/10/96 to 29/1/97). During this time the well was plugged back (19/11/96) to 29.9 m to restrict the operation to the unconfined aquifer. Following this further injection occurred (29/1/97 to 14/5/97, 40 L/s) which resulted in the injection of 250 ML. The well was further plugged back (28/4/97) to 29.8 m. The injected water was left in residence for 70 days (14/5/97 to 23/7/97). A 22 m deep production well was drilled (3/7/97) at a distance of 10 m from the injection well.

Following the end of the 4th injection cycle (total injection of 10 + 40 + 300 + 250 ML), a long term recovery efficiency test was conducted for 55 days (23/7/97 to 16/9/97) at 4 L/s resulting in the recovery of 16 ML. Salinity data is given in Figure 1 (following 8,760 mins) and indicates that the salinity rose in a linear manner (from the initial 560 mg/L) until a time of 51,000 minutes, when 9.5 ML had been pumped and the salinity was 1,000 mg/L.

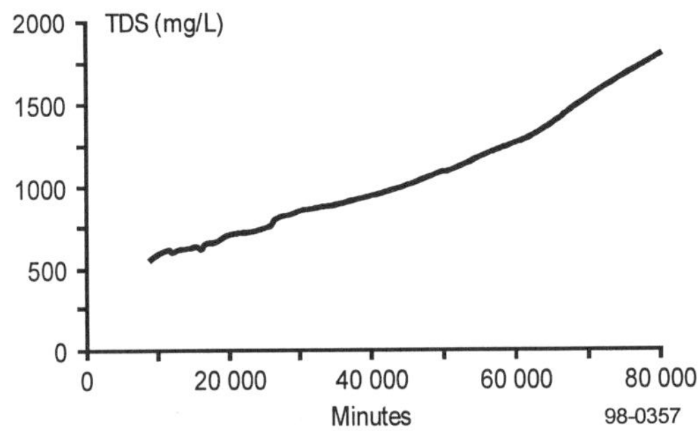

Figure 1. Clayton salinity response during pumping of 9.5 ML.

The final salinity was 1,780 mg/L. Bacteriological sampling indicated the water satisfied potable water quality requirements with no treatment. Pumping resulted in a significant decay in the lens.

2.3 *Trial injection and recovery efficiency*

Due to concern that there was insufficient understanding of the nature and stability of the injected lens, it was decided to operate the system through the 1997/98 summer with continuous injection and pumping until an algal outbreak occurred. Injection occurred (3/10/97 to 17/1/98, rate 40 L/s) which resulted in a volume of 280 ML being injected. Injection ceased due to the need to complete infrastructure, no algal problem occurred in the lake during the summer. The fifth injection cycle may indicate evidence of slight clogging (injection heads rising by 0.5 m) although this was unclear at this stage.

Pumping of 20 ML for the town water supply (UV sterilised) occurred for a period of 69 days (23/12/97 to 2/3/98) at a rate of 4 L/s. Between 21/1/98 and 2/3/98 (40 days), following the end of injection, 14 ML was pumped with a salinity less than 1,000 mg/L. Salinity data is given in Figure 2, which indicates that the salinity rose rapidly from the initial salinity of 530 mg/L to 750 mg/L at 10,000 minutes, and then in a linear manner until 2/3/98 when the salinity was 1,000 mg/L at a time of 58,000 minutes.

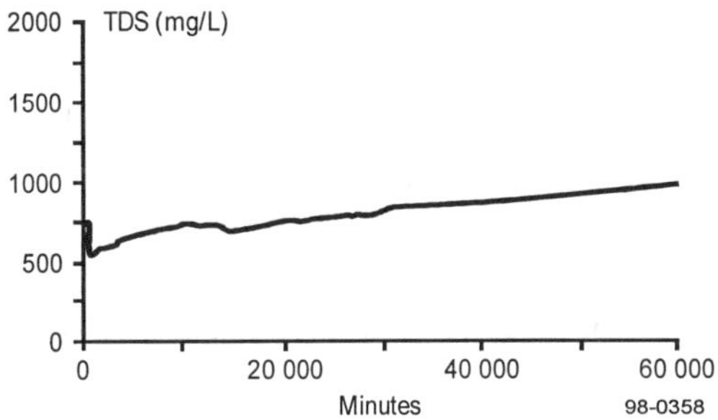

Figure 2. Clayton salinity response during pumping of 14 ML.

2.4 *Conclusion*

The Clayton ASR scheme is an investigation of a novel and non-routine nature. Work to date indicates that a potable water supply can be developed in the highly saline groundwater in the unconfined high transmissivity fractured and karstic Tertiary limestone aquifer. A large 'sacrificial lens' has been injected to create a buffer and transition zone for the decay process to the native groundwater. Inside this initial lens the potable water supply lens is injected. It is believed that the water is being injected into, and pumped from, both the limestone matrix and the large fractures and karstic space of the aquifer.

The salinity data indicates that the injection of the initial 50 ML reduced the salinity from 35,000 mg/L to 4,500 mg/L at the injection well (8 fold reduction). The reduction to a level of 500 mg/L (9 fold reduction) required the injection of a further 250 ML, five times as much water. Residence followed by pumping of operational volumes resulted in significant decay, which may reduce with repeated cycles.

Due to the aquifer filtering and ultra violet (UV) sterilised water it has been proposed the scheme be used throughout the year. It is recommended that injection occurs on an 'as needs basis', determined by either high use periods, or preparation for summer algal outbreaks, which may require the injection of 300 ML. In any event sufficient injection should be ensured to maintain the 'sacrificial lens'. The salinity response during pumping can be expected to improve as the aquifer is further flushed. In addition the injection requirement may reduce as the 'sacrificial lens' further develops.

Ideally calibration of the lens is required. The 'salinity state' that the lens must be injected to be able to undergo residence periods of 1, 2, or 3 months and be capable of producing a supply of 10, 20, or 30 ML of water, of which a salinity of less than 1,000 mg/L is required. This must be determined over a period of years with the use of observation well salinity profiles, and possibly computer simulation. As a result it may be possible to predict the recovery efficiency.

Neither clogging or bio-geochemical problems have been shown to be a problem to date. The injected silt collects on the well and fracture walls and in time will reduce the efficiency of the injection well. The well can then be backflushed, acid treated, or a new injection well drilled. Reducing the permeability of the aquifer may have some advantage as it may increase the stability of the injected lens resulting in the need for less injection.

3 ASR IN A SALINE DEEP CONFINED LOW TRANSMISSIVITY AQUIFER

3.1 *Initial investigation*

A wetland site in Adelaide known as 'The Paddocks', was developed with the aim of conjunctive wetland treatment and aquifer storage of storm water.

A 164 m deep exploration/injection/production well was drilled and completed (7/7/94) with 30 m of open hole within a confined silty sandy limestone. A final depth to water of 9.9 m, a salinity of 1,860 mg/L and an airlift yield of 10 L/s were recorded. Step drawdown (S) test (24/7/94 steps: 5, 10, 15.4 L/s) analysis resulted in the well equation:

$$S_t = 48.81\ Q + 1.22\ Q^2 + 4.76 \log_{10}(t)\ Q \qquad (1)$$

The specific capacity at 100 minutes for step-1 was 0.29 L/s/m of drawdown. Constant discharge test (2/8/94 15 L/s, duration 1,500 mins) indicated a specific capacity at 100 minutes of 0.28 L/s/m of drawdown and a transmissivity of 77 m^2/d.

The initial injection trial using wetland water (6/6/95 duration 1,440 mins) resulted in an injection rate initially starting at 15 L/s and falling to 4.2 L/s by the end of the test. At the end of the test an injection head of -68 m had developed. The specific capacity at 100 minutes was 0.15 L/s/m of injection head. Following the test it was found that there were iron bacteria present in the wetland, and the well was chemically treated and airlifted. A step drawdown test (21/9/95 steps: 5, 8.25 L/s) indicated that the well efficiency was similar to that prior to injection. Further injection testing using mains water (29/9/95 duration 480 minutes) resulted in an initial injection rate of 12.5 L/s falling during the test to 10.6 L/s. At the end of the test an injection head of -58 m had developed. The specific capacity at 100 minutes was 0.2 L/s/m of injection head. This injection test was accepted as a baseline for injection behaviour. Discharge behaviour should be used as a baseline as this is the most efficient hydraulically. The injection testing indicated that the well was not able to readily accept injection of wetland water or mains water, developing injection heads greater than expected, however mains water was more readily accepted. It was apparent that there was a water compatibility problem and a decision was made to increase the well efficiency by acidising.

3.2 *Acidisation*

Acidisation (20/11/95) was conducted using hydrochloric acid. Caliper logging indicated the well diameter was increased from a 200 mm to between 300 and 600 mm. Step drawdown testing (23/11/95 steps: 5, 7.5, 8.5 L/s) resulted in the well equation:

$$S_t = 35.94\ Q + 1.73\ Q^2 + 4.01 \log_{10}(t)\ Q \qquad (2)$$

The specific capacity at 100 minutes for step-1 was 0.37 L/s/m of drawdown. A comparison of the pre and post acidisation well equation constants indicated the aquifer loss 'b' was not seriously affected, the well loss for turbulent flow 'c' increased slightly and the well loss for laminar flow 'a' was significantly reduced, by 25%. Specific capacity values at 100 minutes indicate an improvement in the wells discharge performance by 25%.

Constant rate injection tests using mains water was conducted on three consecutive days (28/11/95 5 L/s, duration 300 mins, 29/11/95 7.5 L/s, duration 300 mins, 30/11/95 10 L/s, duration 1,440 mins). Analysis resulted in the well equation:

$$S_t = -36.12\ Q - 8.82\ Q^2 - 4.67 \log_{10}(t)\ Q \qquad (3)$$

The specific capacity at 100 minutes was constant at 0.33 L/s/m of injection head in each test. At the end of the third test an injection head of -32.5 m had developed. Clogging was indicated within the first 30 minutes which was interpreted as being caused by a combination of density differences and dissolved gasses. The specific capacity values pre and post acidisation indicate that the well was 50% more efficient during injection with mains water following acidisation. The well constants 'a' and 'b' for post acidisation discharge and injection with mains water are similar (when the signs are ignored), however the much higher 'c' value during injection was due to clogging of the well.

Recovery efficiency testing (19/3/96 12.7 L/s, duration 2,880 mins) resulted in the pumping of 2.2 ML. This test followed a residence of 110 days since the mains water injection. The initial salinity was 720 mg/L and rose in an almost linear manner to the final value of 1,760 mg/L, 100 mg/L below the original groundwater salinity.

3.3 *Operational injection and recovery efficiency*

Constant rate injection was conducted for 15 cycles (7/6/96 to 8/10/96) the duration of which varied between 2,000 and 17,000 minutes. Injection rates started at 10 L/s and reduced during the injection cycles to 7 L/s. Backflushing was carried out between the cycles on a weekly basis. Injection heads increased between the first and fifth cycles to a level of -73 m, during the following seven cycles injection heads reduced slightly to -67 m. These injection heads were within the safe operating limit, calculated at -120 m, to prevent a risk of heaving the confining bed. Specific capacity data indicated the specific capacity fell from the start of the first cycle, 0.2 L/s/m of injection head, to 0.1 L/s/m of injection head by the end of the third cycle. In cycles 10, 11, 12 the specific capacity did not fall below 0.13 L/s/m of injection head. These results may be explained by early clogging followed by the effects of dissolution of calcium carbonate and backflushing increasing well efficiency.

If the specific capacity for injection with mains water is taken as 0.33 L/s/m of injection head at 100 minutes, and that for injection with wetland water as 0.2 L/s/m of injection head at 100 minutes, then injection with wetland water is 60% as efficient as injection with mains water.

Recovery efficiency testing (14/1/97 13 L/s, duration 96,000 minutes) resulted in the pumping of 76 ML, salinity data is plotted in Figure 3. This test followed a residence of 97 days since the end of the 1996 injection. The data from this first cycle indicated that 40% was recovered with a salinity similar to that of the injected water (250 mg/L), 90% was recovered with a salinity less than 1,000 mg/L (the limit for drinking water) and 100% was recovered with a salinity less than 1,300 mg/L (good quality for irrigation). It was estimated that 120% of the injected volume would have to be pumped to raise the salinity level to that of the original groundwater.

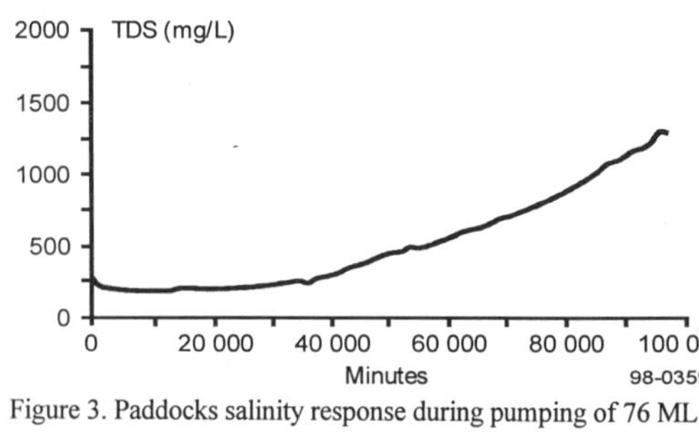

Figure 3. Paddocks salinity response during pumping of 76 ML.

3.4 *Conclusion*

An irrigation water supply can be injected into a low transmissivity confined aquifer and be stored for subsequent pumping in spite of the high injection heads required for injection at rates of up to 10 L/s. Acidisation resulted in an increased well performance. The low transmissivity assists in retarding mixing and movement of water away from the site of the injection well. Clogging processes may be dominated by dissolution of calcium carbonate which may improve well efficiency with time.

4 CONCLUSION

Although limited experience with ASR exists within Australia, South Australia has developed the expertise of aquifer injection in various hydrogeological environments. The results of using ASR in South Australia will be the development of a water resource management tool. This will assist in sustaining groundwater systems, reducing reliance of urban users on imported water, reducing the use of mains water for irrigation; and developing economically viable alternatives where traditional water supply is impractical.

The two extreme examples in carbonate aquifers cited indicate the difference in injecting and recovering water from a saline low transmissivity confined aquifer where the injected 'bubble' remains localised with little decay, and from a highly saline unconfined high transmissivity aquifer where the injected 'lens' may be subject to rapid movement and decay. The dissolution effects noted are a result of injecting water with a lower pH to that of the native groundwater. This has contributed to the sustained and increased well efficiency, and in retarding clogging processes.

5 REFERENCES

Emmett, A., S. Clarke, & S. Howles, 1994. *Conjunctive wetland treatment/aquifer storage and recovery at Regent Gardens residential development, Northfield, South Australia.* Proc. Int. Specialist Conf. on desalination and water reuse, Murdoch University, Perth, Western Australia, December 1994.

Gerges, N.Z. 1992. *A preliminary assessment of artificial recharge potential, with reference to the Northern Adelaide Plains irrigation area.* South Australian Department of Mines and Energy. Report Book, 92/40.

Gerges, N.Z. 1995. *Proposals for injecting effluent for aquifer storage and recovery scheme.* Proc. Int. Symp., Helsinki, Finland, 3-5 June 1996.

Gerges, N.Z. 1996a. *The potential for aquifer storage and recovery in the Northern Adelaide Plains.* South Australian Department of Mines and Energy. Report Book in preparation.

Gerges, N.Z. 1996c. *Proposals for injecting effluent for an aquifer storage and recovery scheme.* Proc. Int. Symp., Helsinki, Finland, 3-5 June 1996.

Gerges, N.Z. & Armstrong, D.A. 1996. *Enfield cemetery storm water recharge investigation stage 1, the shallow sand aquifer.* South Australian Department of Mines and Energy. Report Book in preparation.

Gerges, N.Z. & S.R. Howles, 1996. *Aquifer storage and recovery in a highly saline aquifer as a solution for town water supply.* AWWA Proceedings 17th Federal Convention Melbourne, Australia, 1997.

Gerges, N.Z., X.P. Sibenaler, & D. Armstrong, 1994. *Experience in injected stormwater into aquifers to enhance irrigation water supplies in South Australia.* In: Artificial Recharge of Groundwater 11. (Eds. Johnson, A.I. and Ryne, R.D.G.). Proc. 2nd Int. Symp. on Artificial Recharge of Groundwater, Orlando, Florida, 17-22 July 1994.

Gerges, N.Z., X.P. Sibenaler, & S.R. Howles, 1996. *South Australian experience in aquifer storage and recovery.* Proc. Int. Symp., Helsinki, Finland, 3-5 June 1996.

Howles, S.R. 1989. *Artificial recharge in the Angas - Bremer irrigation area.* Mines and Energy Review No 157.

Howles, S.R. 1992a. *Suspended solids clogging and reversal in recharge well 6727-2310, Langhorne Creek South Australia.* Department of Mines and Energy. Report Book, 92/69.

Howles, S.R. 1992b. *Enhancement of recharge rate of recharge well 6727-1491, Langhorne Creek South Australia.* Department of Mines and Energy. Report Book, 92/35.

Howles, S.R. & N.Z. Gerges, 1993. *Artificial recharge potential at Precinct 1 Northfield residential development, Adelaide, South Australia.* Department of Mines and Energy. Report Book, 93/11.

Pavelic, P., P. Dillon, N.Z. Gerges, 1998. *Restoring heterogeneous brackish aquifer through aquifer storage and recovery, TISAR '98.* Proc. 3rd Int. Symp. on Artificial Recharge of Groundwater, 12-25 September 1998, Amsterdam (this volume).

Artificial Recharge of Groundwater, Peters et al. (eds) © 1998 Balkema, Rotterdam, ISBN 90 5809 017 5

Aquifer storage recovery – Wessex Water's experience using the chalk aquifer

J.C. Eastwood
Wessex Water Services Limited, Bristol, UK

C.M. Bienfait
CH2M Hill United Kingdom, London, UK

R.D.G. Pyne
CH2M Hill International Incorporated, Gainesville, Fla., USA

ABSTRACT: Wessex Water Services Ltd. started a year long trial of Aquifer Storage Recovery (ASR) in December 1997 in which treated drinking water is being injected into a confined Chalk aquifer. The object of the trial is to create a freshwater 'bubble' or reservoir within a non-potable aquifer that can be re-abstracted for public water supply without further treatment. The trial is of particular significance in that it represents the first large-scale ASR trial of its type in the United Kingdom. The results will therefore be of importance, not just to Wessex Water but to other British water companies, especially those who are considering ASR schemes in the Chalk. This paper describes the background and initial test results from the ASR trial and what this may mean in terms of its subsequent development.

1 BACKGROUND AND NEED

Wessex Water Services Ltd., supplies water to more than one million people in central southern England. The region is a popular holiday area and demand for water increases significantly each summer. Peak week demand can reach 140% of the annual average demand which is currently around 400 Ml/d. This region is also an area of growth which will inevitably increase the pressure on existing supplies.

The Wessex Water region supports an extremely rich and diverse range of habitats and there has always been a need to try and maintain a balance between the provision of safe and reliable sources for water supply, and safeguarding the aquatic environment. Groundwater supplies 75% of Wessex Water's resources, the remainder is from springs and surface water sources. Abstraction from some of the long-standing groundwater sources are now thought to be contributing to low river flows by reducing baseflow and the Environment Agency, in conjunction with Wessex Water, is drawing up plans to reduce abstraction from sensitive sites. At the same time concern is growing that changing rainfall patterns, possibly associated with climate change, will lead to more frequent and prolonged periods of drought. Such events could increase the stress on existing sources of supply and the aquatic environment.

Wessex Water's strategy for meeting these challenges is by means of a twin-track approach. On the one hand significant investment is being made in a programme of demand management and leakage control measures. In parallel with this, a range of options are being explored that will enable the Company to shift the balance in terms of future water abstraction by taking more water when the environment permits, normally during the winter, and then storing it. This water will then be made available to meet peak demand during the summer months without damage to the environment. A range of schemes are under consideration including conventional surface water bankside storage.

ASR may have the potential to provide part of the solution. The problem is that ASR has yet to be proven in a dual porosity aquifer such as the Chalk and the overall potential for ASR in the Wessex Water region is unknown. Hence, the first step is to carry out a full scale trial on a Chalk borehole to establish the viability of the technique, following which, depending on results, a phased expansion programme can be considered.

2 CHOICE OF ASR SITE AND BACKGROUND HYDROGEOLOGY

Two main factors need to be considered in selecting an ASR site. Firstly, the site must have suitable hydrogeology. Secondly, to be cost effective it should be located as close as possible to existing infrastructure which has the capacity for off-peak injection.

A feasibility study of potential ASR sites in the Wessex Water region was carried out in 1996. This led to a trial site being chosen to the west of Poole in Dorset on the south coast of the United Kingdom, in an area that had been investigated in 1994 as part of a conventional groundwater investigation, Figure 1. The main aquifer in this locality is the Chalk. Table 1 summarises the thicknesses and depths of the geological strata at the trial site.

Figure 1. Location of Poole, United Kingdom

Table 1. Summary of stratigraphic units penetrated by the ASR well

Lithostratigraphic Unit	Thickness (m)	Depth to Base (mbgl)
Poole Formation	14	14
London Clay	45	59
Upper Chalk	138	197

The Poole Formation comprises fine to coarse grained sand with interbeds of clay and silty clay. The London Clay sediments, which acts as an aquiclude, comprise a stiff fissured silty clay with a basal unit (15 m thick) of clayey sand. The Upper Chalk comprises a fissured fine grained microporous limestone. The matrix typically has a high porosity (34%) but the pore throats are small, giving rise to very low permeabilities (10^{-4} - 10^{-2} m/day). The fissures and fractures provide a secondary component of porosity and the main component of permeability in the Chalk. The permeability at this site is in the order of 2 m/day.

The Chalk at this location is well confined with a piezometric surface of approximately 20 metres above ground level and an artesian yield from the borehole of up to 3Ml/d. The groundwater is essentially fresh in origin although it exhibits slight mineralisation with fluoride being the main parameter that exceeds drinking water standards. The water is estimated to be around 10,000 years old with no history of abstraction to disturb the aquifer.

The effects of a 6 month constant rate pump test, pumping at 5.4 Ml/d, were localised with high drawdowns around the well with no adverse environmental impacts at the surface or in the unconfined Chalk. Values of transmissivity typically ranged from 200 to 500 m^2/day and storage coefficients were typically in the order of 2 x 10^{-4}. Both the field data and the hydraulic modelling indicated that natural recharge is limited and that even seasonal abstraction is unlikely to be sustainable without some form of artificial recharge. These findings make the aquifer at this location particularly suitable for ASR.

3 ASR CYCLE TESTING PROGRAMME

The primary objective of the trial scheme is to determine the feasibility of ASR in a confined Chalk aquifer and to collect sufficient information for the construction of an operational facility at Poole. The testing programme was divided into two parts: initial tests followed by a series of full scale tests.

The initial tests were completed in January 1998 and lasted a period of two months. The information collected from these tests was used to determine the likelihood of ASR being successful on this site prior to embarking with the full scale tests and to help refine the programme for the full scale cycle tests.

The full scale cycle tests commenced in May 1998 and are expected to be completed in December 1998. The objectives of these tests are to evaluate the long term performance and the full operational capability of the ASR well. Specific objectives include determining the degree and nature of mixing between the stored treated drinking water and the native groundwater, determining the potential for well clogging, determining the recovery efficiency and the impact of storage on recovery efficiency, determining hydraulic and /or water quality changes in the ASR well and nearby observation boreholes, determining the maximum operational capability of the ASR system and the maintenance requirements.

4 ASR WELL AND WELLHEAD FACILITIES

The ASR well was originally constructed in April 1993 as a production borehole for direct groundwater abstraction. A 609 mm diameter mild steel casing was installed from ground level to 59 m depth. Below this, the well is open to a depth of 110 m. On completion of its construction, the borehole was developed by acidisation.

In September 1997, work commenced to retrofit the borehole from a production well to an ASR well. The main criteria for the design was to inject potable water at a rate of 2 Ml/d and to recover water at the artesian flow rate (approximately 3 Ml/d). Figure 2 illustrates the schematic wellhead piping configuration and basic components.

Potable water is directed to the ASR well from a 16" trunk main located approximately 150 m from the well. To reduce the risk of non-potable water from the ASR well entering the public supply during the cycle testing programme, a Class A air gap was installed in the delivery line. This was achieved by installing a break pressure tank which provides a 20 minute buffer storage. As the cycle testing programme proceeds and the results demonstrate that a storage zone of drinking water quality has been successfully developed, the air gap should no longer be required and the break tank would be removed.

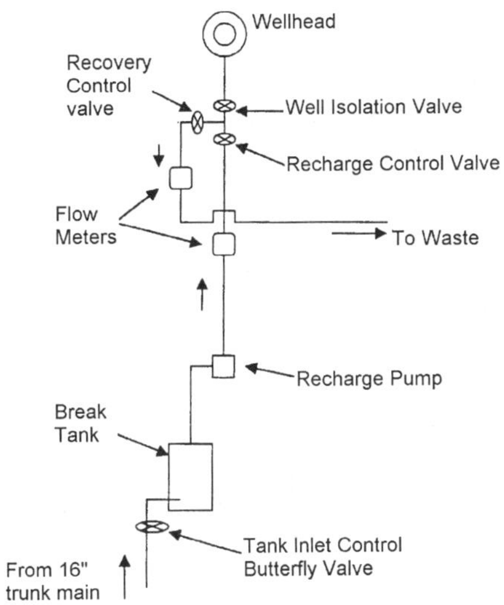

Figure 2. Schematic Diagram of Wellhead

An injection pump with a duty point of 2 Ml/d @ 50 head delivers potable water from the break pressure tank into the ASR well. During the recovery phases of the cycle tests, artesian flow is directed via a weir tank into a nearby stream. Several valves to control injected and recovered flows were installed, along with flow meters, pressure transducers, sampling ports and alarms to monitor and regulate the operation.

Pressure transducers were installed in three observations boreholes that are located approximately 20 m from the ASR well. One observation borehole records groundwater pressure in the Upper Chalk aquifer (OBS), the second records groundwater pressure in the basal sandy unit of the London Clay (P2) and the third records groundwater pressure in an upper unconfined sandy aquifer (P1).

5 WATER QUALITY

The native groundwater of the Chalk aquifer is a calcium bicarbonate water type chemistry with reducing conditions prevailing. A typical geochemical profile is given in Table 2.

Table 2. Typical groundwater and recharge chemistry

Parameter	Unit	Chalk Groundwater	Recharge Water
Temperature	°C	12.5	10
pH		7.2	7.2
Eh	mV	-120	630
Conductivity	µs/cm	690	594
Turbidity	NTU	6.9	0.8
Total Organic Carbon	mg c/l	1.4	-
Alkalinity	mg CaCO3/l	246	187
Total Hardness	mg CaCO3/l	239	299
Ammonical Nitrogen	mg N/l	0.21	0.01
Nitrite	mg N/l	0.003	0.01
Nitrate	mg N/l	0.15	7.12
Orthophosphate	mg P/l	0.02	0.02
Sulphate	mg SO4/l	34	20
Chloride	mg Cl/l	58	28
Fluoirde	mg F/l	3.35	0.18
Silica	mg Si/l	12.4	5.6
Sodium	mg Na/l	46	14.9
Magnesium	mg Mg/l	20	4.2
Potassium	mg K/l	4.0	2.8
Calcium	mg Ca/l	63	113
Dissolved Iron	mg Fe/l	0.34	0.06
Dissolved Manganese	mg Mn/l	0.005	0.005

Fluoride and the dissolved concentrations of iron are the only ionic species to exceed the drinking water quality limits. Hydrogen sulphide is also present, giving rise to strong odours from groundwater samples.

The injected potable water is an alkaline, well buffered calcium bicarbonate type water. The main compositional differences between the native groundwater and the injected water are in the concentration of fluoride ions, which is 19 times greater in the native groundwater, and the concentration of nitrate ions, which is 166 times greater in the injected water.

Water samples from the sampling ports in the wellhead were collected at regular intervals during each injection and recovery cycle. Water samples were also collected from the nearby observation wells to monitor the progress of the storage bubble and its impact in adjacent formations.

6 RESULTS OF THE INITIAL TESTS

6.1 *Aquifer Hydraulics*

A step drawdown test was undertaken to confirm the baseline aquifer hydraulics. These results indicated specific capacities of 240 m^3/day/m and 195 m^3/day/m for discharge rates of 2 Ml/d and 3Ml/d respectively. A step injection test was subsequently undertaken to determine the level of increase in well head pressure at varying injection rates and to determine the maximum injection rate that the pump was capable of delivering. It was anticipated that similar or slightly lower specific capacities would be recorded for comparable rates of flow from the step drawdown test. However, to everyone's surprise higher specific capacities were recorded for similar rates of flow. Similar improvements in specific capacities were also recorded during the cycles tests, confirming that aquifer development had taken place during the step injection test. Overall, the specific capacity of the borehole has increased by approximately 25% resulting in improved well efficiency.

Table 3 summarises the testing schedule of the Initial Tests. The rates of injection and recovery were kept constant during each cycle. The aquifer pressure was allowed to recover to static conditions at the end of each phase of injection and recovery.

The first cycle comprised a short period of injection followed by the recovery of a larger volume of water to define the recharge water / groundwater characteristic mixing curve. Identical volumes were injected in Cycle 2, 3 and 4 to measure the improvement in recovery efficiency with successive cycles. However, less water was recovered than injected to initiate the development of a buffer zone around the storage bubble. This buffer zone will be critical to the success of the final operational scheme.

Monitoring data collected from the ASR well and from nearby observations boreholes did not identify any negative environmental impacts. In particular, no anomalous observations indicating breaches of well integrity have been recorded throughout the cycle tests.

Increases in well pressure during injection were typically around 8 m. Subsequent decreases in well pressure during the recovery phases of the cycles were up to 16.8 m. Changes in aquifer pressure during the cycle tests were recorded in the nearby observation boreholes of OBS and P2. The aquifer pressure in OBS typically rose by 4.7 m during injection and fell by up to 7.5 m during recovery. P2 recorded pressure increases of up to 3.3 m during injection pressure decreases of up to 4.3 m during recovery. P1 installed in the upper unconfined aquifer remained unaffected throughout the duration of the cycle tests.

Aquifer pressures recovered quickly at the end of each injection and recovery phase but they never returned back to pre-cycle testing levels. This is indicative of a closed fissured aquifer system that receives little natural recharge.

The total volume of potable water remaining in the aquifer at the end of Cycle 4 was approximately 30 Ml. Assuming a bulk chalk porosity of 2% and an aquifer thickness of 50 m, the theoretical size of the storage bubble would be in the order of 100 m.

6.2 *Geochemical Observations*

The success of ASR at Poole will be measured on the basis of the recovery efficiency of fluoride since this is the main parameter to exceed the drinking water quality limits in the native groundwater. Recovery efficiency is defined as the percentage of water volume stored that is subsequently recovered while meeting a target water quality criterion in the recovered water, which in the case of fluoride is 1.5 mg/l. Fluoride is therefore the principal tracer during these tests, although chloride and nitrate are also monitored closely. The results of fluoride concentration in the recovered water of Cycles 1, 2, 3, and 4 are given in Figure 3.

Table 3. Cycle Testing Programme of the Initial tests

Cycle	Time (days) Recharge	Storage	Recovery	Total	Cum' tive	Volume (m^3) Recharge	Recovery	Recovery % of volume injected
1	5	0	8	13	13	8780	21422	243
2	11	0	2	13	26	20735	3622	17
3	11	0	5	16	42	20735	12731	61
4	11	0	5	16	58	20735	155551	75

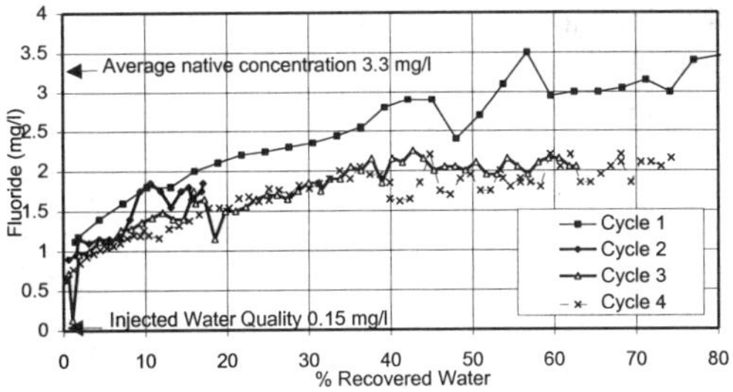

Figure 3. Change in Fluoride Concentration with successive cycles

These results indicate successive improvements in recovery efficiency, particularly from Cycles 1 to 3. In order for this site to become operational, a much greater improvement in recovery efficiency will be required. The strongly convex shape of the recovered water quality curve is indicative of mixing close to the well, an observation which was supported from an analysis of pH and Eh results and the results of analysis of the main ionic species. Such responses are typical when the volumes of recharge are small, but also if a strong hydraulic gradient exists between the productive fissures in the aquifer, giving rise to anisotropic flow conditions during recharge and recovery. Insufficient information is currently available to verify whether such anisotropic conditions exist in this well. To help improve the current knowledge of the behaviour of these productive fissures, it is hoped that flow logging equipment can be installed during the full scale cycle tests.

Initial geochemical analyses from water samples collected during the recovery phases of the four cycles, indicate that geochemical processes other than just mixing are occurring in the storage bubble. The mechanism(s) of these processes are not yet fully understood and will be the subject of further study during the full scale tests. Figure 4 indicates the relative change in concentration of fluoride, chloride, nitrate, sulphate, sodium and calcium during Cycle 4 recovery. On the vertical axis, 1.0 is indicative of native groundwater quality whilst at the base of the axis, zero is indicative of injected water quality. Most of these parameters including chloride, which is normally regarded as a conservative ion, do not reflect a normal mixing profile, indicating that some interesting geochemical reactions are taking place. Oxidation reactions are occurring in which iron (probably from pyrite) is being oxidised by the injected water to form oxyhydroxides. The potential for well plugging as a result of iron precipitation is anticipated, however, to be slight since the chalk is expected to buffer any shift in pH associated with pyrite oxidation. Nevertheless, indications of well plugging from changes in specific capacity will be monitored closely during the full scale cycle tests. The reduction of nitrate in the recovered water in Figure 4 indicates that nitrate is being converted to nitrite and possibly nitrogen. Although this reaction was expected, the scale of change is significant and indicates that ASR has the potential to play a major role in reducing the level of nitrates in drinking water. The reduction in calcium during recovery confirms that no major dissolution of the chalk is occurring.

Free and total chlorine levels in the recovered water have consistently remained below detection limits.

6.3 *Current direction of the full scale cycle tests*

One of the main requirements of the full scale tests will be to improve the recovery efficiency of the well and to determine the target storage volume at this site. These will be addressed in the full scale tests by increasing the injected volumes and by increasing the injection rate from 2 Ml/d to 4 Ml/d.

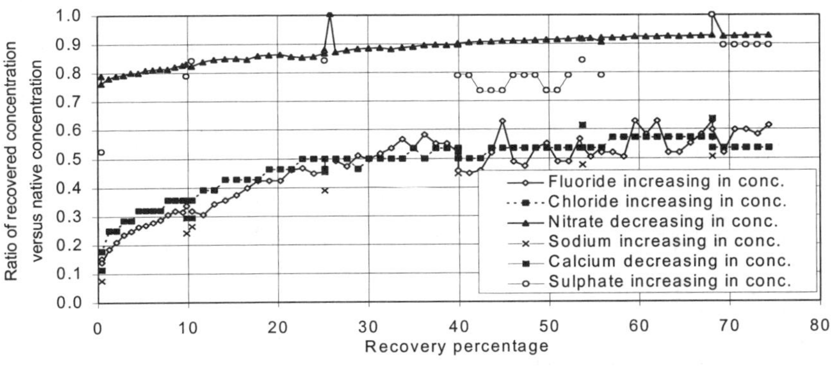

Figure 4. Comparison of change in Fluoride, Chloride, Sodium, Sulphate, Sodium and Calcium Concentration during Cycle 4 recovery.

The recovery rates will also be increased to determine the full operational capability of the well. Towards the end of the cycle testing programme, the recovered water will be directed back into supply. Recovered volumes will continue to be less than the injected volumes to develop further the buffer zone around the storage bubble.

Plans are currently being developed to install a flow logger and conductivity logger in the ASR well to obtain detailed information on the mixing characteristics of recharge water and native groundwater at each of the productive fissures intercepted by the ASR well.

Monitoring of nearby observation wells, both in the Chalk aquifer and the overlying sandy silt unit of the London Clay will continue to be a part of the full scale cycle test programme to track the development of the storage bubble.

A storage phase will also be introduced between injection and recovery, the prime purpose of which will be to assess the impact of storage on the recovery efficiency. Of particular interest will be the potential for diffusion between the connate water in the rock matrix and the injected water. The impact of dual porosity diffusion is being modelled by the British Geological Survey using SWIFT/486 which is a fully transient, three-dimensional, finite difference model. The model allows for a one dimensional subsystem to be attached to the grid blocks which can be used to characterise the secondary porosity element of a dual porosity fractured medium.

Water quality analysis will include the determination of Disinfection-By-Products (DBPs) concentration in the recovered water. This is not anticipated to be a serious issue owing to the low concentrations of organic material in the injected and native groundwater. Results received during the Initial Tests appear to indicate a reduction in the concentration of total haloforms in the recovered water, however, this trend needs to be verified particularly after periods of storage.

7 CONCLUSIONS

The first four cycles of the ASR trial at Poole have produced encouraging results which will be used to refine the programme of full scale cycle tests. The concentration of fluoride in the recovered water has been reduced from an average native groundwater concentration of 3.3 mg/l to 2 mg/l and it is expected that, with the injection of larger volumes of water and at higher rates of flow, this concentration will be reduced further to below 1.5 mg/l.

The initial indications confirm that the success of ASR in the chalk aquifer at this site will be dependant on two main factors:

- The formation of a buffer zone around the storage bubble to minimise the mixing potential between the native groundwater and the injected water;
- Proving that dual porosity diffusion will not deleteriously impact the recovered water quality.

No negative environmental impacts have been identified during the trial, either in near by observation wells or the near by stream.

The initial results of recovered water quality indicate that geochemical reactions in addition to just mixing are occurring within the storage zone. Of particular interest is the process of denitrification which has major potential benefits for the development of ASR schemes in areas where current problems exist with nitrate rich drinking water.

ACKNOWLEDGEMENTS

The authors are grateful to Wessex Water Services Ltd. for their permission to publish this paper and also to staff within Wessex Water and CH2M HILL for their various contributions. We are also particularly grateful to the contribution and support provided throughout this project by Ian Gale and Ann Williams of the British Geological Survey. The views expressed in this paper are solely that of the authors and are not necessarily those of Wessex Water Services Ltd and CH2M HILL.

REFERENCES

Buckley, D.K. 1996. A review and interpretation of geophysical logging performed for the Wareham Groundwater Project. British Geological Survey Technical Report WD/96/45C.

CH2M HILL UK 1996. Report on the ASR Potential in the Wessex Water Service Area. Unpublished.

Edmunds, W.H. 1996. Groundwater Quality in the Chalk - Lulworth to Poole Harbour. British Geological Survey Technical Report WD/96/49/C.

Freeze, A.A. & J.A.Cherry 1979. *Groundwater*. Prentice Hall

Pyne, R.D.G. 1995. *Groundwater Recharge and Wells: A Guide to Aquifer Storage Recovery*. Lewis Publishers.

University of Birmingham School of Civil Engineering 1997. Wareham Groundwater Scheme - Phase III Pumping Test Analysis. Unpublished.

1.12 Chemical reactions and well recharge

Artificial Recharge of Groundwater, Peters et al. (eds) © *1998 Balkema, Rotterdam, ISBN 90 5809 017 5*

Quality changes upon injection into anoxic aquifers in the Netherlands: Evaluation of 11 experiments

Pieter J. Stuyfzand
Kiwa Research and Consultancy, Nieuwegein, Netherlands

ABSTRACT: Since 1973 11 deep well recharge experiments, each at a different site with intensive hydrochemical monitoring, were carried out in the Netherlands. At each test site the influent was oxic and the anoxic target aquifer contained pyrite. All experiments revealed a typical quality evolution in 7 phases due to sequential leaching of aquifer constituents. The main processes were, in mean order of increasing duration: mixing with native groundwater, cation exchange, SiO_2 desorption, PO_4 and F exchange, pyrite oxidation, acid-buffering by calcite dissolution or HCO_3^-, and oxidation of organic matter. Differences in quality changes between the experiments relate to variations in aquifer geochemistry and influent quality. Pyrite oxidation is kinetically hindered, it takes about 10-100 days to deplete O_2 and NO_3^- in the influent. Oxygen is reacting faster with pyrite than with labile organic matter, for nitrate the reverse holds. Pyrite oxidation leads to mobilization of As, Co, Ni and Zn, of which only As may reach the recovery well.

1 INTRODUCTION

In the past 25 years, 11 independent experiments with deep well recharge and intensive hydrochemical monitoring were carried out in the Netherlands (Fig.1). These experiments aimed at getting insight into (a) the feasibility of recharging confined aquifers with surface water through wells, for drinking water supply, and (b) the resulting water quality changes, because these have consequences for pre- and post-treatment and effects on clogging of the recharge and recovery wells. The general driving force was an increasing demand for drinking water, in areas where the government forced the water supply companies to restrict their groundwater pumping, in order to combat the drawdown of groundwater tables and to restore wetlands in selected areas. In most cases oxic water was prepared from surface water (by partial purification or by full treatment to drinking water standards; Table 1), and was subsequently injected into a deep anoxic, sandy aquifer and recovered by a pumping well at 75-200 m distance. In this contribution the hydrochemical results of the experiments are compared and integrated to provide general patterns in the quality evolution downgradient and with time.

2 SITES AND METHODS

The experiments were carried out in various environments (Fig.1): the coastal dunes (1-6), the Rhine fluvial plain (7-9) and Pleistocene uplands (10-11). Some details on the 11 experiments, of which number 6 (with 24 injection wells) is actually part of a current drinking water production system, are listed in Table 1.

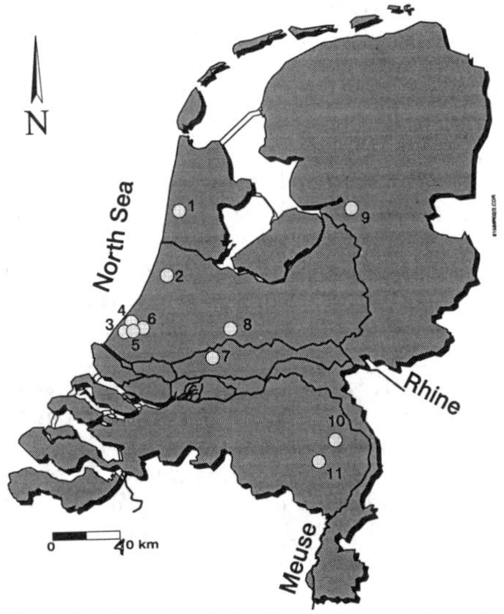

Figure 1. Location of the 11 experimental sites where oxic water was or still is injected into an anoxic aquifer, with intensive hydrochemical monitoring.

Table 1. Some details on the 11 experiments with deep well injection of oxic water into an anoxic aquifer. The locations are shown in Fig.1 (MSL = Mean Sea Level).

site no	site name	Water Supply Company	period of injection	source injection water @	injection depth m-MSL	injection rate m^3/h	target aquifer m-MSL	geologic formation**	number sampled wells	anal. program water#	further data in §
1	Castricum AIP	PWN	75-77	Y_p	81-90	10-30	41-102	En	2	MC	
2	Leiduin WIP	GW	76-77	$R_{d,p}$	22-36	30	14-74	E,D	9	MC	a
3	Scheveningen DIP	DZH	73-80	R_d,M_d	25-45	37-60	20-53	E,U/S	2	MC	b
4	Scheveningen CIP	DZH	74-77	R_p,M_p	23-42	36	11-47	E,U/S	3	MC	
5	Scheveningen F/F	DZH	84-88	M_p	29-42	15-20	27-44	E,U/S	4	All	i
6	Waalsdorp	DZH	90-date	M_p	25-63	20	23-67	E,U/S	3	All	
7	Langerak	WZHO	96-98	G_{d+t}	73-92	35	68-93	K,H	13	MC,TE	c,d,e,j
8	Nieuwegein	WZHO	96-98	R_p	104-132	40	90-136	K,H	10	MC,TE,P	c,j
9	St. Jansklooster	WMO	96-97	G_d	48-85	28-53	22-110	En,H	9	MC,TE,P	f
10	Breehei	WML	94-95	G_{d+t}	15-26	26	5-28	Ki	10	MC,TE	
11	Someren DIZON	WOB	96-98	Z_p	257-284	30	248-287	Ki	22	MC,TE	g,h

@: G = local groundwater; M = river Meuse water; R = river Rhine water; Y = lake IJsselmeer water; Z = canal water from Zuid-Willemsvaart; d = treated to drinking water; p = pretreated (half fabricate); t = NaCl and $NaNO_3$ added.

**: D = Drenthe Formation (glacial); E = Eemian Formation (marine); En = Enschede Formation (fluvial); H = Harderwijk Formation (fluvial); K = Kedichem Formation (fluvial); Ki = Kiezeloolite Formation (fluvial); U/S = Urk or Sterksel formation (fluvial).

#: All = MC + TE + organic micropollutants; MC = Main Constituents; P = Parallel column; TE = Trace Elements.

§: a = Appelo et al. 1979; b = Van Beek & Van Puffelen 1986; c = Timmer & Stuyfzand 1998; d = Saaltink et al. 1998; e = Brun et al. 1998; f = De Ruiter & Stuyfzand 1998; g = Straatman & Brekvoort 1998; h = Stuyfzand 1998; i = Rutte 1990; j = Stuyfzand & Timmer 1998.

Quality monitoring consisted of frequent analysis of: (a) the oxic influent, (b) groundwater from observation wells at various distances from the recharge well, and (c) the percolate of parallel columns filled with aquifer cores and flushed with injection water on site (only sites 8-9). Most monitoring wells were situated along the shortest flow paths between injection and recovery well. Groundwater samples were obtained by suction lift from PVC-piezometers (well screen 1-2 m long). The water content of riser plus well screen was purged 3 times prior to sampling. Temperature, EC, pH and O_2 were measured on site. Alkalinity, pH and EC were measured in the laboratory, on the day of sampling. The portion to be analyzed for most trace elements, Ca^{2+}, Mg^{2+}, Fe, Mn, SiO_2 and PO_4 (both ortho and total) was filtrated in situ over 0.45 μm, directly acidified to pH=1,5 using HNO_3-suprapure and was stored in the dark at 4 °C in polyethylene vessels rinsed out with HNO_3 and the water to be sampled. Main constituents were analysed with conventional, well-standardized analytical methods, most trace elements were analyzed using Atomic Absorption Spectrophotometry (AAS) with graphite furnace.

In 7 experiments sediment cores from the target aquifer were obtained using a core catcher during bailer drilling of the monitoring wells. The sediment cores including their pore water were immediately sealed on site using liquidized paraffin, and kept at 4°C in the dark till analysis. Samples were opened and pretreated in an anoxic glove-box since 1996. The samples were analysed for amongst others: total content of 27 elements by X-Ray Fluorescence, organic matter and calcium carbonate by thermogravimetry (resp. 550 and 1000°C), $CaCO_3$ by HCl-extraction of Ca^{2+} (correcting for exchangeable Ca; ppm level), organic carbon by pyrolysis, exchangeable cations using $SrCl_2$, sulphide species (especially pyrite) with a sequential extraction, non-Si-bound trace elements by aqua regia extraction and amorphous Al-, Fe- and Mn-(hydr)oxides by ammonium oxalate/oxalic acid.

3 THE ORIGINAL SITUATION

3.1 *Hydro(geo)logy*

The target aquifers are composed of Pleistocene sands (1-9) or Miocene sands (10-11), in most cases with a modal grain size of 300-500 μm and 0.5-2% <2 μm (clay fraction) and 0-3% >2 mm (gravel). The resulting permeability is in most cases 30-50 m/d, porosity varies from 29-35%. In all cases the recharged sediments exhibit, within the scale of about 100 m (horizontal) by 30 m (vertical), a significant heterogeneity especially in vertical direction. In all cases the target aquifer is (semi)confined.

3.2 *Hydrochemistry*

In most cases the native groundwater in the target aquifer was fresh (Cl^- 7-70 mg/l), calcite saturated (pH 6.6-7.6, Ca^{2+} 40-100 mg/l) and either anoxic (no oxygen, no nitrate, sulphate (meta)stable) or deep anoxic (no oxygen, no nitrate, no sulphate), with relatively high concentrations of Fe^{2+} (1-10 mg/l), Mn^{2+} (0.1-0.6 mg/l), NH_4^+ (0.3-5 mg/l), SiO_2 (18-40 mg/l) and, exclusively in deep anoxic water (sites 1, 2, 7-10), methane (2-32 mg/l).

Table 2. Survey of mean geochemistry of the various geological formations which were flushed by oxic water during deep well recharge. The experimental site numbers correspond with those in Table 1 and Fig.1.

Geol. Formation code	name[§]	exper. site no.	fraction <2 µm %	mean grain size µm	CEC meq/kg	$CaCO_3$ mg/kg	organic carbon % d.w.	FeS_2 as S mg/kg	Fe_{EXCH} (calc) meq/kg	$NH_{4\text{-}EXCH}$ (calc) meq/kg	$Fe(OH)_3$ as Fe mg/kg
Pleistocene Formations											
E	$Eemian^{Ma}$	2-6	0.9	350	9.0	45,000	0.11	220	0.55	0.05	-
D	$Drenthe^{G}$	2	-	325	-	≤10,000	≤0.1	-	-	-	-
U/S	$Urk^{R}/Sterksel^{RMe}$	3-6	0.8	340	8.0	17,500	0.09	230	0.10	0.10	-
En	$Enschede^{E}$	9	0.5	450	4.5	<50	0.05	440	0.14	0.45	30
K/H	$Kedichem/H^{ER}$:U	7,8	3.9	240	24.0	18,160	0.14	225	0.34	0.20	1,085
H/K	$Harderwijk/K^{E}$:M+L	7	1.4	360	9.5	5,600	0.12	160	0.19	0.07	1,020
H/K	$Harderwijk/K^{E}$:M+L	8	0.6	390	6.7	1,360	0.08	65	0.04	0.03	570
Miocene formations											
Ki	$Kieseloölite^{R}$:U	10-11	1.7	330	68	≤200	0.8	4,520	1.07	0.05	525
	$Kieseloölite^{R}$:M+L	10-11	0.8	350	10	<200	0.4	700	0.30	0.03	≤9

§: L = Lower part; M = Middle part; U = Upper part; $_E$ = fluvial, origin in NE-Germany; $_G$ = glacial; $_{Ma}$ = marine; $_{Me}$ = fluvial, river Meuse; $_R$ = fluvial, river Rhine;

Within each target aquifer some vertical hydrochemical stratification was noticed, often connected to the vertical sedimentary stratification.

3.3 *Geochemistry*

The chemical composition of the 11 target aquifers varied especially in connection with the geological formation (Table 1) and within each formation in connection with the clay fraction (<2 µm) or mean grain size. The mean characteristics of the discerned formations are listed in Table 2.

Most formations contain (very) low amounts of very reactive pyrite. Calcium carbonate is typically present in marine deposits and Pleistocene sediments deposited by the river Rhine, and practically absent in deposits from ancient rivers with an eastern provenance and in Rhine deposits of Miocene age. Calcium was in all cases the dominant exchangeable cation (58-82% of CEC; 37% on site 1), followed by resp. Mg^{2+} (6-26%; 33% on site 1), Na^+ (1-8%; 22% on site 1), K^+ (1-6%; 8% on site 1), Fe^{2+} (1-6%), NH_4^+ (0.2-2%; 11% on site 9) and Mn^{2+} (0.1-0.4%). The content of exchangeable Fe^{2+}, Mn^{2+} and NH_4^+ was calculated (as measurements were always biased by oxygenation) according to a procedure described by Stuyfzand & Timmer 1998.

The maximum trace element content of pyrite was estimated by linear regression of As, Co, Ni and Zn (as extracted with aqua regia) with pyrite. On site 7 and 8 the following stoichiometry was thus deduced for these fluvial sediments:

$Fe_{0.98}Co_{0.0037}Ni_{0.01}Zn_{0.01}S_2As_{0.0053}$.

Several aquifers are suspected to contain manganous siderites ($Fe_{1-x}Mn_xCO_3$) in small amounts. Their quantification and reactivity are matter for future research.

4 QUALITY AND TRACING OF INFLUENT

4.1 *Quality*

The mean quality of the water injected, is shown in Table 3. On sites 2, 3, 4, 7 and 10 there were important changes during the experiment, due to either taking another water source (2, 3, 4, 10) or stopping the addition of NaCl and $NaNO_3$ (7,10). Where drinking water was used (2, 3, 7, 9, 10) only minor quality and temperature fluctuations occurred, contrary to the sites receiving pretreated surface water.

Iron, Mn, PO_4 and CH_4 (not shown in Table 3), were in general <0.05 mg/L. High SiO_2 concentrations (19-21 mg/L) are connected with groundwater, moderate levels with drinking water prepared from surface water by basin recharge and recovery (10-13 mg/L) and low contents with pretreated surface water (2-5 mg/L). The extremely low DOC levels on sites 8 (1.3 mg/L) and 11 (0.8 mg/L) are caused by filtration over activated granular carbon as part of the pretreatment.

Important parameters for behaviour during aquifer passage, are the calcite saturation index ($SI_c = \log [Ca^{2+}][CO_3^{2-}]/K_{CaCO3}$) and the Modified Oxidation Capacity (MOC), a modification of the 'oxidation capacity of water' as introduced by Broers et al. 1991. MOC is explained in footnote '§' of Table 3; the modifications include the assumption of stable behaviour of sulphate (without oxidation capacity), and a correction for oxidizable compounds in the influent itself (100% nitrification of ammonium and, based on numerous observations, 20% oxidation of DOC). It should be noted that MOC is in milli-electrons/L, and that the oxidation of 1 mmol of pyrite and organic matter lower MOC by resp. 15 and 4 me/L.

Table 3. Mean quality of the water injected.

site		type	EC	temp	pH	O_2	Cl	SO_4^{2-}	HCO_3^-	NO_3^-	Ca^{2+}	NH_4^+	DOC	SiO_2	MOC	SI_c
no	name	@	µS/cm	°C	-	mg/L	mg/L	mg/L	mg/L	mg/L	mg/L	mg/L	mg/L	mg/L	me/L	-
1	Castricum	Y_p	1100	9.5	7.95	11	220	135	130	7	96	0.03	5.7	2	1.55	0.26
2	Leiduin	R_d	966	9.5	8.0	7.1	180	75	213	6	92	0.05	2.3	12.9	1.21	0.53
		R_p	840	13	7.6	4.2	150	73	160	22	81	0.06	3.0	5	2.09	0.03
3	Schev. DIP	R_d	910	12.4	7.61	8.1	169	83	191	8.5	96	0.03	2.8	10	1.51	0.16
		M_d	630	11.5	7.76	9.4	89	69	198	7.3	80	0.04	3.6	10	1.52	0.25
4	Schev. CIP	R_p	950	12.0	7.40	8.0	180	80	140	14	80	0.10	3.5	4	1.88	-0.26
		M_p	660	12.0	7.40	9.0	85	95	160	15	80	0.17	4.0	4	2.04	-0.20
5	Schev. F/F	M_p	580	12.3	7.08	9.5	81	75	144	15.0	76	0.03	2.8	3.7	2.21	-0.56
6	Waalsdorp	M_p	586	13.0	7.32	10.0	73	78	146	15.1	72	<0.03	2.3	3.5	2.31	-0.33
7	Langerak	G_{d+t}	700	13.5	7.7	9.5	92	8	298	14.9	60.2	0.02	3.0	20	2.19	0.34
		G_d	470	11.0	7.6	9.5	11	8	298	14.9	60.2	0.02	3.0	20	2.19	0.34
8	Nieuwegein	R_p	746	13.7	7.57	5.7	126	63	158	14.9	74.5	0.03	1.3	4	1.82	-0.02
9	St. Janskl.	G_d	499	11.9	7.51	9.9	38	11	265	9.2	85.2	0.06	7.2	19.1	1.49	0.17
10	Breehei	G_{d+t}	425	11.5	7.65	10.5	51	10	185	11	52	<0.05	4.2	20	1.92	-0.02
		G_d	290	11.5	7.45	8.9	13	6	176	1	54	<0.05	4.2	19	0.91	-0.22
11	Someren	Z_p	557	12.0	7.12	9.9	82	62	117	19.2	68.4	0.07	0.8	3.4	2.72	-0.60

@: G = local groundwater; M = river Meuse water; R = river Rhine water; Y = lake IJsselmeer water; Z = canal water from Zuid-Willemsvaart; d = treated to drinking water; p = pretreated (half fabricate); t = NaCl and $NaNO_3$ added.

§: MOC = $4(O_2 - 2\ NH_4^+) + 5(NO_3^- + NH_4^+) - 0.8$ DOC, with concentrations in mmol/L, MOC in me/L.

The range in SI_c is from -0.60 (calcite agressive; site 11) to +0.53 (calcite supersaturated; site 2 first period). The range in MOC is from 0.9 (site 10 second period) to 2.7 me/L (site 11).

4.2 Tracing

Chloride, electrical conductivity, temperature, sulphate, methane and hydrogencarbonate proved to be very useful tracers of the injection water. Although only chloride behaves as a perfectly conservative tracer, the others were conservative enough for use (see methane in Fig.2).

With temperature a retardation factor of 2.0 ± 0.1 should be taken into account due to heat exchange with the aquifer matrix (Olsthoorn 1981; Stuyfzand 1997). At the Langerak site where sodium chloride was added for a short time (30 days), sulphate (high in influent) and methane (low in influent) were especially useful for the recovery well where dispersion of the tracer added and mixing with native groundwater faded away the initial chloride contrast (Stuyfzand & Timmer 1998).

The break-through at levels in between observation screens was established using (a) 0.01 °C accurate temperature sounding in the deepest piezometer prior to any pumping for sampling, and (b) resistivity measurements (of formation plus groundwater; latter variable) with fixed electrodes, both on a frequent basis.

The mean travel time (t_{50}) is defined as 50% break-through, and is used to calculate for each observation the amount of pore flushes (PFs) with the injection water during the injection period (Δt_{inj}): PF = $\Delta t_{inj}/t_{50}$.

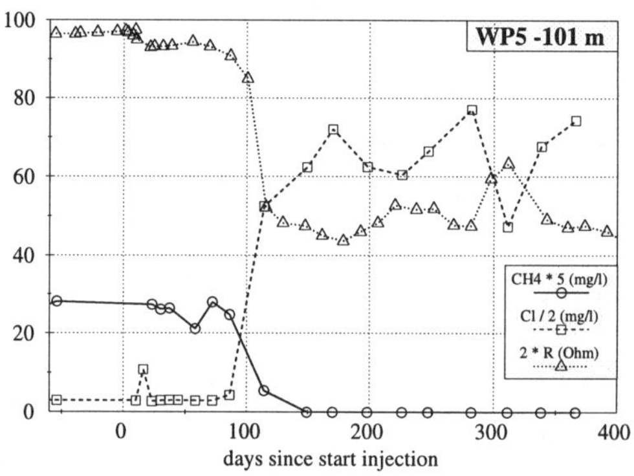

Figure 2. The break-through curves of chloride and methane (in groundwater sampled with a 1 meter long well screen) and of the electrical resistivity (measured with a fixed electrode couple in gravel pack of well) yield about the same mean travel time for observation well WP5-101 (site 8) and longitudinal dispersivity: resp. 105 days and 0.3 m.

5 GENERALIZED QUALITY CHANGES

The water quality changes of oxic water upon anoxic aquifer passage generally evolve in a specific way with time, due to sequential leaching of reactive components of the aquifer matrix. The quality evolution of the main constituents is shown for a calcareous aquifer in Fig.3. The patterns shown are based on numerous observations at test sites 1-8, with emphasis on sites 5-6. The evolution reveals 7 phases, which are discussed below in order of increasing number of pore flushes (PFs). The number of PFs for phases 3-6 is only approximative and holds for water with a travel time of 3-10 days (most data). The first break-through of nitrate and sulphate, initiating resp. phases 3 and 4, requires less PFs for water with a shorter travel time due to the relatively slow kinetics of redox reactions. As the available data contain time series up to 400 PFs, the further evolution is based on calculations using the transport code EASY-LEACHER (Stuyfzand 1998). Of course, the evolution depicted is only valid for a more or less continuous injection without clogging or with an easily curable clogging.

Phase 1: displacement (0.5-1.5 PFs)

During this initial phase the native groundwater is displaced by the water injected. Both fluids mix due to dispersion by both the porous medium and the sampling with 1-2 m long well screens. The longitudinal dispersivity, as derived from breakthrough curves, in fact combines both mixing mechanisms and is about 0.5-1 m for most sands. Close to the injection well this mixing will result in typical reactions like the precipitation of ferrihydroxides and manganous oxides, as well as nitrification. These precipitation reactions do not pose clogging problems because (a) clogging was never observed in the beginning of injection and (b) the reactants oxygen and nitrate in the injection water are soon used up by redox reactions making the front of injection water totally anoxic.

Phase 2: cation exchange and sorption (1.5-7 PFs)

Exchange and sorption reactions between influent and aquifer matrix dominate during the second phase. In most cases Na^+ and especially Ca^{2+} expel part of the adsorbed K^+, Mg^{2+}, NH_4^+, Fe^{2+} and Mn^{2+} from the exchangers (reaction 13 in Table 4), while SiO_2 and PO_4^{3-} desorb (reactions 14-15 in Table 4), and F^- adsorbs (reaction 16 in Table 4).
A general order of increasing duration of the exchange or sorption can be given for several constituents, with the PFs within brackets (see also Table 6): Na^+ (1.5-2) < Mg^{2+} (6) < NH_4^+ (7) < SiO_2 (7-8) < K^+ (8) < PO_4^{3-} (14).
Sodium, K^+ and Mg^{2+} do not have other sources and sinks; the abundant silicate minerals are practically inert due to approximate saturation of the influent and the relatively short travel times (maximum 100-500 days). Desorption is the main source of NH_4^+ and PO_4^{3-}. The oxidation of solid phase Natural Organic Matter (NOM) constitutes in some cases a minor source of both, whereas calcite and calciumphosphates may form an additional phosphate source in some cases. Desorbing NH_4^+ can be nitrified by oxygen, and phosphate may sorb to neoformed ironhydroxides.

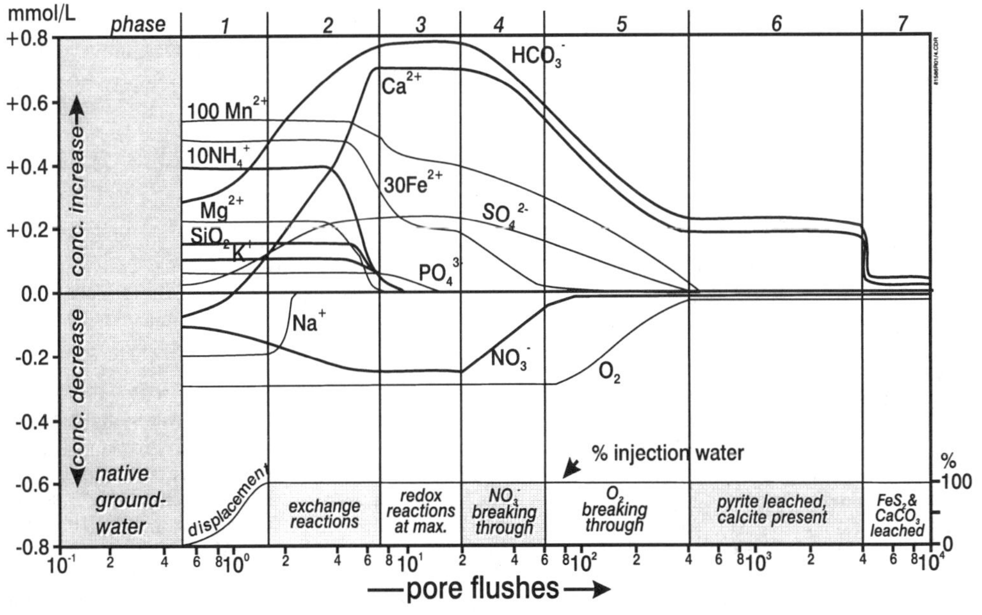

Figure 3. Generalized evolution of the quality changes of oxic injection water in an anoxic, calcareous aquifer, as a function of the number of pore flushes of the soil in between injection and observation well (modified after Stuyfzand 1989). The X-axis also gives the retardation factor for species with a negative change, and the leach factor for species with a positive change.

This apparently speeds up the leaching process. Silica mainly derives from adsorbed SiO_2 (Bassett, in Wood & Signor 1975), but also amorphous SiO_2 (like diatom skeletons) may dissolve especially in marine formations. The behaviour of Fe^{2+} and Mn^{2+} is not dictated by desorption alone but also by redox reactions and possibly the dissolution of a manganous siderite. During phase 2, redox reactions (discussed below) and calcite dissolution rapidly grow to full size. However, in various cases some lag develops probably due to adaptation of the microbiological population, which is needed to catalyze oxidation processes.

Phase 3: redox reactions at maximum (7-20 PFs)

Redox reactions are quantitatively the most important reactions, which also trigger acid buffering reactions like the dissolution of calcite. The maximum size of redox reactions is reached when O_2 and NO_3^- are completely consumed, and when SO_4^{2-}, TIC (= Total Inorganic Carbon = CO_2 + HCO_3^-) and Ca^{2+} (in case $CaCO_3$ is buffering the acid produced) reach their maximum concentration. This moment is in general already reached after 1.5-7 PFs, so that phase 3 may overlap phase 2. The size of both the redox reactions and calcite dissolution declines when nitrate is breaking through in phase 4. Using the mass balance approach the water-aquifer interactions could be simulated very satisfactorily with O_2 mainly reacting with pyrite (mainly reaction 5a, also reaction 5b), and NO_3^- mainly with NOM (largely reaction 8a, reaction 8b in traces). The best lead in quantifying pyrite oxidation is sulphate, which is produced and does not have other sources and sinks. Some nitrate or, only in specific layers low in NOM and high in pyrites, all nitrate is used for pyrite oxidation as well. The main reaction product of NOM-oxidation is HCO_3^-, which is obscured, however, by a much larger HCO_3^- source in case of calcite dissolution (reaction 17 in Table 4), and by a HCO_3^- sink (reaction 21 in Table 4) in case $CaCO_3$ is lacking.

Phase 4: nitrate breaking through (20-60 PFs)

The first break-through of nitrate initiates phase 4, which is characterized by a slowly increasing redox potential in consequence of a steady NO_3^- increase. This is probably connected with the depletion of: (a) the most labile fractions of NOM and, to a smaller degree, pyrite; and (b) mobilized Fe^{2+} (from desorption and pyrite oxidation). As a matter of fact the increasing NO_3^- levels are mirrored by decreasing Fe^{2+} concentrations. This may be partly connected with mixing reaction 4a in Table 4. Also Ca^{2+}, HCO_3^- and Mn^{2+} are declining but less dramatically than Fe^{2+}.

Phase 5: oxygen breaking through (60-400 PFs)

When the front of pyrite leaching is approaching, oxygen will break through which heralds phase 5. The oxygen concentrations do either rise slowly (Fig.3) or more abruptly, probably depending on the width of the pyrite leaching zone. The behaviour of oxygen is typically followed by decreasing SO_4^{2-}, Ca^{2+} and HCO_3^- concentrations. Nitrate removal and iron mobilization may approach zero after about 80 PFs.

Phase 6: pyrite consumed (400-4,000 PFs)

When sulphate stops to increase, pyrite is completely consumed or at least the reactive pyrite particles. This is the start of phase 6. It is calculated for sites 1-8 and also demonstrated by observations, that calcite is still present at this stage. Now the main reactions in the aquifer are dominated by the oxidation of NH_4^+ and DOC in the influent, and by the calcite saturation index of the influent. The latter is assumed in this generalization to be negative, implying a calcite agressive influent. In most cases there still is a less reactive NOM fraction which consumes oxygen or nitrate at a very slow rate.

Phase 7: pyrite/calcite consumed (>4,000 PFs)

At this stage the CO_2 produced by oxidation reactions (NH_4^+ and DOC in the influent and NOM in aquifer matrix) and the negative calcite saturation index of the influent, are not buffered any more by calcite dissolution. This results in minor quality changes during aquifer passage, with a small pH decrease and small increase in TIC.

6 COMPARING SITES 1-11

The water quality changes and evolution patterns at the 11 sites are compared in Table 5. Most conspicuous are the differences in the parameter $\Delta C/\Delta S$, which is the ratio of the mean molar change in HCO_3^- and SO_4^{2-} as observed during phase 3 of aquifer passage. As ΔSO_4 is positive on each site, a positive value indicates HCO_3^- production and a negative value HCO_3^- consumption. Pyrite oxidation by O_2 according to reaction 5a (Table 4) yields in combination with calcite dissolution $\Delta C/\Delta S$ = +2. And pyrite oxidation by NO_3^- according to reaction 6b (Table 4) yields in combination with calcite dissolution $\Delta C/\Delta S$ = +0.5.
Values >+2.5 therefore point at a significant HCO_3^- production by calcite dissolution in connection with a negative calcite saturation index (SI_c) of the influent or denitrification by NOM.

Table 4. Listing of important reactions between on the one hand the injected water and native groundwater (upon mixing), and on the other hand injected water and solid constituents of the target aquifer.

Reactions upon mixing	
$O_2 + 4\ Fe^{2+} + 8\ HCO_3^- + 2\ H_2O$ --> $4\ Fe(OH)_3 + 8\ CO_2$	(1)
$2\ O_2 + NH_4^+ + 2\ HCO_3^-$ --> $NO_3^- + 2\ CO_2 + 3\ H_2O$	(2)
$0.5\ O_2 + Mn^{2+} + 2\ HCO_3^-$ --> $MnO_2 + 2\ CO_2$	(3)
$0,2\ NO_3^- + Fe^{2+} + 1,8\ HCO_3^- + 0,6\ H_2O$ --> $Fe(OH)_3 + 0,1\ N_2 + 1,8\ CO_2$	(4a)
$NO_2^- + 3\ Fe^{2+} + 5\ HCO_3^- + 2\ H_2O$ --> $0,5\ N_2 + 3\ Fe(OH)_3 + 5\ CO_2$	(4b)
Reactions with pyrite and NOM	
$3,75\ O_2 + FeS_2 + 4\ HCO_3^-$ --> $Fe(OH)_3 + 2\ SO_4^{2-} + 4\ CO_2 + 0,5\ H_2O$	(5a)
$3,5\ O_2 + FeS_2 + 2\ HCO_3^-$ --> $Fe^{2+} + 2\ SO_4^{2-} + 2\ CO_2 + H_2O$	(5b)
$2,8\ NO_3^- + FeS_2 + 0,8\ H^+$ --> $Fe^{2+} + 2\ SO_4^{2-} + 1,4\ N_2 + 0,4\ H_2O$	(6a)
$3\ NO_3^- + FeS_2 + HCO_3^- + H_2O$ --> $Fe(OH)_3 + 2\ SO_4^{2-} + 1,5\ N_2 + CO_2$	(6b)
$O_2 + CH_2O$ --> $CO_2 + H_2O$	(7)
$NO_3^- + 1,25\ CH_2O$ --> $0,5\ N_2 + 0,25\ CO_2 + HCO_3^- + 0,75\ H_2O$	(8a)
$NO_3^- + 0,5\ CH_2O$ --> $NO_2^- + 0,5\ CO_2 + 0,5\ H_2O$	(8b)
Other redox reactions	
$MnO_2 + 0,5\ CH_2O + 1,5\ CO_2 + 0,5\ H_2O$ --> $Mn^{2+} + 2\ HCO_3^-$	(9)
$Fe(OH)_3 + 0,25\ CH_2O + 1,75\ CO_2$ --> $Fe^{2+} + 2\ HCO_3^- + 0,75\ H_2O$	(10)
$O_2 + 4\ FeCO_3 + 6\ H_2O$ --> $4\ Fe(OH)_3 + 4\ CO_2$	(11)
$NO_3^- + 5\ FeCO_3 + 8\ H_2O$ --> $5\ Fe(OH)_3 + ½N_2 + HCO_3^- + 4\ CO_2$	(12)
Exchange reactions	
$[Fe,NH_4,Mn,K,Mg]$-EXCH $+ Ca^{2+} + Na^+$ <--> $[Ca,Na]$-EXCH $+ Fe^{2+} + NH_4^+ + Mn^{2+} + K^+ + Mg^{2+}$	(13)
$[H_4SiO_4]$-EXCH + X <--> [X]-EXCH + H_4SiO_4	(14)
$[H_2PO_4]$-EXCH + HCO_3^- <--> $[HCO_3]$-EXCH + $H_2PO_4^-$	(15)
[OH]-EXCH + $F^- + CO_2$ <--> [F]-EXCH + HCO_3^-	(16)
Dissolution/precipitation and pH-buffering	
$CaCO_3 + H_2CO_3$ <--> $Ca^{2+} + 2\ HCO_3^-$	(17)
$Fe_{1-x}Mn_xCO_3 + H_2CO_3$ <--> $1\text{-}x\ Fe^{2+} + x\ Mn^{2+} + 2\ HCO_3^-$	(18)
$SiO_2 + 2H_2O$ --> H_4SiO_4	(19)
$Fe(OH)_3 + x\ H_4SiO_4 + H_2O$ <--> $Si_xFe(OH)_{3+4x}\cdot H_2O$	(20)
$H^+ + HCO_3^-$ <--> H_2CO_3 <--> $CO_2 + H_2O$	(21)

$CaCO_3$ = calcite; CH_2O = organic matter; EXCH = exchanger; $Fe(OH)_3$ = ironhydroxide; $FeCO_3$ = siderite; FeS_2 = pyrite; H_4SiO_4 = dominant silicon species dissolved in water.

Finally, negative ΔC/ΔS values indicate internal acid buffering by HCO_3^- which then converts to CO_2 (reaction 21 in Table 4). Now the ΔC/ΔS values in combination with the SI_c values and calcite content of the aquifer (Table 5) reveal 3 groups:
(a) sites 1-6 and 8 (site 2 with $SI_c = 0.03$) where calcite dissolved in significant amounts;
(b) sites 2 (with $SI_c = 0.53$) and 7 (with $SI_c = 0.34$) where calcite is present but not dissolving due to supersaturation and where ΔC/ΔS is negative; and
(c) sites 9-11 where calcite is lacking and the negative ΔC/ΔS indicates acid buffering by HCO_3^-.
As expected, group a also shows high leach factors (up to >400) for both Ca^{2+} and HCO_3^-, whereas groups b and c display low values (1.5-16).
The parameter $f\Delta SO_4$ is defined as the ratio of the observed change in SO_4^{2-} during phase 3 of aquifer passage and the calculated, potential maximum change of SO_4 (being in mmoles: 2*MOC/15). It is deduced from Table 5 that site 2 has an exceptionally low value (0.1), and that sites 5, 7 (only lower part of aquifer), 9 and 11 (only lower part of aquifer) display a relatively high value (≥0.65), pointing at resp. a low and high capture efficiency of oxidants by pyrite. This is probably connected with differences in the amount and reactivity of labile NOM and labile pyrite. A small size of pyrite crystals and the absence of coatings (especially by NOM) raise $f\Delta SO_4$.
There is an exceptionally fast, low-level breakthrough for nitrate only at sites 2, 10 and 11 ($R_{10\%}$ <2), whereas oxygen is showing there a normal retarded, low-level break-through ($R_{10\%}$ >33 up to >146). This indicates a very low reactivity of the NOM, which in case of sites 10-11 can be attributed to the relatively high age of the Miocene sands as compared to the Pleistocene aquifers on the other sites. The fast nitrate break-through combines with a relatively low leach factor for Fe^{2+} (4-7 compared with 14-≥73).
Sites 2, 9, 10 and probably 7 (not measured) are unique regarding the behaviour of SiO_2: its concentration drops below the relatively high influent level (which is only little lower than in the native groundwater). This excludes the normal sorptive behaviour which is observed elsewhere. It is assumed that SiO_2 coprecipitates with neoformed $Fe(OH)_3$ according to reaction 20 (Table 4). Only at site 2 this SiO_2 removal was overshadowed for a short period by a strong desorption. The sites in the coastal dunes (1-6) display a relatively high leach factor for K^+ (8-27 versus 5-7). This is probably connected with previous salt water flushing of these coastal aquifers, which is also manifested by raised K^+ concentrations in the fresh dune water today.

Table 5. Comparison of the water quality changes c.q. leaching and retardation factors upon injection of oxic water into an anoxic aquifer at 11 sites, for observation points with a travel time of 3-10 days. R_X = retardation factor for $X = t_X/t_{50}$; L_X = leach factor for $X = t_X/t_{50}$, with t_{50} = 50% breakthrough of conservative tracer like Cl^-; t_X = time for 10 or 90% breakthrough of solute X (if influent higher than native groundwater then retardation, otherwise leaching).

site no	site name	SI_c @	$CaCO_3$† %	ΔC/ΔS #	fΔSO_4 ##	R_{O2} 10-90%§	R_{NO3} 10-90%§	L_K 90%	L_{Mg} 90%	L_{NH4} 90%	L_{Fe} 90%	L_{Mn} 90%	L_{SiO2} 90%	L_{PO4} 90%	L_{Ca} 90%	L_{HCO3} 90%
1	Castricum	+0.26	>1	+2.6	0.5	27->102	18-102	14	11	11	32	80	11	15	>140	>140
2	Leiduin	+0.53	5	-3.2	0.1	>14	1->14	>14	9	9	2-13	>14	6	13	5^R	1.5
		+0.03	5	+5.0	0.1	>69	1-69	27	-	-	-	>69	-	-	>69	>69
3	Schev. DIP	+0.20	2.5	+3.1	0.6	140->244	<22-109	-	-	-	38	94	-	-	>244	>244
4	Schev. CIP	-0.23	2.5	+6.3	0.4	<240->400	<40-100	-	-	-	>40	>40	-	-	>400	>400
5	Schev. F/F	-0.56	2.5	+5.7	0.7	49-95	14-83	12	7	5	≥73	≥95	10	24	>95	>95
6	Waalsdorp	-0.33	2.5	+5.0	0.6	>49	>49	8	6	8.5	>49	>49	7	-	>49	>49
7	Langerak	+0.34	0.65	-0.3	0.4/0.9[a]	>61	24->61	$6^{R/L}$	$4^{R/L}$	14	14->57	31->57	-	-	≤3^R	1.5
8	Nieuwegein	-0.02	0.8	+3.8	0.35	24->114	1->114	5	5	7	14	>114	-	-	>114	>114
9	St. Janskl.	+0.17	≤0.002	-1.1	0.7	>95	>26	6^R	5.5^R	5	21	>26	>26^R	14	1.5	1.5
10	Breehei	-0.12	≤0.013	-1.2	0.4	>33	2->15	$7^{R/L}$	$10^{R?L}$	11	4	26	>10^R	-	4^R	1-16
11	Someren	-0.60	<0.02	-1.4	0.4/0.7[a]	>146	1-43	-	≤2	6	7	17	4	-	-	-

@ = calcite saturation index of influent; † = mean calcite content of target aquifer; § = if one value, then for both 10 and 90%.
R = retardation; R/L = retardation and leaching on different levels in aquifer; a: X/Y = (upper + middle)/(lower part of aquifer);
#: ΔC/ΔS = ratio of mean molar change in HCO_3^- and SO_4^{2-} as observed during aquifer passage (ΔSO_4 always positive);
##: fΔSO_4 = ratio of observed SO_4^{2-} change during aquifer passage and the calculated potential maximum SO_4 change (being in mmoles: 2*MOC/15).

7 SOME DETAILS ON PYRITE OXIDATION

Pyrite oxidation which was observed on all test sites, can be considered as the most important water-sediment reaction. This reaction proved to be kinetically hindered on all sites (Fig.4), which corresponds with Nicholson et al. (1988, 1990). It is concluded from Fig.4 that it takes about 10-100 days to use up the O_2 and NO_3^- in the influent for pyrite oxidation, and that 40-100% of the oxidants is actually consumed exclusively by pyrite oxidation. Saaltink et al. (1998) modelled pyrite oxidation for site 7, including reaction kinetics.
Where denitrification by NOM is important, an inverse behaviour of O_2 and ΔSO_4, and of NO_3^- and HCO_3^- was observed.

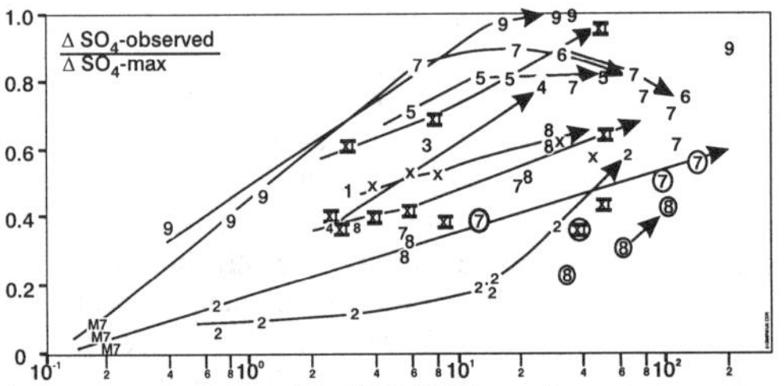

Figure 4. The oxidation of pyrite is a relatively slow process taking about 10-100 days to deplete O_2 and NO_3^- in the influent. The reaction progress variable is the observed sulphate increase (ΔSO_4), which is plotted here for phase 3 (redox reactions at maximum) as the ratio to the maximum potential sulphate increase (ΔSO_4-max = 2*MOC/15). This ratio is denominated fΔSO_4 in Table 5. The numbers indicate the test site, interconnected numbers are situated along a flow path, and encircled numbers are those with an anomalously high denitrification by natural organic matter.

This indicates that oxygen is reacting faster with pyrite than with labile NOM, and that the reverse holds for nitrate. The rate of pyrite oxidation further depends on temperature as demonstrated in Fig.5. When temperature drops below ca. 10°C the reaction slows down considerably.
The oxidation of pyrite leads to some mobilization of As, Co and Ni, as shown in Fig.6. This is consistent with the geochemical composition of pyrite (section 3). Zinc did not mobilize, however, possi-

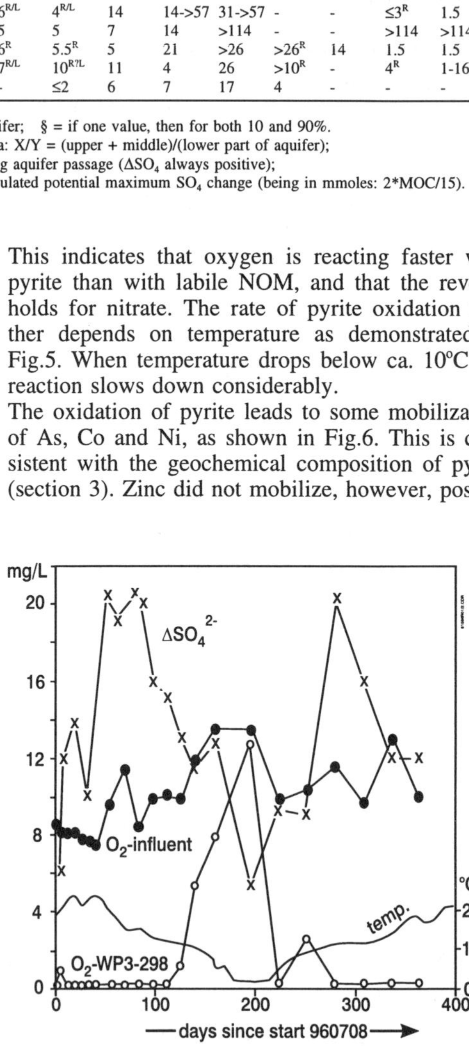

Figure 5. The temperature dependency of pyrite oxidation in observation well WP3-298 at test site 11 (distance to injection well 8 m, travel time 3 days), is manisfested by oxygen and the sulphate increase (ΔSO_4 = $(SO_4^{2-})_{WP3-298}$ -$(SO_4^{2-})_{influent}$).

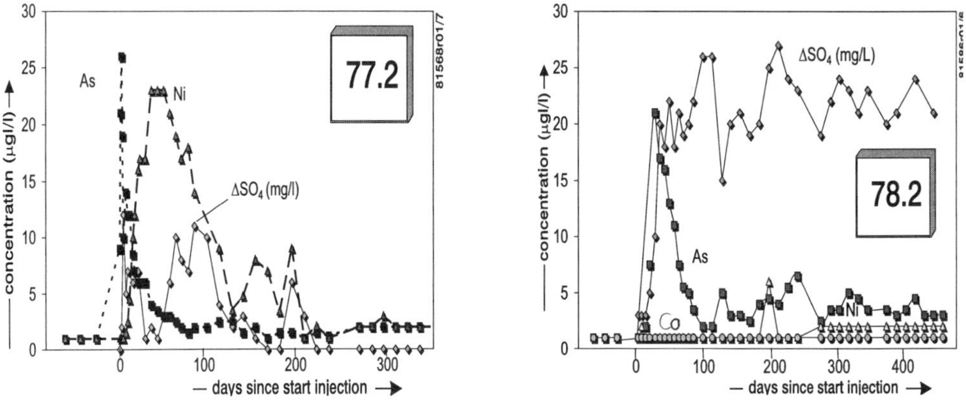

Figure 6. Relation between the observed sulphate increase (ΔSO_4) and the concentrations of As, Co and Ni in observation wells WP.77-60 and WP.78-60 at test site 9 (distance to injection well resp. 3 and 33 m, travel time resp. 0.5 and 21 days).

bly due to a more effective scavenging by neoformed $Fe(OH)_3$. Cobalt and Ni are clearly much less mobile than As which is noticed further away from the injection well. These metals probably coprecipitate with or strongly adsorb to the neoformed $Fe(OH)_3$. Arsenic is initially more mobile due to its reduced state (occurring in water as H_3AsO_3) which prevents it to sorb to $Fe(OH)_3$. After some time the advancing oxygen and nitrate fronts converse the dissolving arsenic into the oxidized state (AsO_4^{3-}) which is much less mobile due to preferential sorption to neoformed $Fe(OH)_3$.

REFERENCES

Appelo, C.A.J., P.J. Stuyfzand & G.B. Engelen 1979. Reactions of Rhine water in the Veluwe and dunes: an experimental study and an injection test. *H_2O* 12 (in dutch), 328-332.

Broers, H.P., J. Griffioen & E.A. Buijs 1994. Relations between groundwater and sediment composition for 5 well sites in N. Brabant. TNO-report OS.94-38-A, 90p.

Brun, A., F.D. Christensen, J.S. Christiansen, P.J. Stuyfzand & H. Timmer 1998. Water quality modelling at the Langerak deep-well recharge site. This volume.

De Ruiter, H. & P.J. Stuyfzand 1998. An experiment on deep well recharge of oxic water into an anoxic aquifer near St. Jansklooster. This volume.

Nicholson, R.V., R.W. Gillham & E.J. Reardon 1988. Pyrite oxidation in carbonate-buffered solution: 1. Experimental kinetics. *Geochim. Cosmochim. Acta* 52, 1077-1085.

Nicholson, R.V., R.W. Gillham & E.J. Reardon 1990. Pyrite oxidation in carbonate-buffered solution: 2. Rate control by oxide-coatings. *Geochim. Cosmochim. Acta* 54, 395-402.

Olsthoorn, T.N. 1982. Clogging of recharge wells. Kiwa-Meded. 71 (in dutch).

Postma, D., C. Boesen, H. Kristiansen & F. Larsen 1991. Nitrate reduction in an unconfined sandy aquifer: water chemistry, reduction processes and geochemical modeling. *Water Resources Research* 27, 2027-2045.

Saaltink, M.W., C. Ayora, P.J. Stuyfzand & H. Timmer 1998. Modelling the effects of deep artificial recharge on groundwater quality. This volume.

Straatman, R. & Y. Brekvoort 1998. Well recharge pilot South-East Netherlands, this volume.

Stuyfzand, P.J. 1989. Comparing artificial recharge using basins and wells on a geohydrochemical basis. *H_2O* 22, 721-728.

Stuyfzand, P.J. 1998. Simple models for reactive transport of pollutants and main constituents during artificial recharge and bank filtration. This volume.

Stuyfzand, P.J. & H. Timmer 1998. Injecting oxic surface water into an anoxic aquifer along the river Rhine; water-rock interactions including redox dynamics. Kiwa-report EU-contract ENV4-CT95-0071, in prep.

Timmer, H. & P.J. Stuyfzand 1998. Deep well recharge in a polder near the Rhine. This volume.

Van Beek, C.G.E.M. & J. Van Puffelen 1987. Changes in the chemical composition of drinking water after well infiltration in an unconsolidated sandy aquifer. *Water Resources Research* 23, 69-76.

Wood, W.W. & D.C. Signor 1975. Geochemical factors affecting artificial groundwater recharge in the unsaturated zone. *Transactions ASAE* 18, 677-683.

Artificial Recharge of Groundwater, Peters et al. (eds) © 1998 Balkema, Rotterdam, ISBN 90 5809 017 5

Well recharge in California: Water quantity and quality considerations

D.Wendell
CH2M Hill, Oakland, Calif., USA

D.Glanzman
CH2M Hill, Denver, USA

ABSTRACT: Artificial recharge projects are of critical importance to California in meeting ever-increasing demands for water. Recharge projects are regulated under strict State standards designed to ensure protection of public health, especially where reclaimed wastewater is used for recharge. Use of ASR wells is relatively new to the state, but is now gaining widespread acceptance. Operation and testing of ASR facilities has produced new information regarding chemical changes in water quality during injection and storage of water. This paper provides an overview of California well recharge projects and related water quality issues.

1 INTRODUCTION

California has the world's seventh largest economy, but several of its major metropolitan centers are located in water-short regions whose people and industries are entirely dependent upon massive statewide water projects for continued prosperity. Conjunctive use projects utilizing artificial recharge are critical in helping the State continue to meet its municipal, industrial, and agricultural water needs. Recharge projects are regulated under strict State standards designed to ensure protection of public health, especially where reclaimed wastewater is used for recharge.

The Los Angeles metropolitan area finds itself in the most severe predicament. This area is home to over 10 million people living in a desert that receives about 28 cm/year of rain. Necessity has required development of innovative water management techniques. Three major canals and conduits import water to Los Angeles: 1) the 500 km long California Aqueduct; 2) the 350 km long Colorado River Aqueduct; and, 3) the twin barrel Los Angeles Aqueducts that reach 250 to 350 km up into the Owens and Mono Valley areas (Fig. 1). Seasonal artificial recharge of import water and intensive reuse and recharge of reclaimed wastewater have become key water management techniques. Artificial recharge of import water via surface spreading basins has been practiced in the Los Angeles area since 1952, and recharge of reclaimed wastewater since 1962.

Beginning in the early 1950's, a series of recharge wells were built along coastal portions of Los Angeles to protect large inland groundwater basins from seawater intrusion. The seawater injection barriers currently consist of about 250 wells distributed along four separate barriers that have a total length of 30 km. These recharge wells originally used treated drinking water for injection, but more recently reclaimed water that has undergone reverse osmosis treatment has been blended with import water and used for injection. This has been the practice at Orange County's Talbert Barrier since 1975, at Los Angeles County's West Coast Basin Barrier since 1995, and is planned to begin at the Los Alamitos Barrier in the year 2000. These projects use percentage blends of reclaimed/potable water that vary between 50/50 and 66/44. Several projects are now under construction to inject reclaimed/RO treated water into closed (i.e. non-coastal) groundwater basins in the state, where the water will be used to augment local supplies and decrease discharges of effluent to sensitive surface water bodies.

Use of dual-purpose Aquifer Storage and Recovery (ASR) wells is relatively new to the state, but is now gaining widespread acceptance (Table 1 and Figure 1). Recent advances in solving well clogging account for this growing acceptance. The Goleta Water District has operated an ASR system since 1975. Since 1990, five major ASR systems have been brought on line. The largest of these is operated by Calleguas Municipal Water District, which currently has one active ASR well that was used for demonstration purposes; eight ASR wells built but not yet outfitted and connected to the distribution system; and another 10 wells under construction. The ultimate project will consist of as many as 40 ASR wells recharging an average of 125

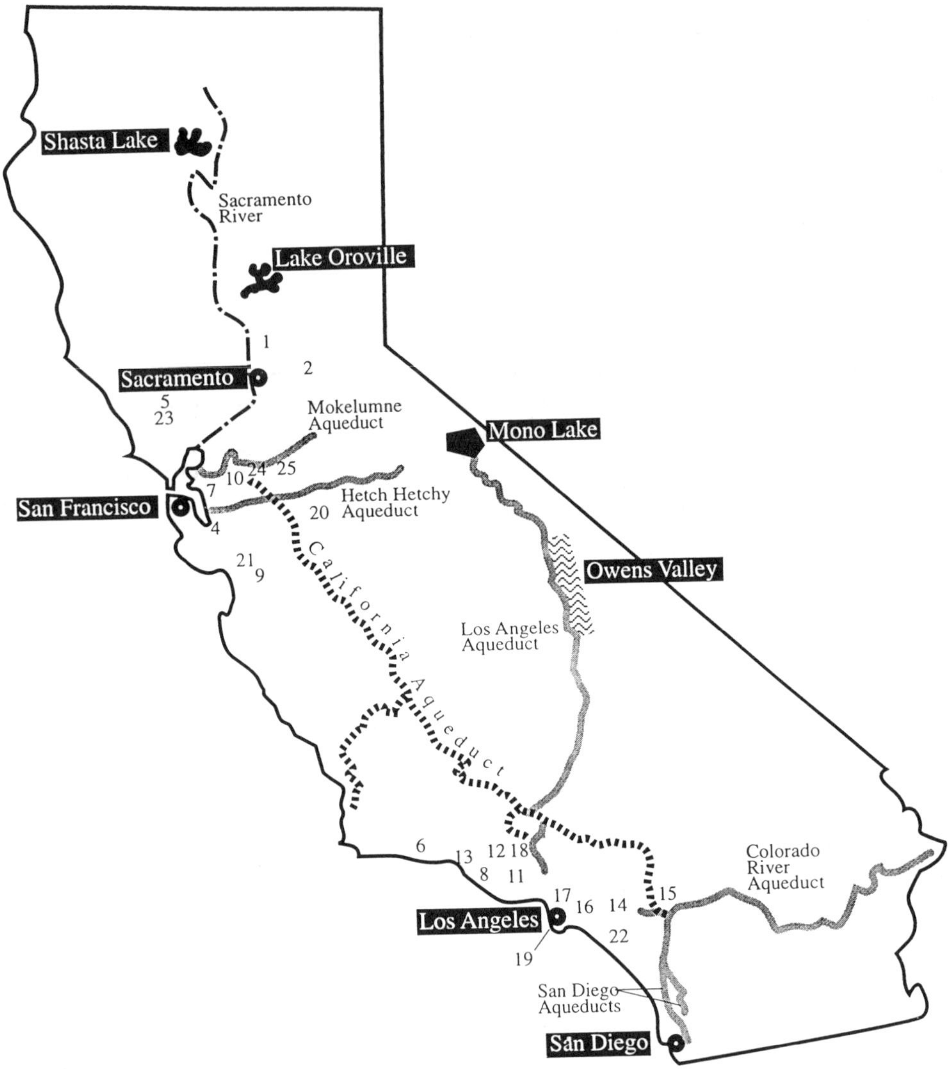

Figure 1. Location Map. Numbers are keyed to Table 1.

million cubic meters of water per year.

The goal of all the above projects is to increase the operational yield of groundwater supplies via artificial recharge of surplus winter runoff and/or reclaimed water while maintaining high quality water supplies that meet public health standards. Operation and testing of these facilities has produced invaluable information regarding chemical changes in water quality during injection and storage of water.

2 WATER QUALITY

2.1 *Trihalomethanes (THMs)*

Water supplies in the U.S. are commonly disinfected using chlorine although water retailers are now switching to chloramine and ozonation disinfection methods. Reaction between free chlorine and naturally-occurring organic compounds in the recharge water lead to formation of THMs, which are classified as probable human carcinogens. Accordingly, response of THMs to ASR and injection operations has received much attention.

Concentrations of THMs (after accounting for dilution) at the Calleguas project were observed to decrease by about 35 to 70 percent. At the Pasadena site, however, the concentrations of THMs were actually observed to increase shortly after injection, indicating continued reaction between the chlorinated water and subsurface organic matter.

Differences in THM response may be due to redox conditions of groundwater, the major form of THM present (chloroform versus bromo-THMs), and the type of disinfectant used (chlorine versus chloramine). For example, the Calleguas project used chlorinated water injected into a confined aquifer that is under highly reducing conditions, whereas the Pasadena project used chloraminated water injected into an unconfined aquifer saturated with oxygen. It is notable that at the Calleguas project the concentration of chloroform actually increased during injection and storage, but this increase was more than offset by a decrease in brominated forms of THM. It is thought that because chlorine is more reactive than chloramine, and has been observed to react and essentially disappear within hours of injection, that it may have less THM formation potential after injection relative to chloramine.

2.2 *Iron*

The concentration of iron in groundwater at the Calleguas project was about 250 ug/L prior to injection, which is slightly below the allowable regulatory level ("maximum concentration level" or "MCL") of 300 ppb. Test injection operations were carried out for 40 days followed by extraction of 150 percent of the volume injected. After accounting for mixing effects the observed concentration of iron in recovered water was about 10 ug/L. Three years later iron concentrations were still only about 50 ug/L after this single injection event.

2.3 *Manganese and Arsenic*

An ASR well has recently been completed in an aquifer containing elevated concentrations of manganese (100 ug/L; MCL = 50 ug/L) and arsenic (20 ug/L; MCL = 50 ug/L), but low concentrations of iron (30 ug/L). In an attempt to provide insitu treatment of these metals super chlorinated water containing 50 mg/L free chlorine was injected into the aquifer to oxidize the aquifer and enhance precipitation of manganese and arsenic. A one-week long ASR cycle was conducted and demonstrated decreases in manganese concentrations to 60 ug/L, but little or no decrease in arsenic. It is thought that if iron concentrations were higher, that arsenic might have been more readily removed via adsorption. Additional future ASR cycles may be conducted to further test benefits of insitu treatment, possibly including addition of stronger oxidizing agents.

2.4 *Mercury*

Apparent elevated concentrations of mercury have been observed at three ASR sites in California (Pasadena, Foothill, and Zone 7) following injection and storage. This is one of the few apparent adverse water quality effects noted to date. Water produced during the initial phases of recovery contained as much as 3.9 mg/L mercury (MCL = 2 ug/L) at the Pasadena site. Mercury concentrations at the other sites remained below MCLs. The reason for these elevated mercury concentrations is not currently understood, but is a transitory event that might be laboratory related. This issue is currently being evaluated.

2.5 *Insitu Blending*

The Foothill project uses two ASR wells and two additional production wells to perform insitu blending to mitigate high concentrations of nitrate and PCE in groundwater. This mode of operation allows the water agency to purchase less expensive water during wet winter months and decrease purchase of higher priced blend water during peak demand summer months. Testing over one year demonstrated the ability to blend down nitrate in the wellfield. A wellfield simulation model was used to predict nitrate concentrations under various operations alternatives. PCE, THMs, and mercury were all elevated beyond concentrations predicted through simple mixing. It is believed that PCE present in "dead" pore space was remobilized in the unique hydraulic regime generated by ASR operations. The increase in THMs and mercury is not clearly understood, but were both below drinking water standards.

3 RECLAIMED WATER

Potential health affects arising from injection of reclaimed water are addressed by stringent regulation of recharge water quality and percent reclaimed water produced by surrounding potable water supply wells (Table 2). All reclaimed water that is injected through wells is required to undergo organics removal (via reverse osmosis or granular activated carbon) prior to injection. Injected water may contain no more than 2 mg/L of total organic carbon (TOC) of reclaimed water origin, and nearby potable wells may produce no more than 50 percent reclaimed water. The intent of these rules is to

Table 1. Summary Data For California ASR and Well Recharge Programs

	Agency	Status**	Number of Wells	Inception Date	Average Recharge Rate (lpm/well)	Average Production Rate (lpm/well)	Specific Capacity of Production (lpm/m)	Specific Capacity of Injection (lpm/m)	Transmissivity (m2/d)		Source of Water
1	City of Roseville	D	0	1995	-	3800	740		370		Local potable water
2	Private Developer	S	0	1995	-	3800					Local potable water
3	Monterey Peninsula WMD	T	1	1996	-	1900					Local potable water
4	Santa Clara Valley WD	T	1	1996	1100	1900			120	310	Local potable water
5	Sonoma CWA	T	1	1993	2200	5700					Local potable water
6	Goleta Water District	O	9	1978	1500	900	190	50	190	500	Import potable water
7	East Bay MUD	D	1	1997	-	1900					Import potable water
8	Camrosa WD	D	0	1995	-	2800					State Water Project
9	San Benito WD	D	1	1997	-	1900					State Water Project
10	Zone 7	O	1	1997	7900	15100	1120	500	1240	3730	State Water Project
11	City of Camarillo	T	1	1994	2300	4200					State Water Project
12	Ventura County WW	T	1	1991	2300	2600	270	120	880	1180	State Water Project
13	City of Oxnard	O	5	1990	5700	7600			2490	3730	State Water Project
14	Chino Basin MWD	S	0	1995	-	3800					SWP & Colorado River
15	Eastern MWD	S	0	1992	-	2800					SWP & Colorado River
16	City of Pasadena	O	2	1992	6800	7600	250	190	930	2490	SWP & Colorado River
17	Foothill Municipal WD	O	2	1992	2600	3000	370	310	820	1860	SWP & Colorado River
18	Calleguas Municipal WD	O	15	1990	2300	3800	250	150	820	1240	SWP & Colorado River
19	Los Angeles Co. DPW*	O	210	1950	900	0			370	3110	Wastewater/import water
20	East Bay MUD	S	0	1994	-	3800					Local non-potable water
21	Pajaro Valley WMD	T	1	1995	1100	3000			340	390	Local non-potable water
22	Orange Co. WD (Talbert)	O	23	1970	800	0					Wastewater/groundwater
23	City of Santa Rosa	S	0	1995	-	1900					Wastewater
24	Dublin-San Ramon SD*	D	2	1998	1000	0			620	1240	Wastewater
25	Livermore	D	1	1998	2800	0			620	1240	Wastewater
26	Monterey Regional WPCA	T	1	1992	-	1900					Wastewater

* All wells are injection wells, not ASR wells.

** Status: S = feasibility study; D = design; T = tested; O = operational facilities

Table 2. Summary of reclaimed water regulatory requirements

	Surface spreading		Direct Injection	
Treatment Process:				
Primary	X	X	X	
Secundary	X	X	X	
Filtration	X	X	X	
Disinfection	X	X	X	
Organic removal	X		X	
Minimum Depth to Groundwater	3	3	-	Meters
Minimum Retention Time	6	6	12	Months
Minimum Horizontal Separation from Domestic Well	150	150	610	Meters
Maximum Amount of Reclaimed Water in Domestic Well	50	20	50	Percent
Maximum Allowable TOC after Organics Removal				
Maximum reclaimed water in domestic well = 0 to 20 percent	20	-	5	
Maximum reclaimed water in domestic well = 46 to 50 percent	6	-	2	
Maximum Nitrogen Concentration in Reclaimed Water Effluent	10	10	10	mg N/l

ensure that potable water supplies contain no more than 1 mg/L TOC of reclaimed water origin. Injected water must have a minimum residence time of one year prior to being extracted. Ensuring compliance with regulations often requires groundwater modeling.

Two projects in Southern California (West Basin and Talbert Barriers) currently utilize blends of reclaimed water and potable water for injection. Two reclaimed water injection projects in Northern California (Livermore and Dublin-San Ramon) are currently under construction. These Northern California projects are unique in that the water is being injected into a closed groundwater basin not affected by seawater intrusion. In all these projects blending at the wellhead is required to ensure that the 50 percent rule is met at nearby domestic production wells. All these projects hope to obtain permission to inject 100 percent reclaimed water in the future.

4 CONCLUSIONS

Operation and testing of well recharge facilities has produced valuable information regarding chemical changes in water quality during injection and storage of water. Recovered water from these projects is typically of very high quality, and water quality changes typically beneficial. Beneficial effects of recharge include decreased concentrations of THMs, iron, manganese, and nitrate. Adverse effects may include local increases in THMs, possible increases in mercury, and flushing of contaminants from pore spaces. In all cases drinking water standards have been met either because adverse effects were minimal, or wellhead treatment was already in place.

Artificial Recharge of Groundwater, Peters et al. (eds) © 1998 Balkema, Rotterdam, ISBN 90 5809 017 5

An experiment on well recharge of oxic water into an anoxic aquifer

Hannie de Ruiter
WMO, Zwolle, Netherlands

Pieter J. Stuyfzand
Kiwa Research and Consultancy, Nieuwegein, Netherlands

ABSTRACT: The Drinking Water Company for the Dutch province of Overijssel (WMO) is investigating well recharge of semi-purified surface water at a future production plant Leeuwterveld. At a nearby existing conventional well field an injection experiment was carried out using one of the pumping wells for injection and another, at a distance of 85 m, as a recovery well. Two observation wells were placed between these two wells, each with 3 observation screens. For a period of 1.5 years oxic drinking water was injected into the anoxic aquifer. Chemical analysis of water samples shows that the injected water underwent reduction due to the oxidation of pyrite. The sequence of the release of heavy metals from pyrite, adsorption processes and cation exchange is demonstrated in graphs. The retardation of several parameters was calculated using natural tracers.

1 INTRODUCTION

The Drinking Water Company for the Dutch province of Overijssel (WMO)is investigating the possibility of implementing well recharge with semi-purified surface water at the future production plant Leeuwterveld (Fig. 1). In order to gain insight into potential problems, we carried out a recharge experiment. The purpose of this experiment was to gain insight in the character, the sequence and the time scale of changes to the water quality of the injection water on its way to the recovery wells. Other aims were to gain insight into the attainable injection rate and in potential problems of clogging. For the experiment we chose the nearby St Jans-klooster production plant which is hydrologically and chemically similar to the future recharge area. The experiment started in April 1996 and is still in progress. This paper presents the information collected up until June 1997 and focuses on hydro-chemical aspects.

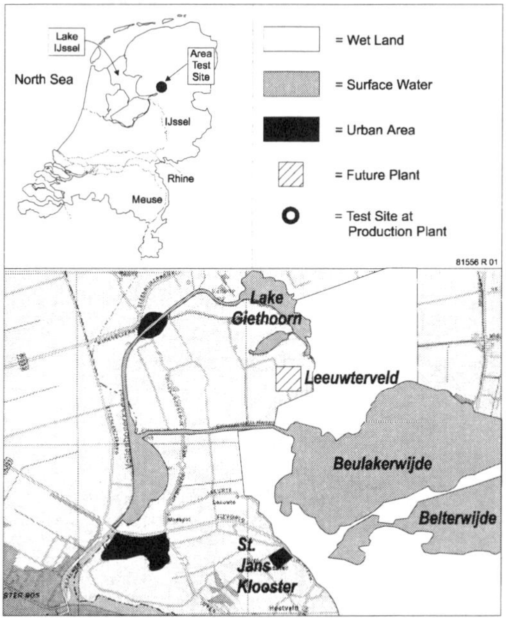

Figure 1. Situation of the present and future drinking water production plants and the test site

2 SET UP OF THE EXPERIMENT

2.1 *St Jansklooster as a test site*

The reasons for selecting St Jansklooster as a test site are: (1) The borehole descriptions of both sites show great hydrological similarity. (2) The quality of the water to be injected at the future production plant is almost identical to that of the drinking water produced by the St Jansklooster production plant. (3) The existing recovery wells and infrastructure of the production plant could be used, keeping the costs of the experiment low. (4) Two wells were especially suitable as they were situated in a favourable part of the flow domain, where the injection water could be recovered almost exclusively by one well.

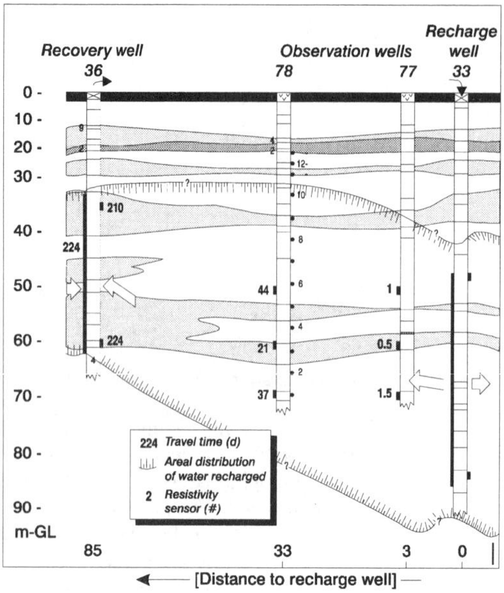

Figure 2. Set up of the test site with hydrogeological schematisation and the extent of the injected drinking water after 14 months (June 1997) with an indication of the mean travel times

2.2 *Set up of the experiment*

Injection started in April 1996 and stopped in December 1997. Drinking water from the St Jansklooster production plant was used. At the beginning of the experiment the injection rate was 30 m^3/h. After 3 months the rate was increased to 60-m^3/h to see what effect this would have. The extraction rate was kept fairly constant at 125 m^3/h throughout the experiment. The set up of the observation facilities is shown in Figure 2. Samples were taken just before the start and during the whole period of injection from the injected water, all 1.0 m screens and the total recovery water. The sampling frequency started off at once a week and was reduced during the experiment to once a fortnight.

The screens of the recovery well will be sampled for another half year after the injection stop. The samples were analysed for the main constituents using conventional techniques and for 7 trace elements using an ICP-MS cq AAS graphite furnace. Other regular measurements throughout the injection period comprised hydraulic heads at several depths and a weekly scan of the electrical resistivity sensors. The measured changes in their resistance are mainly due to the changes in the electrical conductivity of the groundwater which can be used as a multilevel tracer device. Halfway through the experiment we installed 5 parallel filters in a laboratory filled with samples from the aquifer. Two filters were used to simulate the clogging of the well and the other three to obtain insight into water-aquifer reactions during extremely short contact times.

3 HYDRO-GEOLOGICAL, GEO-CHEMICAL AND HYDRO-CHEMICAL ENVIRONMENT

The hydro-geological structure is shown in Figure 2. The recharged aquifer is composed of alternate coarse and fine-grained sand layers. The fine-grained layers are of limited horizontal extent. A peat layer confines the aquifer. Geo-chemically the composition of the aquifer is anoxic, without $CaCO_3$ and low in organic matter content. It contains some pyrite with traces of Co, Ni, Zn and As. At the adsorption complex we find significant amounts of NH_4^+, Fe^{2+} and Mn^{2+} though dominated by Ca^{2+}. CEC is rather low and we find no iron and manganese oxides. The native groundwater consists of surface water that has infiltrated through the anoxic bottom layers of the surrounding lakes and marshes (Fig. 1). As a result

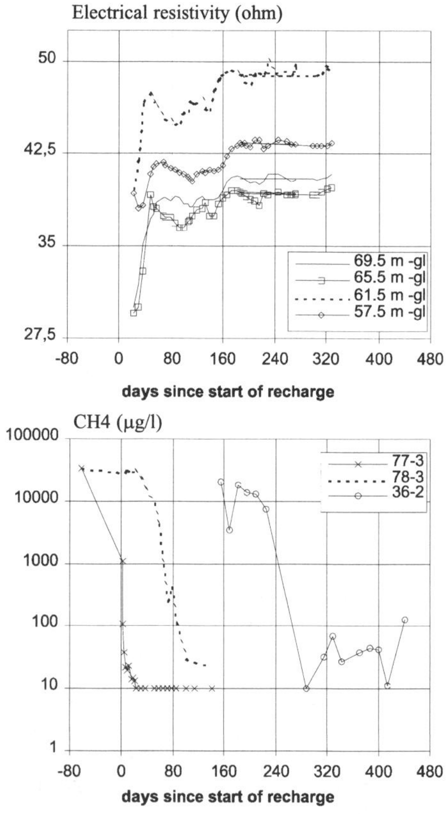

Figure 3. Electrical resistivity using permanent sensors in observation well 78 (at 33 m from the injection well) and methane concentration of the injected water and samples from the deepest filters of the observation wells 77, 78 and recovery well 36

this groundwater is deeply anoxic, high in HCO_3^-, Ca^{2+}, CH_4, NH_4^+ PO_4^{3-}, organic compounds and EC but it lacks SO_4^{2-} and NO_3^-.

4 RECOGNITION AND TRAVEL TIME OF THE INJECTED DRINKING WATER

The treatment of drinking water causes the injected drinking water to differ from the native groundwater (which is the source of drinking water). The treatment consists mainly of aeration in order to remove methane, iron and manganese. As a result we find O_2 and NO_3^- instead of NH_4, lower HCO_3^- concentrations, lower EC and no CH_4, Fe^{2+} and Mn^{2+}. Apart from these general differences we find some hydrochemical stratification in the aquifer. The groundwater in the upper part of the aquifer has lower concentrations of DOC and Cl^- than the injected water, whilst the groundwater in the bottom part has higher Cl^- concentrations. We used these parameters together with measurements of resistivity as tracers, to be able to recognise the drinking water and to calculate travel times (Fig. 2 and 3). From these calculated travel times we could discern the flow rates at different depths.

5 WATER QUALITY CHANGES ALONG THE FLOW PATH

The most important reaction that was seen to affect the quality of the injected water is pyrite oxidation. This reaction influences the redox parameters and trace elements. The changes in these parameters are discussed in this chapter. In table 1 the quality changes of injected drinking water are shown along the deepest flow path. Because the travel times of the injected water vary greatly, both horizontally and vertically, the chemical processes are compared according to the number of 'pore flushes' or bed volumes. This is the number of times that the injected water has passed through the aquifer at a particular point. After 2-6 'pore flushes' the injected drinking water has replaced the original groundwater for 100% but all soil reactions are still active.

5.1 *Redox parameters*

The following parameters play an active role in redox reactions: O_2, NO_3^-, NO_2^-, SO_4^{2-}, CH_4, Fe^{2+}, Mn^{2+} and NH_4^+ (Fig. 4 and 5). Oxidation of pyrite plays the leading role, shown by diminishing O_2 and NO_3^- and increasing SO_4^{2}. At the same time cation exchange restores equilibrium by release of Fe^{2+}, Mn^{2+} and NH_4^+ against capture of Ca^{2+}. Oxidation of organic matter does not seem to play an important role, as there has not been an increase of TIC (Total

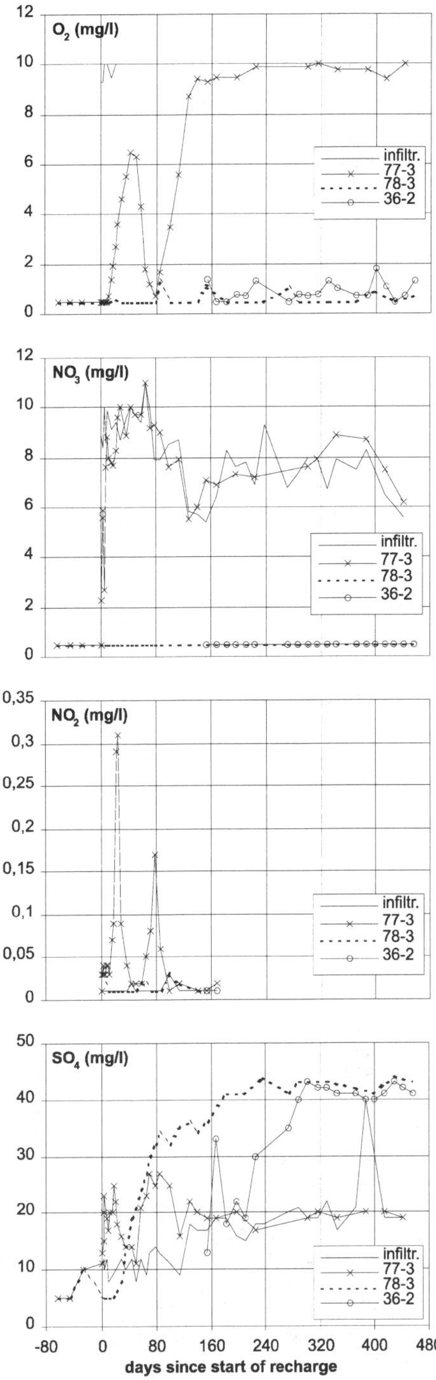

Figure 4. Concentration of 4 redox-parameters (oxygenated species) of the injected water and samples from the deepest observation screen of wells 77, 78 and 36 (at 3, 33 and 85 metres resp. 1, 37 and 224 days) from the injection well

Table 1. Mean quality of injected drinking water and mean quality at the deepest observation screens after 2-6 pore flushes (for well 36 after 2.1 pore flushes)

well nr		33	77	78	36
depth	m-GL	0	-70	-70	-60
distance	m	0	3	33	85
travel time	d	-	1.5	37	224
PF's	-	-	2-6	2-6	2.1
EC	µS/m	499	498	484	437
pH	-	7.51	7.39	7.19	7.27
O_2	mg/l	9.9	0.5	0.7	0.8
CH_4	mg/l	0.01	0.05	0.12	0.13
DOC	mg/l	7.2	7.0	6.1	5.3
Cl^-	mg/l	37.6	33	38.7	39
SO_4^{2-}	mg/l	10.8	19.8	35.7	42
HCO_3^-	mg/l	265	247	232	218
NO_3^-	mg/l	9.2	6.2	0.5	0.5
NO_2^-	mg/l	0.01	0.035	0.015	-
PO_4^{3-}	mg/l	0.09	-	0.32	0.23
Na^+	mg/l	20	19.8	20.2	19.7
K^+	mg/l	3.2	2.2	1.5	1.9
Ca^{2+}	mg/l	85.2	80.8	83.6	80
Mg^{2+}	mg/l	5.9	5.0	5.2	5.7
Fe^{2+}	mg/l	0.10	1.32	4.0	3.1
Mn^{2+}	mg/l	0.01	0.09	0.16	0.13
NH_4^+	mg/l	0.06	0.73	1.3	0.49
As	µg/l	1	7	5	3
Co	µg/l	1.4	1	< 2.3	2
Ni	µg/l	1.4	1.1	< 2.3	2
Zn	µg/l	23	9	9	2
$SI_{calcite}$	-	0.17	-0.02	-0.24	-0.18
TIC	mmol/l	4.69	4.50	4.57	4.12

Inorganic Carbon) while the increase of SO_4^{2-} definitely points to pyrite oxidation. Around the injection well, at short travel distances, the injected water shows for some time a lack of equilibrium. Measurable amounts of O_2 and NO_3^- are found together with Fe^{2+}, Mn^{2+}, NO_2^- and NH_4^+ in all screens of observation well 77 at 3 m from the injection well. This shows that thermodynamic equilibrium at that moment has not yet been reached. It takes about 15 days to reach equilibrium. It took 224 days, according to the concentrations of O_2, NO_3 and SO_4^{2-} to consume all of the pyrite during the first 3 m of aquifer passage (observation well 77). After about 5-11 days all Fe^{2+} and Mn^{2+} has been flushed out at observation well 77, but flushing at the other wells is not yet complete. The flushing of NH_4^+ is faster than that of Fe^{2+} and Mn^{2+} in all wells.

5.2 *$CaCO_3$-CO_2 parameters*

The following parameters play an active role in $CaCO_3$-CO_2 reactions: pH, Ca^{2+}, HCO_3^--, SI (calcite Saturation Index) and TIC (Total Inorganic Carbon). Due to a lack of $CaCO_3$ and readily degradable organic matter in the recharged aquifer, all observation screens show no increase of Ca^{2+}, HCO_3^- and TIC. On the contrary, pH, Ca^{2+}, HCO_3^-, TIC and SI decrease for a long time on their way to the extraction well. At observation well 77 the decrease in Ca^{2+} stops first (after 10 days) whilst the decrease in pH, HCO_3^-, TIC and SI stops when SO_4^{2-} production stops. The decrease in Ca^{2+} is caused by cation exchange with NH_4^+, Fe^{2+} and Mn^{2+},

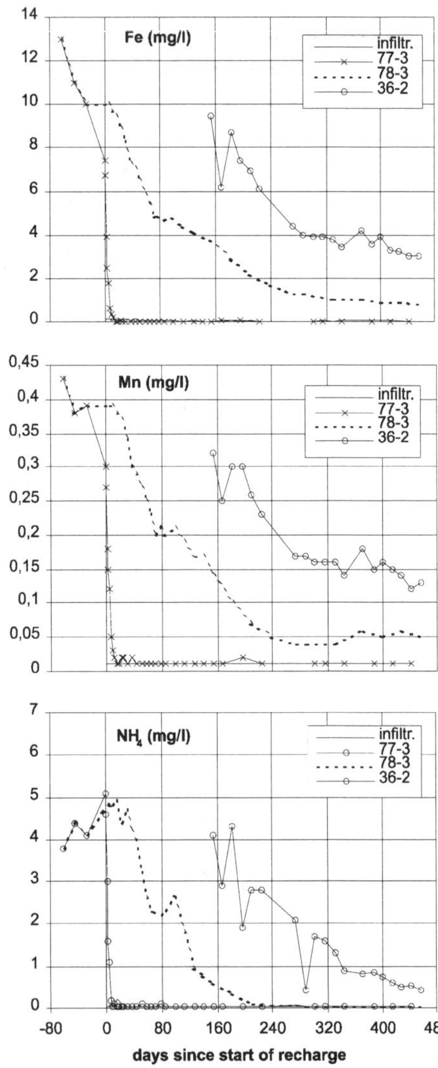

Figure 5. Concentration of 3 redox-parameters (reduced species) of the injected water and samples from the deepest observation screen of wells 77, 78 and 36 (at 3, 33 and 85 metres resp. 1, 37 and 224 days) from the injection well

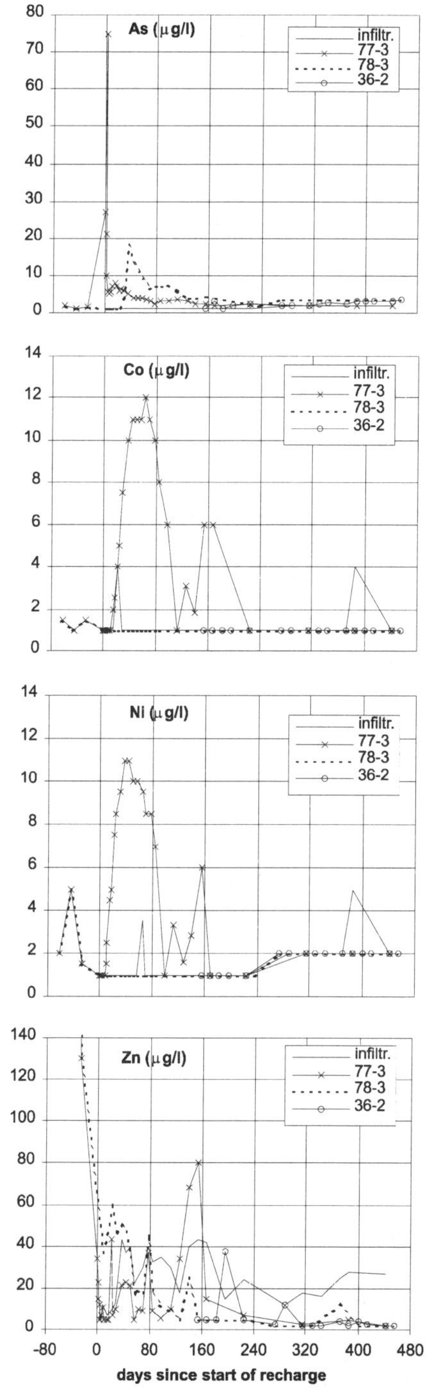

Figure 6. Concentration of 4 trace elements released from pyrite during aquifer passage. Shown are the injected drinking water and the samples from the deepest observation screens of the wells 77, 78 and 36 (at 3, 33 and 85 metres resp. 1, 37 and 224 days) from the injection well.

whilst the decrease in pH, HCO_3^-, TIC and SI is caused by internal pH buffering. Acid produced during pyrite oxidation is neutralised by the conversion of HCO_3^- into CO_2. TIC should not change by this reaction but it clearly does. This is probably caused by the escape of CO_2 during or after sampling.

5.3 *Trace elements connected with pyrite*

Trace elements that were detected in pyrite in significant amounts are: As, Co, Ni and Zn. The concentration of these elements increases to some tens of μg/l during aquifer passage (Fig. 6). They are clearly connected to pyrite as (1) their increase stops as soon as SO_4^{2-} production (a product of pyrite oxidation) stops; (2) their concentration peaks follow SO_4^{2-}; and (3) their concentration peaks approach the calculated pyrite-bound content as analysed from aquifer soil samples. Only arsenic has reached the recovery well up to date with its front flattened. Co, Ni and Zn got stuck between the observation wells 77 and 78, where especially freshly formed iron (hydr)oxides (a product from pyrite oxidation) adsorb them. Also Arsenic as AsO_4^{3-} is adsorbed by these oxides, but 10 % (as H_3AsO_3) escapes adsorption thanks to its lack of charge. That part reaches the recovery well at the same time as SO_4^{2-}. The retardation factor of Co, Ni and Zn is about 20 - 100 times that of SO_4^{2-} and arsenic because of adsorption. The reason is that 2^+ charged cations have stronger adsorption bonds. Therefore Co, Ni and Zn are not expected in this experiment to peak in the extraction well.

5.4 *Other parameters*

Of course Cl^- shows fully conservative behaviour and Na^+ is almost conservative. K^+ and Mg^{2+} however, are subject to cation exchange processes. They are being adsorbed after 5-10 days at observation well 77 and after 110-224 days at observation well 78. DOC (Dissolved Organic Carbon) decreases asymptotically along its flow path. Only about 1,1 mg C/l of the mean of the injected water, 6,4 mg C/l, seems to be decomposable. This means that the decomposition of DOC takes time.

CONCLUSIONS

The results of this recharge experiment have so far yielded many answers to the questions about the changes to the quality of the injection water on its way to the recovery wells. We gained insight into the character and sequence of the changes. As to the time scale, further calculations are planned using the mass balance approach and the model EASY-LEACHER (St, this volume). As for the clogging around the injection well, mixing of oxic and anoxic water might have resulted in precipitation of (hydr)oxides around the well. Its volume can be neglected however, as doubling the injection rate did not influence the injection cone very much. This leads to the conclusion that clogging is not a serious problem around the injection well. Around the recovery well however, we do foresee clogging as a result of mixing. Anoxic and oxic water will be extracted at the same time once oxic water breaks through. We will be able to quantify and to set a time scale when the results from the parallel filters, the flowmodel and EASY LEACHER are available. We are confident that the results of the experiment can be extrapolated to the future production plant Leeuwterveld because of the similarity of the two sites. So far the recharge of semi-purified surface water in the future production plant Leeuwterveld looks promising!

REFERENCES

Stuyfzand, P.J. & J.C. de Ruiter 1997. Kwaliteitsverandering van drinkwater bij diep-infiltratie te St. Jansklooster. Nieuwegein: Kiwa (in dutch)

Horn, P. van den & J.C. de Ruiter 1996. Infiltratieproef Sint Jansklooster: Onderzoeksplan. Zwolle: WMO (in dutch)

Stuyfzand, P.J. 1998. Simple models for reactive transport of pollutants and main constituents during artificial recharge and bankfiltration: infomi 3.1 and easy-leacher. (this volume).

Artificial Recharge of Groundwater, Peters et al. (eds) © *1998 Balkema, Rotterdam, ISBN 90 5809 017 5*

Water quality modelling at the Langerak deep-well recharge site

Adam Brun, Flemming Damgaard Christensen & Jesper Skovdal Christiansen
VKI, Horsholm, Denmark

Pieter J. Stuyfzand
Kiwa Research and Consultancy, Nieuwegein, Netherlands

Harrie Timmer
Water Supply Zuid-Holland Oost, Gouda, Netherlands

ABSTRACT: This paper describes the application of a multi-species reactive transport model to a closely monitored deep-well artificial recharge test site at Langerak, The Netherlands. Using default parameter values for the processes and the measured native and recharge water quality parameters, it is possible to simulate the developments in the water quality very precisely.

1 INTRODUCTION

The water supply company South-Holland East (WZHO) supplies drinking water in the Rhine fluvial plain in the area between Amsterdam, Rotterdam, and Utrecht in the Netherlands. The main source of drinking water supply comes from bank-infiltrated Rhine River water that requires advanced and expensive purification techniques. At Langerak, The Netherlands, WZHO is investigating the use of a natural aquifer as an optional treatment plant for pre-treated river water. The main objectives for WZHO is to investigate clogging effects and changes in the water quality during passage in the aquifer between an injection and a recovery well. The development of the content of trace metals is also in focus. The investigation is based on water quality observations from several multilevel wells and experiments conducted on the aquifer sediment. The data collection and experiments conducted were explicitly designed to enable a quantitative analysis of the change in native water quality due to the injection.

The quality of the water that is withdrawn from the aquifer depends on the quality of the native groundwater, the characteristics of the sediment and the quality and quantity of the injected water. Apart from the transport of solutes owing to the mixing of waters, several reactive processes also have to be accounted for in explaining the development of the water quality. The differences between the two solutions in oxidation level and inorganic cation concentration and the presence of various minerals in the aquifer will strongly affect the final quality of the water withdrawn in the recovery well. The following possible reactions are implicated: Flow and the related transport of solutes, redox processes, mineral dissolution/precipitation, complexation, and ion exchange.

A way to handle these processes quantitatively is in the framework of a reactive transport model, which introduces the need for an accurate process formulation and the geometric distribution of the process parameters. Numerous chemical analyses are required, making it expensive to establish the distribution. The model application in this paper will try to evaluate the use of reactive transport models for describing and predicting the water quality changes at the Langerak site, predominantly using default parameter values known from geochemical model databases.

2 LANGERAK INJECTION SITE

Two wells, an injection and a recovery well, distanced by 190 m constitute the test field, see Figure 1.

Between these wells, three monitoring wells are situated (WP1, WP2, and WP3) at distances of 20, 40, and 91 m from the injection well, respectively. Within each monitoring well, it is possible to conduct multilevel sampling from three to six differently elevated screens.

2.1 *Recharge water*

The injection experiment at the Langerak site took place during the period July 22, 1996 to July 22, 1997, one year in all. The water used for the recharge injection was local groundwater pre-treated by aeration and rapid sand filtering. Since the plan

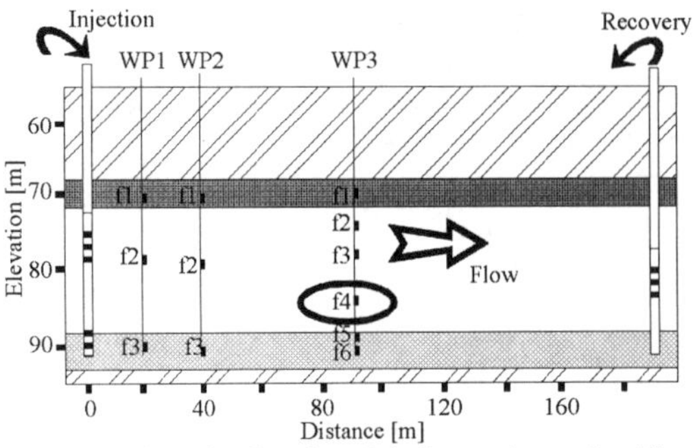

Figure 1. Schematic of section of the Langerak test site. After Stuyfzand & Timmer 1998.

for future production is injection of pre-treated Rhine water, $NaNO_3$ was been added continuously, increasing the oxidation capacity of the injection water to allow it to become comparable to pre-treated Rhine water.

During the first month of the experiment, NaCl was added for tracer purposes in the attempt to determine the groundwater velocity and solute dispersion (Stuyfzand & Timmer 1998). This initial NaCl pulse was injected during the period July 22 - August 22, 1996 and was succeeded by injection of plain water.

The mean of the measured concentrations of the major ions in the recharge water is shown in Table 1, along with the equilibrated concentrations used in the modelling. The recharge water is in an oxidized state, and the inorganic chemistry reflects these conditions. The recharge water is supersaturated with respect to calcite ($CaCO_3$).

2.2 *Aquifer water and sediment*

The native aquifer is in an oxidative state that is more reduced than that of the recharge water. The reduction capacity is mainly associated with pyrite (FeS_2) and other reduced minerals (e.g., siderite ($FeCO_3$)), reduced exchanged cations (NH_4^+, Fe^{2+}, and Mn^{2+}), and methane (CH_4), see Table 2.

2.3 *Controlling reactive processes*

It is expected that the recharge of oxidized water will implicate oxidation of aquifer constituents, and the oxidized species will subsequently take part in other processes, e.g. precipitation of oxidized iron minerals here expressed by goethite (FeOOH). Oxidation of organic matter is here assumed insignificant owing to a low observed change in concentration of organic matter (not shown).

The different concentrations of the cations will influence the ion exchange complex. In particular, the high concentration of Na^+ injected as a pulse is expected to replace Ca^{2+} and Mg^{2+} initially bound in

Table 1. Total aqueous concentrations in the recharge water. (kgw: kg water).

Species	Recharge water (Pulse) [mol/kgw]	Model input [mol/kgw]
pH	7.7	7.7
Pe	-	12.85
Cl^-	$2.58\cdot10^{-3}$ ($3.05\cdot10^{-4}$)	$2.58\cdot10^{-3}$ ($3.05\cdot10^{-4}$)
Ca^{2+}	$1.50\cdot10^{-3}$	$1.50\cdot10^{-3}$
Mg^{2+}	$4.54\cdot10^{-4}$	$4.54\cdot10^{-4}$
Na^+	$3.64\cdot10^{-3}$ ($1.58\cdot10^{-3}$)	$3.64\cdot10^{-3}$ ($1.58\cdot10^{-3}$)
K^+	$1.67\cdot10^{-4}$	$1.67\cdot10^{-4}$
Fe^{3+}	$7.23\cdot10^{-7}$	$7.23\cdot10^{-7}$
Mn^{2+}	$1.65\cdot10^{-7}$	$1.65\cdot10^{-7}$
NH_4^+	$8.36\cdot10^{-7}$	$8.36\cdot10^{-7}$
O_2	$2.53\cdot10^{-4}$	$2.527\cdot10^{-4}$
NO_3^-	$2.40\cdot10^{-4}$	$2.40\cdot10^{-4}$
SO_4^{2-}	$7.96\cdot10^{-5}$	$7.96\cdot10^{-5}$
TIC	$5.11\cdot10^{-3}$	$5.09\cdot10^{-3}$
HCO_3^-	$4.88\cdot10^{-3}$	$4.88\cdot10^{-3}$

Table 2. Total aqueous concentrations in the native groundwater at WP3-f4. The X indicates exchanged species. Mineral concentration and exchanged species are given as their equivalent aqueous concentrations.

Species	Native [mol/kgw]	Model input [mol/kgw]
pH	7.3	7.16
Cl^-	$2.70\cdot10^{-4}$	$2.70\cdot10^{-4}$
Ca^{2+}	$1.87\cdot10^{-3}$	$1.84\cdot10^{-3}$
Mg^{2+}	$6.20\cdot10^{-4}$	$6.03\cdot10^{-4}$
Na^+	$1.82\cdot10^{-4}$	$1.82\cdot10^{-4}$
K^+	$1.90\cdot10^{-4}$	$1.88\cdot10^{-4}$
Fe^{2+}	$8.71\cdot10^{-5}$	$1.41\cdot10^{-5}$
Mn^{2+}	$1.20\cdot10^{-5}$	$1.17\cdot10^{-5}$
NH_4^+	$5.20\cdot10^{-5}$	$4.97\cdot10^{-5}$
CH_4	$3.90\cdot10^{-4}$	$3.98\cdot10^{-4}$
NO_3^-	$< 7.00\cdot10^{-6}$	$7.00\cdot10^{-6}$ as N_2
O_2	$< 3.00\cdot10^{-5}$	0.0
SO_4^{2-}	$\leq 1.20\cdot10^{-5}$	$2.01\cdot10^{-12}$
TIC	$8.17\cdot10^{-3}$	$8.28\cdot10^{-3}$
HCO_3^-	$7.36\cdot10^{-3}$	$6.94\cdot10^{-3}$
NaX	-	$3.57\cdot10^{-4}$
CaX_2	-	$7.34\cdot10^{-3}$
MgX_2	-	$2.82\cdot10^{-3}$
FeX_2	-	$3.35\cdot10^{-5}$
MnX_2	-	$2.79\cdot10^{-5}$
KX	-	$2.47\cdot10^{-4}$
NH_4X	-	$3.27\cdot10^{-5}$
HX	-	$1.52\cdot10^{-7}$
$CEC_{\Sigma X-}$		$2.00\cdot10^{-2}$
Pyrite	$5.73\cdot10^{-3}$	$5.73\cdot10^{-3}$
Calcite	0.227	0.227
Goethite	0.0	0.0

aquifer material. The subsequent normalisation will reverse the exchange. The ion exchanged concentrations listed in Table 2 are based on equilibrium calculations using Gaines-Thomas coefficients from literature (Appelo & Postma 1993). pH is assumed to be controlled by the presence of calcite.

3 REACTIVE TRANSPORT MODEL

The flow and reactive solute transport are calculated using the MIKE SHE modelling system (Abbott et al. 1986a&b). Three modules are used to compute the reactive solute transport:

1. The AD-module, which calculates three-dimensional solute transport.
2. The BM-module, which calculates kinetically controlled microbial-mediated degradation of organic matter under the influence of the surrounding geochemical environment.
3. The GM-module, which calculates reversible thermodynamically controlled equilibrium ion exchange, precipitation/dissolution, complexation, and redox processes. The GM module is based on the geochemical mass transfer model PHREEQC (Parkhurst 1995). It is modified slightly, so it can be coupled to the MIKE SHE system.

Since degradation of organic matter is assumed unimportant in these preliminary simulations, only the modules AD and GM are used when modelling the Langerak deep-well injection site, and hence only these will be discussed here. Several methods are available for solving the reactive transport equations (Engesgaard & Kipp 1992). In the MIKE SHE system, the solution to the coupled transport and the geochemical problem is carried out with a sequential non-iterative approach (SNIA). In SNIA, the AD-module solves the advection-dispersion equations in the first step hereby introducing a non-equilibrium state. In the following and final step, the GM module adjusts the transported concentrations to obtain equilibrium.

4 MODELLING THE LANGERAK SITE

At this preliminary stage, the modelling of the groundwater flow and the related transport will only take place in the framework of a one-dimensional (1D) model aquifer. The advantage of 1D simulations is a highly reduced calculation time. Among disadvantages are that variations in dilution, velocity, and dispersivity through the system, which are due to inhomogeneities in the aquifer and the flow pattern induced from the pumping system, are all neglected. The Langerak data support a fully three-dimensional (3D) model setup, but so far a 3D model has not been applied to the problem.

4.1 *Flow and transport setup*

The pore-velocity and the dispersivity are estimated by means of the analytical 1D model CXTFIT (Toride et al. 1995) for fitting the breakthrough curves for chloride for each screen. The velocity and dispersivity used in the following simulations are 0.9 m/day and 0.5 m, respectively. These parameters represent transport from the injection well to a specific screen WP3-f4, which is situated at well WP3, at level 86 m, see Figure 1. Comparison between measured and simulated concentrations will only be done for screen WP3-f4.

4.2 *Reactions setup*

Using an equilibrium assumption when modelling the reactive processes implies that the input solution and the native solution have to be charge-balanced and in equilibrium. Only the transport of solutes can disturb this equilibrium. The solution used in the model can therefore not be exactly matched since the measured concentration levels of the various species are not exactly charge-balanced or at equilibrium. The geochemical model does not accept, for example, that O_2 and NH_4^+ are present at the same time, since these constituents are not at chemical equilibrium. O_2 will be used to oxidize NH_4^+ until either O_2 or NH_4^+ is totally consumed. The equilibrium assumption explains the difference between the observed and modelled solution composition listed in Table 1 and 2. In addition, it has been necessary to treat NH_4^+ as an individual component not influenced by redox processes. If set at equilibrium, it will be instantaneously oxidized by O_2, NO_3^- or SO_4^{2-}. It is included in the model as a cation exchanger only.

Default process parameters from the PHREEQC database are used with respect to complexation, redox and mineral processes. Partly fitted values are used only with respect to cation exchange reactions.

The mean of the measured solution concentrations in the recharge water is assumed to be at equilibrium with atmospheric $O_2(g)$ and $CO_2(g)$. The partial pressure of CO_2 has been slightly adjusted to make a better estimate of the modelled pH. In general, the calculated composition of the recharge water is comparable to the measured concentration. Only minor differences are observed in Table 1. The oxidizing capacity, associated with O_2 (4 eq./M), NO_3^- (5 eq./M), and SO_4^{2-} (8 eq./M), is approximately $2.85 \cdot 10^{-3}$ eq.

The background water as observed in WP3-f4 is set to be at equilibrium with the exchanger (X^-) and with minerals calcite and pyrite. The modelled reduction capacity, mainly associated with pyrite (15 eq./M), is $8.6 \cdot 10^{-2}$ eq. No direct measurement of

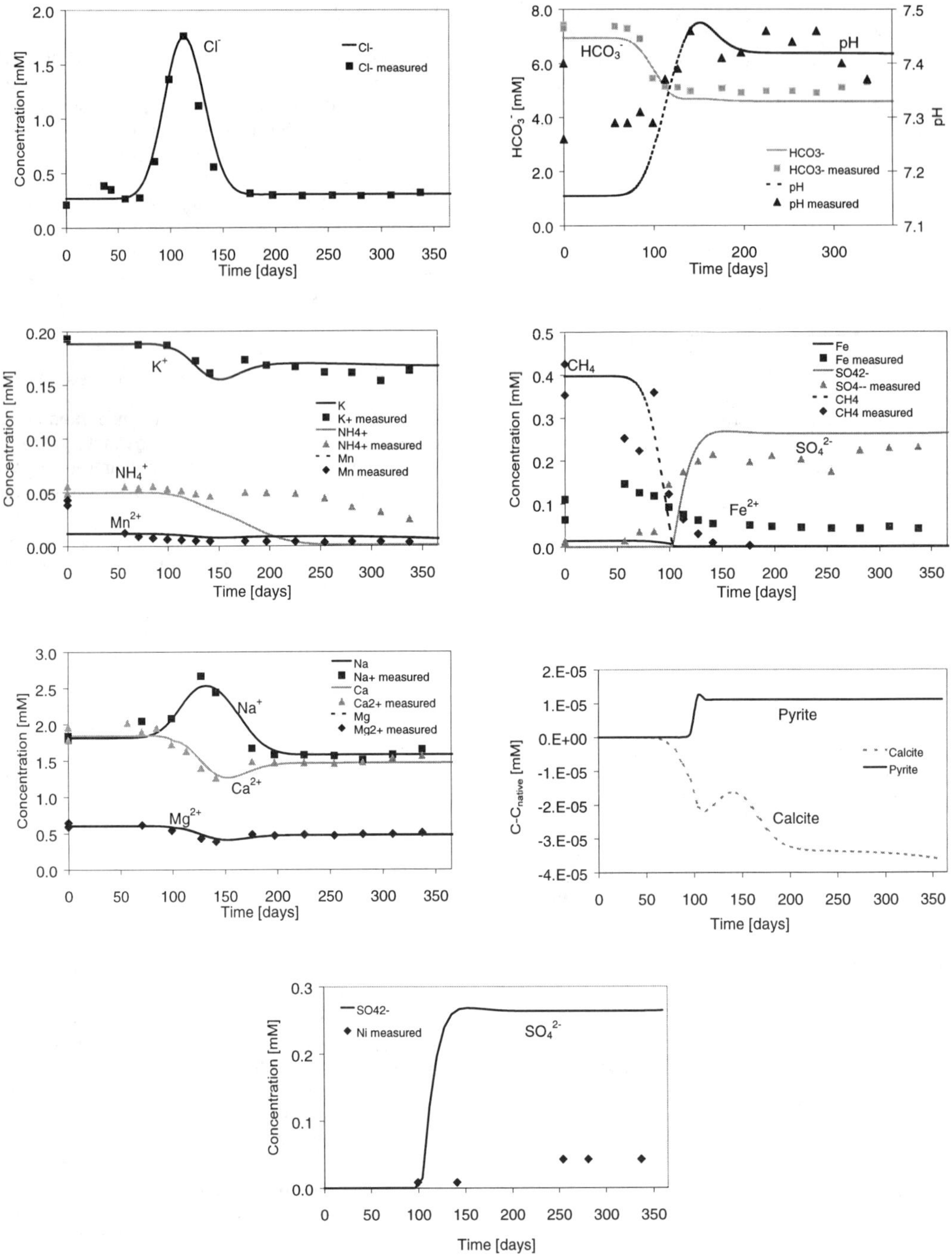

Figure 2. Simulated (lines) and observed (symbols) breakthrough curves of numerous species for screen WP3-f4.

CEC is available, so the exchanged concentrations are calculated on the basis of the concentration of the various ion exchangeable species. The selectivity coefficients used are based on literature values (Appelo & Postma 1993, Stuyfzand & Timmer 1998), see Table 3.

Table 3. Gaines-Thomas selectivity coefficients.

NaX	1
CaX_2	0.45
MgX_2	0.5
FeX_2	0.6
MnX_2	0.55
KX	0.15
HX	1
NH_4X	0.3

4.3 *Results and discussion*

Figure 2 shows the resulting breakthrough curves from the reactive transport simulation for several species. In all seven plots, solid or dashed lines represent simulated results whereas symbols represent temporal measurements from the injection experiment. The location of the breakthrough is for all curves at WP3-f4. Looking first at the breakthrough curve for Cl^-, it is seen that the measured arrival and disappearance of the pulse is very well reflected in the simulation. Recall that the determination of the groundwater velocity and the dispersivity is based on this breakthrough curve. Thus, a good agreement between measured and simulated Cl^- is almost obligatory. The background concentration is followed by the pulse, which is superseded by the recharge water concentration that is slightly higher than the background concentration. Only a few measurements are not exactly on the simulated line. Cl^- broke through after approximately 100 days, and the pulse duration increased from the original 30 days to 40 days at 91 m downstream from the injection. It is worth noticing that approximately 100% Cl^- mass recovery was found at this distance from the injection relative to the total input of Cl^-, indicating that the 1D assumption is acceptable. Both HCO_3^- and pH are well predicted by the simulations. As the concentration of HCO_3^- generally seems to slightly underestimate the measurements, measured and simulated pH are in good agreement, except for the background values of the aquifer at the beginning. The initial saturation of calcite could be said to "lock" initial pH to an erroneous result. With respect to the redox environment, the evolution of CH_4, SO_4^{2+}, and Fe^{2+} is given. For CH_4, there is fair agreement between measured and simulated results, but the simulation seems to be insufficiently dispersed. The SO_4^{2+} simulation tends to overestimate the measured concentration in the longer time scale corresponding to the recharge water. Similarly, it can be hard to recognise the concentration similarities between measured and simulated Fe^{2+}. The poor fit of Fe^{2+} and overestimation of SO_4^{2+} are believed to be caused by an only partial oxidation of pyrite and perhaps a presence of siderite. After one year of travel, oxidized conditions have still not been obtained in the aquifer at this distance. O_2 and NO_3^- are not shown, since no changes in concentration occur. Thus, O_2 and NO_3^- are strongly retarded relative to Cl^-.

The chromatographic behaviour of the cation exchange is illustrated by the breakthrough curves of the cations. In general, there is good agreement between the simulated and measured cation concentrations. As the Cl^- pulse passes, the Na^+ concentration increases as well. Ca^{2+}, Mg^{2+}, and K^+ decrease to a level lower than both background and recharge water. As Na^+ replaces the other cations on the exchange complex, a release is observed. However, as the concentration of Na^+ decreases again, mass from the aqueous phase of the other cations will be taken to cover the loss of the exchange complex. Mn^{2+} is also well described by the simulation, although slightly overestimated. A manganese substitution in siderite was suggested by Stuyfzand & Timmer (1998) as an explanation for the deviations. Unfortunately, the simulation of NH_4^+ does not yield a picture similar to these measurements. All cations break through a little later than Cl^-, indicating a minor retardation. It is seen that Na^+ is slightly retarded, Ca^{2+} and Mg^{2+} are more retarded, and K^+ is even more retarded. According to the measured data, NH_4^+ should be strongly retarded. Because of ion exchange competition, the shape of the retarded species is not perfectly Gaussian. The retarded species recover more slowly to the recharge concentrations characterised by more dispersed breakthrough curves relative to chloride.

The evolution of minerals pyrite and calcite is shown as the difference between the actual concentration and the native mineral concentration. Pyrite starts precipitating at the time of the breakthrough of Cl^-. Near the injection well, pyrite is oxidized, and Fe^{2+} and SO_4^{2-} are dissolved as the aquifer becomes oxidized by the recharge water. Downstream, the increased concentrations of Fe^{2+} and SO_4^{2-} make pyrite precipitate again, because the aquifer at downstream locations still is reduced. After the breakthrough of Cl^-, the concentration of pyrite increases to a constant level at equilibrium with the incoming Fe^{2+} and SO_4^{2-} because of the constant upstream pyrite oxidation rate.

The concentration will remain constant until the retarded oxidation front arrives at the well, but this had not happened within the 365 days of the experiment. Calcite starts to be dissolved when Cl^- breaks through owing to the lower content of HCO_3^- in the recharge water. Because Ca^{2+} contributes to the cation exchange process, the Na^+ pulse is reflected in calcite dissolution as well, which explains the lack of dissolved Ca^{2+} in the cation breakthrough curve.

The measured concentrations of nickel (Ni) and SO_4^{2-} are also illustrated. Ni is probably substituted in the pyrite and consequently dissolved as pyrite is oxidized. The same picture can be observed for other trace minerals like arsenic, cadmium, and zinc (not shown). It is expected that when the pyrite has been totally oxidized, a precipitation of Fe^{3+}-minerals may constitute a base for trace mineral sorption. However, this state of aquifer development had not been reached at WP3 within the test period.

In general, a good agreement between the measured and simulated concentrations is seen despite the simplification by using only 1D and default values from a chemical data base.

A future application of 2D and 3D flow patterns will probably improve the results with respect to fitting the breakthrough curves from all screens. None of the investigated species constitute a danger to human health in the concentrations developed. However, an inclusion of the transport of trace metals associated with pyrite and goethite will justify the modelling in this sense. Clogging effects owing to mineral precipitation should also be analysed. Sorption of organic micropollutants (e.g., pesticides) to precipitated iron minerals will be investigated. In the end, the oxidation of pyrite will finally be complete, and the aquifer will be entirely oxidized. Thus, there is an upper limit of the reduction capacity of the aquifer. Likewise, if the artificial recharge at the Langerak site is halted, a re-reduction of the aquifer will take place. Sorbed trace metals and organic micropollutants will be desorbed, with consequences for a future withdrawal. Reactive transport simulations can, in this respect, be a powerful tool for investigating the future behaviour of the aquifer development independent of human choices.

5 CONCLUSION

Owing to time constraints, not all of the information available at the Langerak deep-well injection site has been incorporated into this model exercise. The results presented here are focused on the observations from a single screen in one of the three wells. The observations concerning the other screens are from a reactive transport viewpoint somewhat analogous to the observations in the screen used in this study. The difference is believed to be caused by different travel times from the injection point to the various screens and the lack of equilibrium in the screens closest to the injection point. Hopefully, the promising results from the one-dimensional model aquifer can be verified when the site is modelled in several dimensions. The setup of the model has been based on observations of the quality of the recharge water and the native aquifer, the observation of the breakthrough of a conservative tracer, and default parameters for the reactive parameters. From these observations, it has been possible to simulate the major geochemical processes occurring in the aquifer and the related change in water quality precisely and quantitatively. On the basis of the work carried out until now at the Langerak site, it is concluded that a reactive model is a valuable tool in the design and operation of an artificial recharge plant, at least for this specific site. Whether this is also the case for other aquifers with other physical and geochemical properties has yet to be proven.

6 ACKNOWLEDGEMENT

This research was supported by VKI, Water Supply South-Holland Oost, Kiwa, and the EC Environment, Climate Research Programme (Contract: ENV4-CT95-0071).

REFERENCES

Abbott, M.B., J.C. Bathurst, P.E. O'Connell & J. Rasmussen 1986a. An introduction to the European Hydrological System - Système Hydrologique Européen, SHE 1: History and philosophy of a physically-based, distributed modelling system. J.Hydro., 87 45-60.

Abbott, M.B., J.C. Bathurst, P.E. O'Connell & J. Rasmussen 1986b. An introduction to the European Hydrological System - Système Hydrologique Européen, SHE 2: Structure of a physically-based, distributed modelling system. J.Hydro., 87 61-77.

Appelo, C.A.J & D. Postma 1993. *Geochemistry, groundwater and pollution*, A.A. Balkema, Rotterdam, Brookfield.

Engesgaard, P. & K.L. Kipp 1992. A Geochemical model for redox-controlled movement of mineral fronts in groundwater flow systems: A case of nitrate removal by oxidation of pyrite, *Water Resour. Res.*, Vol 28 (10), pp 2829-2843.

Parkhurst 1995. D. L. User's Guide to PHREEQC – A Computer program for speciation, reaction-path, advective-transport, and inverse geochemical calculation. U.S. Geological Survey, Water-Resources Investigation Report 95-4227.

Toride, N., F.J. Leij, & M. Th. Van Genuchten 1995. The CXTFIT Code for Estimating Transport Parameters from Laboratory or Field Tracer Experiments (version 2.0). Research Report No. 137, U. S. Salinity Lab., USDA, ARS, Riverside, Ca.

Stuyfzand, P.J. & H. Timmer 1998. Injecting oxic surface water into an anoxic aquifer along the Rhine River. Water-rock interactions including redox dynamics (draft) Nieuwegein.

1.13 Well recharge case studies

Artificial Recharge of Groundwater, Peters et al. (eds) © 1998 Balkema, Rotterdam, ISBN 90 5809 017 5

Environmental guidelines for aquifer storage and recovery: Australian experience

Peter Dillon & Paul Pavelic
Centre for Groundwater Studies and CSIRO Land and Water, Glen Osmond, SA, Australia

ABSTRACT: An essential prerequisite for aquifer storage and recovery (ASR) of stormwater and reclaimed water is to ensure environmental protection from such operations. A set of guidelines was produced in Australia in 1996, based on a review of ASR international experience and guidelines, and reviews of source water quality and passive treatment processes (Dillon and Pavelic, 1996). Laboratory and field studies on in-aquifer physical and biogeochemical processes affecting the chemical and microbial composition of groundwater and the recovered water, and the hydraulic characteristics of the aquifer were also performed. These provided a sound basis for determining the quality of water for injection, and the means of preventing irreversible clogging of wells. Unlike other guidelines, these allow for a range of beneficial uses of recovered water and take into account the sustainable attenuation of injected contaminants within the aquifer. These guidelines have provided a basis for EPA Policies and Codes of Practice for use by regulators, proponents, and operators of such systems. These have in turn acted as a stimulus for appropriate ASR development in this semi-arid area.

1 ASR WITH WATERS OF IMPAIRED QUALITY : SOUTH AUSTRALIA

In semi-arid regions where water resources are highly valued, ASR with stormwater and treated effluent adds a new dimension to urban water management. Much of the development of ASR in the northern hemisphere has concentrated on injection of high quality water or highly treated effluent with the intention of recovering it for potable use. However in southern Australia, the emphasis has been on using otherwise wasted water of lower quality to restore brackish aquifers and thereby develop new irrigation water supplies. A Mediterranean climate, summer irrigation demand, and thick surficial clays overlying moderately transmissive brackish aquifers led to consideration of ASR. Initially, stormwater was the preferred source of water for injection for four reasons.

1. *proliferation of stormwater detention ponds.* As urban areas spread upstream in a surface water catchment detention ponds became mandatory simply for flood control, and as a by-product provided a means of harvesting stormwater for ASR.
2. *development of urban wetlands.* Progressive developers appreciated the aesthetic (and real-estate) value of constructed wetlands in residential areas, and the benefits for water quality downstream. Generally these have been incorporated with stormwater detention, and provide an opportunity for some passive pre-treatment of water prior to injection.
3. *public acceptance of stormwater re-use.* While there was widespread community acceptance of re-use of treated sewage for irrigation, stormwater was perceived to have a lower public health risk, and was preferred, at least initially, for injection into aquifers.
4. *magnitude of the resource.* The annual volume of stormwater is equivalent in size to annual consumption of mains water.

Several experimental licences to inject stormwater were granted by the State environment protection agency and its predecessor, commencing in October 1993, leading to well managed and demonstrably successful stormwater ASR operations, and data on which to found guidelines. These operations were all designed and conducted by Primary Industries and Resources, SA (formerly Mines and Energy, SA), a partner of the Centre for Groundwater Studies. Some of the results at three of these sites are presented in companion papers by (Gerges & Howles 1998), and (Pavelic et al 1998). Currently work is underway to demonstrate the environmental sustainability and technical viability of injection of post-secondary treated effluent (Rinck-Pfeiffer & Ragusa 1998), as part of a major effluent re-use initiative (Boscher & Kracman 1998).

2 ASR WITH WATERS OF IMPAIRED QUALITY : WORLD PERSPECTIVE

In an international context, use of urban stormwater for ASR is relatively unique. A review of international experience in ASR (Pavelic & Dillon 1997) identified 45 case studies, including 70 known sites in 12 countries, with published information on; site characteristics and recharge techniques; operational problems such as clogging, and means to resolve these; and monitoring on impacts on groundwater quality or the quality of recovered water. Of the 45 case studies, 71% use "natural" source waters (rivers, lakes and groundwater), and the remainder use treated sewage effluent (20%) or urban stormwater runoff (9%) (Fig. 1). Highly treated sewage effluent is commonly used in the United States of America. This yields a very consistent, high quality water at a relatively uniform flow rate, making this attractive as a source of water for ASR. The sustained injection of urban stormwater was found in only four documented cases, three of which were in South Australia.

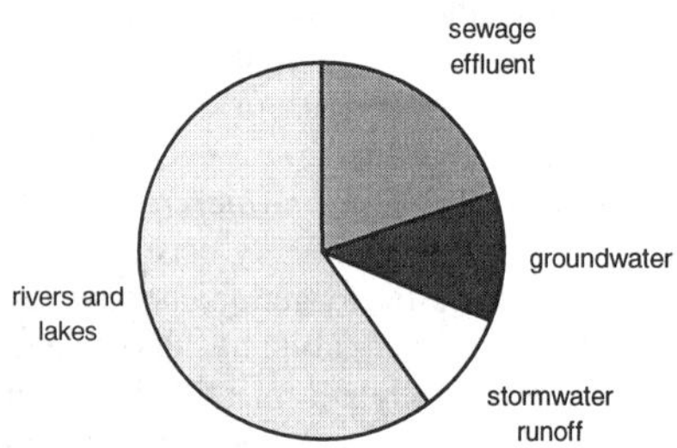

Figure 1. Types of source waters *(n=45)*

The quality of the various sources of water can, of course, be improved by pre-treatment to a level deemed suitable for injection. That level is determined by various factors; the quality of the ambient groundwater, uses of the recovered water and the need to prevent clogging and to avert adverse biogeochemical reactions. The level of risk tolerance should be no different to the normal health and environmental standards applying to the classes of use assigned to the groundwater and recovered water. Consequently, the degree of pre-treatment required may range from a precautionary screen through to advanced tertiary treatment (including nutrient removal, reverse osmosis and disinfection). Passive forms of treatment (eg. wetlands) have been rarely used in comparison with what are considered more predictable engineered treatment processes. The exception is for stormwater runoff, where wetlands also provide an element of detention storage, assisting to buffer the large temporal variations in quantity and quality of flow, which is one of the reasons that stormwater was not considered a prime resource for artificial recharge in the United States of America (National Research Council 1994).

More than half the documented sites (Pavelic & Dillon 1997) involve injection of potable source water into potable groundwater (Fig. 2). A further 13% of cases involve recharge of potable water into non-potable groundwater, generally with the intention of restoring aquifers for recovery of potable water. In 16% of cases non-potable water is injected into non-potable aquifers for the purpose of creating irrigation water supplies, while improving some aspects of groundwater quality. In at least one case (Dillon et al 1995, 1997; Pavelic et al 1998) it would appear that stormwater injected into a brackish aquifer has the potential to create a potable water resource due to the sustainable attenuation of stormwater contaminants within the aquifer. More investigations are needed to substantiate this. The final category, non-potable recharge into potable aquifers, covering 16% of ASR operations, is of concern. This reflects either a reliance on passive treatment within the aquifer or a disregard for protecting groundwater quality.

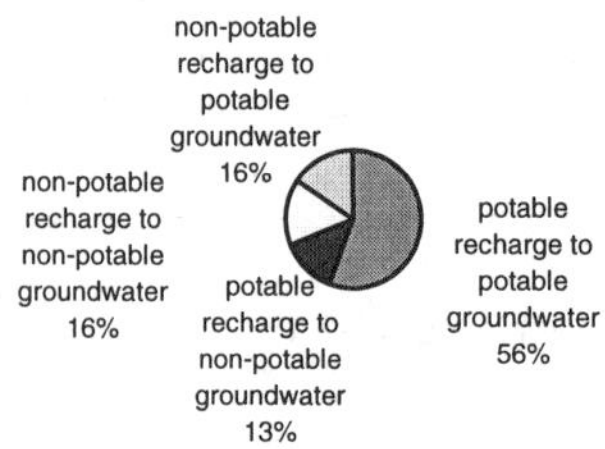

Figure 2. Potability of source waters and ambient groundwaters at ASR sites *(n=45)*

In recent years there has been substantial research on passive attenuation of plumes of industrial pollutants in unconfined aquifers, however there appears to be little quantitative information on attenuation of typical stormwater and treated effluent contaminants in confined systems. Licensing of ASR operations which have potential to deteriorate groundwater quality, should be linked to, and dependent on the outcomes of such research.

3 BASIS FOR AUSTRALIAN ASR GUIDELINES

In 1982 the Australian Water Resources Council (AWRC 1982) published guidelines for the use of reclaimed water for aquifer recharge. These were aimed primarily at recharge via spreading basins (as the preferred method of aquifer recharge), although some guidance was given on water quality and aquifer types suitable for injection of reclaimed water. These included guidance on turbidity, salinity, sodium adsorption ratio, pH, and redox reactions, as well as aquifer types, and recommended site specific assessments, and a national repository for pilot study results.

The Australian Water Resources Council Draft Guidelines for Groundwater Protection (NWQMS, 1992*a*), developed as part of the National Water Quality Management Strategy (NWQMS), established the national objectives and principles for groundwater protection policies to be developed by each State. They recognised that:

- groundwater is a pervasive resource which is often difficult to define, requires skilled management, and may have limited ability to assimilate contaminants.
- groundwater which is polluted is technically difficult and expensive to remediate.
- groundwater is to be managed as a sustainable resource so that the needs of future generations are taken into account in today's decisions.

The goal of groundwater protection is to ensure that the groundwater resources of the State can support their identified environmental values (including beneficial uses) in an economically, socially and environmentally sustainable and acceptable manner. This relies on the identification of existing or potential environmental values for each groundwater resource, which then determine the required level of protection. Once environmental values have been determined, the obligation for protection of groundwater lies with the industry or activity which has the potential to contaminate the groundwater body, such as an ASR operator.

Categories of environmental values include:

- human consumption/food production
- agricultural, industrial and mining (including irrigation and stock watering)
- ecosystem support (particularly adjacent surface water bodies)
- no definable beneficial use/controlled degradation

For each category or sub-category, the *Water Quality Guidelines for Fresh and Marine Waters* (NWQMS 1992*b*) provides schedules of acceptable values of water quality parameters for a wide range of known indicators, inorganic and organic substances, and microbiota.

In summary therefore, *Australia has accepted a differential protection policy whereby the level of protection offered at sites where reclaimed water is injected into aquifers will depend on the potential environmental values of ambient groundwater and therefore on its current water quality.* This takes into account that much of the groundwater in Australia is too saline for potable supplies, and that a simple potable/non-potable differentiation is inadequate for protecting major environmental values. Also, the allowable concentrations of some substances, such as phosphorus, are lower for ecosystem support (say where an aquifer contributes baseflow to a stream), than they are for human consumption. All relevant environmental values which apply to an aquifer are required to be protected.

This policy is being implemented through legislation on land use planning, groundwater management, and environmental protection. The roles of these are described in (Dillon & Pavelic 1996). Other NWQMS guidelines relevant to ASR include:

1. *Draft guidelines for urban stormwater management* (NWQMS 1994*a*) which promotes the valuing of stormwater as a water resource, and
2. *Draft guidelines for the use of reclaimed water* (NWQMS 1995), which allows for recharge by injection into potable aquifers for subsequent potable reuse. This specifies that the injectant would require secondary treatment, filtration, and pathogen reduction, in addition to advanced treatment which may be site-specific. The reclaimed water should have no deleterious effects on groundwater quality, and faecal coliforms should be less than 1000 colony forming units per 100 mL. The draft specifies, but without explanation, and in keeping with Californian guidelines, that injectant should be retained below ground for a minimum of 12 months before withdrawal. The recovered water should meet the quality requirements for sources of raw drinking

water. Recharge to aquifers to prevent saline groundwater intrusion are also included, and any treatment in addition to secondary treatment would depend on water quality requirements which are specific to the site. Injection for other non-potable uses is not considered in the draft guidelines, but some water quality parameters are specified for point of use. Such uses include irrigated food crops, non food crops, garden watering, aquaculture and toilet flushing.

4 AUSTRALIAN ASR GUIDELINES

Guidelines on the quality of stormwater and treated effluent for injection into aquifers for storage and reuse were recently developed (Dillon & Pavelic 1996) based on the NWQMS principles, and including and drawing from reviews of; international experience and guidelines, the quality of urban stormwater and treated effluent, clogging and redevelopment of injection wells, and attenuation of chemical and micriobiological contaminants in aquifers. Preliminary results from a pilot study on ASR with stormwater (Dillon et al 1997) were also included as supporting evidence and to ensure that guidelines were practical.

The water quality requirements for injection of stormwater and treated effluent into aquifers for storage and reuse are based on three objectives:
- to prevent irreversible clogging
- to protect or improve groundwater quality, and
- to ensure that the quality of recovered water is fit for its intended use.

These follow the principles established in the NWQMS and allow for the level of groundwater quality protection to be matched to the potential beneficial uses of the ambient groundwater. In a departure from previous guidelines, these specifically allow for the sustainable treatment of water provided by its passage though the aquifer, via processes such as: microbial attenuation, and biodegradation of organic chemicals.

The precautionary principle applies, and in the absence of reliable information on contaminant attenuation, this must be estimated conservatively. The demonstration licence approach described below is intended to extend our knowledge base while protecting groundwater quality. Unless integrated catchment/ aquifer management principles are applied (which in Australia is becoming more common), the precautionary principle suggests that there should be no allowance for dilution of injectant by natural flow in the aquifer system, where this would relax restrictions on the quality of injectant to meet the objectives above.

The guidelines contain provisions concerning demonstration licensing, environmental management plans, pre-treatment, monitoring, maximum contaminant concentrations, and minimum residence times. Proposed new ASR sites should apply for a demonstration licence, which provides an opportunity for the operator to demonstrate that the licence conditions are met. The demonstration licence would be based on an environmental management plan (EMP) to account for all environmental impacts and associated risks. A checklist of items for the EMP is given in the guidelines, and includes details of monitoring and reporting procedures to demonstrate performance with respect to objectives, as well as operational and contingency plans to manage risks, and projections of performance over the longer term. Appropriate opportunity for public comment should be allowed. Licence renewal would be subject to a performance review.

The overarching principle for pre-treatment should be to remove or reduce whatever contaminants can be viably attenuated at the surface before injection, in particular those contaminants which are resistant to degradation in the aquifer. The minimum level of pre-treatment required for a demonstration licence is that which results in predicted achievement of water quality criteria (eg. NWQMS 1992*a* 1994*b*) relevant to all pre-existing potential beneficial uses for the aquifer. Where the beneficial uses of recovered water are different from those of the ambient groundwater, pre-treatment combined with aquifer treatment should meet the relevant criteria at the point of extraction. This will determine whether disinfection or any other pre-treatment is required during the period of the demonstration licence. Pre-treatment methods may include passive and engineered systems, or a combination of these, at the discretion of the proponents as part of their environmental management plan.

Where any relevant water quality criterion is not initially met by the ambient groundwater, excluding the effects of anthropogenic pollution, the injected water only needs to have a concentration less than that of the ambient groundwater, at least for the period of the demonstration licence. Detention storage as part of pre-treatment is desirable in reducing the variability of the quality of injectant. This has the advantage of increasing confidence in the results of monitoring for any given level of monitoring effort.

At least one observation well is required at each ASR site, and where reliance is placed on water treatment within the aquifer to meet water quality objectives, we recommend at least three observation wells in addition to the injection well. In fractured or heterogeneous aquifers more wells may be needed. Extraction wells may be used as observation wells, if sited appropriately. In areas where existing drainage wells or ASR operations are prolific and have a history of use, a rationalised monitoring program designed on risk management principles may be adopted in the environmental management plan. The guidelines also refer to frequency of sampling, analytes/species, triggers for remediation, and hydraulic effects such as over-pressurisation. Sites where monitoring results prove the achievement of water quality and other objectives during their demonstration licence, may be granted a full licence for a period to be nominated by the regulatory authority, with appropriate terms and conditions, presumably with a modified sampling regime.

Guideline values for individual parameters are determined by water quality objectives, and by the capacity for treatment within the aquifer. Some specific guidance is given for several parameters which are commonly regarded as constraints to ASR with reclaimed waters; suspended solids, total dissolved solids, faecal coliforms, and nitrogen. A minimum residence time for undisinfected injectant of 50 days is recommended to provide an acceptable degree of health protection when recovered water is used for recreation or irrigation. Shorter residence times may be allowed if source water quality and exposure paths provide an equivalent level of public health protection. Protozoa and viruses may have longer survival times in aquifers, and recovered water should not be used for water supplies unless there is either a field-based assessment of the potential for breakthrough of these species, or the injected or recovered water is suitably treated.

A well considered environmental management plan presents the opportunity to trade-off sustainable treatment in the aquifer against pre-treatment of injectant, providing all objectives can be met. Aquifer treatment effects on recovered water may be increased for some contaminants by lengthening the flow path and residence time in the aquifer by the arrangement of injection and recovery wells. The increase in knowledge of the sustainable attenuation capacity of aquifers as a result of information collected at these sites will allow improved design of matching pre-treatment systems, and allow costs to be contained, particularly for non-potable uses of recovered waters from aquifers which are initially non-potable.

There is a very strong linkage between environmental sustainability and economic feasibility. Operations which are marginally economic may be unable to meet the required costs of monitoring, and the level of operational management may be compromised. It is important therefore that proposed ASR facilities are economically viable, taking account of monitoring and maintenance costs.

5 POST-GUIDELINE DEVELOPMENTS AND CONCLUSIONS

Groundwater protection policies are still evolving in the Australian States. The Victorian Government was the first to enact its post-NWQMS policy (Victoria, EPA 1997), and draft policies are being developed into legislation in other states. In some States, codes of practice will be established as regulations, and a Code of Practice for ASR is under development in South Australia. Adhering to the code of practice provides legal protection for ASR operators from environmental damages claims. A national repository for ASR monitoring data is being established, to expedite the growth in understanding of technical, environmental, and economic viability of ASR.

These guidelines, while largely untested, have provided some much-needed goal posts for the injection of water of impaired quality for storage and re-use. They have been a focus for debate by EPA's and Health Departments. It is clear that regulators will need to have, or have access to, a higher level of groundwater expertise to make good licensing decisions. The principles are agreed, and it is hoped that the debate necessary to produce a sound Code of Practice for ASR will lead to Codes and operations which consistently meet the objectives of the guidelines.

6 ACKNOWLEDGEMENTS

These guidelines were developed with the support of CSIRO Urban Water Systems Program, the Urban Water Research Assoc. of Aust. and Hickinbotham Homes. The authors acknowledge our colleagues in Primary Industries and Resources, without whom this work would not have been possible.

REFERENCES

Australian Water Resources Council 1982 Guidelines for the use of reclaimed water for aquifer recharge. Water Management Series No 2. Aust. Govt. Publishing Service, Canberra.

Boscher, C. & Kracman B. 1998. ASR with treated effluent on the Northern Adelaide Plains, South Australia. *ibid*

Dillon, P.J., Pavelic, P., Gerges, N.Z., Armstrong, D. & Emmett, A.J. 1995 Stormwater injection effects on groundwater quality in South Australia. In:"Artificial Recharge of Groundwater II", [Eds. A.I. Johnson and R.D.G. Pyne], *Proc. 2nd Int. Symp. on Artificial Recharge of Groundwater*, 17-22 July, 1994, Orlando, Florida, pp.426-435.

Dillon, P.J & Pavelic, P 1996. Guidelines for the quality of stormwater and treated wastewater for injection into aquifers for storage and reuse. Urban Water Research Assoc. of Australia, Research Report No 109.

Dillon, P., Pavelic, P., Sibenaler, X., Gerges, N., & Clark, R. 1997. Storing stormwater runoff in aquifers. Aust. Water and Wastewater Assoc. J. Water, 24 (4) 7 - 11.

Gerges, N.Z. & Howles, S.R. 1998. ASR, Hydraulic and salinity response in unconfined / confined aquifers. *ibid*

National Research Council, 1994 Groundwater Recharge Using Waters of Impaired Quality. National Academy Press, Washington DC.

National Water Quality Management Strategy 1992*a* Draft guidelines for groundwater protection. Australian Water Resources Council, August 1992.

National Water Quality Management Strategy 1992*b* Australian water quality guidelines for fresh and marine waters. ANZECC, Canberra.

National Water Quality Management Strategy 1994*a* Draft guidelines for urban stormwater management. ARMCANZ and ANZECC, November 1994 (pre-release).

National Water Quality Management Strategy 1994b Australian drinking water guidelines. NHMRC & ANZECC, June 1994 (draft).

National Water Quality Management Strategy 1995 Draft guidelines for the use of reclaimed water. ARMCANZ and ANZECC, April 1995 (pre-release).

Pavelic, P. & Dillon, P.J. 1997 Review of international experience in injecting natural and reclaimed waters into aquifers for storage and reuse. Centre for Groundwater Studies Report No. 74.

Pavelic, P., Dillon P., Barry, K.E., Herczeg, A.L., Rattray, K..J., Hekmeijer, P., & Gerges, N.Z. 1998. Well clogging effects determined from mass balances and hydraulic response at a stormwater ASR site. *ibid*

Pyne, R.D.G. 1995 Groundwater recharge and wells: a guide to aquifer storage and recovery. CRC Press, Florida.

Rinck-Pfeiffer, S., Ragusa, S.R., Vandervelde, T. & Dillon, P.J. 1998. Parameters controlling biochemical clogging in laboratory columns receiving treated sewage effluent. *ibid*

Artificial Recharge of Groundwater, Peters et al. (eds) © 1998 Balkema, Rotterdam, ISBN 90 5809 017 5

Berkheide well recharge system: Design, implementation and initial experience of operation

M.W. Kortleve
Dune Water Company of South-Holland, The Hague, Netherlands

ABSTRACT: The dunes in the Netherlands have been used for more than a century as a source for drinking water. Initially, only groundwater was abstracted. Then, as the demand for drinking water continued to rise, the capacity was increased by means of artificial recharge with spreading ponds. However, because of the impact this method had on the environment, use is being made today of a new recharge and recovery method that employs deep wells below natural layers of clay.
In the Berkheide dunes, this new method is being used to produce 4 million m^3 of drinking water a year. Given the specific local conditions, existing technical ideas have been further elaborated and much thought devoted to the prevention of upconing of salt or brackish groundwater.

1 INTRODUCTION

The Berkheide deep well recharge project is located in the dunes between Katwijk and The Hague (Fig. 2). Because of the presence of a natural reserve of freshwater (Koopman 1995 & Looy 1997), water has been recovered from this area since 1887. This natural reserve was formed by natural rainfall. Due to the difference in density between the fresh water and the salt groundwater, a freshwater lens was formed above the salt groundwater (Fig. 1). The freshwater reserve was ideal for the recovery of water to be used for supplying drinking water.

As a result of increasing demand for water, the available groundwater became insufficient. In 1940, therefore, the first practical experiments of using ponds for recharging were conducted. Gradually, the capacity of the open recharge system expanded to reach its present level of 24 Mm^3 a year. Roughly a further 1 Mm^3 a year is abstracted from the deeper-lying aquifer.

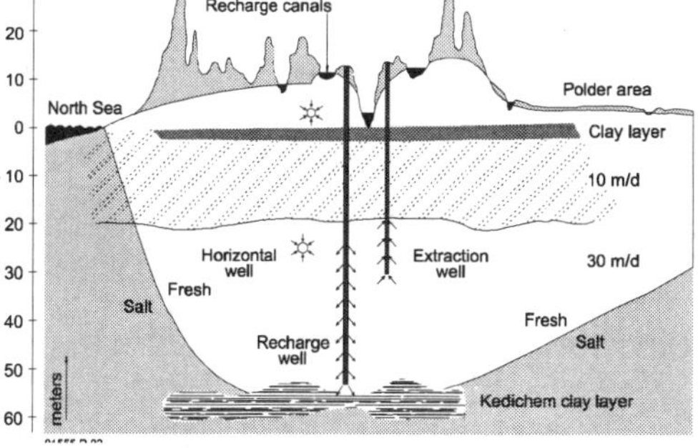

Figure 1. Hydrogeological cross section

The water recovery area lies in a very valuable nature reserve. In the course of time, water recovery and artificial recharge have affected some of the dunes' ecological value. For this reason, expansion of the open recharge system is no longer permitted. An alternative method is deep well recharge. With this method, which also uses aquifer passages, there are no major negative ecological impacts because the recharge and recovery occur at greater depths and below layers of clay. To achieve this, wells are used for recharge and recovery. There is also hardly any impact on the groundwater table or the quality of the shallow groundwater.

2 HYDROGEOLOGICAL STUDY

The hydrogeological study into the possibility of a deep well recharge system in the area started in 1987. Given the specific local hydrogeological conditions, deep well recharge seemed feasible in this location. The aquifer held freshwater, and above and below it there was a layer of clay.

Figure 1 shows a cross section of the area, at right angles to the coast. The geological structure can be viewed as two major aquifers. One, a phreatic aquifer, is an aeolian deposit with permeability of approximately 10 m/d. The open recharge and recovery are

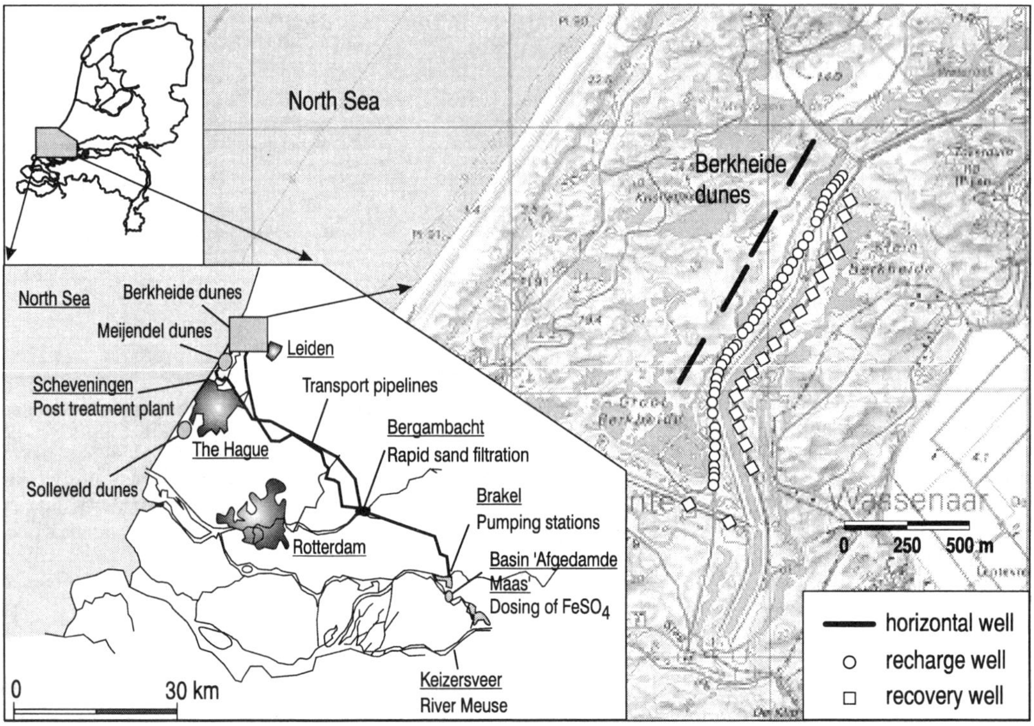

Figure 2. Transport to the recharge systems in the dunes and location of the Berkeheide deep well recharge project

carried out here. Slightly below sea level is a layer with low permeability and very high hydraulic resistance (approximately 100,000 d). This fluviatile deposit of the Rhine extends under almost the entire area. Because of its high resistance, this layer allows hardly any feeding of the underlying aquifer. The freshwater lens is also relatively thin.

Between 20 and 50 m below sea level is another aquifer with permeability of approximately 30 m/d. This aquifer is extremely suitable for deep well recharge and is confined underneath by the Kedichem clay layer. The thickness and hydraulic resistance of this clay layer vary greatly in this area. Drillings carried out at the time of the study show that the clay layer extends throughout the area. The aquifers under the clay layer contain salt groundwater (Zagwijn & Van Staalduinen 1997, Stuyfzand, 1993).

The location of the recharge and recovery wells has been chosen to disturb the area as little as possible, that is, by using existing routes (roads, paths, pipelines, etc). For recovery, use is even made of an existing set of deep wells. The definitive well plan comprises 45 recharge wells and 25 recovery wells (21 vertical and 4 shafts with horizontal wells).

Based on the hydrogeological information, models have been used to calculate the capacity of the system, residence time of water in the ground, and the system's environmental impact.

Environmental impact plays an important role in requests for recharge and recovery permits. Of significance is the ecological impact, as well as that on agriculture (crop damage) and buildings (subsidence). After a long procedure involving objections, the project was finally permitted in 1997. The recovery capacity is 4.3 Mm^3 per annum, with a maximum excess recharge of 10%. This excess recharge is needed to prevent any rising of salt groundwater. However, the higher the percentage the greater the ecological and agricultural impacts.

3 DESIGN, INSTALLATION AND OPERATION

The technical design and installation of the system drew on experience with the Waalsdorp deep well recharge system, that has been running since 1990. Extensive pretreatment and a low entrance rate at the drilled wall of each recharge well were chosen with the intention of practically eliminating well clogging.

Moreover, given the local conditions, it is necessary to control the wells individually. Recovered water, with the recovered water of the open recharge system, is pumped through existing pipelines to the existing posttreatment plant in Katwijk.

3.1 *Pretreatment*

To allow surface spreading, transport pipelines have been laid over a distance of 80 km for the conveyance of (pretreated) surface water from the Afgedamde Maas. After dosing with $FeSO_4$, the settling of $FePO_4$ and other substances adhering to these flocs occurs in this blind arm of the river. The water is then filtrated over rapid sand filters, making it ready for surface spreading.

Deep well recharge places higher standards on the water it uses. Firstly, because entrance rates in the wells are greater than in recharge lakes, the wells become clogged quicker (Olsthoorn 1982). To prevent this, therefore, the water has to be even cleaner. A measure of the clogging potential of water is its MFI index, a value below 3 s/l^2 representing no danger of recharge well clogging by suspended solids.

Secondly, the permit requires that pretreatment includes activated carbon filtration for the removal of organic micropollutants, such as pesticides.

To meet these stringent quality requirements, the supplementary pretreatment has to consist of flocculation, flotation, rapid sand filtration and activated carbon filtration (Fig. 3). This extra pretreatment of the recharge water takes place in the Lindenberg pretreatment plant, adjacent to the dune area.

3.2 *Recharging*

The water is recharged in 45 wells. Forty of the wells have a drilled bore hole diameter of 1000 mm; the remaining five, 500 mm. The purpose of this is to compare two types of well construction operating under otherwise identical conditions. Wells with a diameter of 500 mm are recharged with the same volume flux as the other wells, so that the entrance rate at the drilled hole wall is twice as great. As a result, clogging will occur sooner. To prevent this, these wells are regularly (daily) pumped clean by abstracting water with a high flowrate during a short period. This is possible through the installation of a submersible pump.

For drilling the recharge wells, a method developed by Amsterdam Water Supply and Drilling Contractor De Ruiter in 1993 has been used (Olsthoorn & van Harlingen, 1994). This involves drilling a well with an exceptionally clean drilled bore hole wall. The well is first drilled using standard suction drilling: drilled material is removed from the drill hole by a recirculating flow, and after the larger particles have settled above ground, the water is allowed to flow back into the bore hole. Because pressure in the drill hole is greater than in the aquifer, during this stage of the drilling the wall of the drill hole becomes coated with a skin of fine particles still present in the water.

With the new method, a slightly larger drilling bit is then used to scrape this layer of sludge from the drill hole wall, while, simultaneously, clean water is introduced and water containing the scrapings is abstracted. The effectiveness of the method is shown by the fact that, during scraping, the recharge rate of the drill hole increases from 25 m^3/h to 400 m^3/h for an unchanged pressure differential (Woering, 1995).

To achieve a correct distribution of recharge water

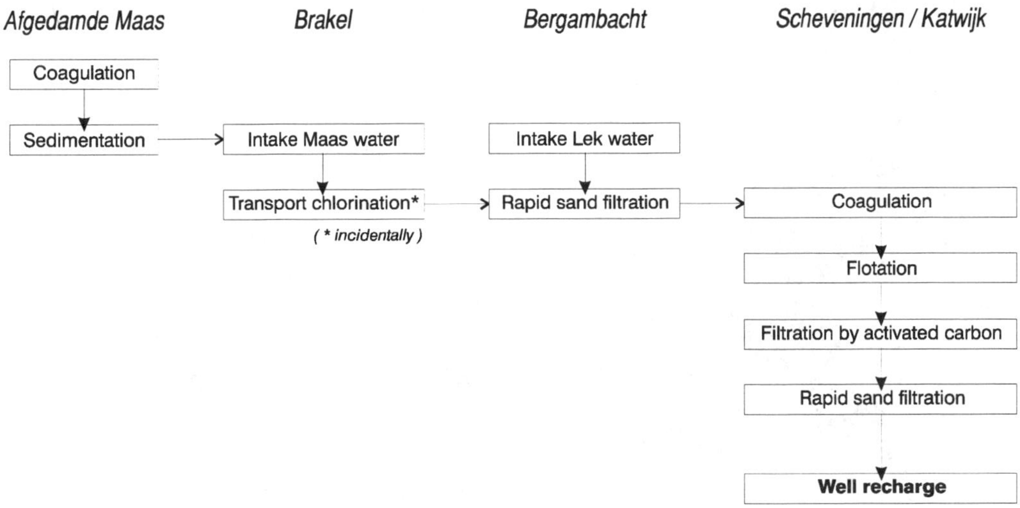

Figure 3. Pretreatment of surface water used for well recharge

over the 45 wells and to prevent underpressure in the transport pipelines of the feeding system, the amount of recharge water for each well is controlled by a specially constructed regulator valve. This valve is mounted on the lower end of the recharge pipeline, at a depth of MSL-5 m (MSL is Mean Sea Level). By placing this valve at the end of the recharge pipeline, recharging can be regulated so that there is always overpressure in the feed pipeline.

During operation, the total recharge volume, MFI index, turbidity and temperature are measured. The MFI value determines whether the water is suitable for recharging. If it is not suitable for well recharge, it is used for surface spreading.

3.3 *Recovery and posttreatment*

Recovery is carried out with 21 deep (vertical) recovery wells and in four vertical shafts, each equipped with two horizontal 90 m long well-screens (Fig. 2). During drilling of the vertical recovery wells, it transpired that the clay layer on the lower side of the aquifer was not continuous, and even absent from one part of the area. Because of the risk of drawing up brackish, salt groundwater, a study was made into the best way of operating the system.

Using model-based calculations that allow for differences in density between fresh and salt groundwater, the percentage excess recharge and the division of recharge and recovery over the different resources have been determined. This optimization was carried out using the calculated value for the leakage through the Kedichem clay layer at about 60 m below sea level. To prevent salt groundwater from leaking to the recovery wells, in the vicinity of the system the leakage must be directed downwards through this clay layer. Figure 4 shows the leakage in both the original and optimized situations. In the optimized situation, the percentage excess recharge has been increased to 14% and the division of recharge and recovery has been changed (Van Gerven, 1997).

The horizontal wells were built earlier, being combined with the installation of a recovery unit in the surface spreading system. Drains have been installed in both the phreatic and deep aquifers (Fig. 5) by drilling horizontally outwards from a central shaft. The maximum length of this type of well is 90 m, permitting in this case a total length of 180 m. This particular method has been used to keep the excavation and ecological damage in the area to a minimum.

Using the flow rates and operating hours of the pumps, the distribution of the extraction over the recovery wells was established. All the wells have

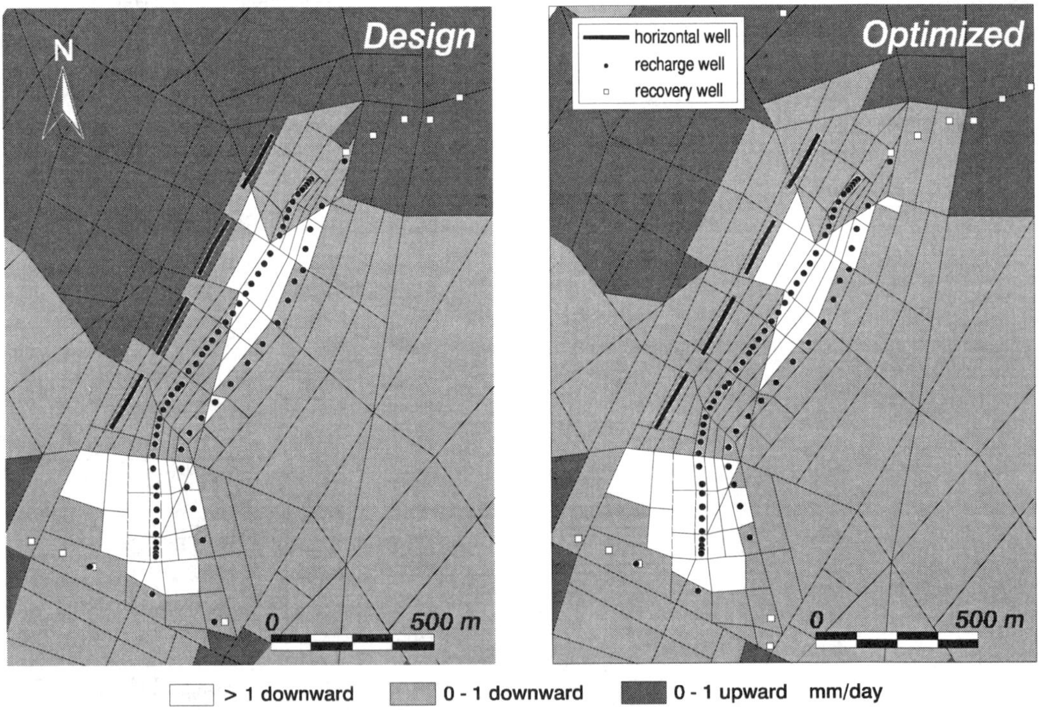

Figure 4. Leakage through the Kedichem clay layer during deep well recharge (Van Gerven, 1997)

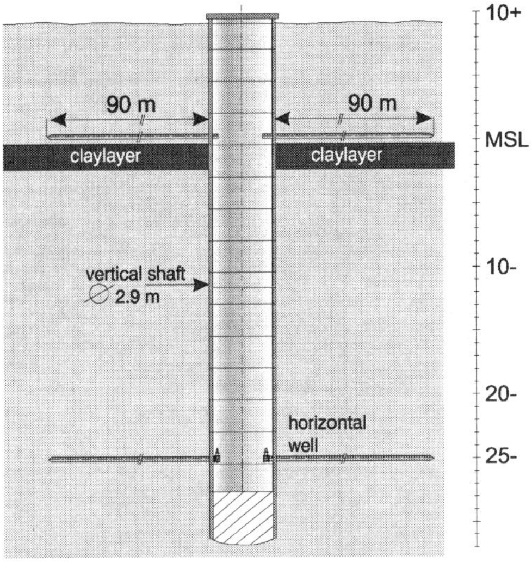

Figure 5. Cross section of a horizontal well

submersible pumps that pump the water through existing transport pipelines directly to the post treatment. This constitutes a final treatment consisting of aeration, rapid sand filtration and slow sand filtration. Because of the aquifer passage, the recovered water is hygienically safe, so that the final treatment need not include chemical disinfection.

3.4 *Monitoring*

Monitoring focuses on the quantity and quality of the recharge water and the recovered groundwater, as well as on the impacts of the project on the groundwater tables and the groundwater quality (Koopman 1995, Looy 1997). This latter aspect mainly concerns possible ecological and agricultural damage.

For the purpose of monitoring the groundwater, an extensive measurement network has been installed. This comprises observation wells with 1 m long screens at different depths. At regular intervals, samples are taken and the hydraulic head measured. Practically all these observation wells are fitted with an geo-resistivity measuring cable. This type of cable allows the measurement, at various depths, of the conductivity of the soil and the groundwater at the same time. Based on these measurements, the shifting of salt groundwater may be deduced.

3.5 *Initial experience*

In June 1997, a trial phase of the system began. The first months concentrated on testing the technical infrastructure and overcoming teething troubles in the system. Due to problems with pretreatment the project ran during the fist half year at 60% of its capacity. In connection with possible salination, the study has devoted much attention to the operation of the recharge and recovery resources. In practice, software controls the distribution of the flow of water over the recharge wells, with the help of the regulator valves. System operation can be so configured that the result is a stable system.

However, the distribution of the abstraction over the recovery means has to be set up manually by starting and stopping pumps individually. For the operators a less ideal situation, leading to a inefficient operation in which some wells are used continuously too long, and others too little. Hence the proposal to control the recovery means by software as well. In this way, it would be possible to precisely regulate the distribution of the abstraction over the different recovery wells.

Concerning the movements of the interface between fresh and salt groundwater, the period of less than a year allows only very general conclusions to be made. Below the recharge wells, the interface moves clearly downwards. In the neighbourhood of the recovery wells, there is virtually no movement at all.

In the future, more attention will be paid to the further movement of the interface between fresh and salt groundwater, and the dispersion of recharge water outside the well recharge system. Problems involving the clogging of recharge wells are not expected.

REFERENCES

Gerven, M.W. 1998. Optimalisatie diepinfiltratie-systeem Berkheide met het oog op verzilting. Geohydrologische modelstudie met dichtheids-stroming, Kiwa, Nieuwegein (in Dutch).

Koopman, K. 1995. Karteren van zoet-zout grondwater in het waterwingebied Berkheide (ZH). Afstudeerverslag Vrije Universiteit Amsterdam, Energie- en Watervoorziening Rijnland (in Dutch).

Looy 1997. Hydrochemie Berkheide; Bepaling nulsituatie diepinfiltratieproject Berkheide. N.V. Duinwaterbedrijf Zuid-Holland, Voorburg (in Dutch).

Olsthoorn, T.N. 1982 Verstopping van persputten. Kiwa mededeling 71 (in Dutch).

Olsthoorn, T.N. & S. van Harlingen 1994. Infiltratieputten: schoon boren door wandschrapen. H_2O, 27,(21), pp 636-639 (in Dutch).

Stuyfzand, P.J. 1993. Hydrochemistry and hydrology of the coastal dune area of the western Netherlands. PhD-thesis, Kiwa, Nieuwegein

Strack, O.D.L. 1994. A Dupuit-Forcheimer model for three-dimensional flow with variable density. Water Resources Research, 31, (12), pp 1309-1315

Woering M. 1995. Evaluatie boren diepinfiltratieputten diepinfiltratieproject Berkheide. Afstudeerverslag Vrije Universiteit Amsterdam, Energie- en Watervoorziening Rijnland (in Dutch).

Zagwijn, W.H. & C.J. van Staalduinen 1975. Toelichting bij de geologische overzichtskaarten van Nederland, Rijks Geologische Dienst, Haarlem, Nederland (in Dutch).

Artificial Recharge of Groundwater, Peters et al. (eds) © *1998 Balkema, Rotterdam, ISBN 90 5809 017 5*

Preliminary study for deep injection experiments at the Cornellà site, Barcelona

A. Pérez-Paricio & J. Carrera
Universitat Politècnica de Catalunya, Barcelona, Spain

ABSTRACT: Several tests are currently under execution at the Cornellà site within the framework of an European Project. One of the existing Aquifer-Storage-and-Recovery (ASR) wells was previously selected to perform specific experiments. The main objective is to study clogging around the ASR well itself, although redox reactions and water-aquifer material interactions are also being examined. Three different tasks were scheduled: analysis of recharge water and groundwater, analysis of soil cores, and completion of field tests. Both water and soil investigations served to obtain the basic parameters, to identify the aquifer characteristics and to assess some clogging-related properties. Still, natural heterogeneity and main hydraulic connectivity can not be derived through laboratory analysis. Therefore, various field tests are being carried out. The approach presented here is innovative as regards two aspects: first, a-priori relevant clogging parameters have already been measured, and, second, specifically derived numerical codes will be applied to obtained data.

1 INTRODUCTION

Aigües de Barcelona (Water Supply Company of Barcelona), acronym *Agbar*, is operating two Artificial Recharge systems in the Llobregat River aquifers, near Barcelona.

In 1950 they started by increasing the natural recharge through ploughing the riverbed sediments in Pallejà, where the aquifer is phreatic (*Agbar* 1987). However, surface recharge is only done when restrictions shown in Table 1 are applicable .

Table 1. Limits on river water for surface recharge.

Flow:	Between 10 and 35 m^3/s
Turbidity:	< 150 NTU
Chlorine Break-point:	< 12 mg/l
Chloride:	< 700 mg/l
Other quality parameters:	Regular monitoring

In 1969 a new methodology was adopted: seven existing wells were adapted for double purpose use, becoming one of the first ASR schemes in the world. These are 950 mm diameter wells, 40 m deep, 10 m screen, with a recharge capacity of 50 l/s (pumping rate up to 200 l/s), and are operating in the Llobregat River Lower Valley. The aquifer splits here into two different units: the upper (phreatic) one and the lower (semi-confined) one, where the ASR wells are screened. The ASR scheme was projected to improve water management, when an additional (higher quality) source of water from the Ter River was available for the Barcelona metropolitan area, in 1966. As a result of this, the water treatment plant of Sant Joan Despi did not operate at full capacity during lower-demand periods. ASR was contemplated as an effective means of increasing water storage at relatively low costs. It proved to be a successful experience, so that five ASR wells were added to the old ones in the late 70's. These have 20 m screen and can inject at 100 l/s, whilst the pumping rate can attain 400 l/s (*Agbar* 1987).

Consequently, the ASR wells inject the surplus of treated water (accomplishing the drinking water standards) whenever it is considered necessary. This allows to store the water in the aquifer, recovering it when peak demands exist. In spite of the treatment, some restrictions are set on the river water quality in order to preserve groundwater quality (Table 2).

Table 2. Restrictions on raw water quality. These are maximum concentrations in Llobregat River limiting Artificial Recharge.

Turbidity:	< 0.6 NTU
Salinity (in [Cl^-]):	< 500 mg/l
Chlorine Break-point:	< 15 mg/l
Detergents:	< 1 mg/l

The effectiveness of such management methodology is evident, as it is nowadays operating (Valdés

1992). A good picture is provided by Figure 1, where annual abstracted and recharged volumes are plotted. Success is due to several factors: good quality of recharge water, appropriate restrictions to preserve groundwater quality, clogging prevention by specific protocol, dynamic operation of the whole system, and constant improvement of facilities.
It is worth commenting that a conservative injection protocol was adopted in order to prevent clogging. It basically consists of recharge cycles of up to 15 days of continuous operation (for each well, irrespective of the differences in recharge flowrate). A shock pumping, at a flowrate which is four times larger than the injection one, is immediately carried out. This helps minimise the impact around the recharge wells. Airlifting is applied when further redevelopment is needed.

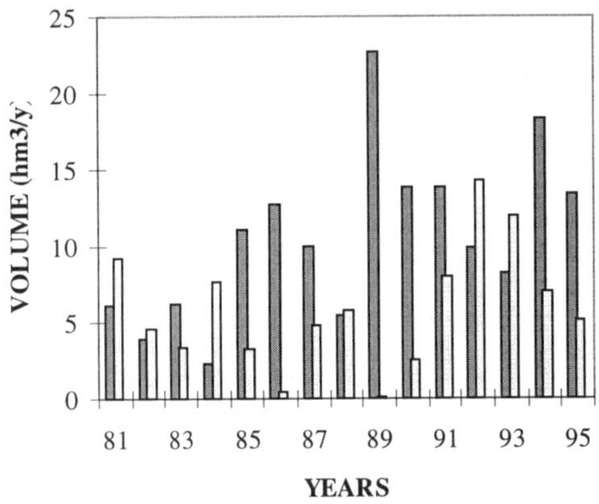

Figure 1. Abstraction and Artificial Recharge by *Agbar*. Dark columns refer to abstraction, while grey ones represent total (surface and deep) recharge. Typical oscillations of Mediterranean climate can be observed.. 1 $hm^3 = 10^6 m^3 = 1 Mm^3$.

This paper concentrates on the preliminary study experiments that has been conducted around one of the ASR wells (P-13). It was selected because several observation wells exist near the recharge well, which is basic both to describe the hydrogeology of the systems and to allow for accurate measurements. Moreover, two new piezometers were drilled in the close vicinity of well P-13 with the aim of achieving a good characterisation of aquifer heterogeneity and obtaining soil cores.

All the investigations are being conducted within the framework of an European Union Project entitled *Artificial Recharge of Groundwater*. The basic objective in the Cornellà site is to investigate the validity of a recently developed clogging model (Pérez-Paricio & Carrera, this volume), which is capable of dealing with physical, biological and chemical processes. Moreover, redox reactions and water-aquifer material interactions are also being considered.

Three main tasks have been carried out to achieve the aforementioned objectives: analysis of recharge water and groundwater, analysis of soil aquifer cores, and field tests.

This paper describes the preliminary study at Cornellà.. A brief description of the site is presented first. Then, all the experimental procedure is explained, together with the ongoing work Finally, a discussion of gathered information and future operations is enclosed.

2 SITE DESCRIPTION

Recharge well P-13 was drilled in 1934. It is located in Cornellà (Barcelona), close to the Sant Joan Despí Treatment Plant. It is 40.10 m deep, with a 950 mm inner diameter iron casing and 9.60 m screen in the lowest part. It was drilled to benefit from the high transmissivity of the semi-confined aquifer unit.

The aquifer is formed by boulders, coarse gravel and sand, with finer-sized embedded layers. Transmissivity values of 20,000 m^2/d are typical. The aquifer is phreatic up to city of Cornellà, where it is 2100 m wide and 40 m deep. From here to the Mediterranean sea (eastwards), the aquifer splits into two different units. Both units are separated by a low-permeable wedge of mud, whose maximum thickness (near the sea) is 40 m. The storage capacity of the semi-confined lower aquifer is estimated at around 114 hm^3, and its thickness varies between 30 and 45 m (*Agbar* 1987).

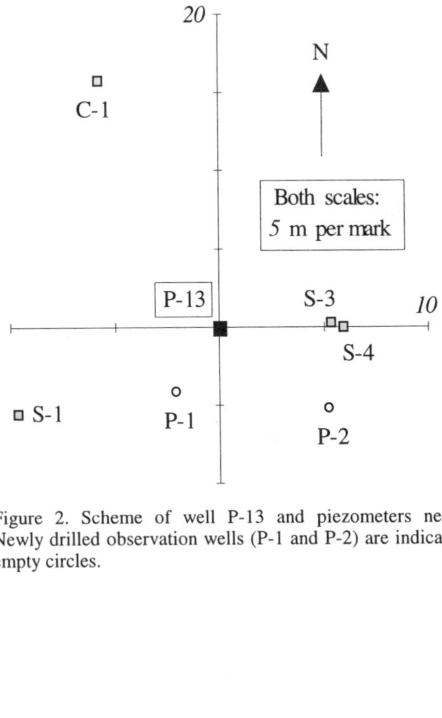

Figure 2. Scheme of well P-13 and piezometers network. Newly drilled observation wells (P-1 and P-2) are indicated by empty circles.

A schematic plot of recharge well P-13 and the existing observation wells is shown in Figure 2. Two groups of observation wells exist: the already existing ones and the specifically-drilled piezometers. Table 3 depicts some of the essential characteristics of all these wells.

On the other hand, groundwater chemistry stands out by notable concentrations of basic ions. This can be observed in Table 4. Also, all concentrations remain fairly constant despite the dynamic evolution of the aquifer.

Table 3. Characteristics of existing wells.

Well	Length (m)	Diameter (mm)	Distance** (m)	Screen (m)
P-13	40.10	950	0.47	9.60
P-1	47.50	200	4.68	15.00
P-2	53.30	200	7.38	18.00
S-1	37.40*	75*	11.15	-
S-3	20.90*	75*	5.35	-
S-4	39.50*	75*	5.94	-
C-1	-	75*	16.84	-

*Estimated values.
**From P-13 well.

Table 4. Groundwater: main parameters.

PARAMETER		1994	Apr-97	Nov-97
Turbidity	(NTU)	1.12	-	-
EC	(µS/cm)	2023	2010	1950
Alkalinity	(mg/l $CaCO_3$)	299	-	307
Hardness	(mg/l $CaCO_3$)	609	-	670
Cl^-	(mg/l)	386	363	348
HCO_3^-	(mg/l)	365	390	374
SO_4^{-2}	(mg/l)	218	323	340
NO_3^-	(mg/l)	17	15.8	18.6
Ca^{+2}	(mg/l)	156	192	184
Mg^{+2}	(mg/l)	44	52	51
Na^+	(mg/l)	213	193	193
K^+	(mg/l)	41	42	41
Fe (total)	(µg/l)	24	38	23
Al	(µg/l)	21	220	-
TOC	(mg/l C)	-	1.9	-

3 PRELIMINARY STUDY

Experiments conducted to obtain a satisfactory description of the system within the EU Project are presented in this section. We have focused on the recharge water parameters, soil aquifer properties and field tests.

3.1 *Recharge water*

Tap water is the only source of recharge water, so that this one satisfies all drinking water standards. In particular, significant doses of chlorine are permanently added to it in order to achieve a biologically stable water. This is why we have not looked at the biological aspects of Artificial Recharge. Instead, two major issues have been investigated: suspended particles and redox reactions.

The former is relevant concerning clogging at the recharge well. The latter is significant due to the oxidising potential of chlorine.

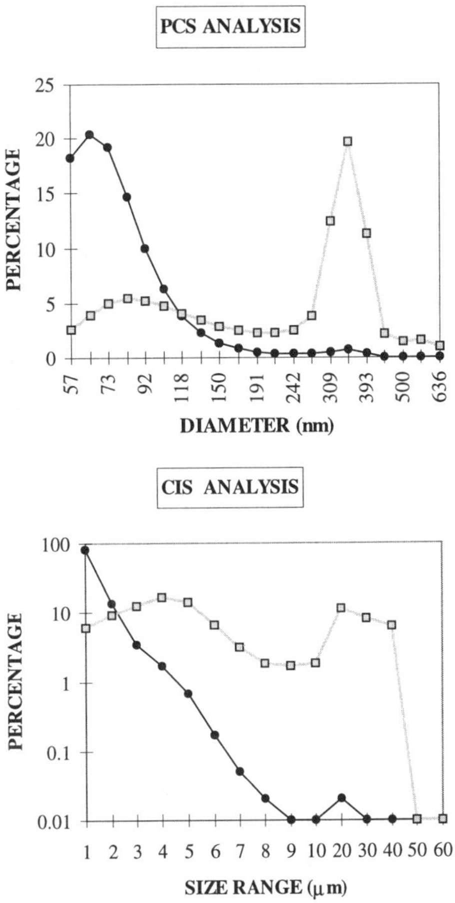

Figure 3. Suspended solids: characterisation by optical methods (T. Hofmann, IfW, Dortmund, Germany). Black: number concentration; grey: volume concentration.

The concentration of suspended particles is obviously low. Conventional techniques, such as membrane filtration (concentration of suspended solids) and turbidity measurements, cannot provide detailed information for high quality water (Pérez-Paricio 1998). Instead, special optical methods have been applied to obtain precise determinations of the particle size distribution (psd) and mean values (Figure

3). Analysis were performed at the IfW (Schwerte, Germany), as a collaboration in the aforementioned ARGW EU Project (Second Annual Progress Report, 1998). Two optical methods were used: Photon correlation spectroscopy (PCS) was applied to characterise the fraction of particles whose diameter is lower than 600 nm (1 nm = 10^{-9} m), whilst Computer image system (CIS) was used for the fraction between 500 nm and 60 μm (1 μm = 10^{-6} m).

This procedure involved transportation and a lag time between sampling and analysis. These factors can affect the psd of the samples, especially through processes such as coagulation and precipitation. They could alter the obtained conclusions in case it affected the distinction between colloids (particles whose diameter is smaller than 1 μm), intermediate particles (size between 1 μm and 30 μm) and large particles (above 30 μm). Given the fact that coagulation processes are employed during water treatment, through addition of aluminium sulphate, it is unlikely that significant changes have been artificially generated during manipulation. Sedimentation is not relevant in view of results presented in Figure 3, apart from a possible artefact at the 20 μm size. However, the number percentage can not be considered representative enough for clogging classification purposes.

This classification is essential for modelling, because the physico-chemical behaviour of suspended particles depends on their size.

- Colloids are only affected by surface forces, such as London-van der Waals attraction and electrokinetic interactions, and by Brownian diffusion.
- Large particles, on the other hand, are only subject to volumetric forces, such as sedimentation and inertia, and to interception mechanisms.

It stems from Figure 3 that the only relevant size of particles lies in the colloidal range, with a predominant class at around 70 μm and a secondary peak at 350 μm, approximately. It is precisely this secondary class that is important from the point of view of clogging. From CIS analysis, it is deduced a concentration of suspended solids of $8.3 \cdot 10^4$ particles/ml water (that is, of particles between 500 nm and 60 μm). The mean diameter value as determined from the volume concentration analysis is 0.69 μm, with a standard deviation of 0.71 μm. As for PCS, the volumetric mean value is set at 250 nm. Therefore, recharge water only carries small colloidal particles, as expected.

Recharge water quality is extensively and continuously controlled by *Agbar*, that publishes annual reports collecting average and extreme concentrations. Most of the key parameters tend to stabilise or even to decrease during the last years (e.g. peak values of turbidity). Average values of some of these are presented in Table 5.

Table 5. Average concentration values of key chemical parameters for groundwater, in 1995 (*Agbar*, 1996).

Parameter	Units	Value
Turbidity	NTU	0.5
pH	-	7.2
Electrical Conductivity	μS/cm	
Chloride	mg/l Cl^-	
Total organic carbon (TOC)	mg/l C	
Iron	μg/l Fe	

3.2 *Soil analysis*

The properties of the aquifer material are fundamental with respect to clogging, water-aquifer interactions and redox processes. The following ones were carefully analysed: granulometry, cation exchange capacity (CEC) and mineralogy.

Granulometric curves of available cores from the newly drilled observation wells (P-1 and P-2) were obtained. It is worth noting that the drilling method (cable tool) is not very precise to perform these determinations, as both the finer and the coarser fractions are generally biased. In any case, results agree with visual inspections. Mean grain diameter of the order of centimetres are deduced from cores.

CEC analysis were performed, after drying the core samples, at the Universitat de Barcelona (Geology Department). Results are not explicitly shown in this paper, but they yield logical higher values for mud and clay layers. Values of up to 17.20 meq/100 g soil were obtained for these, while layers formed by coarse materials have lower values (below 10). It must be insisted that the sampling method is essential to achieve reliable figures.

Semi-quantitative x-ray analysis were executed at UPC and also at CSIC (Institut Jaume Almera, Barcelona). High contents of quartz were detected, in accordance with the known mineralogy of the aquifer. Also, calcite and dolomite were identified, as were minor fractions of the following minerals: clinochlore, microcline ferroan, albite, muscovite and chamosite.

3.3 *Field tests*

The following field tests were conducted at the Cornellà site (Table 6):

1. Development of new observation wells. After drilling piezometers P-1 and P-2, they were pumped at a rate of 15 l/s both to achieve a good development and to control the aquifer response. However, this one was very small (around 5 cm after pumping for 4 hours), in accordance to the high previous estimates of aquifer transmissivity.
2. First step drawdown test. A two-step test was done, by pumping from P-13 well at 45 l/s for 73 minutes and at 190 l/s for 5.5 hours. Heads were manually registered at all available wells.

Interpretation with MariaJ_IV (Carbonell et al. 1997) -code for automatic calibration of pumping tests- was applied at each point individually, that is, without proposing a unique model for all of them. Rather homogeneous transmissivity values, ranging between 15,000 m^2/d and 40,000 m^2/d, were obtained by means of the Hantush model. On contrary, the storage coefficient varied between 10^{-7} and 10^{-2}. An interesting discussion on this effect can be found in Meier et al. (1998).

3. Second step-drawdown test. This time, a four-steps test was done, again by pumping from the P-13 well. Flowrate was 50 l/s for 62 minutes, 120 l/s for 62 minutes, 160 l/s for 89 minutes and 207 l/s for 18 hours. However, during this fourth step it was necessary to detain pumping because of Plant needs. Obtained results were similar to the previous test.

4. Pilot recharge test. A short-term recharge run was executed, injecting at 50 l/s in P-13 for 22.5 hours. This test served to check the performance of both equipment and wells.

5. Pumping test at P-1. Another pumping test was carried out at P-1 in order to obtain quantitative information of aquifer heterogeneity. The objective was to characterise the medium by monitoring the hydraulic response at different piezometers. But only 17 l/s were abstracted, for 8 hours, so that it was not enough to create a big impact on the aquifer heads. A maximum drawdown of 8 cm was recorded.

Table 6. Objective and schedule of field tests.

TEST	DATES	PURPOSE
Develop (o.w.)	February 97	Hydraulics
2-steps drawdown	April 97	Hydraulics, heterogeneity
4-steps drawdown	September 97	Hydraulics
Pilot recharge	September 97	Hydraulics, quality
Pumping (o.w.)	November 97	Heterogeneity
Tracers	*	Interaction, heterogeneity
Recharge	*	Clogging
Analysis	*	Clogging

* Not completed at the time of preparing the paper.
o.w. = observation wells

4 ONGOING WORK

This section provides a simple description of ongoing work at Cornellà, namely the completion of field tests (Table 6) and the application of some new numerical codes.

4.1 *Field tests*

1. Tracer test. Bromide and lithium have been selected, as the groundwater content of other ions is considerable and it is not possible to add undesirable compounds. The test will carried out by introducing the tracers in piezometers P-1 and P-2, and pumping at 50 l/s from well P-13.

Tracer tests are essential to obtain travel times, assess interactions between aquifer material and non-conservative tracers, and monitor breakthrough curves for redox compounds.

2. Long-term recharge test. The purpose of this test is to measure clogging-induced reductions of intrinsic permeability around P-13 well, basically by retention of suspended particles.

3. Analysis of the clogging materials. Just after stopping recharge an immediate pumping test will be done both to redevelop P-13 well and to examine the composition of the clogging layer. Results from mineralogical analysis will suggest the origin of these materials.

4.2 *Numerical modelling*

Should the recharge run cause significant variations in permeability, a new numerical comprehensive code for clogging (CLOGEL) will be applied to the recorded data (Pérez-Paricio & Carrera, this volume). As for the tracer tests, chemical reactions will be modelled too.

5 DISCUSSION

This paper describes the detailed characterisation of an Artificial Recharge site in order to achieve a good understanding of the main processes. Assessing the nature of these processes is essential when dealing with complex inter-related phenomena, as is the case of clogging.

Analysis of water and aquifer material provide evidence of the major issues, but it is only through field tests that a reliable representation of the system can be drawn. The development of more elaborated mathematical models is essential to improve the performance of operating facilities, but it is also necessary from a conceptual point of view. And this is only possible if the basic parameters in actual systems are well controlled.

ACKNOWLEDGEMENTS

This work was funded by the European Union (Project ARGW, Artificial Recharge of Groundwater). Thanks are also due to T. Hofmann (IfW, Dortmund, Germany), for the suspended particles determinations and to C. Ayora (Institut Jaume Almera, CSIC) for his generous collaboration concerning x-ray diffractions.

REFERENCES

Aigües de Barcelona (*Agbar*) 1987. Reàrrega artificial d'aqüífers als deltes dels rius Llobregat i Besòs. *Monograph* No. 3, 16 pp.

Aigües de Barcelona (*Agbar*) 1996. Water quality data, period 1991-1995. *Monographs.*

Carbonell, J.A., A. Pérez-Paricio & J. Carrera 1997. User's guide of MariaJ_IV: an inverse calibration code for pumping tests with analytical and numerical models. *Univesitat Politècnica de Catalunya & ENRESA.*

ARGW EU Project 1998. Second Annual Progress Report, corresponding to year 1997. Not published.

Meier, P., X. Sánchez-Vila & J. Carrera 1998. *Water Resour. Res.*, accepted for publication.

Pérez-Paricio, A., 1998. Clogging and Artificial Recharge of Groundwater: Fundamental aspects. *M.Sc. Thesis at the Universitat Politecnica de Catalunya,* Barcelona (Spain). 170 pp.

Valdés, J.L. (*Agbar*) 1992. Experiencias de recarga artificial en los acuíferos del Río Llobregat y Río Besós. Barcelona.

Artificial Recharge of Groundwater, Peters et al. (eds) © 1998 Balkema, Rotterdam, ISBN 90 5809 017 5

Well recharge pilot in the south-east Netherlands

R. Straatman, Y. Brekvoort & S. M. L. Verheijden
Oost Brabant Water Company (WOB), 's-Hertogenbosch, Netherlands

ABSTRACT: Because of the restrictions placed on the use of groundwater to produce drinking water and industrial water, the WOB, NRE and WML water companies took the initiative in 1994 to conduct a deep well recharge tests with surface water from the Zuid-Willemsvaart, a small canal in the south of the Netherlands. The project, known as DIZON, will be completed at the end of 1998. The DIZON project demonstrates that deep well recharge is a suitable method for preparing drinking water from surface water obtained from the Zuid-Willemsvaart canal. To produce drinking water of equal quality to that prepared from groundwater, a pilot plant for deep well recharge has been set up in the town of Someren. The water is pretreated, deep well recharged and recovered locally at a depth of 300 metres below surface level.
Recharge ensures hygienic reliability. The pretreatment, including, in order, microstraining, flocculation, flotation, rapid sand filtration, active carbon filtration and slow sand filtration, makes the surface water suitable for deep well recharge. Problems such as well clogging are now under control. The DIZON-project shows that operating on a production scale is technically feasible. This paper concentrates on the experience with recharge. The treatment results of the pretreatment, and the above-ground simulation of well clogging are also briefly considered.

1 INTRODUCTION

In 1994, three water companies started a five-year pilot for the well recharge of pretreated surface water from the Zuid-Willemsvaart canal, at a depth of approximately 300 metres below surface level. At the same depth, water was pumped using a separate well. The pumping and recharge wells are 100 metres apart. The pilot is being carried out in order to have all knowledge available by the end of the millennium, in case the use of groundwater for producing extra drinking water is no longer permitted. Loss of wetlands is currently a key issue in the Netherlands. Artificial recharge is studied because it combines the best of both worlds: ground water and surface water. A source that is qualitatively unreliable is turned into a safe one. During recharge, surface water turns into groundwater of constant quality and temperature.

2 PRETREATMENT

2.1 Purpose and design of the study

Surface water that is to be recharged is evaluated against the criteria included in the Soil Conservation Recharge Decree. To prepare this water from the surface water from the Zuid-Willemsvaart canal, a pretreatment plant has been set up on the DIZON site. The flowchart shows the processes used.

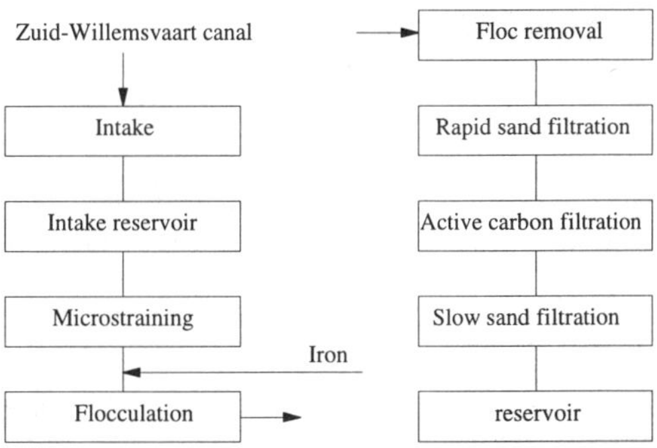

Figure 1. DIZON pretreatment flowchart (Keltjens & Van de Wetering 1997)

The pretreatment consists of the following seven steps, of which the necessity of the last step (slow sand filtration) is questioned:

1. The intake reservoir, for the sedimentation of suspended solids and the smoothing out of extre-

me variations in quality (residence time: 1 day).
2. Microstraining, during which algae and relatively large elements are removed.
3. Flocculation, in which dosage of a flocculant (iron(III)chloride) causes flocs to be formed. The flocs adsorb organic matter, as well as suspended solids and colloids.
4. Floc removal by means of flotation. Through the introduction of supersaturated water, air bubbles are created. The air bubbles attach themselves to the flocs, causing them to float to the surface.
5. Rapid sand filtration, which removes suspended material.
6. Active carbon filtration, which removes organic constituents through adsorption and biological action.
7. Slow sand filtration, which removes small, filterable particles and colloids, and reduces the biological activity of the water.

2.2 Results

The water quality of the Zuid-Willemsvaart canal is comparable to that of the Maas (Supèr et al 1996). Specific features of the Zuid-Willemsvaart canal are its relatively high concentrations of heavy metals (cadmium, arsenic, chromium and lead), cyanide and pesticides (atrazine and simazine). The microbiological quality is better than that of the Maas. The results presented here are based on one year's experience of pretreatment (Keltjens & Van de Wetering 1997).

In broad terms, the chemical composition of the water changes during pretreatment, as described below. Through the dosage of iron(III)chloride, the pH value and hydrogen carbonate content both reduce during flocculation. The chloride content increases, as does the manganese content. During floc removal, an average of 95% of iron and flocs are removed. In relation to the intake pond (1), the total organic carbon (TOC) content decreases by 35%, and the UV extinction by 40%. Following rapid sand filtration, the iron content is less than 0.03 mg/l. At water temperatures above 7°C, denitrification occurs and manganese completely disappears. The active carbon filtration causes UV extinction and TOC content to decrease considerably. The membrane filtration index (MFI) of the filtrate is on average between 1 and 5 s/l^2. The assimilable organic carbon (AOC) content of the filtrate is on average between 5 and 20 µg/l carbon acetate. The slow sand filtration produces a further improvement in the water composition. The MFI decreases on average to between 0.2 and 1.0 s/l^2. The AOC decreases on average to between 2.5 and 11.0 µg/l carbon acetate.

2.3 Conclusions

The pretreated water from the Zuid-Willemsvaart canal completely satisfies the criteria of the Soil Conservation Recharge Decree, and can be recharged at the DIZON site without any structural problems of a technical nature.

3 CHANGE IN WATER QUALITY DURING RECHARGE

3.1 Purpose and design of the study

To understand the treatment necessary for the water after recharge and recovery, the water quality is tracked through monitoring wells from the time of recharge until the time of recovery. The drawings below (plan and cross section) show the locations of the recharge, monitoring and pumping wells.
The pretreated water from the Zuid-Willemsvaart canal is recharged by a recharge well (IP.2) at a depth of between 280 and 310 metres below surface level. The recharged water is recovered through a pumping well (PP.1) at a distance of 98 metres from the recharge well. At distances of 8, 12, 22 and 38 metres from the recharge well are monitoring wells WP.3, WP.2, WP.4 and WP.1 respectively, each with observation filters in their target aquifers (figure 3). In terms of permeability, the upper part of the target aquifer is moderate (transmissivity is 250 - 275 m^2/d) and the lower part is the best (transmissivity is 330 - 370 m^2/d). For all parts, permeability increases in the direction of the pumping well (Castenmiller 1995).
Since the start of the study, samples have been regularly taken from nearly all the filters for extensive anorganic chemical examination. The recharge and recovery was 30 m^3/h. During the recharge experiment, measurements are taken at regular intervals of (Stuyfzand 1996):

- The recharge flow, water level in the recharge well, groundwater levels and heads
- The water temperature in the subsoil at the level of the recharge route
- The composition of the recharged water before recharge, and after recovery.

To determine the initial geochemical status, five sand samples were taken from the target aquifer, at depths ranging from 275 and 312 metres below surface level. The samples came from the Waubach Sands, which are part of the continental Kiezel-oolyt-formation from the Late Miocene epoch (Stuyfzand & Lüers 1994). The samples are low in calcium (0.02–0.30%), highly anoxic (no oxygen, nitrates or sulphates) and basic (pH > 7). For a cation exchange capacity that is low for sand (10-110 meq/kg), with

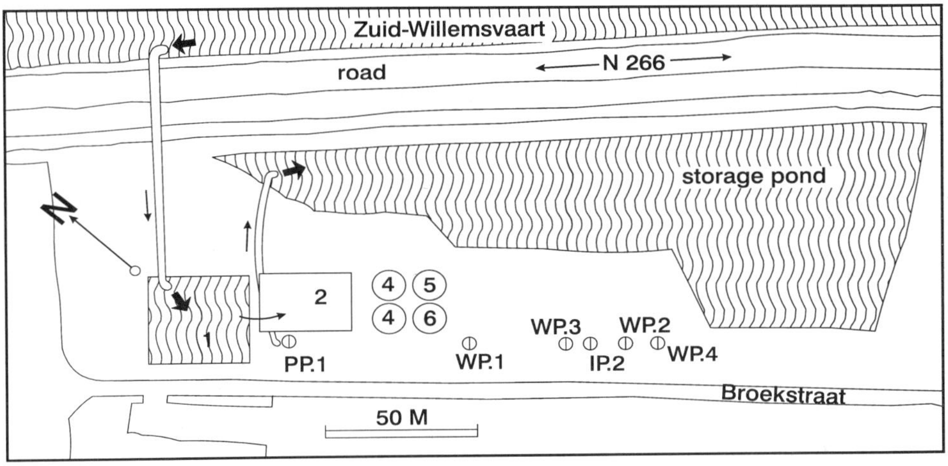

Figure 2. Plan of DIZON site showing locations of recharge well (IP.2), sampling wells (WP.1, WP.2, WP.3 and WP.4) and pumping well (PP.1)
Key: 1= intake reservoir, 2 = pretreatment, 4 = slow sand filters, 5 = recharge /backflush water input buffer, 6 = backflush water output buffer

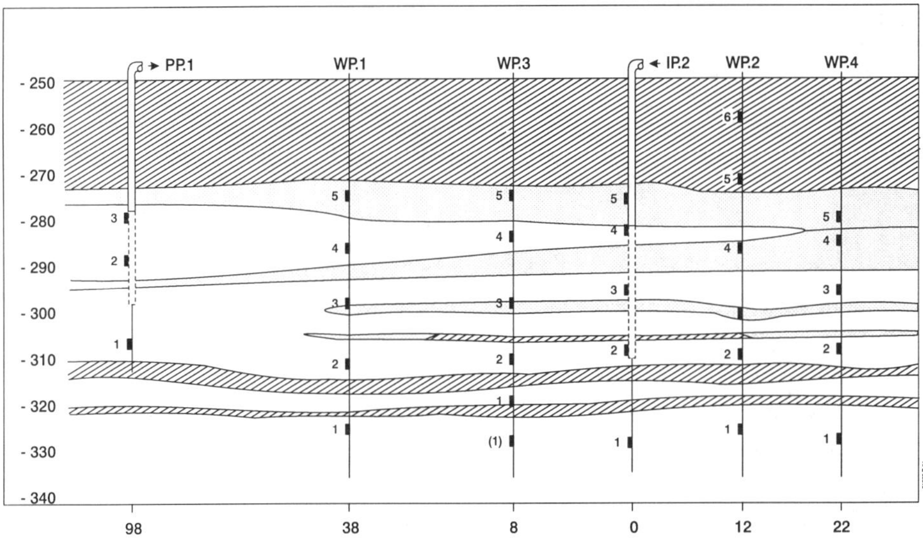

Figure 3. Cross section of the monitoring wells from recharge well (IP.2) to pumping well (PP.1). On vertical axis is depth (m) below surface level. On horizontal axis is distance (m) to IP.2.

an adsorption complex very high in calcium (> 80%), they have a fairly high content of adsorbed Fe^{2+} (and relatively little ammonium), a moderate organic matter content (0.3–2.3%), and a high sulphide content (650–4640 mg S/kg), probably mostly in the form of FeS_2.

The target aquifer contains groundwater of excellent quality. The water at the location of the monitoring wells can be described as highly anoxic, calcium-rich water with noticeably low concentrations of methane and chloride (Stuyfzand 1997). The water contains no organic micro-contaminants and has a very low concentration of trace metals. Arsenic, barium, cobalt and zinc are present in higher concentrations, probably a result of boring the holes with rotary flush using large amounts of working and drilling water.

3.2 Results

The pretreated water from the Zuid-Willemsvaart canal is easily identified in the target aquifer by its higher chloride concentration and abnormal temperature curve. During the first year, the recharge water has hardly spread beyond the target aquifer. The pumping well is reached in approximately 200 days, at least, that is, in the lower part of the aquifer. In one year, approximately 40% of the recharged water has been recovered.

The change in water quality during recharge depends mainly on the oxidation of pyrite, the oxidation of organic matter, and cation exchange (subsequent release of adsorbed iron, manganese and ammonium). Figure 4 shows this change in water quality for oxygen, nitrite, nitrate and sulphate (Stuyfzand 1997).

It is striking that measurable concentrations of nitrate occur at most measuring points in combination with measurable concentrations of iron, manganese and ammonium. In normal groundwater this is inconceivable because iron does not dissolve until all the nitrate has gone. Thermodynamic equilibrium has not been reached yet because the reactions proceed slower than the flux of the groundwater.

During recharge, the total inorganic carbon concentration (TIC = (bi)carbonate + carbon dioxide) continues to increase at all locations by 0.06–0.21 mmol/l, as a result of three processes:

- cation exchange, in which calcium adsorbs under desorption of ammonium, iron and manganese
- neutralization by bicarbonate of the acid that is released during oxidation of pyrite
- oxidation of organic matter by oxygen and/or nitrates.

Sodium exhibits minor adsorption during the beginning. Chloride is an excellent tracer of the recharge water. There is an exchange between potassium and magnesium: the concentration of potassium decreases, while that of magnesium increases, until both reach equilibrium after roughly the same time.

Of the trace metals, only arsenic and barium are regularly detected; while cobalt and zinc are sporadically detected. A link between oxidation of pyrite, and mobilization and occurrence of arsenic, cobalt and zinc cannot be demonstrated. Barium exhibits long-term desorption.

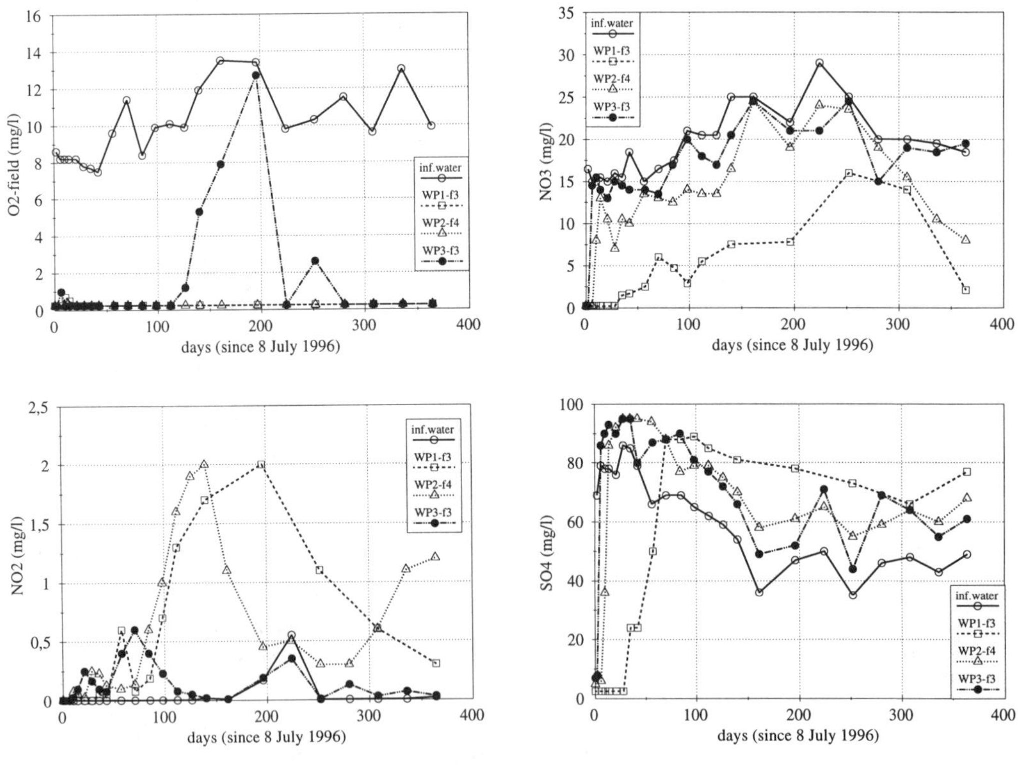

Figure 4. Change in water quality during recharge for oxygen, nitrite, nitrate and sulphate (Stuyfzand 1997).

3.3 Conclusions

If deep well recharge is continued with the same pretreated water from the Zuid-Willemsvaart canal, then, broadly speaking, the same pattern of change in water quality during recharge will emerge (Stuyfzand 1997). Small changes will occur in the middle and lower parts. For the middle part, the exchange reactions will have stopped and equilibrium reached. In the lower part, the picture will probably change because pyrite and weak organic matter will have leached out almost completely. Continuation and adaptation of the monitoring programme are advisable because prognoses are always uncertain, quality changes can occur through well clogging, seasonal influences are not yet sufficiently identified, and because it is still unclear whether or not oxygen and/or nitrates oxidize organic matter that does not degrade easily.

4 SIMULATION OF WELL CLOGGING

4.1 Purpose and design of the study

To predict clogging of the recharge wells and to investigate the effect on well clogging of using slow sand filtration (SSF) as the final pretreatment step, a laboratory simulation using columns has been conducted (Bunnik 1998, Hijnen et al 1998). With these parallel filters, the clogging resistance of the recharge well is compared with the clogging resistance of four columns that are filled with sand typical for the soil sediment of the target aquifer at the DIZON-location.

The study was conducted for 11 months (August 1996–July 1997). Three columns were fed with SSF-effluent, with different fluxes. The values were 3, 6 and 12 l/h, which correspond respectively to one, two and four times the recharge rate at the bore hole wall of the recharge well. To determine the effect of SSF, one column was loaded with active carbon filtrate (ACF). This is water that is subjected to active carbon filtration but not slow sand filtration (SSF). The flux was 6 l/h. At the beginning and end of the study, the clogging of the four columns and the recharge well were measured (kPa at 10°C, corrected for viscosity). In this case, we only briefly mention the results and conclusions relevant to the DIZON-location.

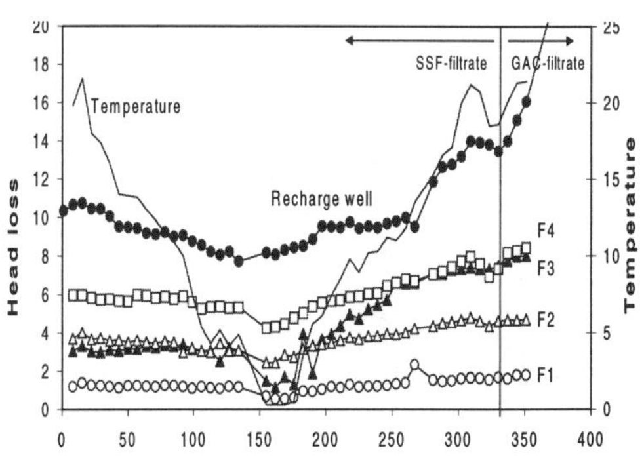

Figure 5 Comparison of parallel filters and recharge well, concerning the rate of clogging (Hijnen et al 1998). Temperature in centigrades celcius. Headloss in kPa.

4.2 Results

Figure 5 shows the increase in clogging for each column that occurred during the study. For comparison purposes, this figure also shows the increase in clogging of the recharge well. The figure illustrates that the clogging of the columns is somewhat less than the clogging of the recharge well.

4.3 Conclusions

The rate of clogging of the recharge well is no cause for concern, although it is somewhat greater than that in the parallel filters. Because of different characteristics (grain size distribution) of the sand in the columns and the sand in the aquifer, it is difficult to predict the well clogging of the recharge well on the basis of the results of the parallel filter study (Hijnen et al 1998). Clearly, however, water not treated with SSF leads more quickly to well clogging.

5 CONCLUSION

The DIZON-project demonstrates that it is possible to prepare drinking water by using deep well recharge and recovery from surface water from the Zuid-Willemsvaart canal. Problems such as well clogging are now under control. If additonal capacity for the production of drinking water is needed, then the DIZON-pilot shows how this can be achieved.

REFERENCES

Bunnik, J. 1998. Meeloopfiltratie-onderzoek DIZON. Rapportage eerste 11 maanden. Kiwa-rapport KOA 97.183. (in Dutch)

Castenmiller, F.J.C. 1995. Basisdocument Geohydrologie; diepinfiltratie Zuid-Oost Nederland. Rapport Werkgroep Diepinfiltratie DIZON (in Dutch)

Hijnen, W.A.M. *et al* 1998. Determining the clogging potential of water used for artificial

recharge in deep sandy aquifers. This volume.
Keltjens, L.L.M. & T.S.C.M. van de Wetering 1997. Ervaring voorzuivering proefproject DIZON.
Stuyfzand, P.J. 1996. Voorstel tot waterkwaliteitsbewaking van diep-geïnfiltreerd Zuid-Willemsvaartwater te Someren, 3e versie. Notitie DIZON 96-03, 15-01-96 (in Dutch)
Stuyfzand, P.J. 1997. Onderzoeksresultaten 1e jaar diepinfiltratie DIZON. Kwaliteitsverandering tijdens bodempassage. Kiwa-rapport KOA 97.220. (in Dutch)
Stuyfzand, P.J. & F. Lüers 1994. Chemische samenstelling van Boven-Miocene zandmonsters uit boring 51H.171 te Someren in relatie tot diepinfiltratie. Kiwa-rapport SWE 94.049. (in Dutch)
Supèr, J. et al 1996. Kleine oppervlaktewateren: een bron voor drinkwater? Verslag van een jaar waterkwaliteitsonderzoek. Kiwa-rapport SWE 96.007 (in Dutch)

2 Poster presentations

2.1 Water management in arid regions

Artificial Recharge of Groundwater, Peters et al. (eds) © 1998 Balkema, Rotterdam, ISBN 90 5809 017 5

Development of a simple tracer method to assess contribution of artificial recharge ponds to augment groundwater resources

B.S.Sukhija, D.V.Reddy & P.Nagabhushanam
National Geophysical Research Institute (NGRI), Hyderabad, India

ABSTRACT: This paper illustrates the development of a new tracer method, which can be applied in artificial recharge studies. The method is simple, reliable and sensitive enough to respond to temporal changes. It is employed to assess quantitatively the contribution of artificial recharge ponds (called Percolation Tanks, in India) to augment groundwater resources. The method, based on the mass balance of environmental chloride in the impounded water, in fact, estimates the efficacy of artificial recharge structures. This simple method has been tested in different tanks situated in various geological terrains. Our measurements show that the percolation rate in granite-gneisses is about 30-45%, basalts about 20-30% and sandstones about 60% of the stored water, as "artificial recharge" to groundwater and the rest evaporates. The results when compared with the values from the water balance method, clearly demonstrated the sensitivity of the developed method to the environmental changes.

1 INTRODUCTION

Continuous declining of groundwater levels due to ever increasing exploitation and limited natural recharge restricted by climatic and hydrogeological constraints (Sukhija et al. 1996) necessitates deployment of artificial recharge structures to augment groundwater resources. Out of various recharge techniques, viz., surface spreading, check dams, subsurface dams, injection wells, percolation tanks; the percolation tanks, especially in hard rock areas, are the most popular in India. A percolation tank is a simple earthen dam constructed across a natural stream (Fig. 1) to impound surface runoff. Due to evaporation losses, only certain fraction of impounded water is subsequently expected to percolate through the tank bed and improve the downstream well yields. Despite huge investments (a few million dollars per year) being made on their construction, detailed scientific studies on their function are very limited (Raju 1985; Sharma 1985; Nair et al. 1978). This state of art led us to develop a simple tracer method of environmental chloride for assessing the contribution of percolation tanks (Sukhija et al. 1997) for different hydrogeological terrains and also evaluate the temporal changes caused by silting and environmental changes in two semi arid/arid regions of India (Fig.1, inset). Here we report the principle of the method, the results from different tanks and their evaluation and interpretation.

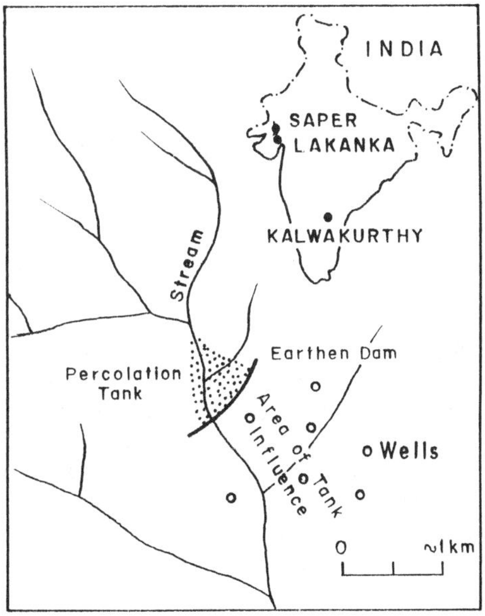

Figure 1. A typical percolation tank and location of studied tanks in India.

2 DEVELOPMENT OF ENVIRONMENTAL CHLORIDE METHOD: A NEW APPROACH

Besides utilisation of environmental chloride for natural groundwater recharge estimates (Allison & Hughes 1978; Sharma & Hughes 1985; Sukhija et al. 1988), Sukhija & Reddy (1987) have demonstrated its potential to detect the presence of percolated water in the downstream wells of a percolation tank.

Environmental chloride has the advantage of being a conservative tracer, and it can be measured using simple analytical techniques. Environmental chloride is deposited on land by rainfall and dry fallout, which is more along the coast and decreases with increasing distance from the coast. The contribution from rainfall is measurable but the dry fall out input, especially in the inland area, is generally negligible.

For developing the method for percolation tanks, we assume, the total chloride input to the tank is from surface runoff only; no seepage through the tank bund; and loss in the tank water is only through evaporation and percolation. Evaporation enriches chloride concentration in the tank water (Fig. 2), and percolated water carries the chloride with it.

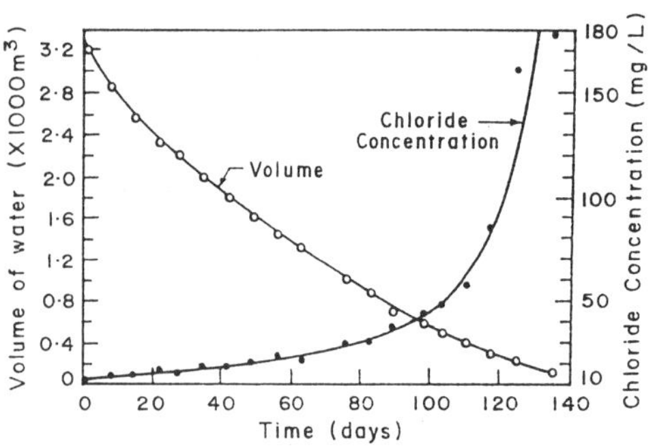

Figure 2. Temporal changes of volume of water in the tank vis-a-vis chloride concentration

Therefore, by measuring both volume of water and the chloride concentration in the tank water at regular intervals (after the monsoon) and by making chloride mass balance in the tank water, assuming no other sources/sinks of chloride in the tank water during the experimental period, the percolated volume can be computed. The chloride balance in the tank water between times t_1 and t_2 can be written as follows:

$$V_1*C_1=V_2*C_2+(1-f)(V_1-V_2)C_p \qquad (1)$$

Where V_1 and V_2 are the volumes of water in the tank at time t_1 and t_2 respectively; C_1 and C_2 chloride concentration in the tank water at time t_1 and t_2 respectively; (V_1-V_2)-loss of water between t_1 and t_2; f-fractional loss of water by evaporation; (1-f)- percolation fraction/efficacy; $f(V_1-V_2)$-loss of water by evaporation; $(1-f)(V_1-V_2)$ - loss of water by percolation; $C_p= \sum C_i*V_i/\ V_i$ is the time weighted average concentration of chloride in percolated water.

The percolated fraction is then,

$$1-f=(V_1*C_1-V_2*C_2)/C_p\,(V_1-V_2) \qquad (2)$$

It is assumed that fractional percolation (and also the fractional evaporation loss) remains constant during the period (t_1 to t_2) i.e. when the volume decreased from V_1 to V_2.

The developed technique is tested in the following tanks situated in various geologic terrains and climatic regions for its validity; 1. Singaram tank, and 2. Kalwakurthy tank, both in Granites, Andhra Pradesh, 3. Lakanka tank in Basalts, and 4. Saper in Sandstone, both in Surendranagar dist., Gujarat (Fig. 1).

3 RESULTS AND DISCUSSIONS

Table 1 demonstrates the successful application of the environmental chloride method to assess the recharge efficacy of the tanks situated in various geological terrains.

Table 1. Performance of different percolation tanks in terms of percentage of impounded water.

	1992-93		1993-94		1994-95		1995-96		1996-97	
T	Cl	WB	Cl	WB	Cl	WB	Cl	WB	Cl	WB
A.	35	49	22	49						
B.					45	29	44	42	58	43
C.							36	57	20	55
D.							64	74	62	63

T Tanks studied, Cl Percolation percentage by chloride method, WB Percolation percentage by water balance method, A. Singaram tank in Granites B. Kalwakurthy tank in Granites, C. Lakanka tank in Basalts and D. Saper tank in Sandstone.

A quick appraisal shows that the geologic fabric under a tank bed has great control on efficacy; fairly good percolation efficacy (30-45%) for granites, very good efficacy (60%) for the sandstones, and poor efficacy (20-30%) for the basalts are measured. Semi-consolidated sandstone with joints and fractures having a higher hydraulic conductivity of 3.1 m/d is ranked over granites with a low hydraulic conductivity of 1.4 m/d for their weathered portion. Thus good characteristics of the underlying aquifer matrix justifies higher percolation efficiencies. The black clayey soil, a weathered product of basalts, with poor hydraulic conductivity of 0.01 m/d is responsible for the poor efficacy. Further, we observe the above results are in tune with natural recharge measurements: 20-25% for sandstones, and 8-10% for granites (Sukhija et al. 1996). The effect of silting, on percolation efficacies is strikingly shown by Singaram tank, which registered a decrease from 35 to 22% during 1992-94. Yearly changes in evaporation do also control the efficacy of the tank, as was demonstrated by the Kalwakurthy tank: during 1996-97 the percolation efficacy (58%) was higher than during 1994-95 and 1995-96 (45%). We relate this variation to relatively lesser pan evaporation (580 mm) during 1996-97 than 1994-95 and 1995-96 (800 mm). The other controlling factor for high percolation is the groundwater abstraction in the downstream, which leads to additional recharge called as "Capture Recharge" (Bredehoeft 1997). The Singaram and Kalwakurthy tanks (efficacies 30 and 45% respectively) situated in similar geological and climatic framework prove this, as there are large number of irrigation wells in the downstream of the Kalwakurthy tank. However, on comparison, we observe that the water balance method does not show temporal changes and climatic effect as clearly as shown by the developed technique (Table 1).

4 CONCLUSIONS

The newly developed environmental chloride method for studying artificial recharge structures like percolation tanks enables a quick appraisal of their efficacies. The role of geology, climate, aging of tank, and geohydrology are evaluated using this method. The developed technique is simple, sensitive, reliable and yet powerful enough to resolve the role of above mentioned controlling factors over the performance of such artificial recharge structures and is therefore recommended for the study of such recharge structures.

5 ACKNOWLEDGEMENTS

Authors express their gratitude to Dr. H.K. Gupta, Director, NGRI for permission to publish the paper, the Department of Science and Technology, Govt. of India, authorities for funding and Aga Khan Rural Support Programme, Ahmedabad for field support.

REFERENCES

Allison,G.B. & M.W. Hughes 1978. The use of environmental chloride and tritium to estimate total recharge to an unconfined aquifer. *Aust. J. of soil Res.*, v.16, p. 181-195.

Bredehoeft, J. 1997. Safe yield and the water budget myth. *Groundwater*, v. 35, no.6.

Nair, A.R., A.S. Pendharkar, S.V. Navada, & S.M Rao 1978. Groundwater recharge studies in Maha- rashtra - development of isotope techniques and field studies. *Isotope Hydrology,* IAEA, v.II, p. 803-825.

Raju, K.C.B. 1985. Recharge through percolation tanks and subsurface dykes, India. *Proc. Semi. on "Artificial recharge of Groundwater"* held at Ahmedabad, India. p 12c-1.

Sharma, S.C. 1985. Artificial recharge in Gujarat state by injection, spreading though percolation tanks, check dams, Bandharas & other methods. *Proc. Semi. on "Artificial Recharge of Groundwater"* held at Ahmedabad, India. p. 16c-1.

Sharma, M.L. & M.W. Hughes 1985. Groundwater recharge estimation using chloride, deuterium and oxygen-18 profiles in the deep coastal sands of Western Australia. *J. of Hydrol*, v. 81, p. 93-109.

Sukhija, B.S. & D.V. Reddy 1987. Study of artificial recharge through Potharam percolation tank by environmental chloride method. *Geophy. Res. Bul.*, v. 25, no.1, p. 27-31.

Sukhija, B.S., D.V. Reddy, P. Nagabhushanam, & R. Chand 1988. Validity of the environmental chloride method for recharge evaluation of coastal aquifers in India. *J. of Hydrology*, v. 99, p. 349-366.

Sukhija B.S., P. Nagabhushanam, & D.V. Reddy 1996. An overview of groundwater recharge in different hydrogeological environs of semi-arid regions of India using tracers. *Hydrogeology Journal*, v.4, no.3, p. 50-71.

Sukhija, B.S., D.V. Reddy, M.V. Nandakumar, & Rama 1997. A method for evaluation of artificial recharge through percolation tanks using environmental chloride. *Groundwater,* v. 35, no.1, p. 161-165.

Artificial Recharge of Groundwater, Peters et al. (eds) © 1998 Balkema, Rotterdam, ISBN 90 5809 017 5

Artificial recharge experiments in Bustan Extension Area, Egypt

Madiha Moustafa Darwish
Research Institute for Groundwater, Kanater, Egypt

ABSTRACT: One of the major challenges for the Egyptian Ministry of Public Works and Water Resources is the need for optimal development and management of its water resources. The increasing population in combination with the need to develop the vast desert area have stressed the need for water conservation. This can be achieved through various means including the application of artificial recharge of groundwater.
The Research Institute for Groundwater has constructed an experimental infiltration basin in the fringes of the Western Delta (Bustan Extension Area). Three infiltration tests in 1996 using Nile Water, have shown that in the specific conditions of this area, moderate infiltration rates (about 30 - 15 cm/day) can be obtained during the first 15-20 days of the infiltration process. If no serious clogging occurs, final infiltration rates are in the range of 10-12 cm/day).
In this paper the experiments at El Bustan Extension Area are evaluated It can be concluded that the first tests with basin recharge showed its technical feasibility, especially when taking into account that the subsoil conditions in Bustan Extension Area are not the most favorable ones due to presence of clay, loam and calcrete layers in the subsoil.

1 GENERAL BACKGROUND AND OBJECTIVES

Groundwater resources management in Egypt aimes at optimal use of the groundwater without creating adverse effects such as large drawdowns, seawater intrusion, and land subsidence. Inaddition the groundwater resources management policy aims at protection of the groundwater resources from industrial, agricultural and domestic pollution.

Egypt's Ministry of Public Works and Water Resources (MPWWR) conciders artificial recharge as an important option in the national water conservation policy. It serves several goals, such as increasing the recharge at the desert fringes, storage of water in case the availability of water does not coincide in time with the demands, reduction of evaporation losses and improvement of water quality due to filtering characteristics of the underground.

In previous studies eleven locations have been indicated as being feasible for an artificial recharge experiment in Egypt. Two areas have been selected for detailed hydrological studies. Depth to water, availability and transport of recharge water, infiltration rate and existence of infrastructures required for the experiments are the major parameter for selecting the areas for Artificial Recharge.

These two areas are Bustan Extension Area located at the desert fringe about 150km northwest of Cairo,and Bahig Hammam Area located at the coastal zone about 50km west of Alexandria. This paper focuses on the work in Bustan Extention Area.

2 HYDROGEOLOGY OF BUSTAN EXTENSION AREA

The area is completely occupied by sedimentary rocks belonging to the quaternary with a maximum thickness of about 200m essentially developed into deltaic sands and gravels. These are underlain by the tertiary clay (Pliocene) and are overlain by shifting sands and calcareous loam. The water in this aquifer has exists under semi-confined or phreatic conditions.

The total transmissivity is ranging from 500 m2/day in southwest to 5000 m2/day in the east and northeast.The depth to the groundwater table is increasing from less than 2 m in the north along Nubariya canal to more than 40 m south of Bustan Extention Area. The depth to groundwater level is between 15 and 30m. Continued increasing of groundwater level is observed due to the extension of land reclamation project by using Nile water (Attia et al, 1998).The salinity of the groundwater is 1000ppm and the water is $Na\text{-}HC^{O}_{3}$,the salinity increases in the northwest direction

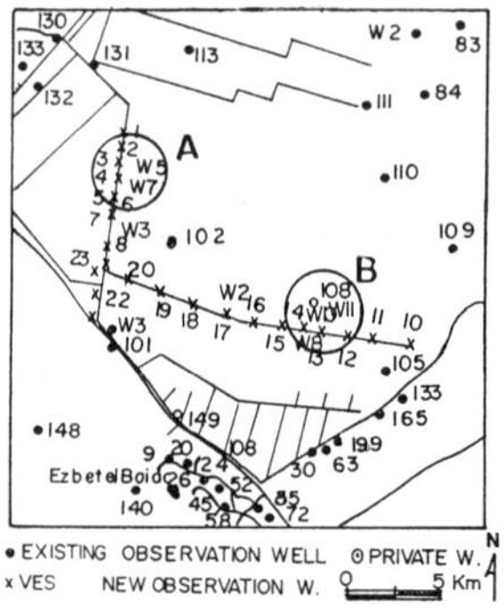

Figure .1 well location and geophysical cross sec (ves) A.B Selected area

3 SITE SELECTION

Intensive field work at Bustan Extension Area shows the complexity and heterogenity of the underground. The new interpretation of the previous studies, additional drillings and geophysics resulted in selection of two areas. Then detailed study resulted in one area for basin recharge (Figure 1). Within the area drilling of 14 wells untill 50m and geophysical investigations (geoelectrical and EM method) were used to find the optimal site.The chosen area is located south of Bustan canal eastern branch no 2 at 15.2 km from its intake .The selection of the final location of the basin was chosen using the following criteria: small amount of low permeability layers in the unsaturated zone, availability of land, and preferably gravity flow from the canal to the basin .

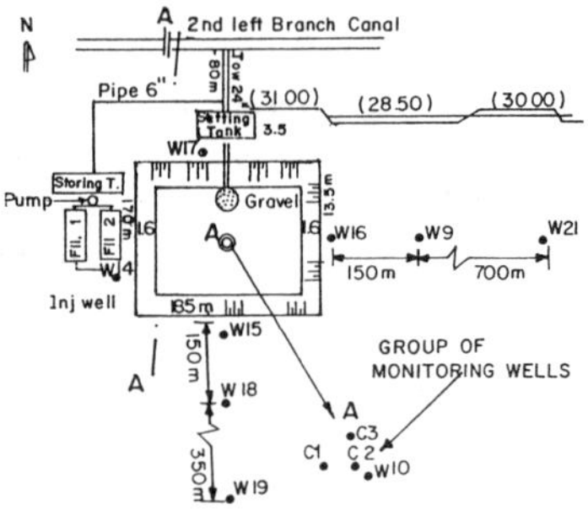

Figure .2 general layout of recharge experiment area

4 DESIGN, CONSTRUCTION OF THE BASIN AND THE MONITORING NETWORK

The surface dimension size of the basin is 185*170m with side slope 1:8,1:6 respectively ,average depth is 2.5m. The dimension of the bottom of the basin is 155*135m Figure 2 presents general layout of the recharge experiment area with the monitoring network.Three deep observation wells are drilled at the basin center with 9 screens(25-140m depth)
- 4 observation wells are at the middle of each side
- 8 observation wells are at regular distances in all directions.Permanent electrode cable is at the basin center with 13 electrodes.
-Automatic recording of groundwater level in 8 locations.
-The evaporation rate were measured inside the basin itself, and at 20 m away from the basin.
-Staff gauges are at the basin and feeding canal .
-Regular manual readings done for levels, water quality and sampling programme.

5 EXPERIMENTS AND RESULTS

Three experiment have been performed in 1996 during January-February for 13 days, March-April for 37 days, June-August for 50days. Water diverted to the recharge basin is 56,000m3,71,000m3 and 145,000m3 respectively.

5.1 Main results

Groundwater levels below the basin started to rise within 4 days after the start of the experiment. Maximum increase of the groundwater level (about 0.4m) ocurred under the center of the basin (Figure 3).Extension of the mound is about 1km.
Intensive leaching of the unsaturated soil resulted in high EC values of the percolated water (Figure4), especially during the first experiment. Leaching diminished during the second and the third experiment. During the three experiments about 0.25million m3 has been infiltrated. Infiltration rates varied from 0.3m/day to 0.1m/day. Figure 5 shows the gradual clogging of basin during the first and the third experiment. Rapid clogging was observed during the second experiment, caused by fine material originating from the dust storms that frequently occur during March- April.
The average evaporation rate during the three experiments was 1.5, 7, 13mm per day respectively.
The average wind speed during the three experiments was 100, 160, 60 km/day respectively (Wadi el Natrun meterological station). A dust storm has been blown with 270 km/day started after scraping of the basin before the second experiment started.

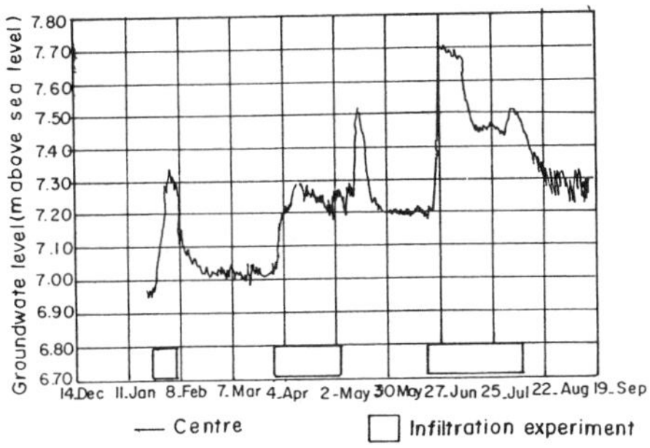

Figure. 3 Groundwater levels during infiltration

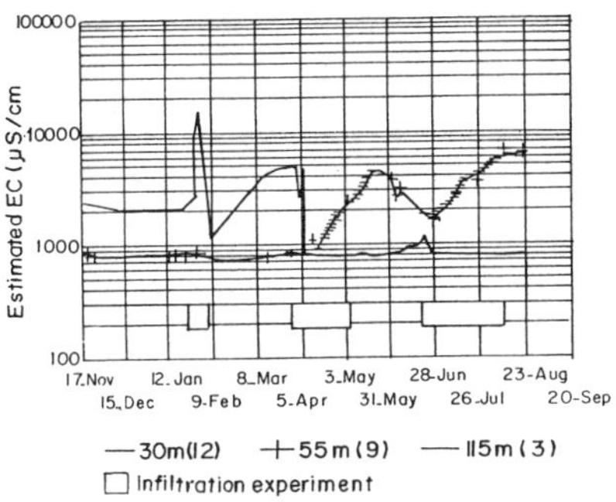

Figure. 4 Salt cable measurements

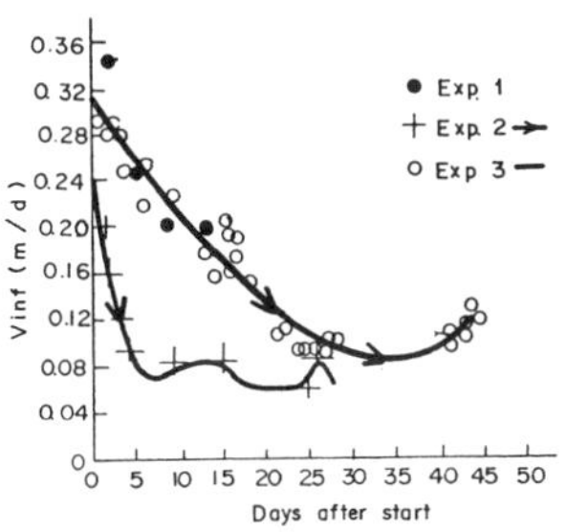

Figure. 5 Change of infiltration rate

6 CONCLUSION AND RECOMMENDATIONS FOR FUTURE WORK

6.1 Conclusions

The hydrogeological situation of El Bustan Extention Area allows for reasonable infiltration rates when using Nile water. Infiltration rates are comparable to values in operating schemes in other countries (e.g The Netherlands). Regular cleaning of the infiltration basin will be needed to prevent clogging, especially after dust storms. Environmental effects of artificial recharge in areas with groundwater depletion are positive.The adverse effect of soil leaching will occur only during the starting phase.

6.2 Recommendations for future work

More exprimental work is needed to obtain the recovery percentage. Also different designs of the infiltration scheme can be tested to minimize clogging.Future experimental work should also include comparison between basin infiltration and deep well injection.

Following technical feasibility studies,economic evaluation is required in order to define the role of artificial recharge in Egypt's water resources management.

When using treated sewage water as a source for artificial recharge experiments appropiate treatment requirements and techniques should be evaluated.

REFERENCES

Attia, F.A.R., A.Fekry & E.Smidt,1995. Artificial recharge of groundwater as a tool in Egypt's future water management IWRA conference on water resources management in Arid countries, p.183-189,Muscat, Oman.

M.Mustafa, A.Abulfutuh, A.Ragab & E.Smidt 1996. Impact of artificial recharge on groundwater potential-A case study from Egypt. ICID 16th congress on Irrigation and Drainage. p 14-16 Cairo, Egypt 1996.

RIGW/IWACO 1997. Hydrogeology of El Bustan Extension area and Bahig-El Hammam area (EMGR) April 1997.

RIGW/IWACO Artificial Recharge Experiment: analysis of first results and advise on future work, June 1996.

Attia,F.A.R., M.Moustafa, Theo Olthoorn,E.Smidt, 1998. The role of Artificial Recharge in Integrated Water Management in Egypt.

Artificial Recharge of Groundwater, Peters et al. (eds) © *1998 Balkema, Rotterdam, ISBN 90 5809 017 5*

Scoping process to determine pilot sites for artificial recharge with wastewater

A.A.Fadlelmawla
Research Institute for Groundwater, Egypt

ABSTRACT: The ambitious plans of Egypt to expand its habitable lands are being challenged by its limited water resources. This fact led the government into considering the wastewater reuse as an important dimension of water resources planning. Artificial recharge with wastewater, due to its advantages over the conventional reuse of wastewater, can be an added dimension to the wastewater reuse policy of the Egyptian government. The preparation of an application plan should clearly identify both locations and methods for the application among other parameters. Determination of optimum locations and suitable methods requires the optimization of a combination of intrinsic characteristics of the wastewater, and the proposed locations in addition to other environmental and economical considerations. Due to the complexity of the involved factors, a scoping process is needed before the final identification of methods and locations of application. The preliminary screening process is described in this paper.

1 INTRODUCTION

In many arid and semi-arid countries, water is becoming an increasingly scarce resource and planners are forced to consider any source of water which might be used economically and effectively to promote further development. At the same time, expansion of urban populations and increased coverage of domestic water supply and sewerage give rise to greater quantities of municipal wastewater. With the current emphasis on environment and water pollution issues, there is an increasing awareness of the need for disposal of wastewater safely and beneficially. All this has led the water resources planners of many countries to consider wastewater reuse into their plans as an added source of water and a mean of safe disposal of the continuously increasing flux of wastewater.

Wastewater reuse in Egypt has been practiced since the beginning of this century. Practices of wastewater reuse have taken two forms; official continuous and organized and unofficial perennial unorganized practices. Wastewater reuse has been officially implemented in Egypt since 1915 at the government farm of El-Gabal El-Asfar near Cairo. Many other farms have recently started such practice in many areas of the country. A number of farms at different locations of the country are using treated and untreated sewage water, for example, farms at Ismaylia, Helwan, Kena, and Abu-Rawash are currently officially operating on treated wastewater from the treatment plants. On the other hand, at many places in the Delta, whenever irrigation water shortages occur, the local farmers unofficially practice wastewater reuse. Many villages (i.e. Mansouria) in the Delta are deprived of a sanitary network, and villagers are depending on septic tanks to drain their domestic wastes to. Such tanks are being drained periodically using specially equipped trucks which, in turn, drain its contents in the nearest agriculture drain. During water shortages, farmers rely on drain water to irrigate their crops. This water is nothing but untreated wastewater mixed with drainage water.

2 THE SCOPING PROCESS

The process of selecting the suitable locations was based on a number of factors: availability of wastewater, the composition of wastewater, hydrogeology of the location, possible benefits, environmental considerations, and availability of data.

As for the treated wastewater availability, numerous locations of wastewater treatment plants are available in the country. However, the suitability of the hydrogeology eliminated the Delta and the Nile Valley due to the presence of a thick clay cap on the top of the aquifer. Such a clay layer makes any attempt to use artificial recharge through infiltrating basins, the most common method, unfeasible.

The western desert fringes represent an ideal situation for artificial recharge with wastewater, suitable hydrogeology, availability of treated wastewater, suitability of wastewater quality for artificial recharge, and present and future plans for development. Additionally, adverse impacts on the groundwater from large and aggressive groundwater-dependent development scheme are already observed in the form of large drawdowns. A number of areas where groundwater drawdowns are severe were identified: Abu-Rawash, West Gabal El-Mansouria, Wadi El-Farigh, and southwest El-Sadat City (Fekry & Attia). As for this study, El-Sadat City (Fig. 1) was selected as one of the potential pilot areas for applying the scheme. The specific selection of the city was based on the availability of properly treated wastewater, the near future availability of monitoring network, and the availability of data.

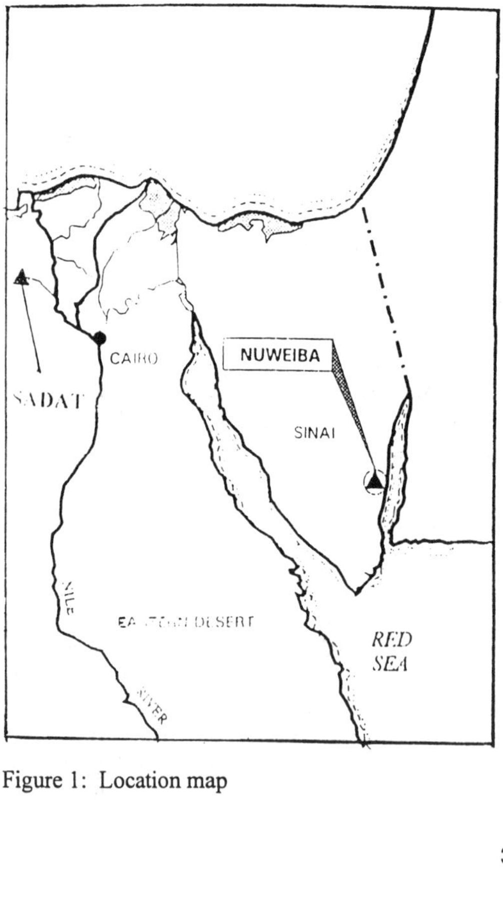

Figure 1: Location map

The city of Nuweiba (Fig. 1) on the east coast of the Sinai peninsula on the Gulf of Aqaba was also selected as a potential location for the scheme. Apart from the suitable hydrogeology and the availability of treated wastewater, the city is totally dependent on groundwater and is suffering from an extreme shortage in fresh water due to the saltwater intrusion. Additionally, Nuweiba is in need for that scheme due to other environmental considerations. The growth of the city is driven by the growth in the tourism industry which is directly related to the Gulf of Aqaba beaches and the coral reefs in the area. Any discharge of wastewater to the Gulf requires advanced treatment levels to meet pollutant concentration limits that will not have any significant effect on the coral reefs and the beaches. For these reasons a wastewater reuse scheme at Nuweiba will enable the reclamation of valuable resources and the abatement of pollution.

3 CONCLUSIONS AND RECOMMENDATIONS

The artificial recharge of groundwater using treated sewage water can be an added dimension to the reuse policies of Egypt. Nevertheless, close monitoring is essential to avoid adverse environmental impacts.

Two locations, El-Sadat City and the City of Nuweiba, were selected as potential candidates for applying the scheme in pilot scales. The locations were selected based on hydrogeological, and environmental considerations, availability of property treated wastewater, and need for groundwater.

Further investigations are needed before the scheme is applied at the selected location.

REFERENCES

DATEX INC, 1997. "Feasibility Study for Waste water Reuse Scheme for Nuweiba." Report pre sented to US Agency for International Develop ment, Cairo, Egypt.

Fekry, A. & F. Attia, 1997. "Preliminary Selection of Suitable Sites for Sewage-Based Irrigation in Egypt." In proceedings for Third Gulf Water Con ference, Muscat, Sultanate of Oman, 1997.

Pescod, M. B., 1992. "Wastewater treatment and use in agriculture." FAO irrigation and drainage paper No. 47, Rome, 1992.

RIGW/IWACO, 1992. "Monitoring and Control of Groundwater Pollution: Local Site Investigation"

Artificial Recharge of Groundwater, Peters et al. (eds) © 1998 Balkema, Rotterdam, ISBN 90 5809 017 5

Need to increase recharge to the aquifer in Sri Lankan hard rock terrain

C. Shanthi de Silva
The Open University, Nawala, Nugegoda, Sri Lanka

ABSTRACT: Water shortage is the primary limiting factor to expanding agriculture in hard rock terrain. Poor maintenance of these surface reservoirs during past several years caused insufficient storage. In addition the siltation in these reservoirs has reduced the recharge to the aquifer and the water stored in the reservoirs lost as evaporation during the first two months after the wet season. Estimated annual average recharge to the aquifer from the case studies conducted in dry and intermediate zones were 200 mm and 250 mm respectively. With the present recharge levels the farmers could cultivate only less than 1 ha. and the income from this cultivation is very marginal to uplift their living standards. Therefore there is a need to increase recharge to the aquifer.

1 INTRODUCTION

Sri Lanka is predominantly an agricultural country consists three major climatological zones namely the dry, intermediate and wet zones. The dry and intermediate zones cover more than two thirds of the country's arable land. The dry season has two major seasons per year and these seasons are strongly influenced by the monsoon circulation. The North east monsoon brings considerable rainfall over the whole of the island from October to January.

Since time immemorial, people in South Asia collected rainfall runoff in small storage reservoirs (called tanks) for later use in agricultural production (Figure 1). During this period, farmers cultivate two season paddy, wet season paddy as rainfed and the dry season paddy with tank water. Over a time, more people were attracted to the command areas of the tanks and many of the systems became unsustainable due to overcrowding. In addition these tanks were not maintained properly during the past several years caused siltation leading to poor storage capacities. As a result now there are several agrowells constructed in lower command area of these tanks and farmers are cultivating rainfed paddy during wet season (October to January) and vegetables in dry season (February to September) using agrowell water.

Basically these agrowells are constructed in the tank command area and mainly fed by the tanks

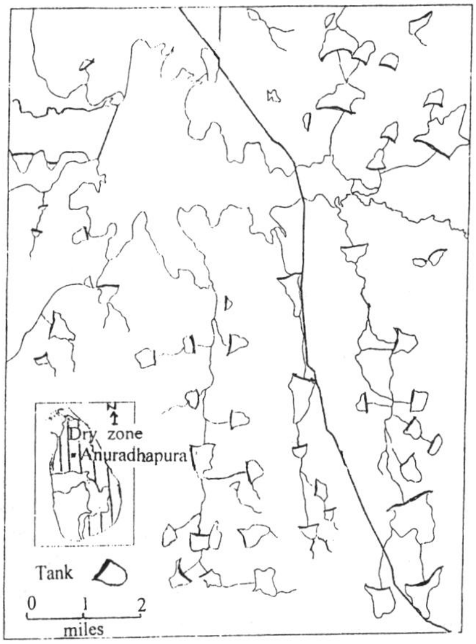

Figure 1. Distribution of small and large storage tanks in the dry zone of Sri Lanka

water. Usually the tanks hold water till end of March and then the agrowell water is used for irrigation. When the tank water is over by March the agrowells can supply water only for other two months till June due to their shallow depth. Therefore the farmers are unable to do cultivation without water till September. Increasing the depth and

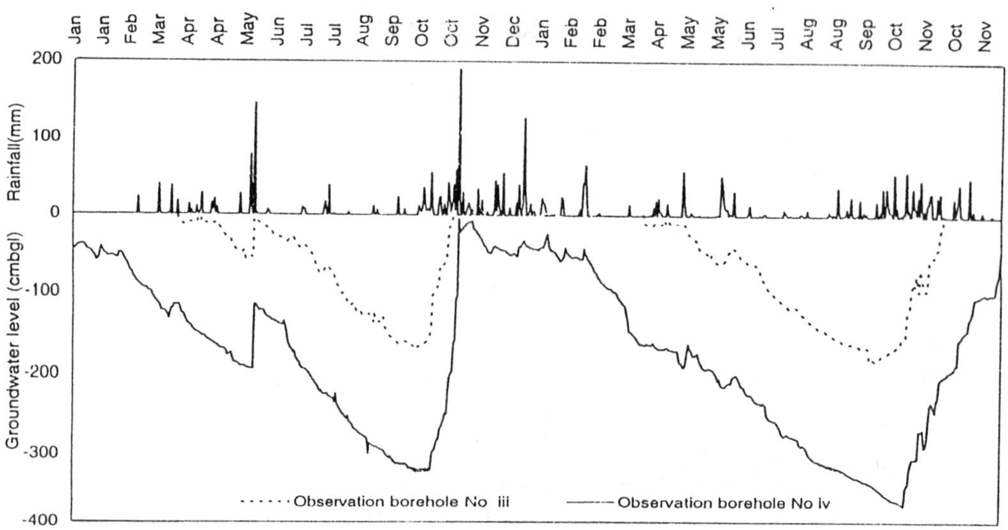

Figure 2. Groundwater level fluctuation in the dry zone agrowell system (From 1994-1995)

capacity of these tanks may help the farmers to do year around cultivation and to uplift their living standards.

2 CASE STUDIES

Case study areas are located in dry and intermediate zones of Sri Lanka and the main aquifer underneath these locations is hard metamorphic rock. Several agrowells in the tank command area were selected and intensively equipped with observation bore holes, raingauges and evaporation pans. Daily monitoring on groundwater level fluctuation in agrowells and observation bore holes before and after pumping activity and rainfall evaporation were carried out for a period of 3 years. In addition crop information, cropping patterns, land use pattern and soil characteristics were also recorded during the study period. Several pumping tests were conducted in most of the selected agrowells during the dry season to study the aquifer flow mechanisms and recovery patterns of the agrowells during early and late dry seasons. Data collected was analysed to understand the agrowell system in a small tank system and the interrelationship between the agrowell and tank water. Daily weather, crop and soil data were used in a computer based soil moisture balance method to estimate the annual average actual recharge to the aquifer.

3 RESULTS AND DISCUSSION

3.1 Aquifer flow mechanisms in hard rock aquifer

The hard rock aquifer consists of very low storage and transmissivity. The estimated hydraulic conductivity and specific yield were 6 m/day and 0.065 respectively (De Silva 1995). Recovery of the well were recorded with the tank water level and the results showed that the recovery rates of the wells were good when there was high tank water level. But the recovery rates decreased with decreasing tank water levels. The large well storage in the agrowells contributed nearly 70%-80% of the total abstraction during pumping when the tank water level was high. Recovery rates varies depending on the aquifer properties (De Silva 1993, De Silva and Weatherhead 1994).

3.2 Recharge characteristics

Basically, the aquifers in the dry and intermediate zones got recharged during October to January with the wet season rainfall. It was observed by the quick groundwater table rise to the ground level soon after the wet season rains, fully saturated soil zone and excessive runoff after filling the surface depressions. From January onwards the groundwater table dropped and in October (prior to the wet season rains) the groundwater table was nearly 3 to 4 m below ground level due to pumping and, evaporation by small and big trees and seepage losses (Figure 2). Abstraction from agrowells during the dry season (February to September) is limited by the wet season recharge and the results indicated that there would be more recharge if the aquifer has the capacity to hold. Estimated average annual recharge by soil moisture balance method for dry and intermediate zones were 200 mm and 250 mm respectively

3.3 Need to increase the aquifer recharge

In the dry and intermediate zones of Sri Lanka, there are thousand of surface reservoirs (tanks) which could be seen easily on an aerial picture. These tanks have been neglected for several years and the average depth of these tanks are only 4 m where as these tanks were about 8m -10 m depth initially. Because the cost of renovating tanks are expensive and farmers are unable to do it with their resources, farmers are constructing 6m diameter and 6m -8 m depth of large diameter wells near to the tank for irrigation purposes.

As the results indicated there is a close relationship between the tank and agrowell water. Therefore increasing the depth of the tanks by desilting them may increase the capacity of water storage and it could decrease the water lost as runoff during wet season. In addition this will help the farmers to abstract more water and deplete the aquifer up to 5m -6m at the end of dry season. This depth could be filled by the following wet season rains and it also help to increase the recharge to the aquifer. In addition there is a need to find artificial recharge methods to increase the recharge.

4 CONCLUSION

The available groundwater resources should be used efficiently by studying the recharge to the aquifer and the ability of the agrowell draw water from the weathered overburden of the hard rock aquifer. In addition there is a need to increase recharge to aquifer as the dry season abstraction is limited by the wet season recharge. Present recharge rates are limited to do successful cultivation during dry season in dry and intermediate zones of Sri Lanka.

REFERENCES

De Silva, C.S. 1993. A preliminary study on Agrowells in Hard rock aquifer, with special reference to well performance. Tropical Agricultural Research, 5:17-28.

De Silva, C.S & K. Weatherhead 1994. Use of large diameter wells (Agrowells) for supplementary irrigation in the dry zone of Sri Lanka. Proc. Inter Conference on 'Water Down Under '94, Adelaide, South Australia, 2B:595-599

De Silva, C.S. 1995. The use of large diameter well (agrowells) for supplementary irrigation from hard rock aquifers of Sri Lanka. Ph.D thesis, Silsoe College, Cranfield University, UK

2.2 Field and desk studies, plans and cases

Artificial Recharge of Groundwater, Peters et al. (eds) © 1998 Balkema, Rotterdam, ISBN 90 5809 017 5

Epe artificial recharge project

B.D.Volkers & G.Roelofs
Gelderland Water Company, Velp, Netherlands

ABSTRACT: The water in the streams known as Klaarbeek and Verloren Beek, has good-quality and serves both nature and agriculture on the Wisselsche and Tongerensche peat moors. Veluwe Water Authority and Gelderland Water Company have developed a plan for reusing the water, once it has performed this function. From September 1998, the groundwater at Epe pumping station will be supplemented by stream water, in order to combat a decline of the groundwater level and loss of wetlands.

1 WATER DEPLETION

Every day, some 375,000 people in the Veluwe region consume drinking water from Gelderland Water Company. Only groundwater is used to produce this drinking water, and up to half a century ago, this groundwater seemed inexhaustible. Since then, however, the increase in the number of homes and businesses has caused the groundwater abstraction to rise dramatically. In addition, drainage of agricultural areas has increased. These factors have led to a drop in the water table, something especially threatening to nature conservation areas. Veluwe Water Authority is responsible for, among other things, the intake and supply of water, an adequate level of water, and the right amount of water wherever nee-

Photo 1. Location of water intake Zuuk.

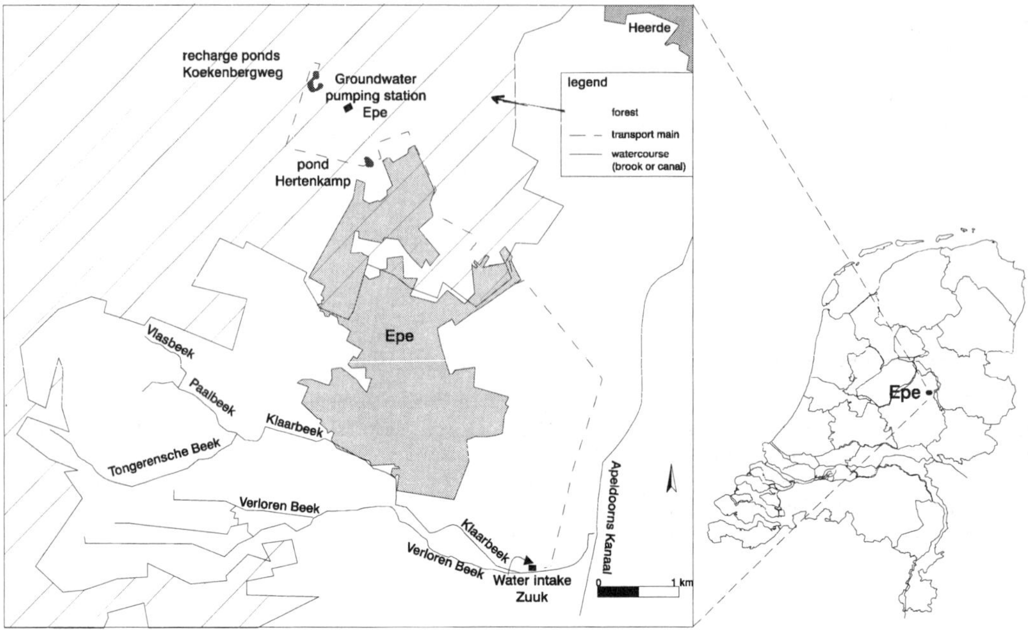

Figure 1. View of the Epe recharge project

ded. Gelderland Water Company and Veluwe Water Authority have a common interest therefore: the prudent use of water.

2 CONSEQUENCES FOR THE DRINKING WATER SUPPLY

As manager of the groundwater, the Province of Gelderland has set the conditions for the abstraction of groundwater. One of these is that, in the Veluwe region, 25% less groundwater is to be pumped than in previous years. The Gelderland Water Company will have to meet this condition by pursuing an active water-conservation policy and by 'importing' drinking water from the Province of Flevoland or the Betuwe region. This aside, the company is looking for other ways to supply its customers in the Veluwe region with sufficient good-quality drinking water at the lowest possible price.

3 RECHARGE

For several years, various studies have looked into ways of combating a declining water table in Oost-Veluwe. Based partly on these studies, Gelderland Water Company and Veluwe Water Authority want to retain the clean surface water of the Veluwe streams and springs for a greater length of time in the region. The plan is, in the winter, to catch the Klaarbeek and Verloren Beek water just before it flows into the Apeldoorns Kanaal (Photo 1), and then use the stream water for recharging a site approximately 600 metres north-west of Epe pumping station (Fig. 1). A pipeline will transport the stream water to the existing Hertenkamp sand pit, which will only be filled with water during the winter months. From here, the water is pumped to the recharge ponds of the water recovery area Koekenbergweg. At this location, the water seeps into the ground. After this, Epe pumping station pumps the water for use in the production of drinking water. Depending on the stream discharge, the amount of surface water abstracted each winter from the stream will probably be 1 to 2.2 million m^3. The quality of the water is so good that treatment is considered unnecessary. In the unlikely event that the water does not meet the desired standard of quality, a contingency system will ensure it is routed back to its original stream.

4 CONSEQUENCES

With this method of recharging, the nett original groundwater abstraction at Epe pumping station will drop by 30%. As a result, the level of groundwater will rise. At the recharge ponds, the rise will be roughly three metres (highest measured water table is presently approximately four metres below ground level). On Tongerensche peat moor, the water table will rise by some 5 to 10 cm. Because of the higher water table, the streams will produce more water. Calculations show that the annual supply of stream water to the Tongerensche and Wisselsche peat moors will increase by approximately 0.7 million m^3.

5 RECREATION

The shallow recharge ponds Koekenbergweg will lie within the existing pine forest. Their construction requires the clearing of roughly 2.2 hectares of forest, which will be compensated for by planting an equal-sized area somewhere else. The existing heath remains largely intact. The banks of the recharge ponds will have very gentle slopes on which common floatgrass, rough-stalked meadow grass, hairy willow-herb and soft rush are expected to take root. Around the ponds in the forest, heather will be planted.

Walkers and cyclists will have access to Hertenkamp pond: in the winter, a secluded wood-land pool; in the summer, a shallow, dry recreation area once again. There will also be new leisure facilities, such as picnic tables, play equipment and a walking route. The recharge ponds Koekenbergweg will not be accessible to the public. However, an outlook hill will be built that will provide a view over the recharge ponds and their surroundings. On this hill, there will be information panels on the relationship between water in the streams and the springs, the recharge with stream water, and the drinking water facilities at Epe.

Artificial Recharge of Groundwater, Peters et al. (eds) © 1998 Balkema, Rotterdam, ISBN 90 5809 017 5

Hydrogeological potential for Aquifer Storage and Recovery (ASR) in England and Wales

H.K.Jones, I.N.Gale & D.M.J.Macdonald
Hydrogeology Group, British Geological Survey, Wallingford, UK

ABSTRACT: A scoping study was initiated in September 1996 to investigate the hydrogeological and regulatory issues constraining the use of ASR in Britain. The major aquifers of Britain were assessed in terms of their hydrogeological potential for ASR, and the regulatory issues addressed. This paper briefly describes the significant hydrogeological issues and assesses the ASR potential of British aquifers in terms of their hydrogeological characteristics.

1 ASSESSMENT OF HYDROGEOLOGICAL POTENTIAL FOR ASR

A scoping study was initiated in September 1996 to investigate the hydrogeological and regulatory issues constraining the use of ASR in Britain. From the hydrogeological perspective, the aim of this study was to identify aquifers with greatest potential for ASR. The second part of the study was to identify the major regulatory issues of concern to both the Environment Agency and water supply companies, and to produce guidelines to facilitate implementation. This paper focuses on the former aim of the study.

One major problem that became apparent early in the study was the difficulty in identifying the hydrogeological factors controlling ASR potential. Examination of existing schemes (largely from the USA) showed no unique factor, or combination of factors, could be identified. This was thought to be partly due to the fact that the siting of many schemes has been based on demand, rather than on hydrogeological suitability; any hydrogeological problems were then overcome by engineering. For instance, if a scheme was needed in an area where the identified aquifer had low transmissivity, more boreholes would be drilled until the required recharge rates were achieved.

After examination of existing schemes and discussion, several factors were identified as being hydrogeologically significant:

- aquifer properties
- degree of confinement
- depth to aquifer
- groundwater quality.

Maps, accompanied by a text description, were produced for each Environment Agency region. The maps showed:

- location of major centres of population as an indicator of potential demand
- areas of outcrop for each aquifer
- 200m and 500m depth to top of aquifer contours
- transmissivity values, grouped into less than 100 m^2/d, 100 - 1000 m^2/d, and more than 1000 m^2/d.

An example, for the South West Region, is shown in Figure 1. Accompanying the maps were descriptive assessments of the ASR potential for each major aquifer in each region.

The hydrogeological factors and sources of data used in compilation of the maps and descriptive assessments were:

- Transmissivity data, largely obtained from the Aquifer Properties Manual (Allen et al. 1997). The data were grouped into the classes shown above, and plotted as different size symbols on the maps to illustrate variation across the area. Although diffusivity would have been a better parameter to map than transmissivity, values of storage coefficient were found to be too sparse to use this parameter, and hence transmissivity was used. Diffusivity was incorporated within the text by considering aquifer type. As anticipated, aquifer properties data were generally sparse for the major aquifers away from outcrop, apart from areas that had been investigated for specific projects, such as geothermal or oil exploration work.
- The contours of depth to the tops of each formation were derived by combining two datasets within ArcInfo. Digital contours of depth to the formation, relative to sea level, were available

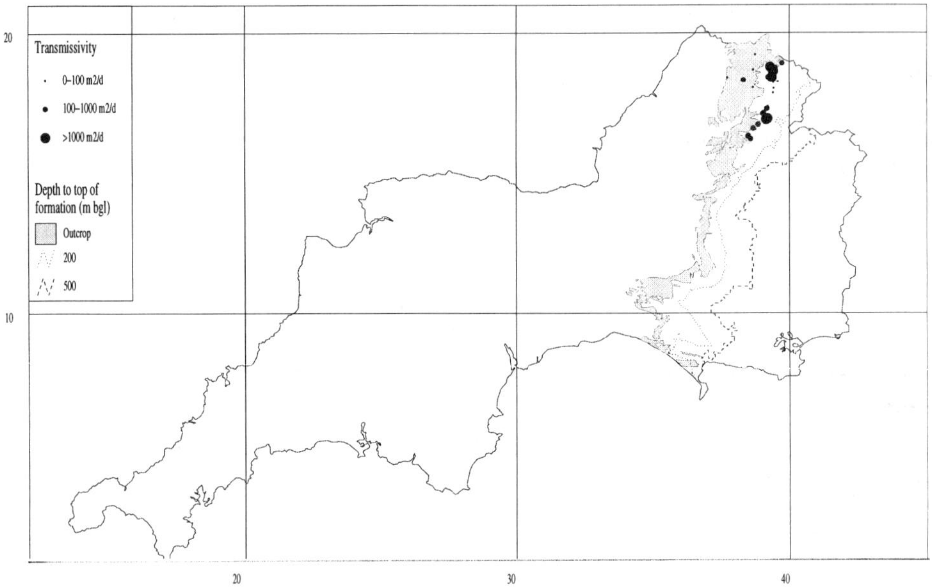

Figure 1. Transmissivity, outcrop and depth to the Great Oolite Series in the South West Region

(British Geological Survey). These contours had been digitised from the Atlas of Onshore Sedimentary Basins (Whittaker 1985). They were imported into ArcInfo and gridded on a 500 m grid. These depths were added to a topographic grid to produce depth below ground level. Thus elevation data on a 1 km grid were imported into ArcInfo, and two sets of data combined to give a depth to top of formation below ground level. The grid was then converted to contours for the purpose of producing the final map.

- Areas of outcrop were obtained from the digital 1:625000 geological map of Great Britain (BGS), and imported into ArcInfo.
- Generally there were little data on water quality that could be incorporated onto the maps. Where available, information was included within the text description.

2 HYDROGEOLOGICAL POTENTIAL OF THE MAJOR BRITISH AQUIFERS

There is a diverse range of geology, topography and climate across England and Wales. This results in different degrees of reliance on groundwater, and also varying importance of different aquifers across the country. In addition to the major aquifers (illustrated in Figure 2), there may be as many as 35 minor aquifers within each region. It was not possible to assess the potential for ASR in each of these, and therefore the evaluation focused on the major aquifers.

2.1 Chalk and Upper Greensand

The Chalk, hydraulically connected with the Upper Greensand in some areas, is an important aquifer over much of southern and central England. The Chalk's ability to function as an aquifer stems from fracture flow, along dissolution-enhanced fractures. As a result of its high transmissivity but low storage, the Chalk frequently has high diffusivity, and the impact of ASR schemes on the aquifer at outcrop would have to be carefully assessed, even at distances of several kilometres from outcrop.

The confined chalk frequently has low transmissivity due to the lack of groundwater movement, and hence the lack of development of fractures. Furthermore, water quality may be poor in the deeply confined chalk due to the lack of groundwater flow. These factors could result in reduced recovery efficiency due to diffusion between the saline porewater and injected fresh water in the fractures. Initial modelling has suggested this could be a significant process (Bennett 1997); ongoing modelling and field investigations should provide valuable data to resolve this issue.

The opportunity for ASR in the confined chalk aquifer would therefore appear to be limited. However, the apparent limited potential of the confined chalk regionally is offset by the success of the North London artificial recharge scheme, and also by recent work in the Wareham Area, both of which utilise confined Chalk. Locally the confined chalk aquifer has good potential in suitable hydrogeological settings.

The Upper Greensand may also have potential due to its high storage. Impermeable horizons within the Upper Greensand may be useful in creating a multi-layered aquifer, and offering the opportunity for stacking. Where there is hydraulic continuity between the Chalk and Upper Greensand, the combined aquifer may have greater potential.

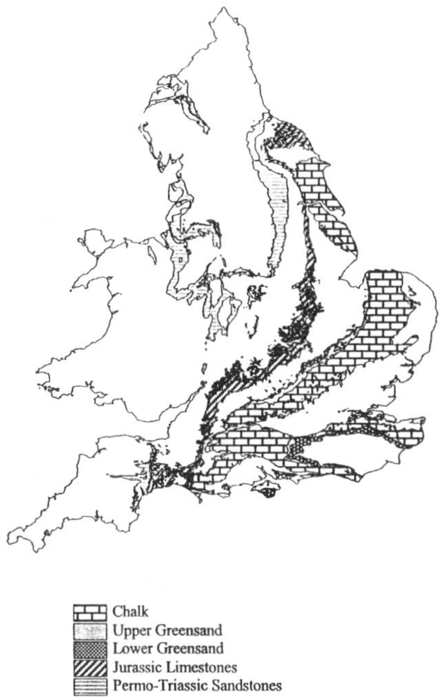

Figure 2. Major aquifers of England and Wales

2.2 Lower Greensand

The Lower Greensand comprises an important aquifer in parts of southern England. It has limited area of outcrop, although its high storage and generally good quality water render it locally important. The Lower Greensand Group includes a complex series of intercalated clays and sands, with varying degrees of cementation, which results in differing reliance on fracture flow. For example, intergranular flow is more dominant in the western part of the Weald in southeast England, with fracture flow becoming increasingly significant in the more cemented sands of the east and south.

The Lower Greensand may present problems for ASR in terms of borehole competence due to the poorly cemented nature of the sandstones.

Generally, the Lower Greensand has good potential for ASR. Transmissivity, even where confined is generally reasonable, and storage coefficients are high. It is most likely to be of use where it attains maximum thickness, and where intergranular flow is dominant, for example in the west of the Weald.

2.3 Jurassic Limestones

The Jurassic sediments outcrop in a broad band from the southwest to the northeast of England. The sediments comprise a very variable sequence of limestones, clays, siltstones and sandstones. The limestones are replaced by clays in some areas.

The two main limestones are the Great Oolite Limestone and the Inferior Oolite. Intergranular permeabilities are generally low, and the dominant flow mechanism is through solution-enlarged fractures, with local karstic flow having been observed in all the main aquifers.

The limestones are dominated by fracture flow, and have high diffusivity. Hence pressure changes in the confined aquifer are rapidly transmitted to outcrop, and the impact of injection or abstraction, even several kilometres from outcrop would have to be carefully assessed. The fast response times mean the unconfined aquifer is unsuitable for ASR, and even the confined parts of the aquifer could have limited potential within a few kilometres of the outcrop.

Water quality is good close to outcrop, but gradually becomes more saline down dip, beneath confining horizons.

Therefore the limestone may have some potential for ASR, although this may be limited by their reliance on fracture flow and deterioration of water quality with increasing confinement.

2.4 Permo-Triassic Sandstones

The Permo-Triassic sandstones outcrop across southwest, central, northeast and northwest England, and the Welsh Vale of Clwyd. They comprise a thick (up to 1000 m), variable sequence of predominantly sandstone, with interlayered fine grained horizons. Although intergranular porosity and permeability are fairly high, the aquifer frequently relies on fractures for much of the groundwater flow. The importance of fracture flow varies both between areas and between different horizons within the sequence.

One problem with utilising a thick, homogeneous, non-potable aquifer is mixing if relatively small volumes of water are injected, resulting in low recovery efficiencies. However, all the Permo-Triassic sediments possess some degree of inter-layering. The degree to which these less permeable layers effectively sub-divide the aquifer would need to be carefully assessed as they may enable "stacking" if they are sufficiently impermeable. The potential problems caused by partial penetration of the thick sandstones, and the potential for vertical subdivision of the sandstone by less permeable layers, also require further investi-gation.

Water quality deteriorates away from outcrop, in many areas, becoming non-potable at distances of 5 to 10 km down dip. In some areas, ASR may not be viable due to the presence of gypsum or anhydrite within the Permo-Triassic sandstones themselves, as these highly soluble salts would cause deterioration of the quality of the injected water.

3 CONCLUSIONS

Maps were produced for each major aquifer in each Environment Agency region, showing transmissivity, outcrop, and depth to the top of the formation, the maps being accompanied by text descriptions and a discussion of ASR potential. In this way, it was possible to produce generalised maps, identifying areas that were likely to have the highest hydrogeological potential for ASR in relation to centres of population. Each of the major British aquifers has individual characteristics that could affect the application of ASR, and such possible constraints were also examined. For example, one such issue is the potential reduction in recovery efficiency in the Chalk due to diffusion between native, saline pore water, and injected fresh water in the fractures. The need to address these issues through a programme of research and careful development of trial schemes was highlighted, in order to reduce the perceived risk, and promote use of ASR in Britain.

Another aspect that was addressed by the scoping study was the regulatory aspects of development of ASR schemes. Although not discussed in this paper, extended debate has resulted in draft guidelines being formulated to aid both developer and regulator.

REFERENCES

Allen, D. J., L. J. Brewerton, L. M. Coleby, B. R. Gibbs, M. A. Lewis, A. M. MacDonald, S. J. Wagstaff & A. T. Williams 1997. *The physical properties of major aquifers in England and Wales*. British Geological Survey Technical Report No WD/97/34.

Bennett, J. 1977. An analytical model of salt diffusion in a fissured aquifer. BGS Technical Rpt. No WD/97/7C.

Whittaker, A. (Editor) 1985. *Atlas of onshore sedimentary basins in England and Wales*. British Geological Survey, Keyworth.

Artificial Recharge of Groundwater, Peters et al. (eds) © 1998 Balkema, Rotterdam, ISBN 90 5809 017 5

Groundwater management and water quality assessment in the City of Vienna

H. H. Hauck
Verbundplan GmbH, Vienna, Austria

ABSTRACT: Due to the construction of the Freudenau hydro power plant on the Danube river in Vienna a groundwater management system was designed and implemented. Control of groundwater level is done by operating 21 pairs of wells and 4 additional wells for surface water supply. Control of water quality is carried out by using on-line monitoring stations and biomonitoring. It is supported by means of a hydrological information system.

1 BACKGROUND

Due to the construction of the Freudenau hydro power plant in Vienna by damming the Danube the elevated water level in the river would change drastically the groundwater flow as well as flood a large amount of house basements (more than 4000) in the 2nd and 20th metropolitan districts of the city. As a precautionary measure, a system of sealing walls was constructed along the right river bank. To avoid stagnation of groundwater a field proof groundwater management and monitoring system was developed to keep the groundwater flow in the affected areas as nearly natural in a predefined state, also preventing any adverse changes in water quality. The computer controlled and fully automated groundwater management system developed, installed and operated by Verbundplan guarantees that after completion of the powerplant there will be no negative effects in the impacted area. A first report on the experience of three years operation is given by Dreher & Gunatilaka (1998).

2 WELL SYSTEM

Along the right river bank 21 pairs of wells (one abstraction and one infiltration well each) were constructed. Another 4 wells were installed for the supply of the surface waters in the riparian woods of the Prater Park area. Between infiltration and abstraction well a control room is located which houses the technical equipment and the measuring instruments. The controlling piezometers located between two adjacent infiltration wells enable a continuous measurement of the groundwater level.

All information collected from the management system, including water levels, pumping rates and on-line water quality data, are continuous transmitted to the computer at the control center. Based

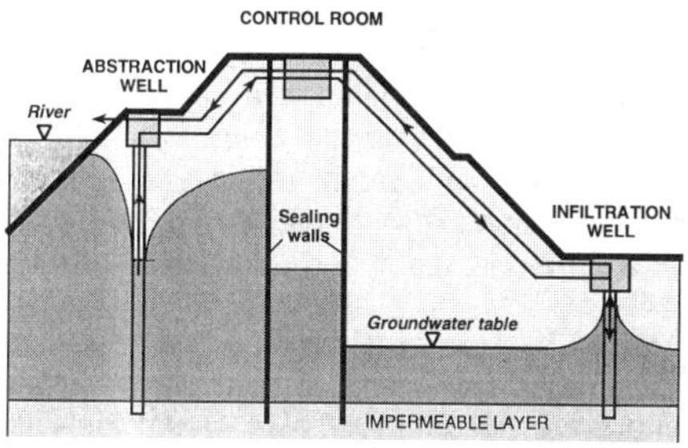

Figure 1. Well system

on this data and supported by the groundwater model the pumping rates are calculated and the operation program is set up every day.

3 MANAGEMENT CONCEPT

To simulate the groundwater flow in the whole project area a 2-dimensional mathematical groundwater model has been developed. This model for transient groundwater flow is based on a representative data set derived from hydrological records over a long period. Using the model the management concept was developed and optimized as well as the number, dimensions and location of the wells could be fixed. The model is also used to operate the system, whereby the amount of infiltration water is calculated, based on the actual flow of the river Danube. Water can also be pumped out from the infiltration well and released into the river to simulate falling groundwater levels.

4 WATER QUALITY ASSESSMENT – CONTINUOUS MONITORING

4.1 *Water quality monitoring around the clock*

The continuous control of the water quality of both the river water and the bank filtered groundwater is carried out through on-line water quality monitoring stations. The following parameters are continuously analysed and recorded: Temperature, electrical conductivity, pH-value, redoxpotential, spectral adsorption coefficient, oxygen concentration, dissolved organic carbon (DOC), dissolved ammonia and oil (petrol, diesel, heating oil, lubricants and other petro products) as total hydrocarbons (THC).

As soon as the threshold value of a given value overshoots, the well functioning is stopped and the abstraction well is flushed automatically. The same happens in case of an alarm due to an accident on the Danube river.

The system can be extended easily to measure a wide range of further parameters as phoshate, nitrate, nitrites, silicate, chloride, chlorine, sulphate, sulphites, cyanides, calcium, magnesium, alkalinity, hardness and metal ions like iron, chromium, copper and zink.

4.2 *Biomonitoring*

Biomonitoring is an alternative to chemical tests which is capable of rapidly detecting acutely toxic conditions in water. „Dynamic-Daphnia-Test" which is installed is a sensitive, automatic biomonitoring system for continuous water quality control. Swimming behavior of 24 hours old healthy water fleas (Daphnia magna) kept in situ river water is continuously recorded through a battery of infra red sensors. The test animals are observed for a week, the records are stored in a data base.

Chemical contaminants in test water tend to cause deviations from the normal behavior pattern of the animals which lead to dynamic alarm levels (upper or lower) that in turn control the well function.

4.3 *Hydrological Information System*

The groundwater management and surveillance system is supported by a hydrological information system (HIS), a modern data base system that provides a full pallet of tools and functions necessary for the capturing, storing and analysing of aquatic environment related data. The HIS concept is based on the client-server principle with ORACLE as data base platform using tools like ORACLE-CDE, ARCInfo, AutoCAD or other external programs with embedded SQL-functions. A Windows based program serves as user interface. Additionally to the available functions it offers a set of predefined DDE (OLE) calls supporting direct data exchange between HIS and other Windows programs like Excel, Winword, Powerpoint, ARCView, COREL Draw).

REFERENCES

Dreher, J.E. & A.Gunatilaka 1996. Groundwater Management in the City of Vienna. Modelling, Testing and Monitoring for Hydropower Plants. II. Int.J. *Hydropower and Dams 2*: 545-554.

Gunatilaka, A. 1997. Online-Meßsystem zur Güteüberwachung des Oberflächen- und Grundwassers der Donau im Stauraum Freudenau. *Schriftenreihe des Bundesamtes für Wasserwirtschaft*, Band 4. Wien.

Dreher, J.E. & A.Gunatilaka 1998. Groundwater management system – an evaluation after three years of operation. TISAR 98 – Third International Symposium on Artificial Recharge of Groundwater, Amsterdam.

Artificial Recharge of Groundwater, Peters et al. (eds) © 1998 Balkema, Rotterdam, ISBN 90 5809 017 5

Use of tracer tests in artificial recharge

L. Maxe
Division of Land and Water Resources, KTH, Stockholm, Sweden

L. Tilly
VAI VA-Projekt AB, Stockholm, Sweden

P.-O. Johansson
VBB VIAK, Stockholm, Sweden

ABSTRACT: A basic requirement of artificial recharge is the ability to recover the water. This implies knowledge of the hydraulic connection between recharge area and abstraction area. The residence time in the aquifer is important for the transition to the groundwater quality. Tracer tests were carried out in four water supply areas in glaciofluvial deposits. Different tracers were used in parallel in order to evaluate their applicability for use in artificial recharge. The best results were obtained with chloride, while Na-Naphthionate exhibited degradation or adsorption. Nitrate was also tested and may under certain conditions be an alternative. The results were evaluated for the rate of water recovery, residence times and flow pattern in the aquifer. The results from one of the water supply areas, Jägarbacken, are presented here. The tracer study altered the earlier conception of the hydraulics of the area.

1 INTRODUCTION

An increasing part of the municipal water distributed to consumers in Sweden is based on artificial recharge of groundwater. Current discussions concerning a higher quality of drinking water, reduced use of chemicals in drinking water production and the need for barriers against pollution indicate that the method has a potential for increased use in the future. About 25% of municipal drinking water is produced with artificial recharge. This is approximately the same amount that is produced from groundwater, while the rest is supplied from surface water plants. Most Swedish artificial recharge plants are located in glaciofluvial deposits. Typically, these are elongated formations with a width of ten to several hundreds of meters.

For proper functioning of a water supply using artificial recharge a good hydraulic control is essential. The hydraulic contact between the infiltration area and the abstraction area is, of course, of utmost importance. The contact determines how much of the infiltrated water that will be recovered. If the infiltrated water undergoes a chemical pre-treatment before infiltration, losses of infiltration water will result in an increased use of chemicals. Loss of water also brings other costs. Exfiltration of water may have negative consequences. The residence time below ground is important as it influences the degradation of organic material and decay of micro-organisms in the water. A long residence time tend to decrease the temperature fluctuations of the water. The contact between infiltration ponds and abstraction wells has been studied using tracer tests. One main objective was to test different tracers and evaluate their suitability for use in water plants with artificial recharge in open basins.

The Swedish experience of adding tracers for this purpose has been limited to chloride, iodide and bromide. Iodide is convenient to use as the natural concentrations are very low. Bromide occurs naturally in much higher concentrations than iodide. Both iodide and bromide are usually considered to be conservative tracers but at low pH values they may sorb to mineral phases. Neither iodide nor bromide should be used in water that will be chlorinated as substances giving taste to the water and toxic substances may form.

2 TRACER STUDIES

The design of tracer studies is dependent on the purpose of the investigation. Important considerations when choosing what tracer to use is the underground transport properties of the tracer, background levels and the detection limit and precision in the analyses. The cost of a tracer might be prohibitive if large volumes of water need to be marked, and the cost of the analyses must also be taken into account. If the water will be used as drinking water, the tracer and its possible metabolites must not be harmful for human health. Further, the tracer should not present any problems regarding taste, odour or colour and it should be environmentally safe. If the tracer is added to a recharge basin, volatile substances are unsuitable.

In this study chloride was used as a reference tracer in combination with Na-Naphthionate (NN) in the Jägarbacken water supply area in Örebro (Fig. 1). This is one of the city's major water supply areas producing 60-100 l/s. The Swedish National Food Administration did not object to the use of NN in the water supply area. The experiment was carried out in co-operation with the Örebro waterworks.

Jägarbacken is a part of a large esker running in a north-south direction. The esker material is mostly highly permeable gravel and sand but local inclusions of fine grained material have been noted.

Only one basin was active during the test (basin 2), recharging 60 l/s of water from the river Svartån. Before recharge, organic substances in the water are reduced through flocculation and sedimentation. Six of the wells in use are arranged more or less linearly to the north; the distance to the first (J1) is 160 m and to the last (J7) 360 m. Approximately 300 m to the south there are two wells that are part of the adjacent water supply area but these were also shown to receive water from Jägarbacken.

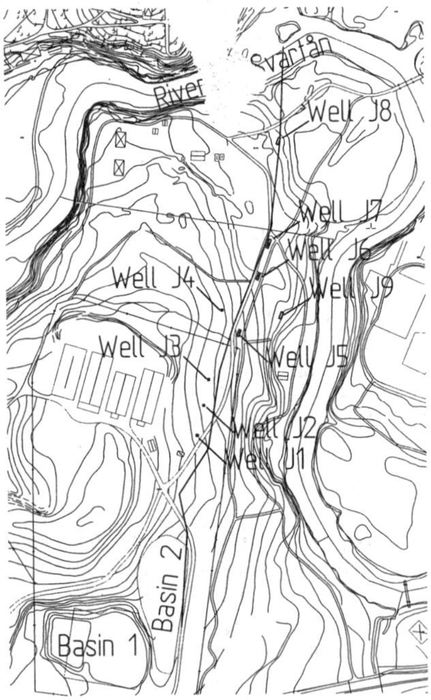

Figure 1. Water supply area Jägarbacken, Örebro

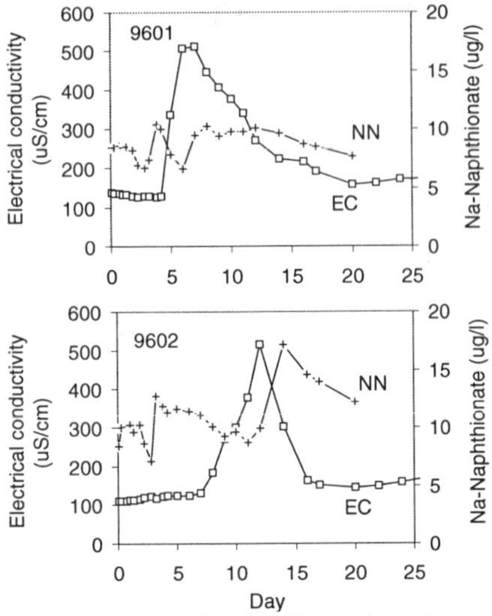

Figure 2. Tracer pulses in observation tubes near the infiltration basin. The curve given for Na-Naphthionate also includes the natural fluorescence of the water.

In preparation for the tracer experiments observation wells were installed close to the basin and along the assumed main flow directions. The open section (3-6 m) was installed immediately below the groundwater table. Dissolved sodium chloride (3 000 kg) and NN (3.9 kg) were added to the inlet of the basin over approximately 40 h giving concentrations of 0.4 mg NN/l and 270 mg Cl/l. Sampling started immediately and was continued for 3 months.

The samples were analysed without filtering. Chloride was mainly detected by careful measurement of the electric conductivity but a large number of chloride measurements were also performed. A few series were also analysed for cations. NN was analysed on a fluorescence spectrometer using an excitation wavelength of 325 nm and measuring the emission at 420 nm (Wernli 1986).

3 RESULTS

The chloride pulse, detected through measurements of the electrical conductivity (EC), rapidly reached the first three wells. Well J3 at the distance of 220 m was affected 2.5 days after the beginning of the addition. Well J5 however, situated only 50 m further north, did not react until day 50. The tracer was not found in the northernmost well. The chloride pulse was also found in the observation wells, even in those not in direct line with the well. This indicates that some water is lost to the sides of the esker. However, it is probable that the slow response of well 5 is due to the infiltrating water making a detour around the water flow that supplies the first three wells.

The tracer NN was not found in any of the wells and only in three of the observation wells (9602, 9603, and 9608) 5-50 m from the basin. A comparison with the chloride response showed that most NN was adsorbed or degraded. Only 8.1 to 3.6 % of the expected response was achieved. In figure 2 the breakthrough curves for observation wells 9601 and 9602 is presented. Well 9602 is situated approximately 25 m from the eastern side of the basin while 9601 is situated between 9602 and the basin. It can be seen that although chloride, as could be expected, was detected first in 9601, no clear pulse of NN was found in this well, while the slightly deeper well 9602 showed such a response. This implies that NN may be selectively retained in the aquifer material. In figure 2, it can also be seen that the natural fluorescence is quite high. The variation is due to disturbances in the pre-treatment of the infiltration water.

The large addition of NaCl caused a substantial cation exchange (Fig. 3). Sodium, although a weak competitor for ion exchange sites, displaced calcium and magnesium and to a minor degree potassium. It may be noted that the response of calcium is quicker than the sodium response.

4 DISCUSSION AND CONCLUSION

Na-Naphthionate exhibited very large degradation and adsorption when added to the infiltration basin. This implies that it cannot be used for quantitative evaluations. Indications of selective adsorption implies that it may also be unsuitable to use for showing the hydraulic connection between infiltration basin and wells even over small distances.

In the adjacent water supply area, Bista, 5.8 kg NN was injected into the aquifer. By this procedure, comparable to German tracer tests (Ptak & Teutsch 1994), NN was easily detected in a well 110 m away.

In another water supply area nitrate was tried in combination with chloride. No degradation was found but it was only possible to follow the tracer for about 20 meters or 54 days. It is possible that nitrate may be an alternative tracer in some aquifers but it is necessary to consider the possibility for nitrate reduction. Substantial reduction is reported by Vanek (1990) who injected nitrate into an aquifer in order to reduce high iron contents.

The results so far indicate that chloride still may be the most practical tracer. Due to the large amount needed for a distinct response it is inconvenient to handle but this is balanced with the very simple detection by EC measurement. The chemical analysis of chloride is available with sufficient precision at all water laboratories.

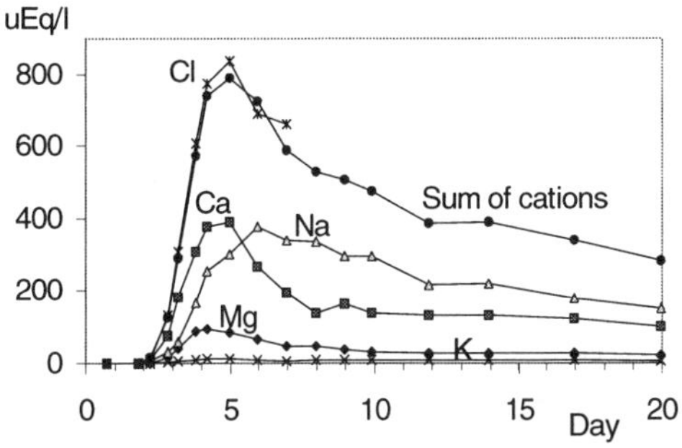

Figure 3. Increase in concentrations of chloride and base cations in well 3 following the addition of NaCl tracer.

Large additions of salt will however increase the density of the water. Experiments have shown that a total salt content of 890 mg/l is enough to cause a sinking to the bottom of the aquifer if injected (Le Blanc et al. 1991). The evaluation of the tracer test would have benefited from a larger salt addition. This should be achieved by keeping up the addition over a longer period of time rather than increasing the concentration. In this water supply area it is not possible to achieve a plateau in the breakthrough curve unless the addition continues for a very extended period. The cation exchange effect should allow an evaluation of the water-solid material interactions if more data of this sort was collected.

5 ACKNOWLEDGEMENT

This research was supported by the Swedish Research Foundation VA-forsk and the participating municipalities. The full results will shortly be published in a research report in the VA-forsk serie.

6 REFERENCES

LeBlanc, D.R., S.P. Garabedian, K.M. Hess, L.W. Gelhar, R.D. Quadri, K.G. Stollenwerk & W.W. Wood 1991. Large scale natural gradient tracer test in sand and gravel, Cape Cod Massachusetts. *Water Resources Research* 27:5: 895-910

Ptak, T. & G. Teutsch 1994. Forced and natural gradient tracer tests in a highly heterogenous porous aquifer: instrumentation and measurements. *J. of Hydrology*. 159:79-104.

Vanek, V. 1990. In situ treatment of iron-rich groundwater by the addition of nitrate. *Limnology*, Dept of Ecology, University of Lund.

Wernli, H.R. 1986. Naphthionat - ein neuer Fluoreszenztracer zur Wassermarkierung. *Deutsche Gewässerkundliche Mitteilungen* 30(1) :16-19.

Artificial Recharge of Groundwater, Peters et al. (eds) © *1998 Balkema, Rotterdam, ISBN 90 5809 017 5*

New-style open recharge in the Oostduinen: An ecological engineering approach

J.H.Peters
Kiwa Research and Consultancy, Nieuwegein, Netherlands

J.S.Rijk, C.Lafort & M.Annema
DELTA Waterworks, Middelburg, Netherlands

ABSTRACT: Based on experience in other parts of the country and on indices measuring terrain occupancy and capacity, the expectation was that optimization of the existing open recharge system in the Oostduinen would benefit both nature and water production. Since the end of 1996, this optimization has been tested in part of the recharge system. Monitoring the ecological impacts and investigating the increase of capacity have also begun.
The redevelopment has resulted in a more diverse and rolling landscape in the dunes. Another result is that valuable plant species, typical of new dune slack, are returning very rapidly. It also leads to a 70% increase in water production. Translating these findings to the rest of the system leads to the conclusion that overall growth of water production by 40% is certainly feasible, even without an increase in occupied terrain. DELTA Waterworks (DELTA) plans to introduce ecologically engineered optimization (upgreening) and to expand the production of drinking water by means of open recharge in the Oostduinen.

1 INTRODUCTION

The dunes in the Netherlands cover some 40,000 ha, and represent an area of great ecological value. Of the total Dutch flora, two-thirds occur in the dunes, with almost 10% occurring only here. There is an enormous variety of plant species concentrated in a relatively small area. Many species of birds, butterflies and reptiles live exclusively in the dunes. The same holds for the Oost- and Middelduinen in the south-west of the Netherlands (fig. 1). These dunes are some of the most valuable, and like all the dunes in the Netherlands, they form part of the ecological infrastructure, the foundation of the country's ecology.
The Oostduinen cover an area of roughly 100 ha. To the north and the east, the dunes become high defences against the sea; to the west, they border on the Middelduinen.
The Oost- en Middelduinen also play a role in the production of drinking water. Every year, DELTA supplies around 50 million m^3 of water to over 200,000 consumers. The water is produced from surface water, groundwater and water from the Haringvliet. This Haringvliet water, a mix from the rivers Rhine and Maas, is taken in and then transported to Ouddorp.
Production facilities in Ouddorp comprise pretreatment units, recharge basins, for recharging the water, and drains for recovering it. In addition, natural dune water is recovered in the Oost- en Middelduinen. This water, together with the recovered recharge water, is then treated to produce drinking water.
In the future recharge and recovery of water will only take place in the Oostduinen, following which there will be possibilities in the Middelduinen to restore the landscape there, which dates from around 1900.

2 GREEN BOTTLENECKS IN THE OOSTDUINEN

Dune water is recovered in the Oostduinen. Pretreated surface water is recharged through basins and recovered by means of drains. The vegetation is prone to be dominated by ruderal species, the wet areas by wild, tall-growing plants and flowers. Even the characteristic form of the dune landscape has been altered through the construction of basins and wells. The result has been the disappearance of natural water gradients. Various species of plants have suffered therefore. We have in mind pioneering species that grow in wet open dune slack (such as Lesser centaury), species that grow in wet, organic soil of dune slack (such as Grass of Parnassus and Early Marsh- orchid), and those that grow in moderately wet, nutrient-poor grasslands (such as Autumn gentian).

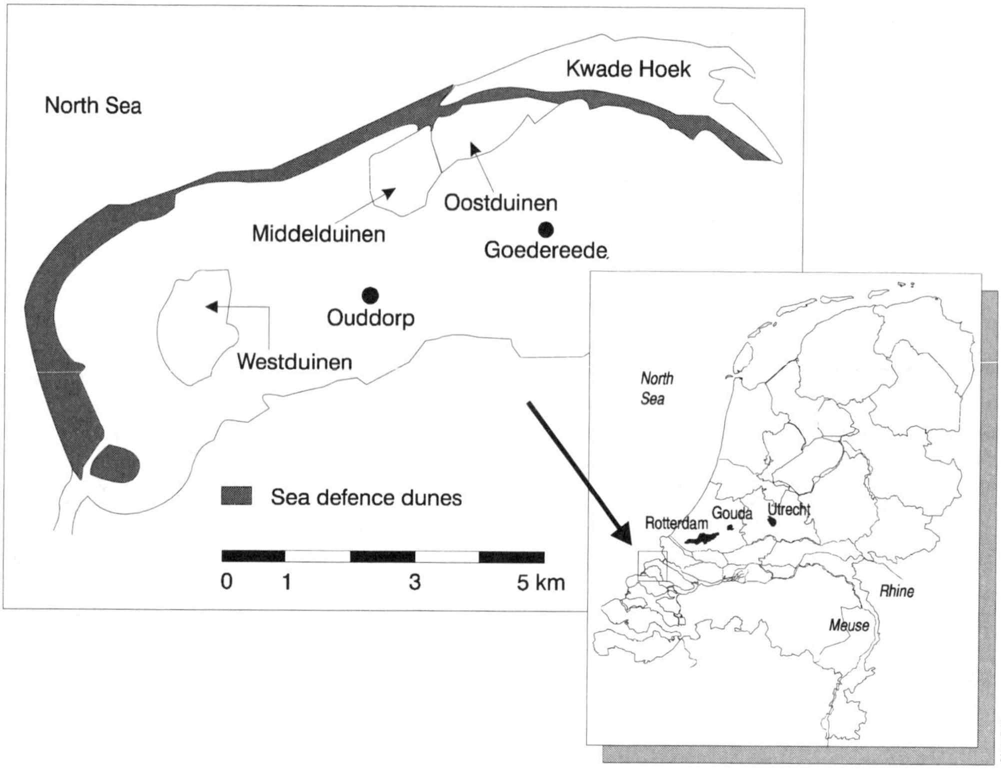

Figure 1. View of dune areas at the island of Goeree

An important cause of the decline of plant species is the change in the water table, because wetland vegetation places great demands on this. Moreover, the annual fluctuation often changes. This is no longer dictated naturally by rainfall and by the seasons, but by the operation of the recharge system.

As well as the water regime, the supply of nutrients is also important. The dunes have by nature a nutrient-poor condition, which contributes to the diversity of the plant growth. Owing to recharge, the vegetation began to be dominated by ruderal species, particularly in dune slack and next to recharge basins. In addition to changes in water tables, vegetation growing wild, and dispersion of transported water, certain forms of disturbance, splitting up of landscape and excavation of soil also occur. As do disturbances from construction work, maintenance, sampling and observation activities and fragmentation from fencing, because some areas are used for grazing and because it is not always possible to manage adjacent areas identically. Excavation occurs due to the construction of basins, drains and pipelines for the transportation of water, energy and signals.

3 BLUE BOTTLENECKS

In the recharge basins, sludge accumulated, which hampered the recharge process. From time to time, the sludge had to be removed. Since the beginning of 1996, pretreatment of recharge water has dramatically reduced the intake of suspended solids and thereby the problem of clogging. Deterioration of wells has made phased renovation necessary, something which is now taking place. Moreover, DELTA also intends overhauling the posttreatment at the Ouddorp pumping station. This work will begin at the end of 1998.

4 ASPECTS OF THE SOLUTION

In the fifties, when most of the recharge systems in the Netherlands were constructed, the environment was not receiving much attention yet. In the nineties, design, construction and renovation follow a new approach. Characteristic of this approach, known as new-style open recharge, is that from the ecological standpoint, the emphasis is on objectives rather than restrictions. This idea, or philosophy, has already

been successfully applied at several locations (see references). New-style open recharge is simple and based on the ecological principle that nature benefits if the following two conditions are satisfied.
First, it is important when carrying out the construction and implementation to create spatial gradients that produce variations in habitat for the vegetation. The second is to have a type of operation that, as far as possible, only allows natural disturbances and modifications. As we know, these are dictated by the seasons, which bring changes in rainfall, temperature and light.

New-style open recharge can thus be defined as a recharge process for which nature determines the implementation and operation. The process breaks down as follows:
- *pretreatment* to limit the intake of nutrients and other substances critical to the environment;
- *concentration* to minimize the terrain occupied by the recharge system;
- *implementation* that creates or restores spatial gradients;
- *operation* that limits non-natural modifications to water levels and recurring human activities (concerning maintenance, sampling, etc.);
- *adaptations for the landscape*; and
- *terrain management* to make and keep fit the habitat for the natural vegetation (mowing, grazing and sod cutting).

The aim of pretreatment is to ensure a nutrient-poor habitat. For concentration, the principal aim is to limit terrain occupancy so that natural processes are dominant in as large an area as possible. Construction and re-sectioning of recharge basins bring new gradients with sloping, natural banks, gentle inclines, lakes and wetlands, and coves. This means a basin does not have a uniform depth. It is even possible to divide a basin into different sections, have some parts deeper than others, or create small islands. Mowing might have to be organized in order to remove nutrients. The grazing, too, ensures adequate light for short herbaceous vegetation. Where a lot of organic material has accumulated in the soil, sod is cut at first. The top layer is removed to reduce the abundance of nutrients and expose the base-rich earth once more.

5 THE PILOT PROJECT AT BASIN 7

By comparison with other recharge systems in the dunes, the territory of the recharge system in the Oostduinen is large. Since future recharge will remain restricted to the Oostduinen, the idea of using new-style open recharge there has been considered. Accordingly, the goal of this in the Oostduinen is to upgreen the system and simultaneously produce more water, with less operational attention and less maintenance.

Figure 2. Adapted basin 7

To demonstrate this on a small scale, it was decided in 1996 to redevelop and modify the operation of the area surrounding basin 7, the most southerly recharge basin close to the pumping station. Since mid-1996, the following work has been carried out (Fig. 2):

- roughly 400 m^3 of sludge has been removed from the basin;
- vegetation alongside the basin has been removed;
- from the area surrounding the basin, roughly 2 ha of the organic top layer has been cut, and roughly 5,600 m^3 of this has been removed;
- existing banks of the basin have been made smooth or broken up;
- the water level in the basin has been raised, especially in the east section; and
- on the east side, the pumping level in a recovery drain 5 East has been lowered.

To determine the effectiveness of these steps, a monitoring programme has been established for measuring flow rates and water levels in pumps and observation wells. Groundwater and recharge water is chemically analysed and an inventory of the vegetation made.

6 RESULTS

Thanks to the optimization, basin 7 has changed from a straight-lined recharge basin into a natural-like body of water with a more natural shape (Fig. 2). The redesigned basin blends in perfectly with the dune landscape immediately surrounding it.

In the basin, target species such as Stonewort occur, and along the banks, Brackish Water-crowfoot. Following the sod cutting work on the low banks, different members of the midget rush family have appeared, rare pioneer plants indicative of a base-rich, nutrient-poor, wetland environment. According to expectations, further succession will lead to the growth of rare plant communities with Black Bog-rush.

The conclusion to be drawn is that redevelopment very quickly leads to the establishment of plant species of the natural target types. Particularly along banks and in dune slack depressions fed with base-rich water, many unusual species of plants occur.

Concerning the fauna, the following species have been seen in the water: green frog, natterjack, small salamander and nine-spined stickleback. Two unusual species of ground beetle have been spotted, as well as the blue-winged grasshopper. In and around the basin live different species of dragonfly (emperor dragonfly, emerald damselfly, darter, lantern fly, large dragonfly, damselfly, black-tailed skimmer). Some of the distinctive and unusual species of birds observed after the redevelopment are the spoonbill, little egret, greenshank, green sandpiper and common sandpiper.

The capacity of the part of the recharge system that has been redeveloped was some 1,250 m^3 a day. Following the changes, this has become 2,100 m^3 a day. In relation to the initial situation, the capacity of this part has increased by approximately 70%.

7 FUTURE TRENDS

The results of the redevelopment represent an unequivocal success. In less than a year, many pioneer species of natural target types have been observed. In terms of landscape and fauna, too, there has been significant improvement.

Since the outcome for the water production is also favourable, the next obvious step is to optimize and redevelop the entire recharge system. Looking at the pilot results, a 40%-increase in water production at Ouddorp is feasible. The permitted water production is currently 3.6 million m^3 a year. DELTA wants to pursue a course that leads to a permit for an annual production of 4.5 million m^3 of drinking water, an amount needed within the next few decades. DELTA will draw up a master plan for the redevelopment of the entire recharge system in the Oostduinen.

REFERENCES

Peters, J.H., A. Stakelbeek, J. Demarteau, Q.L. Slings & J. Mostert 1990. Optimalisatie en renovatie infiltratiegebied Castricum-masterplan-. Ecohydrologische optimalisatie en renovatie. SWO 90.256. Kiwa, Nieuwegein (in Dutch).

Peters, J.H., Q.L. Slings & A. Stakelbeek 1992. Open Infiltratie Nieuwe Stijl. Integrale ontwikkeling van natuur en techniek bij renovatie van een open infiltratiesysteem. *H2O* (25) no.19: 532-537 (in Dutch).

Peters, J.H., M.W. Kortleve, G.M. Leltz, K.J. Wardenaar, E. van der Mark, C.L.H.J. Bessems, C.M. de Jong & W.J. Zaadnoordijk. 1994. Optimalisatie Groot Berkheide. Uitgewerkte ideeën voor natuurontwikkeling, herinrichting en Open Infiltratie Nieuwe Stijl. SWO 93.351. Kiwa & EWR (in Dutch).

Peters, J.H. & R.A. Visser 1994. Ecological Engineering and Artificial Recharge at Groot Berkheide. In: A.I. Johnson & R.D.G. Pyne (1994). Artificial recharge of groundwater. *Proceedings of the Second International Symposium on Artificial Recharge of Ground Water.* A.S.C.E. New York.

Peters, J.H., A. Stakelbeek, Q.L.Slings & H. Eken 1995. Masterplan infiltratiegebied Kieftenvlak te Wijk aan Zee. Uitgewerkte ideeën voor renovatie en voor ecologische aanpassingen. SWO 94.359, Kiwa & PWN, Nieuwegein & Bloemendaal (in Dutch).

Artificial Recharge of Groundwater, Peters et al. (eds) © *1998 Balkema, Rotterdam, ISBN 90 5809 017 5*

Possibilities of enlarging groundwater exploitation reserves of Peć town, Yugoslavia

M. Komatina
Geozavod, Belgrade, Yugoslavia

ABSTRACT: Groundwater sources Crna and Bela voda near Peć have been a subject of various explorations. Abstraction was carried out with a long gallery in limestones and horizontal drainage within alluvial gravel. In such way, double enlarging of the source capacity is possible. Solutions of flow regulation and exploitation reserves enlarging of the both sources were also planned, in order to catch even 1500 l/s, instead of 190 l/s.

1 INTRODUCTION

The first condition for an explorer wishing to get the essence of some solution is, certainly, that a concept of an exploration and giving the solution of some practical task based on the nature itself - at concrete geological conditions of the interesting area, that ideas of an explorer are successfully composed into the network consisting of (numerous) geological elements of the terrain. From the other side, in the previous practice, a solution of groundwater catchment in an area of some resource was based on exploitation minimal flow which is very often a non-rational way, if we consider increasing demands for water and significant possibilities of exploitation reserves enlarging through application of artificial recharge. That is particularly valid for aquifers of karst terrains, where levelling the groundwater flow now becomes one of the primary conditions in the plan of rational use of water wealth. About different possibilities in solving that actual task, there is a great number of papers (Perić 1963, Komatina 1968, 1990). In this poster, sources Crna and Bela voda near Peć town (Yugoslavia), as an example outlining the fact that capabilities of development of a source might be practically unlimited, are discussed.

2 HYDROGEOLOGIC BASE OF GROUND WATER CATCHMENT

Crna and Bela voda karst springs are located just in front of the entrance to the famous Rugovska klisura gorge, upstream from Peć town, in the vicinity of the Pećka Bistrica river (Fig.1). Watershed areas of the occurrences are built of very karstified Triassic limestones insulated by impermeable formations. Areas are about 17.5 km^2 (Crna voda) and 7 km^2 (Bela voda).

Groundwater discharge in the area of Crna voda takes place in two points - diffuse into alluvial deposit and river flow along the wide front over 200 m and relatively concentrated - through the spring. In the second case, from the practical point of view, beside karst aquifer, widened part of the Pećka Bistrica alluvial plain is particularly interesting especially because a close hydraulic link between the two aquifers exists. Conductance of the alluvial plain is very high, and yield of exploratory wells varies from 34 to 38 l/s.

Because of steep gradient of the limestone footwall, within the Bela voda watershed, groundwater flows very fast to the drainage base - the river flow. Concentrated groundwater recharge takes place by percolation of the Milešveska reka river in the part where the river engraved its bed within the limestone, and discharge - through the Bela voda spring completely.

2.1 *Crna voda source*

The place where the Crna voda spring appears is a camouflaged seepage karst source covered by alluvial deposit of the Pećka Bistrica river. Before Holocene, rising of the water probably took place concentrated and directly to the river flow because

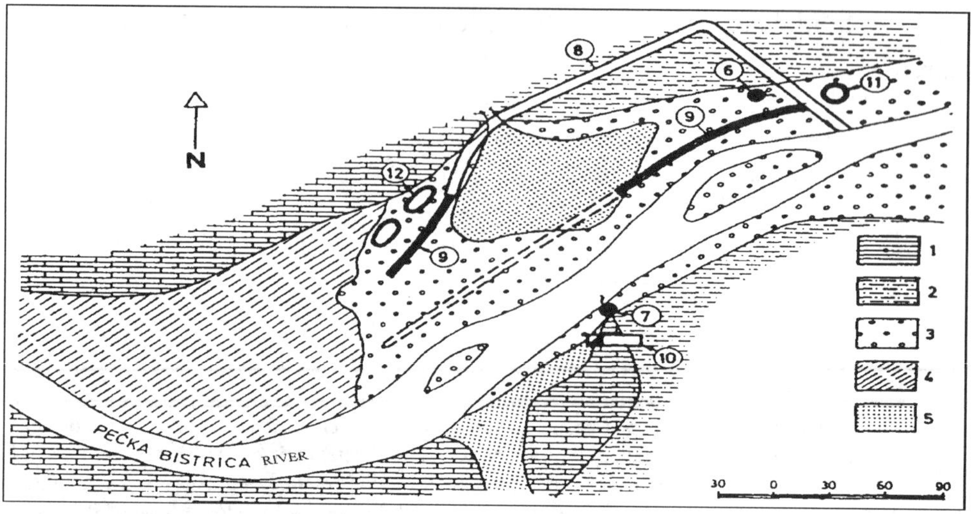

Figure 1. Hydrogeologic map in the area of the Crna and Bela voda springs near Peć. 1. limestone - permeable; 2. diabase - chert formation - impermeable; 3. alluvial deposit - permeable; 4. river terrace - low-permeable; 5. vortex fan; 6. Crna voda old structure; 7. Bela voda old structure; 8. Crna voda gallery; 9. horizontal drainage in the alluvial plain (hatched - planned part); 10. Bela voda shaft and gallery; 11. water-collecting basin; 12. planned infiltration basins (Scale: in meters).

the river intersected the limestone mass. By depositing, original primary spring evoluted into seepage secondary source. Before intake work, it manifested through presence of numerous springs during high and middle water levels, in the area of alluvial plain and, less, along the bank of the river flow.

By the first intake work, performed in 1959, only concentrated groundwater flow was abstracted, supplying the town Peć with 150 l/s drinking water. However, it was clear that the old intake structure was not enough to tap all groundwater inflow and the significant amount disappeared without control. That is the reason why in the period 1979-1982 the territory was an object of voluminous detailed exploratory and intake studies (Komatina 1982).

Results of hydrogeologic exploration showed that there are several possibilities to develop (expand) the source, especially because here two permeable formations with different flow conditions and groundwater regime are in touch: limestone and alluvial, with close surface flow, characterized by high-quality water. That made a specific creative approach possible to defining a solution for ground water management in general. The result is one complex system of intake structures, presented by installation of the spring, and a long gallery within limestones, as well as horizontal drainages and drilled wells in the alluvial plain (Fig.1). For the next stage, objects were planned, because almost ideal hydrological and hydrogeological conditions prevail.

Exploration - water-collecting gallery was planned to cut the major part of the water of seepage source along the wide front and in the primary (limestone) formation. The object, 251 m long, was performed by contact of Triassic limestones and products of diabase - horn formations in the hanging-wall. Open tectonic contact is generally subhorizontal, with expressed rough spots strengthened by small faults. In that range, about 20 stronger or weaker water occurrences from karst channels, small caverns and open fractures were registered, with discharge between 2 and 20 l/s. From the far end of the gallery, thrashing of the 60 m long horizontal drainage was continued. So, water of the alluvial aquifer was directed to the gallery. In that way, additional 220 l/s of water was provided. Later, near to the river flow, a 100 m long horizontal drainage was made, providing additional 100 l/s.

The gallery testing performed during the extreme water levels in 1982 gave very significant results. First of all, capacity of the object was determined as well as defined presence of positive hydraulic connection between the new and old intake structure. Namely, minimal capacity of the old structure is, practically, unchanged (150 l/s), but each attempt to accumulate water in the gallery during periods of lower or medium levels caused increase of the old structure capacity and level rising in all piezometers within the alluvial plain. Several-years observations of piezometers showed, besides, that at the edge of the alluvial plain, narrow belt operating as estavelle exists - namely, during high water, privileged channels within limestone formation are not able to convey all amount of water to places of discharge cut by the gallery, so - they pour into the alluvial deposit and opposite - during the low level, percolating from the alluvial plain occurs and, hence - aquifer recharge. This favorability was used for planning an appropriate system of artificial recharge, estimating the increase of the source capacity up to 1100 l/s.

2.2 *Bela voda source*

In the area of the Bela voda spring, during geological evolution of the river valley, process of subrecent and recent karstification followed the peak elevation of the bottom of the river bed, precisely peak elevation of the bed in the narrow rayon of permeable (limestone) and impermeable (diabase - chert) formation contact, where privileged groundwater collector was modelled. That is clear considering the fact that the river flow is perpendicular to extension of geological formations and so - that is reason why in all moments of the river bed deepening just the mentioned contact was the lowest possible zone for groundwater discharge. As one evidence, beside presence of the spring, serves the Markova pećina cave very near - markant speleological object at about ten meters higher elevation, which presented the main groundwater outcrop during the earlier stages. Note that by exploratory drilling and making an exploratory well, up to 25 m deep, presence of several forms of paleokarst were determined, but completely filled by terra rossa and clay.

Important hydrogeological role around the spring has a Quaternary deposit within a deep depression modelled earlier by water from the cave. Dipping under the river bed peak elevation, and extending approximately 100 m parallel to the river flow, so low-permeable medium serves as a barrier. For groundwater discharge served only a place of cut of permeable limestone medium, 5-6 m wide (Fig.1), and at the spring, 40 l/s occurs as a minimum.

Within mentioned hydrogeological conditions, enlarging of exploitation reserves can be reached by overexploitation of the reserves during two-three arid months, when a problem with lack of water is present. In order to test effects of the overexploitation, a 25 m deep well was made. However, to depth of 13.40 m, the object was drilled through mentioned low-permeable Quaternary deposit, and in the rest through compact limestones. Yield of the object was inevitable 1 l/s.

Object for overexploitation was made near to the rear of the spring (Fig.1). It consists of a shaft and a gallery, which peak elevation was determined so to locate the gallery not only within a limestone mass with filled paleokarst collectors, from one side, and to make maximum water level lowering possible (more than 8 m in relation to minimal water level at the spring) from the other. Length of the gallery coincides with a width of the limestone cross-section at the appropriate depth (Fig.2). The shaft was made to 13 m depth, and from the planned 35 m of the gallery, only 11 m was carried out. Unfortunately, because of unsuitable social situation in 1982, works had to be stopped and the object conserved. So, realizing a modern approach to regulating subsurface flows in karst was temporarily put out of action.

For further development of the source, two measures are possible: 1) groundwater accumulation forming, by injecting very narrow limestone cross-section, parallel to the river flow and to the desired (possible) height and 2) regulation of the Mileševska reka river and discharging of significant quantities of the water through ponors within its bed. Similar artificial groundwater recharge by the river water filtration may be performed in the case of the Crna voda source, too.

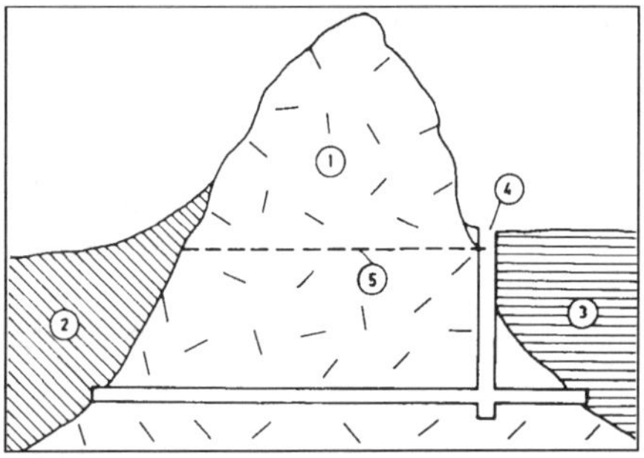

Figure 2. Sketch map of the shaft and gallery in vicinity of the rear of the Bela voda spring (1. limestone; 2. diabase - chert formation; 3. impermeable deposit; 4. shaft with a gallery; 5. groundwater level in the natural conditions).

3 CONCLUSIONS

Practically each aquifer in karst possesses its own characteristics in geological structure, enabling the explorer to perform forming of the source in an unique and creative way. A lot of possibilities exist, particularly when karst and alluvial aquifer are hydraulically connected to each other.

As an example the Crna and Bela voda sources with the alluvial plain of the Pećka Bistrica river are discussed in the poster. Here, both springs were pumped in a classical way and tapped 190 l/s. The capacity was enlarged approximately three times by establishing the gallery within limestones, and by the drainage system within the alluvial plain. Structures for artificial recharge were also planned. Significant potential sources of enlarging the capacity are also available by regulation of ponors in the catchment area and by river infiltration into the limestones.

REFERENCES

Komatina, M. 1968. Karst and hydrogeological capabilities of more rational groundwater catching. *Vesnik*, Geozavod, Belgrade, VIII, Ser.B: 83-120.

Komatina, M. 1980. Hydrogeological approach to exploration of karst territories in order to evaluate possibilities changing the groundwater regime and enlarging their exploitation reserves. *Vesnik*, Geozavod, Belgrade, XVI/XVII, Ser.B: 5-20.

Komatina, M. 1982. Exploratory and catching works in the area of Crna and Bela voda springs near Peć. Fund of Geozavod, Belgrade.

Komatina, M. 1990. Hydrogeological Approach to Investigation in Karst for Possible Modification of Groundwater Regime and Increase of Recoverable Reserves. *Environ.Geol.Water Sci.* 6(2): 149-153.

Perić, J. 1963. Subsurface accumulations as hydrotechnical objects for purpose of regime regulation of groundwater of karst and other types. *Vesnik*, Geozavod, Belgrade, III, Ser.B: 67-96.

2.3 Artificial groundwater recharge in Stockholm, Sweden

Artificial Recharge of Groundwater, Peters et al. (eds) © 1998 Balkema, Rotterdam, ISBN 90 5809 017 5

Artificial groundwater recharge in Stockholm
I. The project and its general aim

J.O. Hjort
Stockholm Water Company, Sweden

P. Ericsson
The Municipal Association Norrvatten, Järfälla, Sweden

ABSTRACT: Stockholm Water Co and The Municipal Association Norrvatten has started a project to investigate if artificial recharge of raw water from Lake Mälaren can be a long term sustainable water supply system for the Greater Stockholm area. Existing water supply system and expectations on the future system are presented. Actual areas in the Uppsala Esker are presented together with a first plan for the formation of the future supply system. A second phase of the project in progress, comprises strategic environment assessment and pilot plant column tests. These investigations are reported separately.

1 GENERAL AIM OF THE PROJECT

In 1995 Stockholm Water Co and The Municipal Association Norrvatten started the project "Artificial ground water as the future water supply for Greater Stockholm ". The idea came from the Swedish consultant firms J&W and Sweco. The basic idea is to infiltrate raw water from Lake Mälaren into the Uppsala Esker situated west of Greater Stockholm and connected with Lake Mälaren. See figure 1.

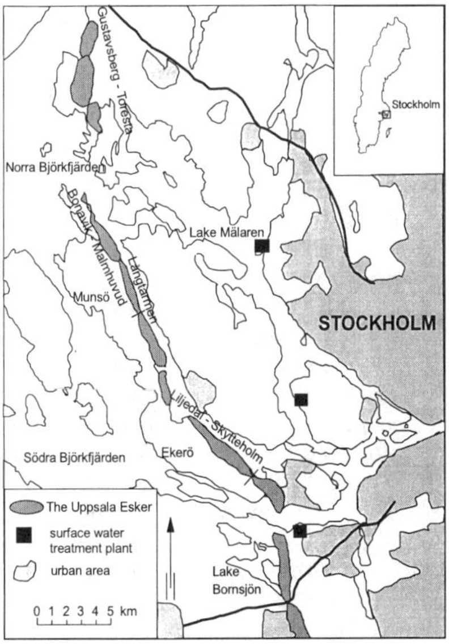

Figure 1. Map of investigated area (Johansson 1996).

The general aim of the project is to investigate if artificial recharge is possible for this area and if it can be sustainable according to Agenda 21. It is of interest to eliminate the use of the flocculant, aluminium sulphate, for the chemical treatment and in that way reduce the sludge problem. It is also of interest to reduce or eliminate the use of chlorine and in that way reduce the formation of chlororganic compounds in the drinking water and the environment.

The most important questions in the project are: Is it possible to get a high quality water with artificial recharge in existing eskers? Will this water supply system be better from an environmental point of view than the existing system? If so, the water suppliers will work to obtain legal protection of the land necessary for the artificial recharge. A possible decision to introduce artificial recharge in the Greater Stockholm area will probably have to wait for at least another ten years.

2 BACKGROUND

Stockholm Water Company and Norrvatten are the main water suppliers of Greater Stockholm. They deliver around 500 000 m^3 of high quality water per day to 1.5 million persons from three water works: Norsborg and Lovö water works belonging to Stockholm Water and Görväln water works belonging to Norrvatten. Raw water from Lake Mälaren is treated by chemical precipitation (aluminium sulphate), sedimentation and rapid sand filtration. The water works of Stockholm Water have slow sand filters as final treatment to improve odour and get a better filtration. Norrvatten adds powdered activated carbon into the flocculation step, in case of disturbing odour to the water due to algae production. In the near future Norrvatten will introduce granulated activated carbon filters to improve taste and odour. The water is pH-adjusted and chlorinated before distribution.

The produced water in the existing water works is of good quality but there is a wish to make some improvements. The content of organic material measured as COD_{Mn} is a little bit high. There is a wish to reduce the existing content, that is around 3 mg/l to less than 2 mg/l. The content of easy degradable organic material, 30 - 60 µg/l measured as AOC, is too high to get a biostable water.

The disinfection with chlorine gas and ammonium sulphate to form monochloramine in the water also results in a formation of 30 to 60 µg/l of adsorbable organic halogen compounds, AOX and around 5 µg/l of trihalomethanes, THM.

The use of aluminium sulphate results in large quantities of sludge, mainly aluminium hydroxide polluted by organic material such as humic substances. With artificial recharge, there is a much smaller sludge production.

3 THE UPPSALA ESKER

In the Stockholm area, there are several eskers. In this work three parts of the Uppsala Esker have been studied. The esker is the largest glaciofluvial deposit in the area. Another reason for choosing the Uppsala Esker is that it is situated near the central part of Lake Mälaren. This means minimum transport distances and access to the best raw water in the region. See figure 1.

4 PROJECT ORGANISATION

An artificial recharge of ground water for the future water supply of Greater Stockholm will result in restrictions in the use of some land areas. For that reason it is important to have all authorities engaged in the project at an early stage. A Steering Committee for the whole project is made up of representatives from: The National Environment Protection Agency, The County Administrative Board, The Office of Regional Planning and Urban Transportation, The Stockholm County Association of Local Authorities, The National Swedish Sand, Gravel and Crushed Stone Association, The City of Stockholm, The affected municipalities Botkyrka, Ekerö and Upplands-Bro and the water suppliers.

For every part project in the project, there is a separate project organisation with representatives from the water suppliers and the engaged consultants and scientists.

5 PHASE ONE

The first phase of the project was a feasibility study made by the water suppliers, the above mentioned consulting companies and The Royal Institute of Technology, Division of Land and Water. The following questions were analysed:

- Future need of water in Greater Stockholm area;
- Experience from existing water works using artificial recharge;
- Water Quality in relevant parts of Lake Mälaren;
- Competing use of land;
- Hydro-geology and ground water quality;
- Proposal for future water works using artificial recharge;
- Co-ordinated use of land, and
- Necessary research and development.

In a prognosis for the year 2100, made in phase one, the population in the water supply area was calculated to increase up to between 2.7 and 3.1 million. The water consumption was calculated to maximum 10 m^3 per second or 860 000 m^3 a day. Today the expectations for the future water demand is lower.

There were a great number of conclusions from the first phase of the investigations. The most important ones were, that a future water supply for the Greater Stockholm area, based on artificial recharge, was judged realistic. The main problem is the relatively short detention time in the esker, estimated to only one or a few days in some parts of the eskers. For that reason there is a need to use the best available technique. There is also a need to further develop existing techniques.

Three main infiltration areas are proposed. These are Bonavik-Malmhuvud for the Görväln water works, Liljedal-Skytteholm for the Lovö water works and Sturehov-Bornsjön for the Norsborg water works. For each infiltration area a separate raw water intake from the central part of Lake Mälaren is proposed. It will be possible to take raw water from different depths in order to optimise the raw water quality.

In each infiltration area there will be a number of infiltration basins. The infiltration rate is preliminary calculated to be 1 meter per day. This is a low value compared to conventional infiltration. The rate is based on the low detention time in the ground, one day up to less than one week. The necessary infiltration area is calculated to be around 1 km^2. If the investigations show that a higher infiltration rate is possible, the infiltration area will of course be reduced.

From the three infiltration areas, the artificial

ground water will be delivered through a rock tunnel to one of the three existing water works in the area. There the water will undergo a final treatment such as pH adjustment and possible disinfection before the water is distributed through the existing distribution system. (Stockholm Water Co et al, Hjort et al, Hjort and Johansson 1996).

6 PHASE TWO

From the first phase of the investigations a number of key issues were identified and have to be investigated more carefully. To get an early answer to the questions around the environmental effects of the proposed future water supply system, a strategic environmental assessment, SEA, was made. The SEA has been remitted to a number of organisations. Their answers have just arrived and will be summed up. The SEA is reported as a separate poster to the TISAR symposium (Blomberg, et al).

There is a lack of knowledge about the different processes in the infiltration zone. Pilot plant column tests have been started up to study the chemical and microbiological processes in the unsaturated zone. These studies are reported as three separate posters at the TISAR symposium (Berggren et al; Johansson et al 1998; Szewzyk, U. et al).

In the original plans the building of a small scale artificial recharge plant was included to follow up the experiences from the pilot plant column tests. Today it is of interest to make these investigations in an existing small water works using artificial recharge of Lake Mälaren raw water into the actual esker.

7 FURTHER INVESTIGATIONS

The land suited for the future artificial recharge is in many parts used for recreational purposes. Some parts are used for excavation of sand and gravel. The introduction of artificial recharge means that there will be a trespass on the recreational use of the land. Artificial recharge means introduction of a great number of infiltration basins, which will stand in the way of outdoor life. These basins have to be operated in a way so that they normally contain water almost all the time. The basins have to be designed in a way so that they merge into the natural surroundings. These problems will be a challenge for the project. There is also a need of safety protection zones around the infiltration basins and the wells. The effect on the recreational use of the land has to be minimised.

In the first stage a principal evaluation was made of the eskers according to the possibility to recharge and take out water. The future work includes a verification of this work.

REFERENCES

Berggren, D. et al, this volume

Blomberg, J, et al, this volume

Hjort, J.O. et al, ARG 96, Helsinki, *Nordic Hydrological Programme*, NHP Report No 38 (1996)

Hjort, J.O. *Stockholm Water Symposium*, 1996.

Johansson, P-O. et al, ARG 96, Helsinki, *Nordic hydrological Programme*, NHP Report No 38 (1996)

Johansson, P-O. et al, this volume

Szewzyk, U. et al, this volume

Stockholm Water Co et al, Storstockholms framtida vattenförsörjning, Undersökning av möjligheterna till konstgjord grundvattenbildning, Etapp 1. (In Swedish)

Artificial Recharge of Groundwater, Peters et al. (eds) © 1998 Balkema, Rotterdam, ISBN 90 5809 017 5

Artificial groundwater recharge in Stockholm
II. Column test design and tracer tests

P.-O. Johansson
VBB VIAK, Stockholm, Sweden

B. Espeby & B. Nilsson
Division of Land and Water Resources, KTH, Stockholm, Sweden

G. Allestam
Swedish Institute for Infectious Disease Control, Stockholm, Sweden

ABSTRACT: Column tests were performed to study water quality changes in the unsaturated zone as a part of a project concerning the future water supply of Stockholm by artificial recharge. Time-domain reflectometry (TDR) probes, water and gas sampling devices, and devices for microbiological studies were installed in the columns. The hydraulic loads were 1 $m^3/m^2/d$ in three of the columns and 2 $m^3/m^2/d$ in the fourth. Free water surfaces appeared after 14-22 weeks in three of the columns. Unsaturated conditions prevailed below the sand surface in all 4 columns. According to a chloride tracer test the mean residence times were 16-18 hours in the columns with a hydraulic load of 1 $m^3/m^2/d$, and 10 hours in the column with 2 $m^3/m^2/d$. A test with Salmonella bacteriophages gave 2-5 hours shorter first arrival times than the chloride test while the arrival times of the peak concentrations were just 1-2 hours shorter.

1 INTRODUCTION

The water supply of the Greater Stockholm Region is today based on water from Lake Mälaren, treated in three conventional water treatment plants. A feasibility study indicated that artificial recharge in the Uppsala Esker is a realistic alternative for the future water supply. In the feasibility study column tests were recommended 1) to increase the knowledge about processes controlling water chemistry and microbial activity in the unsaturated zone, and 2) to investigate how these processes can be optimised in order to achieve and maintain a good water quality. The main project is described in Hjort & Ericsson (1998) and the chemical and microbiological results from the columns tests are presented in Berggren et al. (1998) and Szewzyk et al. (1998).

2 COLUMN TEST DESIGN

Four large columns were constructed (height: 4 m, diameter: 0.3 m). The columns were filled with sand (grain size: 0.1-4 mm, d_{60}=0.9 mm). In column 3 a more fine-grained filter sand (0.1-1 mm, d_{60}=0.5 mm) was used for the uppermost meter. The sand consisted of about 70% silica. The specific surfaces of the sands were 0.8 and 1.0 m^2/g respectively. The average total porosities were 40-42%. The raw water from one of the present surface water plants (Norsborg) was used in the study. The columns were placed in a room at the plant where the air temperature followed that of the raw water.

Time-domain reflectometry (TDR) probes for measurements of the water content and apparent bulk electrical resistivity, suction cups (pore size 1-5 μm) for water sampling, and gas sampling devices were installed at different depths. Furthermore, holes were drilled for access for soil sampling and insertion of glass sheets for microbial analysis (Fig. 1). All devices were made of stainless steel, teflon or silicon and all tubings were shielded from daylight to minimise microbial and algal growth. The columns were exposed to artificial daylight for 6-12 hours/day depending on season.

3 HYDRAULIC LOAD AND WATER CONTENT

The hydraulic load was set to 1 $m^3/m^2/d$ in columns 1, 3 and 4 and to 2 $m^3/m^2/d$ in column 2.

Free surfaces were obtained at the top of columns 2, 3 and 4 after 14, 17 and 22 weeks respectively.

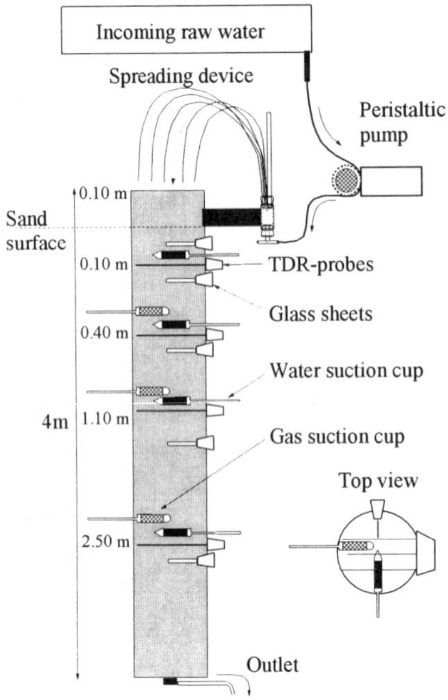

Figure 1. Column setup

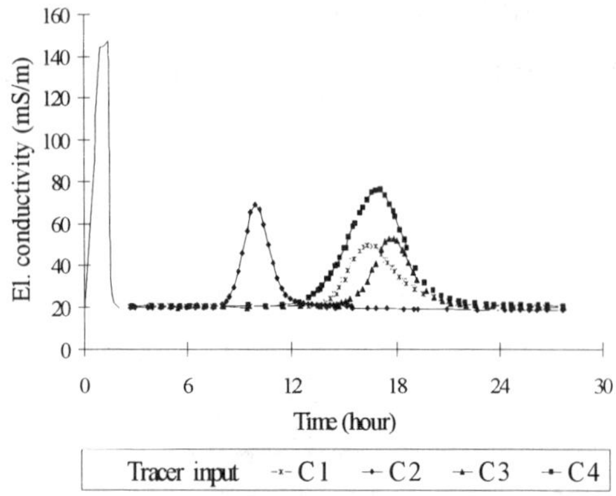

Figure 2.Breakthrough of chloride in column outlets.

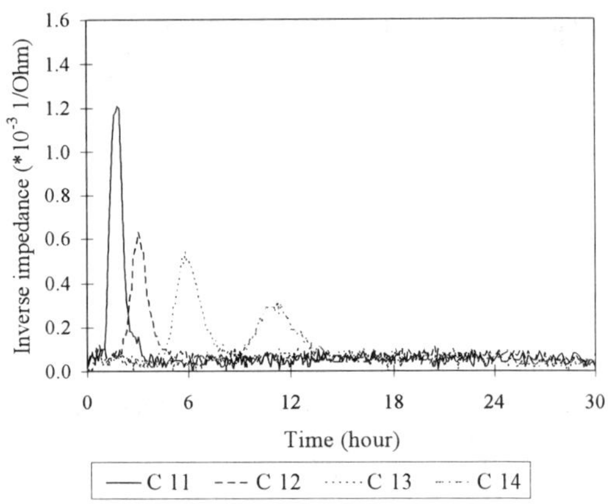

Figure 3. TDR breakthrough curves in column 1, depth: C11=0.1 m, C12=0.4 m, C13=1.1m, C14=2.5 m.

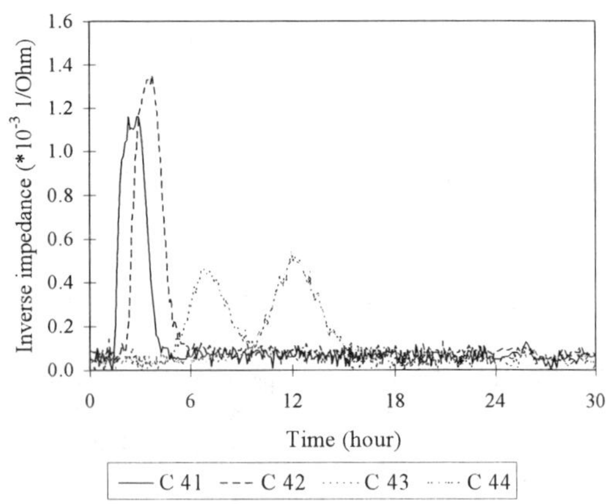

Figure 4. TDR breakthrough curves in column 4, depth: C41=0.1 m, C42=0.4 m, C43=1.1m, C44=2.5 m.

Column 1 had still not obtained a free water surface after 35 weeks.

TDR-measurements after 1 and 25 weeks gave water contents of 13-26% with the exception for the uppermost meter of column 3, where the values were 23-31%.

4 TRACER TESTS

Tracer tests were performed with chloride and bacteriophages. The chloride tracer was added as NaCl in the raw water distribution tank during one hour. The peak concentration in the infiltrating water was 400 mg/l Cl (150 mS/m). Water samples from the outlets of the columns were analysed for conductivity. The first arrival times at the outlets were 12-15 hours for columns 1, 3 and 4, and 8 hours for column 2 while the peak conductivities were reached after 16-18 and 10 hours respectively (Fig. 2). The tracer movement was also followed by TDR-measurements (Fig. 3 and 4). The breakthrough curves of columns 1-3 were in good agreement with a matrix dominated advection-dispersion flow while the results from column 4 indicated preferential flowpaths at some levels.

Bacteriophages (virus that multiply in bacterial cells) have been shown to be valuable tracers in groundwater and also act as a hygiene indicator of animal virus. A rounded phage, Salmonella typhimurium phage 28B, with the size of 60 nm and known to be stable for several years in the environment at hand, was propagated. In total 1.5 x 10^{10} phage particles per column were added in volumes of 100 ml and rinsed down by 1 litre of raw water. Sampling of the effluent (500 ml) was done each hour during the first 24 hours followed by every 6th hour day 2-4. Thereafter control samples were taken weekly. Compared to the chloride test the first arrival times were 5, 2.5, 2 and 5.5 hours shorter in the 4 columns respectively while the arrival times for the peak concentrations were 1-2 hours shorter (Fig. 5).

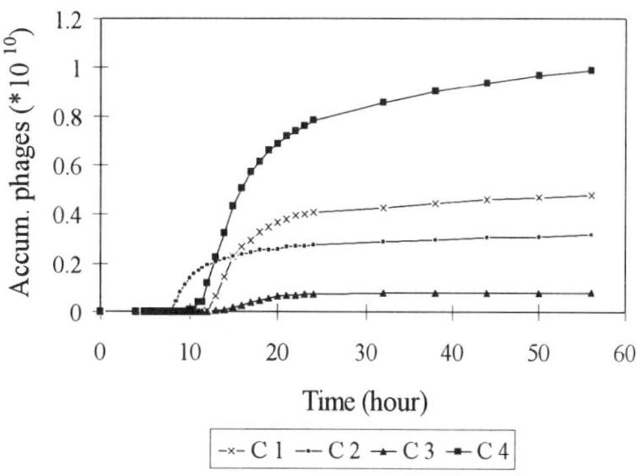

Figure 6. Accumulated breakthrough of phages in the column outlets (totally added 1.5*10^{10}).

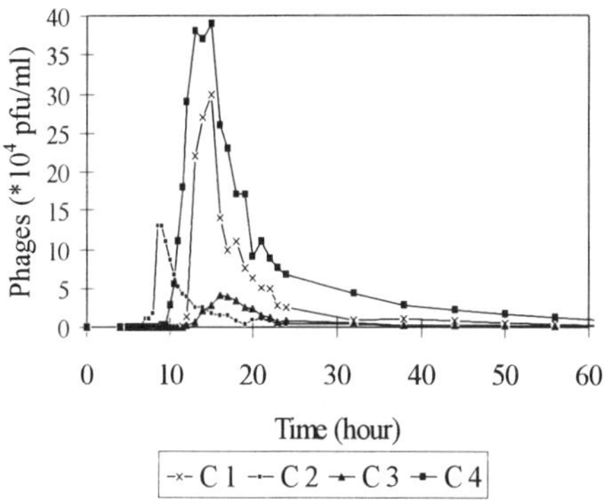

Figure 5. Breakthrough of phages in the column outlets (pfu = plaque forming units).

The most pronounced retention of phages occurred in column 3 with the fine grained top sand. Accumulated recovered number of phages in the effluent after 48 hours were 31, 21, 5 and 64 % for columns 1-4 respectively (Fig. 6). Still after 100 days only 8% were recovered in column 3 compared to 78% in column 4 and phages were found in the outlet water from all of the columns.

5 DISCUSSION AND CONCLUSION

Unsaturated conditions prevailed in all columns; only about half of the pore volume contained water. The free water surfaces at the top were caused by clogging at the sand surface. A advection-dispersion flow in the soil matrix dominated in columns 1-3. The preferential flowpaths indicated at some levels in column 4 was probably caused by an accidental loss of sand under the construction of the column, through the access holes for the microbiological investigation. The shorter first arrival times of the phages, compared with the chloride, are probably due to some phages finding a fast way mainly through large diameter pores. The relatively high retention of phages in column 3 may be caused by the more fine-grained top sand, while the low retention in column 4 may be due to preferential flowpaths. The investigations of the physical properties of the sand, water contents and residence times form an important basis for the accompanying chemical and microbiological studies and enable a coupling of these parameters to the water treatment effects. The column tests are planned to continue for at least one year.

REFERENCES

Berggren, D., J. Blomberg, U.Larsson & I. Nilsson 1998. Artificial groundwater recharge in Stockholm, III An experimental study of chemical processes in the unsaturated zone. This volume.

Hjort, J. & P. Ericsson 1998. Artificial groundwater recharge in Stockholm, I The project and its general aim. This volume.

Szewzyk, U., J. Långmark, S. Kalmbach, W. Manz & T.-A. Stenström 1998. Artificial groundwater recharge in Stockholm, IV Microbiology of the unsaturated zones of experimental sand filter columns. This volume.

Artificial Recharge of Groundwater, Peters et al. (eds) © 1998 Balkema, Rotterdam, ISBN 90 5809 017 5

Artificial groundwater recharge in Stockholm
III. An experimental study of chemical processes in the unsaturated zone

D. Berggren & S. I. Nilsson
Department of Soil Sciences, Swedish University of Agricultural Sciences, Uppsala, Sweden

J. I. Blomberg
Stockholm Water Company, Sweden

U. Larsson
Norrvatten Municipal Association, Solna, Sweden

ABSTRACT: Untreated water from a eutrophic lake (pH 7.0 - 7.5) was infiltrated into columns (diameters 0.3 m; height 4 m) filled with coarse soil material, to test the retention capacity with respect to dissolved organic matter. After 8 months the retention was not more than 15% (initial TOC concentration 5.7 - 6.5 mg C/l). There are indications of a selective retention of hydrophobic and hydrophilic acids.

1 INTRODUCTION

The general aim of the main project, of which the present study forms a part, is to investigate the possibility of using artificial basin recharge of ground-water as water supply for the Greater Stockholm region during the next century. More information about the main project is found in Hjort & Ericsson (1998), this proceedings.

In this article we present some preliminary results from an experimental study, in which water from lake Mälaren is infiltrated into large soil columns. The experiment started in June 1997 and will continue for at least one year, probably longer, to cover seasonal variations in temperature and chemical and microbial composition of the water. The microbiological and hydraulic studies are presented in this proceedings by Szewzyk et al. (1998) and Johansson et al. (1998), respectively.

The objectives of the column experiment are (1) to increase the knowledge about processes controlling water chemistry and microbial activity in the unsaturated zone, and (2) to investigate how these processes can be optimised in order to achieve and maintain a good drinking water quality. We are particularly interested in those processes which control the reduction of the dissolved organic matter (DOM) content of the water.

2 MATERIAL AND METHODS

2.1 *The experimental setup*

Four columns are used in the experiment. Each column consists of a plastic tube with a diameter of 0.3 m and a height of 4 m. They are filled with coarse-grained mineral soil (diameter 0.1-4.0 mm), except for one column (column 3), in which the uppermost 1 m is filled with mineral soil, having a grain size of 0.1-1.0 mm. A pump and a spreading device with 13 tubes distribute the water uniformly on top of each column. The hydraulic load in column 2 is 2 $m^3/m^2/d$ and in the other three columns 1 $m^3/m^2/d$. A typical value for Swedish infiltration plants is about 2 $m^3/m^2/d$. The experimental data of the columns are presented in Table 1. More detailed information about the setup is found in Johansson et al. (1998).

Table 1. Experimental data of the four columns in the study.

Parameter	C1	C2	C3	C4
Hydraulic load ($m^3/m^2/d$)	1	2	1	1
Grain size (mm)	0.1-4	0.1-4	0.1-1 0.1-4	0.1-4
Cleaning of filterskin	scraping	scraping	raking	raking
Other	-	addition of nutrients	-	20 cm soil from B-horizon

2.2 *Sampling and analyses*

In order to follow changes in the chemical composition within the columns, water from the 10, 40, 110 and 250 cm depth is sampled using installed suction cups made of teflon and quartz (Prenart, pore size 1-5 µm, Prenart Equipment Aps, Denmark). Chemical analyses are also performed on the untreated water and on water leaving the columns. Sampling is made about every second week. The main chemical analyses are TOC (Total Organic Carbon, which is a measure of DOM), pH, turbidity, alkalinity (mainly HCO_3^-), total Fe, electrolytical conductivity, and UV light absorbance at 254 nm (UV_{254}). DOM was fractionated into hydrophobic acids (HoA) and neutrals (HoN), as well as hydrophilic acids (HiA), neutrals (HiN) and bases (HiB) according to Leenheer (1981). The soil material in the columns and soil from the illuvial horizon (B horizon) of a near-by podzol were characterized with respect to secondary inorganic Al, Fe and Si compounds and Al and Fe complexed with organic C compounds, by combining ammonium oxalate and sodium pyrophoshate extractions according to Simonsson & Berggren (1998).

3 RESULTS

3.1 *Untreated water*

The water source used in this study (the eutrophic Lake Mälaren) is characterised by a low turbidity (3 FNU as a maximum). The TOC concentrations are normally in the range of 5.5-7.0 mg C/l. The quality of the untreated water is quite stable throughout the year.

3.2 *Infiltrated water*

Up to February 1998 pH increased from 7 - 7.5 in the untreated water to > 8 in the water leaving the columns. The total reduction of TOC during the same time period was rather small. However, the retention tended to increase from 5-10% at the start of the experiment to about 15% after two months of infiltration, as shown in Figure 1. The figure also shows that the retention seemed to decrease with decreasing water temperature.

The reduction of TOC took place, mainly in the upper 10 cm soil layer in the columns (Figure 2). The reduction of turbidity in the water leaving the columns has been quite moderate during the test period, at most about 50 % (column 3) but usually lower (in the range 20-30 %).

3.3 *Fractionation of DOM*

The first fractionation of DOM showed that the hydrophilic compounds (particularly neutrals and bases) seemed to be more prominent in the water from lake Mälaren than in soil water collected from the organic toplayer of podzols in south-central Sweden. In the lake water, hydrophilic compounds accounted for > 65% of the DOM. Even if the retention of DOM was low, a later second fractionation showed a significant (pairwise t-test; $p<0.05$) decrease in HoA, HiA and HiB in the water leaving the columns compared to the lake water (Table 2).

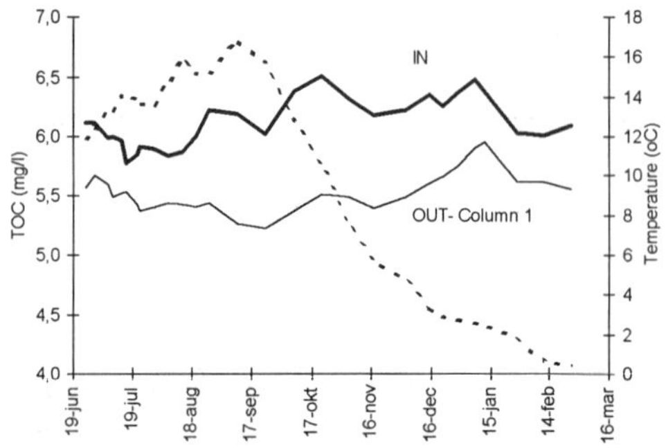

Figure 1. Concentration of TOC in untreated water (thick solid line) and water leaving column 1 (thin solid line). Also shown is the water temperature (hatched line).

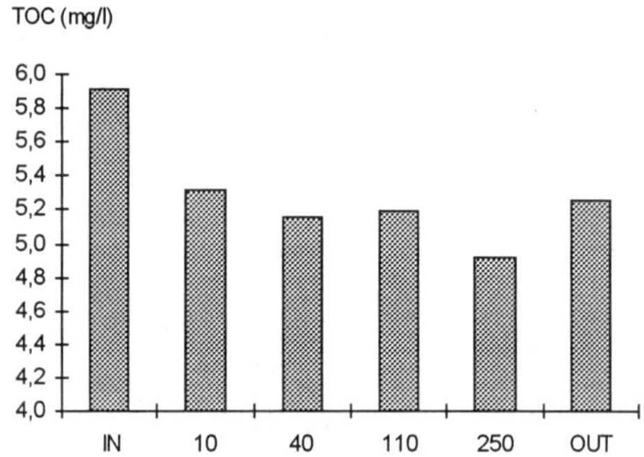

Figure 2. Concentration of TOC in untreated water (IN), water collected at four depths in column 3 and water leaving the column (OUT).

Table 2. DOM composition in untreated water (IN) and in water leaving the columns (OUT) with respect to HoA, HoN, HiA, HiN and HiB (mg C/L).

Sample	HoA	HoN	HiA	HiN	HiB
C1 IN	2.06	0.43	3.11	0.23	0.37
C1 OUT	2.00	0.10	2.80	0.22	0.27
C2 IN	2.11	0.30	2.90	0.40	0.36
C2 OUT	1.98	0.60	2.67	0.28	0.31
C3 IN	2.14	0.40	3.07	0.21	0.41
C3 OUT	1.94	0.61	2.67	0.21	0.28
C4 IN	2.23	0.26	3.11	0.21	0.48
C4 OUT	1.99	0.40	2.86	0.14	0.25

3.4 *Fe and Al content*

The oxalate and pyrophosphate extractions of the soil material in the columns, showed that it contained approximately the same amounts of inorganic Al

Table 3. Al, Fe, Si and organic C in soil material collected from the B horizon of a podzol (B horizon) and from three depths in column 1. The indices ox and pyp denote oxalate extractable and pyrophosphate extractable amounts, respectively.

Sample	Depth (cm)	Fe_{ox} (mmol/kg)	Al_{ox} (mmol/kg)	Si_{ox} (mmol/kg)	Fe_{pyp} (mmol/kg)	Al_{pyp} (mmol/kg)	C_{pyp} (mmol/kg)	Inorganic Fe+Al (mmol/kg)
B-horizon	50-60	16.0	88.5	61.5	3.3	14.6	79.2	96.5
Column 1	5	16.7	163.0	72.8	0.4	5.1	<50	179.7
"	15	10.1	83.7	52.6	0.4	4.9	<50	93.8
"	40	11.0	87.7	41.9	0.1	4.3	<50	98.7

and Fe compounds as an illuvial horizon (B horizon) from a near-by podzol (Table 3). Inorganic Al and Fe were estimated according to the empirical formula $Fe_{ox} + Al_{ox} - 0.1*C_{pyp}$ where the indices ox and pyp denote oxalate and pyrophosphate extraction, respectively. Most of the secondary compounds seemed to be of the imogolite type, as the stoichiometric Al/Si ratio (estimated as $(Al_{ox} - Al_{pyp})/Si_{ox}$) was approximately 2.0 at two column depths.

3.5 *Experimental manipulations*

As the reduction of DOM was low during the first five months of the experiment some further manipulations were carried out in December 1997. Nutrients (R2A-medium) were added to column 2, to stimulate the microbial growth and activity. Column 4 was equipped with a 20 cm top layer of soil collected from the illuvial horizon (B horizon) of a near-by podzol to study to what extent there would be an increased chemical adsorption of DOM due to the presence of aluminium and iron oxides, which were supposed to carry a net positive surface charge. During the first 12 weeks after the manipulations, the DOM "disappearance" (through chemical adsorption and/or microbial assimilation) did not change in either of the two columns. TOC even increased in the water leaving column 4.

4 DISCUSSION

The increase in water pH in the columns indicated calcium carbonate buffering in the soil material. Karltun (1997) showed that the adsorption of fulvic acids on the surfaces of secondary Al and Fe compounds in a B horizon was close to zero at pH 8 (pH 8 about zero point of net charge). However, he also showed that there was an adsorption of about 80% already at pH 6. The fulvic acids are approximately equivalent to the hydrophobic acids obtained in the DOM fractionations. Our conclusion is that the high pH was probably the main reason for the low DOM adsorption in the columns. The high pH might also explain the increase in DOM in column 4 after the addition of B horizon material on top of the column. So, even if the oxalate and pyrophosphate extractions indicated a fair amount of adsorbing material, the high pH was clearly suboptimal for an efficient DOM adsorption, although we are quite confident that the noticed "disappearance" of the hydrophobic and hydrophilic acids indicated by the DOM fractionation was mainly attributed to chemical adsorption rather than microbial assimilation due to the refractory nature of these organic compounds (Qualls & Haines 1992). We intend to test the hypothesis of a suboptimal pH in the columns, by adding hydrochloric acid to the "untreated" water to be added to column 4 (target pH about 6).

REFERENCES

Hjort, J. & P. Ericsson, 1998. Artificial groundwater recharge in Stockholm, I The project and its general aim. This volume.

Johansson, P.O, B. Espeby, B. Nilsson, G. Allestam, 1998. Artificial groundwater recharge in Stockholm, II Column test design and tracer tests. This volume.

Karltun, E. 1997. Modelling SO_4^{2-} surface complexation on variable charge minerals. II. Competiton between SO_4^{2-}, oxalate and fulvate. *Eur. J. of Soil Sci.* 48 (in press).

Leenheer, J.A. 1981. Comprehensive approach to preparative isolation and fractionation of dissolved organic carbon from natural waters and wastewaters. *Environ. Sci. & Tech.* 15 (5), 578-587.

Qualls, R.G. & B.L. Haines, 1992. Biodegradability of dissolved organic matter in forest throughfall, soil solution and stream water. *Soil Sci. Soc. of Am. J.* 56, 578-586.

Simonsson, M. & D. Berggren, 1998. Aluminium solubility related to secondary solid phases in upper B horizons with spodic characteristics. *Eur. J. of Soil Sci.* (in press).

Szewzyk, U., J. Långmark, S. Kalmbach, W. Manz & T.-A. Stenström, 1998. Artificial groundwater recharge in Stockholm, IV Microbiology of the saturated zones of experimental sand filter columns. This volume.

Artificial Recharge of Groundwater, Peters et al. (eds) © *1998 Balkema, Rotterdam, ISBN 90 5809 017 5*

Artificial groundwater as the future water supply of greater Stockholm IV. Phylogenetic diversity of the unsaturated zones of experimental sand filter columns

U. Szewzyk, S. Kalmbach & W. Manz
Microbial Ecology Group, Technical University Berlin, Germany

J. Långmark & T.-A. Stenström
Swedish Institute for Infectious Disease Control, Stockholm, Sweden

ABSTRACT:
The intention and experimental set-up of the pilot sand filter, the treatment of rawwater from Lake Mälaren, is described in detail by Hjort et al. (1998). In this part of the project, the intention was to determine the phylogenetic diversity, density and activity of the microbial cells in different zones of pilot plant sand filters fed with rawwater from Lake Mälaren. The phylogenetic diversity was determined by in situ hybridization, using oligonucleotide probes specific for the α-, ß- and γ-subclass of proteobacteria. The phylogenetic staining revealed shifts in the population composition from ß-Proteobacteria dominated lake water populations to communities representing members of all subclasses of Proteobacteria, as well as the Cytophaga-Flavobacterium group.

1 INTRODUCTION

The lake water was characterized by a high content of organic compounds, composed of a mixture of easily degradable substances, mostly released from photosynthetic algae, and a high proportion of recalcitrant humic substances. To a certain extent, even algae and other plankton organisms were fed to the columns and deposited in the upper zone of each column. Due to the high diversity of organic substrates added to the columns a corresponding diversity in the population structure as well as a clear vertical stratification in the columns was to be expected.

A polyphasic approach is being used, combining cultivation techniques with molecular detection and identification methods to characterize the bacterial population in the sand filters. A similar strategy had been applied to isolate and identify the dominating bacteria of the drinking water distribution systems in Berlin (Kalmbach et al. 1997a/b), and is presently also applied to the drinking water distribution system of Stockholm. The newly isolated and described bacteria from the Berlin drinking water were not only dominant in many drinking water distribution systems, but could also be detected in groundwater and closely related species were even detected in surface water.

In the present study the microbial populations in the sand filter columns were superficially characterized by phylogenetic staining. Based on this first information on the phylogenetic composition of the population, bacteria from the sand filter columns were isolated using R2A-based agar plates. Representative isolates were chosen based on cell and colony morphology and further characterized phylogenetically.

2 MATERIAL AND METHODS

Since the major part of the microorganisms in the sand columns live associated with the surfaces of the sand grains, glass slides were introduced in the column, such that they could be removed after different time intervals. The biofilms on the glass surfaces were further examined using vital staining (DAPI), phylogenetic staining and the PAC (probe active counts) method (Kalmbach et al. 1997b). In situ hybridization with oligonucleotide probes directed to 16S- or 23S-rRNA of bacterial ribosomes was used to identify respectively assign individual cells to kingdoms, domains, or sub-groups of the phylum Bacteria. Probes targeting the following groups were applied: Bacteria, α-, ß-, γ-subgroup of Proteobacteria and *Cytophaga-Flavobacterium*-group.

In addition to the direct microscopic examination of biofilm communities, scrape samples from biofilms were plated on different agar media, e.g. R2A-medium, to obtain pure cultures of as many of the microorganisms as possible. So far about 230 isolates were obtained through the cultivation campaign. From selected isolates genomic DNA was isolated and 16S rDNA sequences were amplified. The amplified and purified DNA was sequenced and phylogenetic analysis of the sequences was performed with the ARB software package. Strain-specific oligonucleotide probes for the isolated and sequenced strains were designed using the Probe_Design tool of the ARB software package.

The oligonucleotide probes were Cy3-labeled and used for in situ detection of the respective strains by in situ hybridization. By combining the in situ hybridization and the method for determination of direct viable counts (Kogure et al. 1984), a new method (PAC = probe active counts) was applied to determine the metabolic potential of individual cells in situ (Kalmbach 1997a).

3 RESULTS AND DISCUSSION

Detection of the bacteria in biofilms from the sand columns by direct staining methods like DAPI-staining for all viable bacteria and in situ probing with 16S rRNA directed oligonucleotides revealed a fast colonization of glass surfaces within 2 to 4 weeks. The activity of cells in the biofilms, indicated by the intensity of emitted light after in situ probing with an eubacterial probe, changed markedly during the first weeks of biofilm development. During the first week approximately 75 to 85% of all viable bacteria could be detected by an eubacterial probe. indicating their high content of ribosomes Already after 3 weeks the proportion of bacteria yielding strong signals after in situ probing decreased to values around 40 to 50%. A similar behavior had been observed for bacterial populations forming biofilms in drinking water distribution systems, indicating that also the bacterial species in the sand columns partly switch to another physiological state.

The results of the phylogenetic staining of 1 week old biofilms extracted from sand material indicate four dominant phylogenetic groups in the different levels of the sand filters, namely representatives of the α-, ß- and γ-subgroup of Proteobacteria and members of the *Cytophaga-Flavobacterium* group, representing approx. 5, 15, 40 and 15% of the viable counts, respectively. The population structure of biofilms developed on exposed glass slides in the sand columns was similar to that on biofilms on the sand grains. However, the proportion of bacteria of the ß-subgroup was slightly higher, reaching values of 20-25% of total viable bacteria. The loss of activity of the biofilm bacteria was comparable to that of bacteria on the sand grains. These proportions of phylogenetic groups are significantly different from surface water systems where bacteria from the ß-subclass of' Proteobacteria dominate with percentages around 70-80% of the DAPI stained cells. The ratio in the sand columns resembles much more situations in terrestrial ecosystems (Felske et al. 1998). This is especially interesting, because the constantly inflowing water from Lake Mälaren continuously introduces the bacterial population from the lake. Obviously the lake-adapted population is not able to survive and grow under the conditions in the sand filter, and already after the first centimeters a new population established, probably specific for the situation in the sand column.

Examinations with the PAC method indicated, that a substantial part of the population reduced their ribosome content what was deducted from weak signals after in situ hybridization. However, many of these bacteria started synthesizing ribosomes after addition of excess substrate. These experiments indicate. that the populations in the sand columns had utilized the easily degradable organic compounds and while adapting to utilize more recalcitrant organic compounds reduced their growth rate and subsequently their protein synthesis. If these bacteria were limited by access to carbon sources or other factors, like for instance phosphate, is one of the questions addressed presently. However, biofilm populations in drinking water distribution systems showed a very similar behavior and yet it is not clear what physiological mechanism is responsible for this phenomenon. Bacteria of the dominating strains in drinking water entered a physiological state, where they were fully viable but lost their culturability on R2A media. Up to 70 to 90 percent of the members of one species entered this VBNC (viable but nonculturable) state after 2-4 weeks biofilm development in drinking water (Kalmbach et al. 1997b). Since members of these in drinking water dominating species were also detected in the sand filters and in surface water, it is possible that they undergo similar physiological changes in the sand columns. Consequently, the approximately 99% of the DAPI-stained bacteria which can not be cultivated on plates do not necessarily represent new species but subpopulations in a less active state. This conclusion might also hold for hygienically relevant species.

REFERENCES

Felske, A., A. Wolterink, R. van Lis & A.D.L. Akkermans 1998. Phylogeny of the main bacterial 16S rRNA sequences in Drentse a grassland soil. Appl. Environ. Microbiol. 64:871-879.

Hjort. J. et al. This conference (1998)

Kalmbach, S. W. Manz & U. Szewzyk 1997a.

Dynamics of biofilm formation in drinking water: phylogenetic affiliation and metabolic potential of single cells assessed by formazan reduction and *in situ* hybridization. FEMS Microbiol. Ecol. 22:265-279.
Kalmbach, S., W. Manz & U. Szewzyk 1997b. Isolation of new bacterial species from drinking water biofilms and proof of their in situ dominance with highly specific 16S rRNA probes. Appl. Environ. Microbiol. 63:4164-4170.
Kogure et al. 1984. Arch. Hydrobiol. 102:117-122.

Artificial Recharge of Groundwater, Peters et al. (eds) © *1998 Balkema, Rotterdam, ISBN 90 5809 017 5*

Artificial groundwater recharge in Stockholm
V. Strategic environmental impact assessment

J.I. Blomberg
Stockholm Water Company, Sweden

U. Larsson
Norrvatten Municipal Association, Solna, Sweden

M. Andersson, H. Djurberg & M. Westerdahl
SWECO, Stockholm, Sweden

ABSTRACT: This strategic environmental impact assessment is the initial part of investigations for the future water supply of the Greater Stockholm Region, Sweden. It describes the differences in environmental impact between using artificial groundwater recharge through basin infiltration and conventional water treatment techniques. Furthermore, the study identifies the key issues for further investigations.

1 INTRODUCTION

As part of the ambitions to achieve a long-term sustainable water supply in accordance with the environmental goals of Agenda 21, the drinking water producers in the Greater Stockholm region, Stockholm Vatten AB and Norrvatten, have commissioned this study. The study was conducted by the consulting company SWECO supported by J&W (consulting company), Royal Institute of Technology and the principals.

Today, the Greater Stockholm area is supplied with high-quality drinking water purified in three water supply plants using the processes chemical precipitation, sedimentation, rapid sand filtration and PAC (powdered activated carbon) or alternatively slow sand filtration. The water is pH adjusted and chlorinated before distribution. Altogether a total of about 500 000 m^3 of drinking water is produced daily. With the ambition to achieve an even longer-term sustainable water supply and to further increase the quality of the water, the possibility to replace currently used technology with artificial groundwater recharge through basin infiltration in eskers (using untreated surface water) where the only additional treatment required is pH adjustment before distribution, has been studied.

2 STRATEGIC ENVIRONMENTAL IMPACT ASSESSMENT

By means of a strategic environmental impact assessment, the impacts of using different techniques of producing drinking water have been described and compared from environmental viewpoints. The following alternatives were studied: existing technology (the zero alternative), further development of existing technology (zero-plus alternative) and the artificial groundwater recharge alternative). Key questions facing a possible expansion of artificial groundwater recharge have also been more precisely defined. In all alternatives use is made of Lake Mälaren as raw water supply. The time horizon selected for the comparison has been placed at year 2010.

A strategic environmental impact assessment is a suitable method in early planning stages where different alternatives are studied on a general basis and where strategic decisions on choice of technology are to be made. Dialogues between different authorities and pressure groups are also achieved in an early stage. In a later stage, when decisions are made on technology, the project environmental impact assessment is made for each individual water supply plant.

3 PREVIOUS INVESTIGATIONS

In an earlier phase of the project, three areas along the Uppsala esker were identified as being of particular interest for artificial groundwater recharge. These areas are located close to Lake Mälaren and west of central Stockholm. The areas of interest are generally classified as of particular national interest for preservation of cultural or natural values, or open-air recreational activities. According to Swedish legislation water production plants can be classified as of national interest for protection. This opportunity is however rarely applied. Groundwater resources can not be classified as such. More information about the complete project can be found in Hjort J. & Ericsson P, this proceedings.

4 DIFFERENT ALTERNATIVES AND THEIR ENVIRONMENTAL IMPACTS

Anticipated environmental impacts are either linked to the actual *treatment process* or to the *geographical area where the treatment takes place*. In the environmental impact assessment interest is thus focused on questions of strategic importance for choice of method used to produce drinking water.

The *zero* and the *zero-plus alternatives* were assessed to imply a fairly minor change in relation to the current situation both as regards the treatment process and the place of treatment. The *artificial groundwater recharge alternative* is assessed to result in both positive and negative consequences in both respects. The major positive consequences are expected to occur with regard to drinking water quality and a decreased use of chemicals. Minor negative consequences will occur for, e.g., energy consumption in the water treatment process. At the recharge sites, small negative consequences will occur in connection with the necessity to encroach on areas of value for protection with regard to historical contexts, nature and recreation.

However, if abandoned gravel pits are utilised, then the consequences may be positive. In a long-term perspective, preservation interests may be strengthened by the formation of water protection areas around the infiltration plants. The assessments of the environmental impacts are presented in Table 1.

In a comparison between alternatives, it is thus found that the *alternative with artificial ground-water recharge* is clearly the most favourable with regard to environmental aspects linked with the treatment process. In the short-term, the *zero alternative* and the *zero-plus alternative* are the most favourable with regard to environmental aspects related to the place of treatment. In a longer perspective, however, the *artificial groundwater*

Table 1. Illustration of environmental impacts as of year 2010

STRATEGIC ENVIRONMENTAL IMPACT ASPECTS	Zero alternative	Zero-plus alternative	Artificial groundwater recharge alternative
Environmental impacts aspects linked to treatment process			
• Water quality-health aspects	⇨	⇨	⇗
• Energy consumption in water processing	⇨	↘	↘
• Accidents and sabotage	⇨	⇨	↘
• Sludge disposal	↗	⇗	⇗
• Use and handling of chemicals	↘	↗	⇗
Environmental impacts aspects linked to place of treatment			
• Recreation and open-air activities	⇨	⇨	(↘)
• Historical environment	⇨	⇨	(↘)
• Natural environment	⇨	⇨	(↘)

Symbols

⇨ No change in relation to present

↗ ⇗ Small or large positive impact (change) in relation to present

↘ ⇘ Small or large negative impact (change) in relation to present

◯ In the long-term preservation interests may be strengthened by means of water protection areas around infiltration plants *recharge alternative* may be equivalent to or even better than the other alternatives also from these viewpoints. If this alternative is to be achieved it will probably require that the areas in question needed for artificial groundwater recharge must be defined as areas of national interest for protection for drinking water production.

5 KEY QUESTIONS

When anticipating an expansion of artificial groundwater recharge on a large scale the work should be focused on two key questions:

- *Development of the basin infiltration technique* with regard to the location, design and surface load. The possibility to integrate different interests at the water production plants, e.g., gravel pits, and recreational activities, should be studied. Fulfilment of the water quality target at the production plants is another question that should be studied.
- *Society's evaluation of protection of water production plants and groundwater resources*. By defining these areas and plants as being in the national interest to protect it will be possible to achieve balance between water supply interests and preservation interests in municipal planning. In their planning activities, the municipalities are obliged to consider national interests.

2.4 Behaviour of contaminants

Artificial Recharge of Groundwater, Peters et al. (eds) © 1998 Balkema, Rotterdam, ISBN 90 5809 017 5

Experimental and numerical analysis of atrazine behaviour in an artificial recharge site

M. Van Hoorick & J. Feyen
Institute for Land and Water Management, Katholieke Universiteit Leuven, Belgium

L. Pussemier
Centre for Veterinary and Agrochemical Research, Ministry of Agriculture, Belgium

ABSTRACT: Laboratory studies of pesticide sorption during transport can provide insight into the basic processes affecting atrazine behaviour. This paper presents the results of miscible displacement experiments performed on undisturbed soil cores in order to determine the governing processes and to assess the retention capacity for atrazine of the sediments and aquifer underlying an artificial recharge basin. The saturation degree of adsorbed atrazine is found to be a possible key factor in the adsorption capacity of the sediment. It is also shown that non-linear adsorption may be a controlling process in atrazine transport when dealing with low pesticide concentrations. Non-linear adsorption should therefore be taken into consideration when deriving the isotherm. Further research to be done in this area is presented.

1 INTRODUCTION

An evaluation of the potential for pesticides to contaminate groundwater in an artificial recharge site requires an understanding of the transport mechanisms that occur during the infiltration process, and an accurate representation of these mechanisms in simulation models. The aim of this study is to conduct laboratory experiments to gain insight in atrazine transport. In a next step this information will be implemented in the numerical code MIKE.SHE to assess the risk for groundwater pollution.

2 MATERIALS AND METHODS

2.1 *Column miscible displacement experiments*

Column displacement experiments were performed to identify the governing physical and chemical processes of atrazine transport at core-scale. Four undisturbed soil cores with a length of 30cm and a diameter of 15cm were taken at the Grobbendonk artificial recharge site, Belgium. Stainless steel columns were used to avoid atrazine adsorption on the walls. Two types of sediments were sampled: the aquifer material (Diestian sands at 4m below surface; Exp. I, II) and the top layer of the existing infiltration basin (calibrated sandfilter of 500μ; Exp. III, IV).

Prior to the application of the pulse each column was slowly saturated from the bottom and was equilibrated to steady state flow conditions applying a constant head of 1.5cm on top of the column, which was maintained throughout the experiment. A pulse of 200ml aqueous solution containing both the 'inert' tracer ^3H_2O (25μg l^{-1}; NEN^{TM} Life Science Products, Boston) and the 'reactive' tracer ^{14}C-labeled atrazine (20μg l^{-1}; 95% pure; Sigma-Aldrich NV) was applied. The simultaneous application of both tracers allowed us to examine inert and reactive transport under the same experimental flow conditions. Column-effluent data were collected at regular time intervals and analysed with a Tricarb 460cd Liquid Scintillation System (Hewlett-Packard). Throughout the experiment the water content (saturated) of the soil cores, which was monitored using TDR (Time Reflectory Domain), remained relatively constant for the four columns at a value ranging between 0.33 to 0.36.

2.2 *Estimation of process parameters*

Soil-column effluent data were used to determine the parameters of the following transport models: the conventional convection dispersion equation (CDE), the physical nonequilibrium (PNE), both with instantaneous adsorption, and the chemical (CNE) nonequilibrium CDE with rate-limited adsorption. The parameters were derived using the non-linear least-squares parameter optimisation code CXTFIT (Toride et al. 1995). All models assume a linear isotherm. Comprehensive description of the transport models tested in this study are discussed in detail by Rao et al. (1979) and van Genuchten & Wagenet (1989), among others.

Tritium was used as a non-sorbing conservative tracer to investigate the hydrodynamic properties of the porous media. Breakthrough curves (BTCs) of tritium were used to determine the physical transport parameters v (pore water velocity) and D (dispersion coefficient) for the equilibrium CDE. For the PNE model the parameters v, D, f (mobile water content) and α (first-order mass transfer coefficient) were determined.

Subsequently, parameters describing the sorption process of atrazine, namely K_d, the distribution coefficient in the linear isotherm, f, the fraction of type-I (instantaneous equilibrium) sorption sites and α (first-order adsorption rate constant) were derived by fits of the CDE and CNE to the atrazine BTCs. In order to reduce the number of parameters, the 'hydrodynamic' parameters v and D were kept constant and set equal to the values derived from BTCs of the inert tracer.

3 RESULTS AND DISCUSSION

Normalised tritium and atrazine BTCs and the model predictions are presented in Figures 1 A-B, respectively.

A major difference between BTCs in the two types of sediments was observed. In Exp. III and IV tritium and atrazine are more or less eluted simultaneously. The BTCs also exhibited no or little tailing. On the other hand, for the aquifer material (Exp. I, II) a large difference between the tritium and atrazine BTCs can be noticed. BTCs of atrazine show a small peak of early breakthrough and a very long tail, indicating strong adsorption. The relative peak values are a factor 10 and 7, for respectively Exp. I and II, smaller than for the tritium BTCs.

Breakthrough curves of tritium were successfully described using the conventional CDE. The results of the curve fitting are presented in Table 1. The goodness of fit was quantified in terms of the

Table 1. Estimated transport parameters for the conventional CDE and the chemical nonequilibrium CDE model

	Aquifer sediment		Infiltration basin	
	Exp. I	Exp. II	Exp. III	Exp. IV
v (cm min^{-1})	1.71	1.37	1.42	1.86
D (cm^2 min^{-1})	1.02	0.30	2.20	0.33
r^2	0.99	0.99	0.93	0.99
K_d (cm^3 g^{-1})	9.5	11.9	0.024	0.036
f	0.001	0.001	0.011	0.41
α (min^{-1})	0.005	0.004	0.978	0.210
r^2	0.76	0.96	0.96	0.98

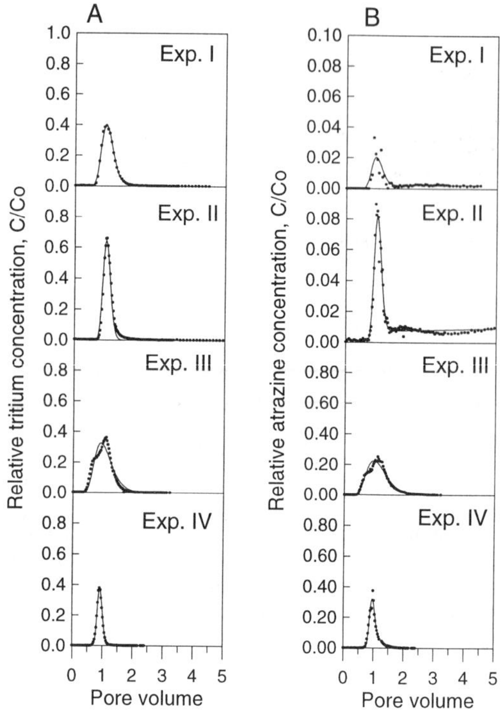

Figure1. Observed (•) and predicted (—) BTCs for tritium (A) and atrazine (B).

coefficient of determination (r^2) that ranged from 0.93 to 0.99. The estimated values for the pore velocity v are consistent with measurements of water content and flow rate. Although the PNE also provided good fits, the ω values (dimensionless form of α) obtained were larger than 100, indicating local equilibrium assumption validity and lack of PNE (Vallochi 1985, Gaber et al. 1992).

A relative accurate prediction of the observed atrazine breakthrough curves was obtained using the chemical nonequilibrium CDE. The coefficient of determination (r^2) ranged from 0.76 to 0.98. Estimated values for K_d, f and α for the different columns are summarised in Table 1. For the aquifer sediments small values for α can be noticed. This in combination with the small values of f may indicate highly nonequilibrium adsorption. In addition, the K_d values strongly differ for both the sediments with much higher values for the aquifer sediments denoting higher atrazine retention. This could already be seen from the BTCs. However, the higher adsorption can not be explained by a higher organic matter content, which is an ideal adsorbent for pesticides, of the aquifer material as it is much the same for both the sediments (app 0.1-0.2 %). Moreover, such high K_d values do not correspond with most values found in literature (Montgomery 1993).

The foregoing indicates that atrazine adsorption may be influenced by another process than rate-limited adsorption, e.g. non-linear adsorption. According to Brusseau et al. (1997) identification of the controlling process based only on the shape of BTCs obtained under steady state flow, is often difficult. Both rate-limited sorption and non-linear sorption cause BTCs to exhibit tailing, but only the former induces a concentration perturbation upon imposition of flow interruption. Therefore, a flow interruption was applied during the experiments. Since the foregoing did not cause a disturbance of the BTC, the results do indicate the presence of non-linear adsorption, rather than rate-limited adsorption (not shown). In addition to the tailing of the atrazine BTCs, non-linear adsorption may also explain the difference between K_d values for the two sediments. When the sorption isotherm is non-linear, the affinity of atrazine to the solid phase and the distribution coefficient depend on the amount of sorption sites that are occupied by sorbed atrazine molecules. K_d observed in the leaching experiments depends on the concentration of the applied solute pulse and the initial atrazine concentration in the sediments. In contrast to the samples I and II (Diestian sands), the sediments of the samples of III and IV were already contaminated with atrazine (7.97 μg/kg). This might explain the smaller K_d values observed in samples III and IV.

Most atrazine adsorption isotherms reported in literature cover the concentration range of 0-20,000 μg $l^{-1}$and are nearly linear over this concentration range. However, these isotherms do not provide detailed information on the shape of the isotherm in the lower range, which is very often obtained from extrapolation of K_d determined at much higher concentrations. The results of this study may indicate that extrapolation can not be applied when low concentrations are involved, as it is often the case in artificial recharge of contaminated surface water to the groundwater.

4 CONCLUSIONS

1. An increasing atrazine concentration in the soil-water system is a key factor in a decreasing retention capacity. Therefore, it is likely that with time the infiltration basin and underlying aquifer might fall out of use due to the percolation of pesticides.
2. The results of the leaching experiments show that atrazine transport may be controlled by non-linear adsorption when the concentration of pesticide in the infiltration water is low. This should be taken into consideration when deriving the isotherm. An appropriate model concept ought to be used.

5 FURTHER RESEARCH

1. At this moment the adsorption isotherm is being determined by means of batch experiments to confirm the non-linear behaviour of the isotherm in the lowest concentration range.
2. The water flow and contaminant transport in the infiltration site will be modelled using the numerical model MIKE.SHE taking into account the insight gained in the laboratory experiments.

6 ACKNOWLEDGMENT

This study was made possible by the financial support of the European Community (contract ENV4-CT95-0071). The first author also wishes to thank the Centre for Veterinary and Agrochemical Research for the use of their laboratory for the radioactive studies. The author is also embedded to S. Goux and I. Joris for their help with the laboratory experiments.

REFERENCES

Brusseau, M.L., H. Qinhong & R. Srivastava 1997 Using flow interruption to identify factors causing nonideal contaminant transport. *J. Contam. Hydrol.* 24:205-219.

Gaber, H.M., S.D. Comfort, W.P. Inskeep & H.A. El-Attar 1992. A test of the local equilibrium assumption for the adsorption and transport of picloram. *Soil Sci. Soc. Am. J.* 56:1392-1400.

Montgomery, J.H. 1993. *Agrochemicals desk reference: environmental data.* Lewis Publishers, Chelsea, USA: 31-34.

Rao, P.S.C., J.M. Davidson, R.E. Jessup & H.M. Selim 1979. Evaluation of conceptual models for describing nonequilibrium adsorption-desorption of pesticide during steady flow in soils. *Soil Sci. Soc. Am. J.* 43:22-28.

Toride, N., F.J. Leij & M.Th. van Genuchten 1995. *The CXTFIT code for estimating transport parameters from laboratory or field tracer experiments.* Research report No. 137, U.S. Salinity Lab., USDA-ARS, Riverside, CA.

Vallochi, A.J. 1985. Validity of the local equilibrium assumption for modelling sorbing solute transport through homogenous soils. *Water Resour. Res.* 21:808-820.

van Genuchten, M.Th. & R.J. Wagenet 1989. Two-site/two-region models for pesticide transport and degradation: 1. *Theoretical Soil Sci. Soc. Am. J.* 53:1303-1310.

Your daily 'drugs' in drinking water? State of the art for artificial groundwater recharge

N. Zullei-Seibert
Institute for Water Research, Schwerte, Germany

ABSTRACT: Pharmaceutical chemicals are discussed under drinking water quality aspects. Most of the detected substances occur in raw water in a concentration range comparable to pesticides. Especially surface water with high waste water loads is contaminated. Natural filtration processes like bank filtration or artificial groundwater recharge in principle are vulnerable to water soluble pharmaceutics. Results show that trace amounts for example of lipid reducing agents, antirheumatics and hormons may enter the ground or drinking water.

1 INTRODUCTION

For human and veterinary health care pharmaceutical chemicals are indispensable. Although the value of these substances is undoubted, anxiety rose about unintentional effects on the environment. Especially hormones gain increasingly interest in the debate about endocrine disruptors. This term summarizes substances influencing the hormonal system of aquatic and terrestric animals or birds as well as humans. Substances discussed in this context are natural sterols (e.g. ß-sitosterol), pesticides (DDT, lindane and other), and industrial chemicals (PCB, phthalates, nonylphenoles or tinorganics). Synthetic hormones like ethinylestradiol, which intend therapeutical effectiviness, link endocrine disruptors with pharmaceutical chemicals. Because in public mind both terms often are inadmissible mixed, public relation work for a correct differentiation becomes necessary (DVGW 1998).

Behaviour of lipophilic pesticides and industrial components during drinking water treatment is comparatively well known. Due to their physicochemical properties these substances mostly are sufficiently reduced by sorption processes during artificial groundwater recharge. In contrast, comprehensive information for risk assessement of pharmaceutics is lacking.

Nevertheless fourty years ago first scientific papers discussed the possible effects of contraceptives and antibiotics on drinking water quality (Demoll 1954). Basing on the improvement of analytical methods twenty years later occasional investigations about the occurence of natural and synthetic sterols (Tabak & Bunch 1970, Rurainski et al. 1977) and other pharmaceutical chemicals (Richardson & Bowron 1985) in waste water, surface and drinking water followed. By coincidence clofibric acid, a metabolite of lipid reducing agents, could be detected during herbicide routine analysis (Stan et al. 1994). Basing on this results research work into antirheumatics, analgetics, hormones and other pharmaceutics (Hirsch et al. 1996, Stumpf et al. 1996) and water works monitoring programs started.

2 USE OF PHARMACEUTICS

In Germany around 8000 pharmaceutical products for human use are regularly approved (Bundesverband der Pharmazeutischen Industrie 1994). The main groups of medical prescribed human pharmaceutics are against pain, rheumatism or for reduction of dangerous blood lipids (Table 1). Besides that a remarkable amount of products, e.g. against common cold or vein infirmities, is in use by domestic self-medicamentation. Information about the hospital turnover is not available, but concentration of antibiotics, X-ray contrast chemicals and cancer chemotherapeutics can be expected.

For the veterinary sector no reliable data are available also. Similar to human application there exists the medical indication as well as the self treatment. In addition the black market for animal breeding and food additive products must be considered an important pharmaceutic source. This is because antibiotics used in this way do not underly the verterinary pharmaceutics legacy in Germany.

Table 1. Estimation of human pharmaceutics prescribed in Germany (after Stumpf et al. 1996).

Substance	estimated for 1993 tons/year
Analgetics	
Acetylsalicylic acid	23 - 116
Antirheumatics	
Diclofenac	48 - 72
Ibuprofen	48 - 96
Ketoprofen	0,7
Lipid regulatorss	
Benzafibrate	38 - 57
Clofibrate	15 - 21
Fenofibrate	11 - 15
Gemfibrocil	14

3 OCCURENCE OF PHARMACEUTICS IN WATER

3.1 *Contamination routes*

Pharmaceutical chemicals enter the water cycle via four main routes. In combination with production or formulations processes point discharge of industrial waste water takes place. The second route refers to domestic or hospital consumption of pharmaceutics, followed by excretion either in form of the active ingredients or metabolites. Because of limited elimination during sewage treatment these substances can be transported into surface waters. Veterinary use of pharmaceutics and antibiotics or hormone containing food additives in combination with the agricultural use of liquid manure or sewage sludge may lead to groundwater contamination or runoff into surface waters as a third route. The last pathway is characterized by disposal of leftovers. The improper way via toilets cause additional surface water contamination. The relevance of impact into the groundwater after disposal on waste deposits cannot be judged due to missing information.

3.2 *Occurence of pharmaceutics in various water types*

Information about the occurence of approximately 60 individual pharmaceutics in various water types is available in literature. Table 2 refers to substances investigated in drinking or groundwater. From this list 29 substances were detected in sewage plant effluents and 22 in surface water. Only 5 are reported in groundwater and 9 in drinking water. Under the viewpoint of artificial groundwater recharge and drinking water quality especially diclofenac, ibuprofen, bleomycin, ethinylestradiol, benzafibrate, clofibric acid, fenofibrate and diazepham are relevant.

Table 2. Occurence of pharmaceutical chemicals in various water types (literature data 1977 - 1996).

Substance	sewage plant effluen	surface water	ground water	drinking water
Analgetics				
Acetylsalicylic acid	◆	□		□
Antirheumatics				
Diclofenac	◆	◆		◆
Fenoprofen	□	□		□
Ibuprofen	◆	◆		◆
Indometacin	◆	◆		□
Ketoprofen	◆	□		□
Antiasthmatics				
Clenbuterol	◆	□		□
Fenoterol	◆	◆		□
Salbutamol	◆	□	□	□
Terbutalin	◆	◆		□
Cancer therapeutics				
Bleomycin	◆	◆		◆
Hormones				
Diethylstilberol			◆	
Estradiol	◆		◆	□
Estradiol-17-valerat	□	□		□
Estriol	◆	□		□
Estron	◆	□		□
Ethinylestradiol	◆	◆	◆	◆
Mestranol	◆	◆		□
Norethisteron	◆	□		□
ß-Sitosterol *	◆	◆		◆
Lipid regulators				
Benzafibrate	◆	◆		◆
Clofibric acide	◆	◆	◆	◆
Fenofibrate		◆	◆	◆
Fenofibrinc acid	◆	◆		□
Gemfibrozil	◆	◆		□
Psychopharmaceutics				
Diazepham	◆	◆		◆
ß-Sympatholytics				
Betaxolol	◆	◆	□	□
Bisoprolol	◆	◆		□
Carazolol	◆	◆		□
Metoprolol	◆	◆	□	□
Nadolol	◆	◆		□
Propranolol	◆	◆		□
Timolol	◆	◆		□

◆ positive findings, □ negative findings, *phytohormone

Concentrations in sewage plant effluents range between the detection limits of 0,004 µg/L and approximately 30 µg/L with single samples of 240 µg/L (Fig. 1). Surface water contamination was detected approximately more than one log-unit less. Especially rivers influenced by industrial or municipal sewage effluents are affected by pharmaceutical chemicals.

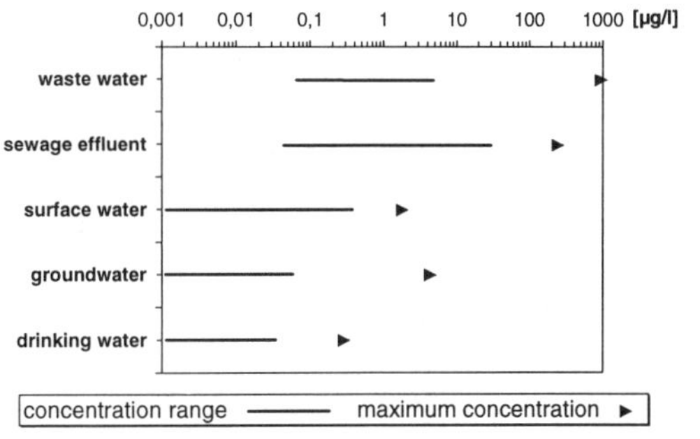

Figure 1. Average and maximum concentration of various pharmaceutical chemicals - literature data.

Periods with low discharge and therefore high loads of sewage effluents differ in their concentration characteristic from flood periods. As can be seen for clofibric acid in the middle part of the River Ruhr this substance could be detected during low flow events only (Fig. 2). Floods with discharge larger 100 m^3/s showed no chlofibric acid.

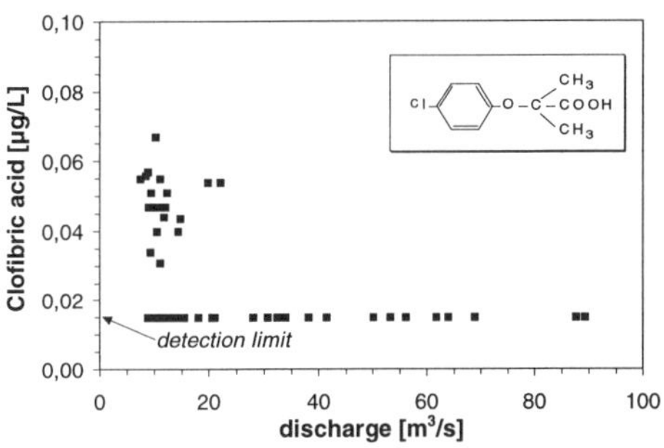

Figure 2. Relation between clofibric acid and discharge of the River Ruhr - data Dortmunder Energie und Wasser.

Positive findings in groundwater are mainly caused by surface water influences (Heberer & Stan 1996, Mendel & Zullei-Seibert 1997). This indicates a vulnerability of artificial groundwater recharge as well as bank filtration against pharmaceutical chemicals. Certainly the detected concentration in ground- and drinking water are low. Trace amounts cover the range between the most sensitive detection limit of 0,001 µg/L and 0,05µg/L with single samples reaching 4 µg/L (Fig. 1). In case of the Dortmund catchment area no pharmaceutics occured in the artificially recharged groundwater. Compared to the prescribed pharmaceutical dose, the reported findings in drinking water are of no relevance for human health but should be avoided in principle.

4 CONCLUSIONS

Various pharmaceutical chemicals have been detected in surface waters. A few of these also appeared in artificially recharged groundwater and bank filtrate in trace concentrations. Knowledge about this substance group by now is unsufficient for comprehensive risk assessement. Therefore further monitoring and research is required under the general aspect of water quality requirements.

REFERENCES

Bundesverband der Pharmazeutischen Industrie e.V. 1994. Arzneimittelverzeichnis des BPI (Rote Liste). *ECV Editor Cantor für Medizin und Naturwissenschaften*

Demoll, R. 1954. Viren, Hormone, Wuchsstoffe und Antibiotika im Abwasser. *gwf Wasser Abwasser* 95 (4): 97-100

DVGW-Arbeitskreis Arzneimittel 1998. Relevanz von Arzneimittelm, hormonell wirksamen Substanzen und Kosmetika in Gewässern. *DVGW-Fachinformation* (in press): 1-8

Heberer, T. & H.J. Stan 1996. Vorkommen von polaren organischen Kontaminanten im Berliner Trinkwasser. *Vom Wasser* 86: 19-31

Hirsch, R., T.A. Ternes, K. Haberer & K.L. Kratz 1996. Nachweis von Betablockern und Bronchospasmolytika in der aquatischen Umwelt. *Vom Wasser* 87: 263-274

Mendel, B. & N. Zullei-Seibert 1997. Sind Arzneimittel relevant für die Trinkwasserversorgung? In DVGW, *Informationsveranstaltung Trinkwasserhygiene, Bonn, 2-3 September 1997*

Richardson, M.L. & J.M. Bowron 1985. The fate of pharmaceutical chemicals in the aquatic environment. *J. Pharm. Pharmacol.* 37: 1-12

Rurainski, R.D., H.J. Theiss & W. Zimmermann 1977. Über das Vorkommen von natürlichen und synthetischen Östrogenen im Trinkwasser. gwf Wasser Abwasser 118: 287-291

Stan, H.J., T. Heberer, & M. Linkerhägner 1994. Vorkommen von Clofibrinsäure im aquatischen System - Führt die therapeutische Anwendung zu einer Belastung von Oberflächen-, Grund- und Trinkwasser? *Vom Wasser* 83: 57-68

Stumpf, M., T.A. Ternes, K. Haberer & W. Baumann 1996. Nachweis von natürlichen und synthetischen Östrogenen in Kläranlagen und Fließgewässern. *Vom Wasser* 87: 251-261

Stumpf, M., T.A. Ternes, K. Haberer, P. Seel & W. Baumann 1996. Nachweis von Arzneimittelrückständen in Kläranlagen und Fließgewässern. *Vom Wasser* 86: 291-303

Tabak, H. & H.L. Bunch 1970. Steroid hormones as water pollutants I. *Dev. Ind. Microbiol.* 11: 367-376

Artificial Recharge of Groundwater, Peters et al. (eds) © 1998 Balkema, Rotterdam, ISBN 90 5809 017 5

Pesticide monitoring: Measuring the right substances at the right place and time

B. Kuhlmann & N. Zullei-Seibert
Institute for Water Research, Schwerte, Germany

ABSTRACT: In order to develop an accurate monitoring program, data on topics influencing the entrance of pesticides in water bodies must be known. Based on information concerning pesticide application and land use as well as on the local conditions (including hydrogeology, climate), a program can be worked out to measure the relevant pesticides at the right time and place. The more precise and direct the information is the more exact can be the measuring program. Starting with this data collection, a schedule for the whole procedure can be developed.

1 INTRODUCTION

Contamination of surface and ground water with pesticides is a fact many water works have to face. Today, looking for pesticides is almost a routine task for every water laboratory. Nevertheless sometimes it seems to be the search for the needle in a haystack.

In Germany, nowadays more than 250 organic substances are registered for use as pesticides (Biologische Bundesanstalt 1997). For only about half of them analytical methods exist. Including the pesticides used in the past, there are about 280 active substances that can be analysed and may be detected in water bodies (Kuhlmann 1998). In a study on the pesticide situation in ten European countries, over 150 pesticides were mentioned which have been found at least once in water bodies in concentrations exceeding the 0,1 µg/l level set by the EU drinking water standard (Heinz et al. 1995).

Therefore it is necessary to develop a monitoring program, that means, a concept which helps to search and analyse for the right substance at the appropriate time and place.

2 INFORMATION NEEDED FOR A MONITORING PROGRAM

Depending on the objectives of the investigation, sampling strategies and substances analysed will differ widely. Waterworks wanting to evaluate the entire pesticide situation in their water catchment area will use a different concept than waterworks that centre the investigations on the product drinking water.

The most crucial point for all kind of measuring is a good knowledge of the factors affecting the entry of pesticides into the aquatic environment. To assess the relevant pesticides in a water catchment area as well as suitable sampling points and frequency, detailed information is needed on:

- the pesticide situation in the catchment area:
agricultural and non agricultural use (i.e. substances, amounts, application period and technique, behaviour), production, handling, transport
- the land use:
agriculture (cultivated crops, crop rotation), industry, estates (e.g. gardens, cemeteries, parks), traffic (e.g. railway tracks, parking lots)
- the hydrogeological situation:
catchment area, type of raw water, climate (e.g. rainfall)
- the water work practice:
capacity, treatment

The quality and accuracy of the measuring program depend basically on the information used to develop it. A specific monitoring is guaranteed by direct and comprehensive information on the topics mentioned above.

Indirect data, for example derived from maps or statistics, lead to a less precise result.

3 DEVELOPING A MONITORING PROGRAM

Most important for the development of a monitoring program is knowledge about the pesticides used in the catchment area including those applied in the

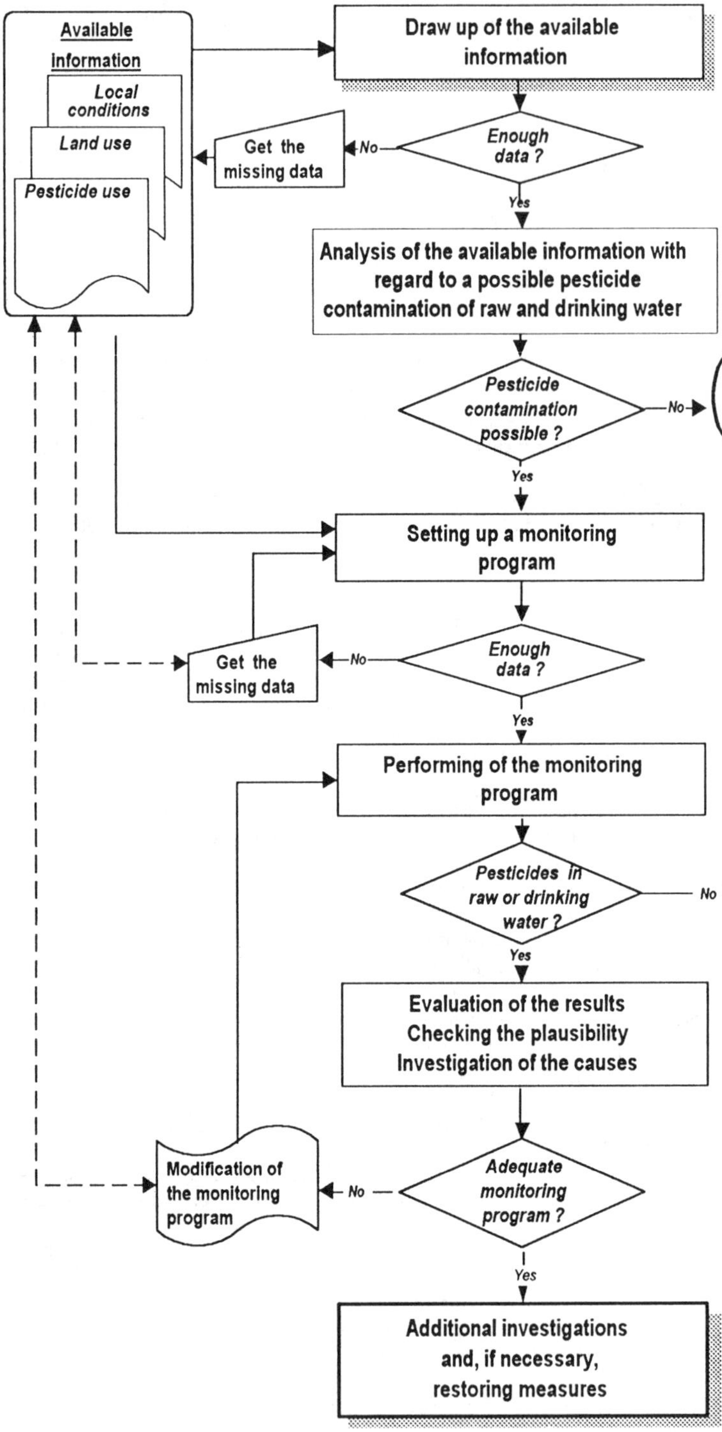

Figure 1. Looking for a possible pesticide problem

past. With these data in hand, together with at least some information on metabolites, a selection must be made which ones to analyse. Of course, those that can not be measured by a well-established analytical method, can be excluded. The potential of a given active substance to enter the aquatic environment depends in particular on its physical and chemical properties that determine its environmental behaviour. In addition, the amount applied, the application technique and the time of use must be considered. For example pesticide application in autumn or winter on the naked ground poses a special risk. As well certain hydrogeological situations or other local conditions may decrease or increase the contamination risk. (Müller-Wegener 1994, Nordmeyer 1994, Seel et al. 1995, Skark & Zullei-Seibert 1995)

By evaluation of all the facts on the pesticide use, the land use and the local situation, appropriate substances and endangered zones can be selected as well as suitable sampling frequencies.

4 PESTICIDE MONITORING

Figure 1 shows a schematic schedule of the whole procedure. Based on the local situation, waterworks must decide if any danger of contamination with pesticides exists. If this can be excluded, no further action is required. If not, the available information must be analysed focusing on a risk estimation of relevant substances, sampling time and place. This leads to a monitoring program and a first measuring campaign.

If no pesticides are found and the monitoring program is really adequate, no further action must be taken.

If there are any pesticides found, they should be assessed in view of the substance and its concentration, together with the place and time of occurrence. Detection of pesticides should also be checked with regard to the analytical methods and its limitations.

Regarding the results of the first campaign, the monitoring program should be reviewed and revised if necessary.

Crucial for this approach is the link between data quality and monitoring strategy. Therefore it is essential that the underlying information is thoroughly checked and rechecked at each stage of the investigation.

REFERENCES

Biologische Bundesanstalt (BBA) 1997. *Pflanzenschutzmittelverzeichnis 11/1997*. Ribbesbüttel: Saphir-Verlag.

Heinz, I., A. Flessau, N. Zullei-Seibert, B. Kuhlmann, U. Schulte-Ebbert, M. Michels, J. Simbrey & G. Fleischer 1995. *Economic Efficiency Calculations in Conjunction with the Drinking Water Directive (Directive 80/778/EEC); Part III. The Parameter for Pesticides and Related Products*. Final Report for the European Commission - DG XI, Brüssel.

Kuhlmann, B. 1998. Leitlinie zur Überwachung von Pflanzenbehandlungs- und Schädlingsbekämpfungsmitteln (PBSM) bei der Wasserversorgung. *Veröffentlichungen des Instituts für Wasserforschung GmbH Dortmund und der Dortmunder Energie- und Wasserversorgung GmbH* Nr. 55.

Müller-Wegener, U. 1994. Eintrag von Pflanzenschutzmittelwirkstoffen aus landwirtschaftlicher Anwendung in Oberflächengewässer. *Bundesgesundhbl.* 4:158-163.

Nordmeyer, H 1994. Bodenvariabilität und Verhalten von Pflanzenschutzmitteln. *Z.Pflanzenern. Bodenkd.* 157:283-288.

Seel, P., S. Lang & N. Zullei-Seibert 1995. Eintragspfade von Pflanzenschutzmitteln in Oberflächengewässer. *Wasserwirtschaft* 85:28-33.

Skark, C. & N. Zullei-Seibert 1995. The occurrence of pesticides in groundwater - Results of case-studies. *Int.J.Environ.Anal.Chem.* 58:207-213.

Artificial Recharge of Groundwater, Peters et al. (eds) © 1998 Balkema, Rotterdam, ISBN 90 5809 017 5

The behaviour of fluorine in a lake-water recharge plant in Southeastern Finland

A. Vuorinen
University of Helsinki, Finland

ABSTRACT: The behaviour of F is studied in the Kuivala plant area where high F contents (over 1.5 mg/l) are sporadically measured in water. The introduced lake water and waters from 11 observation tubes are analyzed since June 1994. The observation tubes are separated into 4 groups according to the increasing groundwater character, decreasing temperature and COD and increasing silica content. Samples from 5 deep-soil profiles, 3 top-soil profiles (down to 2 m) and additional topsoil samples were studied. Mineral studies, total F determinations and leaching experiments were undertaken. The behaviour of F in top-soil profiles from natural soil and sprinkler infiltration area as well as in a deep-soil profile are presented in this paper. F is scavenged by the A and B horizons of the podzolic soils especially in the sprinkler infiltration area. The weakly weathered deep-soils contain the largest amounts of leachable and total fluorine.

1 INTRODUCTION

An infiltration plant has been operated since 1992 at Kuivala in the Southeastern Finland (Vuorinen 1994, 1996). The podsolic soils in the area consist mainly of different glaciofluvial sediments mostly from local rapakivi bedrock ranging from fine fraction to rock fragments and there are also till layers. The main minerals in the soil are quartz and feldspars (90-95 per cent). The main fluorine bearing minerals are hornblende, biotite, chlorite and occasionally fluorite.

2 SOIL PROFILES

In this paper, two top-soil profiles, one from natural soil (C3) and another from sprinkler infiltration area (C3S) are presented.The contents of F, Ca, Mg, K and silica leached from the A horizons (fraction <0.7 mm) in a sequential leaching experiment are presented in Table 1.

The contents of F and silica in the B and lower horizons of natural soil and sprinkler area soil are presented in Figures 1-4. Ca, Mg and K behave similar to F. Water soluble K and exchangeable Ca, Mg, as well as potentially soluble silica and F (ammonium acetate, pH 4.8) (non-mineral-bound forms) have been enriched in the A horizon. More F and silica and potentially soluble Fe (C3 10 mg/kg C3S 770 mg/kg) and Al (C3 85 mg/kg C3S 835 mg/kg) are dissolved from the sprinkler area than

Table 1. Contents (mg/kg dry weight) of F, Ca, Mg, K and silica leached from A horizons of the soil profiles C3 and C3S in: 1) pure water, 2) 1 M ammonium acetate and 3) 1 M ammonium acetate at pH 4.8. The leached contents are summarized in the leaching order.

natural soil				sprinkler recharge area soil			
C3	1.	2.	3.	C3S	1.	2.	3.
F	16	32	50	F	20	57	160
Ca	425	8490	10570	Ca	485	8735	10795
Mg	55	735	815	Mg	65	765	840
K	160	240	243	K	200	310	320
SiO_2	130	305	400	SiO_2	230	315	800

from the natural soil. The humic acid content is also highest. The soil pH is 5.5 for C3 and 6.55 for C3S samples. Exchangeable K, Ca, Mg and potentially soluble F, Al, Fe and silica have been enriched in the B horizons (depths 3-40 cm). More F is dissolved from the sprinkler area than from the natural soil. The soil pH of the upper part is 6.0. In general, there are low contents of sequentially leached K, Ca, Mg, F and silica in the below the A and B horizons (at least down to 2 m) and the soil pH is 5.0-5.5. Also the total contents of F, Ca, Mg and Fe are lower in top-soil samples than in deep-soil samples.

A soil profile taken from a well site is selected to represent features in deep-soil profiles.The sequential leaching results for F and Ca are presented in Figure 5. The contents of Mg and K behave like Ca.

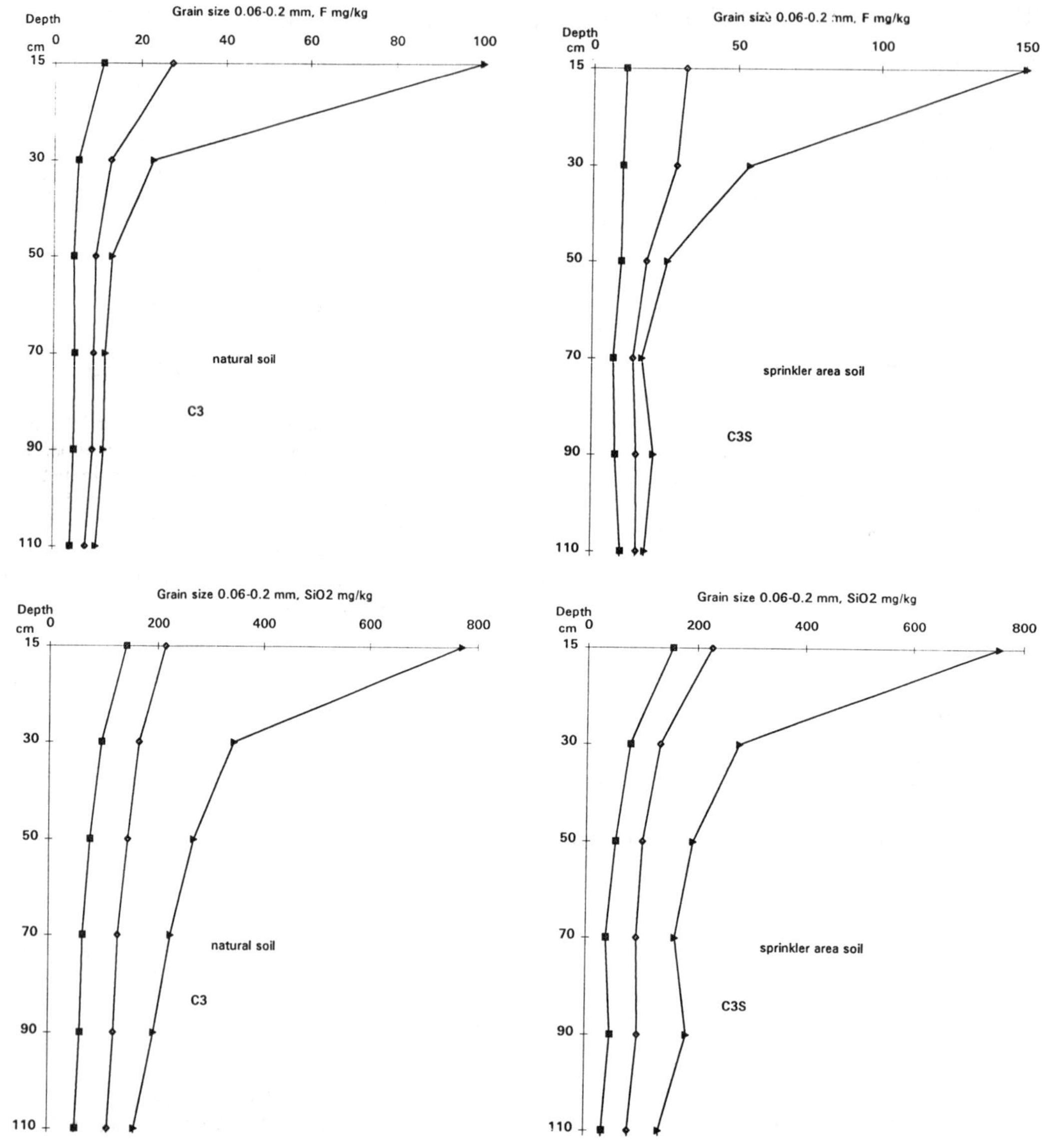

Figures 1-4. The contents of fluorine and silica leached sequentially from soil profiles C3 and C3S.

Potentially soluble silica is enriched in the top part of the profile (1.5 and 2.5 m). The sequentially soluble and total F, Ca, Mg and K are depleted down to the depth of 7.5-8.5 m due to weathering.

3 CONCLUSIONS

1. The organic-rich A horizon materials to the depth of 3 cm and the accumulation zone B horizon materials to the depths of 40 cm scavenge effectively fluorine especially from lake water in the sprinkler infiltration area.
2. The contents of non-mineral-bound fluorine in the weathered horizons below the A and B horizons are very low (at least to the depth of 2 m).
3. Generally, the weakly weathered deep-soil horizons contain the largest amount of leachable and total fluorine and seem to form the main source of fluorine into the infiltering water.
4. Further, the soil profiles under the infiltration basins should be studied in order to elucidate the

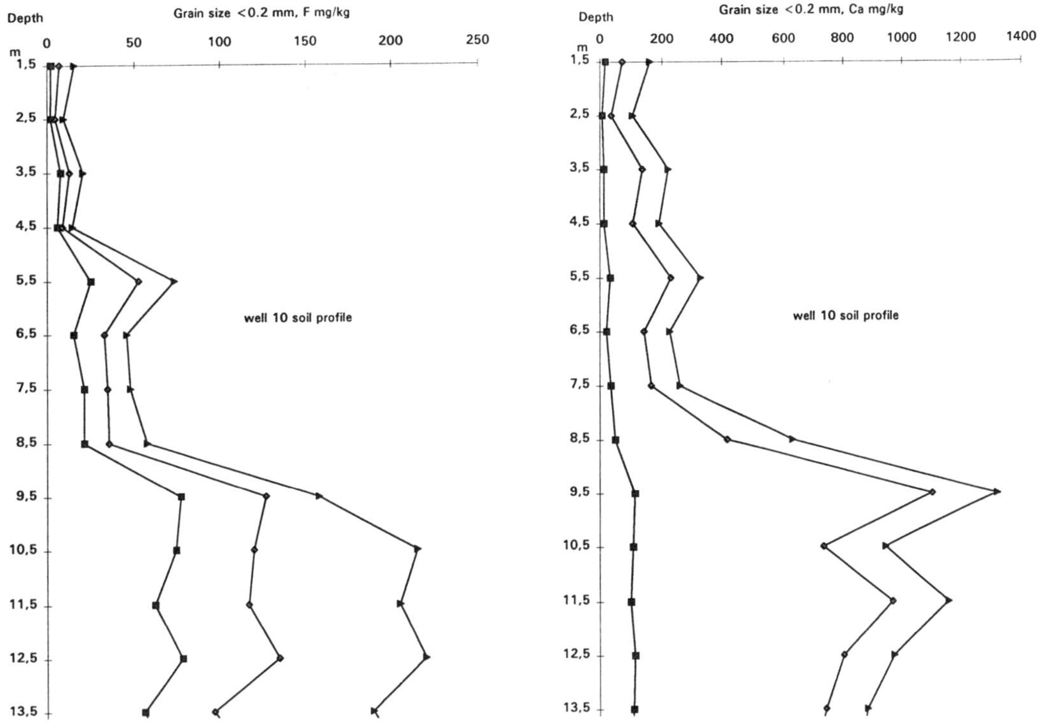

Figures 5 and 6. The contents of fluorine and calcium leached sequentially from the well 10 site soil profile.

differences in behaviour of fluorine between basin and sprinkler infiltration.

REFERENCES

Vuorinen, A. 1994. Geochemistry of fluoride in a lake-water infiltration plant, and prevention of "fluoride-booms" into domestic water? *6th Int. Conf. Environ. Contamination, Delphi, Greece, October 1994: 347-349, Edingburgh: CEP Consultants.*

Vuorinen, A. 1996. The behaviour of fluorine in a lake-water infiltration plant in the Viborg Rapakivi area. *Artificial Recharge of Groundwater. Nordic hydrological Programme NHP. Report No. 38. The Nordic Coordination Commitee for Hydrology*: 239-245.

2.5 Predicting and modelling groundwater quality

Artificial Recharge of Groundwater, Peters et al. (eds) © 1998 Balkema, Rotterdam, ISBN 90 5809 017 5

An integrated model for predicting and assessing the development of groundwater quality

Jasper Griffioen, Aris L. Lourens, Peter Venema, Chris B. M. te Stroet & Bennie Minnema
Netherlands Institute of Applied Geoscience TNO, Delft, Netherlands

Marc P. Laeven, Pieter J. Stuyfzand, Cees G. E. M. van Beek & Wouter Beekman
Kiwa Research and Consultancy, Nieuwegein, Netherlands

ABSTRACT: Development of the groundwater quality at artificial groundwater recharge systems is important for reasons of clogging, etc. We present a general groundwater quality transport model that has an integrated, hierarchical set-up. Both the major groundwater composition and individual trace compounds can be modelled and the transport models are coupled with a data-information system. The approach aims at minimizing the computation time and incorporating mechanistically oriented model parameters as much as possible. The major groundwater chemistry can be modelled with a multiple solute model, when one is solely interested in kinetically controlled biodegradation reactions, or with a multicomponent geochemical transport model, when the entire major composition has to be modelled. The process parameters for the microspecies are considered to be determined by the major chemistry. The actual values for sorption parameters are calculated from spatial data about the contents of reactive phases and typical mathematical functions that refer to the sorption affinity. Degradation constants for microorganics are obtained from literature for each specific redox environment.

1 INTRODUCTION

The development of groundwater quality is one criterium that is considered for designing artificial groundwater recharge systems. Mixing of anaerobic and aerobic groundwater at an abstraction well should be avoided as much as possible in order to prevent clogging of the wells by precipitation of Fe-oxides upon mixing. Numerical reactive transport models can be of use for assessing the development of groundwater quality.

We present a flexible modelling approach that can be used for prediction of groundwater quality at individual drinking water abstraction locations. The location can be any type of groundwater pumping location: either an artificial groundwater recharge site, river bank infiltration site or (phreatic) groundwater pumping site. The model approach is currently being developed. We present here the general outline of the approach together with some details.

2 MODELLING APPROACH

Numerical transport models become attractive for modelling the development of groundwater quality when the subsurface is heterogeneous, interaction of various physico-chemical processes occurs and/or the chemical processes occuring are non-linear. Analytical solutions for transport are no longer available in such cases. The use of transport models is however hampered by relatively long computation time. Optimal choice of the model code becomes therefore of interest. This holds even more when one realises that data about the reactivity of the subsurface is usually limitedly available. Oppositely, the prediction of future groundwater quality asks for a mechanistic approach of the ongoing chemical processes. Non-mechanistic, empirical approaches do not guarantee that the future status of the subsurface system is well described, especially when the status is strongly modified as often happens during artificial groundwater recharge. Modifications happen for example when oxic recharge water is infiltrated in an aquifer that is anoxic at pristine conditions.

The currently developed approach uses as much as possible default process parameters for the chemical processes. The default process parameters are obtained from simplifications of more complex chemical processes, that are valid for a limited range of conditions of the system studied. This approach is justified when one realises that reactive transport modelling within the context of groundwater pumping for drinking water purposes, is solely of interest for fresh groundwater and the solutes of interest have concentrations ranges around the maximum tolerable concentration: no modelling is necessary for lower concentration ranges and drinking water pumping is unlikely for much higher concentration ranges.

Both the major groundwater composition and individual microspecies can be of interest in practice. We have distinguished three types of modelling scenarios. First, kinetic biodegradation is the relevant issue. Here, one is solely interested in the fate of redox-sensitive reactants such as dissolved oxygen, nitrate and dissolved organic carbon. Denitrification is a typical example of this class of processes. A multiple-solute transport model will be sufficient. A multiple-solute transport model is a model in which the dissolved species are described as individual species and a limited set of chemical reactions is included in the model. No speciation calculations, calculations of the individual complexes and free species that make up the total concentration of a species, are performed and the condition of electroneutrality is not taken into account.

Second, the major groundwater quality is the relevant issue, for example hardness, pH, dissolved Fe concentration, etc. Both reactants and products need to be considered. A multicomponent, geochemical transport model is required. Aqueous speciation, ionic strength effects and aqueous complexing are taken into account and various types of aquatic chemical reactions are incorporated. Kinetics may be included.

Third, the fate of individual microspecies is of interest. A single-solute transport model may tackle this problem when sorption processes together with degradation reactions are controlling the mobility. One should assure here that other, classical reactions are irrelevant. The microspecies can be both organic contaminants as pesticides, and trace metal(loid)s. The amount of sorption of the latter should be negligible compared to that of the major ions, otherwise one should use the multicomponent geochemical transport model.

The model set-up is based on the finite difference models MODFLOW (McDonald & Harbaugh 1988) and MT3D96 (Zheng 1996). The set of individual models are embedded in a Graphical User Interface, which takes care of pre- and postprocessing and invokes the models. The integrity of the whole system and its (intermediate) results is controlled and visualised in a decision support fashion.

The single-solute transport model MT3D96 is used for simulations of the microspecies. The preparation of process parameters is described in the next section. A multiple-solute version of MT3D96, named MODBIO, is established for the multiple-solute model. The implicit coupling between transport and chemistry in MT3D96 is set aside in MODBIO. Advective/dispersive transport is calculated with the Eulerian/Lagrangian approach as is done within MT3D96. Degradation is calculated as a redox process among the species involved according to reaction stochiometry. One can define an unlimited amount of oxidants and reductants. Transport and chemistry are explicitly coupled by the operator split method.

The amount of degradation per time step per grid cell, is calculated by explicit analytical solutions of kinetic differential equations and superposition of oxidants and reductants (either dissolved or solid) concentration fields. One can choose either first-order degradation with respect to oxidants (oxidant-limited) or zero-order degradation with respect to organic matter (carbon-limited), for the relevant oxidant/reductant couples. The dissolved oxidants O_2, NO_3 and SO_4 are consumed in order of decreasing energy yield.

A multicomponent geochemical transport model will be established that also uses the operator split method for coupling transport and chemistry. The model uses MT3D96 for advective/dispersive transport, the biodegradation module of MODBIO for kinetic biodegradation and PHREEQC for other aquatic chemical processes (Parkhurst 1995). The biodegradation module will be called before the PHREEQC module in order to let kinetic biodegradation determine the redox state of the system (Griffioen et al. 1995).

3 HIERARCHICAL STRUCTURE OF THE INTEGRATED MODEL

Chemical model parameters should preferably be based on mechanistic concepts for the geochemical processes operative, instead of using empirical concepts. We use the concept of hydrogeochemical environments to establish the relationship between chemical process parameters for microspecies and state of the major chemistry of the system. A hydrogeochemical environment is defined as an underground spatial entity with a unique pore water composition and a unique assemblage of reactive components as the porous matrix. The distinct characteristics refer to both the intensity and capacity factors for geochemical processes during groundwater flow. Intensity factors are associated with the strength and affinity of chemical processes and capacity factors are associated with the extent in which the chemical reactions take place.

Each hydrogeochemical environment has its own values for the process parameters of the microspecies. The values of the sorption parameters are calculated by mathematical functions that combine capacity parameters (concentration, contents), with literature values for the intensity parameters (thermodynamic constants, sorption affinity). The capacity parameters are obtained by combining laboratory determinations of reactive compounds (e.g. CEC, extractable iron oxides) for field collected samples and default values expressed per unit weight of reactive compound for the process parameters, such as sorption capacity.

The values for degradation constants of microorganics are assumed to be controlled by the redox status of the system. In many cases no data are available for degradation in anaerobic conditions, since kinetic studies of anaerobic systems are scarcely performed. In such cases one either has to use values for aerobic systems or assuming no degradation at all.

4 MODELLING SORPTION OF TRACE METALS

Sorption of trace metals is described by a single-solute linear, Freundlich or Langmuir sorption isotherm. The isotherm is constructed from data about the groundwater composition and sorbents present in the porous medium. A sorption isotherm is calculated for each hydrogeochemical environment in the following way.

The activity of the free trace metal (e.g. Zn^{2+}) is calculated from the total dissolved concentration in three non-iterative steps by formulas that take into account the major inorganic aqueous complexes and complexation with dissolved organic acids, including competition of Ca, Mg, Fe, Al and other trace metals. The calculated activity of the free ion is subsequently used to calculate the related sorbed amount for each sorbent present. We consider three types of sorbents and assume that sorption to the three individual sorbents is additive. The three sorbents are clay minerals, iron oxides and organic matter. General isotherm type curves are established for these three sorbents for fresh groundwater systems. Sorption to iron oxides and organic matter is for example described by a Freundlich isotherm:

$$\beta_{Me} = k\,[Me^{2+}]^n \qquad (1)$$

with

$$\log k = a + b\,pH + c\,pH^2 + d\,pH^3$$
$$n = a' + b'\,pH + c'\,pH^2 + d'\,pH^3$$

where β_{Me} is the fraction of the sorbent occupied with the metal, k is the sorption affinity and n is an exponential term. The values for the third-order polynomial pH-function are obtained from speciation calculations using PHREEQC for a range of fresh water compositions from pH is 4 to 7, from non-polluted (soft, slightly mineralised water) to hard, mineralised but still fresh groundwater, and from one tenth to ten times the maximum tolerable concentration of the trace metal of interest. The total sorbed amount is calculated as the sum of the products of the fraction sorbed and the amount of sorbent present:

$$s_{Me} = s_{clay}\,\beta_{Me,clay} + s_{ox}\,\beta_{Me,ox} + s_{OM}\,\beta_{Me,OM} \qquad (2)$$

where s_i refers to the amount of sorbent present. The isotherm that is used in the single-solute transport model, is obtained from an empirical fit of the total sorbed amount versus the total concentration.

5 DISCUSSION AND CONCLUSIONS

A flexible groundwater quality transport model is constructed that can be used for predicting the development of the groundwater quality at artificial groundwater recharge systems. The approach is strongly directed to practical field studies where a limited set of data is available about the geochemical properties of the subsurface.

Many assumptions become included in the hierarchical, integrated model approach, especially in the module that establishes a single-solute sorption isotherm for a trace metal. The alternative for modelling sorption of trace metals as a single-solute sorption process, is using the multicomponent geochemical transport model that includes cation-exchange and surface complexation modules. The degree of accuracy is presumably near-identical for that approach since the major uncertainty lies in the availibility of field data. Comparable assumptions will be necessary for the multicomponent geochemical transport model as we have implicitly incorporated in our sorption module.

6 ACKNOWLEDGMENTS

This research project is financed by nine Dutch drinking water companies (NUON, WG, WLF, WML, WMN, WMO, WNWB, WOB, WZHO) and cofinanced by the Ministry of Economic Affairs.

REFERENCES

Griffioen, J., P. Engesgaard, A. Brun, D. Rodak, I. Mucha & J.C. Refsgaard 1995. Nitrate and Mn-chemistry in the alluvial Danubian Lowland aquifer, Slovakia. In *"Groundwater quality: Remediation and protection"*. IAHS Publ. No. 225, p. 87-96.

McDonald, M.G. & A.W. Harbaugh 1988. A modular three-dimensional finite-difference ground-water flow model. *U.S.G.S. Tech. Water-Res. Inv., Book 6, Chapter A1.*

Morel, F.M. & J.G. Hering 1993. *Principles and applications of aquatic chemistry*. John Wiley & Sons, New York, 588 pp.

Parkhurst, D.L. 1995. User's guide to PHREEQC - a computer program for speciation, reaction-path, advective-transport, and inverse geochemical calculations. *U.S. Geol. Surv. Water Res. Inv.* 95-4227, 143 pp.

Zheng, C. 1996. MT3D96. A modular three-dimensional transport model for simulation of advection, dispersion and chemical reactions of contaminants in ground-water systems. *S.S. Papadopulos & Ass., Inc.*, Bethesda, Maryland, U.S.A.

Artificial Recharge of Groundwater, Peters et al. (eds) © 1998 Balkema, Rotterdam, ISBN 90 5809 017 5

Modelling the effects of deep artificial recharge on groundwater quality

M.W. Saaltink – *Dep. d'Enginyeria del Terreny i Cartogràfica, UPC, Barcelona, Spain*

C. Ayora – *Institut de Ciències de la Terra (Jaume Almera), CSIC, Barcelona, Spain*

P.J. Stuyfzand – *Kiwa Research and Consultancy, Nieuwegein, Netherlands*

H. Timmer – *Watermaatschappij Zuid-Holland Oost, Gouda, Netherlands*

ABSTRACT: We modelled the hydrogeochemical effects of deep well injection of oxidised water into an anoxic sandy. Kinetic expressions have been used for mineral dissolution/precipitation rates and organic matter oxidation, which appear to control the evolution of the redox state of the system. Reasonable matching of calculations with experimental data was obtained by adjusting the kinetic rate constant for the reactions.

1 INTRODUCTION

Surface water may be artificially recharged into deep aquifers to improve the quality of the recharged water for use by water supply companies. To study the feasibility and usefulness, a field test was carried out at Langerak (the Netherlands).

In this paper a reactive transport model of this field test is presented. Such modelling can help to understand the complex relationship between solute transport processes and the chemical reactions.

2 REACTIVE TRANSPORT MODEL

We used a model code for general use that couples solute transport and chemical reactions (Saaltink et al. 1997). It can handle complexation (including redox and acid/base reactions), adsorption (including cation-exchange) and precipitation/dissolution of minerals. The last reaction can be assumed equilibrium or kinetically controlled. The model code uses a one-dimensional or two-dimensional finite element discretization and a Direct Substitution Approach (DSA) for coupling solute transport and chemical equations.

3 EXPERIMENT DESCRIPTION

The experiment consisted of injecting oxidised water into a sandy aquifer with small amounts of calcite and pyrite and with a low cation exchange capacity, which initially contains reduced water of good quality with low concentrations of chloride and nitrate and in equilibrium with calcite. Water was pumped out of the aquifer at a distance of 190 m from the injection well. Monitoring wells were placed at 20m, 40m and 91 m from the injection well. During one year samples for chemical analysis were taken from the monitoring wells, the pumping well and of the injected water (Stuyfzand and Timmer 1998; Timmer and Stuyfzand 1998).

4 CONCEPTUALIZATION

A homogeneous one-dimensional domain was assumed with larger width at the centre to represent the shape of the hydrological system induced by the injection and the pumping.

The chemical composition of both the injected and the initial water were analysed by means of the program EQ3. Together with information on mineral composition of the sand, this led to a geochemical conceptual model of 35 aqueous chemical species, 21 complexation reactions and 7 cation exchange reactions, which we assumed to be at equilibrium. Moreover, we considered kinetically controlled precipitation and dissolution of amorphous ferric hydroxide ($Fe(OH)_3$), siderite ($FeCO_3$) and rhodocrosite ($MnCO_3$) as well as oxidation by oxygen and nitrate of both pyrite (FeS_2) and organic matter (CH_2O):

$$Fe(OH)_3 \rightarrow Fe^{3+} + 3OH^-$$

$$FeCO_3 \rightarrow Fe^{2+} + CO_3^{2-}$$

$$MnCO_3 \rightarrow Mn^{2+} + CO_3^{2-}$$

$$FeS_2 + 3.75O_2 + 0.5H_2O \rightarrow Fe^{3+} + 2\ SO_4^{2-} + H^+$$

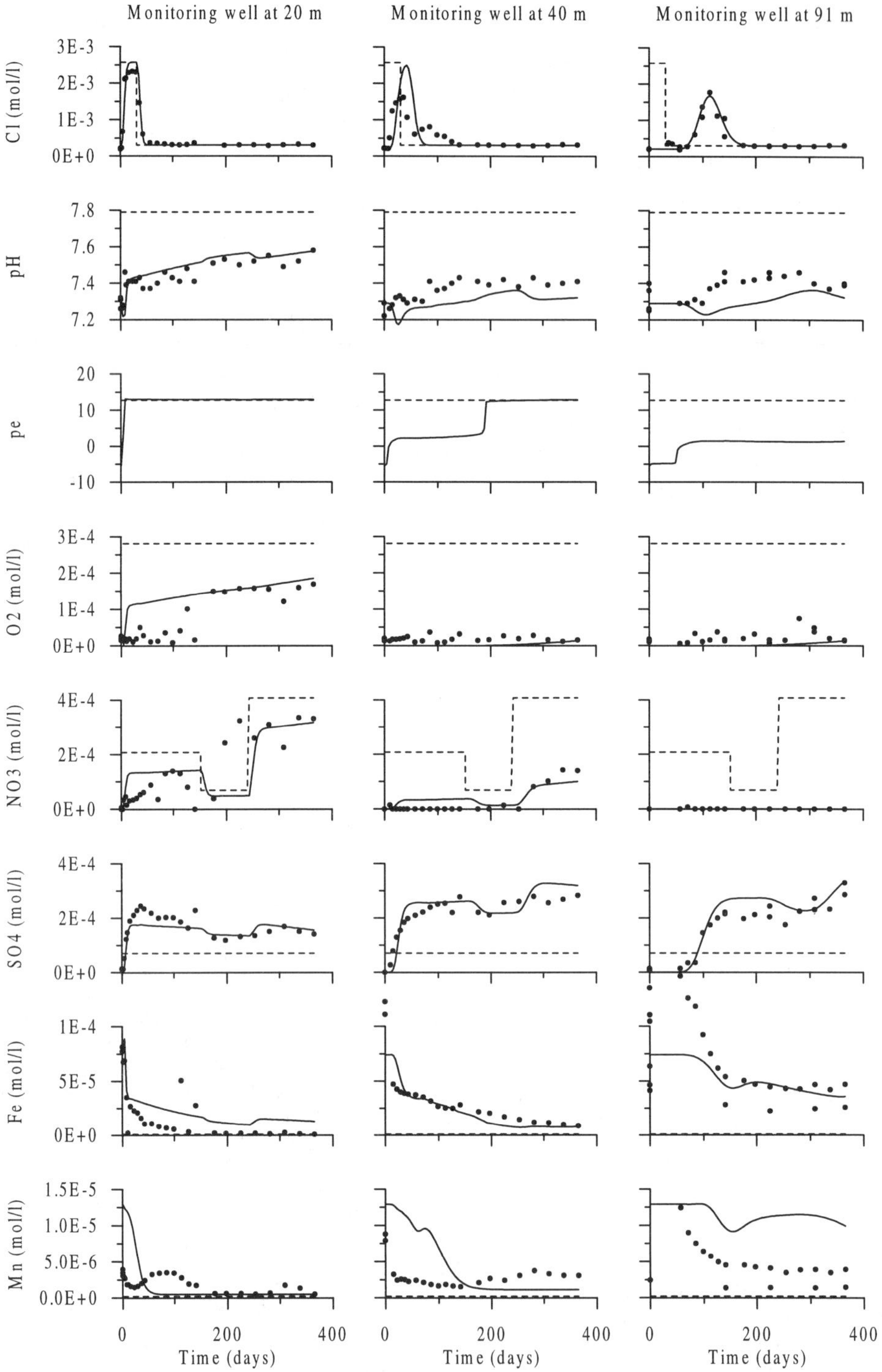

Figure 1. Breakthrough curves of measured (dots) and calculated values (continuous lines). Values of the recharge water are indicated by discontinuous lines.

$$CH_2O + O_2 \rightarrow HCO_3^- + H^+$$

$$FeS_2 + 3NO_3^- + 2H^+ \rightarrow Fe^{3+} + 2SO_4^{2-} + 1.5N_2 + H_2O$$

$$CH_2O + 0.8NO_3^- \rightarrow HCO_3^- + 0.4N_2 + 0.2H^+ + 0.4H_2O$$

For these kinetically controlled reactions one has to express the reaction rates. In theory, they may depend on the reactive surface of the mineral, concentration of (a) bacteria, (b) chemical species involved and (c) catalysers. However, exact expressions are hardly known. We used the following:

$$r = \kappa(\Omega - 1) \quad (1)$$

$$r = \kappa_1 (O_2)^{0.5} (H^+)^{-0.11} \quad (2)$$

$$r = \kappa(O_2) \quad (3)$$

$$r = \kappa(NO_3^-)\left[1 - (O_2)(O_2)_{max}^{-1}\right] \quad (4)$$

where r are the reaction rates, κ the rate constants, which also include the values for reactive surface areas, and Ω the saturation indices of the minerals. Equation (1) is the standard formulation for dissolution or precipitation of a mineral and is used for ferric hydroxide, siderite and rhodocrosite. Equation (2) has been taken from Nicholson (1994) and is used for the oxidation of pyrite by oxygen. Equation (4) is a linear expressions with a factor that represents inhibition by oxygen and is used for denitrification by pyrite and organic matter.

5 RESULTS AND DISCUSSION

In this article we only present chloride and the species that control the redox potential (pH, pe, O_2, NO_3^-, SO_4^{2-}, Fe, Mn), although concentrations of more species have been calculated too. The concentrations in the monitoring wells were calibrated by estimating the dispersivity, the width of the hydrological system and the rate constants of equations (1) to (5). Best fits are shown in figure 1.

Good fits were obtained for chloride except for the monitoring well at 40 m. The measured data show two peaks probably due to heterogeneity of the permeability.

The pH is rather constant. The rise to values of the recharge water is inhibited by proton production of pyrite and organic matter oxidation.

The pe (that is, -log of the electron concentration) and oxygen concentration rise to the values of the recharge water at the monitoring well closest to the injecting well. At the other wells it rises to pe values between 3 and 5, which means practically zero oxygen concentrations.

Nitrate is consumed through denitrification of pyrite and organic matter. Therefore, concentrations reach lower values than those of the recharge water, giving practically zero values at the most distant monitoring well. Concentrations of the recharge water varied considerably in time. For simplicity we represented this by means of three periods, although the measured data do not show such brusque fronts. These fronts can still be seen for the calculated values, especially at the nearest monitoring well.

Due to oxidation of pyrite sulfate concentrations rise to values higher than those of the recharge water. At the calculated values the influence of the varying nitrate input can still be seen.

Initial concentrations of both iron and manganese show high spatial variation. Therefore, it is difficult to reproduce the measured values. However, both measured and calculated values show the influence of siderite and rhodocrosite dissolution

As a conclusion, acceptable but not perfect fits could be obtained. Comparison with both the expressions and rates obtained in similar empirical studies is required. Moreover, the homogeneous 1D model may not represent the spatial heterogeneity of both transport and chemical parameters. So, 2D heterogeneous representations of the problems will contribute to a more accurate understanding of the system.

ACKNOWLEDGEMENT

This work was supported by the Watermaatschappij Zuid-Holland Oost and the European Union (contract: ENV-CT95-0071).

REFERENCES

Nicholson, R.V., 1994, *Iron-sulfide oxidation mechanisms: laboratory studies*, In: J.L. Jambor and D.W. Blowes (eds.): Short course on Environmental geochemistry of Sulfide Mine-wastes. Minerological Association of Canada, p.163-183.

Saaltink, M., I. Benet, & C. Ayora, 1997, *RETRASO, Fortran Code for Solving 2D Reactive Transport Solutes, Users´s guide*, E.T.S.I. Caminos, Canales y Puertos, UPC and Instituto de Ciencias de la Tierra, CSIC.

Stuyfzand, P.J. & H. Timmer, 1998, *Injecting oxic surface water into an anoxic aquifer along the river Rhine; water-rock interactions including redox dynamics*, Kiwa-report, EU-contract ENV4-CT95-0071, in preperation.

Timmer , H. & P.J. Stuyfzand, 1998, *Experiments with deep well infiltration along the river Rhine for drinking water supply*, this volume.

Artificial Recharge of Groundwater, Peters et al. (eds) © 1998 Balkema, Rotterdam, ISBN 90 5809 017 5

Simple models for reactive transport of pollutants and main constituents during artificial recharge and bank filtration

Pieter J. Stuyfzand
Kiwa Research and Consultancy, Nieuwegein, Netherlands

ABSTRACT: As an alternative to reactive transport codes which require a high skill and tedious data input, 2 simple models are presented for use in artificial recharge and bank filtration. INFOMI is a finite element 1D-model using 'mixing cells in series'. It incorporates: a variable input signal; volatilization and breakdown in the infiltration basin; filtration of pollutants attached to suspended particles; and advection, dispersion, sorption and breakdown in the saturated zone. EASY-LEACHER is a 2D spread-sheet model, based on: the mass balance approach; travel times derived from a separate hydrological model; and empirical rules on reaction sequence and kinetics. The model simulates: the displacement of native groundwater; the leaching of exchangeable cations, calcite, organic matter and iron sulphides; the breakthrough of reactive solutes; and sorption and breakdown of micropollutants. The behaviour of 4 radionuclides and trihalomethanes upon artificial recharge is predicted using INFOMI. Quality changes of injected surface water when recollected by a distant well, are calculated by EASY-LEACHER.

1 INTRODUCTION

Modelling the water quality changes of surface water during detention in spreading basins and/or during aquifer passage can contribute in various ways to the design, optimization and monitoring of systems with artificial recharge (AR) or bank infiltration (BI). Important questions to be solved are: (a) which system is to be preferred from a chemical point of view (BI or AR using either basins or deep wells)?; (b) which is the minimum detention time required for removal of specific pollutants?; (c) which aquifer is best suited to eliminate for instance nitrate, nickel, atrazin and trihalomethanes?; (d) which parameters can be used to screen the analytical program?; and (e) on which time scale is oxygen in the influent reaching the recovery well and then triggers the clogging process?

The current generation of complex reactive transport models like PHREEQC (for an application in MIKE-SHE see Brun et al. 1998) and the future generation which will be integrated with geographical information systems (Griffioen et al. 1998), are in a way handicapped. They require a very detailed data entry, long computation times (or powerful computers) and an extremely high skill of the performer. This provoked the development of 2 simple reactive transport models at Kiwa, INFOMI 3.1 (Dijkhuis & Stuyfzand 1996) and EASY-LEACHER (Stuyfzand in prep).

2 INFOMI

INFOMI is the acronym for 'INFiltration Of MIcropollutants'. It is a finite element 1D-model using the 'mixing cells in series' concept, incorporating (Fig.1): (a) a highly variable input signal for the pollutants in surface water; (b) in the infiltration basin or in a specific river segment: volatilization, photochemical and microbiological first order breakdown; (c) at the water/sediment interface: filtration of pollutants attached to suspended particles; and (d) in the saturated zone: advection, dispersion (plus diffusion), linear sorption (including DOC as a carrier), and first order breakdown (organics and radionuclides).

Some general specifications are given in Table 1, and the most important transport equations are collected in Table 2. Advection and dispersion (+diffusion) in the aquifer system are modelled using algorithms of Appelo & Postma (1993). A flow path is essentially subdivided into at least 3 parts (Fig.1): (1) the infiltrating water course, (2) the water/sediment interface, and (3) the aquifer system, which may be further subdivided, for instance into bottom sludge or an upper aquiclude and the aquifer. The concentration which is calculated for the exit of each compartment, forms the input for the next compartment.

Standard the model generates the concentration in the aquifer along the flow path (distance travelled), and the concentration with time (the breakthrough curve) on each desired point along the flow path within the aquifer. Variables which need to be known, are specified in Table 2. The most important assumptions or conditions of the model are: (1) suspended solids are filtrated at the water/sediment interface and do not play a further role, because of inert behaviour or removal by either erosion or sludge scraping; (2) steady hydrology (travel times, water depth in ponds, flow path, etc.); (3) there is no unsaturated zone; (4) each soil compartment is homogeneous and geochemically in steady state (no changes in redox environment and in leaching or accumulation of reacting components); (5) no DOC-accelerated transport of organic pollutants (see also 8); (6) trace metals are only influenced by sorption (no precipitation/dissolution); (7) each decay process is first-order (one T½); (8) the flow velocity and longitudinal dispersivity (α_L) are constant within each soil compartment; (9) linear, reversible sorption without competition; and (10) there is no temperature dependency, except for volatilization.

Table 1. Comparison of characteristics and performance of two simple models dveloped at Kiwa.

item	INFOMI 3.1	EASY-LEACHER
water quality[1]	OMPs, TMs	MCs, OMPs
dimension	1-D (2-D)	2-D (3-D)
input signal	all	step[@]
processes considered:		
•in water course[2]	vol, phot, bio	-
•at water/sed. interface	filtration	filtration
•within aquifer system[3]	advect., disp, sorp., decay	as INFOMI 3.1 + leaching[4]
hardware	PC, ≥486	PC, ≥486
software	Turbo-Pascal	spread-sheet QPRO

1: OMPs = Organic MicroPollutants; TMs = Trace Metals; MCs = Main Constituents
2: vol = volatilization; phot = photolysis; bio = biodegradation
3: advect = advection; disp = dispersion (incl. diffusion); sorp = sorption
4: including: cation exchange, oxidation of pyrite and organic matter, and calcite and MnO_2 dissolution.
@: stepwise change from native situation to artificial recharge or bank filtration (with mean recharge composition).

3 EASY-LEACHER

EASY-LEACHER is an analytical 2D-spread sheet model (Fig.2). It is based on: (a) the mass balance approach, for both reactive solutes in the water recharged and solid reactive phases in the aquifer system; (b) the cumulative frequency curve of detention times for a well or well field as derived from a separately run hydrological model; and (c) some empirical rules regarding the sequence of reactions and reaction kinetics. The model simulates for as many strata as desired: (a) displacement of the native groundwater by the water recharged; (b) the leaching of reactive soil constituents: exchangeable cations, calcite, organic matter and iron sulphides; (c) the breakthrough of reactive solutes; and (d) sorption and breakdown of organic micropollutants.

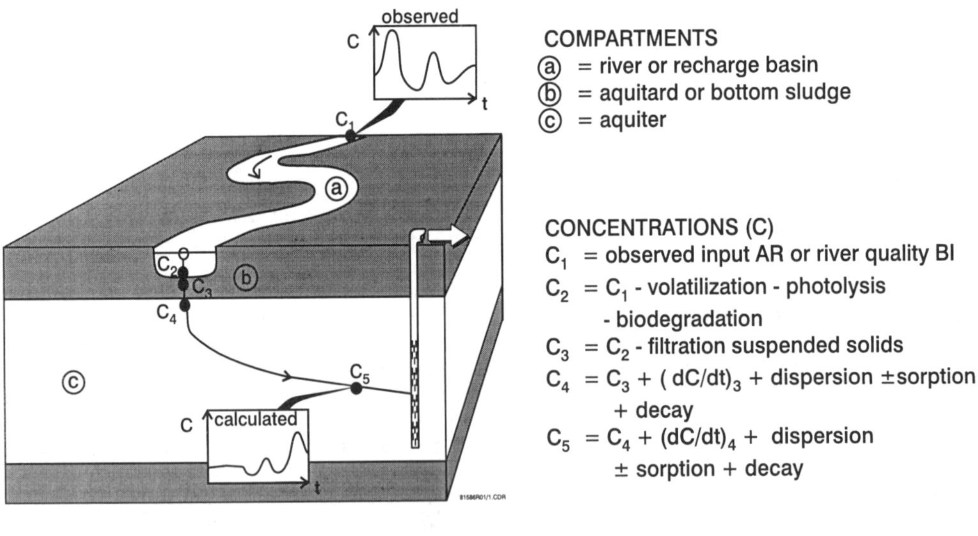

Figure 1. Essential features of INFOMI 3.1.

Table 2. The most important transport equations (hydraulics excluded) used in INFOMI 3.1

Volatilization, photolysis and biodegradation in a recharge basin or river segment

Assuming simultaneous decay and first order decay rates for all 3 processes, we obtain:

$$C_{t_i} = C_0 e^{-(\lambda_v + \lambda_p + \lambda_b)t_i} \quad (1)$$

with:

λ_v, λ_p, λ_b = ln(2)/T½ = decay constant of OMP during resp. volatilization, photolysis and biodegradation [1/d]; (T½) = half life during resp. volatilization, photolysis and biodegradation [d]; t_i = travel time in infiltrating water course [d]; C_0 = initial concentration [mol/L]; C_{ti} = concentration after t_i days [mol/L].

Volatilization is calculated using the approach of Southworth (1979) with estimation of transport coefficients according to McKay & Shiu (1981), and with correction for TOC-binding. Combination yields:

$$\lambda_v = \frac{8.995\,10^{-3} K_H}{(1.68\,10^{-3} K_H + 6.18\,10^{-5} T)(1 + K_{oc} TOC\,10^{-6}) D_{sw} \sqrt{M}} \quad (2)$$

with:

K_H = constant of Henry [Pa.m^3/mol]; T = absolute temperature [K]; D_{sw} = mean water depth of infiltrating water course [m]; M = Molecular weight of organic micropollutant [g/mol]; TOC = Total Organic Carbon in surface water [mg C/L]; K_{oc} = distribution coefficient between organic carbon and water [L/kg C].

If K_{oc} is unknown, calculate by:

$K_{oc} = 10^b K^a_{ow}$, with K_{ow} = octanol water distribution coefficient, a and b = regression parameters (see Appelo & Postma 1993); or $K_{oc} = 10^{3.64}\, S^{-0.55}$ with S = solubility OMP in water [mg/L].

Filtration at the water/sediment interface

Analytical results of OMPs in surface water generally refer to total concentrations. Assuming that all sorption to suspended particles occurs to particulate organic carbon (POC = TOC - DOC), and that POC-binding forces equal TOC-binding forces, yields (Stuyfzand & Lüers 1996):

$$OMP_{diss} = OMP_{tot}\left(\frac{DOC}{TOC} + \frac{1 - (DOC/TOC)}{1 + K_{OC} TOC\,10^{-6}}\right) \quad (3)$$

with:

DOC = Dissolved Organic Carbon in surface water [mg/L]; OMP_{tot} = total concentration of Organic Micro-Pollutant [μg/L]; OMP_{diss} = concentration of dissolved OMP = free + DOC-bound [μg/L].

The fraction removed by filtration then equals:

$(OMP_{tot} - OMP_{diss})/OMP_{tot}$

Sorption of organic micropollutants within the aquifer system

Retardation due to sorption is calculated using the classical approach (see Appelo & Postma 1993) with addition of the effects of DOC-binding according to Kan & Tomson (1990) and of dissociation according to Schellenberg (1984):

$$R = \left(1 + \frac{\rho_b f_{oc} f_{nd} K_{oc}}{n(1 + f_{nd} TOC\,10^{-6} K_{oc})}\right) \quad (4)$$

with:

ρ_b = bulk density of aquifer matrix [kg/L] = $(1-n)\rho_s$ with ρ_s = density of solids in aquifer matrix = 2.65 kg/L; n = porosity [L/L]; f_{oc} = fraction organic carbon in aquifer matrix [-]; $f_{nd} = 1/(1 + 10^{pH-pKa})$ = correction for dissociation, which is important only if K_{oc} or K_{ow} were measured at $pH_{exp} < pKa$ and pH (= pH_{sample}) > pH_{exp}; $pK_a = -\log K_a$, with K_a = first dissociation constant of H-OMP.

Sorption of trace metals within the aquifer system

If calcium is the main cation sorbed to the aquifer matrix, then the retardation of trace metal M can be calculated as follows:

$$R = 1 + \frac{\rho_b CEC}{nm}\left(K_{M\backslash Ca} \frac{\beta_{Ca}^{0.5}}{[Ca^{2+}]^{0.5}}\right)^m \quad (5)$$

with:

CEC = actual Cation Exchange Capacity = $Na_{EXCH} + K_{EXCH} + Ca_{EXCH} + Mg_{EXCH} + NH_{4EXCH}$ + ... [meq/kg dry weight]; $K_{M\backslash Ca}$ = exchange constant (using the Gaines-Thomas convention) for metal M^{m+} in equilibrium with Ca^{2+} [-], according to the reaction: 1/m M^{m+} + ½Ca-$EXCH_2$ ↔ 1/m M-$EXCH_m$ + ½Ca^{2+} . This constant is calculated from $K_{Na/M}$-values listed in Appelo & Postma (1993), by taking: $K_{M\backslash Ca} = (K_{Na/Ca} / K_{Na/M})$; β_{Ca} = fraction of calcium in CEC of aquifer matrix = Ca_{EXCH}/CEC [-]; m = charge of metal M; $[Ca^{2+}]$ = total calcium concentration infiltration water [mmol/L].

Degradation or radio-active decay is calculated using:

$$C_t = C_0 e^{-(\lambda_b t R)} = C_0 / 2^{\frac{tR}{(T_{1/2})_b}} \quad (6)$$

with:

t = travel time of water in aquifer system [d]; T½ = half life due to biodegradation (depends on redox etc.) or radio-active decay [d]; R = retardation factor, Eq.4 for OMPs and Eq.5 for trace metals.

Contrary to most transport codes, the situation after each time step is directly calculated with analytical formulae, without iterations, finite elements and equilibrium calculations (calcite excluded).

In contrast to INFOMI 3.1. the composition of the recharge water is assumed constant, and the leaching of reactive solid phases is accounted for. The latter is essential for both the major constituents and organic pollutants (where the advance of redox zones determines the overall breakdown).

Some general specifications are given in Table 1, and some of the most important transport equations regarding inorganics are collected in Table 3. The equations regarding organic micropollutants are equal to those listed in Table 2. The transport equations involving calcite dissolution and the oxidation of organic matter strongly depend on the leaching sequence.

As shown in Fig.2 there are essentially 2 options for running EASY-LEACHER. When flow parallels the discerned aquifer layers, as in case of deep well injection, the model is built up by the individual layers. If flow crosses the discerned aquifer layers, as in case of basin recharge, the model is built up by selected flow paths. In the latter case some problems may arise and information on the exact position of leaching fronts is lost, due to the necessity to average the geochemistry along each flow path (excluding sludge at the bottom of recharge basins, which is treated separately).

The assumptions or conditions of the model are the same as those mentioned for INFOMI 3.1, excluding number 4 (each soil compartment can be leached here). Additional assumptions are: (a) bottom sludge in a recharge basin remains in steady state leading to constant O_2 and NO_3^- losses by oxidizing organic material; (b) DOC and NH_4^+ in the water recharged are first partly oxidized, the remaining O_2 and NO_3^- can oxidize aquifer materials; (c) exchange reactions and the dissolution of inorganic aquifer phases are without kinetic hinderance (sharp fronts); (d) the leachable phases are all homogeneously distributed within each layer or along each flow path; (e) organic material in the aquifer matrix is composed of a labile (highly reactive), tough (little reactive) and inert fraction, with the tough fraction being oxidized only after total consumption of the labile fraction; (f) the oxidation rate of organic material depends on the presence of O_2 and NO_3^-, and the original amount of the labile c.q. tough fraction; and

(g) the leaching sequence is such that when pyrite is completely leached, tough organic material is still present.

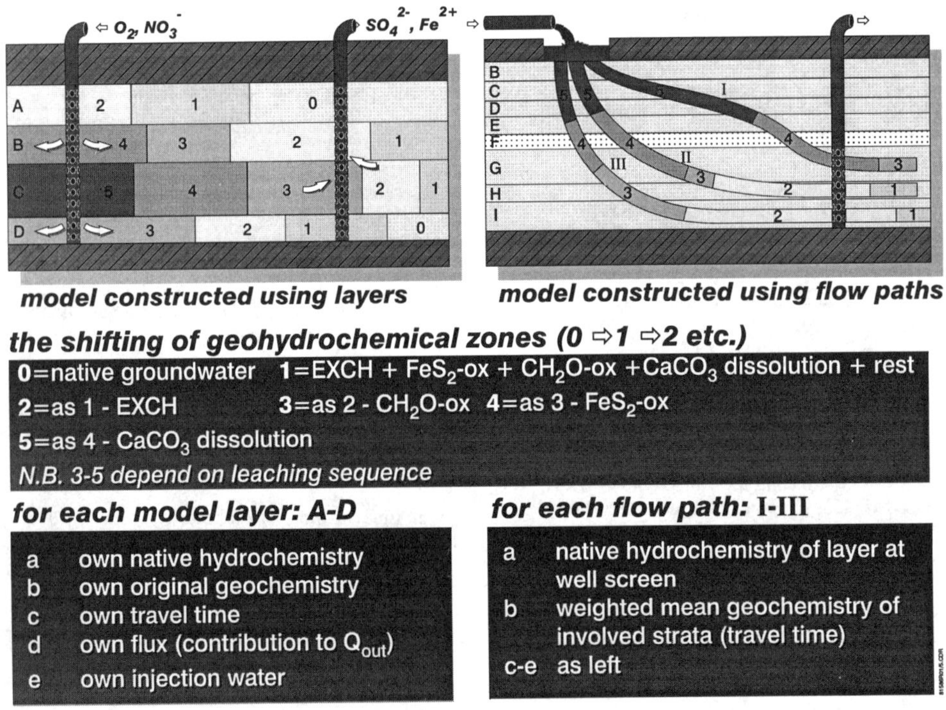

Figure 2. Essential features of EASY-LEACHER in case flow parallels the stratified medium (left) or crosses it (right).

Table 3. Some of the most important transport equations used in EASY-LEACHER.

The retardation and leach factor in general

Sorbing and oxidizing solutes are retarded during aquifer passage, and similarly desorbing and dissolving compounds are delayed by leaching. In the latter case raised concentrations drop to the influent level long after passage of the chloride - front. These delays are quantified by respectively the retardation factor R_i and leach factor L_i:

$$R_i = \frac{t_i}{t_{H2O}} \quad (7.1) \qquad L_i = \frac{t_L}{t_{H2O}} \quad (7.2)$$

with: t_i = time for 90% breakthrough of solute "i" [days]; t_L = time required for 90% leaching or till equilibrium is attained with the injection water [days]; t_{H2O} = mean travel time for water or a conservative tracer like chloride [days].

For practical reasons 90 in stead of 100% breakthrough or leaching has been chosen.

The general formula for a stationary and fast leaching of a homogeneously distributed reactive phase in the aquifer, is:

$$L = 1 + \frac{\rho_b(solid)}{n(reac)r_r} \quad (8)$$

with: (solid) = content of reactive phase in aquifer [mmol/kg dry weight]; (reac) = concentration of reactant in flushing fluid [mmol/L]; r_r = reaction coefficient, i.e. number of mmoles of solid phase leached by 1 mmol of reactant [-].

In stead of the concentration of the reactant (reac) and its reaction coefficient r_r, also the concentration of a reaction product (prod) and its reaction coefficient r_p can be taken, so that Eq.8 becomes:

$$L = 1 + \frac{\rho_b(solid)}{n(prod)r_p} \quad (9)$$

Of course, (reac) or (prod) should have no other sinks or sources, unless these can be properly accounted for.

Cation exchange

The period of cation exchange reactions (here Base EXchange) can be estimated by:

$$R_{BEX} = L_{BEX} = 1 + \frac{\rho_b CEC}{n\Sigma c} \quad (10)$$

with: CEC = cation exchange capacity [meq/kg d.w.]; Σc = sum of cations in flushing liquid, the influent [meq/L].

For the individual, adsorbed cations like Fe^{2+}, Mn^{2+} and NH_4^+ the following approximation holds:

$$L_{X\text{-}EXCH} = 1 + \frac{\rho_b X_{EXCH}}{0.75 n X_O} \quad (11)$$

with: X_{EXCH} = content of X in cation exchange complex [meq/kg d.w.]; X_O = concentration of X in native groundwater prior to displacement [meq/L]. The factor 0.75 is empirical and assumes the water recharged to be without X.

Pyrite oxidation

The leach factor for pyrite (FeS_2) is the result of oxidation by O_2 and/or NO_3^-:

$$L_{FeS2} = 1 + \frac{\rho_b(FeS_2)}{0.5n(\Delta SO_4)} \quad (12)$$

with: (FeS_2) = pyrite content [mmol/kg d.s.]; ΔSO_4 = sulphate-increase due to pyrite oxidation [mmol/L].

4 INFOMI APPLIED TO RADIONUCLIDES

4.1 *Definition of the problem*

The behaviour of ^{90}Sr, ^{60}Co, ^{106}Ru and ^{137}Cs during dune sand passage was modelled after basin recharge without interfering sludge layer. Hypothetical, relatively short pulses (of 1-28 days) with 100% activity were chosen as input signal, in order to simulate a calamity at the recharge facility of The Hague (using spreading basins).

System parameters for the aquifer were:

n = 0.38, ρ_s = 2.65 [kg/L], α_L = 1 m; D = 4 10^{-5} m^2/d; CEC = 20 meq/kg; Ca_{EXCH} = 17 meq/kg (β_{Ca} = 0.85); length of flow path = 70 m; cell size = 1 m; v_{H2O} = 1 m/d; Ca-infiltrate = 2 mmol/L.

Characteristics of the radionuclides are (partly after Haberer 1989): $K_{Co/Ca}$ = 0.67; $K_{Sr/Ca}$ = 1.14; $K_{Ru/Ca}$ = 1.14; $K_{Cs/Ca}$ = 3.43;

$(T½)_{Co}$ = 1925 d; $(T½)_{Sr}$ = 10483 d; $(T½)_{Ru}$ = 368 d; $(T½)_{Cs}$ = 10958 d.

4.2 *Strontium-90*

The propagation of the ^{90}Sr-pulses along a representative 70 m long flow path in dune sand is shown in Fig.3. The duration of the calamity has a strong impact on the final radio-activity along the flow path. Due to a relatively slow radio-active decay and little retardation, strontium activities remain significant in the recovery well. Radiostrontium breaks through there after about 2 years and it takes 2-2.5 years to flush it from the aquifer. A maximum activity of 0.5% is calculated for a pulse of ½ week and of 6% for a pulse of 4 weeks. Mixing in the recovery well is expected to reduce the activities by about 50%.

As radiostrontium is hardly removed by post-treatment, radiostrontium poses a health hazard which requires a save fluvial alarm system and cautious selective intake system.

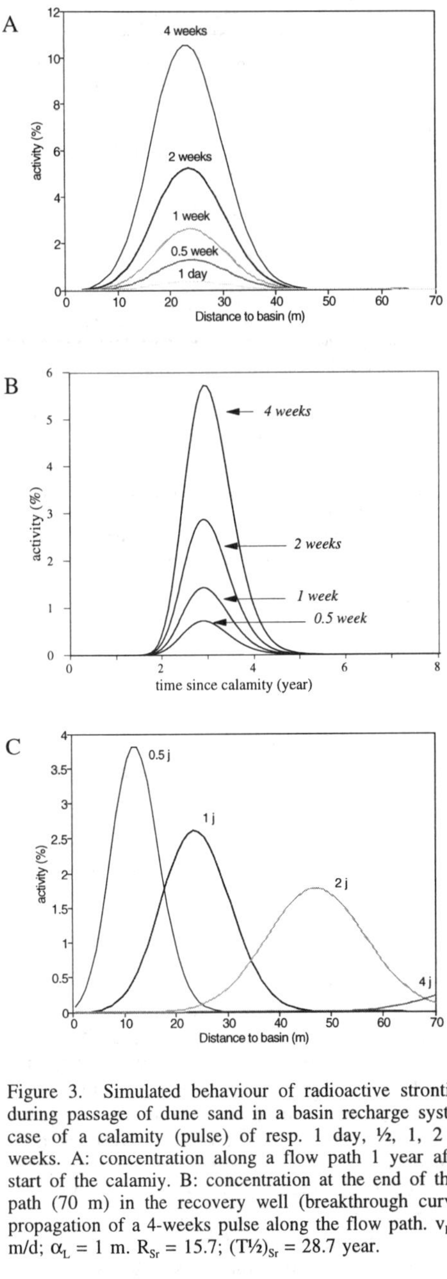

Figure 3. Simulated behaviour of radioactive strontium-90 during passage of dune sand in a basin recharge system, in case of a calamity (pulse) of resp. 1 day, ½, 1, 2 and 4 weeks. A: concentration along a flow path 1 year after the start of the calamiy. B: concentration at the end of the flow path (70 m) in the recovery well (breakthrough curve). C: propagation of a 4-weeks pulse along the flow path. v_{H2O} = 1 m/d; α_L = 1 m. R_{Sr} = 15.7; $(T½)_{Sr}$ = 28.7 year.

4.3 ^{60}Co, ^{90}Sr, ^{106}Ru and ^{137}Cs

The breakthrough of a combined ^{60}Co-, ^{90}Sr-, ^{106}Ru- and ^{137}Cs-puls of 1 week at the end of a 70 m long flow path in dune sand is shown in Fig.4. Differences in retardation clearly lead to peak separation: ^{60}Co (R = 6.1) arrives first after 0.5 year, after about 2 years ^{90}Sr and ^{106}Ru (R = 15.7) follow, and ultimately after 10 years ^{137}Cs (R = 133.8) appears. Again, 100% was the starting activity for each radionuclide upon infiltration. Filtration and precipitation were not considered, which makes the approach for cobalt, ruthenium and especially cesium a true worst-case scenario. Strontium, however, is hardly subject to these processes so that its results may approach reality.

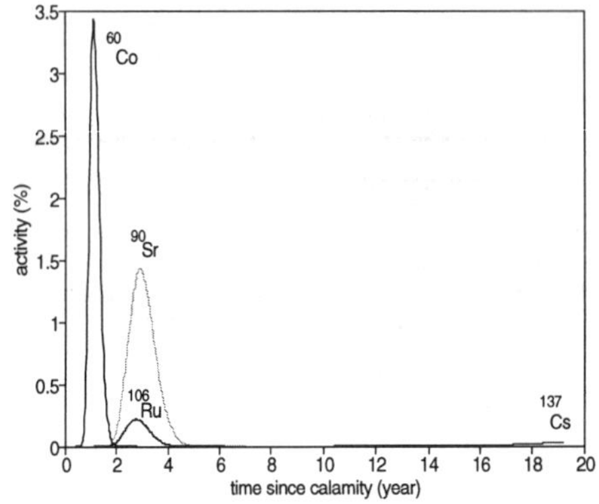

Figure 4. Breakthrough curves for radio-active cobalt-60, strontium-90, ruthenium-106 and cesium-137 after 70 m passage of dune sand in a basin recharge facility, in case of a calamity (pulse) of 1 week. v_{H2O} = 1 m/d; α_L = 1 m. R_{Co} = 6.1; R_{Sr} = R_{Ru} = 15.7; R_{Cs} = 133.8; $(T½)_{Co}$ = 1925 d; $(T½)_{Sr}$ = 10483 d; $(T½)_{Ru}$ = 368 d; $(T½)_{Cs}$ = 10958 d.

5 INFOMI APPLIED TO TRIHALOMETHANES

5.1 *Definition of the problem*

The behaviour of chloroform ($CHCl_3$), bromodichloromethane ($CHCl_2Br$), dibromochloromethane ($CHBr_2Cl$) and bromoform ($CHBr_3$) was simulated for a basin recharge system. The system implies subsequently: (a) 2 days detention in a recharge basin (volatilization and biodegradation in oxic environment); and (b) 70 m transport in dune sand (convective transport, dispersion, sorption and degradation in (sub)oxic environment).

The 2 years input signal is composed of a pulse during the period April - October (214 days), when chlorination for transport occurred. Peak and background levels were resp. 9 and 1 µg/l for chloroform, 10 and 0.5 µg/l for $CHCl_2Br$, 6 and 0.4 µg/l for $CHBr_2Cl$ and 2 and 0.1 µg/l for bromoform.

System parameters for the recharge basin are: residence time = 2 d; water temperature = 17°C; water depth = 1.5 m. System parameters for the sand aquifer are: n = 0.38, ρ_s = 2.65 [kg/L], α_L = 1 m; D = 4 · 10^{-5} m^2/d; f_{OC} = 0.001; TOC-infiltration water = 4.2 mg/l; (DOC/TOC)-infiltration water = 0.85; DOC-infiltrate = 2.8 mg/l; length of flow path = 70 m; cell size = 1 m; v_{H2O} = 1 m/d.

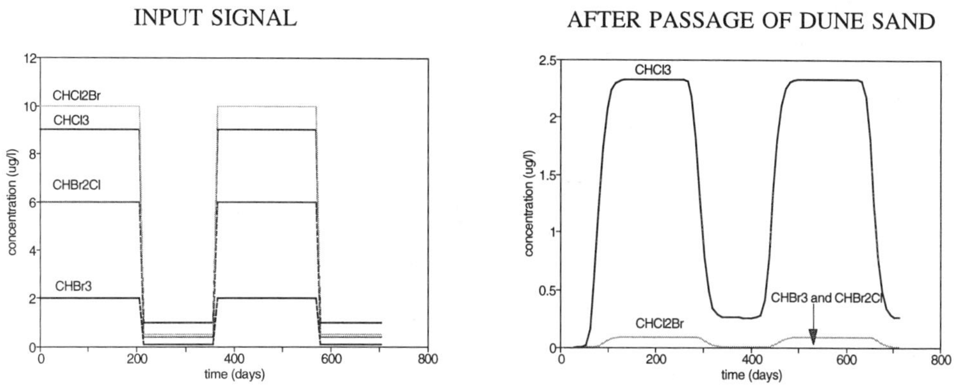

Figure 5. Input signal for trihalomethanes in pretreated water, and the calculated concentrations after a 2 days detention in the recharge basin and after passage of 70 m of dune sand. The input signal contains recurring pulses of 7 months due to transport chlorination in the period April - October. v_{H2O} = 1 m/d; α_L = 1 m. Concentration in the basin is lowered by about 50% for each THM. In dune sand: R_{CHCl3} = 1.17; $R_{CHCl2Br}$ = 1.5; $R_{CHBr2Cl}$ = 1.77; R_{CHBr3} = 2.19.

Characteristics of the OMPs are:
for chloroform: K_H = 304 Pa m^3/mol; $(T½)_b$ = 80 d; log K_{OC} = 1.6; M = 119.4 g/mol.
for $CHCl_2Br$: K_H = 162 Pa m^3/mol; $(T½)_b$ = 17 d; log K_{OC} = 2.07; M = 163.8 g/mol.
for $CHBr_2Cl$: K_H = 88 Pa m^3/mol; $(T½)_b$ = 4.5 d; log K_{OC} = 2.26; M = 208.3 g/mol.
for bromoform: K_H = 44 Pa m^3/mol; $(T½)_b$ = 3 d; log K_{OC} = 2.44; M = 252.7 g/mol.

5.2 *Results*

A 2 days detention in a recharge basin leads to a reduction of each THM by about 50%, mainly due to volatilization and little by biodegradation. The passage of 70 m of dune sand in (sub)oxic environment raises the retardation and degradation in the following order: chloroform < dichlorobromomethane < dibromochloromethane < bromoform. Only chloroform and some dichlorobromomethane 'survive' the aquifer passage (Fig.5). These calculations correspond well with observations (Stuyfzand 1989).

6 EASY-LEACHER APPLIED TO DEEP WELL INJECTION

6.1 *Definition of the problem*

The water quality evolution in the recovery well of deep well injection experiment number 11 (Stuyfzand 1998) was simulated using EASY-LEACHER. Pretreated oxic water from the Zuid-Willemsvaart (a canal containing mainly river Meuse water) is injected into an anoxic Miocene sandy aquifer. Mean characteristics of the experiment are listed in Tables 1-2 of Stuyfzand (1998). The aquifer is schematized by 5 different horizontal strata (A-E), each with a different travel time, flux (contribution to the recovery well) and geochemistry (Table 4), and minor differences in native hydrochemistry (not shown). During the first year of the experiment the injection water reached the recovery well (at 98 m) in layers D and E only.
The most important reactions of the injected water with the aquifer matrix were, in decreasing order (Stuyfzand 1997): (a) pyrite oxidation in layers A-C exclusively by O_2, and in layers D-E by both O_2 and NO_3^-; (b) acid buffering by the injected HCO_3^- (converting to CO_2); (c) oxidation of labile organic material by nitrate (more in A-C than in D-E); (d) cation exchange, with especially Ca^{2+} and Na^+ expelling adsorbed Fe^{2+}, Mn^{2+} and NH_4+; (e) desorption of SiO_2; and (f) adsorption of F^-. These reactions were included in EASY-LEACHER.

6.2 *Results*

Some results of modelling the recharge experiment for 100 years are shown in Fig.6. Although the experiment had been running for only 1 year during our evaluation, measurements and model predictions can be compared over a much longer 'apparent' period, because the amount of pore flushes (similar to bed volumes) in monitoring wells closer to the injection well approach the amount of pore flushes after 100 years for the recovery well. In fact, there was satisfactory agreement (Stuyfzand 1997).
From Fig.6 some conclusions of practical interest were drawn. First, oxygen and nitrate are not ex-

pected to reach the recovery well within resp. 100 and 20 years, which reduces the risk on clogging of the recovery well by $Fe(OH)_3$ and bacteria due to mixing of aerobic and anaerobic water in and around the well screen. Second, a future water softening in the pretreatment is not expected to be counteracted by dissolution of $CaCO_3$, but a significant lowering of HCO_3^- should be avoided because this would reduce the internal buffering of acid produced by pyrite oxidation (and then lead to the mobilization of metals).

Table 4. Geohydrological and geochemical stratification of the Miocene target aquifer of deep well experiment 11 near Someren, the Netherlands.

model layer	unit	A	B	C	D	E
GEOHYDROLOGY						
depth top	m-MSL	248	253	266	270	280
depth base	m-MSL	253	266	270	280	287
permeability	m/d	4	12.5	5	17	25
travel time	d	1250	500	1000	300	200
flux#	%	3.2	27	3.2	27.8	28.8
GEOCHEMISTRY						
Fe_{EXCH}	meq/kg	1.61	0.26	1.12	0.15	0.06
Mn_{EXCH}	meq/kg	0.10	0.02	0.06	0.01	0.005
$NH4_{EXCH}$	meq/kg	0.04	0.02	0.03	0.02	0.01
$CaCO_3$	mmol/kg	0	0	0	0	0
C-organic	mmol/kg	1075	458	1150	167	113
FeS_2	mmol/kg	55.2	10.2	25.5	12.6	8.6

= permanent admixing of authochtonous groundwater is 10%

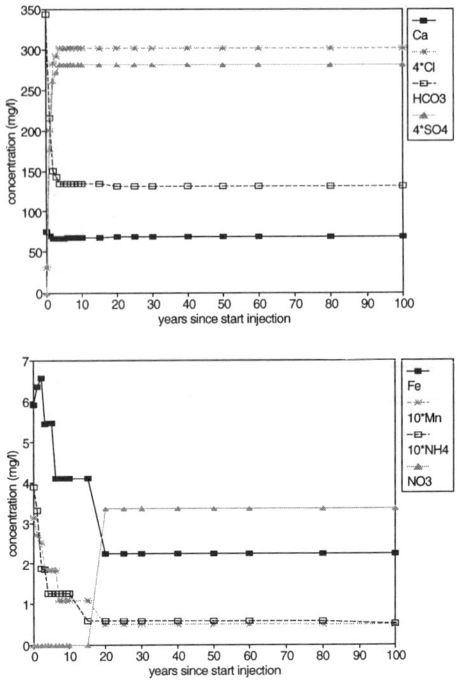

Figure 6. Quality development in the groundwater recovered at 98 m from the injection well of experiment 11, as predicted by EASY-LEACHER. Important input data for the model are listed in Table 4.

REFERENCES

Appelo, C.A.J. & D. Postma 1993. *Geochemistry, groundwater and pollution.* Balkema, 536p.

Brun, A., F.D. Christensen, J.S. Christiansen, P.J. Stuyfzand & H. Timmer 1998. Water quality modelling at the Langerak deep-well recharge site. This volume.

Dijkhuis, L.G. & P.J. Stuyfzand 1996. INFOMI 3.1: a 1-D transport model for micropollutants in a water course and in the underground after infiltration. Kiwa-report SWI 96.219 (in dutch).

Griffioen, J., A.L. Lourens, Ch.B.M. te Stroet, B. Minnema, M.P. Laeven, P.J. Stuyfzand, C.G.-E.M. van Beek & W. Beekman 1998. An integrated groundwater quality model based on hydrogeochemical environments. This volume.

Haberer, K. 1989. *Umweltradioaktivität und Trinkwasserversorgung.* R. Oldenbourg Verlag, München, 142p.

Kan, A.T. & M.B. Thomson 1990. Ground water transport of hydrophobic organic compounds in the presence of dissolved organic matter. *Env. Toxicol. and Chem.* 9, 253-263.

MacKay, D. & W.Y. Shiu 1981. *J. Phys. Chem. Ref. Data* 10, 1175-1199.

Schellenberg, K., C. Leuenberger & R.P. Schwarzenbach 1984. Sorption of chlorinated phenols by natural sediments and aquifer materials. *Environ. Sci. Technol.* 18.

Southworth G.S. 1979. The role of volatilization in removing PAH from aquatic environments. *Bull. Environ. Contam. Toxicol.* 21, 507-514.

Stuyfzand, P.J. 1997. Results of the first year of deep well recharge DIZON; quality changes upon aquifer passage. Kiwa-report (in dutch), 76p.

Stuyfzand, P.J. 1998. Quality changes upon injection into anoxic aquifers in the Netherlands: evaluation of 11 experiments. This volume.

Stuyfzand, P.J. in prep. Manual of EASY-LEACHER, a simple model for reactive transport of main constituents and pollutants, especially for artificial recharge and bank filtration. Kiwa-report, in prep.

Stuyfzand, P.J. & F. Lüers 1996. Behaviour of environmental pollutants upon bank filtration and artificial recharge; effects of aquifer passage as measured along flow paths. Kiwa-Meded. 125 (in dutch), 272p.

2.6 Clogging of recharge systems

Artificial Recharge of Groundwater, Peters et al. (eds) © 1998 Balkema, Rotterdam, ISBN 90 5809 017 5

Determining the clogging potential of water used for artificial recharge in deep sandy aquifers

W.A.M. Hijnen, J. Bunnik & J.C. Schippers
Kiwa Research and Consultancy, Nieuwegein, Netherlands

R. Straatman
Water Supply Company Oost-Brabant, 's-Hertogenbosch, Netherlands

H.C. Folmer
Water Supply Company of North-Holland, Bloemendaal, Netherlands

ABSTRACT: A set of tools for the determination of the clogging potential of water used for artificial recharge in deep sandy aquifers was used at two different locations. The results revealed that with the Parallel Filtration Device (PFD) the prediction of the clogging rate in a recharge well was not possible. MFI and the AOC-concentration of water proved to be valuable parameters to estimate the clogging potential of water. In combination with the PFD using a standard filter medium the clogging potential of water can be assessed more accurately. Further research is necessary to elucidate the relationship between the clogging in recharge wells and in the PFD columns with a suitable standard filter medium.

1 INTRODUCTION

The interest in the use of artificial recharge of pretreated surface water in deep sandy aquifers in the Netherlands for drinking water production has increased. It is an effective way of removing pathogenic micro-organisms and of replenishing the ground water resources. To obtain an economically interesting operating process clogging of the recharge well should be avoided as much as possible. For assessment of the clogging potential of water used for artificial recharge in deep sandy aquifers a set of tools was developed.

2 WATER QUALITY MONITORING

Clogging of recharge wells is the result of deposition of suspended matter and bacterial growth. The Modified Fouling Index (MFI; Schippers & Verdouw 1980) and the assimilable organic carbon content (AOC, Van der Kooij et al. 1982, Hijnen & Van der Kooij 1992) in the water proved to be valuable parameters in practice for both clogging phenomena (Table 1). Guide line values for water used for artificial recharge in deep sandy aquifers were defined to limit the clogging of recharge wells.

Table 1. Water quality parameters

Parameters	Clogging process; guide line value
MFI	Suspended matter; <3 s/l^2
AOC	Bacterial growth; <10 µg ac-C eq/l
BFP	Biofilm formation; not yet defined
IAR	Iron accumulation rate; not yet defined

A new and valuable tool is the biofilm monitor developed for the determination of the biofilm formation potential (BFP, Van der Kooij & Veenendaal 1992). With the biofilm monitor important fouling parameters as the iron accumulation rate of water can also be assessed (Van der Kooij et al. 1994). Guidelines values for the clogging potential of water have not yet been defined.

3 MONITORING THE CLOGGING PROCESS

From the MFI and AOC of water the clogging rate of a recharge well can not be predicted. For this purpose an experimental filtration device was developed, the Parallel Filtration Device (PFD, Schippers et al. 1995). The PFD consists of sand columns supplied with the surface water pretreated before infiltration

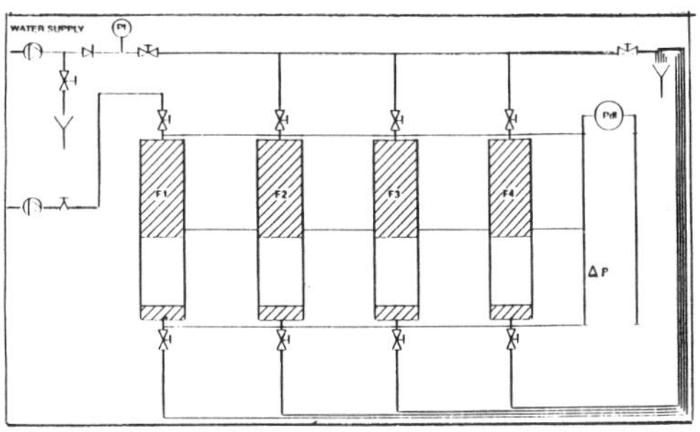

Figure 1. The Parallel Filtration Device (PFD) for simulating and measuring the clogging rate of a recharge well

(Fig. 1). To simulate the clogging process in the aquifer the sand columns are filled with material derived from the aquifer of the borehole or composed sand material with the same hydraulic characters (grain size distribution) as this soil material. The head loss over the sand column is measured during operation at a constant filtration rate (flow rate control). One column is operated with the filtration rate as occuring on the borehole of the recharge well. The other columns are operated at a higher flow rate. The clogging process in the columns is described using an appropriate filtration model (Schippers et al. 1995) and from the estimated filtration constant the clogging rate in the aquifer is predicted. For a reliable prediction the water quality should be constant and the sand columns must have the same hydraulic characteristics as the sandy aquifer that is recharged.

4 TO PREDICT THE CLOGGING RATE OF A DEEP SANDY AQUIFER

At a demonstration plant (30-40 m^3/h) in the south-east of the Netherlands the Water Supply Oost Brabant Ltd. and Water Supply Limburg Ltd. started a research with a recharge well. The project is known as DIZON, the Dutch acronym for Well Recharge in South East Netherlands. Surface water pretreated successively with micro-straining, coagulation/filtration, granular activated carbon filtration (GAC) and slow sand filtration (SSF) was infiltrated in a recharge well (Fig. 2) with an average flow rate of 0.5 m/h. The PFD was used to predict the clogging rate in the recharge well. The four columns were filled with river sand with a grain size distribution equal to the grain size distribution of the soil material as assumed

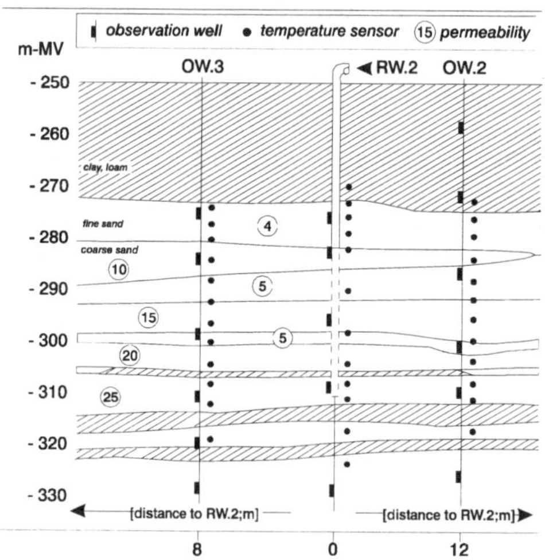

Figure 2. A cross-section of the underground and the sandy aquifer with the recharge well RW2. (permeability in m/day)

in the aquifer. Three columns were supplied with the SSF-filtrate at the flow rate of the aquifer (0.5 m/h; F1) and at a flow rate of 1 and 2 m/h (F2 and F4, respectively). Column F3 was supplied with the GAC-filtrate at a flow rate of 1 m/h. The head loss development in the recharge well and the PFD columns was monitored (kPa_{10} at 10°C; correction for the viscosity). For the recharge well the head loss was calculated from the water heads monitored in observation wells located in the gravel pack of the borehole and in the aquifer at 8 and 12 m distance from the borehole (Fig. 2).

The MFI and the AOC content of the SSF-filtrate were below the guide line values for infiltration water (Table 2). The average MFI of the GAC-filtrate however exceeded the guide line value significantly. Both parameters MFI and AOC in this water as well as the temperature (Fig. 3) showed a strong fluctuation.

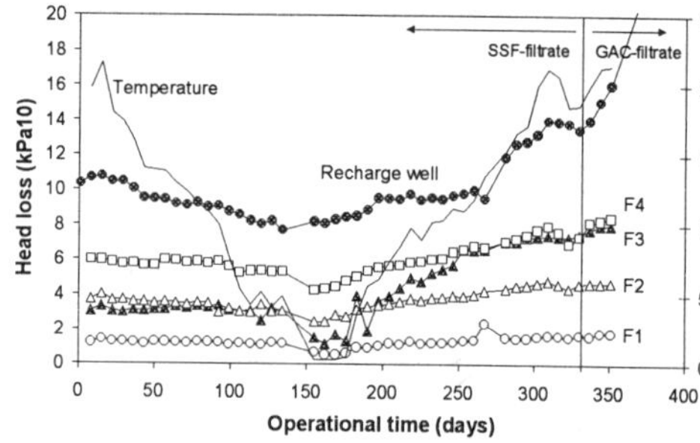

Figure 3. The head loss development in the recharge well (●) and the PFD columns F1 (0.5 m/h;○), F2 (1 m/h;Δ) and F4 (2 m/h;□) supplied with SSF-filtrate and after 340 days with GAC-filtrate; PFD column F3 (1 m/h; ▲) was supplied with GAC-filtrate from the start of the experiment

The results revealed that the clogging rate in the recharge well could not be predicted by the PFD; the columns showed a significant smaller clogging rate than observed in the recharge well (Fig. 3). In the first 140 days of operation the head loss in the recharge well decreased 3 kPa. In the following period the head loss in the well increased; first slowly but when the temperature of the water increased (Fig. 3) the clogging rate increased as well. In the PFD column

Table 2. MFI and AOC-concentrations in the SSF and GAC-filtrate of the pretreatment DIZON

		Average (±Std)	Min.	Max.
SSF-filtrate	MFI (s/l^2)	1.8 (±1)	1	5.2
	AOC (mg/l)	6.7 (±1.5)	4.3	9.2
GAC-filtrate	MFI (s/l^2)	7.3 (±6.6)	0.5	25.6
	AOC (mg/l)	9.5 (±5.7)	4.8	32

with a filtration rate of 0.5 m/h (as in the recharge well) hardly any clogging was observed. Even the clogging rate in the PFD columns with a flow rate of 1 and 2 m/h was not as fast as the clogging rate in the recharge well.
The permeability (k-value) of the sand in the PFD columns, calculated from the initial head loss over the 0.2 m of sand, ranged from 13 to 26 m/d and was somewhat higher than the k-values observed in the direct surrounding of the recharge well (Fig. 2). The heterogeneity of the sandy aquifer in the surrounding of the recharge well might also explain the dissimilarity of clogging in the PFD columns and in the well.
There was a significant effect of an additional treatment with the SSF on the clogging potential of the water. During the short period that the recharge well was supplied with GAC-filtrate (higher MFI and AOC, Table 2) the head loss increased rapidly to 44 kPa_{10} and the well was regenerated (Fig. 3). Also in the PFD column with a flow rate of 1 m/h supplied with the GAC filtrate (F3) the clogging rate was significantly higher compared to column F2 supplied with SSF-filtrate.

5 DETERMINING THE CLOGGING POTENTIAL OF WATER

The PFD was used by Water Supply of North-Holland Ltd. at location Andijk for filtration experiments with pretreated surface water to be used for artificial recharge. The pretreatment consisted of a full scale coagulation/filtration process (COA/FIL) followed by a pilot plant with advanced oxidation (ozone and peroxide; AOP) and a postfiltration either filled with anthracite-sand (ASF) or with GAC. Oxidation of water (AOP) is used for the removal of persistent micro-pollutants. It results in an increase of the AOC-concentration in the water (Table 3). The objective of the study was to verify the influence of the increase of the AOC concentration after oxidation on the clogging potential of the water. The water after each process of the treatment was filtered through 2 sand (Ø 0.4-0.62 mm) columns of the PFD with a constant flow rate of 5 and 1 m/h (low and high flux, respectively). The permeability of the sand column (k-value) was calculated at 120 m/d. The development of the head loss in the sand columns was monitored (kPa_{10} at 10°C). The temperature of the water varied between 5 and 15°C and the range of AOC and MFI values are summarized in Table 3.
The results of the PFD experiments were compared with the results of a laboratory experiment with the PFD operated at similar conditions as described above (filter medium and flow rate). These PFD columns were supplied with drinking water with a low MFI-value (Table 3) supplemented with 25 µg/l acetate-C as a model substrate for AOC. The temperature was 15°C.
The following results were observed:
- The clogging potential of the water treated otherwise could be discriminated by the PFD (Fig. 4a/b).
- The enhanced effect of a high MFI on the clogging rate was clearly demonstrated in the PFD culumns supplied with the water after COA/FIL. Both MFI and AOC-values were above the guide line values (Table 3). The clogging rate in these PFD columns was higher compared to the clogging rate in the PFD columns supplied with drinking water with a low MFI and a higher AOC-concentration (25 µg/l acetate-C/l; Fig. 4).
- The role of biological processes in the clogging process was demonstrated by the retarding effect of (1) the remaining H_2O_2 in the water after the AOP-process and (2) the low temperature (Fig. 4). The AOC concentration in the water was significantly increased by the AOP-process, but the clogging potential was decreased due to the presence of H_2O_2 (Table 3). The effect of H_2O_2 on the head loss was clearly demonstrated in the column with low flux supplied with AOP water; the head loss decreased after the H_2O_2 concentration was raised (Fig. 4b).

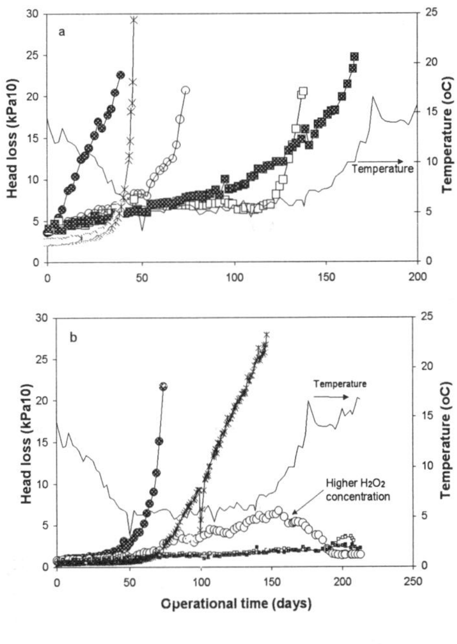

Figure 4. The head loss development in the PFD-columns with a standard filter medium (Flow rate: a = 5 m/h and b = 1 m/h) supplied with water after COA/FIL (●), AOP (O), ASF-filter (□) and GAC-filter (■) of the pilot plant Andijk and of the PFD-columns of a laboratory experiment (X) with drinking water supplemented with 25 µg acetate-C/l

Table 3. MFI, AOC and H_2O_2 concentrations in the water after the different treatment stages of the pilot plant Andijk

Water after Process	AOC-range (µg ac-C eq/l)	MFI (s/l^2)	H_2O_2-range (mg/l)
COA/FIL	5-22	6.5	0
AOP	120-223	8.0	7.5-11
ASF	42-129	nd	0.1-5.4
GAC	30-40	4.2	<0.02
Drinking water[a]	25[b]	<2	0

[a]laboratory experiment; [b]acetate

The ASF-filtrate also contained H_2O_2 (Table 3) which most likely retarded the clogging of the PFD-columns.

- AOC and MFI of the GAC-filtrate was higher compared to the water of the laboratory experiment with 25 µg acetate-C/l (Table 3). Nevertheless, the clogging rate in both PFD-columns supplied with the GAC-filtrate was smaller. From these results it was concluded that the rest AOC-fraction after oxidation and filtration was mainly composed of organic compounds which were less easily biodegradable than acetate used as a model substrate in the laboratory experiment.
- In the upper part of the PFD columns supplied with the water after COA/FIL and with the GAC-filtrate deposits of iron and carbon fines, respectively were observed.

6 CONCLUSIONS AND DISCUSSION

On two different locations the Parallel Filtration Device (PFD) was used to determine the clogging potential of pretreated surface water to be used for artificial recharge in deep sandy aquifers. These studies revealed the following conclusions:

- Prediction of the clogging rate of a recharge well with the Parallel Filtration Device (PFD) is difficult. Variations in the water quality and dissimilarity in the hydraulic characteristics of the sand in the aquifer and in the PFD columns can not be avoided.
- The MFI and the AOC-concentration of water are valuable parameters to estimate the clogging potential of water.
- The clogging potential of water can be assessed more accurately when MFI and AOC are used in combination with the PFD using a standard filter medium. Sand with a grain size distribution of 0.42-0.6 mm used in the described study probably is too coarse (high k-value). A finer and less uniform filter medium corresponds more with the sand in deep sandy aquifers. With analytical procedures the deposits in the sand of the PFD column can be characterized at the end of the operational time and the main cause of clogging can be determined.
- Further research is necessary to select a filter medium and to elucidate the relationship between the clogging process in these PFD columns and in recharge wells.

REFERENCES

Hijnen, W.A.M. & D. van der Kooij 1992. The effect of low concentrations of assimilable organic carbon (AOC) in water on biological clogging of sand beds. *Wat. Res.* 26:963-972.

Schippers, J.C. & J. Verdouw 1980. The Modified Fouling Index, a method of determining the fouling characteristics of water. *Desalination* 32:137-148.

Schippers, J.C., J. Verdouw & G.J. Zweere 1995. Predicting the clogging rate of artificial recharge wells. *J. Water SRT-Aqua* 44, 1:18-28.

Van der Kooij, D., A. Visser & W.A.M. Hijnen 1982. Determining the concentration of easily assimilable organic carbon in drinking water. *J.Am. Wat. Wks Ass.* 74:540-545.

Van der Kooij, D. & H.R. Veenendaal 1992. Assessment of the biofilm formation characteristics of drinking water. *Proceedings AWWA Water Quality Technology Conference*, Toronto, November 15-19, 1992.

Van der Kooij, D., J.S. Vrouwenvelder, H.R. Veenendaal & M.J.C. van Raalte-Drewes 1994. Multiplication of Aeromonads in ground-water supplies in relation with the biofilm formation characteristics of drinking water. *Proceedings AWWA Water Quality Technology Conference,* San Francisco, November 6-10, 1994.

Artificial Recharge of Groundwater, Peters et al. (eds) © 1998 Balkema, Rotterdam, ISBN 90 5809 017 5

Operational guidelines regarding clogging

A. Pérez-Paricio & J. Carrera
Universitat Politècnica de Catalunya, Barcelona, Spain

ABSTRACT: Clogging of artificial recharge systems is a ubiquitous problem. It consists of porosity or infiltration area diminution induced by inter-related physical, biological and chemical processes, resulting in a decrease of recharge effectiveness. Clogging is a highly site-dependent phenomenon. As a consequence, clogging prevention and redevelopment are commonly addressed by resorting to previously reported practices. However, serious reductions can occur if there is a lack of experience or if special problems exist. Based both on a thorough literature review and on field investigations from certain European countries, several recommendations and operating guidelines are presented. These include identification of basic parameters, preventive and re-development techniques, specifically defined clogging-tools and limitations on a few magnitudes. Such recommendations should represent a good starting point in order to prevent clogging, although it is evident that pilot field tests are always unavoidable when accurate clogging estimates are needed.

1 INTRODUCTION

The amount of references on clogging during Artificial Recharge operations is impressive. Interesting reviews exist (Olsthoorn 1982, Custodio 1986, Frycklund 1992, Dillon et al. 1995, Pérez-Paricio 1998). This is a natural consequence of the technological relevance of clogging. Clogging, as well as its obvious scientific impact (Figure 1), may give rise to significant economic losses.

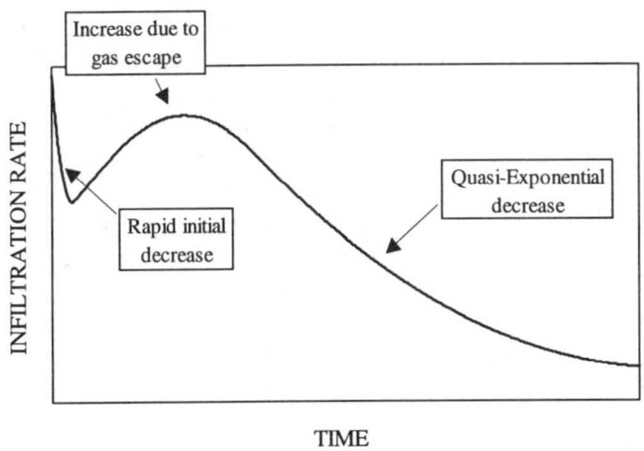

Figure 1. Qualitative time evolution of infiltration rate for a basin. The final asymptotic value is close to hydraulic conductivity at saturation in the absence of clogging, but can be very low (eventually zero) if clogging takes place.

In fact, Artificial Recharge can be performed by using three different facilities or systems:

1. Surface systems. These include constructed devices -such as basins, trenches, pits or artificial wetlands- and natural grounds (like extension grounds, flooding areas). They require large field sites to achieve good recharge yields, although recharge water quality requirements are relatively low as compared to deep systems.
2. Deep systems. Three types of wells can be distinguished: dry wells, normal wells and dual purpose or Aquifer Storage and Recovery (ASR) wells.
The former are only useful if very thick non-saturated zones exist, but are highly susceptible to clogging. ASR wells (Pyne 1995) have proven to be very effective to minimise clogging by adopting periodic pumping schemes, as commented below.
Deep systems do not require large areas, but are more expensive and must be carefully monitored to prevent irreversible clogging.
3. Bank filtration. It defines the intentional increase of natural recharge from surface reservoirs or rivers by pumping at nearby wells. Clogging can be very fast, due to the content of suspended solids, if natural flooding is not allowed (especially when upstream dams exert a tight control on stream flow).

Five clogging-related processes can be identified (Pérez-Paricio 1998): attachment/detachment of suspended particles ('physical clogging'), biological growths ('bioclogging'), precipitation/dissolution of minerals ('chemical clogging'), gas generation, and compaction of the clogging layer. In general terms, processes affecting Artificial Recharge systems are essentially identical. However, the importance of such processes can be different for each system. For instance, algal blooms are only relevant for surface systems and bank filtration. A general treatment is adopted here, while the particularities of recharge systems are expressly remarked when necessary.

It is well known that clogging is a highly site-dependent phenomenon. Climatic conditions, natural aquifer heterogeneity, chemistry of recharge water and groundwater, operational practices, among others, have shown to be very influential. Type of infiltration and construction (or drilling) must also be taken into account. Usual management techniques rely on reported experiences and on analogies with other field-sites. This proves to be enough in general terms, providing satisfactory results for most recharge facilities. Nevertheless, clogging is a complex phenomenon. While countries that possess a strong background on Artificial Recharge have come to excellent practical recommendations, the situation can be completely different when field examples are scarce -that is, most of the countries. Irremediable damage of recharge wells due to clogging is the extreme consequence.

An evident conclusion can be drawn from the former discussion: some general guidelines should be given to address clogging prevention and redevelopment, irrespective of the site-dependent factors. Based both on a thorough literature review and on experiences collected from certain European countries, this paper concentrates on the practical (operational) aspects of clogging.

The structure of the paper is as follows: first, some comments on the basic parameters for clogging are presented. Second, preventive and redevelopment techniques are briefly outlined. Third, a review of specific clogging tools, such as especial indexes and graphs, is done. Finally, operational guidelines or recommendations -to minimise the effects of clogging- follow.

Most of these sections are based on Pérez-Paricio (1998), whose work is a prelude of the Clogging Handbook that will result from the European Union joint Project entitled Artificial Recharge of Groundwater.

2 BASIC PARAMETERS

Apart from the basic physical parameters (temperature, pressure of both liquid and gas phases, viscosity and density of the liquid), it is possible to find some parameters in order to characterise clogging. This is presented here by associating a few parameters to each clogging process (Table 1).

Table 1. Clogging processes and associated parameters.

Physical	d_g, d_p, q
Biological	Bacteria: e^- donor & acceptor, nutrients, i Algae: P & N, light
Chemical	SI, a_m, catalysts
Gas generation	Bacteria & algae, design
Compaction	Water height

*T, P_l, P_g, μ_l, ρ_l are obviously included.
Note: see Notation at the end of the paper.

1. Physical clogging: mean grain (aquifer) size, suspended particles diameter, flow velocity.
2. Biological clogging. For bacteria: electron donors (mainly organic matter), electron acceptor (i.e. oxygen, nitrate, ferric ion) and nutrients (organic carbon). For algae: light, nutrients (phosphorous, nitrogen), infiltration rate.
3. Chemical clogging: saturation index, availability of catalysts, reactive surface area.
4. Gas generation: metabolic products of bacteria and algae, design of deep recharge systems.
5. Compaction: height of water in surface systems.

3 PREVENTION AND REDEVELOPMENT

Prevention is essential in order to reduce the negative effects of clogging, so that it is advisable to summarise which measures are most beneficial. However, after a certain recharge period -whose duration depends on design, water quality and environmental and hydrogeological conditions- it is often necessary to apply a redevelopment method. Both topics are addressed in this section.

3.1 *Prevention*

Needless to say, the best preventive tactics consists of ideal design of systems together with high water quality and optimum hydrogeological characteristics.

Actual recharge facilities are not ideal, though. This is why various measures must be adopted to maximise benefits. As shown in Table 2, some of these are strictly necessary to obtain a satisfactory groundwater quality, preserving it from pollution. Details about numerical constraints on certain parameters are presented in section 4.

Table 2 includes different possibilities, but they are not mutually excluding. In general, it has been observed that surface systems can maintain high yields if recharge cycles are applied, that is, if the clogging material is allowed to dry and crack naturally. This tends to break clogging deposits, as well as impairing algal growth and enhancing biodegradation within the non-saturated zone. As for deep systems, the best management option consists of

Table 2. Processes and Preventive measures.

Physical clogging	Flocculants / Reverse osmosis / Prior filtration / Backwashing Surface systems → Desiccation
Bioclogging	Bacteria: Disinfectants / Reduction of nutrients / Constraints Algae: Filtration / Reduction of nutrients / Chemicals / Prevent light
Chemical clogging	Acidification / Daily pumping / Separate strata / Avoid oxygen
Gas generation	Adequate design (valves, positive pressures) / Recharge water T larger than groundwater one
Compaction	Limit water height
Corrosion	Avoid: Work-hardened materials / Stressed parts / High T & q
Generic rules	1. Appropriate water treatment 2. Minimise aeration (wells) 3. Proper design is essential 4. Install high-water alarms

Table 3. Redevelopment techniques.

Surface systems	During runs:	1. Filtration: mats, filters 2. Chemicals (calcium) 3. Natural: aeration, no light 4. Others: underwater robot
	Between runs:	1. Mechanical methods: tillage, scrapping 2. Chemicals (calcium) 3. Natural: drying, freezing 4. Acid baths (deposits)
Deep systems	General methods	1. Pumping 2. Chemicals 3. Fluid-based methods 4. Protocols 5. Others (less frequent)
	Physical clogging	1. Pumping 2. Chemicals 3. Fluid-based methods 4. Protocols
	Bio-clogging	1. Pumping 2. Chemicals 3. Physically-based methods 4. Combinations
	Chemical clogging	1. Chemicals 2. Hydro-mechanical 3. TV surveys

periodic backflushing (reversing flow direction by pumping). In both cases, prefiltration can be very helpful. It is highly recommended to consult the work of Wegelin et al. (1991) on horizontal roughing filters, which are simple, effective and inexpensive.

3.2 *Redevelopment*

Prior to the occurrence of notable clogging impacts it is necessary to redevelop the recharge system. Redevelopment and prevention are not separable in many cases, as it is not feasible to distinguish between them. This is the case of management alternatives such as desiccation or periodic backflushing, which can be indistinctly viewed either as preventive or redeveloping measures.

Table 3 presents some redevelopment techniques (Olsthoorn 1982, Driscoll 1986) but, contrary to preventive measures, a distinction is made between surface and deep systems. The main reason to do this is based on the technological differences between them, rather than on conceptual dissimilarities.

In summary, it can be affirmed that the following methods are preferable (Pérez-Paricio 1998): polyphosphates for physical clogging, oxidants for bioclogging and chemicals for chemical clogging. However, these are only completely effective when combined with hydraulic methods, such as pumping, surging or juttering. This one consists of intermittent pumping at the damaged well, as denoted by Olsthoorn (1982).

4 USEFUL TOOLS

Given the significance of clogging, some authors have concentrated on generating applicable tools. These include specific parameters, relationships and generalised charts. The most interesting ones are briefly described below, insisting on their applicability range and main drawbacks.

4.1 *Specific parameters*

Conventional parameters, such as turbidity and concentration of suspended solids (physical clogging) or dissolved organic carbon (bioclogging) are certainly valid, but they are hardly interesting when recharging high quality water. Two parameters have been developed in the Netherlands:

1. Modified fouling index (MFI). It is proportional to the concentration of both colloidal and suspended matter, and is determined by membrane measurements. Physically, it represents the slope of inverse filtered flow versus cumulative filtered volume. Although it is used as a guideline in the Netherlands (assuming a maximum MFI value of 3 to 5 s/l^2 for well recharge in fine sand aquifers), some limitations arise: it is only reliable when clogging is strictly physical, and threshold values can only be established after doing field research, taking into account the aquifer characteristics.
2. Assimilable organic carbon (AOC). It was introduced to quantify the biostability of drinking water (not subject to disinfection). AOC represents the

fraction of organic carbon that can be rapidly assimilated by microorganisms. This is a problem, as it can over-simplify actual conditions.
In order to overcome these and other limitations, another parameter is currently under study (van der Kooij et al. 1995): the biofilm formation rate, or BFR. This accounts for all biomass in water in order to provide better results (Hijnen et al., this volume).

Despite all limitations, both MFI and AOC (or, alternatively, BFR) should be considered as excellent clogging-indicators. It is evident that prediction can not be adequately addressed (Schippers et al. 1995), but they can indicate whether clogging is to be relevant or not.

4.2 *Simple mathematical relationships*

1. Normalised build-up. Olsthoorn (1982) proposed a standardised equation to compare the clogging effects in recharge wells under different circumstances. This equation adopts several reference conditions: 20 °C water temperature, infiltration velocity of 1 m/h, injection time of 1 year.
2. Simple filtration formulae. Different variations on Darcy's equation are available to express the time evolution of filtration rate, where a given time-dependence for hydraulic resistance (inverse of conductivity) must be adopted.
3. Parallel filters. Due to the problems associated to previous parameters, a new approach applies simple semi-empirical equations to data recorded during laboratory experiments with parallel filters. These are columns filled-in with aquifer material, dosing recharge water under field-like hydraulic conditions (Schippers et al. 1995).

Parallel filters can provide very valuable estimates of clogging evolution, but it is difficult to extrapolate laboratory conclusions to field conditions. This is especially questionable when natural heterogeneity in the aquifer is high, what is very usual.

4.3 *Especial charts*

We will only refer here to the Hutchinson's plot (Hutchinson & Randall 1995), because it is particularly important. It incorporates three different magnitudes, each one related to basic properties of a recharge system: hydraulic conductivity of the aquifer, concentration of suspended particles and clogging (normalised) build-up. Therefore, at least three essential features are considered.

Figure 2 presents an example from Hutchinson & Randall. It is worth remarking that some cases (see the *Garfield Well*) exhibit anomalous behaviours. This is a consequence, for instance, of the interaction of other clogging sources, namely biochemical reactions and related processes.

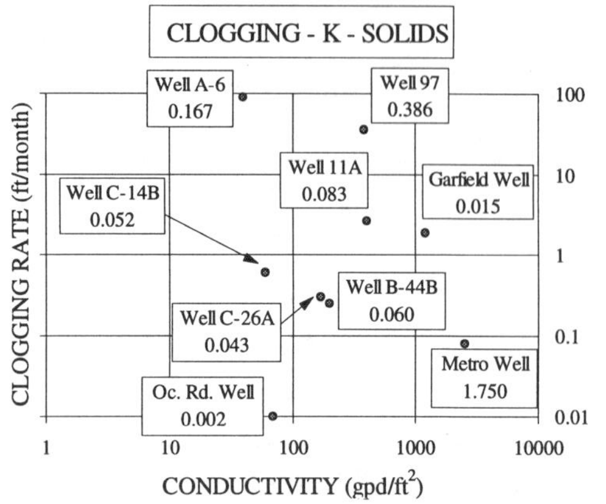

Figure 2. Hutchinson's graph: normalised build-up, hydraulic conductivity and suspended solids (mg/l). Note that only physical clogging is considered.

Table 4. Recommended values for basic parameters.

CONCEPT	LIMITATIONS	COMMENTS
Physical clogging	Deep systems: [9][TSS] < 2 mg/l [1,9][TSS] < 0.1 mg/l [2]MFI < 3 to 5 s/l2 [1,3]Tur < 1 NTU Surface systems: [TSS] < 10 mg/l Tur < 5 NTU	[9]K > 40 m/d [9]4 m/d<K<40 m/d
Biological clogging	Deep systems: [4]pH > 7.2 (Fe bacteria) [4]Avoid CO_2 [4]Eh > 10 mV [10]DOC < 2 mg/l Surface systems: TOC < 10 mg/l	[10]Swiss standards
Chemical clogging	Deep systems: [6]$[Fe^{2+}]_{GW}$ < 11.2 mg/l [7]Saturation index [8]Ryznar Index > 7 [8]pH < 7.5 [8]Low $[Ca^{2+}]$ & $[Mg^{2+}]$ [8]TDS < 150 mg/l	More important for abstraction wells than for recharge ones. Saturation indexes are always essential.
Corrosion	[8][Cl-] < 500 mg/l [8]DO < 2 mg/l [8]Ryznar Index < 9 [8]pH > 7 [8]CO_2 < 50 mg/l [8][TSS] < 1000 mg/l [8]Avoid H_2S	Applicable to wells and pipes

[1]The Netherlands [2]Olsthoorn [3]Germany [4]Degalier [5]van der Kooij&Hijnen [6]Lucas et al. [7]Hills et al. [8]Ford [9]Pyne [10]Hurni
See Pérez-Paricio (1998) for non-included references

5 RECOMMENDATIONS / GUIDELINES

Recommendations to minimise clogging follow. Values coming from the literature or personal communications are expressly indicated, while the rest have been suggested by the present work.

Table 4 displays a set of threshold values, where differences between surface and deep systems are explicitly remarked. Notice that some of these values are rather exigent, especially for recharge wells. However, they can be relaxed if other conditions apply (limitations used in the Netherlands are very tight, as there are many fine sand aquifers).

Finally, Table 5 shows simplified criteria to determine the extent of clogging based on a few simple parameters. Such criteria are only indicative, as not all basic magnitudes are included (light, temperature, other nutrients, hydraulic conductivity, recharge regime). However, suggested values should be accounted for to improve management.

Table 5. Guidelines for clogging, based on simple parameters.

	CLOGGING	RECHARGE WATER	REDEVELOPMENT
SURFACE SYSTEMS	Slight	TSS < 10 mg/l Tur < 5 NTU TOC < 10 mg/l	Natural drying+cracking Once a year: mechanical
	Notable	10 mg/l < TSS < 20 mg/l 10 mg/l < TOC < 25 mg/l	Frequent drying+cracking Twice a year: mechanical
	Strong	TSS > 20 mg/l TOC > 25 mg/l	Prefiltration recommended
DEEP SYSTEMS	Slight	TSS < 1 mg/l Tur < 1 NTU TOC < 5 mg/l	Frequent pumping Surging/Jetting: once a month
	Notable	1 mg/l < TSS < 10 mg/l 1 NTU < Tur < 10 NTU 5 mg/l < TOC < 15 mg/l	Pumping: once a day Surging/Jetting: once a week
	Severe	TSS > 10 mg/l Tur > 10 NTU TOC > 15 mg/l	Daily pumping Adapted protocol

NOTATION

a_m	Reactive surface area
d_g	Mean grain diameter
d_g	Mean suspended particles diameter
e^-	Electron
i	Infiltration rate
q	Darcy's flow velocity
AOC	Assimilable organic carbon
BFR	Biofilm formation rate
DO	Dissolved oxygen
DOC	Dissolved organic carbon
MFI	Modified fouling index
N	Nitrogen
P	Phosphorous
P_l	Liquid pressure
P_g	Gas pressure
SI	Saturation index
T	Temperature
TOC	Total organic carbon
TSS	Concentration (total) of suspended solids
Tur	Turbidity
μ_l	Liquid dynamic viscosity
ρ_l	Liquid density

ACKNOWLEDGEMENTS

This work was funded by the European Union (Artificial Recharge of Groundwater Project). Thanks are also due to P.J. Stuyfzand and J.H. Peters (Kiwa, the Netherlands) for their comments.

REFERENCES

Custodio, E. 1986. Recarga Artificial de acuíferos. *Boletín de Informaciones y Estudios, No. 45,* Servicio Geológico, Ministerio de Obras Públicas y Urbanismo (MOPU, Madrid). 148 pp.

Dillon, P., M.R. Hickinbotham & P. Pavelic 1995. Review of international experience in injecting water into aquifers for storage and reuse. *Proc. of Water Down Under 1994 Conf.* (Adelaide, Australia). Part 2-A, No. 94/14: 13-19.

Driscoll, F.G. 1986. Groundwater and wells. *Johnson Filtration Sys. Inc.* (St Paul, Michigan, USA).

Frycklund, C. 1992. Artificial Groundwater Recharge -State of the Art. *VA*VA-FORSK, Report 1992-04.* 55 pp.

Hutchinson, A. & R. Randall 1995. Estimation of injection well clogging with the Modified fouling index (MFI). Proc. 2nd Int. Symp. on Artificial Recharge of Groundwater, 1994 (Orlando, FL, USA). Publ. by the ASCE in 1995 710:719.

Olsthoorn, T.N. 1982. The clogging of recharge wells -Main subjects. *Kiwa communications* 72, Working Group on recharge wells (the Netherlands). 136 pp.

Pérez-Paricio, A. 1998. Clogging and Artificial Recharge of Groundwater: Fundamental aspects. *M.Sc. Thesis at the Universitat Politècnica de Catalunya,* Barcelona (Spain). 170 pp.

Pyne, R.D.G. 1995. Groundwater recharge and wells: a guide to Aquifer Storage and Recovery. *CRC Press, Inc. (Lewis Publ.),* Boca Ratón (FL, USA).

Schippers, J.C., J. Verdouw & G.J. Zweere 1995. Predicting the clogging rate of Artificial Recharge wells. *J. of Water Supply Res. & Technol.* (*Aqua,* Oxford). 44(1): 18-28.

Van der Kooij, D., H.S. Vrouwenvelder & H.R. Veenendaal 1995. Kinetic aspects of biofilm formation on surfaces exposed to drinking water. Water Sci. & Technol. 32(8): 61-65.

Wegelin, M., R. Schertenleib & M. Boller 1991. The decade of roughing filters -Development of a rural water treatment process for developing countries. *J. of Water Supply Res. & Technol.* (*Aqua,* Oxford). 40(5): 304-316.

Artificial Recharge of Groundwater, Peters et al. (eds) © 1998 Balkema, Rotterdam, ISBN 90 5809 017 5

Analysis of a well recharge experiment

P.H.Giao, N.Phien-wej & P.Nutalaya
Asian Institute of Technology, Pathumthani, Thailand

ABSTRACT: A correct understanding of recharge well behaviour has both theoretical and practical significance as a recharge well may behave quite differently from an extraction one due to clogging effects. This paper presents an analysis of a well recharge experiment made at the Asian Institute of Technology (AIT) from 1993 to 1995. The field test data could be used to define a parameter, called Recharge Clogging Factor (RCF), to characterize the time-dependent reduction of hydraulic conductivity of the clogging zone around the well. FEM analyses using this parameter were done and simulated successfully the recharge test curves.

1 INTRODUCTION

Apparently, recharging an aquifer is just a reverse process to pumping. But, in practice, the difference in design, set up and data analysis is quite significant between the recharging and pumping tests. In the past, a number of laboratory and field experiments were carried out to investigate artificial recharge through wells, a summary can be found in Giao (1997). In general, these experiments provided corroborative evidences on possible clogging types as well as their adverse effects on recharge well efficiency. Main causes for the clogging were identified such as : air bubbles, suspended solids, bacteria, chemical incompatibility between recharge water and native groundwater. Clogging causes alteration of soil medium immediately adjacent to the well and reduces its hydraulic conductivity. Because the ability to understand clogging mechanism, to simulate and predict its rate are needed for a rational design of a recharge structure (Wood 1976), evaluation of clogging by just a qualitative description is not sufficient. For well recharging to become a tool for an effective water management one must have ability to assess reliably the extent of clogged zone from the recharge well, the rate of permeability reduction around the well, and to simulate the recharge curves. In this study, an approach to assess quantitatively clogging around a recharge well and its associated behaviour is presented based on field test data and FEM analysis.

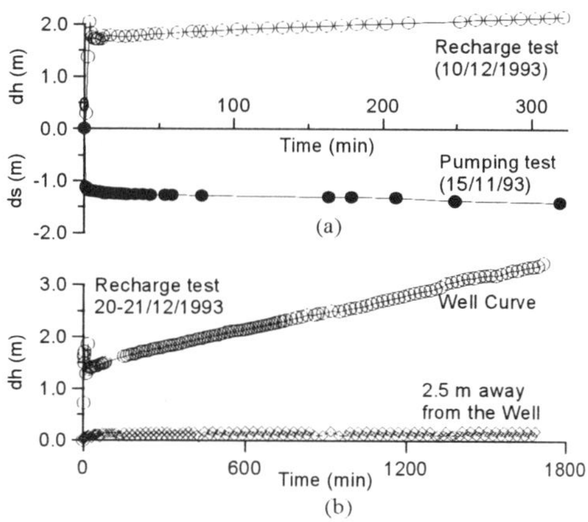

Figure 1. Field observations

2 FIELD OBSERVATIONS

Artificial recharge has been proposed as a supplemental measure of mitigation for Bangkok land subsidence (AIT 1981). As the first step, an experiment with more than 50 field tests was carried out from 1993 to 1995 to investigate technical aspects of well recharging into the uppermost aquifer of the Bangkok aquifer system which include design, construction, operation of a recharge well, instrumentation, monitoring and data analysis (Phien-wej et al. 1998). Some field observations related to the recharge well behaviour are presented below.

1. With reference to the well, at the beginning of the recharge experiment, a recharge curve shape was somehow similar to that of a discharge curve which was leveled-off with time (Fig. 1a). But in the later tests it became steeper with a continuously increased head inside the well as seen in Figure 1b.

2. A curve measured at a distance from the well had not changed its shape with recharging time, e.g., 2.5 m in Figure 1b, which means that clogging was developed only in proximity of the recharge well.

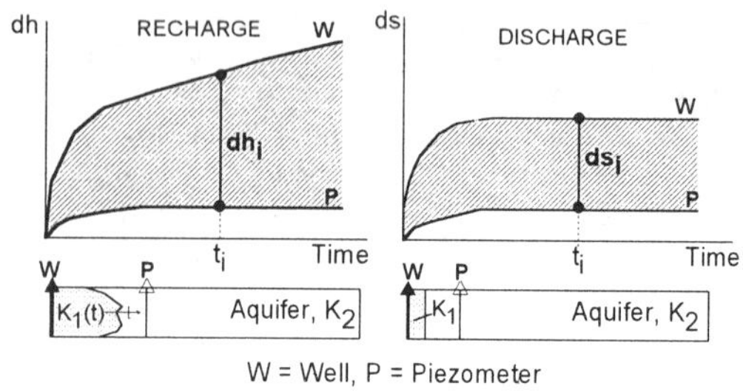

Figure 2. Basis for RCF derivation

3 RECHARGE CLOGGING FACTOR (RCF)

3.1 *Derivation of RCF*

With recharge progressing, a clogged zone is developed around the well, over which the head loss becomes larger and larger due to gradual reduction in permeability. One can determine head differences between a pair of curves, measured at the well and at a piezometer, respectively, for each case of recharge or discharge, e.g., dh and ds in Figure 2. Applying Darcy's law for the assumed clogged zone one can obtain the following equations :

$$Q_R = k_R \frac{dh}{L} 2\pi rb \qquad (1a)$$

$$Q_D = k_D \frac{ds}{L} 2\pi rb \qquad (1b)$$

Where Q_R, Q_D = recharge/discharge rate; dh, ds = head loss ; k_R, k_D = hydraulic conductivity of the considered zone in the recharge/discharge test, respectively ; L = width of the zone over which the hydraulic gradient is considered, e.g. the width of the clogged zone in the recharge test ; r = the well radius; b = the confined aquifer thickness.

For two recharging and discharging tests of the same pumping rate, i.e., $Q_R=Q_D$, one can deduce :

$$RCF = \frac{k_R}{k_D} = \frac{ds}{dh} \qquad (2)$$

RCF, therefore, indicates how much the hydraulic conductivity of the clogged zone (k_R) is reduced comparing to the value which is not affected by recharge-induced clogging (k_D).

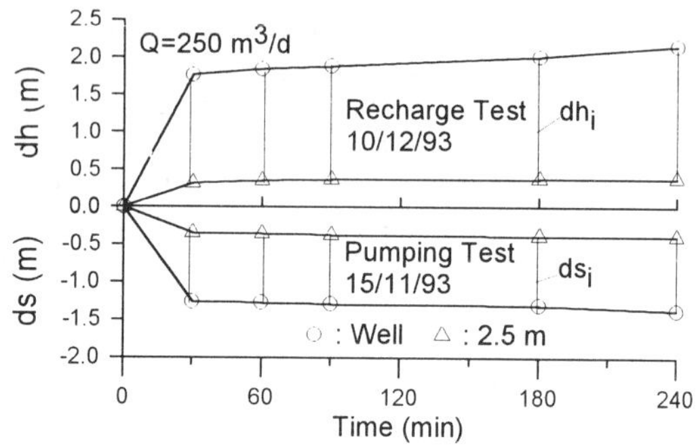

Figure 3. Working example of determination of RCF

3.2 *Procedure of determination of RCF*

A procedure of determination of RCF in a field experiment is presented in the following :

1. After the recharge well was developed and ready to be operated, a series of pumping tests are recommended with different pumping rates from small to large values. These "reference pumping tests" should precede any recharge test to avoid being affected by recharge-induced clogging.

2. The recharge rate is selected based on the specific recharge well design and after some initial tests in practice. To determine RCF it should be equal to one of the reference pumping rates.

3. As a working example, two sets of data monitored at the well location and 2.5 m were used to calculate RCF (Fig. 3). Firstly, dh_i, ds_i at a number of discrete times, i.e., 0, 30, 60, 180 and 240 min were calculated. Then, values of RCF can be determined by the ratio ds_i/dh_i at the corresponding times. It is noted that at t = 0, although $dh_i=ds_i=0$, the RCF is considered to be 1 which means no reduction occurred yet. Some empirical rules of RCF change were found based on the curve fitting methods as shown in Figure 4.

4 FEM ANALYSIS OF RECHARGE CURVES

In the actual experiment, a 150mm-diameter well injected reservoir water into a confined aquifer located at depths 16-28 m. To simulate recharge test curves, a 2-D groundwater model with some modifications to treat the clogging zone was employed.

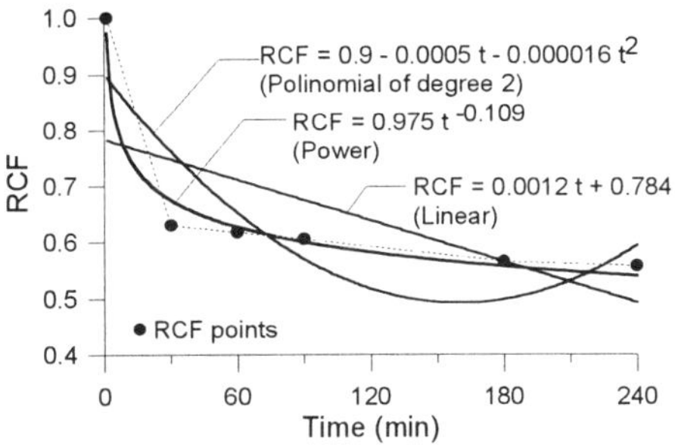

Figure 4. Empirical rules of RCF change

A FEM program was developed for this purpose (Giao 1997). In the FEM analysis, the aquifer was discretized by a rather simple mesh (Fig. 5), a narrow zone near the well representing the clogging zone was assigned with a reduced hydraulic conductivity, k_1. The extent of this clogged zone was varied in the analysis and it is called the Clogging Development Radius (CDR). The remained zone had k_2 which was the virgin hydraulic conductivity of the aquifer. A storage coefficient S = 0.001 was used. Two types of FEM analyses were performed.

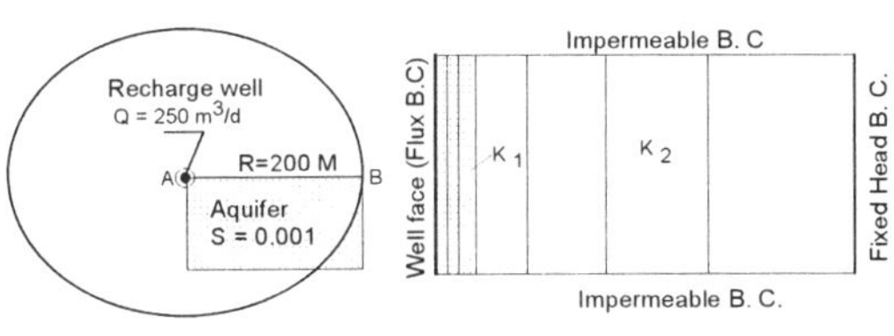

Figure 5. FEM mesh of the recharged aquifer

1. *Analysis 1* : The recharge well curve of the test on 10 December, 1993 was chosen to be simulated for different sets of parameters k_1 , CDR. No satisfactory results were obtained as seen in Figure 6. Thus, one can conclude that the analysis with k_1 and CDR being constant could not

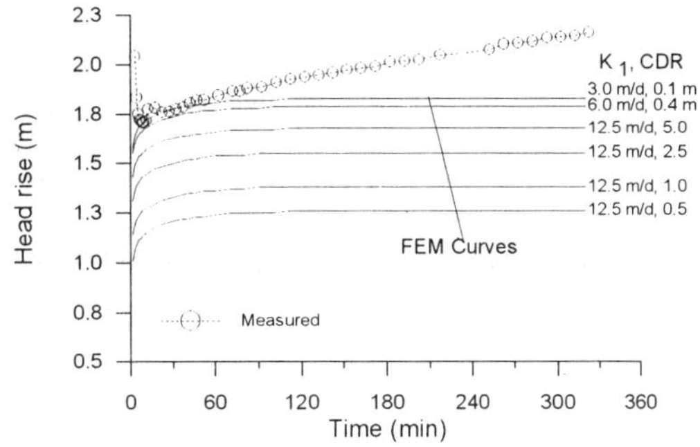

Figure 6. Analysis 1

simulate properly the feature of continuously increased shape of the curves recorded in the well.

2. *Analysis 2* : In this analysis, k_1 varied with time, i.e., $k_1(t) = k_1 \times RCF(t)$, where k_1 was the initial value for the clogging zone. A power rule, i.e., RCF = a t^{-b} (a=0.98, b=0.11) was put in the FEM program at the part where the element stiffness matrix was formed to simulate the time-dependent change of the hydraulic conductivity of the clogged zone. The results of analyses which successfully simulated the well curve are presented in Figure 7. As this recharge test was performed at the early stage, the clogged zone corresponding to the second parameter set with k_1=6.0 m/d, CDR=0.15 m, somewhere near the gravel pack/sand interface, seems to be the suited scenario.

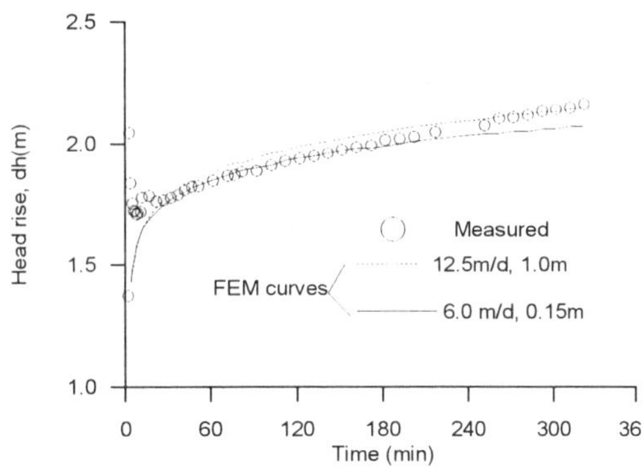

Figure 7. Analysis 2

5 CONCLUSIONS

RCF can be a useful parameter in analysis of clogging around the recharge well. The procedure to determine it from field test data is simple, straightforward and it can be applied for any well recharge experiment in the other places. The empirical rule of RCF can be integrated into a FEM program to simulate successfully a behaviour specific to the recharge well, i.e., the continuous head rise in the well caused by recharge-induced clogging effects.

REFERENCES

AIT 1981. Investigation of land subsidence caused by deep well pumping in the Bangkok area, comprehensive report 1978-1981, Research Report No. 91, AIT, Bangkok

Giao, P. H. 1997. Artificial recharge of the Bangkok aquifer system for mitigation of land subsidence, D. Eng. Dissertation, No. GE/96-2, AIT, Bangkok.

Phien-wej, N., P. H. Giao & P. Nutalaya 1998. Field experiment of artificial recharge through a well with reference to land subsidence control, to be published in *Engineering Geology*, vol. 51(2).

Wood W. W. 1976. Toward a rational development of artificial-recharge research, *Advances in Groundwater Hydrology*, The American Water Resources Association.

2.7 Biological treatment and removal of organic matter

Artificial Recharge of Groundwater, Peters et al. (eds) © 1998 Balkema, Rotterdam, ISBN 90 5809 017 5

Biological treatment and re-infiltration for denitrification of groundwater

J. Lindberg
Division of Land and Water Resources, KTH, Stockholm, Sweden

P.-O. Johansson
VBB VIAK, Stockholm, Sweden

ABSTRACT: A biological reactor for denitrification above ground combined with re-infiltration for removal of organic matter was tested for treatment of nitrate contaminated groundwater. In a column test pulp mixed with sand was used for the biological reactor. A sand column, where unsaturated conditions prevailed, simulated the re-infiltration. The test study was run for 255 days with a hydraulic load of 0.22 m/h. After a start up phase of 30 days a nitrate reduction of 70-100% was obtained with exception for a break-through period probably caused by a short stop in the nitrate input. After about 200 days the reduction declined. The TOC concentrations after the biological reactor was about 4.5 mg/dm^3 and no significant reduction was obtained in the sand column. However an efficient reduction in substances causing taste and odour in the water was observed.

1 INTRODUCTION

Increasing concentrations of nitrate in groundwater have become a problem in many areas worldwide. Leakage from agricultural land and from sewage water facilities often causes nitrate concentrations exceeding drinking water standards. Existing techniques for treatment of nitrate contaminated water include blending, anion exchange, membrane techniques (reverse osmosis and electrodialysis) and biological methods. Most methods are rather complex and expensive to implement. There is a need for less complicated and inexpensive treatment methods.

In the current project a system based on biological denitrification in a biological reactor above ground combined with re-infiltration is tested. The biological denitrification is based on the shift of electron acceptor from oxygen to nitrate by microorganisms. A prerequisite for the process is thus that anoxic conditions prevail. Furthermore the microorganisms are in need of an organic carbon source, phosphorous and micronutrients. The biological activity is dependent on pH and temperature and the efficiency is also ruled by the retention time in the system.

Pulp was chosen as a solid phase carbon source. Among earlier investigations with solid phase carbon sources can be mentioned Connesson et al. (1986) /straw/, Hansen (1993) /straw/ and Volokita et al. (1996a, b) /cotton and newspaper/.

The purpose of the re-infiltration treatment step is to reduce organic carbon and to remove taste and odour. The need for post-treatment by e.g. re-infiltration has been stated e.g. by Mercado et al. (1988), Hiscock et al. (1991) and Dahab (1993). In Sweden re-infiltration is a common practice to treat water with high concentrations of iron and manganese (Agerstrand 1982).

2 MATERIAL AND METHODS

A column study was performed using PVC columns (height 100 cm, diameter 30 cm). A first column was used as a biological reactor and a second column filled with sand was used to simulate the re-infiltration (Fig. 1). The biological reactor was from the bottom filled with sand (16.6 kg) mixed with lime (4.5 kg) and phosphorous (0.35 kg), sand (17.7 kg) mixed with pulp (3.3 kg) and water (17.3 dm^3) and sand (15.4 kg). Tap water was used and mixed with $NaNO_3$ to a concentration of about 15 mg NO_3-N/dm^3. The flow through the system was about 250 ml/min (0.22 m/h) and directed upward in the biological reactor. The biological reactor was water saturated while unsaturated conditions prevailed in the sand column.

Water samples were taken out every week for analysis (Table 1). The nitrogen parameters were analysed using a fluid injection analysis instrument and total organic carbon (TOC) was analysed using a

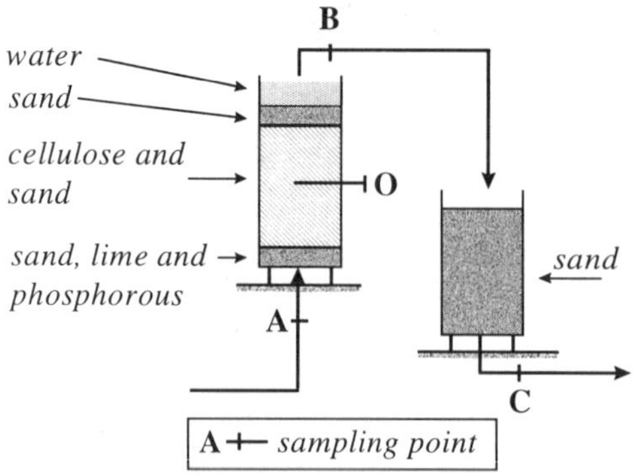

Figure 1. Set-up of the column study.

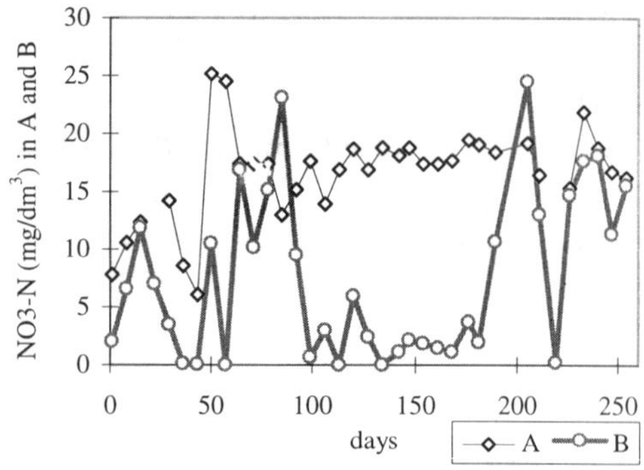

Figure 2. Concentrations of NO_3-N in A and B.

TOC-analysator. The taste and odour was detected manually and the colour was determined by a comparative method. Dissolved oxygen (DO) was determined by the method according to Winkler. The column study was run for 255 days (June 1997 - February 1998).

Table 1. Scheme for analysis of water samples.

	A	*B*	*C*	*O*
pH	×	×	×	×
temperature	×	×	×	×
NO_3-N	×	×	×	
NO_2-N	×	×	×	
NH_4-N	×	×	×	
TOC		×	×	
taste		×	×	
odour		×	×	
colour		×	×	
dissolved oxygen (DO)				×

3 RESULTS

The nitrate concentrations at the inlet (A) and the outlet (B) of the biological reactor are shown in Figure 2. Five different phases can be differentiated; (i) day 1-30 (the start up time for the system with a gradually increasing removal efficiency), (ii) day 30-70 (the system works with a high degree of efficiency, between 60 and 100% of nitrate reduction), (iii) day 70-100 (a complete break-through), (iv) day 100-200 (the system works with a high degree of efficiency, 70 to 100% nitrate reduction) and (v) day 200-255 (a break-through).

The nitrate concentrations at the outlet of the sand column (C) was almost the same as at the outlet of the biological reactor. The total amount of NO_3-N added during the experiment was 1808 g and the total amount of NO_3-N removed during the experiment was 950 g.

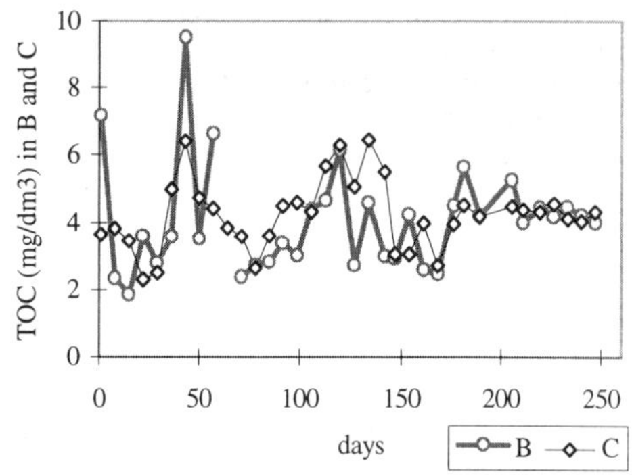

Figure 3. Concentrations of TOC in B and C.

The total volume of treated water during the experiment was 118 m^3 which corresponds to 4370 pore volumes. The temperature varied during the study period generally between 20-22°C but a maximum value of 26°C was recorded during August. pH has varied between 7.0-7.5 during the study period, DO between 0-9 mg/dm^3 (normally about 4 mg/dm^3) and the colour in the water samples has been in the range 0-5 mg Pt/dm^3 for samples from sampling point B. For the samples from sampling point C no colour has been detected. A slight smell and odour was sometimes detected from B but never from C. The detected concentrations of nitrite and ammonium was generally below 0.05 mg/dm^3 and significantly lower in C compared to B.

The TOC concentration in A is about 4.0 mg/dm^3. Figure 3 shows the TOC concentration in B and C. No accumulation in TOC takes place in the biological reactor and no significant reduction of TOC takes place in the sand filter.

4 DISCUSSION AND CONCLUSION

The results from the study are promising. Except from the start-up phase and the period of break-through after about 70 days the system worked with a high degree of efficiency until the decrease in efficiency after about 200 days. The break-through after about 70 days was preceded by an unintentional stop in the dosage of nitrate to the system. This may be the explanation for the break-through as a new start-up phase had to follow. Stable conditions should be favourable for the microbiological processes. The operation of the system is favoured by the upward directed flow of water in the biological reactor as entrapment of gas is avoided that otherwise will reduce the effective porosity and cause partly clogging of the system and a following reduction in the hydraulic retention time.

Based on the assumption that all organic carbon available for the microorganisms is consumed the C:N-ratio for the experiment is calculated to be 1.3 (3.3 kg added pulp corresponds to 1.2 kg organic carbon and 0.95 kg NO_3-N is removed). This value corresponds with values from earlier investigations (Chalupa 1988 /C:N=1.5/, Dahab & Lee 1992 /C:N=1.0/).

No substantial reduction in TOC was obtained in the sand column. However, Swedish experiences from artificial recharge plants indicate that a substantial reduction can be obtained in full-scale plants even with very short residence times. The substances causing taste and odour in the water are effectively removed in the sand column.

A pilot-scale experiment is now conducted at a municipal water work. The set-up for the experiment is the same as for the column study. The volume of the filters is 4.5 m^3 (compared to 55 dm^3 for the column study). The hydraulic load is 0.2 l/s corresponding to 0.29 m/h (0.22 m/h for the column study). The residence time for the water in the biological reactor is longer in the pilot scale experiment to compensate for the lower water temperature (3.4 h compared to 1.4 h).

5 ACKNOWLEDGEMENT

This research was (in part) supported by the EC Environment and Climate Research Programme (contract: ENV4-CT95-0071) and in part by the Swedish Association of Graduate Engineers.

REFERENCES

Agerstrand, T. 1982. Re-infiltration, a method for removing iron and manganese and for reducing organic matters in groundwater recharged by bank infiltration. In: *DVWK Bulletin 13, Artificial groundwater recharge. International Symposium on Artificial Groundwater Recharge*, Dortmund, 1979. 253-266. Verlag Paul Parey, Hamburg/Berlin.

Chalupa, M. 1988. Problems in Czechoslovakia regarding methods of removal of nitrates from drinking water. *Wat. Sci. Tech.* Vol.20, No.3, pp. 211-213.

Connesson, C., M. Gillet, I. Morvan, G. Martin, M. Tulbo & M. Landream 1986. Denitrification in Situ a l'aide d'un substrat carbon solide. *BRGM-report, 1986*. (BRGM: Avenue de Concyr, BP 6009, 45060 Orleans Cedex).

Dahab, M.F. & P.Y. Lee 1992. Nitrate reduction by in-situ bio-denitrification in groundwater. *Wat.* Sci. Tech. Vol.26, No.7-8, pp. 1493-1502.

Dahab, M.F. 1993. Comparison and evaluation of in-situ bio-denitrification systems for nitrate reduction in groundwater. *Wat. Sci. Tech.* Vol.28, No.3-5, pp. 359-368.

Hansen, H.O. 1993. *Grundvandsdannelse - kvantitet og kvalitet.* Danmarks Tekniske Højskole, 1993. (In Danish)

Hiscock, K.M., J.W. Lloyd & D.N. Lerner 1991. Review of natural and artificial denitrification of groundwater. *Wat. Res.* Vol.25, No.9, pp. 1099-1111.

Mercado, A., M. Libhaber & M.I.M. Soares 1988. In situ biological groundwater denitrification: concepts and preliminary field tests. *Wat. Sci.* Tech. Vol.20, No.3, pp. 197-209.

Volokita, M., S. Belkin, A. Abeliovich & M.I.M. Soares 1996. Biological denitrification of drinking water using newspaper. *Wat. Res.* Vol.30, No.4, pp. 965-971.

Volokita, M., A. Abeliovich & M.I.M. Soares 1996. Denitrification of groundwater using cotton as energy source. *Wat. Sci. Tech.* Vol.34, No.1-2, pp. 379-385.

Artificial Recharge of Groundwater, Peters et al. (eds) © 1998 Balkema, Rotterdam, ISBN 90 5809 017 5

Biological purification of recharge water by forested soils

D.Rüetschi, U.Geissbühler, M.Schmid & Ch.Wüthrich
Department of Geography, University of Basel, Switzerland

ABSTRACT: In this paper we present first results of a research project on the artificial groundwater recharge plant in the „Langen Erlen", a former floodplain area near Basel, Switzerland. There, forested alluvial soils are used for water purification. We investigated the biological purification processes in those soils by measuring soil microbial activity and biomass as indirect parameters of purification performance. In contradiction to ordinary recharge plants where the biofilm on the (sand-) filter surface is very important, we found, that in these forested systems also deeper soil layers are crucial for producing a high groundwater quality. In connection with improved purification performance, we also try to increase the ecological significance of the recharge basins as they are functioning as the last remnants of the former riparian ecosystem. We found that diversified reed vegetation at open sites showed unchanged or even better purification performance than forested sites.

1 INTRODUCTION

The city of Basel (200,000 inhabitants) is supplied by 95% with drinking water gained through artificial groundwater recharge. Half of it is produced in the area „Langen Erlen", a floodplain of the now channelized river „Wiese", located northeast of the town (47°35 N, 7°38'E).

Figure 1. View of a recharge basin.

The raw water is taken in from the river Rhine and prefiltered through a 90 cm thick sand layer to remove suspended solids. The prefiltered water is pumped into 11 forested recharge basins, mainly covered with poplars, with a total area of 22 ha (Fig. 1). As this is done intermittently (10 days watering, 20 days aeration period), only a third of the basins are in operation at a certain time. 60,000 m^3 Rhine water/day is pumped into the basins.

The water seeps through a humus layer of approximately 40 cm and a 2.5-3.5 m thick natural sand and gravel layer before it reaches the groundwater level. Hereby the water is mechanically and biologically purified in the soil. An aeration period of 20 days has shown to maintain the infiltration capacity of the forest soils and to provide good conditions for microbial decomposition processes and for forest growth.

The system is in operation since 1920 and is strongly based on empirical experiences. It seems to be unique worldwide and only very few scientific knowledge is available on the purification processes in the soils of this recharge plant. The objectives of our research are to better understand these processes, to improve the purification performance and to increase the ecological significance of the recharge basins.

All of the basins are normally covered by a dense forest with few underwood to provide good microclimatic conditions. One of our topics was to gain data on change of purification efficiency if these forests were converted into reed habitat to increase biodiversity. This research is conducted by the Department of Geography (sections Physiogeography and NLU-Biogeography) and is partly financed by the Industrielle Werke Basel (IWB).

2 METHODS

As soil microorganisms are very important for the water purification we measured soil microbial activity (CO_2-emission measured by Infrared-Gas-Analysis; Wüthrich 1994) and microbial biomass (measured by CFE-method; Jörgensen 1996, Vance et al. 1987) in three basins of the recharge plant in different depths and within the course of the year. These parameters function as indirect indicators of purification performance. Also oxygen concentration of the seeping water and soil temperature were monitored as they are important ecological factors for microbial purification. Special consideration was taken to compare purification performance and groundwater quality of forested basins with basins dominated by a diverse reed vegetation.

3 RESULTS

As a consequence of the high infiltration capacity (during flooding period 1-2 m^3 per m^2 and day) of the prevailing floodplain soils with undersoils composed by alluvial sand and gravel, no sites showed anoxic conditions. The soils reached field capacity no later than 2 days after finished watering. Aeration was thereby sufficient to bring enough oxygen into the soil macropores. This provides good conditions for fast microbial decomposition of organic material filling the soil pore system.

Microbial soil activity remains high even in deeper soil layers (down to 90 cm depth; Fig. 2). Figure 3 shows, that microbial activity was clearly temperature dependent: in summer, especially the upper 10 cm of the soil profile showed the highest activity (35% higher than in 30-40 cm depth). In winter, however, microbial activity between 30-40 cm was 10% higher than in the surface layer.

Our respiration data are supported by the data on microbial biomass, measured independently by the CFE-method (Fig. 4). Although the surface layer showed no great activity in winter, the groundwater quality in general was as good as throughout the year (Geissbühler 1998, Schmid 1997, Rüetschi, in prep.).

The forests in the basins were not decisive for groundwater quality: More open sites in the basins with reed vegetation (mainly *Phalaris arundinacea*) which have a thick fine root zone in the upper 10 cm of the soil showed even significantly higher microbial activities (Fig. 5). In dry summer periods, particularly the reed vegetation keeps the soil moist which is favourable for microbial decomposition. In deeper soil layers there is no difference between forested and open sites (data not shown). Therefore the total purification performance at open sites is better than at forested sites.

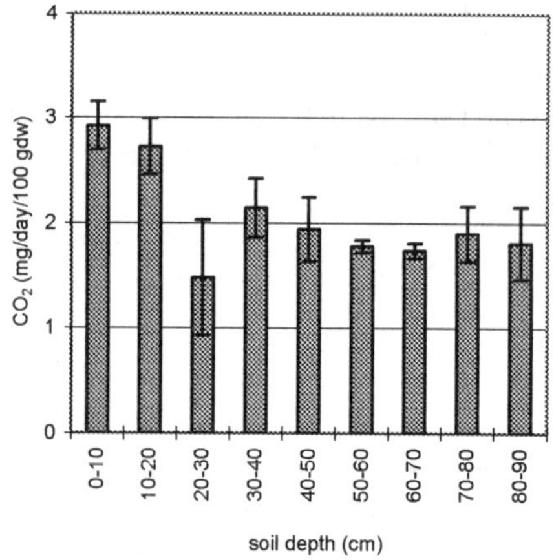

Figure 2. Soil microbial activity in a groundwater recharge basin at 9/4/97, expressed as emitted CO_2 in mg per day and dry weight of fine soil (n=3).

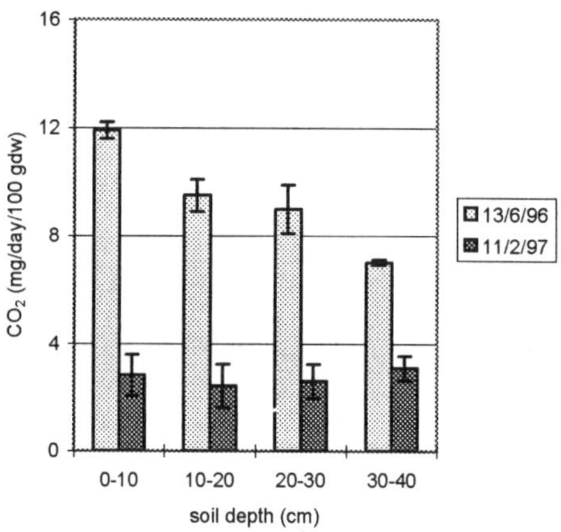

Figure 3. Seasonal variation of soil microbial activity in the same basin as Figure 2 (n=3).

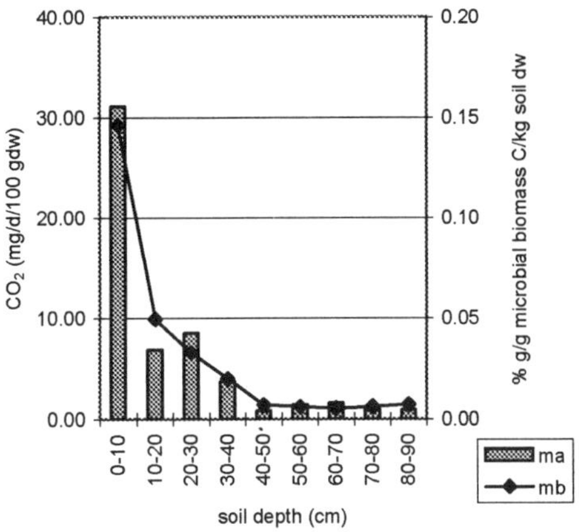

Figure 4. Comparison of mean microbial activity (ma; n=3) and biomass (mb; n=2; expressed as percentage of microbial biomass carbon per kg soil dry weight) in another recharge basin at 03/10/97.

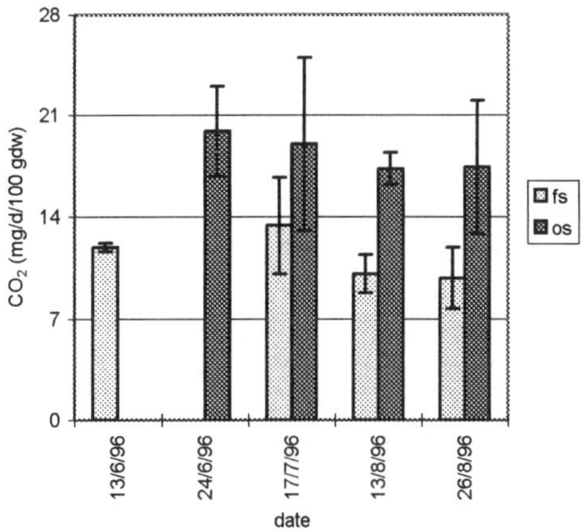

Figure 5. Comparison of microbial activity between soil surface and 10 cm depth at a forested site (fs) and an open site (os), dominated by reed vegetation during summer 1996.

4 CONCLUSIONS

Our results show that in the presented groundwater recharge system also deeper soil layers are important for water purification, since the surface layers become too cold (especially in winter) or too dry (especially in summer) to allow sufficient microbial activity. Though purification processes are generally less intensive in winter, a high groundwater quality during the whole year is guaranteed by the thick soil profile of the investigated floodplain.

Open sites with reed vegetation showed a slightly better purification performance than forested sites. We therefore postulate that parts of the forests should be invaded by reed plants after clearcutting the poplars. The clearcutted area (about 0.5 ha in average, normally every 35 years in the same basin) could be covered temporarily by reed plants until the growing trees are shading them too much (as it would happen in a natural floodplain forest because of vegetation succession after a flood event). In this way the artificial recharge basins in the „Langen Erlen" could not only provide high quality drinking water but could also become more significant as compensatory living space for endangered riparian organisms in an urban floodplain.

REFERENCES

Geissbühler, U. 1998. Changes of the biological filtering process in the infiltration basins of the „Langen Erlen" during winter season. Diploma thesis, Department of Geography, University of Basel.

Jörgensen, R.G. 1996. The fumigation-extraction method to estimate soil biomass: calibration of the k „sub ec" value. *Soil Biol. Biochem.* 28:25-31.

Rüetschi, D. (in prep.). The infiltration cycle of the artificial groundwater recharge in the „Langen Erlen" near Basel: Effects on groundwater quality, microbial soil activity and filtration capacity. (Draft title). Doctoral thesis, Department of Geography, University of Basel.

Schmid, M. 1997. Suitability of a reed meadow basin for artificial groundwater recharge in the „Langen Erlen". Diploma thesis, Department of Geography, University of Basel.

Vance, E.D., P.C. Brookes, & D.S. Jenkinson 1987. An extraction method for measuring soil microbial biomass C. *Soil Biol. Biochem.* 19:703-707.

Wüthrich, Ch. 1994. The biological activity of arctic soils with special consideration of ornithogenic soils. *Physiogeographica* 17:1-222.

Artificial Recharge of Groundwater, Peters et al. (eds) © 1998 Balkema, Rotterdam, ISBN 90 5809 017 5

Impact of drinking water sources on refractory DOC in water reuse systems

J.E. Drewes & P. Fox
Department of Civil and Environmental Engineering, Arizona State University, Tempe, Ariz., USA

D. Ziegler
Department of Water Quality Control, Technical University of Berlin, Germany

ABSTRACT: The scope of this study was to investigate the impact of drinking water sources on the refractory DOC (dissolved organic carbon) in treated effluents used for groundwater recharge. Because the City of Mesa uses different surface waters and local groundwater for its water supply, drinking water DOC varies from approximately 1.3 to 3.2 mg/L. The variations of this DOC of natural origin significantly influences DOC concentrations of the treated wastewater. Drinking water DOC represents probably a 30 to 50 percent of the refractory DOC in the Mesa effluent. Refractory DOC was determined with an adapted soil-column system which simulates the degradation of DOC in aquifers. The impact of drinking water quality on treated wastewater was also confirmed by high correlations of chloride and sulfate concentration. This correlation allowed an estimation of the increase of salt concentrations from the use of water.

1 INTRODUCTION

Residual organic compounds are of concern in groundwater recharge systems, infiltrating surface water or treated domestic effluents. Regarding potable and non-potable reuse, quality issues for these compounds originate from their persistence to both sorption and biodegradation during wastewater treatment and subsequent soil-aquifer treatment (SAT). There are three groups of organic chemicals that require attention:

1. refractory synthetic organic compounds added to the sewage by households, hospitals, industry and other sources (Drewes & Jekel in press),
2. disinfection byproducts (DBP) formed by chlorination of the water before SAT, and
3. refractory natural organic compounds already present in the drinking water or formed in the wastewater treatment plants due to decomposition of organic material (Chudoba 1985; Hejzlar & Chudoba 1986; Link, Gilbert & Eberle 1989).

The bulk of refractory DOC in groundwater recharge systems consists of compounds like humic and fulvic acids. Further removal by biodegradation of these compounds is only possible over a long period of time (years, decades), and it is currently believed that these compounds might significantly influence the DOC of groundwater at the point of recovery. With regard to the character of organics in treated wastewater, this is determined by the character of drinking water sources (background groundwater or surface water quality), the effluent type (portion of households versus industry), and the specific local wastewater treatment.

The scope of this study was to investigate the impact of applied drinking water sources on the refractory DOC in tertiary effluents, and to estimate the influence of refractory DOC from drinking water on aquifers subject to SAT.

2 METHODS

The study was conducted at the City of Mesa AZ, Northwest Reclamation Plant (30,282 m^3/d), a nitrifying/denitrifiying plant using tertiary filtration and subsequent infiltration basins for groundwater recharge (13,200 m^3/d). Organic matter was characterized using DOC-analysis, UV-absorption and ultrafiltration for molecular weight distributions. The DOC was measured using a Dohrmann DC-180 TOC-analyzer. UV-absorbance was measured at a wavelength of 254 nm using a Hewlett Packard 8452A spectrophotometer. Ultrafiltration tests were carried out using Amicon 50 ml stirred cells and membranes of nominal molecular weight cut-off of 1, 3 and 10 kDalton.

The portion of refractory DOC in different waters was determined using an adapted soil-column system which simulated aquifer conditions (saturated, anoxic redox conditions, retention times of 25 days) (Drewes 1997), and an adapted batch-experiment to determine the biodegradable dissolved organic carbon (BDOC).

3 RESULTS

The drinking water supply of the City of Mesa is based on three different water resources: surface water from the Colorado River (Central Arizona Project), the Salt and Verde Rivers (reservoirs operated by Salt River Project - SRP), and local groundwater. The portions of all three sources vary nearly daily due to demand and technical requirements (City of Mesa 1997; SRP 1997) (Fig. 1).

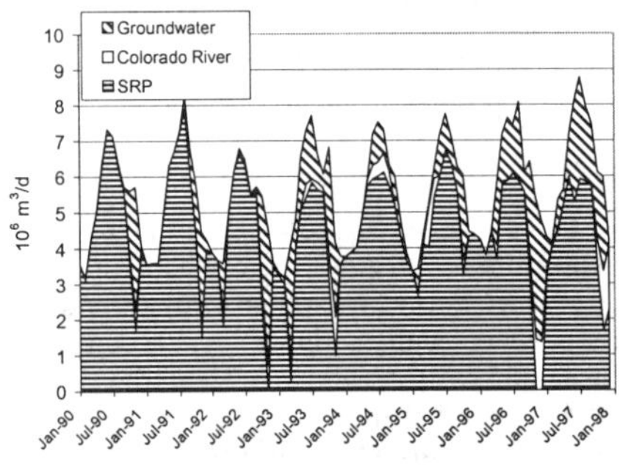

Figure 1. Drinking water sources in 10^6 m^3/d for water supply City of Mesa AZ during 1990 - 1997

The fact that the three water sources showed a different character regarding the natural organic matter content was considered to influence the portion of refractory dissolved organic carbon in the Mesa tertiary effluent. This in turn would have effects on the character and level of groundwater DOC influenced by wastewater recharge.

The DOC of the local groundwater (well depth more than 100 m) was less than 0.5 mg/L. It can be described as persistent regarding biodegradation. Groundwater is pumped after disinfection directly into the drinking water supply system. The mean DOC concentration of Colorado River water was 3.0 mg/L, and 2.5 mg/L after water treatment. These concentrations are relatively constant. SRP is using Salt River and Verde River water in different proportions during the year. SRP water provides the majority of drinking water for the City of Mesa. DOC concentrations in this source were measured daily at the water treatment plant.

Testing surface waters with a BDOC test indicated a high amount of refractory DOC. According to ultrafiltration test results, more than 50 % of the surface water had molecular weight fractions less than 3,000 Dalton. Similar results were shown by Malcolm (1990) for the distribution of surface and groundwater DOC in rivers and aquifers of the Unites States.

City of Mesa Operation Utilities provided daily data of different water volumes and portions used for Mesa drinking water supply. Based on this composition of drinking water sources, the DOC of the drinking water in the distribution system could be estimated by a mixing calculation. Since the drinking water DOC is of natural origin, no significant change should occur on the way to the wastewater treatment plant. Considering the local water supply and sewer system, and estimated water travel times through these systems to the Mesa wastewater treatment plant, a correlation between calculated drinking water DOC and tertiary effluent DOC could be made (Fig. 2).

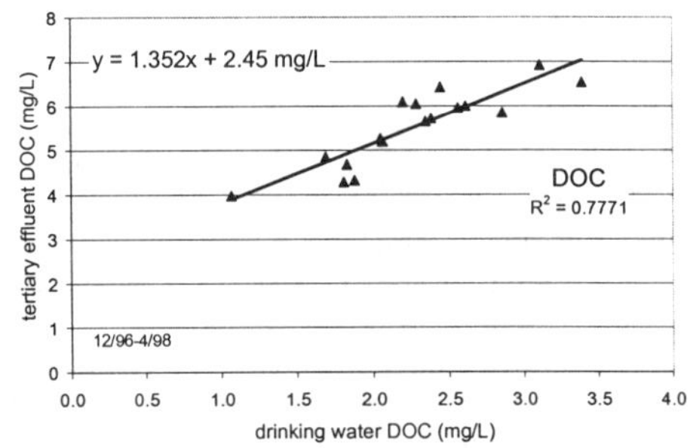

Figure 2. Correlation between tertiary effluent and drinking water DOC, City of Mesa AZ

Obviously, the change of the Mesa drinking water DOC, which varies in a range from 1.3 to 3.2 mg/L (UV/DOC-ratio average 1.45 L/mg * m), seems to cause the variations of tertiary effluent DOC. Decreasing drinking water DOC also leads to an increase of tertiary effluent UV/DOC-ratio (Fig. 3), which confirms that tertiary effluent DOC consists partly of drinking water DOC.

High correlations between drinking water and effluents were also found for chloride (R^2 = 0.89) and sulfate (R^2 = 0.98). These correlations can be used for mass balances of salts, and thus allow to

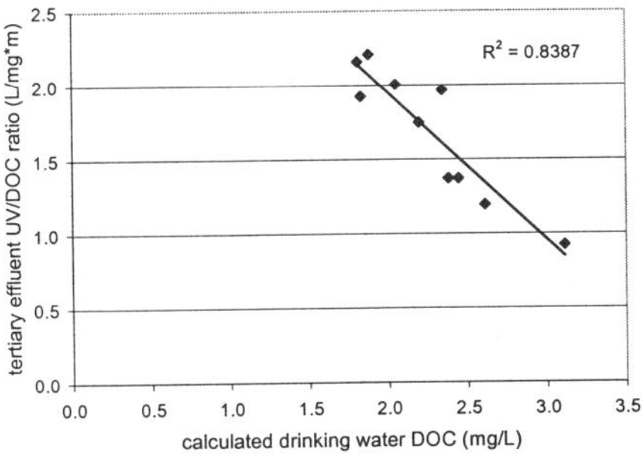

Figure 3. Correlation between drinking water DOC and tertiary effluent UV/DOC ratio.

estimate the increase of salts due to the use of drinking water in a local water cycle.

To determine the refractory DOC of different effluents used for groundwater recharge, an adapted soil-column system was used which simulated aquifer conditions with retention times of 21 days. According to these experiments, refractory DOC of the Mesa effluent consists of approximately 70 percent of drinking water DOC. Therefore, infiltrating effluents consisting of drinking water with high DOC might result in an increase of groundwater DOC concentration. Refractory DOC added by consumers and formed in the wastewater treatment process could be estimated to 1 mg/L at the Mesa field site. Since effluent affected groundwater at the Mesa site showed a little lower DOC concentration than the determined refractory DOC, long-term sorption processes might be responsible for a partly further removal.

4 CONCLUSIONS

The DOC variations of tertiary effluents were found to be influenced by sources of drinking water. The results of this study allows to estimate the portion of additional formed refractory DOC in wastewater treatment plants, when organic matter is decomposed. In infiltrating surface water directly or via effluents, the DOC of groundwater in recharge systems will eventually increase due to refractory DOC of different sources. Therefore, the management of water resources should be included in pre-studies for groundwater recharge systems. Also, refractory DOC of source water has to be considered when regulating SAT systems legally.

REFERENCES

City of Mesa 1997. Data of drinking water sources for public water supply 1990-1998. Mesa, Arizona (unpublished).

Chudoba, J. 1985. Quantitative estimation in COD units of refractory organic produced by activated sludge microorganisms. *Water Research* 19, 37-47. Elsevier: London.

Drewes, J. E. 1997. Behavior of organic compounds using tertiary effluents for groundwater recharge. Ph.D.-thesis Technical University of Berlin, VDI Fortschritt-Berichte, Series 15, No. 174, Duesseldorf: VDI-Verlag (in German).

Drewes, J. E. & M. Jekel in press. Behavior of DOC and AOX using advanced treated wastewater for groundwater recharge. *Water Research*.

Hejzlar, J. & J. Chudoba 1986. Microbial polymers in the aquatic environment – I-III. *Water Research* 20, 1209-1227.

Link, J., E. Gilbert & S. H. Eberle 1989. Untersuchungen über die Art und Menge der Reststoffe in den Abläufen biologischer Kläranlagen. *Vom Wasser* 72, 349-370.

Malcolm, R. L. 1990. Factors to be considered in the isolation and characterization of aquatic humic substances. Meeting on Humic Substances in the Environment, Proceedings of Linköping. 390-417. Linköping University, Sweden.

Salt River Project (SRP) 1997. 1996 Annual Water Quality Report. Phoenix, Arizona.

Artificial Recharge of Groundwater, Peters et al. (eds) © *1998 Balkema, Rotterdam, ISBN 90 5809 017 5*

Artificial recharge of ground water, organic carbon and microbes

I.T. Miettinen & M.J. Lehtola
National Public Health Institute, Division of Environmental Health, Kuopio, Finland

T. Vartiainen & P.J. Martikainen
National Public Health Institute, Division of Environmental Health / University of Kuopio, Department of Environmental Sciences, Finland

ABSTRACT: Artificial recharge is increasingly used in Finland to produce drinking waters from surface waters, which often have high organic carbon content. In this research we studied the changes in organic carbon (NPOC, AOC), microbial numbers and microbial growth during artificial recharge of ground water. Slow sand filtration decreased the content of NPOC only slightly, the content of AOC decreased on average by 74 μg AOC/l. Chlorination of artificially recharged ground waters increased the concentration of assimilable organic carbon on average by 56 μg AOC/l. Chlorination decreased the number of microbes in filtrated waters. However, chlorination halted microbial growth only temporarily. Surprisingly, microbial growth was enhanced in chlorinated samples compared to unchlorinated drinking waters.

1 INTRODUCTION

A major problem in Finnish drinking waters is the high organic carbon content (Vartiainen et al. 1987, Miettinen et al. 1995). The reason for this is that 45% of the drinking water is produced from surface waters (National Board of Water and Environment 1994), which typically have high organic carbon (humus) content (Ryhänen 1968). Content of organic carbon has traditionally been reduced by chemical coagulation/flotation techniques. Unfortunately, the odor and taste problems and also the formation of halogenated organic by-products often deteriorates the quality of drinking waters produced from surface waters. Quality of drinking waters is nowadays increasingly improved in Finland by either using ground waters or by using artificial ground water recharge techniques (Kivimäki 1992, Kivimäki 1995).

2 MATERIALS AND METHODS

2.1 Samples

Samples of raw (surface waters) and finished drinking waters were taken. Qualitative changes in water quality were studied in five different artificial recharge ground water works. All water works (A, B, C, D and E) were located in esker area consisting mainly of sand and gravel layers. Surface water (raw water) was infiltrated into the ground by basin infiltration, bank filtration or irrigation techniques. 75% of the produced drinking water in the water works B and 20% in water works C, was produced also by bank filtration. Of the produced drinking waters 10 - 40% consists of natural ground water. Water works A and B did not disinfect the finished drinking waters. Water works C, D and E had chlorine disinfection (post-chlorination) (0.05 - 0.2 mg Cl_2/l).

2.2 Analyses of organic carbon

The contents of total organic carbon (NPOC) was analysed by high-temperature combustion method using Shimadzu TOC 5000-analyser.

Assimilable organic carbon (AOC) was analyzed by a modified method originally presented by van der Kooij (1982). Inorganic salts were added to the samples before analysis to ensure that organic carbon was the only factor limiting the bacterial growth (Miettinen et al. 1996). AOC was analyzed by using *Pseudomonas fluorescens strain P17 biotype 7.2* (ATCC 49642) and *Spirillum NOX* (ATCC 49643) test strains.

2.3 Analysis of microbes

Heterotrophic plate counts (HPC) were analyzed by R2A agar plate counting (Reasoner and Geldreich 1985).

Microbial growth potential in the samples were tested by *in vitro* laboratory test. Samples were incubated in acid washed and heat treated (250°C) glass bottles in dark at 15°C for 14 days. Microbial growth was followed as HPC.

3 RESULTS AND DISCUSSION

3.1 Organic carbon

The content of NPOC decreased only slightly during filtration (on average 0.1 mg/l) (Table 1). The AOC concentration in raw waters of water works, which had no disinfection, varied from 130 to 150 µg AOC/l. The AOC concentration decreased during filtration. The content of AOC in these undisinfected drinking waters ranged from 60 to 75 µg AOC/l. In chlorinated drinking waters the AOC content ranged from 120 to 130, while the AOC content in raw waters varied from 50 to 100 µg AOC/l . This agrees with the earlier results showing that the chlorination increases AOC content (Van der Kooij et al. 1989).

3.2 Microbial numbers

Heterotrophic plate counts increased during the underground passage as shown by the HPC values in drinking water obtained from water works A and B (Table 1). HPC in drinking waters was lower than in raw waters only in chlorinated drinking waters (Table 1).

Microbial growth occurred in all drinking water samples. Surprisingly, the microbial growth was highest in chlorinated drinking waters (Fig. 1).

Obviously the increase in microbial nutrients enhanced microbial growth. The low doses of chlorine (0.05-0.2 mg Cl_2/l) were inadequate to permanently prevent microbial growth. These results show that microbes recovered after chlorination and that they were able to use the nutrients made available by chlorination.

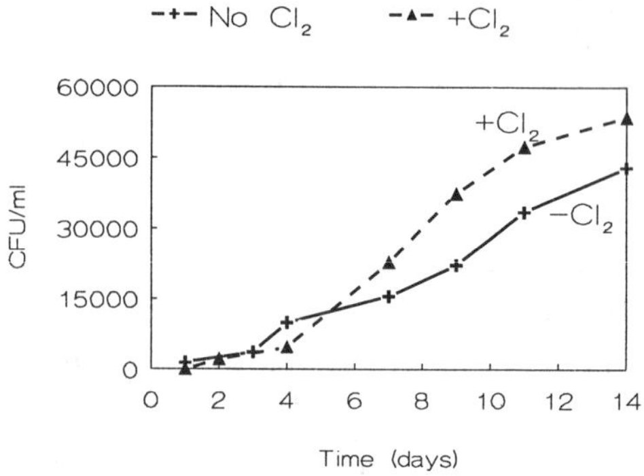

Figure 1. Microbial growth as heterotrophic plate counts in undisinfected (No Cl_2) and disinfected ($+Cl_2$) drinking waters.

4 CONCLUSIONS

Sand filtration decreased the content of organic carbon (NPOC/AOC), but chlorination increased the AOC concentration.

Chlorination decreased microbial numbers temporarily but the microbial growth was enhanced compared to unchlorinated drinking waters.

Table 1. Means of non-purgeable organic carbon (NPOC), assimilable organic carbon (AOC) and heterotrophic plate counts (HPC) in raw and drinking waters of chlorinating and non-chlorinating artificial recharge ground water works. Range in parenthesis.

	NPOC (mg/l)	AOC (µg/l)	HPC (CFU/ml)
Without chlorination (n=2)			
Raw waters	1.8 (1.5-2.1)	140 (132-147)	17 (14-20)
Drinking waters	1.7 (1.5-1.8)	66 (57-74)	1,410 (14-2,800)
Chlorination (n=3)			
Raw waters	2.1 (1.5-2.8)	67 (47-97)	200 (100-300)
Drinking waters	2.0 (1.3-2.7)	123 (119-128)	20 (10-30)

5 ACKNOWLEDGEMENTS

The research was financed by National Public Health Institute and the Academy of Finland project no. 34538.

REFERENCES

Kivimäki, A-L. 1992. *The articial ground water plants in Finland.* Publications of the Water and Environment Administration - series A98, Helsinki. *Abstract in English.*

Kivimäki A-L. 1995. *Production of articially recharged ground water using bank filtration.* Publications of the Water and Environment Administration - 573, Helsinki. *In Finnish.*

Miettinen, I, P.J. Martikainen, E. Iivanainen & T. Vartiainen. 1995. *Mikrobial regrowth, by-products of chlorination and mutagenicity in different Finnish drinking waters in 1994.* Publications of National Public Health Institute. B11/1995. Kuopio (*Summary in English*)

Miettinen I, T. Vartiainen & P.J. Martikainen. 1996. *Bacterial enzyme activities in ground water during bank filtration of lake water.* Water Research. 30:2495-2501.

National Board of Water and Environment. 1994. *Water supply and sewer systems in Finland 1993.* National Board of Water and Environment. Helsinki. Abstract in English

Reasoner, D.J. & E.E. Geldreich. 1985. *A new medium for the enumeration and subculture of bacteria from potable water.* Appl. Environ. Microb. 49: 1-7.

Ryhänen, R. 1968. *Die bedeutung der humus-substanzen im stoffhaushalt der gewässer Finnlands.* Mitt. Internat. Verein. Limnol.14:168-178. *In German.*

Van der Kooij, D., W.A.M. Hijnen & A.Visser. 1982. *Determinating the Concentration of Easily Assimilable Organic Carbon.* J. Amer. Water Assoc. 74:540-545.

Van der Kooij, D., W.A.M. Hijnen & J.C. Kruithof. 1989. *The effects of ozonation, biological filtration and distribution on the concentration if easily assimilable organic carbon (AOC) in drinking water.* Ozone Sci. Tech. 11:297-311.

Vartiainen, T, A. Liimatainen & P. Kauranen. 1987. *The Use of Size Exclusion Columns in Determination of the Quality and Quantity of Humus in Raw Waters and Drinking Waters.* Sci.Total Environ. 62:75-84.

Artificial Recharge of Groundwater, Peters et al. (eds) © 1998 Balkema, Rotterdam, ISBN 90 5809 017 5

Current research into removal of natural organic matter at Baltezers artificial groundwater recharge plant, Latvia

T. Juhna
Riga Technical University, Latvia and Luleå University of Technology, Sweden

J. Sprogis
Riga Technical University, Latvia

ABSTRACT: Column experiments, adsorption studies and microbial analyses of aquifer material were found to be useful at the estimation of retention capacity and function of aquifer material in removing NOM from the infiltrated water at Baltezers artificial groundwater recharge (AGR) plant (Latvia). Natural organic matter (NOM) was used as a limiting factor at the determination of necessary water retention time in an aquifer during artificial recharge of groundwater. Biodegradation and adsorption of NOM during water passages throughout all the aquifer were considered to be important removal processes.

1 BACKGROUND

At present, the City of Riga uses about 320,000 m^3 of water per day for its drinking water supply. About 51% of this water comes from either natural or artificially recharged groundwater (AGR) and 49% from surface water.

Baltezers AGR plant yields about 65000 m^3/d of drinking water. A favourable geological formation (good-sized permeable 30-40 m thick quaternary deposits) makes it possible to replenish natural groundwater with about 50,000 m^3 water per day. Water for replenishing is pumped from Lake Mazais Baltezers and without pre-treatment distributed among 17 ponds excavated along the south bank of the lake. The distance between the ponds is 50-60 m and they are situated in a coniferous forest some 500–900 m from the lake. In the ponds water percolates to the groundwater level with an average entrance velocity of 1 m/day. The groundwater is at a level of about 6 m below the bottom of the ponds, which allows the water to pass a relatively long vadose zone. Infiltrated water is collected by production wells 150-250 m distant from the infiltration ponds, permitting underground travel times of 30-180 d. As a final step desinfection is used after which water is pumped to the city water distribution net. The basins are cleaned once or twice a year by mechanically removing (scraping) the surface layer (4-10 cm) from the bottom of the infiltration ponds.

1.1 *Lake Mazais Baltezers water quality*

Lake M. Baltezers is eutrophic with a high amount of humic substances. The concentration of organic matter in the lake water is about 10 mg O_2/L measured as COD_{Mn}. The surges of seawater from the Gulf of Riga into the Lake M. Baltezers contribute to a high electrical conductivity, around 800 µS/cm, in the water. Due to anthropogenic impact, the water quality in Lake M. Baltezers deteriorates every year.

2 FUTURE STRATEGY

Today, the Enterprises of Riga's Municipality Riga Water assess priorities on the development of groundwater sources till year 2010. An expansion of Baltezers AGR plant is considered as being one of the alternatives for water supply in the future.

Baltezers AGR plant is located closer to the city than most of the groundwater sources that can potentially be used for water supply. However, the capacity of the aquifer available for artificial groundwater recharge is limited. More efficient exploitation of the aquifer can be achieved by reducing the distance between the infiltration ponds and the production wells. When Baltezers AGR plant was designed, this distance was determined by hydrogeological and physical considerations. A reduction demands that the distance becomes determined by considering all ingredient present in the water. At planning the measures, focus has been set at natural organic matter (NOM), which appears to be one of the main limiting factors regarding reduction of the distance. Presence of natural organic mater in drinking water is not desired, firstly because it gives the water colour, a bad taste and odour, secondly, because NOM may form potentially carcinogenic compounds at chlorination.

3 OBJECTIVES

The overall objective was to elucidate removal mechanisms for NOM at artificial groundwater recharge. The intentions also were to acquire preliminary data about the retention capacity of the aquifer material at Baltezers AGR plant regarding NOM, and to propose modes how to study removal of NOM in the course of artificial groundwater recharge. It was considered important to understand the role and functions of the filter cake established (Schmutzdecke) as well as the roles of the unsaturated zone and the saturated zone in removing NOM by artificial groundwater recharge.

4 RESEARCH METHODS

4.1 *Column experiments*

The retention capacity of an aquifer material can be estimated by taking water samples from different depths directly from the aquifer and then by measuring changes in the concentration of organic matter. The reliability of such results may be low taking into account that changes in already relatively low concentrations of organic matter are to be detected as water passes through the aquifer. Moreover, it is difficult to precisely predict the actual water passage through different zones from the infiltration ponds. For the sampling in the unsaturated zone lysimeters may be used (Frycklund 1995), but this process is time consuming.

McCarty et al. (1993) estimated retardation coefficients in field-scale. Water with a high amount of humic substances was injected into the limited aquifer (known volume) and captured by a collecting well at some distance. This method may give very precise results despite of a heterogeneous aquifer material. However, this method requires sophisticated equipment and was considered too expensive for acquiring the primary data.

Thus, column experiments were preferred. The columns (PVC tubes with diameter of 5 cm and length of 42 cm) were filled with aquifer material and water from Lake M. Baltezers was filtered through the columns by gravity with a surface loading of about 1 m/d. The water was sampled before entering the columns and when leaving each column. The turbidity and the NOM of the water before and after the columns were determined during two months until breakthrough occurred. The COD_{Mn} and the colour were used as indicators of NOM in the water. The conditions in the columns were as close as possible reflecting natural conditions in the aquifer. In a few columns unsaturated flow was simulated whereas some other columns simulated saturated conditions. It was assumed that NOM was removed by adsorption and by biodegradation.

The influence of the Schmutzdecke on the removal of NOM was studied in three columns. The efficiency of removing NOM (measured as absorbance at 410 nm) and turbidity was considered possible to estimate by comparing two columns filled with the aquifer material. The Schmutzdecke layer removed from the bottom of the infiltration ponds was added to the top of the first column, while the second column was tested without any Schmutzdecke. The values obtained were compared with results from the third column where the Schmutzdecke was allowed to develop in a natural way.

4.2 *Adsorption studies*

Adsorption of NOM onto the natural filtersand material was considered useful to study in a laboratory scale. The bacterial activity was assumed to be insignificant. Commercially available aquatic humic substance as well as natural humus, isolated from the lake, were used as representatives of refractionary NOM. To identify the main constituents of the aquifer material responsible for adsorption of NOM and to elucidate the nature of adsorption process in aquifers, the humic substances were sorbed onto different model sorbents (e.g. Al_2O_3, Goethite, clay mineral precoated with organic matter and bare clay mineral). The tested solution of humic substances was filtrated through small glass columns filled with aquifer material and its concentration at each outlet was detected.

From dynamic adsorption experiments the maximum adsorption capacity, was acquired. This value was compared to the retention value obtained from the column experiments. The difference between these values was attributed to biodegradation.

4.3 *Microbiological sampling*

Samples of the aquifer material were collected from different depths and positions below the infiltration ponds. Different species of bacteria, algae and single-cells were analysed at different stages of vegetation.

5 RESULTS AND DISCUSSIONS

The column experiments at Baltezers ARG plant were simple to accomplish and they have so far given primary data about the adsorption capacity of the aquifer material. In addition, it was possible to manipulate different parameters. By simulating unsaturated and saturated flow conditions results were achieved that provided useful information about processes in the aquifer, and allowed a comparison of the removal of NOM in different zones. The main limitation of the method was that materials used for filling the columns were recovered from upper layers of aquifer and thus might not be the representative of the aquifer material as a whole. The studies of NOM were restricted to the information obtained by analyses of COD_{Mn} and colour.

There was no significant statistical difference regarding the removal of NOM between columns with and without Schmutzdecke respectively. Probably, the retention time of water in the Schmutzdecke layer was too short for significant biodegradation to take place.

Adsorption mechanisms onto heterogeneous material were described applying Initial Mass Method, proposed by Nodvin et al (1986). It was possible to compare samples from different depth and to calculate an adsorption capacity for the filtersand. Results from the dynamic experiments were compared with results from column experiments and a relative role of biodegradation was estimated. A number of assumptions were made, especially concerning organic matter, but the results were sufficient to satisfy the purpose of the study.

Microbial analyses of aquifer material at different depths have contributed to the understanding of biodegradation processes in the unsaturated zone as well as to the role of the Schmutzdecke at the removal of organic matter. So far, the results indicate that biodegradation occurs not only in the uppermost layer of the aquifer material but also in the unsaturated zone below the infiltration ponds.

6 CONCLUSIONS

Preliminary values of the retention capacity of aquifer material were possible to estimate by column experiments, but field-studies are needed to confirm this value. Field studies followed by development of a mathematical model for the removal of NOM will be useful at determination of optimum distance between the infiltration ponds and the water production wells. Biodegradation has been found to be an important process at the removal NOM in saturated as well as unsaturated columns. Column experiments may well be used as a first step in determination of retardation factor of aquifer for removal of NOM. The existence of a schmutzdecke did not contribute essentially to the removal of NOM.

Future research on the subject should focus more upon biological removal of NOM during AGR.

7 ACKNOWLEDGEMENT

We express our appreciation to Professor Jörgen Hanæus and the staff of the Division of Sanitary Engineering, Luleå University Technology for academic and practical support. We acknowledge Riga Water for their financial support. We are grateful to our colleges Maris Klavins and Gunta Springe, University of Latvia, for they contribution to this research.

REFERENCES

Frycklund, C. 1995. Total organic carbon retention by filtersand in an infiltration pond for artificial groundwater recharge. *Aqua Fennica* 25,1: 5-14

McCarthy, J. F., T. M. Williams, L, Liang, P. M. Jardine, L. W. Joiley, D. L. Taylor, A. V. Palumbo & L. W. Cooper 1993. Mobility of Natural Organic Matter in a Sandy Aquifer. *Environ. Sci. Technol.* 27: 667-676

Nodvin, C. S., C. T. Driscoll, & G. E. Likens 1986. Simple partitioning of anions and dissolved organic carbon in a forest soil. *Soil Sci.* 141: 27-35

Author index